Earth

Then and Now

Earth

Then and Now
Third Edition

Carla W. Montgomery
Northern Illinois University

David Dathe
Alverno College

WCB **Wm. C. Brown Publishers**

Dubuque, IA   Bogotá   Buenos Aires   Caracas   Chicago   Guilford, CT   London
Madrid   Mexico City   Seoul   Singapore   Sydney   Taipei   Tokyo   Toronto

Project Team

Editor *Lynne M. Meyers*
Developmental Editor *Daryl Bruflodt*
Production Editor *Cheryl R. Horch*
Marketing Manager *Keri L. Witman*
Designer *Barb Hodgson*
Art Editor *Brenda A. Ernzen*
Photo Editor *Nicole Widmyer*
Permissions Coordinator *Gail I. Wheatley*

Wm. C. Brown Publishers

President and Chief Executive Officer *Beverly Kolz*
Vice President, Director of Editorial *Kevin Kane*
Vice President, Sales and Market Expansion *Virginia S. Moffat*
Vice President, Director of Production *Colleen A. Yonda*
Director of Marketing *Craig S. Marty*
National Sales Manager *Douglas J. DiNardo*
Advertising Manager *Janelle Keeffer*
Production Editorial Manager *Renée Menne*
Publishing Services Manager *Karen J. Slaght*
Royalty/Permissions Manager *Connie Allendorf*

A Times Mirror Company

Copyedited by Laura Beaudoin

Cover/Interior design by Rokusek Design

Cover photograph: Bridalveil Falls, Yosemite National Park, © Kerrick James Photography

All photographs in chapters 1–18 and 27, unless otherwise indicated, provided by Carla W. Montgomery

New art (figures 8.11, 8.12b, 9.11a, 9.11b, 10.3, 10.8b) and revisions prepared by Bensen Studios

Library of Congress Catalog Card Number: 96–83506

ISBN 0–697–28281–3

Printed in the United States of America by Times Mirror Higher Education Group, Inc.,
2460 Kerper Boulevard, Dubuque, IA 52001

10 9 8 7 6 5 4 3 2 1

CONTENTS

16 LANDSLIDES AND MASS WASTING   284

17 WIND AND DESERTS   303

18 ICE AND CLIMATE   317

19 FOSSILS AND EVOLUTION   340

PREFACE

eology, more so than most other sciences, is much concerned with the passage of time. Many geologic processes are so slow or occur in such minute stages that their effects become significant only over long time periods. Over the course of the earth's history, certain irreversible changes have altered the nature of geologic processes. For example, the earth is slowly cooling, which means that the extent and nature of modern volcanic processes and other temperature-related phenomena are somewhat different from their ancient counterparts; in addition, biological evolution has fundamentally altered the composition of the atmosphere and, consequently, the nature of many surface processes. Therefore, a general introduction to geology appropriately includes both physical geology and historical geology.

Physical geology comprises the study of the various geologic features and materials of the earth, as well as the nature of the processes by which they are formed and modified. It therefore gives us a greater understanding of the planet upon which we live. As humans have come increasingly to depend on geologic resources, knowledge of geology has been important in realizing the origins and the limitations of those resources. We have also come to recognize that we can anticipate—and therefore avoid—some of the disasters that are a natural consequence of certain geologic processes; in a few cases, we may even be able to modify those processes for our benefit.

Historical geology, as the name suggests, is concerned mainly with the history of the earth. Beyond the inherent interest we might naturally have in the development of our home planet and the life-forms on it, historical geology adds to our understanding of the processes that have shaped our world and gives us a particular appreciation for the role of time in geology.

In practice, most students who study geology fall into one of two groups. Some are prospective geology majors, for whom the broad perspectives of physical and historical geology are the foundation on which more advanced coursework builds. A much larger number are nonmajors, and indeed not prospective scientists of any kind, who are prompted to take the course by some mixture of interest in the subject and the need to satisfy a science distribution requirement. Although the needs of these two groups are not identical, every effort has been made to devise a text with sufficient versatility and appropriate learning aids to serve both kinds of students.

ABOUT THE BOOK

This book is intended for an introductory-level college course or course sequence in physical and historical geology. It assumes no prior exposure to college-level mathematics or science. For the most part, metric units are used throughout, except where other units are conventional within a discipline. For the convenience of students not yet comfortable with the metric system, a unit-conversion table is given in appendix A, and English-unit equivalents are frequently given within the text also.

In the early chapters, the student is introduced to the nature of geology as a science; to a brief outline of the history of the earth by way of historical perspective; and to minerals and rocks, the building blocks of geologic features, and the processes by which they form. Included in this section are the closely related topics of volcanic activity, soils, and weathering. This group of chapters concludes with the chapter on geologic time, concepts from which are fundamental to understanding many aspects of geologic process rates.

Once this groundwork has been laid, the next several chapters explore the major features and processes of the earth's interior. This exploration begins with plate tectonics, which provides the conceptual framework for understanding much about seismicity and volcanism. The data of seismology are also the major source of information about the earth's interior. Moving from the interior toward the surface, the next two chapters focus, respectively, on the continental crust, including crustal structures and mountain building, and on the ocean basins and oceanic crust.

Superimposed on these large-scale features are the effects of various surface processes, the subjects of the next several chapters. These processes include the various ways in which water, ice, wind, and gravity act to modify the earth's surface features and forms. It is at the surface that we see the interaction between the internal heat that drives the internal processes of volcanism, seismicity, and tectonics, and solar energy, which drives the winds and the water cycle.

Explaining the geologic and biologic history of the world in chronologic order for the last 4.5 billion years is difficult, if not impossible, for two reasons. First, the amount of information is immense. To circumvent this problem, the historical-geology chapters concentrate exclusively on North America, the United States in particular. Not only is the intended reader already somewhat familiar with the geology—or at least the geography—of this part of the world, but that reader is also most likely to travel within this area in the future. A second potential difficulty is the tedium of presenting geologic history by listing events and saying "this happened here at this time" over and over again. To address this problem, details have been kept to a minimum. Major rock types, styles and patterns of sedimentation, and deformational and mountain-building events for each period are summarized. The goal is to describe the rocks we see for a given period and show what their character and distribution allow us to infer about the earth of that time.

The last chapter, in a sense, looks ahead to the future. As populations grow and technology advances, there is growing dependence upon the availability of the supporting resources. Mineral and many energy sources are formed or concentrated by geologic processes, and issues of resource occurrence and availability are explored in chapter 27.

Appendix A shows the common unit conversions. Appendix B provides an identification chart for common minerals and some guidelines and a key for recognizing different rock types. Appendix C offers a brief introduction to topographic and geologic maps and satellite imagery, which should be particularly helpful to students whose courses do not include a laboratory. Appendix D is a guide to the major characteristics of important fossil groups. Appendix E is a listing of state, province, and territory geological surveys. Appendix F describes and illustrates localities in which certain rock units are particularly well represented.

TO THE INSTRUCTOR

The text organization just described places internal processes ahead of surface processes, with the latter followed by historical-geology chapters. This organization puts discussion of the large-scale processes ahead of that of the more localized surface processes that act on the resulting features, and it aims to help the reader achieve some understanding of current processes before reaching back in time to review the historical development of the planet and its organisms. However, the various chapters are relatively independent overall, so an instructor who prefers to do so can cover surface processes first with minimal difficulty. In addition, individual chapters are as self-contained as possible (without

undue repetition), so that the order can be adjusted by the instructor if desired. The glossary should help to bridge any gaps arising from reordering of blocks of the text.

The historical-geology chapters follow a consistent organization, both for the convenience of the reader and to facilitate comparisons between different portions of the earth's history. Each chapter begins with a few general comments: What was the regional setting of North America during this interval? Were there any events going on in the world that particularly affected North America? What was the tectonic situation? Included in this section is a paleogeographic map showing the inferred distribution of land and sea for the periods under consideration. Next, each part of the continent is considered—the craton, the eastern margin, and the western margin. Major rock types found and inferred tectonics are reviewed. Cross sections of the regions are included, and some of the more significant geologic formations are identified. Finally, the life of that interval of time is discussed. Index fossils are listed, and important evolutionary and extinction events are noted.

A discussion of environmental geology has been added to some texts as a separate chapter. However, there are environmental aspects to many of the topics in the text, from volcanoes to streams to resources. Therefore, in this book, environmental and human-impact considerations are woven into many physical-geology chapters as appropriate. This helps students to appreciate the current relevance of the subject matter while mastering the corresponding facts, theories, and vocabulary.

A variety of pedagogical aids and features are included. Each chapter begins with an outline of the subject headings to follow, by way of overview. Key terms are printed in boldface and defined at first encounter; these boldfaced terms are collected as "Terms to Remember" at the end of each chapter and are defined in the glossary for quick reference. At the end of each chapter are "Questions for Review" to assist students' study efforts, and a small number of questions or problems "For Further Thought" that go beyond basic review of text material. Certain of the latter might serve as starting points for class discussions or for short outside-of-class research projects. There are also "Suggestions for Further Reading" that include several kinds of material: up-to-date (but sometimes relatively sophisticated) references in the subject area of the chapter; materials that may be more readable for the nonspecialist (including some older but fundamentally accurate works); and, occasionally, "classic" works by prominent geologists.

All of the chapters contain one or more boxed inserts. These are of several types. Some describe tools of the geologic trade (for example, thin sections, mass spectrometry). Some present case studies or regional close-ups related to chapter material (flood recurrence-interval projection for a particular stream, groundwater depletion in the Ogallala aquifer system, geologic highlights of particular national parks). A few present somewhat more advanced concepts that might be appreciated most by better-prepared students (for example, an introduction to simple binary phase diagrams). In all cases, the material is included for enrichment or information without disrupting the flow of the main body of the text and the presentation of fundamental concepts. Individual boxes may be included or omitted at the instructor's discretion. Occasional "miniboxes" also appear within the text. These could be viewed somewhat as long parenthetical remarks, minor digressions not lengthy enough to justify a major boxed insert and usually lacking associated figures.

Appendices B, D, and F will probably be of most use to those students whose geology course does not include a required laboratory. They may also be helpful in cases in which lecture and laboratory sections proceed independently, so that the lecture may get ahead of the corresponding subject in the lab.

Users of the second edition will be interested in changes made in the third. The increase in number of chapters from twenty-three to twenty-seven does not reflect a major increase in the relative emphasis placed on physical geology, but rather the splitting of several chapters into more manageable and, perhaps, more logically constructed shorter chapters, as suggested by reviewers and adopters. Certain of the physical-geology chapters have significant new additions—for example, earthquake-cycle theory in chapter 10, and expanded discussion of global climate change in chapter 18. Data in the resources chapter (27) have been brought more up-to-date, and a number of figures and photos throughout the text have been improved or replaced. The map appendix (C), dropped in the second edition, has been returned by popular demand. And, of course, numerous smaller corrections and improvements have been made.

ACKNOWLEDGMENTS

A great many individuals have contributed to this project. The first edition of this text owes its existence particularly to the persistence and energy of Jeff Hahn, who instigated it, and Lynne Meyers, who has perfected the gentle art of nudging busy authors into producing their manuscripts more or less on time, without exhausting her seemingly infinite supply of patience. The second edition was thoughtfully shepherded to completion through the very supportive guidance of Bob Fenchel; Jeff Hahn elected to reentangle himself in the third edition.

Our sincere appreciation goes to the reviewers of this material in the various stages of its development. The first edition took shape with the valuable assistance of Richard Bonnett, Marshall University; Joanne Danielson, Shasta College; David Darby, University of Minnesota–Duluth; John Howe, Bowling Green State University; Larry Middleton, Northern Arizona University; Ed Ruggiero, Eastfield College of the Dallas County Community Colleges; and Monte D. Wilson, Boise State University. The second edition was refined through the careful reviews and thoughtful suggestions of Kenneth F. Griffin, Tarrant County Junior College, NW Campus; Monte D. Wilson, Boise State University, Department of Geosciences; and Robert C. Melchior, Bemidji State University. The third edition has been further improved by the insights and advice of Hobart King, Mansfield University; Monte D. Wilson, Boise State University; Lisa C. Sloan, University of California—Santa Cruz; and Dewey D. Sanderson, Marshall University. The perceptive comments and suggestions of all these individuals have contributed to many improvements in the text. Any remaining shortcomings in this book are, of course, the responsibility of the authors.

Thanks also go to Carol Edwards and Joe MacGregor at the U.S. Geological Survey Photographic Library, for their energetic help with the photo research (in spite of the exigencies imposed by the Gramm-Rudman-Hollings amendment and the disorganizing effects of a relocation); to Donald M. Davidson, Jr., for approving the loan of several departmental rock and mineral specimens for preparation of illustrations, and Ross D. Powell, for the use of his personal library; and to Jerrold H. Zar, for his continuing interest in other authors' grapplings with the muse. Thanks, also, to Anne Pach, Alverno College Interlibrary Loan. Her tireless effort in obtaining pertinent geology texts is gratefully appreciated. Much of the credit for the quality of the finished book goes to the whole book team at WCB, a group of dedicated, talented, and hardworking professionals. And last, but certainly not least, we would like to thank our spouses: Warren Montgomery, for his remarkable tolerance of yet another book, and Cindy Dathe, who suffered as all spouses do when such projects monopolize the authors.

INTERNET RESOURCES

ORGANIZATIONS

AGI American Geological Institute

http://agi.umd.edu/agi/agi.html

GSA Geological Society of America

http://www.aescon.com/geosociety/index.htm

NASA National Aeronautics and Space Administration

http://hypatia.gsfc.nasa.gov/NASA_homepage.html

The Paleontological Society

http://www.uic.edu/orgs/paleo/homepage.html

SEPM Society for Sedimentary Geology

http://www.ngdc.noaa.gov/mgg/sepm/sepm.html

DOE U.S. Department of Energy

http://www.em.doe.gov

USGS United States Geological Survey

http://www.usgs.gov/

MINERALOGY

Mineralogical Society of America teaching resource site

http://geology.smith.edu/msa/Teaching.html

USGS Minerals Page

http://minerals.er.usgs.gov

EARTHQUAKES AND VOLCANOES

The National Earthquake Information Center

http://wwwneic.cr.usgs.gov/

The National Geophysical Data Center

http://www.ngdc.noaa.gov/whatsnew.html

PALEONTOLOGY

The Burgess Shale

http://www.geo.ucalgary.ca/~macrae/Burgess_Shale

Paleo Net Pages

http://www.nhm.ac.uk/paleonet/

The Trilobite Page

http://www.ualberta.ca/~kbrett/Trilobites.html

MUSEUMS

The Museum of Paleontology—University of California, Berkeley

http://ucmpl.berkeley.edu/

The Royal Tyrrell Museum

http://www.tyrrell.com/

The Smithsonian Institution gem and mineral collection

http://galaxy.einet.net/images/gems/gems-icons/html

Internet addresses for geological surveys can be found in appendix E.

1

INTRODUCING THE EARTH

From massive granite crystallized deep in the crust, ice, water, and wind carve mountains. El Capitan and the Merced River, Yosemite National Park, CA. © Doug Sherman/Geofile

INTRODUCTION

Geology is the study of the earth and the processes that shape it. **Physical geology,** in particular, is concerned with the materials and physical features of the earth, changes in those features, and the processes that bring them about. **Historical geology,** as the term suggests, examines the development of the earth and the organisms on it over time. Physical and historical geology together, then, provide a basis for understanding much about the earth and its evolution.

Intellectual curiosity about the way the earth works is one reason for the study of geology. Piecing together the history of a mountain range or even a single rock can be exciting. The nonspecialist can better appreciate the physical environment—from distinctive features like the granite domes of Yosemite or the geysers of Yellowstone (figure 1.1) to the rocks exposed in a roadcut or found in one's own backyard.

There are also practical aspects to the study of geology. Certain geologic processes and events can be hazardous (figure 1.2), and a better understanding of such phenomena may help us to minimize their risks. We have also come to depend heavily on certain earth materials for energy or as raw materials for manufacturing (figure 1.3), and knowing how, where, and when those resources formed can be very useful to modern society.

FIGURE 1.2 Fourth Avenue, Anchorage, Alaska, after the 1964 earthquake. Note how far the shops and street on the right have dropped relative to the left side of the street.
Photograph courtesy of USGS Photo Library, Denver, CO.

FIGURE 1.1 Geyser eruption. Riverside Geyser, Yellowstone National Park.

FIGURE 1.3 Bingham Canyon copper mine, the world's largest open-pit mine, from which over $6 billion worth of minerals has been extracted.
Photograph courtesy of Kennecott.

Before entering into the detailed study of geology, we briefly survey geology as a discipline, and the history of the earth that is its subject.

Geology as a Discipline

Geology is a particularly broad-based discipline, for it draws on many other sciences. Knowledge of physics contributes to an understanding of rock structures and of deformation into folds and faults and supplies tools with which to investigate the earth's deep interior indirectly. The chemistry of geologic materials—rocks, minerals, fossils, fuels—provides clues to their origins and history. Modern biological principles are important in studying ancient life-forms. Mathematics provides a quantitative framework within which geologic processes can be described and analyzed. Physical geographers study the earth's surface features much as some geologists do. What makes geology a distinctive discipline, in part, is that it focuses all these approaches, and others, on the study of the earth. Moreover, having the earth as a subject also introduces some special complexities not common to most other sciences.

The Issue of Time

As noted below, the modern earth has been over 4½ billion years in the making. Moreover, many of the processes shaping the earth are extremely slow on a human timescale, some barely detectable even with sensitive instruments. It is therefore difficult to observe or to demonstrate directly, in detail, how certain materials or features have formed.

Furthermore, materials may respond differently to the same forces, depending on how those forces are applied. This can be seen, for example, in the phenomenon of fatigue of machine parts, in which a material quite strong enough to withstand a certain level of sustained stress fails during the application of a much smaller stress that has been repeated many times. As a practical matter, then, it may be impossible to duplicate some geologic processes in the laboratory because the human lifetime is simply too short.

The Matter of Scale

Likewise, some natural systems are just too large to duplicate in the laboratory. A single crystal or small piece of volcanic rock can be studied in great detail. But it is hardly possible to build a volcano (figure 1.4) or a whole continent in a laboratory to conduct experiments on them. Scale-model experiments, in which the materials and the forces applied to them are scaled down proportionately, are one compromise. For example, in studies of designs for earthquake-resistant buildings, model structures are shaken by small vibrations in the hope that the results will mimic the response of large buildings to great earthquakes. As another example, a large tank of water can be agitated artificially to study wave motion or the

Figure 1.4 Mount St. Helens in eruption, May 1980.
Photograph by P. W. Lipman, USGS Photo Library, Denver, CO.

effects of currents. Scale modeling is an inexact science, however, and not all natural systems lend themselves to such studies. The alternative, for the study of large features, may be extensive travel and field work, and perhaps even use of a satellite's broad perspective from space.

Complexity in Natural Systems

A laboratory scientist about to conduct an experiment tries to minimize the number of variables, so as to obtain as clear a picture as possible of the effect of any one change—whether of temperature, pressure, or the quantity of some particular substance present. Typically, the experimental materials are kept simple: a single rock type or mineral, chemically quite pure. Natural geologic systems, however, are rarely so simple. Natural rocks and minerals invariably contain chemical impurities and physical imperfections; many different rocks and minerals may be mixed together; and temperature and pressure may change simultaneously while gases, water, or other chemicals flow in and out. Extrapolating from carefully controlled laboratory experiments to the real world becomes correspondingly difficult.

Also, the laboratory scientist can perform an experiment in stages, examining the results after each step. The geologist, however, may be confronted by rocks or other materials or structures that have been altered or recycled several times, perhaps dozens of times, each time under different conditions, over millions or even billions of years. That history can be difficult to decipher from the present end product, particularly because the same end product can often be formed from several different possible starting materials, via different combinations of geologic processes, just as one can arrive at a given spot by traveling in various ways from various starting points.

Geology and the Scientific Method

The **scientific method** is a means of discovering basic scientific principles. One begins with a set of observations and/or a body of data that are based on measurements of natural phenomena or on experiments. One or more **hypotheses**—possible explanations, models, or relationships—are formulated to explain the observations or data. It is also possible that no systematic relationship exists among the observations (*null hypothesis*). A hypothesis can take many forms, ranging from a general conceptual framework or physical model describing the functioning of a natural system, to a very precise mathematical formula relating several kinds of numerical data. What all hypotheses have in common is that they are unproven and must be susceptible to testing.

In the classical conception of the scientific method, one uses a hypothesis to make a set of predictions and then devises and conducts experiments to test each hypothesis to determine whether experimental results agree with predictions based on the hypothesis. If they do, the hypothesis gains credibility. If not, if the results are unexpected, the hypothesis must be modified to account for the new data. Several cycles of modifying and retesting of hypotheses may be required before a hypothesis consistent with all the observations and experiments that one can conceive is developed. A hypothesis that is repeatedly supported by new experiments advances, in time, to the status of a **theory,** a generally accepted explanation for a set of data or observations.

This approach is not strictly applicable to many geologic phenomena because of the difficulty of experimenting with natural systems. In such cases, hypotheses are often tested entirely through further observations and modified as necessary until they accommodate all the relevant observations. This broader conception of the scientific method is well illustrated by the development of the theory of plate tectonics, discussed in chapter 9.

Some people casually use the term "theory" to describe what is really a hypothesis or even an educated guess, creating the impression that there is much uncertainty about theories. A genuine scientific theory, however, has been rigorously tested, often challenged by a number of researchers, and found to be a solidly convincing explanation for a body of facts or observations. Even a well-accepted theory, however, may ultimately be found to require extensive modification. In the case of geology, a common cause of this is the development of new analytical or observational techniques, which make available wholly new kinds of data that were unknown at the time the original theory was formulated.

In addition to hypotheses and theories, there is a smaller body of scientific **laws:** fundamental, typically simple principles or formulas that are invariably found to be true. In this category are Newton's law of gravity and the principle of physical chemistry that states that heat always flows from a warmer body to a colder one, never the reverse.

Key Concepts in the History of Geology

Humans have wondered about the earth in some way for thousands of years. The ancient Greeks measured it and recognized fossils preserved in its rocks as remains of ancient life-forms. Theologians, philosophers, and scientists have speculated on its age for centuries. The systematic study of the earth that constitutes the science of geology, however, has existed as an organized discipline for only about 250 years. It was first developed formally in Europe. In its early years, it was predominantly a descriptive subject. Two principal opposing schools of thought emerged in the eighteenth and nineteenth centuries to explain geologic observations.

One, popularized by James Hutton and later named by Charles Lyell, was the concept of **uniformitarianism.** Sometimes condensed to the phrase, "The present is the key to the past," uniformitarianism comprises the ideas that the earth's surface has been continuously and gradually changed and modified over the immense span of geologic time, and that by studying the geologic processes now active in shaping the earth, we can understand how it has evolved through time. It is not assumed that the *rates* of all processes have been the same throughout time, but rather that the *nature* of the processes is similar—that the same physical principles operating on the earth in the past are also operating in the present.

Scotsman James Hutton was a remarkably versatile individual—physician, farmer, and only part-time geologist. In the early days of a developing science, many advances in understanding are made by nonspecialists capable of careful observation and logical deduction. As various disciplines become more advanced and more sophisticated, however, the amateur is far less able to play a major role. Too much accumulated knowledge must be assimilated to arrive at the forefront of research.

The second, contrasting theory was **catastrophism.** The catastrophists, led by French scientist Georges Cuvier, believed that a series of immense, worldwide upheavals were the agents of change and that, between catastrophes, the earth was static. Violent volcanic eruptions followed by torrential rains and floods were invoked to explain mountains

FIGURE 1.5 Meteor Crater, Arizona.
Photograph by D. R. Roddy, USGS Photo Library, Denver, CO.

and valleys and the burial of animal populations that later became fossilized. In between those episodic global devastations, the earth's surface did not change, according to catastrophist theory. Catastrophists also believed that entire plant and animal populations were created anew after each such event, to be wholly destroyed by the next.

Is the Present the Key to the Past?

More detailed observations and calculations, greater use of the allied sciences, and development of increasingly sophisticated instruments have collectively provided overwhelming evidence in support of the uniformitarian view. The great length of the earth's history makes it entirely plausible that processes that seem gradual and even insignificant on a human timescale could, over those long spans of time, create the modern earth in all its geologic complexity.

This is not to say that the earth's history has not been punctuated occasionally by sudden, violent events that have had a substantial impact on a regional or global scale. Geologists see evidence of past collisions of large meteorites with the earth (figure 1.5) and find volcanic debris from ancient eruptions that would dwarf that of Mount St. Helens. But these events are unusual and, for the most part, of only temporary significance in the context of the earth's long history. By and large, the modern earth is the product of uncounted small and gradual changes, repeated or continued over very long spans of time.

The same physical and chemical laws can be presumed to have operated throughout the earth's history. Thus, by observing modern geologic processes, geologists can learn much about how those same processes might have shaped the earth in the past. The relative importance of each process, however, may not always have been just the same as it is now; nor can it be assumed that all processes operated in detail just as they do now. Certain irreversible changes in the earth have no doubt changed the nature and intensity of corresponding geologic processes. For example, the earth has been slowly losing heat since it formed. Present internal temperatures must be substantially lower than they were several billion years ago, especially near the surface. The earth's internal heat plays a key role in melting rocks and thus in volcanic activity, as well as in plate tectonics (see box 1.1). It is reasonable to infer that melting in the interior was more extensive in the past than it is now, that the products of that more extensive melting might be somewhat different from modern volcanic rocks, and also that volcanic activity was more extensive in the past. The earth's atmosphere, too, has undergone profound changes, as is described later in the text. Briefly, the atmosphere has gone from oxygen-poor to oxygen-rich, and this change, in turn, has necessarily affected the chemical details of such processes as weathering of rocks by interaction with air and water. However, geologists can determine, both experimentally and theoretically, how rocks would react with the kind of atmosphere deduced for the early earth, and thus they can characterize ancient weathering processes, even though they cannot observe the exact equivalents in nature today. The *earth* may change; the *physical laws* do not. This concept is fundamental to modern uniformitarianism.

E volution of organisms provides a biologic analogy. Such evolution is, for the most part, a process of many small, individual, adaptive changes that collectively amount, over time, to a significant change in the basic characteristics of the species. Occasionally, however, there are much more rapid changes, caused perhaps by unusual mutations or extraordinary environmental stresses; sometimes, whole plant or animal groups rapidly decline and become extinct.

EARTH PAST AND PRESENT

The sun and its system of circling planets began to form from a *nebula,* a rotating gas cloud, nearly five billion years ago. Most of the mass of the hydrogen-rich nebula coalesced at its center to create the sun. As the balance of the gas cooled, solid minerals condensed from it, and the condensates were assembled by gravity into planets, Earth among them. The earth has changed considerably since its formation, however, undergoing some particularly profound changes in its early history.

Box 1.1

Revolution in Earth's Evolution . . . Plate Tectonics

Rarely has a new geologic theory come so rapidly to dominate the discipline and, at the same time, to capture public attention. Plate tectonics, which became generally accepted only two decades ago, now provides the conceptual framework that allows geologists to understand much about the nature of mountain building and other processes that shape our planet's surface.

Tectonics is the study of large-scale movement and deformation of the earth's crust. The basic concept of *plate tectonics* is simple: that the earth has an outermost rigid, rocky shell, consisting of the crust and uppermost mantle, broken up into a series of *plates* that can move on an underlying soft or plastic layer in the upper mantle (figure 1). Plates can collide or be split apart; as two plates converge, one can override the other.

Continents shift in position on the globe as the plates of which they are a part move. The stresses associated with plate movements and deformation can cause earthquakes, especially at the boundaries between plates. Where one plate plunges below another into the hot mantle, melting occurs, and the melt may rise up into the plate above to erupt from a volcano (figure 2).

The theory of plate tectonics was synthesized from hypotheses devised to explain such diverse observations as the occurrence of volcanoes and earthquakes in linear belts; the topography of the sea floor; patterns of magnetism of continental and seafloor rocks; and global distributions of fossil forms and glacial debris. It illustrates well the interdisciplinary na-

FIGURE 1 Plates and plate boundaries (boundaries in bright highlights).
Source: National Oceanic and Atmospheric Administration.

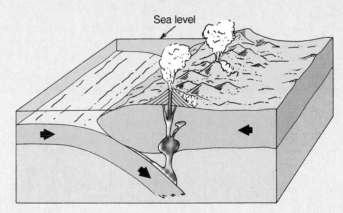

FIGURE 2 Ocean–continent convergence. The melting accompanying sinking of the down-going plate produces volcanic mountains on the continent.

ture of geology, for the evidence supporting it is drawn from many sciences—biology, chemistry, physics, geography, and

mathematics. This will become more apparent in chapter 9, where plate tectonics is discussed in detail.

The Early Earth: Formation, Heating, and Differentiation

Like most of the planets, the earth is believed to have begun as a sort of "dust ball" of small bits of condensed material collected together by gravity, with no free water, no atmosphere, and a very different surface from the present one. The dust ball was heated in several ways. The impact of the colliding particles as they came together to form the earth provided some heat. Much of this heat was radiated out into space, but some was trapped in the interior of the accumulating earth. As

the dust ball grew, compression of the interior by gravity also heated it. (That materials heat up when compressed can be demonstrated by pumping up a bicycle tire and then feeling the barrel of the pump.) Furthermore, the earth contains small amounts of several naturally radioactive elements that decay, releasing energy (see chapter 7). These three heat sources combined to raise the earth's internal temperature enough that parts of it, eventually most of it, melted. At one time, the earth's surface may have been a layer of molten rock.

When accretion of cosmic dust into planets was largely complete, slow cooling of the newly formed earth set in.

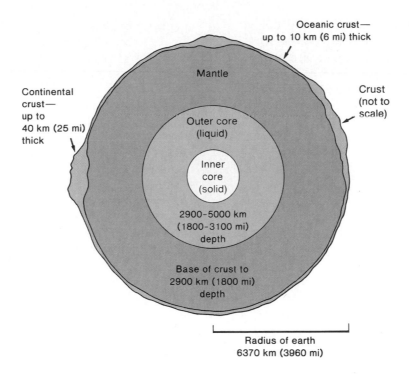

Oceanic crust—
up to 10 km (6 mi) thick

Mantle

Crust
(not to
scale)

Continental
crust—
up to
40 km (25 mi)
thick

Outer core
(liquid)

Inner
core
(solid)

2900–5000 km
(1800–3100 mi)
depth

Base of crust to
2900 km (1800 mi)
depth

Radius of earth
6370 km (3960 mi)

FIGURE 1.6 A chemically differentiated earth (crust not to scale). Oceanic crust, which forms the sea floor, has a composition somewhat like that of mantle but is richer in silicon. Continental crust is thicker and contains more of the low-density minerals rich in calcium, sodium, potassium, and aluminum. It rises above both the sea floor and the ocean surface.

Metallic iron, being very dense, sank toward the middle of the earth. As cooling progressed and the remaining melt began to crystallize, lighter, low-density minerals floated toward the surface. The eventual result was an earth differentiated into several major compositional zones: a large, iron-rich **core;** a thick, surrounding **mantle;** and at the surface, a thin, low-density **crust** (see figure 1.6).

Chapter 10 examines how geologists know the structure and composition of the earth's interior. It can be shown that the core consists mostly of iron, with some nickel and a few minor elements, and that the mantle is made up principally of the elements iron, magnesium, silicon, and oxygen combined, in varying proportions, into several different minerals. The crust is much more varied in composition and very different chemically from the average composition of the earth, as is also discussed in chapter 10. The continental crust is thicker, less dense, and more buoyant relative to the mantle than is the thinner, denser oceanic crust, which accounts for elevation of the continents above the sea floor. The differentiation process that produced an earth of several compositionally distinct zones was complete at least four billion years ago.

The heating and subsequent differentiation of the early earth led to another important result: formation of the atmosphere and oceans. Many minerals that had contained water or gases locked in their crystals released them during the heating and melting. It is not clear whether the earth was ever wholly molten at the surface. At the very least, the early earth was much hotter than at present and subject to more extensive volcanic activity, with water among the gases thus released. As the earth's surface cooled, the water could condense to form the oceans. Without this abundant surface water, which in the solar system is unique to earth, most life as we know it could not exist.

The earth's early atmosphere was quite different from the modern one, even disregarding the effects of modern pollution. The first atmosphere had little or no free oxygen. Humans could not have survived in it. Oxygen-breathing life could not exist before the first simple plants—the single-celled blue-green algae—appeared in large numbers to modify the atmosphere. Their remains are found in rocks more than three billion years old. They manufacture food by photosynthesis, using sunlight for energy and releasing oxygen as a by-product. In time, enough oxygen accumulated that the atmosphere could support oxygen-breathing organisms.

Subsequent History: The Changing Face of Earth

After the early differentiation, the earth's crust with its continents and ocean basins did not look the way we know it today. For one thing, the continents have moved (see chapter 9). They have not always been the same size and shape, either. Rocks that must have been formed in ocean basins can now be found high and dry on land, revealing the former presence of inland seas followed by great uplift. New pieces have been added to the edges of the continents. Volcanoes once erupted where none now exist, leaving behind evidence of their earlier activity in ancient volcanic rocks. Tall mountains have been built up and then eroded away, sometimes several times in the same place, over billions of years.

Geologists can to some extent reconstruct the distribution of land, water, and surface features as they were at times in the past and identify geologically active areas, such as developing mountain ranges, on the basis of the kinds of rocks or fossils of each age found and what is known of how such rocks formed or in what setting the fossilized creatures lived. Such reconstruction becomes more difficult the farther back geologists try to go in time, for the oldest rocks have often been covered by younger ones or disrupted by more recent geologic events. Aside from its academic interest, reconstruction of ancient geography and geology can be of practical value. If certain kinds of needed mineral or energy resources are known to have formed in particular geologic settings, geologists can look for those resources not only in the appropriate modern geologic environments but also in rocks that formed in similar environments in the past.

The rock record also shows when different plant and animal groups appeared on the earth. Some are represented schematically in figure 1.7. Such information has practical applications: certain of our energy sources have been formed from plant or animal remains (see chapter on mineral and energy resources). Knowing the times at which particular groups of organisms appeared and flourished is helpful in assessing the probable amounts of these energy sources available and in concentrating the search for these fuels in rocks of appropriate ages.

On a time scale of billions of years, human beings have just arrived. The most primitive human-type remains are no more than three to four million years old, and modern,

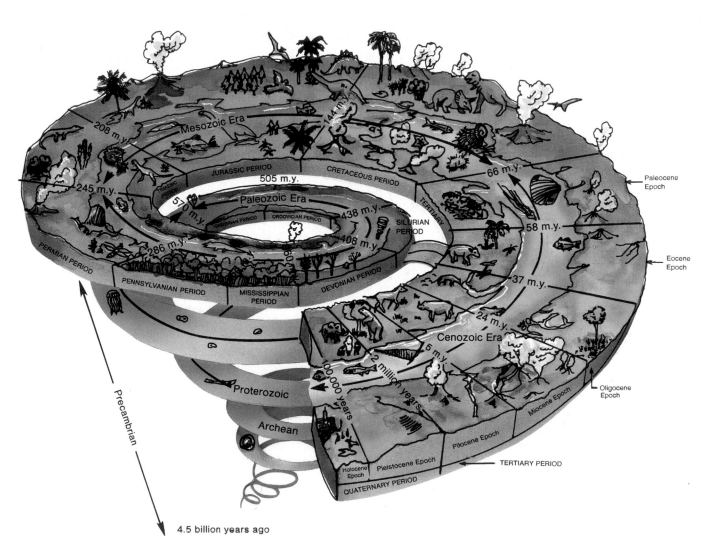

FIGURE 1.7 The so-called geologic clock. Important plant and animal groups are shown where they first appear in significant numbers. All complex organisms—especially humans—have developed relatively recently in the geologic sense.

Source: After U.S. Geological Survey publication, "Geologic Time."

rational humans (*Homo sapiens*) developed only about half a million years ago. Half a million years may sound like a long time, and it is if compared to a single human lifetime. In a geologic sense, though, it is a very short time. Nevertheless, humans have had an enormous impact on the earth—at least at its surface—an impact far out of proportion to the length of time we have occupied the planet.

The Modern Earth: Systems, Cycles, and Change

Although it has been cooling for billions of years, the earth still retains enough internal heat to drive large-scale mountain-building processes, to produce volcanic eruptions, to make continents mobile and indirectly to trigger earthquakes. At the same time, the continual supply of solar energy to the surface drives many of the surface processes: water is evaporated from the oceans to descend again as rain and snow to feed the rivers and glaciers that sculpture the surface; differential heating of the surface leads to formation of warmer and colder air masses above it, which, in turn, produce atmospheric instability and wind. The earth is a dynamic, constantly changing planet—its crust shifting to build mountains; lava spewing out of its warm interior; ice and water and windblown sand and gravity reshaping its surface, over and over. Many of the processes are cyclic in nature.

Consider, for example, such basic materials as water or rocks. Streams drain into oceans, and would soon run dry if not replenished; but water evaporates from oceans, to make the rain and snow that feed the streams to keep them flowing. This describes just a part of the *hydrologic* (water) *cycle,* explored more fully in chapters 14 and 15. Rocks, despite their appearance of permanence in the short term of a human life, participate in the *rock cycle,* introduced in chapter 2 and revisited in chapter 9. The kinds of evolutionary paths rocks may follow through this cycle are many, but consider this illustration: A volcano's lava (figure 1.8) hardens into rock; the rock is weathered into sand and dissolved chemicals; the debris, deposited in an ocean basin, is solidified into a new rock of quite different type; and that new rock is carried into the mantle via plate tectonics, to be melted into a new lava. The time frame over which this process occurs is generally much longer than that over which water cycles through atmosphere and oceans, but the principle is similar. The Rocky Mountains as we see them today (figure 1.9) are not the Rocky Mountains as they formed, tens of millions of years ago; they are much eroded from their

FIGURE 1.8 Molten lava pours from Kilauea and solidifies to make new rock.
Photograph by Hawaii Volcano Observatory, USGS Photo Library, Denver, CO.

FIGURE 1.9 The Beartooth Mountains near Yellowstone National Park, a part of the U.S. Rocky Mountains.

original height, and in turn contain rocks formed in water-filled basins and deserts from material eroded from more ancient mountains before them.

Chemicals, too, cycle through the environment. The carbon dioxide that we exhale into the atmosphere is taken up by plants through photosynthesis, and when we eat those plants for food energy, we release CO_2 again. The same exhaled CO_2 may also dissolve in the oceans from the atmosphere, to make dissolved carbonate, which then contributes to the formation of carbonate shells and carbonate rocks in the ocean basins; those rocks may later be exposed and weathered by rain, releasing CO_2 back into the atmosphere or dissolved carbonate into streams that carry it back to the ocean. The cycling of chemicals and materials in the environment may be complex.

Furthermore, these processes and cycles are often interrelated, and seemingly local actions can have distant consequences. We dam a river to create a reservoir as a source of irrigation water and hydroelectric power, inadvertently trapping stream-borne sediment at the same time; downstream, patterns of erosion and deposition in the stream channel change, and at the coast, where the stream pours into the ocean, coastal erosion of beaches increases because a part of their sediment supply, from the stream, has been cut off. The volcano that erupts the lava to make the volcanic rock also releases water vapor into the atmosphere, and sulfur-rich gases and dust that influence the amount of sunlight reaching earth's surface to heat it, which in turn can alter the extent of evaporation and resultant rainfall, which will affect the intensity of landscape erosion and weathering of rocks by water. . . . So although we divide the great variety and complexity of geologic processes and phenomena into more manageable, chapter-sized units for purposes of discussion, it is important to keep such interrelationships in mind. We will note a number of examples throughout the text. And superimposed on, influenced by, and subject to all these natural processes are humans and human activities. Table 1.1 presents the time spans required for a selection of geologic processes, to give a sense of the rates at which those processes occur relative to human activities.

Dynamic Equilibrium in Geologic Processes

Natural systems tend toward a balance or equilibrium among opposing forces. Internal forces push up a mountain; gravity, wind, water, and ice collectively act to tear it down again. Minerals are dissolved and their component chemicals washed into the sea by rivers; fresh sediments are precipitated in the ocean basins, removing dissolved chemicals from solution. When one factor changes, other compensating changes occur in response. If the disruption of a system is relatively small and temporary, the system may, in time, return to its original condition, and evidence of the disturbance is erased.

TABLE 1.1 *Some Representative Illustrations of Geologic Process Rates*

Process	Occurs Over a Time Span of About This Magnitude
Rising and falling of tides	1 day
"Drift" of a continent by 2–3 centimeters (about 1 inch)	1 year
Accumulation of energy between large earthquakes on a major fault zone	10–100 years
Rebound (rising) by 1 meter of a continent depressed by ice sheets during the Ice Age	100 years
Flow of heat through 1 meter of rock	1000 years
Deposition of 1 centimeter of fine sediment on the deep-sea floor	1000–10,000 years
Life span of a small volcano	100,000 years
Ice sheet advance and retreat during an ice age	100,000 years
Life span of a large volcanic center	1–10 million years
Creation of an ocean basin such as the Atlantic	100 million years
Duration of a major mountain-building episode	100 million years
History of life on earth	Over 1 billion years

A coastal storm may wash away beach vegetation and destroy colonies of marine organisms living in a tidal flat, but when the storm has passed, new organisms start to move back into the area and new grasses begin to establish roots in the dunes. The violent eruption of a volcano like Krakatoa or El Chichón may spew ash high into the atmosphere, partially blocking sunlight and causing the earth to cool, but within a few years, the ash settles back to the ground, and normal temperatures are restored.

This is not to say that permanent changes never occur in natural systems. The size of a river's channel depends, in part, on the maximum amount of water it normally carries. If long-term climatic or other conditions change so that the volume of water regularly reaching the stream increases, the larger quantity of water will, in time, carve out a correspondingly larger channel to accommodate it. The soil carried downhill by a landslide certainly does not begin moving back upslope after the landslide is over; the face of the land is irreversibly changed. Even so, a hillside forest uprooted and destroyed by the slide may within decades be replaced by fresh growth in the soil newly deposited at the bottom of the hill.

FIGURE 1.10 Severe erosion of a steep bank exposed during construction.
Photograph courtesy of USDA Soil Conservation Service.

FIGURE 1.11 Aftermath of the Big Thompson Canyon flood, Colorado.
Photograph courtesy of USGS Photo Library, Denver, CO.

The Impact of Human Activities; Earth as a Closed System

For all practical purposes, the earth is a **closed system** with respect to matter, meaning that the amount of matter in and on the earth is fixed. No new elements are being added. There is, therefore, an ultimate limit to how much of any metal we can exploit. There is also only so much land to live on. Conversely, any harmful elements we create remain in our geologic environment, unless we take some extraordinary step, such as expending the funds, materials, and energy required to cast matter out into space via spacecraft.

Human activities can cause or accelerate permanent changes in natural systems. The impact of humans on the global environment is broadly proportional to the size of the population, as well as to the level of technological development achieved. This can be illustrated especially readily within the context of pollution. The smoke from one campfire pollutes only the air in the immediate vicinity; by the time that smoke is dispersed through the atmosphere, its global impact is negligible. The collective smoke from a century of industrialized society, on the other hand, has caused measurable increases in several atmospheric pollutants worldwide, and those pollutants continue to pour into the air from many sources.

Likewise, in the context of resources, recent human impact has been dramatic. The world's population has soared from two billion to over five billion in little more than fifty years. That fact, combined with desires for ever-higher standards of living around the world, has created voracious demand for mineral and energy resources. Yet, although the population is growing, the earth is not. Therefore, our resources are in finite supply, and some could be exhausted within our lifetimes. This problem is considered in the chapter on mineral and energy resources. The very presence of humans, along with their building, mining, farming, and other activities, alters the landscape, and does so increasingly as the population grows (figure 1.10). Superimposed on the natural geologic processes of change on the earth, then, are further changes, deliberate or unconscious, caused by human activities. Many of the latter are occurring more rapidly than the compensating natural processes.

The principal focus of this text is natural geologic processes. However, from time to time, particularly significant human impacts on geologic systems, as well as the reverse—geologic impacts on human activities (such as the hazards associated with floods, earthquakes, and other geologic processes)—are highlighted (figure 1.11).

SUMMARY

Geology is the study of the earth. It is not an isolated discipline, but rather draws on the principles of many other sciences. The earth is a challenging subject, for it is old, complex in composition and structure, and large in scale. Physical geology focuses particularly on the physical features of the earth and how they have formed. Observations suggest that, for the most part, those features are the result of many individually small, gradual changes continuing over long periods of time, punctuated by occasional, unusual cataclysmic events. Historical geology concentrates on the details of the history and development of the earth and its organic occupants.

Shortly after its formation, the earth underwent melting and compositional differentiation into core, mantle, and crust. The early atmosphere and oceans formed at the same time. Heat from within the earth and from the sun together drives many of the internal and surface processes that have shaped and modified the earth throughout its history and that continue to do so. Earth materials cycle through the various geologic processes and systems, often in complex or interrelated cycles.

Humans only appeared three to four million years ago, but their large and growing numbers and technological advances have had significant impacts on natural systems, some of which may not readily be erased by slower-paced geologic processes.

TERMS TO REMEMBER

catastrophism 4
closed system 11
core 7
crust 7

historical geology 2
hypothesis 4
law 4
mantle 7

physical geology 2
scientific method 4
theory 4
uniformitarianism 4

QUESTIONS FOR REVIEW

1. What is *physical geology? historical geology?*
2. Describe how the time factor complicates human attempts to understand geologic processes.
3. What is the *scientific method?* To what extent is it applicable to geology?
4. Compare and contrast the concepts of uniformitarianism and catastrophism.
5. What are the principal compositional zones of the earth?
6. How were the earth's atmosphere and oceans first formed? How and why has the atmosphere changed over time?
7. The earth is a closed system. Explain this concept in the context of resource use or of pollution.

FOR FURTHER THOUGHT

1. If the whole 4½-billion-year span of the earth's history were represented by one 24-hour day, how much time would correspond to (a) the 600 million years of existence of complex organisms with hard shells or skeletons and (b) the half-million years that *Homo sapiens* has existed?
2. Many geologic processes proceed at rates on the order of 1 centimeter per year (1 inch = 2.54 centimeters). At that rate, how long would it take you to travel from home to school or from your room to class?

SUGGESTIONS FOR FURTHER READING

Cloud, P. 1988. *Oasis in space: Earth history from the beginning.* New York: W. W. Norton.

Eicher, D. L., McAlester, A. L., and Rottman, M. L. 1984. *The history of the earth's crust.* Englewood Cliffs, N.J.: Prentice-Hall.

Head, J. W., Wood, C. A., and Mutch, T. 1976. Geological evolution of the terrestrial planets. *American Scientist* 65: 21–29.

Pilbeam, D. 1984. The descent of hominoids and hominids. *Scientific American* 250 (March): 84–96.

Siever, R. 1983. The dynamic earth. *Scientific American* 249 (September): 46–65.

World Resources Institute. 1994. *World Resources 1994–95.* New York: Oxford University Press.

York, D. 1975. *Planet earth.* New York: McGraw-Hill.

MINERALS

All minerals are crystalline; given time and space enough, they can develop well-shaped crystal forms. Here, brassy metallic pyrite crystals—in a distinctive shape with five-sided faces known as a pyritohedron—occur with quartz. © Doug Sherman/Geofile

INTRODUCTION

It is difficult to talk at length about geology without talking about rocks or the minerals of which they are composed. There are several reasons for studying minerals, aside from the fact that they are all around us. Some have become economically valuable to our societies. Minerals form in a variety of ways—by crystallization from melts, by precipitation from solution, and in other ways—and each mineral forms within a restricted range of temperature, pressure, and other conditions. Thus each mineral and rock contains clues to its origins and history. These clues can be studied not only for the intellectual challenge, but also from the practical standpoint of helping us locate useful mineral deposits, or understand better such potentially hazardous geologic processes as volcanic eruptions.

Just eight elements account for over 98% of the earth's crust, yet there are dozens of common minerals and more than 3000 recognized minerals in all, with a corresponding diversity of physical properties. Fortunately for the beginning student, it is necessary to become familiar with only a small number of the more common minerals and mineral groups to discuss a wide range of rock types and geologic processes. This chapter introduces those minerals and some underlying chemical and mineralogical concepts, and then finishes with a brief look at rocks considered as assemblages of minerals.

ATOMS, ELEMENTS, AND ISOTOPES

All natural and most synthetic substances on earth are made from the ninety naturally occurring chemical elements. An **element** is the simplest kind of chemical; it cannot be broken down further by ordinary chemical or physical processes, such as heating or reaction with other chemicals. An **atom** is the smallest particle into which an element can be subdivided while still retaining its distinctive chemical characteristics.

Basic Atomic Structure and Elements

A simplified sketch of the structure of an atom is shown in figure 2.1. Atoms contain several types of subatomic particles: **protons,** which have a positive electrical charge; **electrons,** which have a negative electrical charge; and **neutrons,** which, as their name implies, are electrically neutral (have no charge). The **nucleus,** at the center of the atom, contains the protons and neutrons; the electrons move around the nucleus.

The number of protons in the nucleus is unique for each element and determines what chemical element that atom is. Every atom of hydrogen contains one proton in its nucleus; every carbon atom, six protons; every oxygen atom, eight protons; every uranium atom, ninety-two protons. The characteristic number of protons is the **atomic number** of the element.

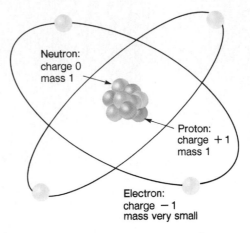

FIGURE 2.1 Basic atomic structure: protons and any neutrons in the nucleus, electrons circling outside the nucleus.

Isotopes

Except for the simplest hydrogen nucleus, all nuclei contain neutrons as well as protons, and the number of neutrons is similar to or somewhat greater than the number of protons. The number of neutrons in different atoms of the same element is not always constant. The sum of the number of protons and the number of neutrons in a nucleus is that atom's **atomic mass number.** (Protons and neutrons have similar masses; electrons are much lighter and contribute little to the mass of the whole atom.) Atoms of a given element with different atomic mass numbers—in other words, atoms with the same number of protons but different numbers of neutrons—are distinct **isotopes** of that element (see figure 2.2). Some elements have only a single isotope, while others may have ten or more. (The reasons for this are complex, involving principles of nuclear physics, and we will not pursue them here.)

I n the lightest nuclei, the numbers of protons and neutrons tend to be approximately equal. In heavier elements, neutrons generally outnumber protons. A possible explanation for this is that the additional neutrons in heavier elements help to stabilize a large nucleus by separating the positively charged protons, which tend to repel each other.

A particular isotope is designated by the element name (which, by definition, specifies the atomic number, or number of protons) and the atomic mass number (protons plus neutrons). Carbon, for example, has three natural isotopes. By far the most abundant is carbon-12, the isotope with six neutrons in the nucleus in addition to the six protons common to all carbon atoms. The two rarer isotopes are carbon-13 (six protons plus seven neutrons) and carbon-14

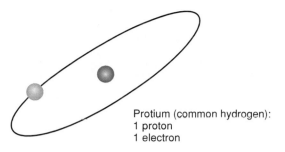

Protium (common hydrogen):
1 proton
1 electron

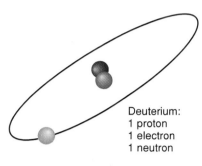

Deuterium:
1 proton
1 electron
1 neutron

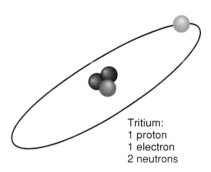

Tritium:
1 proton
1 electron
2 neutrons

FIGURE 2.2 Isotopes of hydrogen: each has one proton in the nucleus, but there may be zero, one, or two neutrons. The isotope with two neutrons—tritium—is unstable and naturally radioactive.

(six protons plus eight neutrons). For most applications, we need be concerned only with the elements involved, not with specific isotopes. Chemically, all isotopes of one element behave alike. The human body cannot, for instance, distinguish between sugar containing carbon-12 and sugar containing carbon-13.

> The principal practical difference between isotopes is in the stability of the different nuclei. Certain combinations of protons and neutrons are inherently unstable, and in time, those unstable nuclei decay, or break down, spontaneously. This is the nature of **radioactivity.** The existence of naturally occurring radioactive isotopes makes accurate dating of many materials possible, as will be seen in chapter 8. Differences in the nuclear properties of two uranium isotopes are relevant to our nuclear-power options.

IONS AND BONDING

In an electrically neutral atom, the number of protons equals the number of electrons. The negative charge of one electron just equals the positive charge of one proton. Most atoms, however, can gain or lose electrons. When this happens, the atom acquires a positive or negative electrical charge and is termed an **ion.** If it loses electrons, it becomes a positively charged **cation,** as the number of protons exceeds the number of electrons. If it gains electrons, the resulting ion has a negative electrical charge and is termed an **anion.**

Causes of Ion Formation

The tendency to gain or lose electrons is related to the number of electrons present in the neutral atom. Electrons are organized into shells, or orbitals, each of which can accommodate up to a specific maximum number of electrons (figure 2.3; note that in figure 2.3B and C the orbitals making up the outer shells are not shown individually, for simplicity). Electrons fill these shells in order, beginning with the shell closest to the nucleus, which can hold up to two electrons. For most atoms, the last shell is only partially filled. Added chemical stability is associated with having all shells exactly filled. Therefore, atoms whose outermost electron shells are only partially filled tend to gain, lose, or share electrons with other atoms to achieve evenly filled shells. The arrangement of elements in the periodic table (see box 2.1 on page 18) reflects the electron-shell structures of the elements.

Atoms that have only a few "leftover" electrons in their outermost shell tend to lose them. A neutral sodium atom, for example, has one electron in its outermost shell, so the simplest way to achieve a configuration consisting of all filled shells is to lose that one electron, forming a sodium cation with a +1 charge (Na^+). A neutral chlorine atom, by contrast, has an outermost shell containing seven electrons, though that shell can hold eight. Chlorine, then, tends to gain one electron, to form the chloride anion (Cl^-). Carbon has four electrons in its outermost shell, which can hold eight electrons. It also has six protons in the nucleus. Gain or loss of four electrons creates, proportionately, a large imbalance between protons and electrons. However, carbon can also share electrons with neighboring atoms to achieve filled shells for all. Indeed, this is what occurs in diamond, a mineral consisting of pure carbon in which each carbon atom shares electrons with adjacent atoms (figure 2.4).

Bonding and Compounds

Opposite electrical charges attract. Cations and anions may become chemically bonded together by this attraction, forming an **ionic bond.** Sodium and chloride ions in table salt (sodium chloride) display ionic bonding. The sharing of electrons between atoms creates another kind of bond, a **covalent bond.** The carbon atoms in diamond display covalent bonding. Molecules of gaseous elements that form diatomic (two-atom)

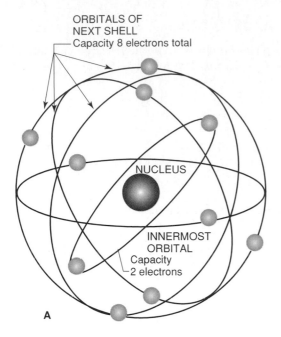

ORBITALS OF
NEXT SHELL
Capacity 8 electrons total

NUCLEUS

INNERMOST
ORBITAL
Capacity
2 electrons

A

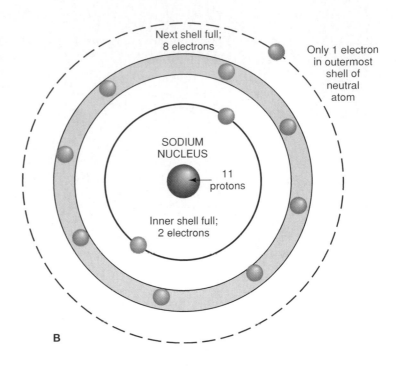

Next shell full;
8 electrons

Only 1 electron
in outermost
shell of
neutral
atom

SODIUM
NUCLEUS

11
protons

Inner shell full;
2 electrons

B

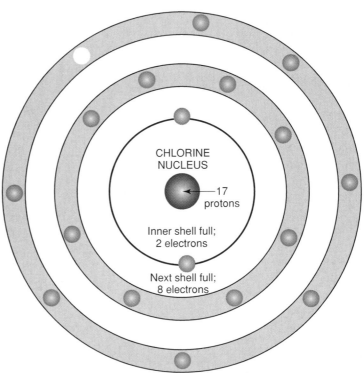

CHLORINE
NUCLEUS

17
protons

Inner shell full;
2 electrons

Next shell full;
8 electrons

C

Outer shell (capacity 8)
lacks 1 electron in
neutral atom.

FIGURE 2.3   Electronic structure and ion formation.
(A) The electron-shell structure of atoms.
(B) Sodium can best achieve a filled outermost shell
by losing one electron, making the Na$^+$ ion. (C) A
neutral chlorine atom lacks one electron in its
outermost shell, so it forms an ion (Cl$^-$) by gaining
one electron.

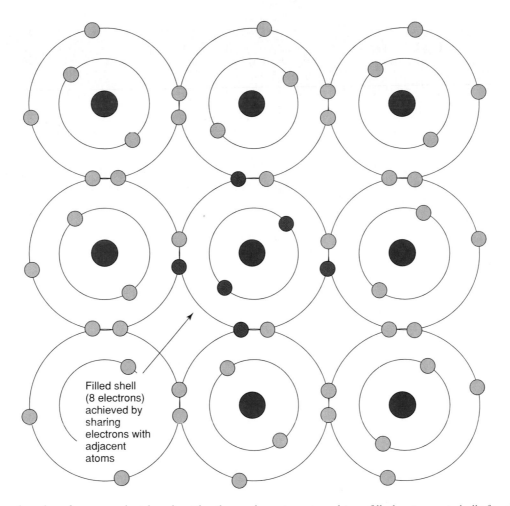

FIGURE 2.4 In diamond, carbon forms covalent bonds with other carbon atoms to achieve filled outermost shells for all.

Filled shell (8 electrons) achieved by sharing electrons with adjacent atoms

molecules, such as nitrogen (N₂) and oxygen (O₂), also show covalent bonding. Ionic and covalent bonding are the two most common kinds of bonding in minerals. Actually, few, if any, chemical bonds in nature are either wholly ionic or wholly covalent, but many are predominantly one or the other kind, and it is convenient to think of them that way.

Other, less common kinds of bonding are important in some materials. One of these is *metallic bonding,* in which the outer shell electrons are not firmly associated with particular atoms but instead tend to travel freely from atom to atom. As the name suggests, metallic bonding is common in metals, and the resultant mobility of the electrons contributes to the ability of metals to conduct electricity well.

A few elements have neutral atoms with all electron shells exactly filled already. They have, therefore, no need to gain, lose, or share electrons to achieve chemical stability. These elements are described as chemically **inert,** for they do not generally bond with other elements. They exist on earth as gases, with individual atoms floating freely and independently. Helium and neon are among the inert gases.

When atoms or ions of different elements bond together, they form a **compound.** A compound is a chemical combination of two or more elements, in particular proportions, that has a distinct set of physical properties, usually very different from those of the individual elements in it. Hydrogen and oxygen are colorless gases at the earth's surface, but combined in a compound with two atoms of hydrogen per oxygen atom, H₂O, they make water, an essential prerequisite to life on earth that (depending on temperature and pressure) can exist as solid, liquid, or vapor at the earth's surface. Sodium is a silver-colored metal; chlorine is a greenish gas that is poisonous in large doses. When equal numbers of sodium and chlorine atoms combine to form table salt (halite)—sodium chloride (NaCl)—the resulting compound consists of colorless crystals and has properties that do not resemble those of either of the constituent elements.

Whatever the nature of the bonding in a solid, the solid is, overall, electrically neutral, the positive charges of the cations just balanced by the negative charges of the anions. Any ions present are firmly bonded in place, not free to drift through the solid. Liquids, too, are electrically neutral overall, but within a liquid, compounds can break up into individual free ions that can move independently. When table salt goes into solution, it breaks up into sodium and chloride ions.

Box 2.1

The Periodic Table

Some idea of the probable chemical behavior of elements can be gained from a knowledge of the **periodic table** (figure 1). The Russian scientist Dmitri Mendeleyev first observed that certain groups of elements showed similar chemical properties, which seemed to be related in a regular way to their atomic masses. At the time (1869) that Mendeleyev published the first periodic table, in which elements were arranged so as to reflect these similarities of behavior, not all the elements had even been discovered, so there were some gaps. Subsequent additions of elements identified later confirmed the basic concept, and, in fact, some of the missing elements were found more easily because their properties could to some extent be anticipated from their expected position in the periodic table.

We now can relate the periodicity of chemical behavior to elements' electronic structures, which, in turn, are a function of their atomic numbers. For example, those elements in the first column of figure 1, known as the alkali metals, have one electron in the outermost shell of the neutral atom. Thus, they all tend to form cations of +1 charge by losing that odd electron. Outermost electron shells become increasingly full from left to right across a row. The next-to-last column, the halogens, are those elements lacking only one electron in the outermost shell, and they thus tend to gain one electron to form anions of charge −1. In the right-hand column are the inert gases, whose neutral atoms contain all fully filled electron shells. Those filled shells are the cause of these gases' lack of chemical reactivity: they do not readily gain or lose electrons to form ions, nor do they share electrons to make covalent bonds.

| 12 ← Atomic number |
| Mg ← Chemical symbol |
| 24.31 ← Atomic weight |

* Elements heavier than uranium synthesized experimentally

1a																	0
1 H 1.008	IIa											IIIa	IVa	Va	VIa	VIIa	2 He 4.00
3 Li 6.94	4 Be 9.01	IIIb	IVb	Vb	VIb	VIIb	\~ VIII \~		Ib	IIb		5 B 10.81	6 C 12.01	7 N 14.00	8 O 15.99	9 F 18.99	10 Ne 20.18
11 Na 22.99	12 Mg 24.31											13 Al 26.98	14 Si 28.09	15 P 30.97	16 S 32.06	17 Cl 35.45	18 Ar 39.95

19 K 39.10	20 Ca 40.08	21 Sc 44.96	22 Ti 47.90	23 V 50.94	24 Cr 51.99	25 Mn 54.94	26 Fe 55.85	27 Co 58.93	28 Ni 58.71	29 Cu 63.54	30 Zn 65.37	31 Ga 69.72	32 Ge 72.59	33 As 74.92	34 Se 78.96	35 Br 79.91	36 Kr 83.80
37 Rb 85.47	38 Sr 87.62	39 Y 88.91	40 Zr 91.22	41 Nb 92.91	42 Mo 95.94	43 Tc (99)	44 Ru 101.97	45 Rh 102.91	46 Pd 106.4	47 Ag 107.87	48 Cd 112.40	49 In 114.82	50 Sn 118.69	51 Sb 121.75	52 Te 127.60	53 I 126.90	54 Xe 131.30
55 Cs 132.91	56 Ba 137.34	57–71 see below	72 Hf 178.49	73 Ta 180.95	74 W 183.85	75 Re 186.2	76 Os 190.2	77 Ir 192.2	78 Pt 195.09	79 Au 196.97	80 Hg 200.59	81 Tl 204.37	82 Pb 207.19	83 Bi 208.98	84 Po (210)	85 At (210)	86 Rn (222)
87 Fr (223)	88 Ra (226)	89–103 see below	104* Rf (261)	105* Ha (260)	106* 263												

57 La 138.91	58 Ce 140.12	59 Pr 140.91	60 Nd 144.24	61 Pm (147)	62 Sm 150.35	63 Eu 151.96	64 Gd 157.25	65 Tb 158.92	66 Dy 162.50	67 Ho 164.93	68 Er 167.26	69 Tm 168.93	70 Yb 173.04	71 Lu 174.97
89 Ac (227)	90 Th 232.04	91 Pa (231)	92 U 238.03	93* Np (237)	94* Pu (242)	95* Am (243)	96* Cm (247)	97* Bk (247)	98* Cf (251)	99* Es (254)	100* Fm (253)	101* Md (256)	102* No (254)	103* Lw (257)

FIGURE 1 The periodic table.

MINERALS—GENERAL

A **mineral,** by definition, is a naturally occurring, inorganic, solid element or compound with a definite composition and a regular internal crystal structure. *Naturally occurring,* as distinguished from synthetic, means that minerals do not include the thousands of chemical substances invented by humans (silicon carbide or Teflon, for instance). *Inorganic* means not produced solely by living organisms, by biological processes. (Pearls are not minerals, for example.) That minerals must be *solid* means that the ice of a glacier is a mineral, but liquid water is not.

Chemically, minerals either consist of one element—for example, diamonds, which are pure carbon—or are compounds of two or more elements. Some mineral compositions are very complex, consisting of ten elements or more. The presence of certain elements in certain proportions is one of the defining characteristics of a mineral. Each mineral has a definite chemical composition or at least a compositional range within which it falls. The reason for the qualifying phrase "or a compositional range" is the phenomenon of **solid solution,** observed in many minerals, in which two or more ions of similar size and charge may substitute more or less interchangeably for each other in the same mineral structure. For example, magnesium (Mg^{2+}) and iron (Fe^{2+}) ions are nearly the same size (figure 2.5), so any mineral that contains one tends to contain some of the other, with the proportions varying from sample to sample. Although they are less similar in size, sodium (Na^+) and potassium (K^+) can exhibit similar behavior, as can many other sets of elements. The effect of this solid solution on the mineral's chemical formula can be seen in some of the silicates, described later in the chapter.

Finally, minerals are crystalline, at least on the microscopic scale. **Crystalline** materials are solids in which the atoms are arranged in regular, repeating patterns. Crystals may or may not show planar faces reflecting their internal symmetry, but most solid elements and compounds are in fact crystalline, and their crystal structures can be recognized and studied using X rays and other techniques.

Crystals

Crystal structures form because they are the most stable arrangement of atoms in a solid. Consider, for example, a solid containing various cations and anions. Opposite electrical charges attract, but like charges repel, and the repulsion is minimized when positive and negative charges are symmetrically distributed through the solid. A crystal structure can be thought of as a single structural and compositional unit, repeated over and over in three dimensions. A basic unit of the sodium chloride crystal is shown in figure 2.6A. The more complex the mineral, the larger and more complex the repeating unit (see, for ex-

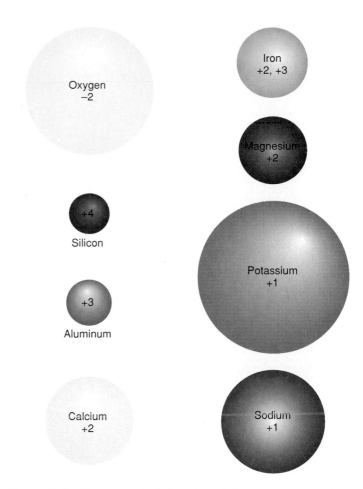

FIGURE 2.5 Charges and relative sizes of the most common ions in the earth's crust.

ample, figure 2.7A). When minerals exhibit their symmetric crystal forms, those forms are determined by, and are evidence of, the regular internal structure on the atomic level (figures 2.6 and 2.7).

In order for crystals of a given mineral to form, the requisite elements must be present in appropriate proportions, and the atoms must have enough time to arrange themselves into the proper regular pattern. A simple analogy illustrates the importance of time. Imagine a roomful of people milling about, who are suddenly told to arrange themselves in rows of six. If they are given ten seconds to do this, some will succeed in forming small, organized groups, but many will still be out of position when the time is up. If they are given half a minute, results will be more orderly; given several minutes, the entire group can probably arrange itself properly. Likewise with crystals growing in a melt, such as volcanic lava: if the liquid cools and solidifies very quickly, there may not be time for regular crystals to grow. The resulting solid, in which atoms are randomly arranged, is a noncrystalline **glass.** With slower cooling, orderly crystals have more time to grow.

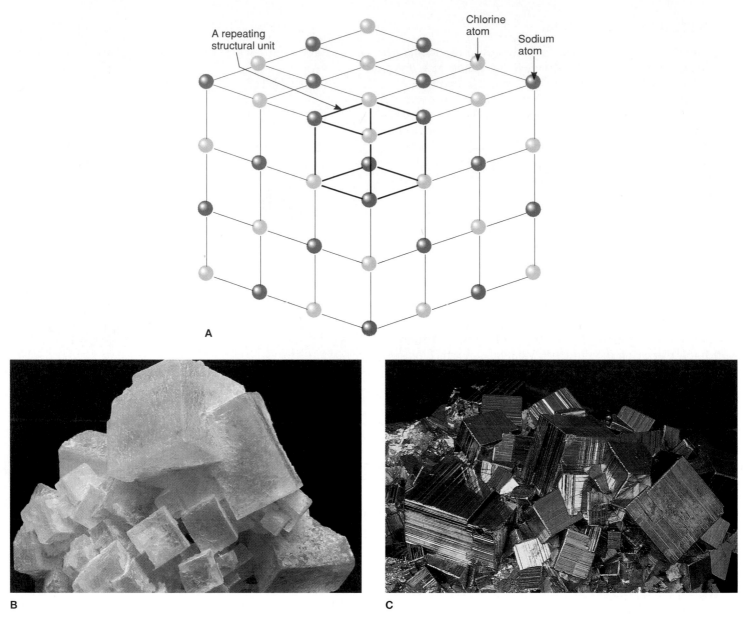

FIGURE 2.6 A crystal can be thought of as a structural unit repeated in three dimensions. (A) The basic structural unit of a crystal of halite (sodium chloride, NaCl). (B) The crystal form of halite reflects its cube-shaped building block. (C) Pyrite, like halite, has a cubic internal structure. Sometimes, it forms obvious cubic crystals, as here; sometimes not.
(B) and (C) photographs © Times Mirror Higher Education Group, Inc./Doug Sherman, photographer.

Some solids satisfy all the conditions of the definition of a mineral except that they lack a regular crystal structure. Such solids are termed **mineraloids.** Opal is an example of a mineraloid. Also, some true minerals can be produced artificially (synthetic diamonds, for instance). Others can be produced by organic as well as inorganic means. For example, seashells are obviously produced by organisms, but the materials that make up seashells are nevertheless minerals, for they are formed inorganically in nature also.

True Identifying Characteristics of Minerals

The two fundamental characteristics of a mineral that together distinguish it from all other minerals are its chemical composition and its crystal structure. That is, each mineral is defined by a unique combination of composition and crystal structure. No two minerals are identical in both respects, though they may be alike in one. For example, diamond and graphite (the "lead" in a lead pencil) are chemically the same: both are made up of pure carbon. Their physical

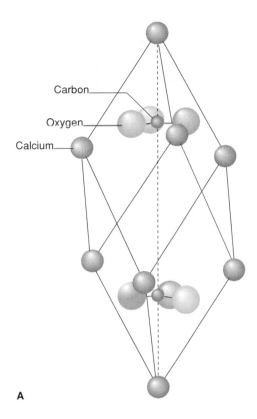

A

B

FIGURE 2.7 The more complex structural unit of calcite (calcium carbonate, $CaCO_3$), shown in (A), is mimicked in the "dogtooth" calcite crystals (B).
© *Times Mirror Higher Education Group, Inc. /Bob Coyle, photographer.*

In the diagram, labels point to: Carbon, Oxygen, Calcium.

---

properties, however, are vastly different because of differences in their internal crystalline structures. In a diamond, each carbon atom is firmly bonded to every adjacent carbon atom in three dimensions. In graphite, the carbon atoms are bonded strongly in two dimensions, into sheets, but the sheets are only weakly held together in the third dimension. Diamond is clear, transparent, colorless, and very hard, and a jeweler can cut it into brilliant, precious gemstones. Graphite is black, opaque, and soft, and its sheets of carbon atoms tend to slide apart as the weak bonds between them are broken; this makes it a useful lubricant, as in motor oil.

D iamond and graphite are examples of the phenomenon of *polymorphism* in minerals. **Polymorphs** are minerals having the same composition but distinctly different crystal structures. There may be more than two polymorphs of a single composition. Silica (SiO_2), which most often occurs as the mineral quartz, has at least half a dozen known polymorphs, though most are geologically rare, formed only under unusual conditions. Under a given set of physical and chemical conditions, one polymorph of a set will be the stable one; under different conditions, other atomic arrangements may be more stable. The very compact crystal structure of diamond is stable under high-pressure conditions; at low pressure, graphite forms instead. As pressures increase with depth in the earth, low-pressure minerals give way to denser, higher-pressure forms (see chapter 10).

A mineral's composition and crystal structure usually can be determined only by using sophisticated laboratory equipment. When a mineral has large crystals with well-developed shapes, a trained mineralogist may be able to infer some characteristics of the mineral's internal atomic arrangement, but most minerals do not show large, symmetric crystal forms by which they can be recognized with the naked eye. Certainly, no one can look at a lump of some mineral and know its chemical composition without first recognizing what mineral it is. Thus, when scientific instruments are not at hand, mineral identification must be based on a variety of other physical properties that in some way reflect the mineral's composition and structure. These other properties are often what make the mineral commercially valuable. However, they are rarely unique to one mineral and often are deceptive. A few examples of such diagnostic properties are outlined in the next section.

Other Physical Properties of Minerals

The most immediately obvious characteristic of most mineral samples is *color*. Is color a good way to identify minerals? While some minerals always appear the same color, many vary from specimen to specimen. Variation in color is usually due to the presence of small amounts of chemical impurities that have nothing to do with the mineral's basic, characteristic composition. Such color variation is especially common when the pure mineral is light-colored or colorless. The mineral quartz, for instance, is colorless in its pure form. However, quartz also occurs in other colors, among them rose pink, golden yellow, smoky brown, purple, and commonly, milky white. Clearly, quartz cannot always be recognized by its color (or lack of it).

U nusual coloring may make a particular mineral specimen especially valuable. Many colored quartz samples are used as semiprecious gems. The purple variety is amethyst. Other valued colors include citrine (a golden quartz named from the Latin *citrus*), rose quartz, and smoky quartz.

Another example of this phenomenon of color variation is the mineral corundum, a simple compound of aluminum and oxygen. In pure form, it too is colorless and also quite hard, which makes it a good abrasive. It is often used for the grit on sandpaper. Yet, a little color from trace impurities can transform this common, utilitarian material into highly prized gems: blue-tinted corundum is also known as sapphire, and red corundum is ruby. Even when the color shown by a mineral sample *is* the true color of the pure mineral, it is probably not unique. With thousands of minerals known to occur on earth, there are usually many of any one particular color.

Perhaps surprisingly, **streak,** the color of the powdered mineral, is more consistent from sample to sample than the color of the bulk mineral. Streak is conventionally tested by scraping the sample across a piece of unglazed tile or porcelain and then examining the color of the mark made. There are two principal limitations to the usefulness of streak as an identification tool. First, like color, it is not unique. Many different minerals leave a brown streak, or white, or gray. Second, minerals that are harder than ceramic simply gouge the test surface and are not powdered onto it. Streak is a particularly useful identification aid for many metallic ores.

Hardness, the ability to resist scratching, is another physical property that can be of help in mineral identification. It is related to the strength of the bonds holding the mineral together. Classically, hardness is measured using the Mohs hardness scale (table 2.1), on which ten minerals are arranged in order of hardness, from talc (the softest, assigned a hardness of 1) to diamond (10). Unknown minerals are assigned a hardness on the basis of which minerals they can scratch and which minerals scratch them. A mineral that can both scratch and be scratched by quartz would be assigned a hardness of 7. One that scratches gypsum but is scratched by calcite has a hardness of 2.5 (the hardness of an average fingernail). Because diamond is the hardest natural substance known, and corundum is the second-hardest mineral, these might be readily identified by their hardnesses. However, among the thousands of "softer" (more easily scratched) minerals, there are many of any particular hardness, just as there are many of any particular color.

TABLE 2.1 *The Mohs Hardness Scale*

Mineral	Assigned Hardness	
Talc	1	
Gypsum	2	
Calcite	3	← Fingernail 2.5
Fluorite	4	← Copper penny 3.5
Apatite	5	
Orthoclase	6	← Glass 5 to 6
Quartz	7	
Topaz	8	
Corundum	9	
Diamond	10	

For comparison, the approximate hardnesses of some common objects, measured on the same scale, are shown.

T he exceptional hardness of diamond makes it not only a particularly durable gemstone but also a valuable industrial commodity. Large, clear, gem-quality diamonds are relatively rare. The smaller and less attractive industrial-grade diamonds are useful in abrasives for grinding and polishing and in saw blades and drill bits, where the material being shaped is itself hard and the tools used must be correspondingly durable.

Crystal form, the shape of well-developed crystals of a mineral, is a very useful clue to mineral identification because it is related to the (invisible) internal geometric arrangement of atoms in the crystal structure. In some cases, the relationship is readily apparent (recall figures 2.6B and C, and 2.7): the ions in sodium chloride are arranged in a cubic structure, and table salt (halite) tends to crystallize in cubes; so-called dogtooth crystals of the mineral calcite are similar in shape to the rhombohedral atomic units from which they are built. However, variations in the conditions under which crystals grow cause different planes of atoms to be more or less prominent in the overall crystal form. A given mineral with a single internal structure can grow in several quite distinctive crystal shapes. Each one necessarily reflects in some way the internal symmetry of the crystal structure, but all crystals of a specific mineral do not necessarily look alike in form. Conversely, different minerals may show the same crystal form, so crystal form is often not uniquely diagnostic of minerals. (For example, many common minerals form cubes as halite does.) Moreover, as noted earlier, well-developed crystal forms are relatively uncommon.

Another property controlled by internal crystal structure is **cleavage,** the tendency of minerals to break preferentially in certain directions, corresponding to zones of weakness in the bonding of the crystal structure. Cleavage can be investigated simply by striking a mineral sample with a hammer. It

FIGURE 2.8 The cleavage directions of halite parallel the cube faces of its crystals.
© *Times Mirror Higher Education Group, Inc. /Bob Coyle, photographer.*

FIGURE 2.10 Calcite cleaves into rhombohedral fragments.
© *Times Mirror Higher Education Group, Inc., Robert Rutford, and James Zumberge/James Carter, photographer.*

FIGURE 2.9 Fluorite forms cubic crystals also but cleaves into octahedra.
© *Times Mirror Higher Education Group, Inc. /Doug Sherman, photographer.*

FIGURE 2.11 Example of conchoidal fracture in volcanic glass.
© *Times Mirror Higher Education Group, Inc. /Bob Coyle, photographer.*

can be tested even on irregular chunks of mineral that show no well-developed crystals. In some cases, the prominent cleavage directions are the same as the prominent faces of well-formed crystals. Halite (sodium chloride) cleaves well in three mutually perpendicular directions that correspond to the faces of its cubic crystals (figure 2.8). The cleavage fragments of other minerals do not necessarily resemble their well-grown crystals (figures 2.9 and 2.10). Invariably, however, cleavage directions are directly related to the arrangement of atoms/ions in the crystal structure. The structure tends to break along planes where bonding is weakest.

Many minerals have no well-defined planes along which the crystals break cleanly. These minerals exhibit more irregular **fracture,** rather than cleavage. A distinctive type of fracture is the *conchoidal,* or shell-like, fracture shown by volcanic glass, quartz, and a few other minerals (figure 2.11).

The surface sheen, or **luster,** of minerals is another diagnostic property. Very few terms are used to describe this quality, and they are, for the most part, self-explanatory. Examples include *metallic* (bright and shiny, like metal; recall figure 2.6C), *pearly* (softly iridescent, like pearl), *vitreous* (glassy; figure 2.11), and *earthy* (figure 2.12).

FIGURE 2.12 Example of earthy luster in clay.
© *Times Mirror Higher Education Group, Inc./Bob Coyle, photographer.*

A mineral's **specific gravity** is related to its density. Specific gravity is the ratio of the mass of a given volume of the mineral to the mass of an equal volume of water. By definition, a mineral having the same density as liquid water has a specific gravity of 1. The higher the specific gravity, the denser the mineral. For comparison, sodium chloride has a specific gravity of about 2.16; garnet, 3.1 to 4.2, depending on its exact composition; metallic copper, 8.9; and gold, 19.3. Specific gravity is related to the atomic weights of the elements in the mineral (gold, for example, is one of the heaviest elements) and to the closeness of atomic packing in the crystal structure. The precise determination of specific gravity requires specialized equipment, and many of the most common minerals fall in a narrow range of specific gravity, from about 2.5 to 4. For the nonspecialist, specific gravity is a qualitative tool, most useful for identifying those minerals of unusually high or low specific gravity (density).

Many other properties are useful mainly in identifying those few minerals that possess them. The iron mineral magnetite, as its name suggests, is strongly magnetic, which is very rare among minerals. Sodium chloride, known mineralogically as halite, naturally tastes like table salt; the similar salt potassium chloride (sometimes used in salt substitutes for those on low-sodium diets) also tastes salty but is more bitter. A few minerals effervesce (fizz) when acid is dripped onto them. These are carbonate minerals (discussed later in the chapter), which react with the acid and release carbon dioxide gas in the process. Uranium-bearing minerals are radioactive, which can be detected with a Geiger counter. Some minerals *fluoresce* (glow) in ultraviolet light; calcite and fluorite are examples. Graphite may feel slippery.

Without exact knowledge of a mineral sample's composition and crystal structure, one must rely on other, less distinctive properties, such as color, hardness, and cleavage. While one physical property rarely identifies a mineral

uniquely, considering a set of properties collectively may narrow the possibilities down to one or a very few minerals. That is, while there may be many green minerals, very few are green *and* show conchoidal fracture *and* exhibit a glassy luster *and* have a hardness of 6.5 to 7, and so on. The more properties that are considered at once, the fewer minerals fit them all, and with experience, it is often possible to recognize particular minerals quickly on the basis of several of these physical properties.

U nique or not, the physical properties arising from minerals' compositions and crystal structures are often what give minerals value from a human perspective—the slickness of talc (main ingredient of talcum powder), the malleability and conductivity of copper, the durability of diamond, and the rich colors of tourmaline gemstones are all examples. Some minerals have several useful properties: table salt (halite), a necessary nutrient, also imparts a taste we find pleasant, dissolves readily to flavor liquids but is soft enough not to damage our teeth if we munch on crystals of it sprinkled on solid food, and will serve as a food preservative in high concentrations, among other helpful qualities.

TYPES OF MINERALS

As was indicated earlier, minerals can be grouped or subdivided on the basis of their two fundamental characteristics—composition and crystal structure. In this section, some of the basic mineral groups are introduced briefly, along with common examples of each. A comprehensive survey of minerals is beyond the scope of this book, and the interested reader is referred to standard mineralogy texts for more information.

Silicates

The two most common elements in the earth's crust, by far, are silicon and oxygen, which together make up over 70% of the crust (table 2.2). It comes as no surprise, therefore, that by

TABLE 2.2 *Average Composition of the Crust*

Element	Weight Percent
Oxygen	46.6
Silicon	27.7
Aluminum	8.1
Iron	5.0
Calcium	3.6
Sodium	2.8
Potassium	2.6
Magnesium	2.1
Total	98.5

Source: Data from B. Mason and C. B. Moore, 1982, *Principles of Geochemistry,* 4th ed., John Wiley & Sons, Inc.

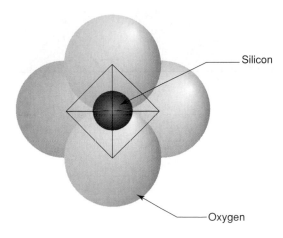

Silicon

Oxygen

FIGURE 2.13 The basic silica tetrahedron, building block of all silicate minerals.

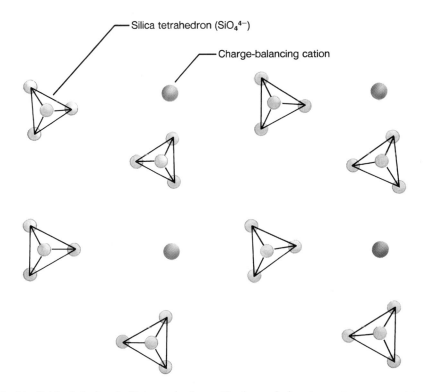

Silica tetrahedron (SiO_4^{4-})

Charge-balancing cation

FIGURE 2.14 Silicate formed of individual, isolated silica tetrahedra, with charge-balancing cations symmetrically distributed among the tetrahedra.

far the largest group of minerals is the **silicate** group, all of which are compounds containing silicon and oxygen, and most of which contain other elements as well. Because this group of minerals is so large, it is subdivided on the basis of crystal structure, by the ways in which the silicon and oxygen atoms are linked together.

The basic structural unit of all the silicates is the *silica tetrahedron,* a compact building block formed by a silicon cation (Si^{4+}) closely surrounded by four oxygen anions (O^{2-}) (figure 2.13). The different silicate mineral groups are distinguished by the way in which these tetrahedra are assembled. A number of specific minerals of various compositions fall into each structural group. Only some of the more common representatives of each structural type are noted here.

The simplest arrangement of tetrahedra is as isolated units (figure 2.14). It can be seen arithmetically that an individual SiO_4 tetrahedron has a net negative charge of -4: $[1(+4) + 4(-2)] = -4$. Thus, in such a silicate, additional cations must balance the electrical charge in order for the solid to be electrically neutral overall and to bond the negatively charged tetrahedra together. For each tetrahedron, cations totaling four positive charge units are required. In the most common single-tetrahedron silicate, **olivine,** the charges are balanced by iron (Fe^{2+}) and/or magnesium (Mg^{2+}), which requires a total of two +2 ions per formula unit: $(Fe,Mg)_2SiO_4$. Another fairly common single-tetrahedron silicate is *garnet,* for which a variety of chemical compositions are possible; a generalized formula, illustrating the effects of solid solution, is $(Ca,Mg,Fe)_3(Al,Fe)_2Si_3O_{12}$.

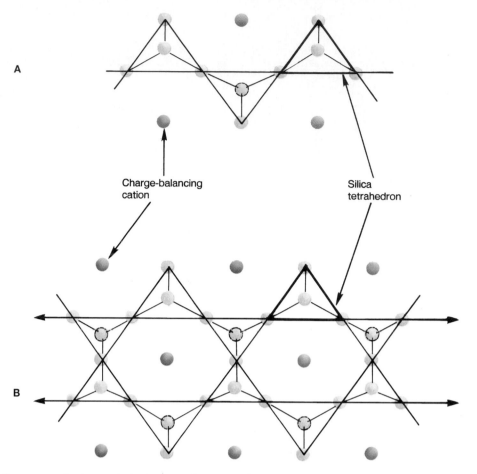

FIGURE 2.15 In chain silicates, tetrahedra are linked in one dimension by shared oxygen atoms. (A) Single chains, characteristic of pyroxenes. (B) Double chains, characteristic of amphiboles.

Other silicates are known as **chain silicates** because their silica tetrahedra are arranged in chains formed by the sharing of oxygen atoms between adjacent tetrahedra in one dimension. The most common arrangements are single chains (figure 2.15A) and double chains (figure 2.15B). When tetrahedra share oxygen atoms, the net effect is less total negative charge from the silica tetrahedra to be balanced by additional cations.

Consider one tetrahedron in a single-chain silicate (figure 2.16). Two of its oxygen atoms are associated only with that tetrahedron, while two are shared with other tetrahedra in the chain, so that, effectively, only half of each of the shared atoms is associated with that tetrahedron. The net ratio of Si to O is then 1: [2(1) + 2(½)], or 1:3 rather than 1:4. The net negative charge on an SiO_3 unit is $[1(+4) + 3(-2)] = -2$, so that cations totaling only two positive charges suffice to balance each tetrahedron.

One large group of single-chain silicates is collectively known as **pyroxenes.** This group comprises many specific minerals, differing in the cations involved in the charge-balancing (and in details of crystal structure). A compositionally simple example in which charges are again balanced by

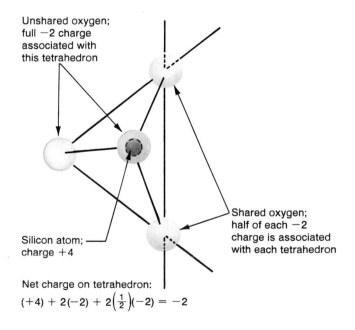

Unshared oxygen; full -2 charge associated with this tetrahedron

Shared oxygen; half of each -2 charge is associated with each tetrahedron

Silicon atom; charge $+4$

Net charge on tetrahedron:
$$(+4) + 2(-2) + 2\left(\tfrac{1}{2}\right)(-2) = -2$$

FIGURE 2.16 The net negative charge on a single tetrahedron in a single chain is -2.

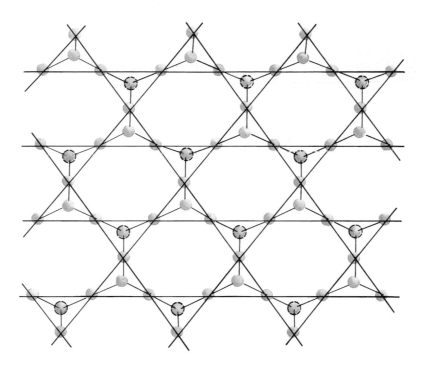

FIGURE 2.17 The two-dimensional linking of tetrahedra in a sheet silicate.

FIGURE 2.18 Excellent cleavage in micas results from splitting of the crystals between weakly bonded sheets.
© *Times Mirror Higher Education Group, Inc. /Bob Coyle, photographer.*

The chain silicates represent sharing of oxygen between adjacent tetrahedra in one dimension. In the **sheet silicates,** tetrahedra are linked by shared oxygen atoms in two dimensions (figure 2.17). The net charge on each tetrahedral unit is −1, so still fewer charge-balancing cations are required. In fact, most sheet silicates contain some water (as hydroxyl ions) also, substituting for oxygen as in the amphiboles, which further reduces the need for charge-balancing cations. The cations commonly fit between stacked sheets, bonding the sheets together. Typically, however, there are not enough cations to hold the sheets very tightly, which is reflected in the macroscopic properties of many sheet silicates. The **micas** are a compositionally diverse group of sheet silicates that have in common excellent cleavage parallel to the weakly bonded sheets of tetrahedra (figure 2.18). Common examples are the light colored mica *muscovite,* often found in granite, and the dark mica *biotite,* rich in iron and magnesium. **Clay minerals** are also sheet silicates, and their often slippery feeling can likewise be attributed to the sliding apart of such sheets of atoms.

When tetrahedra are firmly linked in all three dimensions by shared oxygen atoms, the result is a **framework silicate** (figure 2.19). If the framework consists entirely of silicon-oxygen tetrahedra, the net charge on each tetrahedron is $[1(+4) + 4(\frac{1}{2})(-2)] = 0$, and there is no need for further charge-balancing cations in the structure. This is the case with the mineral *quartz,* compositionally the simplest silicate (SiO_2). The uniform, three-dimensional linkage of tetrahedra in quartz means no planes of particularly weak bonds, and as a result, quartz has no cleavage.

There are other framework silicates besides quartz and its polymorphs. This is possible because, in silicates of virtually all

iron and magnesium would have the formula $(Fe,Mg)SiO_3$. The common pyroxene *augite* is compositionally more complex, containing calcium and aluminum in addition to iron and magnesium.

The chemical formulas for the double-chain **amphiboles** are somewhat more complex, in part because the amphiboles contain some water in their crystal structures in the form of hydroxyl ions (OH^-) substituting for oxygen in the tetrahedra. The structural geometry and nature of the charge-balancing are correspondingly more complex. The general formula for a very common amphibole, *hornblende,* illustrates this well: $(Ca,Na)_3(Mg,Fe,Al,Ti)_5(Si,Al)_8O_{22}(OH,F)_2$.

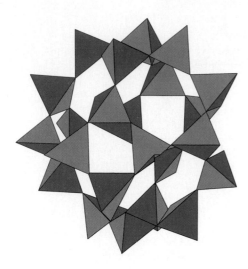

Figure 2.19 When silica tetrahedra are linked in three dimensions, the result is a framework silicate.

structural types, some aluminum ions (Al^{3+}) can substitute in tetrahedra for silicon (Si^{4+}). The resulting charge imbalance must be compensated by additional cations. The most abundant minerals in the crust, the **feldspars,** are such framework silicates in which the charge-balancing cations are sodium (Na^+), potassium (K^+), or calcium (Ca^{2+}).

The feldspars illustrate some of the limitations of solid solution in minerals. Sodium and calcium are very close in size, though slightly different in charge, so they can substitute effectively for each other in a series of sodium-calcium feldspars known as the *plagioclase* feldspars, which have a generalized formula of $(Na,Ca)(Al,Si)_2Si_2O_8$. Potassium is so much larger that it does not substitute in plagioclase to any great extent. Conversely, some sodium may substitute in potassium feldspar ($KAlSi_3O_8$) because the charges of sodium and potassium ions are the same (both +1), although the ions differ somewhat in size. But calcium differs significantly in both size and charge from potassium, and thus these two elements are not readily interchangeable in the potassium-feldspar structure.

Other structural variants occur in the silicates, but less commonly—for example, structures in which tetrahedra are linked into rings or those in which pairs of tetrahedra are joined. In some contexts, it may be useful to refer collectively to a set of silicate minerals of different structures that share a compositional characteristic. For example, the **ferromagnesian** silicates, as the name suggests, comprise those silicates that contain significant amounts of iron (Fe) and/or magnesium (Mg). (They may or may not contain other cations also.) The group includes olivines, pyroxenes, amphiboles, some micas, and other silicates as well. The ferromagnesians are commonly characterized by dark colors, most often black, brown, or green. The **hydrous** (water-bearing) silicates, which include micas, clays, amphiboles, and others, are another pos-

sible compositional group. A silicate mineral may thus fall within one or more particular compositional groupings in addition to its structural category.

Nonsilicates

Just as the silicates, by definition, all contain silicon plus oxygen as part of their chemical compositions, each nonsilicate mineral group is defined by some chemical constituent or characteristic that all members of the group have in common. Most often, the common component is the same negatively charged ion, or group of atoms (somewhat analogous to a silica tetrahedron; a structural unit). Discussion of some of the nonsilicate mineral groups, with examples of more common or familiar members of each, follows. (See also table 2.3.)

The **carbonates** all contain carbon and oxygen combined in the proportions of one atom of carbon to three atoms of oxygen (CO_3). The carbonate minerals all dissolve relatively easily, particularly in acids. The oceans also contain a great deal of dissolved carbonate, from which many marine organisms build their shells. Geologically, the most important and most abundant carbonate mineral is *calcite,* which is calcium carbonate ($CaCO_3$). Another common carbonate mineral is *dolomite,* which contains both calcium and magnesium in approximately equal proportions ($CaMg(CO_3)_2$). Carbonates may contain many other elements—iron, manganese, or lead, for example.

The **sulfates** all contain sulfur and oxygen in a ratio of 1:4 (SO_4). A calcium sulfate—*gypsum*—is the most important, for it is both relatively abundant and commercially useful as a construction material. Sulfates of many other elements, including barium, lead, and strontium, are also found.

When sulfur is present without oxygen, the resultant minerals are called **sulfides.** A common and well-known sulfide mineral is the iron sulfide *pyrite* (FeS_2). Pyrite has also been called "fool's gold" because its metallic golden color often deceived early prospectors into thinking they had struck it rich.

H ad the prospectors who struck pyrite tested its specific gravity, of course, they would have known better: the specific gravity of pyrite is about 5, while that of gold is 19.3! Pyrite and gold also have different crystal structures, they streak differently (gold has a golden streak, while that of pyrite is dark greenish black), and pyrite scratches glass, while the softer gold does not.

Pyrite is not a commercially useful source of iron because there are richer ores of this metal. Nonetheless, the sulfide group comprises many economically important metallic ore minerals. An example that may be familiar is the lead sulfide mineral *galena* (PbS), which often forms in silver-colored

TABLE 2.3 *Some Nonsilicate Mineral Groups*

Compositional Class	Compositional Characteristic	Examples
Carbonates	Metal(s) plus carbonate (1 carbon + 3 oxygen atoms, CO_3)	Calcite (calcium carbonate, $CaCO_3$); dolomite (calcium-magnesium carbonate, $CaMg(CO_3)_2$)
Sulfates	Metal(s) plus sulfate (1 sulfur + 4 oxygen atoms, SO_4)	Gypsum (calcium sulfate, with water, $CaSO_4 \cdot 2\ H_2O$); barite (barium sulfate, $BaSO_4$)
Sulfides	Metal(s) plus sulfur, without oxygen	Pyrite (iron sulfide, FeS_2); galena (lead sulfide, PbS); cinnabar (mercury sulfide, HgS)
Oxides	Metal(s) plus oxygen	Magnetite (iron oxide, Fe_3O_4); hematite (iron oxide, Fe_2O_3); corundum (aluminum oxide, Al_2O_3); spinel (magnesium-aluminum oxide, $MgAl_2O_4$)
Hydroxides	Metal(s) plus hydroxyl (1 oxygen + 1 hydrogen atom, OH)	Gibbsite (aluminum hydroxide, $Al(OH)_3$, found in aluminum ore); brucite (magnesium hydroxide, $Mg(OH)_2$, an ore of magnesium)
Halides	Metal(s) plus halogen element (fluorine, chlorine, bromine, or iodine)	Halite (sodium chloride, $NaCl$); fluorite (calcium fluoride, CaF_2)
Phosphates	Metal(s) plus phosphate group (1 phosphorus + 4 oxygen atoms, PO_4)	Apatite ($Ca_5(PO_4)_3F$)
Native elements	Mineral consists of a single element	Gold (Au), silver (Ag), copper (Cu), sulfur (S), graphite (C)

Other groups exist, and some complex minerals contain components of several groups (carbonate and hydroxyl groups, for example).

cubes. The rich lead ore deposits near Galena, Illinois, Galena, Kansas, and Galena Park, Texas, gave these towns their names. Sulfides of copper, zinc, and numerous other metals may also form valuable ore deposits.

The **halides** are the minerals composed of metal(s) plus one or more of the halogen elements (the gaseous elements in the next-to-last column of the periodic table, box 2.1: fluorine, chlorine, iodine, and bromine). The most common halide is *halite* (NaCl), the most abundant salt dissolved in the oceans.

Minerals that contain just one or more metals combined with oxygen and that lack the other elements necessary for them to be classified as silicates, sulfates, carbonates, and so on are the **oxides.** Iron combines with oxygen in different proportions to form more than one oxide mineral. *Magnetite* (Fe_3O_4) is one of these. Another iron oxide, *hematite* (Fe_2O_3), is sometimes silvery black but often has a red color and gives a reddish tint to many rocks and soils. (All colors of hematite have the same characteristic reddish brown streak, however, illustrating again the usefulness of streak as a diagnostic property in mineral identification.) Many other oxide minerals are also known, including *corundum,* the aluminum oxide mineral (Al_2O_3) mentioned earlier.

The **native elements,** as shown in table 2.3, are even simpler chemically than the other nonsilicates. Native elements are minerals that consist of a single chemical element. The minerals are usually named for the corresponding elements. Not all elements can be found, even rarely, as native elements. However, some of the most highly prized materials, such as gold, silver, and platinum, often occur as native elements. Diamond and graphite are both examples of native carbon; here, two mineral names are needed to distinguish these two very different crystalline forms of the same element. Sulfur may occur as a native element, either with or without associated sulfide minerals. Some of the richest copper ores contain native copper. Other metals that can occur as native elements include tin, iron, and antimony.

MINERALS, ROCKS, AND THE ROCK CYCLE

A **rock** is a solid, cohesive aggregate of one or more minerals (or mineral materials, including volcanic glass and small fragments of other rocks). This means that a rock consists of many individual mineral grains—not necessarily all of the same mineral—or of mineral grains plus glass, all firmly held together in a solid mass. Because the many mineral grains of beach sand fall apart when handled, sand is not a rock, although, in time, sand grains may be cemented together to make a rock. The physical properties of rocks are important in determining their suitability for particular applications, such as for construction materials or for the base of a building foundation. Each rock also contains within it a record of at least a part of its history, in the nature of its minerals, in the sizes and shapes of the mineral grains, and in the way the mineral grains fit together.

In chapter 1, it was noted that the earth is a constantly changing body. Mountains come and go; seas advance and retreat over the faces of continents; surface processes and processes occurring deep in the crust or mantle are constantly

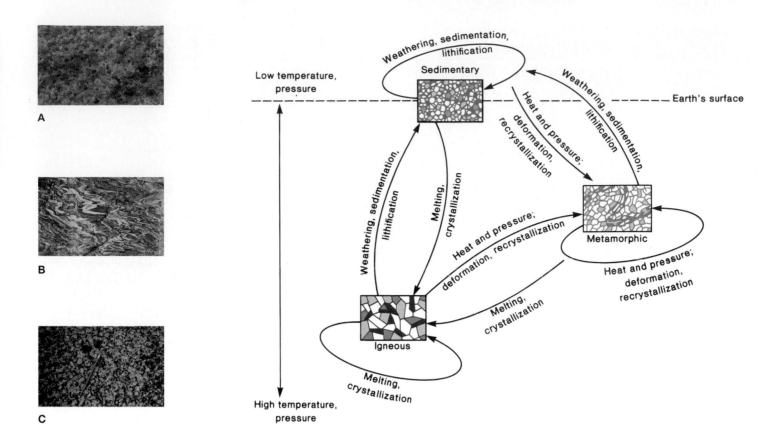

A

B

C

FIGURE 2.20 The rock cycle (schematic). Geologic processes act continuously to produce new rocks from old.
(A) and (B) © Times Mirror Higher Education Group, Inc. /Doug Sherman, photographer.

altering the planet. One aspect of this continual change is that rocks, too, are always subject to change. We do not have a single sample of rock that has remained unchanged since the earth formed; the oldest known crustal rocks are nearly half a billion years younger than the earth. Many rocks have been changed many times.

Rocks are divided into three classes on the basis of the way in which they form. *Igneous* rocks (chapter 3) crystallize from hot silicate melts. They are common in the deep crust and make up the whole of the mantle. *Sedimentary* rocks (chapter 6) form at the very low temperatures at and near the earth's surface, as a result of the weathering of preexisting rocks. *Metamorphic* rocks (chapter 7) are those that have been changed (deformed, recrystallized) by the application of moderate heat and/or pressure in the crust or by changes in pressure, temperature, or chemistry. A rock of one type may be transformed into a rock of another type (or a different rock of the same type) by various geologic processes. Pressures and temperatures increase with depth in the earth. A sedimentary rock, for example, may be deeply buried, squeezed, heated, and changed into a metamorphic rock. It may even be heated so much that it melts; when the melt cools and crystallizes, the result is a new igneous rock. If that igneous rock is uplifted on a continent, it can be weathered and redeposited as sediment, to make new sedimentary rock; and so on.

The idea that rocks are continually subject to change through time is the essence of the **rock cycle** (figure 2.20). The chapters that follow treat individual rock types in some detail. Keep in mind, however, that the labels "igneous," "sedimentary," and "metamorphic" necessarily describe a rock only as it has emerged from its last cycle of formation or change. Many earlier stages through which that material has passed have been wholly or partially obliterated by later changes—hence, some of the difficulty in piecing together the 4½-billion-year history of the earth from rocks collected today.

SUMMARY

Chemical elements consist of atoms, which are in turn composed of protons, neutrons, and electrons. Isotopes are atoms of one element (having, therefore, the same number of protons) with different numbers of neutrons; chemically, isotopes of one element are indistinguishable. Ions are atoms that have gained or lost electrons and thus acquired a positive or negative charge. Atoms of the same or different elements may bond together. The most common kinds of bonding in minerals are ionic (resulting from the attraction between oppositely charged ions) and covalent (involving sharing of electrons between atoms). When atoms of two or more different elements bond together, they form a compound.

A mineral is a naturally occurring, inorganic, solid element or compound with a definite composition (or range in composition) and a regular internal crystal structure. When appropriate instruments for determining composition and crystal structure are unavailable, minerals can be identified from a set of physical properties including color, crystal form, cleavage or fracture, hardness, luster, specific gravity, and others. Minerals are broadly divided into silicates and nonsilicates. The silicates are subdivided into structural types (for example, chain silicates, sheet silicates, framework silicates) on the basis of how the silica tetrahedra are linked in each mineral. Silicates may alternatively be grouped by compositional characteristics. The nonsilicates are subdivided into several groups, each of which has some compositional characteristic in common. Examples include the carbonates (each containing the CO_3 group), the sulfates (SO_4), and the sulfides (S).

Rocks are cohesive mineral aggregates. Certain of their physical properties are consequences of the ways in which their constituent mineral grains are assembled. All rocks are part of the rock cycle, through which old rocks are continually being transformed into new ones. A consequence of this is that no rocks have been preserved throughout the earth's history, and many early stages in the development of one rock may have been erased by subsequent events. Still, all minerals and rocks provide information about the conditions under which they have formed.

TERMS TO REMEMBER

amphibole 27
anion 15
atom 14
atomic mass number 14
atomic number 14
carbonates 28
cation 15
chain silicates 26
clay minerals 27
cleavage 22
compound 17
covalent bond 15
crystal form 22
crystalline 19
electron 14
element 14
feldspars 20

ferromagnesian 28
fracture 23
framework silicates 27
glass 19
halides 29
hardness 22
hydrous 28
inert 17
ion 15
ionic bond 15
isotope 14
luster 23
mica 27
mineral 19
mineraloid 20
native element 29
neutron 14

nucleus 14
olivine 25
oxide 29
periodic table 18
polymorph 21
proton 14
pyroxene 26
radioactivity 15
rock 29
rock cycle 30
sheet silicates 27
silicates 25
solid solution 19
specific gravity 24
streak 22
sulfate 28
sulfide 28

QUESTIONS FOR REVIEW

1. What are *elements?* What are *isotopes?*
2. Compare and contrast ionic and covalent bonding.
3. How is a *mineral* defined? What are the two key identifying characteristics of a mineral?
4. What is the phenomenon of solid solution, and how does it affect the definition of a mineral?
5. Explain the limitations of color as a tool in mineral identification. Cite and explain any three other physical characteristics that might aid in mineral identification.
6. What is the basic structural unit of all silicate minerals? Describe the basic structural arrangements of chain silicates, sheet silicates, and framework silicates.
7. Give the compositional characteristic common to each of these nonsilicate mineral groups: carbonates, sulfides, oxides.
8. What is a *rock?*
9. Describe the basic concept of the rock cycle.

FOR FURTHER THOUGHT

1. More ancient rocks generally are less widely found and more difficult to interpret than are younger rocks. Explain why, in the context of the rock cycle.

2. Choose one of the following minerals or mineral groups, and investigate its range of physical properties and its uses: quartz, calcite, garnet, clay.

SUGGESTIONS FOR FURTHER READING

Berry, L. G., Mason, B., and Dietrich, R. V. 1983. *Mineralogy*. 2d ed. San Francisco: W. H. Freeman.

Dana, J. D. 1985. *Manual of mineralogy*. 20th ed. Revised by Hurlbut, C. S., Jr., and Klein, C. New York: John Wiley and Sons.

Deer, W. A., Howie, R. A., and Zussman, J. 1978. *Rock-forming minerals*. 2d ed. New York: Halstead.

Dietrich, R. V., and Skinner, B. J. 1979. *Rocks and rock minerals*. New York: John Wiley and Sons.

Ernst, W. G. 1969. *Earth materials*. Englewood Cliffs, N.J.: Prentice-Hall.

Hurlbut, C. S., Jr. 1968. *Minerals and man*. New York: Random House.

Kirklady, J. F. 1972. *Minerals and rocks in color*. New York: Hippocrene Books.

O'Donoghue, M. 1976. *VNR color dictionary of minerals and gemstones*. New York: Van Nostrand Reinhold.

Philips, W. J., and Philips, N. 1980. *An introduction to mineralogy for geologists*. New York: John Wiley and Sons.

Sinkankas, J. 1964. *Mineralogy for amateurs*. Princeton, N.J.: Van Nostrand.

Watson, J. 1979. *Rocks and minerals*. 2d ed. Boston: Allen and Unwin.

Zoltai, T., and Stout, J. H. 1984. *Mineralogy: Concepts and principles*. Minneapolis, Minn.: Burgess.

Much of the continent crust consists of granite and related igneous rocks. East Mojave National Scenic area near Barstow, CA. © Doug Sherman/Geofile

The term **igneous** comes from the Latin *ignis,* meaning "fire." The name is given to rocks formed at very high temperatures, crystallized from a molten silicate material known as **magma.** This chapter first examines magmas and aspects of their formation and crystallization, basic characteristics and classification of igneous rocks, and some of the structures magmas form within the crust. In the next chapter, we go on to investigate the surface manifestations of magmatic activity, lava flows and volcanoes.

ORIGIN OF MAGMAS

High temperatures are required to melt rock. Within the earth, temperatures increase naturally with depth, since the earth still retains much of the heat associated with its formation. The rate of temperature increase with depth, called the *geothermal gradient,* averages about 30°C per kilometer (90°F per mile) of depth in the crust. Magmas originate at depths where temperatures are high enough to melt rock, usually in the upper mantle at depths between about 50 and 250 kilometers (30 and 150 miles). However, the required temperatures (and depths) vary considerably, depending on several other factors considered in the sections that follow.

Effects of Pressure

As the temperature of a solid is increased, its individual atoms vibrate ever more vigorously, until their energy is sufficient to break the bonds holding them in place in the solid structure. They then flow freely in a disordered liquid.

Most substances are more dense in their crystalline solid form than in the liquid state, and thus an increase of pressure favors the more compact, solid arrangement of atoms. At high pressure, then, correspondingly higher temperatures are required to impart enough energy to the atoms to cause melting. This explains why most of the earth's interior is not molten. Temperatures are indeed high enough through most of the earth to melt the rocks if they were at atmospheric (surface) pressures, but the weight of overlying rock puts enough pressure on the rocks below that most remain solid even at temperatures of thousands of degrees. If that pressure is lessened—for example, by the opening of fractures in rocks above—the solid may begin to melt.

The effects of pressure can be illustrated by several commonplace examples involving water. The first is the difference in boiling temperature with altitude. Vaporization of water involves a transition from a liquid to a less dense, disorderly gas. At sea level, water boils at 100°C (212°F). On a high mountain, where the air is thinner and pressure reduced, water boils at temperatures several degrees lower; less heat is required for the molecules to break away from the liquid at lower pressure. A second example is a pressure cooker: when the pressure on the water is raised, the water can be heated well above its (sea-level) boiling temperature without boiling. Ice skating provides a third example of the effects of pressure. Ice is unusual in that, at least close to its melting point,

FIGURE 3.1 The magma from which this basalt crystallized trapped bubbles of gas as it solidified.
© *Times Mirror Higher Education Group, Inc. /Bob Coyle, photographer.*

it is *less* dense than water (note that ice cubes and icebergs float). An increase in pressure favors, as usual, the denser form—in this case, water. This is what makes ice skating possible: concentrating the weight of the body on a narrow skate blade puts tremendous pressure on the ice below, and a little of it melts, providing a watery layer on which the skate can glide smoothly.

Effects of Volatiles

Natural magmas contain dissolved water and various gases (oxygen, carbon dioxide, hydrogen sulfide, and others). Sometimes, this is obvious from the resultant rock, which may preserve bubbles formed by gases in the magma (figure 3.1). The general effect of dissolved volatiles is to lower the melting temperatures of silicate minerals. This is illustrated in figure 3.2. Increasing the water pressure drives more water into the melt and lowers the melting temperature of sodic plagioclase (albite), most dramatically for small additions of water; the effect decreases for high water pressure. Exactly how much melting temperatures are lowered in the presence of volatiles varies with the nature of the volatiles and the minerals being melted, but the general principle holds in any case: more volatiles, lower melting temperatures.

Effects of Other Solids Present

When two or more different minerals are in contact, mixed together, the presence of each lowers the melting temperature of the other, to a point. This is why salt can be used to melt ice on a sidewalk in winter. The presence of the salt lowers the melting temperature of the ice so that the ice melts below the freezing point of pure water (0°C, or 32°F). (Presumably,

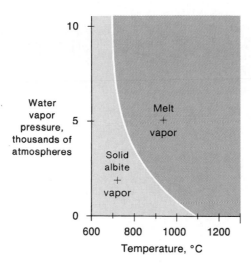

FIGURE 3.2 The effect of increasing water pressure on the melting temperature of sodic plagioclase (albite).

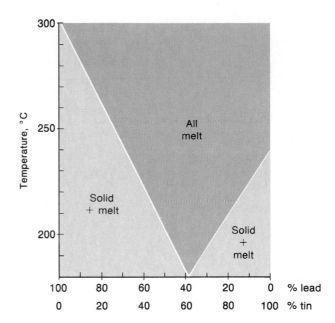

FIGURE 3.3 Mixing of tin and lead in solder lowers melting temperatures.

the ice lowers the melting temperature of the salt, too, but salt has such an extremely high melting point to begin with that it remains solid unless, of course, it simply dissolves in the melted ice.) Another common example is provided by lead-tin solders of various compositions (figure 3.3). Nearly pure lead and tin melt at close to 300°C and 240°C, respectively. Adding some tin to the lead lowers the latter's melting point, and vice versa. As molten solder cools, the temperature at which it begins to solidify varies with composition. For a mix of 63% tin, 37% lead, crystallization does not occur at temperatures above 182°C. Similar effects are observed with minerals; see box 3.1.

Since the vast majority of rocks contain many minerals, it is safe to expect melting of the rocks at temperatures somewhat below those at which the pure minerals would melt, but the relationships are likely to be complex, and it may not be possible to predict each rock's exact melting temperature. Furthermore, melting in natural rocks typically occurs over a range of temperatures, with different minerals melting at different temperatures. The transition from a wholly solid to a completely molten material may thus span several hundred degrees, and, in fact, complete melting may not be necessary for the magma to mobilize and flow. Many magmas are a "mush" of crystals suspended in silicate liquid, somewhat like the slush of ice and water that forms in snowy areas at near-freezing temperatures. (Keep this in mind when examining diagrams in this chapter: a volume of magma may be shown as mostly or wholly liquid for simplicity, but in fact, the melt may be thick with suspended crystals.)

Effects of Solid Solution

Many solid solutions can be regarded as a mix of two or more endmember, or limiting, compositions. For example, olivine ((Fe,Mg)$_2$SiO$_4$) can be thought of as intermediate in composition between the compounds Fe$_2$SiO$_4$ and Mg$_2$SiO$_4$. These endmembers do not melt at the same temperatures: the pure iron olivine melts at 1205°C, while the pure magnesium

olivine melts at 1890°C. An intermediate composition melts between these extremes, at temperatures determined by its exact composition (proportion of iron to magnesium), along with pressure and other factors already noted.

Other Sources of Heat

As mentioned earlier, the most basic source of elevated temperatures is depth. Locally, other factors may also be important in producing heat. Radioactive decay of naturally radioactive elements produces heat, and in rocks containing unusually high concentrations of radioactive elements, this supplementary heat can be significant. Friction produces heat, too. Rubbing two rocks together does not cause sufficient heating to initiate melting, but rubbing two continents together (figuratively speaking) may add enough extra heat to warm rocks at depth to start them melting. Even the movement of existing magma can be a factor: if a hot mass of magma rises up in the crust, it heats surrounding rocks (originally cooler because they are at shallower depths) and under some conditions may begin to melt them.

CRYSTALLIZATION OF MAGMAS

Once an appropriate combination of factors has produced a quantity of melt, the melt tends to flow away from where it was produced. Usually, it flows upward, into rocks at lower temperatures and pressures. As it moves upward, it cools. The cooling, in turn, leads to crystallization, as the atoms slow down and eventually settle into the orderly arrays of crystals. The details of the rock thus produced vary with the composition of the melt and the conditions under which it crystallized. However, some generalizations about the resulting mineralogy and texture are possible.

Box 3.1

Melting Relations in Graphical Form

Figures 3.2 and 3.3 are examples of simple *phase diagrams,* graphs that show a range of conditions under which particular minerals, or minerals plus fluids, are stable. These figures plot temperature against composition. Other sets of variables, such as pressure against temperature, also can be used. Let us consider the interpretation of a simple mineralogic phase diagram in greater detail (figure 1).

There are two kinds of lines on the graph in figure 1. A line separating conditions under which the system is entirely molten (liquid) from conditions under which some solid and some liquid are present is called a *liquidus.* Analogously, a line separating conditions under which only solids are present from conditions under which solids and melt are present is a *solidus.* (The terms can be kept straight by remembering that liquid is present on both sides of a liquidus, solids on both sides of a solidus.) In figure 1, the liquidus lines are the curved lines; the solidus is the horizontal line at about 1060°C.

Above 1713°C, a system with a composition that is a mix of quartz and albite is entirely molten. Below 1060°C, it is entirely solid, consisting of crystals of albite and quartz. At intermediate temperatures, whether solids will be present, and which ones, depends on the bulk composition of the system. If, for example, the composition is as represented by point *A* (about 70% quartz, 30% albite), the system is molten above the liquidus temperature at that composition (about 1470°C). A melt of that composition begins to crystallize quartz at that liquidus temperature. With more cooling, more

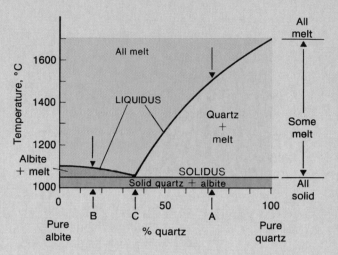

FIGURE 1   The albite-quartz phase diagram.

quartz crystallizes, until the solidus temperature (1060°C) is reached. The system then crystallizes quartz and albite, staying at the solidus temperature until all the melt has crystallized.

A melt of composition *B* (about 83% albite, 17% quartz) cools to about 1075°C before any crystals form. These first crystals are albite rather than quartz, and albite continues to crystallize as the melt cools to the solidus temperature. Then quartz and albite crystallize as before until all the melt has solidified.

Melting relationships are the reverse; that is, a mix of quartz and albite crystals is entirely solid until the solidus temperature is reached. With continued heating, quartz and albite both melt, in the proportions indicated by point *C* (about 35% quartz, 65% albite), until one or the other mineral is used up. If the

starting composition corresponds to *A* (or to any point on the quartz-rich side of *C*), the albite is used up first during melting. With still more heating, the remaining quartz gradually melts until, at the liquidus temperature for that composition (1470°C for composition *A*), all the solid has melted. With an albite-rich composition such as *B,* melting of quartz plus albite at the solidus proceeds until the quartz is used up. Then, with more heating, the remaining albite melts until, at about 1075°C (for composition *B*), the whole sample has melted.

This is actually quite a simple example compared to a natural rock system. Such diagrams, and the corresponding melting and crystallization relationships, become more complex when more than two minerals are involved or if one or more of the minerals is a solid solution.

Sequence of Crystallization

As already observed, a mix of minerals melts over a range of temperatures. Likewise, a magma crystallizes over a range of temperatures, or, in other words, over some period of time during cooling, as different minerals begin to crystallize at different temperatures/times. Moreover, because most magmas originate in the upper mantle, they are in a very general sense similar in composition initially, consisting predominantly of silica (SiO_2), with varying lesser proportions of aluminum, iron, magnesium, calcium, sodium, potassium, and additional minor elements. Magmas therefore tend to follow

a predictable crystallization sequence in terms of the principal minerals appearing. However, the proportions of these minerals in the ultimate rock vary, and igneous rocks, overall, span a broad range of compositions.

The basic crystallization sequence was recognized more than half a century ago by geologist Norman L. Bowen, who combined careful laboratory studies of compositionally simple silicate systems with wide-ranging field observations of the more complex natural rocks. The result, known as **Bowen's Reaction Series,** is illustrated in figure 3.4. Those minerals that tend to crystallize at

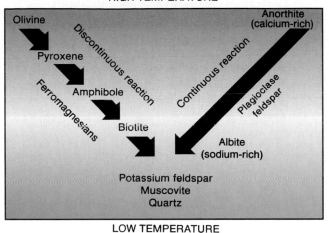

Olivine

Discontinuous reaction

Anorthite
(calcium-rich)

Pyroxene

Continuous reaction

Ferromagnesians

Amphibole

Plagioclase
feldspar

Biotite

Albite
(sodium-rich)

Potassium feldspar
Muscovite
Quartz

LOW TEMPERATURE

FIGURE 3.4 Bowen's Reaction Series.

higher temperatures (olivine, pyroxene, and calcium-rich plagioclase) are shown nearer the top of the series; the later, lower-temperature minerals are near the bottom. In general, the earliest minerals to crystallize are relatively low in silica, so that the residual magma remaining after their crystallization is more enriched in silica relative to its starting composition. The high-temperature portion of the sequence is also subdivided into a ferromagnesian branch and a branch involving plagioclase feldspar.

The plagioclase branch is a **continuous reaction series.** This refers to interaction between crystals already formed and the remaining melt. Recall from chapter 2 that plagioclase is a solid solution between a calcium-rich endmember (anorthite, $CaAl_2Si_2O_8$) and a sodium-rich endmember (albite, $NaAlSi_3O_8$). The more calcic compositions are the higher-temperature end of the series: pure anorthite melts at about 1550°C, pure albite at 1100°C. Sodium and calcium are freely interchangeable in the plagioclase crystal structure. The first plagioclase to crystallize from a magma, at high temperatures, is a rather calcic one; as the magma cools, the crystals react continuously with the melt, with more and more sodium entering into the plagioclase, but no changes in basic crystal structure. (Note that sodic plagioclase also contains a higher proportion of silicon, so the later plagioclases are more silica-rich, too.) If cooling is too rapid for complete reaction between crystals and melt during cooling, the resultant crystals show concentric compositional zones, with calcium-rich cores grading outward through increasingly sodium-rich compositions.

The ferromagnesian side of the crystallization sequence is a **discontinuous reaction series.** Olivine is the first of the ferromagnesians to crystallize. After a period of crystallization, the olivine is so out of balance chemically with the increasingly silica-rich residual melt that the olivine and melt react to form pyroxene. (The ratio of [iron plus magnesium] to silicon in olivine is 2:1, while in pyroxene it is about 1:1.) Assuming that there is sufficient silica available, all of the olivine is converted to pyroxene at that point. After an inter-

val of pyroxene crystallization, the pyroxene, again out of chemical balance with the remaining melt, reacts with it, and pyroxenes are converted to amphiboles, and so on. The last of the common ferromagnesians to crystallize is biotite mica. There is a progression in structural complexity, too, from the simplest silicate structure (olivine), through chains of tetrahedra, to a sheet structure. This discontinuous reaction series, then, is marked by several changes in mineralogy/crystal structure during cooling and crystallization.

At the end of the crystallization sequence, at the lowest temperatures, any remaining, silica-rich residual melt crystallizes to form potassium feldspar, muscovite mica, and quartz. Note that the hydrous silicates—amphiboles and micas—are relatively late in the sequence. At very high temperatures, hydrous minerals are unstable, and any water stays in the melt. Also, not every magma progresses through the whole sequence. A more **mafic** magma (rich in magnesium and iron, poorer in silicon) is completely crystallized before the lattermost stages of the sequence are reached; it does not start with sufficient silica to form quartz at the end. A very **silicic** (silica-rich, iron- and magnesium-poor) magma reaches the final stages, by which time reactions with the melt have eliminated early-formed olivine, pyroxene, and calcic plagioclase. Mafic magmas, then, produce rocks rich in the minerals near the top of the diagram in figure 3.4; silicic magmas produce rocks that are dominated by the minerals near the bottom and are poor in ferromagnesians. The latter rocks are typically rich in feldspar and quartz (silica) and are therefore also termed **felsic.**

Modifying Melt Composition

The previous discussion tacitly assumes that each crystallizing magma behaves as a *closed system,* neither gaining nor losing matter. This is often not the case in natural systems. Magma compositions may be modified after the melt is formed. The result is a product somewhat different from what would be expected on the basis of the original melt composition.

Fractional crystallization is the primary way in which melt composition can be changed. In this process, early-formed crystals are physically removed from the remaining magma and so are prevented from reacting with it. One way in which this can happen is if the early-formed crystals settle out of the crystallizing magma and are isolated from the melt by later crystals settling above them (figure 3.5). The remaining melt may even move away through zones of weakness in surrounding rocks, leaving the early crystals behind altogether. The result is that the average composition of the melt, minus its early, low-silica minerals, is shifted toward a more silica-rich, iron- and magnesium-poor composition. That melt can then progress further down the crystallization sequence than would have been possible if all crystals had remained in the melt as it cooled. A great variety of rock compositions can be produced from a single starting magma composition by varying the extent and timing of fractional crystallization during cooling.

Overall magma composition also can be changed if a magma **assimilates** the rock around it, incorporating pieces

Igneous Rocks and Processes   37

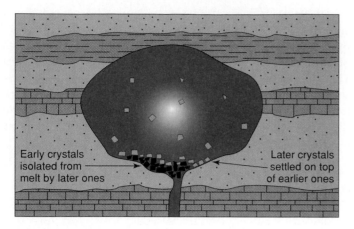

FIGURE 3.5 Crystal settling in a magma chamber as a mechanism of fractional crystallization. Early crystals are isolated from the residual melt by later ones.

of it, melting and mixing it in (figure 3.6). Given the relative temperatures at which mafic and silicic rocks crystallize, assimilation occurs most readily when the initial melt is a (hotter) mafic magma and the assimilated material is more silicic. The result is a modified magma of somewhat more silicic composition, which could move correspondingly further down the crystallization sequence.

Magma mixing, in which two melts combine to produce a hybrid melt intermediate in composition between them, is another way in which melt composition can be changed. Because unusual geologic conditions are needed to produce two distinctly different magmas in nearly the same place at the same time, however, magma mixing is rather rare.

Textures of Igneous Rocks

The most noticeable textural feature of most igneous rocks is *grain size,* the size of the individual mineral crystals. An important control on grain size, as indicated in chapter 2, is cooling rate. If a magma cools slowly, there is more time for atoms to move through the melt and attach themselves at suitable points on growing crystals. The resulting rock is more coarsely crystalline (figure 3.7). In a rapidly cooled magma, there is less time for crystal growth, so the rock is finer-grained (figure 3.8). In extreme cases, the result is a *glassy* rock, with no obvious crystals (recall figure 2.11). Some quickly cooled magmas also trap bubbles or pockets of gas, which are termed **vesicles;** the resulting texture is described as *vesicular* (recall figure 3.1).

Some igneous rocks have a two-stage cooling history, with an initial stage of slow cooling that allows some large crystals to form, followed by a stage of rapid cooling that leaves the rest of the rock **(groundmass)** finer grained. This might happen, for example, if a magma began to crystallize

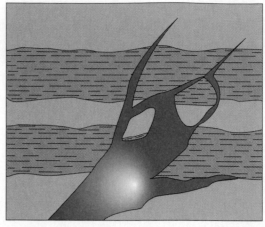

A

B

FIGURE 3.6 Assimilation by magma. (A) The piece of wallrock is partially melted by and mixed into the magma (schematic). (B) Assimilation was in progress here when the melt solidified.

FIGURE 3.7 A coarse-grained igneous rock (granite), formed from a slowly cooled magma. (This granite is also porphyritic, with feldspar phenocrysts.)

FIGURE 3.8 A fine-grained igneous rock (basalt), formed from a rapidly cooled magma.

slowly at depth and then was suddenly erupted from a volcano. The resulting rock is called a **porphyry;** the texture is described as *porphyritic* (figure 3.9; box 3.2). The coarse crystals embedded in the finer groundmass are termed **phenocrysts,** the prefix *pheno-* coming from the Greek for "to show." In other words, they are very obvious crystals.

Grain size can also be affected by melt composition. Silicic melts are more viscous—thicker, more resistant to flow—than mafic ones. In all silicate melts, silica tetrahedra exist in the melt even before actual crystallization. In a mafic melt, most of these tetrahedra float independently. In more silica-rich melts, the tetrahedra are more extensively linked into clumps and chains. As a result, atoms move less freely through the melt, and more time is required for atoms to move into position on growing crystals. Most volcanic glasses are silicic in composition for this reason (even when dark-colored). Such melts are so stiff and viscous that rapid cooling produces not small crystals, but virtually no crystals at all.

On the other hand, some late-stage residual melts left over after much of a magma has crystallized have accumulated high concentrations of dissolved volatiles, so they are quite

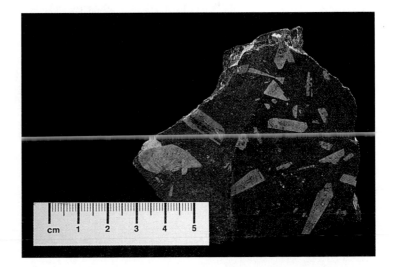

FIGURE 3.9 The phenocrysts in this porphyry formed early, during a period of slow cooling. Later, rapid cooling caused the remaining melt to crystallize in a mass of smaller crystals.
© *Times Mirror Higher Education Group, Inc. /Bob Coyle, photographer.*

BOX 3.2

Looking at Rocks Another Way

THIN SECTIONS

FIGURE 1 Thin section of andesite porphyry. Note how transparent it is.
© *Times Mirror Higher Education, Inc. /Bob Coyle, photographer.*

Volcanic rocks are so fine-grained that it is commonly impossible to distinguish individual crystals with the naked eye. With the majority of rocks of all types, even if individual mineral grains can be seen, it is hard to examine in detail the grain shapes, internal characteristics of individual crystals, and so forth. Geologists therefore rely on microscopic examination of **thin sections,** paper-thin slices of rock mounted on glass slides, to learn more than they can from looking at a chunk of rock unaided.

A thin section is typically ground down to a thickness of about 0.03 millimeters (0.0012 inches), at which thickness most minerals are transparent or translucent (figure 1). The special microscopes for examining thin sections use polarized light (light oriented so that all rays vibrate in parallel). The light rays are deflected or rotated in passing through crystals, and different minerals produce different deflections. If the thin section is sandwiched between two polarizers at right angles to each other, *interference colors* diagnostic of the different minerals

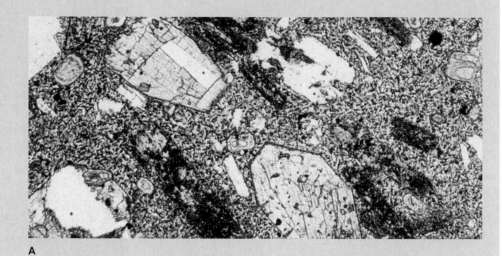

A

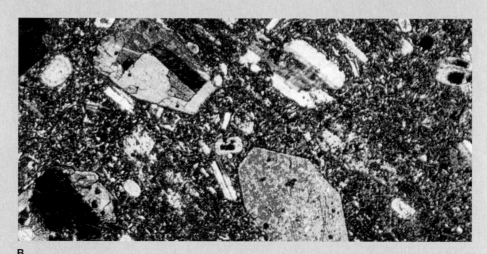

B

FIGURE 2 (A) Microscopic view of the thin section, in plane-polarized light. (B) The same thin section viewed between crossed polarizers.

are seen. (Compare figures 2A and 2B.) A suitably trained geologist is able to recognize plagioclase (gray-and-white striped) and pyroxene (colored) phenocrysts in this volcanic rock, as well as smaller crystals of both, plus some glass and magnetite (the latter opaque even in plain light), in the groundmass.

fluid, and atoms move easily through them. Even with moderate cooling rates, very large crystals can grow in a volatile-rich, fluid magma. The resultant extremely coarse-grained rock is termed a **pegmatite** (figure 3.10). Pegmatites may also concentrate unusual elements that don't fit into the structures of the common silicates, so they may contain rare and valuable minerals; see chapter on mineral and energy resources.

CLASSIFICATION OF IGNEOUS ROCKS

The most fundamental division of igneous rocks is made on the basis of depth of crystallization, as reflected in texture. **Plutonic** igneous rocks are those crystallized at some depth below the surface. They take their name from Pluto, the Greek god of the lower world. Their exposure at the surface

FIGURE 3.10 Pegmatite, a very coarse-grained igneous rock.

FIGURE 3.12 An ultramafic rock (*dunite*), made up of olivine (light green) and some pyroxene (black).
© *Times Mirror Higher Education Group, Inc./Bob Coyle, photographer.*

PLUTONIC

| Granite | Granodiorite | Diorite | Gabbro | Ultramafic |

Quartz

Potassium feldspar

Plagioclase

Ferromagnesians

Proportions of minerals

| Rhyolite | Andesite | Basalt |

VOLCANIC

FIGURE 3.11 A simplified classification of igneous rocks, based on mineralogy (related to chemical composition) and depth of crystallization (plutonic versus volcanic).

implies the erosion of the rocks that overlay them at the time of crystallization. Rocks are poor heat conductors, so magmas at depth are well insulated and cool slowly. Plutonic rocks, then, are generally recognized on the basis of their coarser grain sizes, with individual crystals readily visible to the naked eye. Such a texture is termed **phaneritic. Volcanic** rocks are those formed from magmas cooled at or near the surface. They are typically fine-grained or glassy, with individual crystals not easily seen with the naked eye; this is **aphanitic** texture. Porphyritic rocks with very fine-grained groundmass are also considered volcanic. Further subdivisions within the textural classes are made primarily on the basis of composition, which permits a specific rock name to be assigned. A summary of principal igneous rock types and their corresponding compositions is shown in figure 3.11 and explained in the sections that follow.

Plutonic Rocks

The coarse crystals of plutonic rocks make their preliminary identification relatively simple even without special equipment. They are classified on the basis of relative proportions of certain light and dark (ferromagnesian) minerals. A rock consisting entirely of ferromagnesians and dark, calcic plagioclase feldspar is a **gabbro.** In the extreme case where feldspar is virtually absent and the rock consists almost wholly of olivine and pyroxene, it is termed **ultramafic** (figure 3.12). A rock that is somewhat more silica-rich than gabbro contains some lighter-colored sodic plagioclase and potassium feldspar, with the mix of light and dark minerals giving it a salt-and-pepper appearance. This is **diorite** (figure 3.13). If the rock is sufficiently silica-rich that appreciable quartz is present, and the proportion of ferromagnesians correspondingly less, the rock is a **granite** (recall figure 3.7).

Although plutonic rocks are put in distinct categories, they actually span a continuum of chemical and mineralogical compositions. The boundaries between categories are somewhat arbitrary. One can indicate that a rock is of an intermediate composition by using a hybrid name for it: for example, a dioritic rock containing just a little quartz could be termed a *granodiorite* to indicate that its composition lies between the quartz-rich granite and quartz-free diorite classes.

FIGURE 3.13 Close-up view of diorite, an intermediate plutonic rock with salt-and-pepper coloring.
© *Times Mirror Higher Education Group, Inc. /Doug Sherman, photographer.*

Many rock types were named before the development of sophisticated chemical analytical methods and before there was general agreement on nomenclature. This has left some confusing contradictions among the rock names. The name *gabbro* comes from the locality Gabbro, Italy, but by modern classification, the plutonic rocks found there would now be called diorite, not gabbro!

Volcanic Rocks

Most plutonic rock types have aphanitic volcanic compositional equivalents, as can be seen in figure 3.11. Volcanic rocks are more difficult to identify definitively in handsample, for the individual crystals are, by definition, tiny. Color is the most frequently used means of identification in the absence of magnifying lenses or other equipment, but as with mineral identification, it is not an entirely reliable guide.

The most common volcanic rock is **basalt** (recall figures 3.1 and 3.8), the fine-grained equivalent of gabbro.

FIGURE 3.14 Rhyolite, a silicic volcanic rock chemically equivalent to granite. (Sample on left is porphyritic.)
© *Times Mirror Higher Education Group, Inc. /Bob Coyle, photographer.*

This dark-colored (usually black) rock makes up the sea floor, and basaltic lava is erupted by many volcanoes on the continents as well. Mafic magma cooling quickly at shallow depths in the crust may also form basalt. **Andesite** is the name given to volcanic rocks of intermediate composition. They are typically lighter in color than basalt, often green or gray (the porphyry in figure 3.9 is an andesite porphyry). The volcanic equivalent of granite is **rhyolite** (figure 3.14). Rhyolites may be light gray to pink or pinkish gray in color. Distinguishing among the light-colored andesites and rhyolites can be difficult unless the rhyolite is porphyritic and quartz is visible among the phenocrysts.

A few volcanic rock names lack compositional implications. An example is **obsidian** (volcanic glass): the term *obsidian* can be applied to glassy rock of any composition, although, in fact, most obsidians are rhyolitic in composition, for reasons already explained. If the lava is very gassy, the resultant rock may be so full of vesicles that it will float on water; such a rock is called *pumice.*

As with plutonic rocks, mixed names can denote rocks with compositions overlapping two compositional classes (for example, basaltic andesite). A textural and a compositional term also can be combined to provide a more complete description of the rock (for example, porphyritic andesite, vesicular basalt).

The reader should be warned that nongeologists who work with geologic materials have appropriated some of these terms but given them different meanings. It is common engineering practice, for instance, to call all phaneritic igneous rocks "granite" and all aphanitic ones "basalt," regardless of composition, because for many engineering problems, the texture is the critical factor in determining the rock properties. Engineers may even describe as "rock" soil that is so hard as to require blasting.

FIGURE 3.15 Pyroclastics: cinders from Sunset Crater, Arizona.

FIGURE 3.16 Rock formed from pyroclastic material: the Bandelier Tuff, New Mexico.

There are other volcanic rocks formed not by crystallization of lava, but from hot fragmented material. **Pyroclastic** is a general term, derived from the Greek *pyros* ("fire") and *klastos* ("broken"), which describes hot fragments forcibly ejected from a volcano. These fragments can vary considerably in size, from gritty volcanic **ash** and **cinders** (figure 3.15), which range up to golf-ball size, to the larger volcanic *blocks* that can be as large as a house. If volcanic ash is hot enough when it falls, the fragments may weld together to form a rock known as **tuff** (figure 3.16). A rock consisting of large angular fragments in a finer matrix of ash or cinders is a *volcanic breccia.*

INTRUSIVE ROCK STRUCTURES

Magma moving through the crust is intruding the surrounding rock, which is commonly termed the **country rock,** or **wallrock.** The rising of the magma may be aided by buoyancy if the magma is less dense than the country rock. If it is more dense, it has to be under pressure to force its way upward through the crust. If the magma erupts at the earth's surface, it has become *extrusive.* Extrusive (volcanic) rocks and structures (for example, volcanoes) are discussed in detail in chapter 4. Here, we briefly review the more common forms and features of *intrusions,* igneous rock masses formed by magma crystallizing below the earth's surface.

Factors Controlling the Geometry of Intrusions

Both the properties of the magma and the properties of the surrounding rocks play a role in determining the shape of the intrusion formed. Magmas vary in density, with the iron-rich mafic magmas being the more dense. Denser magmas are less buoyant with respect to the country rock, and their extra mass may even cause the country rocks to sag around the magma body. The viscosity of the magma influences how readily it flows through narrow cracks or other openings in the country rock. This can be demonstrated in the kitchen using a sieve: water pours rapidly through the sieve; honey or molasses, which are more viscous, seep through more slowly; and honey that has partially crystallized (analogous to the crystal-liquid mush of a partly solidified magma) may not pass through the sieve at all, instead remaining as a coherent mass within it. Similarly, very fluid magmas can move through quite narrow cracks in rocks, while thick, viscous magmas are more likely to remain in a compact mass. Sometimes, too, the magma is under unusual pressure from trapped gases within it, which allows it to intrude forcibly wallrocks that it might ordinarily be unable to penetrate.

The strength of the wallrock, and whether or not it is fractured, also influences the shape of intrusions. Fractures in rock are zones of weakness through which magmas can pass more easily. Zones of weakness may also exist at a contact between dissimilar rock types in the wallrock. The shape of an intrusive body may be controlled by the geometry of zones of weakness in the wallrock through which the magma flows preferentially.

Forms of Intrusive Bodies

Pluton is a general term for any body of plutonic rock—that is, any igneous rock mass crystallized below the earth's surface. The term has no particular geometric significance. Plutons are classified according to their shapes and their relationship to structures in the country rock. A pluton is said to be **concordant** if its contacts are approximately parallel to any structure (such as compositional layering or folds) in the country rock; it is said to be **discordant** if its contacts cut across the structure of the country rock. A few examples should clarify the distinction.

Cylindrical plutons, elongated in one dimension (usually the vertical), are **pipes** or **necks** (figure 3.17). Many are believed to be the remains of feeders or conduits that once carried magma up to a volcano above them; erosion has since

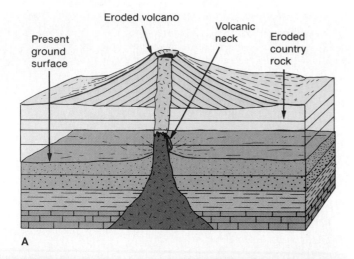

A

B

B

FIGURE 3.17 A volcanic neck or pipe—all that remains of a long-eroded volcano. (A) (schematic) Note cylindrical shape and discordant character. (B) Devil's Tower, Wyoming, is an example of a pipe, though no volcano has been proved to have existed above it.

removed the volcano and much of the country rock around the pipe. Pipes are typically discordant.

Tabular, relatively two-dimensional plutons commonly result from magma emplacement along planar cracks or zones of weakness. They are termed **dikes** if they are discordant, **sills** if they are concordant (figure 3.18).

Concordant plutons that are more nearly equidimensional are less common. Those that have flat floors, or bottoms, and that have caused doming or arching of the rocks above them, are termed **laccoliths.** Those that have floors

FIGURE 3.18 Tabular plutons. (A) The Ferrar Dolerite in Antarctica (dark layer) is a concordant sill emplaced between rock strata. (B) Multiple dikes (lighter-colored) in Black Canyon of the Gunnison River, Colorado.

(A) Photograph by W. B. Hamilton, USGS Photo Library, Denver, CO.

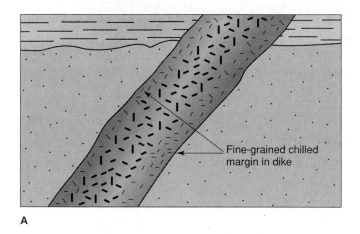

A

B

FIGURE 3.19 Chilled margins. (A) Rapid cooling at the edges of a pluton results in a fine-grained chilled margin. (B) Example of a natural chilled margin: country rock (lower left) has chilled invading magma that formed the rock at upper right; chilled margin is the dark, finer-grained zone at margin.

(B) Photograph by W. B. Hamilton, USGS Photo Library, Denver, CO.

that are concave upward (their tops may not be visible) are called **lopoliths.** Laccoliths are commonly formed by silicic rocks (magmas), lopoliths by mafic ones. This suggests that the density of the magma plays a significant role in shaping the pluton, the lopolith perhaps showing sagging of the country rock under the weight of dense mafic rocks.

A discordant equidimensional pluton is generally described as a **batholith.** Sometimes, a batholith of relatively small areal exposure may also be called a *stock.* Very large batholiths, which can cover thousands of square kilometers of outcrop area and extend to depths of 5 kilometers (3 miles) or more, are multiple intrusions. Many batches of magma were emplaced to form them, and often, many smaller plutons can be distinguished within the large batholith.

Other Features of Intrusions

A pluton may exhibit various textural features, regardless of its overall geometry.

When a hot magma intrudes much cooler country rock, the melt near the contacts is quenched, or rapidly cooled. The resultant *chilled margins* are recognizable because they are less coarsely crystalline than the interior of the pluton (figure 3.19). A gabbroic pluton, for instance, may have a basaltic margin—same composition, different grain size. Chilled margins are more commonly found in shallower plutons (and volcanic rocks) because the temperature contrast between magma and country rock is greater at shallower depths in the crust.

Plutons also sometimes show compositional layering. It usually develops as a result of gravitational settling-out of denser crystals to the floor of the magma chamber as crystallization proceeds, as was shown in figure 3.5. Layering is more frequently observed in mafic plutons, probably for two

FIGURE 3.20 Flow texture may be shown by parallel alignment of elongated crystals in an igneous rock.

reasons. First, in the early stages of crystallization, mafic magmas crystallize abundant, dense ferromagnesian minerals, which are especially prone to settling. Second, settling occurs more readily in a less viscous liquid, and mafic magmas are generally less viscous. (To illustrate the effect of viscosity, drop a pebble into a glass of water and a glass of molasses, and compare the results.)

If some crystals have formed before the magma stops flowing actively, evidence of that flow may be found in the parallel alignment of platy or elongated crystals (figure 3.20). Near the edges of a pluton, at least, the flow direction indicated is generally parallel to the contacts. Bubbles, like crystals, may also be aligned during flow.

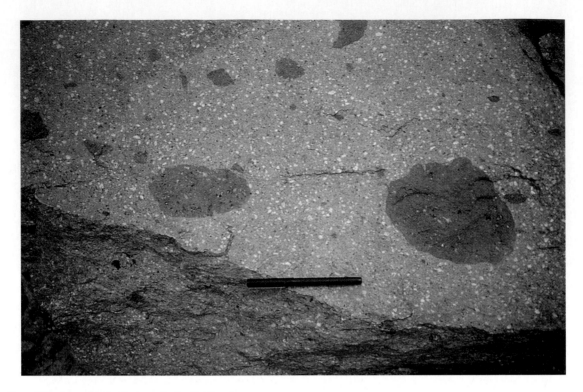

FIGURE 3.21 Xenoliths in andesite. Lassen Park, California.

As magma intrudes country rock, bits of that rock may be broken off by the force of the invading magma and caught up in the molten mass. Some are assimilated through melting, but some may be preserved as **xenoliths** (figure 3.21). Xenoliths take their name from the Greek *xenos,* meaning "stranger," and *lithos,* meaning "stone" or "rock"; they are alien bits of rock caught up in a genetically unrelated magma.

Batholiths and the Origin of Granite

Most batholiths are granitic to granodioritic in composition. What is the source of such large volumes of granitic magma?

As mentioned earlier, most magma originates in the upper mantle, where temperatures are high enough and pressures simultaneously low enough to permit melting. What occurs is actually **partial melting,** in which those minerals with the lowest melting points do melt, while higher-temperature minerals do not. Inspection of Bowen's Reaction Series (figure 3.4) indicates that the first minerals to melt with increasing temperature are the major constituents of typical granite. However, the mantle is ultramafic in overall composition. Even the very earliest melt of ultramafic mantle would not be granitic; by the time a significant fraction of the mantle had melted, the melt would be basaltic in composition through the addition of melted ferromagnesians and plagioclase. If one assumes that only a very small percentage of melt forms, in order to maintain a more silicic composition, there is a twofold problem: first, that a huge volume of mantle would then have to be involved to account for the immense volume of granite in a major batholith, and second, that it is difficult, mechanically, to squeeze a percent or two of melt out of almost completely solid mantle, in order to emplace it into the overlying crust.

An alternative way to make granitic magma would be to start with the basaltic magma expected with significant melting of an ultramafic mantle and then subject it to fractional crystallization at depth. This could proceed, removing the more mafic constituents, until the residual melt was granitic, at which point the melt could intrude the crust. Again, there is a volume problem. To end up with the large volumes of granitic melt represented by batholiths, it would be necessary to start with enormous volumes of basaltic melt. There is little evidence of such extensive melting in the upper mantle.

A third possibility for generating granitic magma is to involve some crust in the melt. For instance, the continental crust is, on average, granodioritic in composition. Partial or complete melting of continental crust could produce magma of the required composition. So would assimilation of considerable continental crust by a rising basaltic, mantle-derived magma. In either case, the problem is one of heat budget, for the continental crust is not normally hot enough to melt, even at lower-crustal depths. Granitic rocks do melt at lower temperatures than does basalt, but whether rising hot basaltic magma could cause enough heating and melting of continental crust to create a granitic batholith is unclear.

Geochemical and other evidence suggests that no single, simple model accounts for all granitic batholiths. The mechanism(s) involved vary from batholith to batholith and sometimes from pluton to pluton within one batholith. We return to the question of the origin of batholiths in connection with the subject of the growth of continents (chapter 11).

SUMMARY

Igneous rocks are those crystallized from magma, a silicate melt. Most magmas are produced in the upper mantle, where temperatures are high enough, and pressures low enough, to allow melting. The amount of melting is increased, and the melting temperatures of minerals reduced, by the presence of volatiles and by the presence of several different minerals in one rock.

Once formed, a cooling magma normally crystallizes principal silicate minerals in a sequence predicted by Bowen's Reaction Series, beginning with olivine and calcic plagioclase and ending with quartz and potassium feldspar. The composition of a magma can be changed by various processes, including fractional crystallization, assimilation, and magma mixing. The crystal size of an igneous rock is determined fundamentally by cooling rate: all other factors being equal, slower cooling means larger crystals. Very rapid cooling produces aphanitic or even glassy rocks. Melt composition and the presence or absence of dissolved volatiles also influence grain size. Typically, plutonic rocks, which crystallize at depth, are more coarsely crystalline than volcanic rocks.

Igneous rocks are classified on the basis of crystal size and chemical composition. For most compositions, there are plutonic and volcanic equivalents, with different names reflecting the depths of crystallization. The mafic volcanic rock basalt is the principal rock type of the sea floor; continental crust is mainly granitic or granodioritic. Intrusive igneous rocks form plutons that may be pipelike, sheetlike, or fairly equidimensional. Plutons are classified on the basis of their shape and whether they are concordant or discordant with respect to the country rock. The shape of a pluton is controlled by the physical properties of both the magma and the country rock. Plutons may exhibit such additional features as chilled margins, compositional layering, and flow textures, and may contain xenoliths of the country rocks through which their magmas have flowed. The origin of the large plutonic complexes called batholiths is problematic, for no single mechanism (partial melting of mantle, fractional crystallization of basalt, melting of continental crust) readily accounts for the production of the necessary volume of granitic or granodioritic magma.

TERMS TO REMEMBER

andesite 42
aphanitic 41
ash 43
assimilate 37
basalt 42
batholith 45
Bowen's Reaction Series 36
cinder 43
concordant 44
continuous reaction series 37
country rock 43
dike 44
diorite 41
discontinuous reaction series 37
discordant 43
felsic 37

fractional crystallization 37
gabbro 41
granite 41
groundmass 38
igneous 34
laccolith 44
lopolith 45
mafic 37
magma 34
magma mixing 38
neck 43
obsidian 42
partial melting 46
pegmatite 40
phaneritic 41
phenocrysts 39

pipe 43
pluton 43
plutonic 40
porphyry 39
pyroclastics 43
rhyolite 42
silicic 37
sill 44
thin section 40
tuff 43
ultramafic 41
vesicles 38
volcanic 41
wallrock 43
xenolith 46

QUESTIONS FOR REVIEW

1. What is an *igneous* rock?
2. Explain briefly the effect of each of the following on the melting temperature of rock: (a) changes in pressure, (b) presence of water vapor.
3. Cite three possible heat sources that might contribute to melting.
4. What is *Bowen's Reaction Series?* How do the continuous and discontinuous branches differ?
5. Describe two ways in which magma composition can be modified.
6. Assimilation of felsic rocks by mafic magmas is more common than the reverse. Why?
7. How is the grain size of an igneous rock related to its cooling rate? What does a porphyritic texture indicate?
8. Plutonic rocks may be more readily identifiable in handsample than volcanic rocks. Why?
9. Most volcanic glasses are rhyolitic in composition. Compositional layering is more often observed in mafic than in felsic plutons. What property of a magma may have a bearing on both of these observations? Explain.
10. What is a *discordant* pluton? Give two examples.
11. What is a *batholith?* Suggest two ways in which the necessary granitic magma might be formed, and discuss any problem with each idea.

FOR FURTHER THOUGHT

1. Considering magma viscosity and density, would you expect volcanic rocks more often to be mafic or silicic in composition? Why?
2. The minerals quartz, albite, and orthoclase (a potassium feldspar) are sometimes called, collectively, the "residua system" in the study of igneous rocks. How do you suppose this name arose?

SUGGESTIONS FOR FURTHER READING

Barker, D. S. 1983. *Igneous rocks.* Englewood Cliffs, N.J.: Prentice-Hall.

Bowen, N. L. 1956. *The evolution of the igneous rocks.* New York: Dover.

Carmichael, I. S., Turner, F. J., and Verhoogen, J. 1974. *Igneous petrology.* New York: McGraw-Hill.

Cox, K. G., Bell, J. D., and Paukhurst, R. J. 1979. *The interpretation of igneous rocks.* London: Allen and Unwin.

Ehlers, E. G., and Blatt, H. 1982. *Petrology: Igneous, sedimentary, and metamorphic.* San Francisco: W. H. Freeman.

Hughes, C. J. 1982. *Igneous petrology.* New York: Elsevier.

Maaloe, S. 1985. *Principles of igneous petrology.* New York: Springer-Verlag.

MacKenzie, W. S., Donaldson, C. H., and Guilford, C. 1982. *Atlas of igneous rocks and their textures.* New York: Halstead Press.

4

VOLCANOES

Snowy cap gives serene appearance to Mt. Rainier, but this volcano in the Cascade Range of the U.S. has the potential for violent eruptions.
© Doug Sherman/Geofile

Chapter 3 introduced igneous rocks and processes. This chapter explores in more detail the surface effects of magmatic activity. Both the form of a volcano and the characteristic style of its eruptions can be related to the kind of magma supplying it and, in turn, to the tectonic setting in which the volcano has developed. The extent to which volcanoes pose hazards is directly related to their eruptive style, some volcanoes erupting relatively quietly, others capable of violent explosions.

MAGMATISM AND TECTONICS

Chapter 1 introduced the concept of plate tectonics, the idea that the earth's crust and upper mantle are broken up into a series of rigid, mobile plates. The plates move over a partly molten zone in the mantle that is the source of most of the magma that accounts for volcanic activity. Magmas are typically generated in one of three plate-tectonic settings: (1) at divergent plate boundaries, where plates split and move apart; (2) over *subduction zones,* a type of plate boundary at which two plates converge and one plate is thrust beneath the other; and (3) at **"hot spots,"** isolated areas of volcanic activity that are not associated with plate boundaries and that are the cause of lone volcanoes away from plate boundaries (*intraplate volcanism*) (figure 4.1).

Magmatism at Divergent Plate Boundaries

Most divergent boundaries are seafloor spreading ridges, described more fully in chapters 9 and 12. The magma produced there is derived by partial melting of mantle material below the rift. The melting may be facilitated by the reduction in pressure from the overlying plate associated with rupture and spreading (figure 4.2). Since the upper mantle is ultramafic and must be fairly extensively melted to produce the volume of magma needed to make the amount of new rock generated at spreading ridges, the resulting melt would be expected to be rather mafic, and indeed it is. The dominant rock type is basalt, which forms the sea floor.

Rifts in continents are less common. Much of the volcanism along continental rift zones is also basaltic, with voluminous lava flows formed from mantle melting as at seafloor-spreading ridges. In addition, rising hot, basaltic magma may warm the more granitic crustal rocks sufficiently to cause melting. Some silicic volcanism is thus found in continental rift zones also, although volumetrically it is usually less important than the basaltic volcanism. The East African rift, which includes dormant Mount Kilimanjaro, is an example of a continental rift zone.

Subduction-Zone Volcanism

The complexity of a subduction zone leads to a variety of possible magma sources. The down-going plate typically comprises seafloor basalt, mafic to ultramafic upper-mantle

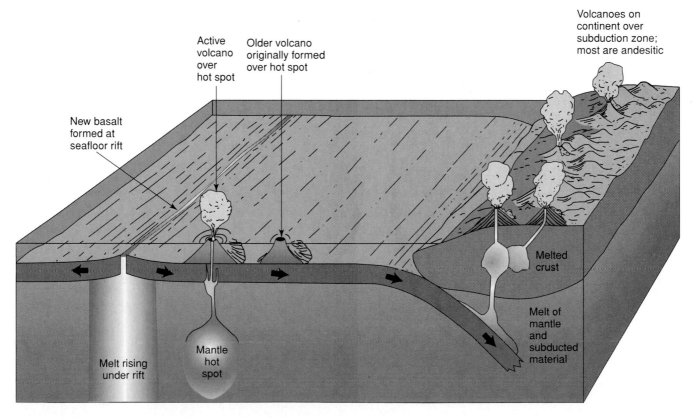

FIGURE 4.1 Volcanism and plate tectonics.

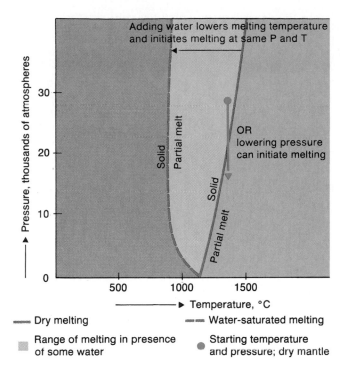

FIGURE 4.2 These melting curves for ultramafic material similar in composition to the mantle show that the addition of water lowers the melting temperature. Reducing the pressure on the material can also cause melting, even with no change in temperature.

rocks, and some seafloor sediments. Any or all of these can melt, wholly or in part. Where the overriding plate is continental, the subducted sediments are, in part, material weathered from the continent. Those sediments are relatively more silicic and tend to melt (producing small quantities of granitic melt) before the more mafic rocks melt appreciably. Ultimately, at least the basaltic crustal rocks melt also.

The down-going sediments and, to a lesser extent, the underlying basalt contain water. During subduction, *dewatering* occurs: the water is released and rises into the portion of the upper mantle between the subducted and overriding plates. The water lowers the melting temperature of the mantle material and induces magma formation. The composition of the resulting magma is andesitic to basaltic, depending on the extent of melting.

Interaction with and assimilation or melting of the overriding plate by rising magma can further modify the magma composition before it reaches the surface. Where the overriding plate is continental, assimilation of granitic or granodioritic crustal material by mafic magma often produces an intermediate-composition, andesitic melt; hence the common observation that volcanoes on continental edges above subduction zones are andesitic. As will be seen later in this chapter, this has implications for the eruptive style of these volcanoes.

Intraplate Volcanism, Hot Spots

Geologists are not in complete agreement about the nature of hot spots. Isolated hot-spot volcanoes are usually attributed to the presence of mantle **plumes,** columns of warm material rising in the upper mantle. As the mantle material rises, the pressure on it is reduced, and melting may begin. If the overlying plate is sufficiently weak, some of the magma breaks through to form a volcano.

The composition of the material erupted depends largely on the composition of the plate through which the magma rises. That is, the melt derived from the plume is expected to be basaltic, like the magma at a spreading ridge. If it rises up through the (mafic) sea floor, it is still basaltic when it erupts, whether or not some sea floor is assimilated. Hot-spot volcanoes in the ocean basins—the Hawaiian Islands, for instance—are commonly built of many thin layers of fluid basaltic lavas. If the magma must make its way up through continental crust, there is more potential for assimilation of granitic material and production of a more silicic final magma.

What causes plumes is not known for certain. They may form over regions of locally high concentrations of (heat-producing) radioactive elements in the mantle, with the extra heat causing extra melting. Some researchers have suggested that plumes rise over anomalous regions of the outer core. Whatever their cause, hot spots are long-lived features. For example, we will see in chapter 9 that the path of the Pacific Plate over the now-Hawaiian hot spot can be traced over the last seventy-five million years.

EXTRUSIVE ROCK STRUCTURES: VOLCANOES AND FISSURE ERUPTIONS

If volcanic rock is crystallized from magma at or near the earth's surface, then most volcanic rock is formed as new sea floor is produced at seafloor spreading ridges, described in detail in chapter 9. However, when most people think of volcanic activity, they think in terms of **volcanoes,** individual mountains built around discrete vents through which magma can erupt at the surface. **Lava** is simply magma that reaches the surface, and many volcanoes are built of layer upon layer of lava. Not all volcanoes erupt only lava, however. Differences in the materials that make up a volcano contribute to differences in both form and eruptive style.

Fissure Eruptions

The outpouring of magma forming volcanic rock at rifts on the sea floor is an example of **fissure eruption,** the eruption of lava out of a long crack rather than from a single pipe or vent (figure 4.3). There are also examples of fissure eruptions on the continents, in which many layers of lava are erupted in succession. One example in the United States is the Columbia Plateau, an area of about 50,000 square kilometers (20,000 square miles) in Washington, Oregon, and Idaho, covered by layer upon layer of basalt, piled up over 1.5 kilometers deep in places (figure 4.4). This area may represent the ancient beginning of a continental rift that ultimately stopped spreading. It is no longer active, but it serves as a reminder of how large a volume of magma can come welling up from the mantle where zones of weakness or fractures in the plate above provide suitable openings. Even larger examples of such *flood basalts* on continents, covering up to 750,000 square kilometers, are found in India and Brazil.

Extensive lava flows often develop a distinctive structure upon cooling. Like most materials, lava and volcanic rock contract as they cool. This contraction can fracture the flow into a mass of polygonal columns (figure 4.5). Such **columnar jointing** is especially common in basaltic flows, perhaps because they start at higher temperatures and therefore undergo more contraction in cooling to surface temperatures.

Shield Volcanoes

Basaltic lavas, relatively low in silica and high in iron and magnesium, are comparatively fluid, so they flow very freely and far when erupted. Consequently, the kind of volcano they build is very flat and low in relation to its diameter. This low, shieldlike shape has led to the use of the term **shield volcano** for such a structure (figure 4.6). Though the individual lava flows may be thin—perhaps less than a meter in thickness—the buildup of hundreds or thousands of flows over time can produce quite large volcanic structures. The Hawaiian Islands

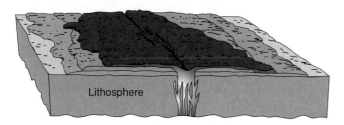

FIGURE 4.3 Fissure eruption (schematic).

FIGURE 4.4 Multiple lava flows from fissure eruption. (A) The Columbia River flood basalts (map extent). (B) Several separate lava flows, piled one atop another, can be seen in this photograph taken near Coulee City, Washington.

(B) © Bruce F. Molnia, Terraphotographics, BPS.

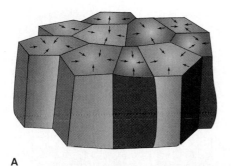

A

B

FIGURE 4.5 Columnar jointing. (A) Formation of fractures by contraction in a cooling lava flow. (B) Columnar jointing at Devil's Postpile National Monument, California.

are all shield volcanoes. Mauna Loa, the largest peak on the still-active island of Hawaii, rises 3½ kilometers (about 2½ miles) above sea level. If measured properly from its true base, the sea floor, it is even more impressive: about 10 kilometers high—higher than Mount Everest rises above sea level—and with a base diameter of 100 kilometers. With their broad, flat shapes, the Hawaiian Islands do not necessarily even look like one's concept of a volcano from sea level.

E ven with limited variation in magma composition, there can be obvious variation in the appearance of the resultant lava flow. Especially fluid lavas that form a smooth, quenched, hardened skin as they cool develop a ropy appearance as they flow (figure 4.7A); this ropy lava is termed *pahoehoe* (pronounced "pa-hoy-hoy"). Other lavas flow less readily and produce jumbled, blocky flows (figure 4.7B); this material is called *aa* ("ah-ah") and is reputed to have been so named from the pained cries of persons attempting to walk barefoot across its jagged surface. When lavas are extruded under water, rapid quenching of the hot flow surfaces leads to development of bulbous flow forms resembling pillows (figure 4.7C). In these **pillow lavas,** each pillow has a glassy rind and a coarser-grained, more slowly cooled interior.

Volcanic Domes

The less mafic, more silicic lavas, andesitic and rhyolitic in composition, tend to be more viscous and flow less readily. They ooze out at the surface like thick toothpaste from an upright tube, piling up close to the volcanic vent rather than spreading freely. The resulting structure is a more compact and steep-sided **volcanic dome.** Modern eruptions of Mount

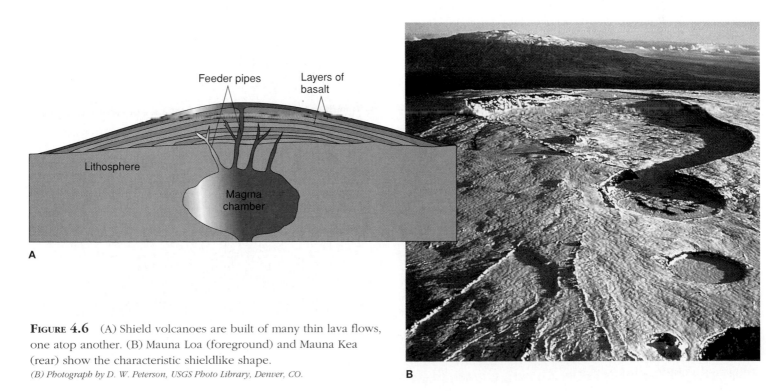

A

FIGURE 4.6 (A) Shield volcanoes are built of many thin lava flows, one atop another. (B) Mauna Loa (foreground) and Mauna Kea (rear) show the characteristic shieldlike shape.
(B) Photograph by D. W. Peterson, USGS Photo Library, Denver, CO.

B

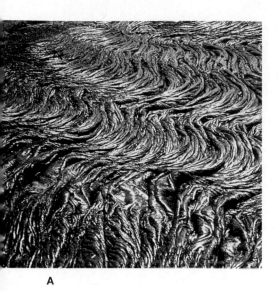

A

B

C

FIGURE 4.7 Types of lava flows. (A) Pahoehoe, with a smooth, ropy surface. Hawaii Volcanoes National Park. (B) Aa, rough and blocky. Snake River Plain, Idaho. (C) Pillow lavas formed underwater.
(C) Photograph by W. R. Normark, USGS Photo Library, Denver, CO.

St. Helens are characterized by this kind of stiff, viscous lava, and a volcanic dome several hundred meters high has formed in the crater left by the 1980 explosion (figure 4.8). Such thick, slowly flowing lavas also seem to solidify and stop up the vent from which they are erupted before much material has emerged. Volcanic domes, then, tend to be relatively small in areal extent compared to shield volcanoes, although through repeated eruptions over time, such volcanoes can build quite high peaks.

Cinder Cones

As noted earlier, magmas do not consist only of melted silicates; they also contain dissolved water and gases, which are trapped under high pressures while the magma is at depth. As the magma wells up toward the surface, the pressure on it is reduced, and the dissolved gases try to bubble out and escape. The effect is much like popping the cap off a soda bottle: the soda contains carbon dioxide gas under pressure, and when the pressure is released by removal of the cap, the gas comes bubbling out.

Sometimes, the built-up gas pressure in a rising magma is released suddenly and forcefully by an explosion that flings bits of magma and rock out of the volcano, as the pyroclastic material described in the last chapter. The most energetic pyroclastic eruptions are more typical of volcanoes with the more viscous andesitic or rhyolitic lavas because these thicker lavas tend to trap more gases. Gas usually escapes more readily and quietly from the fluid basaltic lavas, though even basaltic volcanoes may sometimes emit quantities of finer pyroclastics. Block-sized blobs of still molten lava may also be thrown from a volcano; these volcanic *bombs* commonly develop a streamlined shape as they deform in flight before solidifying completely. When pyroclastics fall close to the vent from which they were thrown, they may pile up into a very symmetric cone-shaped heap known as a **cinder cone** (figure 4.9).

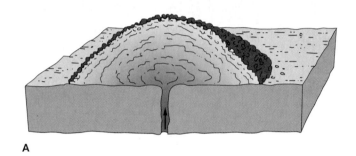

A

B

FIGURE 4.8 Volcanic domes. (A) Formation of a dome (schematic). (B) Dome built in the crater of Mount St. Helens.
(B) Photograph by R. E. Wallace, USGS Photo Library, Denver, CO.

A

FIGURE 4.10 A composite volcano in the Aleutian Islands.
Photograph by R. E. Wilcox, USGS Photo Library, Denver, CO.

Composite Volcanoes

Many volcanoes, andesitic ones especially, erupt different materials at different times. They may emit some pyroclastics, then some lava, then more pyroclastics, and so on. Volcanoes built up in this layer-cake fashion are called **stratovolcanoes** or, alternatively, **composite volcanoes,** because they are built up of more than one kind of material. Most of the potentially dangerous volcanoes of the western United States, in the Cascade Range, are composite volcanoes. They typically have fairly stiff, gas-charged lavas that sometimes flow and sometimes trap enough gases to erupt explosively with a rain of pyroclastic material. The combination of pyroclastics and viscous lavas tends to produce rather steep-sided cones of relatively symmetric shape (figure 4.10).

Calderas

An eruption from a volcanic vent is fed from a magma chamber below, sometimes a very large one. When much of the magma has erupted or, perhaps, has drained back down to deeper levels, the volcano is left partially unsupported. If the overlying rocks are too weak, they may collapse into the hole, forming a depressed **caldera** much larger than the original summit crater from which the lava emerged. Alternatively,

B

FIGURE 4.9 Cinder cone. Parícutin in Mexico. (A) Night eruption shows ejection of hot pyroclastics, piling up in a cone-shaped mound. (B) Parícutin by day, a classic cinder cone.
(A) Photograph by R. E. Wilcox, USGS Photo Library, Denver, CO. (B) Photograph by K. Segerstrom, USGS Photo Library, Denver, CO.

Figure 4.11 Crater Lake, Oregon. Wizard Island, within the lake, is a cinder cone.

a caldera may be formed by a violent explosion, like that of Mount St. Helens in 1980, which greatly enlarges the vent. Major calderas can cover tens or even hundreds of square kilometers. The best-known example in the United States is the misnamed Crater Lake (figure 4.11). The lake fills the depression formed by caldera collapse of the ancient volcano Mount Mazama in southern Oregon; at 600 meters depth, it is the deepest freshwater lake in the United States.

VOLCANIC HAZARDS

Direct Hazards: Materials and Eruptive Style

Primary volcanic hazards include both the products of the eruptions (such as lava, gas, and pyroclastics) and, sometimes, the nature of the eruption itself.

Until the eruption of Mount St. Helens, most people in the United States, if they had thought about volcanoes at all, would have regarded lava as the principal hazard. Actually, lava is not generally life-threatening. Most lava flows advance at speeds of a few kilometers an hour at most, so one can evade the advancing lava readily even on foot. The lava will,

of course, destroy or bury any property over which it flows. Lava temperatures are typically over 500°C (over 950°F) and may be over 1400°C (2550°F). Combustible materials—houses and forests, for example—burn at such temperatures. Other property may simply be engulfed in lava, which then solidifies into solid rock (figure 4.12).

Lavas, like all liquids, flow downhill, so one way to protect property is simply not to build close to a volcano's slopes. Throughout history, however, people *have* built on or near volcanoes, for many reasons. They may simply not expect the volcano to erupt again. Also, soil formed from the weathering of volcanic rock forms slowly but is often very fertile. The Romans cultivated the slopes of Vesuvius and other volcanoes for that reason. Sometimes, too, a volcano is the only land available, as in the Hawaiian Islands or Iceland.

Some strategies do exist for reducing property damage from lava. In Iceland, in 1973, flow-quenching operations saved a crucial harbor when the volcano Heimaey erupted. As lava cools, it becomes thicker and more viscous and flows more slowly; if chilled to the point of solidification, it may stop flowing altogether. Heimaey is an island, surrounded by plen-

FIGURE 4.12 Development below the east rift of Kilauea has been devastated by lava. Roads so new that center lines were still fresh were set aflame by advancing flows: Royal Gardens subdivision, Hawaii, 1983.
Photograph by J. D. Griggs, USGS Photo Library, Denver, CO.

FIGURE 4.13 The 1973 eruption of Heimaey, Iceland: lava flow is controlled by quenching the hot melt.
Photograph courtesy of USGS Photo Library, Denver, CO.

tiful cooling water. Boats sprayed water on flows encroaching on the harbor (figure 4.13), and eventually the harbor—essential to the island's fishing-centered economy—was saved.

Where it is not practical to arrest a lava flow altogether, it may be possible to divert it from a course along which a great deal of damage may be done, to an area where less valuable property is at risk. Sometimes, a lava flow is slowed or halted temporarily during an eruption because the volcano's output has lessened or because the flow has encountered a natural or artificial barrier. A solid crust develops on the flow; the interior remains molten for days, weeks, or months thereafter. If a hole is then punched in this crust by explosives, the remaining fluid magma inside can flow out and away. Careful placement of the explosives can divert the flow in a chosen direction. This was tried in Italy in early 1983, when Mount Etna began another in an intermittent series of eruptions. Unfortunately, the effort was only briefly successful. Part of the flow was deflected, but within four days, the lava had abandoned the planned alternate channel and resumed its original flow path. Later, new flows threatened further destruction of inhabited areas.

Lava flows may be hazardous, but in one sense, they are at least predictable: like other fluids, lavas flow downhill. Their possible flow paths can be anticipated, and once the lavas have flowed into a relatively flat area, they tend to stop. Other kinds of volcanic hazards can be more challenging to deal with and affect much broader areas.

Pyroclastics are often more dangerous than lava flows. They may erupt more suddenly and explosively, and spread faster and farther. The largest blocks and volcanic bombs present an obvious danger because of their size and weight. For the same reasons, however, they usually fall quite close to the volcanic vent, so they affect a relatively small area.

The sheer volume of the finer ash and dust particles can make them as severe a problem, and they can be carried over a much larger area. Also, ashfalls are not confined to valleys and low places. Instead, like snow, they can blanket the countryside. The 18 May 1980 eruption of Mount St. Helens was by no means the largest such eruption ever recorded, but the ash from it blackened the midday skies more than 150 kilometers away, and measurable ashfall was detected halfway across the United States. Even in areas where only a few millimeters of ash fell, transportation ground to a halt as drivers skidded on the slippery roads and engines choked on the dust in the air. Homes, cars, and land were buried under the hot ash. Volcanic ash is also a health hazard that makes breathing both uncomfortable and difficult. The cleanup effort required to clear the debris strewn about by Mount St. Helens was enormous. In the 1991 eruption of Mount Pinatubo in the Philippines, the combination of thick ashfalls and soaking monsoon rains caused widespread collapse of homes under the weight of the sodden debris, and this was the primary reason for the casualties.

Past explosive eruptions of other violent volcanoes have been equally damaging. When the city of Pompeii was destroyed by Mount Vesuvius in A.D. 79, it was buried not by lava but by ash. (This is why extensive excavation of the ruins has been possible.) Contemporary accounts suggest that most residents had ample time to escape as the ash fell, though many chose to stay, and died, but the town was ultimately obliterated. Its very existence had been forgotten until some

A

B

FIGURE 4.14 (A) Mudlfow and flood damage from eruption of Mount St. Helens, 1980: damaged homes along the north fork of the Toutle River. (B) Aerial view of Abacan River channel in Angeles City near Clark Air Force Base, Philippines: mudflow has taken out the main bridge, and pedestrians cross on makeshift bridges.

(A) Photograph by C. D. Miller, USGS Photo Library, Denver, CO. (B) Photograph by T. J. Casadevall, U.S. Geological Survey.

ruins were discovered in the late 1600s. The 1815 explosion of Tambora, in Indonesia, ejected an estimated 30 cubic kilometers of debris—about thirty times the volume of the pyroclastics from Mount St. Helens in 1980!

Pyroclastic materials are a special hazard with snow-capped volcanoes like Mount St. Helens. The heat of the falling ash melts the snow and ice on the mountain, producing a mudflow of meltwater and volcanic ash called a **lahar.** Such mudflows, like lava flows, flow downhill. They may follow stream channels, choking them with mud and causing floods of stream waters. Flooding produced in this way was a major source of damage near Mount St. Helens (figure 4.14A). In A.D. 79, the city of Herculaneum, closer to Vesuvius than was Pompeii, was partially invaded by volcanic mudflows. This made possible the preservation, as mud casts, of the bodies of some of the volcano's victims. The 1985 eruption of Nevado del Ruíz in Colombia is a more recent example of the devastation possible from such mudflows. The swift and sudden mudflows triggered as its snowy cap was melted by hot

ash were the major cause of the more than 20,000 deaths in towns below the volcano. In the 1991 eruption of Pinatubo, mudflows occurred when rain-soaked ash on the mountain's slopes suddenly slid downhill (figure 4.14B).

Another special kind of pyroclastic outburst is a deadly, denser-than-air mixture of hot gases and fine ash known as a **pyroclastic flow,** or a *nuée ardente,* from the French for "glowing cloud." A pyroclastic flow is very hot—temperatures can be over 1000°C in the interior—and it can rush down the slopes of the volcano at more than 100 kilometers per hour (60 miles per hour), charring everything in its path and flattening trees and weak buildings. A pyroclastic flow accompanied the major eruption of Mount St. Helens in 1980 (figure 4.15).

Perhaps the most famous such event in recent history occurred during the 1902 eruption of Mont Pelée, on the Caribbean island of Martinique. The volcano had begun erupting weeks before, emitting both ash and lava, but was believed by many to pose no imminent threat to surrounding towns. Then, on the morning of May 8, with no immediate

FIGURE 4.15 Pyroclastic flow from Mount St. Helens, 18 May 1980.
Photograph by P. W. Lipman, USGS Photo Library, Denver, CO.

advance warning, a nuée ardente emerged from that volcano and swept through the nearby town of St. Pierre and its harbor. In a period of about three minutes, an estimated 25,000 to 40,000 people were fatally injured, burned or suffocated (figure 4.16). The single reported survivor in the town was a convicted murderer who had been imprisoned underground in the town dungeon, where he was shielded from the intense heat.

Just as andesitic volcanoes more often have a history of explosive eruptions, so do many of them have a history of eruption of pyroclastic flows. Again, the composition of the lava is linked to the likelihood of such an eruption. While the emergence of a pyroclastic flow may be sudden and unheralded by special warning signs, it is not generally the first activity shown by the volcano during an eruptive stage. Steam had issued from Mont Pelée for several weeks before the day St. Pierre was destroyed, and lava had been flowing out for over a week. This suggests one possible strategy for avoiding the dangers of a pyroclastic flow: When a volcano known or believed to be capable of such eruptions shows signs of activity, leave.

FIGURE 4.16 Devastation from a nuée ardente: St. Pierre destroyed by Mont Pelée, 1902.
Photograph by Underwood and Underwood, courtesy of Library of Congress.

Evacuations can themselves be disruptive, however. If, in retrospect, the danger turns out to have been a false alarm, people may be less willing to heed a later call to clear the area.

In addition to lava and pyroclastics, volcanoes emit a variety of gases. Many of these, such as water vapor and carbon dioxide, are nontoxic but may nevertheless kill, by suffocation. Others, including carbon monoxide, various sulfur gases, and hydrochloric and hydrofluoric acids, are poisonous. Many people have been killed by volcanic gases even before they realized the danger. During the A.D. 79 eruption of Vesuvius, fumes overcame and killed many unwary observers. Again, the best defense against hazardous volcanic gases is the commonsense one: to get well away from the erupting volcano or escaping gases as quickly as possible.

Some volcanoes are deadly not so much because of any characteristic inherent in the particular volcano but because of where it is located. In the case of a volcanic island, large quantities of seawater may seep down into the rock, come close to the hot magma below, turn to steam, and blow up the volcano like an overheated steam boiler. This is termed a **phreatic eruption.** Phreatic eruptions can also occur when any water—groundwater, lake water, snowmelt, and so on—seeps into the crust close to a hot magma body.

The classic example is Krakatoa, Indonesia, which exploded in this fashion in 1883. The force of its explosion was estimated to be comparable to that of 100 million tons of dynamite, and the sound was heard 3000 kilometers away in Australia. Some of the dust was shot 80 kilometers into the air, causing red sunsets for years afterward. Ash was detected over an area of 750,000 square kilometers. The shock of the explosion generated a fast-moving sea wave that produced breakers over 40 meters high! Krakatoa itself was an uninhabited island, yet its 1883 eruption killed an estimated 36,000 people, mostly in low-lying coastal regions inundated by the sea.

Of course, it is not only island volcanoes that explode. When viscous rhyolitic or andesitic lavas plug the vent, the pressure of gases associated with the magma may build until it rips the volcano apart—and such explosions, too, are often unpredictable in the short term. When Mount St. Helens exploded in May of 1980, the volcano had shown signs of activ-

FIGURE 4.17 Nearly a decade after the 1980 eruption of Mount St. Helens, the force of the blast and the resultant devastation were still evident.

ity for some time, and its north slope was bulging, a sign of potential explosion. Authorities had evacuated the area, leaving only scientific personnel and a small number of commercial loggers, turning away droves of sightseers. But the moment of explosion was recognized only seconds before, too late to rescue those remaining from the searing blast that cropped over 400 meters from the mountain's elevation, cost an estimated $1 billion in damages, killed twenty-five people, and left another thirty-seven people unaccounted for, presumed dead (figure 4.17). The suddenness of that event confirmed the wisdom of the evacuations.

Secondary Effects: Climate

A single volcanic eruption can have a global impact on climate, although the effect may be only brief. Intense explosive eruptions put large quantities of volcanic dust high into the atmosphere, from which it may take years to settle. In the interim, the dust partially blocks out incoming sunlight, thus causing measurable cooling. After Krakatoa's 1883 eruption, worldwide temperatures dropped nearly half a degree centigrade, and the cooling effects persisted for almost ten years. The larger 1815 eruption of Tambora (in Indonesia) caused still more dramatic cooling; 1816 became known in the Northern Hemisphere as the "year without a summer."

FIGURE 4.18 Satellites tracked the path of the airborne sulfuric-acid mist formed by SO$_2$ from Mount Pinatubo; winds slowly spread it into a belt encircling the earth.
Image by G. J. Orme, Department of the Army, Ft. Bragg, NC.

The climatic impacts of volcanoes are not confined to the effects of volcanic dust. The 1982 eruption of the Mexican volcano El Chichón did not produce a particularly large quantity of dust, but it did shoot volumes of unusually sulfur-rich gases into the atmosphere. These gases produced clouds of sulfuric acid droplets that spread around the earth. Not only do such acid droplets block some sunlight, as does dust, but in time they also settle back to the earth as acid rain. The 1991 eruptions of Mount Pinatubo, in the Philippines, likewise involved extensive output of both ash and sulfur gases, and the resultant sulfuric-acid mist circled the globe (figure 4.18).

As climate change becomes a subject of increasing scientific research, increased scrutiny is being given to the role of volcanic eruptions. The unusually cool summer of 1992 in the Northern Hemisphere was attributed to the explosive summer 1991 eruption of Pinatubo. Some researchers would propose a further link with other anomalies, such as El Niño effects (see chapter 12).

Prediction of Volcanic Eruptions

In terms of their activity, volcanoes are divided into three categories: **active; dormant,** or "sleeping"; and **extinct,** or "dead." Unfortunately, the rules for assigning a particular volcano to one category or another are not precise. A volcano is generally considered *active* if it has erupted within recent (recorded) history. When the volcano has not erupted recently but is fresh-looking and not too eroded or worn down, it is regarded as *dormant,* inactive for the present but with the potential to become active again. Historically, a volcano that not only has no recent eruptive history but also appears very much eroded has been considered *extinct,* very unlikely ever to erupt again.

As volcanologists learn more about the frequency with which volcanoes erupt, however, it has become clear that these guidelines are too simplistic. Volcanoes differ widely in their normal patterns of activity. Statistically, a "typical" volcano erupts once every 220 years, but 20% of all volcanoes erupt less than once every 1000 years, and 2% erupt less than once in 10,000 years. Long quiescence, then, is no guarantee of extinction.

The first step in predicting volcanic eruptions is monitoring, keeping an instrumental eye on the volcano. However, there are not nearly enough personnel or instruments available to monitor every volcano all the time. There are an estimated 300 to 500 active volcanoes in the world (the uncertainty arises from not knowing whether some are truly active, or dormant). Their locations are shown in figure 4.19. Note that most are located over subduction zones. The so-called Ring of Fire, the collection of volcanoes rimming the Pacific Ocean, is really a ring of subduction zones, above which rising magmas have produced the Andes, the Cascades, and the Aleutians, among other mountain ranges.

Monitoring those several hundred volcanoes alone would be a large task. Dormant volcanoes might become active at any time, so they also should be watched. Theoretically, extinct volcanoes can safely be ignored, but that assumes that extinct volcanoes can be distinguished from long-dormant ones. The detailed eruptive history of many volcanoes is imperfectly known over the longer term. Some volcanoes, such as Mount St. Helens, seem to erupt at more or less regular intervals (in the case of Mount St. Helens, every 150 years or so), and such information can serve as a general guide to the likelihood of an eruptive phase in the near future.

There are several advance warnings of volcanic activity. A common one is seismic activity. The rising of a volume of magma and gas up through the crust beneath a volcano puts stress on the rocks of the crust, and the process may produce months of small (and occasionally large) earthquakes. The eruptions of Mount St. Helens have been preceded by increased frequency and intensity of seismic activity. Sometimes, the shock of a larger earthquake may itself unleash the eruption. The major explosion of Mount St. Helens in 1980 is believed to have been set off indirectly in this way: an earthquake shook loose a landslide from the bulging north slope of the volcano, which lessened the weight confining the trapped gases inside and allowed them to blast forth. Detailed studies of the seismic activity at Mount St. Helens over the past few years have also suggested that its major eruptions may be preceded by characteristic *harmonic tremors*—distinctive, rhythmic patterns of seismicity. This provides an additional tool in eruption prediction, though not an infallible one.

Bulging, tilt, or uplift of the volcano's surface is also a warning sign. It often indicates the presence of a rising magma mass, the buildup of gas pressure, or both. Figure 4.20,

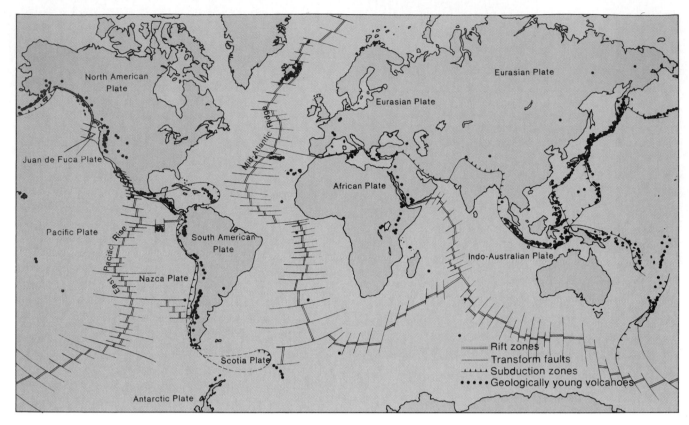

FIGURE 4.19 Volcanic areas of the modern world.

Source: After R. Decker and B. Decker, Volcanoes, 1981, W. H. Freeman and Company, New York, NY.

which is a record of tilt and of eruptions on Kilauea in Hawaii, shows that eruptions are preceded by an inflating of the volcano as magma rises up from the mantle to fill the shallow magma chamber. Unfortunately, it is not possible to anticipate exactly when the swollen volcano will crack to release its contents. That varies from eruption to eruption with the pressures and stresses involved and the strength of the overlying rocks. This limitation is not unique to Kilauea. Uplift, tilt, and seismic activity may indicate that an eruption is approaching, but geologists do not yet know how to predict its exact timing.

Other possible predictors of volcanic eruptions are being evaluated. Changes in the mix of gases coming out of a volcano may give clues to impending eruptions. For example, increases in sulfur-gas emissions may reflect the approach of magma toward the surface, with the associated gases escaping from the magma as it rises. Surveys of ground-surface temperatures may reveal especially warm areas where magma is particularly close to the surface and about to break through. There have been reports that volcanic eruptions have been anticipated by animals, which have behaved strangely for some hours or days before the event. Perhaps they are sensitive to some changes in the earth that scientists have not thought to measure.

Certainly, much more work is needed before the precise timing and nature of major volcanic eruptions can be anticipated consistently. When data indicate an impending eruption that might threaten populated areas, the safest course is evac-uation until the activity subsides. However, a given volcano may remain more or less dangerous for a long time. An active phase, consisting of a series of intermittent eruptions, may continue for months or years. In these instances, either property must be abandoned for prolonged periods, or inhabitants must be prepared to move out not once but many times. Given the uncertainty of eruption prediction at present, some precautionary evacuations will continue to be shown, in retrospect, to have been unnecessary, or unnecessarily early.

Current and Future Volcanic Hazards in the United States

Several of the Hawaiian volcanoes are active or dormant, and eruptions are frequent in some places. It is, of course, not practical to evacuate the islands indefinitely, but a first, commonsense step toward reducing losses from eruptions might be to prevent any new development in recently active areas. Even this is not being done, however; box 4.1 (on pages 64 and 65) shows some of the results of recent unwise development near Kilauea's East Rift Zone.

Mount St. Helens is only one of a set of volcanoes that threaten the western United States and southwestern Canada. Subduction beneath the Pacific Northwest is the underlying cause of continuing volcanism there (figure 4.21). Several more of the Cascade Range volcanoes have shown signs of reawakening. Lassen Peak last erupted in 1914 and 1921, not

FIGURE 4.20 Tilt record of Kilauea in Hawaii. Increasing tilt of volcano's slopes reflects bulging due to rising magma below. Note the deflation after eruptions.

Source: "The Pu'u O'o Eruption of Kilauea Volcano" in Volcanoes in Hawaii, *U.S. Geological Survey.*

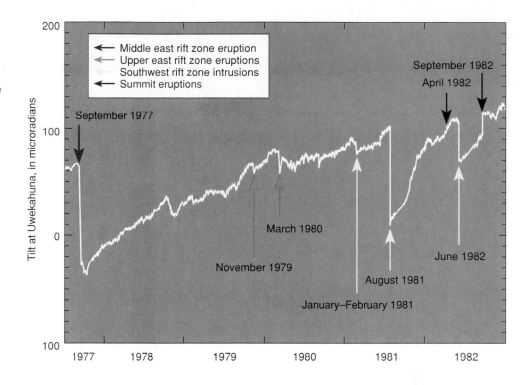

FIGURE 4.21 The Cascade Range volcanoes and their spatial relationship to the subduction zone and to major cities.

Source: R. Decker and B. Decker, "The Eruption of Mount St. Helens," Scientific American, *March 1981; data from U.S. Geological Survey.*

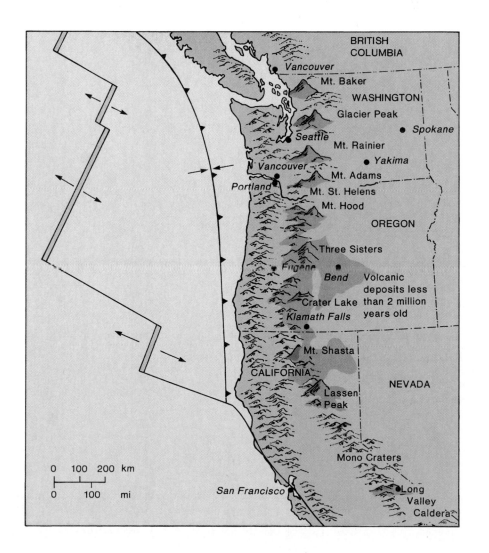

Box 4.1

Life on a Volcano's Flanks

The island of Hawaii is really a complex structure built of five shield volcanoes (figure 1).

The so-called East Rift of Kilauea in Hawaii has been particularly active in recent years, with major eruptions in the 1970s. Less than ten years later, a new subdivision—Royal Gardens—was begun below the East Rift. In 1982–83, lava flows from a renewed round of eruptions reached down the slopes of the volcano and quietly obliterated new roads and several houses. The eruptions have continued intermittently over more than a decade since then. The photographs in figure 2 illustrate some of the results.

When asked why anyone would have built in such a spot, where the risks were so obvious, one native shrugged and replied, "Well, the land was cheap." It will probably remain so, too—a continuing temptation to unwise development in the absence of tighter zoning restrictions.

Of course, when one lives on an active or dormant volcano, no spot may be altogether safe. Still, relative levels of risk can be identified on the basis of past eruptive and seismic activity, and topography (figure 3). While the risks to life are generally low everywhere because of the quiet character of Hawaiian eruptions, property damage could be greatly reduced by greater attention to the level of probable vulnerability to lava flows of areas under consideration for further development.

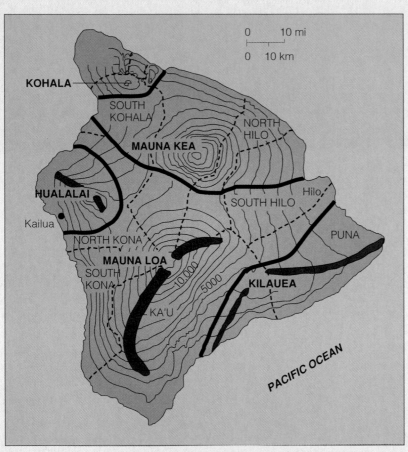

FIGURE 1 Map of Hawaii showing five volcanoes and major associated rift zones.
Source: After C. Heliker, Volcanoes and Seismic Hazards of Hawaii, *U.S. Geological Survey.*

A

FIGURE 2 (A) Lavas invade Royal Gardens subdivision, Hawaii, 1983. Flows from March through July 1983 alone covered 330 lots and destroyed sixteen homes. (B) House aflame, touched off by hot lava. (C) Ruins of million-dollar visitors' center in Hawaii Volcanoes National Park, engulfed by lava in 1989.
Photograph (A) by J. D. Griggs; photographs (A, B) and (C) courtesy of USGS Photo Library, Denver, CO.

B

C

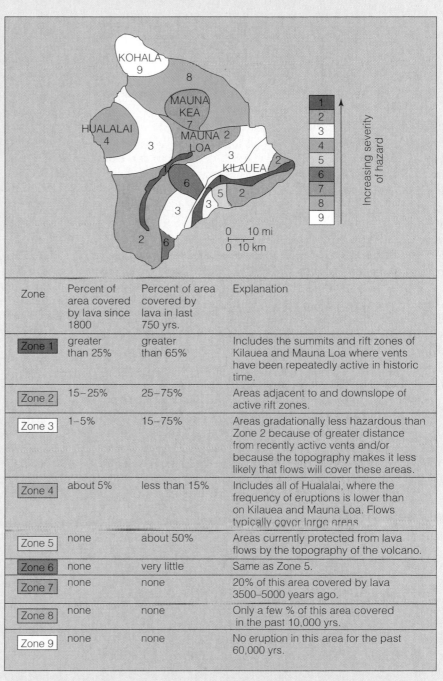

Zone	Percent of area covered by lava since 1800	Percent of area covered by lava in last 750 yrs.	Explanation
Zone 1	greater than 25%	greater than 65%	Includes the summits and rift zones of Kilauea and Mauna Loa where vents have been repeatedly active in historic time.
Zone 2	15–25%	25–75%	Areas adjacent to and downslope of active rift zones.
Zone 3	1–5%	15–75%	Areas gradationally less hazardous than Zone 2 because of greater distance from recently active vents and/or because the topography makes it less likely that flows will cover these areas.
Zone 4	about 5%	less than 15%	Includes all of Hualalai, where the frequency of eruptions is lower than on Kilauea and Mauna Loa. Flows typically cover large areas.
Zone 5	none	about 50%	Areas currently protected from lava flows by the topography of the volcano.
Zone 6	none	very little	Same as Zone 5.
Zone 7	none	none	20% of this area covered by lava 3500–5000 years ago.
Zone 8	none	none	Only a few % of this area covered in the past 10,000 yrs.
Zone 9	none	none	No eruption in this area for the past 60,000 yrs.

FIGURE 3 The Island of Hawaii is divided into zones of relative hazard from lava flows; 1 = highest degree of risk.

Source: After C. Heliker, Volcanic and Seismic Hazards of Hawaii, *U.S. Geological Survey, 1991.*

so very long ago geologically. Its products are very similar to those of Mount St. Helens; violent eruptions are certainly possible. Mount Baker (last eruption, 1870), Mount Hood (last eruption, 1865), and Mount Shasta have shown seismic activity, and steam is escaping from these and from Mount Rainier, which last erupted in 1882. In fact, at least nine of the Cascade peaks are presently showing thermal activity of some kind (steam emission, hot springs). The eruption of Mount St. Helens has in one sense been useful: it has focused attention on this threat. Scientists are watching many of the Cascade Range volcanoes very closely now. With skill, and perhaps some luck, they may be able to recognize when others of these volcanoes are close to eruption.

In 1980, the Mammoth Lakes area of California suddenly began experiencing earthquakes. Within one forty-eight-hour period in May 1980, four earthquakes of magnitude 6 rattled the region, interspersed with hundreds of lesser shocks. Mammoth Lakes lies within the Long Valley Caldera, a 13-kilometer-long oval depression formed during violent pyroclastic eruptions 700,000 years ago. Smaller eruptions occurred around the area as recently as 50,000 years ago. Geophysical studies have shown that a partly molten mass of magma close to 4 kilometers across still lies below the caldera. Furthermore, since 1975, the center of the caldera has bulged upward more than 25 centimeters (10 inches); at least a portion of the magma appears to be rising in the crust. In May 1982, the director of the U.S. Geological Survey issued a formal notice of potential volcanic hazard for the Mammoth Lakes/Long Valley area. Seismicity is continuing. So far, there has been no eruption, but scientists continue to monitor developments very closely, seeking to understand what is happening below the surface so as to anticipate what volcanic activity may develop, and when.

Another area of uncertain future is Yellowstone National Park. Yellowstone, at present, is notable for its geothermal features—geysers, hot springs, fumaroles. All of these features reflect the presence of hot rocks at shallow depths in the crust below the park. Until recently, it was believed that the heat was just left over from the last cycle of volcanic activity in the area, which began 600,000 years ago. Recent studies, however, suggest that considerable magma may remain beneath the park, which is also still a seismically active area. One reason for concern is the scale of past eruptions: in the last cycle, 1000 cubic kilometers (over 200 cubic miles) of pyroclastics were ejected! Activity on such an impressive scale is not unusual for calderas, like Yellowstone or Long Valley, that become active after long dormancy. Whether activity on that scale is likely in the foreseeable future, no one presently knows. The research goes on.

The Aleutian Islands and the Alaskan peninsula together represent the most concentrated region of volcanic activity in the United States. Located above a subduction zone, the region is certainly one in which continuing volcanic activity can be expected. The area receives relatively little attention from the perspective of volcanic hazards, however, because it is sparsely populated.

One of the more widely publicized recent eruptions in this area was that of Mount Augustine, a volcano on Augustine Island in Cook Inlet of the Gulf of Alaska, in 1986. The eruption of Mount Augustine was the first to draw attention to a hazard of volcanic eruptions that is unique to modern technological society. It seems that the volcano's andesitic ash melts at temperatures comparable to the operating temperatures of jet engines. Even unmelted, the ash interferes with fuel/air mixing and proper combustion, and glassy shards grind bearings and turbine blades. In 1989, all four engines of a Boeing 747 aircraft choked on the ash from an eruption of Mount Redoubt, another Alaskan volcano, and the jet plunged thousands of feet before the crew succeeded in restarting the engines. Pilots now fly clear of the ash-clouded air near such volcanoes during eruptions, aided by a satellite-based warning system that monitors ash plumes.

SUMMARY

Most volcanic activity is concentrated near plate boundaries. Volcanoes differ widely in eruptive style, and thus in the kinds of dangers they represent. These differences in eruptive style are related to compositional differences among magmas and, in turn, to tectonic setting. Seafloor rift zones and hot spots are characterized by the more fluid, basaltic lavas. Subduction-zone volcanoes typically produce much more viscous, silica-rich, gas-charged andesitic magma, so, in addition to lava, they may emit large quantities of pyroclastics and other deadly products like nuées ardentes. Lava is perhaps the least serious hazard associated with volcanoes: it moves slowly, it can sometimes be diverted, and its path can be predicted. The results of explosive eruptions are less predictable and the eruptions themselves more sudden. One secondary effect of volcanic eruptions, especially explosive ones, is global cooling, which occurs as a result of dust and gases being thrown into the atmosphere and blocking incoming sunlight.

Early signs of potential volcanic activity include bulging and warming of the ground surface and increased seismic activity. Volcanologists cannot yet predict the exact timing or type of eruption very precisely, except insofar as they can anticipate eruptive style on the basis of historic records, the nature of the products of previous eruptions, and tectonic setting.

TERMS TO REMEMBER

active volcano 61
caldera 55
cinder cone 54
columnar jointing 52
composite volcano 55
dormant volcano 61

extinct volcano 61
fissure eruption 51
hot spot 50
lahar 58
lava 51
phreatic eruption 60
pillow lava 53

plume 51
pyroclastic flow 58
shield volcano 52
stratovolcano 55
volcanic dome 53
volcano 51

QUESTIONS FOR REVIEW

1. Describe the nature of fissure eruptions, and give an example.
2. What is a *shield volcano,* and from what kind of lava are shield volcanoes usually built?
3. Compare and contrast a volcanic dome with a shield volcano. What causes the differences in form between these two kinds of volcanic structures?
4. Describe the processes leading to magma production at subduction zones.
5. Hot-spot volcanoes commonly erupt basaltic magma; volcanoes formed on continents above subduction zones (for example, in the Andes) often erupt more silicic lavas. Suggest why this may be so.
6. What causes the eruption of pyroclastics, and with what type of lava are they more often associated?
7. Lava is usually more of a hazard to property than to lives. Why? Describe one strategy for protecting property from an advancing lava flow.
8. On what kind of volcano is a volcanic mudflow most likely to develop? Name a recent example.
9. What is a *pyroclastic flow?* What strategy can minimize the loss of life from eruption of pyroclastic flow?
10. What causes a phreatic eruption? Cite an example.
11. Volcanic eruptions may alter climate. How?
12. Describe two kinds of precursors used in the prediction of volcanic eruptions.
13. What is the underlying cause of volcanic activity in the Cascade Range of the western United States?

FOR FURTHER THOUGHT

1. Look up the projections, made prior to the May 1980 eruption of Mount St. Helens or the 1991 eruptions of Mount Pinatubo, of the areas that might be affected by these eruptions, including secondary effects such as mudflows. Compare the actual consequences with the predictions.
2. Find out if there is any history of volcanic eruptions in your area (a) in historic times or (b) within the last one to two million years. If so, what was the nature of the activity? What is the explanation for it? Might future activity be expected?

SUGGESTIONS FOR FURTHER READING

Alaska Volcano Observatory Staff. 1990. The 1989–1990 eruption of Redoubt Volcano. *EOS* (Trans. Amer. Geophys. Union) 13 (February): 265–75.

Axelrod, D. I. 1981. *Role of volcanism in climate and evolution.* Geological Society of America Special Paper 185.

Bailey, R. A. 1983. Mammoth Lakes earthquakes and ground uplift: Precursors to possible volcanic activity? *Earthquake Information Bulletin* 15 (3): 88–102.

Blong, R. J. 1984. *Volcanic hazards: A source book on the effects of eruptions.* New York: Academic Press.

Bullard, F. M. 1984. *Volcanoes of the earth.* 2d ed. Austin, Tex.: University of Texas Press.

Decker, R., and Decker, B. 1981. *Volcanoes.* San Francisco: W. H. Freeman.

Delaney, P. T., et al. 1990. Deep magma body beneath the summit and rift zones of Kilauea Volcano, Hawaii. *Science* 247 (10 March): 1311–16.

Foxworthy, B. L., and Hill, M. 1982. *Volcanic eruptions of Mount St. Helens: The first 100 days.* U.S. Geological Survey Professional Paper 1249.

Francis, P. I., and Self, S. 1987. Collapsing volcanoes. *Scientific American* 256 (June): 90–97.

Garesche, W. A. 1902. *Complete story of the Martinique and St. Vincent horrors.* L. G. Stahl.

Gore, R. 1984. A prayer for Pozzuoli. *National Geographic* 165: 614–25.

Heliker, C. 1991. *Volcanoes and seismic hazards on the Island of Hawaii.* U.S. Geological Survey.

MacDonald, G. A. 1983. *Volcanoes.* 2d ed. Englewood Cliffs, N.J.: Prentice-Hall.

National Research Council. 1984. *Explosive volcanism: Inception, evolution, and hazards.* Washington, D.C.: National Academy Press.

Pinatubo Volcano Observatory Team. 1991. Lessons from a major eruption: Mt. Pinatubo, Philippines. *EOS* (Trans. Amer. Geophys. Union) 72 (3 December) 545–55.

Rampino, M. R., and Self, S. 1984. The atmospheric effects of El Chichón. *Scientific American* 250 (January): 48–57.

Tilling, R. I. *Eruptions of Mount St. Helens: Past, present, and future.* Washington, D.C.: U.S. Government Printing Office.

Tilling, R. I. (ed.) 1988. *How volcanoes work.* (A collection of readings from the *Journal of Geophysical Research*.) Washington, D.C.: American Geophysical Union.

Volcanoes and the earth's interior. 1983. San Francisco: W. H. Freeman. (A selection of readings from *Scientific American,* 1975–1982.)

Williams, R. S., Jr., and Moore, J. G. 1973. Iceland chills a lava flow. *Geotimes* 18 (August): 14–17.

5

WEATHERING AND SOIL

Weathering eats away rocks exposed at earth's surface, a little at a time; these feldspar crystals are just a little more resistant than their granite matrix. Yosemite National Park, CA. © Doug Sherman/Geofile.

Broadly defined, **soil** is the surface accumulation of rock and mineral fragments and, usually, some organic matter that covers the underlying solid rock of most areas. Soil scientists distinguish between true soil, which they define as capable of supporting plant growth, and **regolith,** which is a more general term for all surface sediment accumulations in place, regardless of their agricultural promise. By and large, the same weathering processes are involved in the formation of the material in either case, and most geologists and many engineers use the terms *soil* and *regolith* interchangeably. This chapter begins by surveying principal weathering processes, and controls on the composition of the resulting soil. It then moves on to consider some special soil-related problems: the difficult nature of lateritic soil, and aspects of soil erosion and its minimization.

Once the products of weathering are transported and redeposited by wind, water, or ice, they fall into the broader category of **sediment.** Sediment and the rocks produced from it are explored more fully in the next chapter.

WEATHERING PROCESSES

The term **weathering** encompasses a variety of chemical, physical, and biological processes that act to break down rocks in place. The relative importance of different kinds of weathering processes is, in turn, largely determined by climate. Climate, topography, the composition of the bedrock or sediment on which the soil is formed, the nature of added organic matter, and time determine a soil's final composition.

Mechanical Weathering

Mechanical weathering is the physical breakup or disintegration of rocks without changes in their composition. It requires that some physical force or stress be applied. Often, the stress is caused by water, as when water in cracks repeatedly freezes and expands, forcing rocks apart, then thaws and contracts or flows away. This process is termed **frost wedging.** Crystallizing salts in cracks may have the same wedging effect, as may plant roots forcing their way into crevices (figure 5.1). In extreme climates like those of deserts, where the contrast between day and night temperatures is very large, many cycles of the daily thermal expansion of rocks baked by sunlight and contraction as they cool at night might cause enough stress to break up the rocks, although this phenomenon has not been reproduced in the laboratory.

Removal of stress can also break up rocks (figure 5.2). When a rock once deeply buried and under pressure is uplifted and the overlying rock eroded away, the rock is *unloaded*. It tends to expand and may fracture in concentric or parallel sheets in the process. The breakup of the rock mass as these sheetlike layers of rock flake off is called **exfoliation.** Granites, which would necessarily have crystallized at some depth initially, often weather by exfoliation when exposed at the surface.

FIGURE 5.1 Cracks in granite widened by the growth of tree roots. Yosemite National Park.

Whatever the cause, the principal effect of mechanical weathering is the breakup of large chunks of rock into smaller ones. In the process, the total exposed surface area of the particles is increased (figure 5.3). Fragments may further break up as sediment is transported from the site of the source bedrock.

Chemical Weathering

Chemical weathering involves the breakdown or decomposition of minerals by chemical reaction with water, with other chemicals dissolved in water, or with gases in the air. Minerals differ in the kinds of chemical reactions they undergo, as well as in how readily they weather chemically.

Calcite (calcium carbonate) dissolves completely, leaving no other minerals behind in its place. This is a contributing factor to the formation of underground limestone caverns. Calcite dissolves rather slowly in plain water but more rapidly in acidic water, such as is produced by acid rainfall. Calcite's susceptibility to dissolution in acid is becoming a serious problem where limestone and its chemical equivalent—marble—are widely used for outdoor sculptures and building stones.

A   B

FIGURE 5.2 Unweighting, and the associated pressure release, causes exfoliation. (A) Exfoliation (schematic): The pluton was originally deeply buried and under pressure. As erosion exposes the pluton, it is unloaded and uncompressed and may break apart in sheets. (B) Granite weathering by exfoliation, Yosemite National Park.

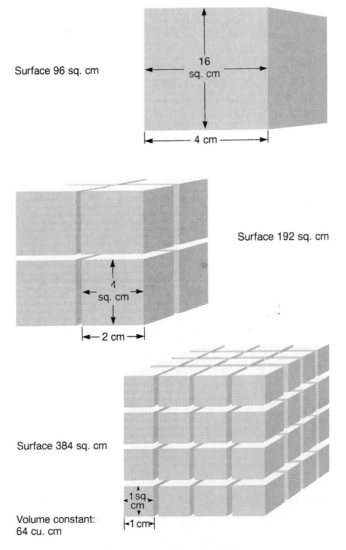

Surface 96 sq. cm

16 sq. cm

4 cm

Surface 192 sq. cm

1 sq. cm

2 cm

Surface 384 sq. cm

1 sq. cm

Volume constant: 64 cu. cm

1 cm

FIGURE 5.3 Mechanical breakup increases both surface area and surface-to-volume ratio.

Box 5.1

Acid Rain and Weathering

Interaction of CO_2 and H_2O in the atmosphere to produce carbonic acid causes rain to be somewhat acidic naturally. As the amount of CO_2 in the air is increased by the burning of fossil fuels, acidity of rainfall tends to increase. Nitrogen gases, produced in internal-combustion engines, also react to form some nitric acid (HNO_3). What concerns those who focus on so-called **acid rain,** however, is the acidity resulting not from these but from sulfur in the air.

Sulfur released into the air reacts with oxygen and water vapor to produce sulfuric acid (H_2SO_4), a strong acid that is both corrosive and highly irritating to eyes and lungs. Human activities contribute sulfur to the air principally through the burning of coal (see chapter on mineral and energy resources). Downwind of

coal-burning furnaces, rainfall is often significantly more acidic than normal; this is acid rain. It causes more rapid chemical weathering of rocks and buildings, acidifies lakes and streams (sometimes to the detriment of wildlife), and may more readily leach toxic metals, such as lead and cadmium, out of soils and into water supplies. (In places, measured rainfall acidity has been comparable to that of the stomach acid that digests our food!)

Local geology may either moderate or aggravate the problems. Limestone tends to neutralize acid, while waters in granitic rocks and soils tend already to be somewhat acidic. Unfortunately, many areas of North America that lie downwind (northeast) of major sulfur sources have granitic bedrock (figure 1). This means that, in these areas, the effects of

acid rain are not moderated by interaction with the rock or soil after the rain falls, and acid-rain problems may be especially severe.

The Canadian government has expressed increased concern over the effects of acid rain in various parts of that country, particularly in the south. What makes this an international political issue is the possibility that the acid-making compounds may be air pollutants generated in the United States that subsequently drift over Canada. Maps of both bedrock (figure 2A) and glacial sediments deposited over that bedrock (figure 2B) reveal broad areas of high geologic sensitivity to acid precipitation in southeastern Canada.

FIGURE 1 Regions downwind of major sources of atmospheric sulfur (dots) are, in many cases, regions underlain by granitic bedrock (shaded).

From Walter W. Roberts, "Rains of Troubles," Science 80: 1 (5): 74–79. Used by permission of the cartographer, Walter R. Roberts.

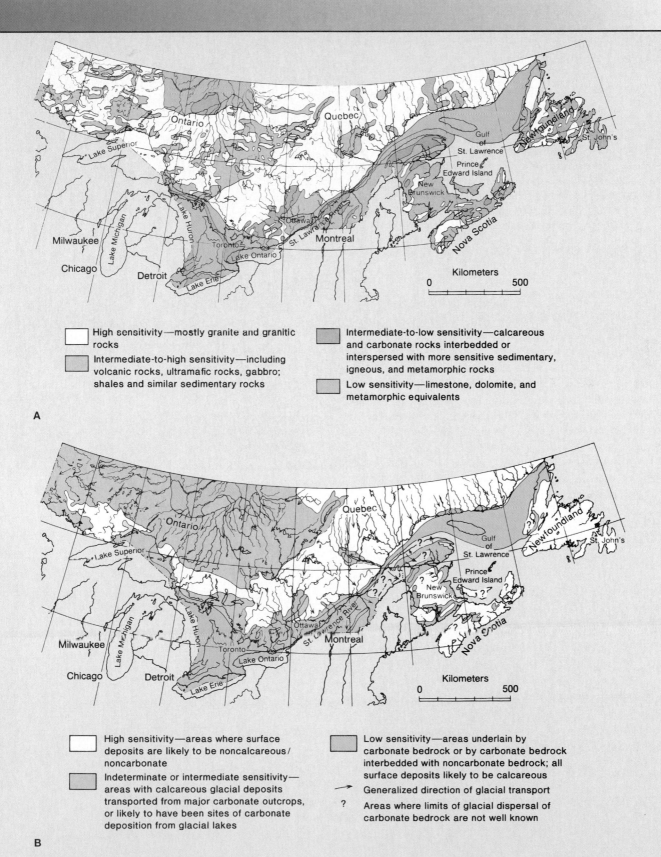

A

High sensitivity—mostly granite and granitic rocks

Intermediate-to-high sensitivity—including volcanic rocks, ultramafic rocks, gabbro; shales and similar sedimentary rocks

Intermediate-to-low sensitivity—calcareous and carbonate rocks interbedded or interspersed with more sensitive sedimentary, igneous, and metamorphic rocks

Low sensitivity—limestone, dolomite, and metamorphic equivalents

B

High sensitivity—areas where surface deposits are likely to be noncalcareous/ noncarbonate

Indeterminate or intermediate sensitivity— areas with calcareous glacial deposits transported from major carbonate outcrops, or likely to have been sites of carbonate deposition from glacial lakes

Low sensitivity—areas underlain by carbonate bedrock or by carbonate bedrock interbedded with noncarbonate bedrock; all surface deposits likely to be calcareous

→ Generalized direction of glacial transport

? Areas where limits of glacial dispersal of carbonate bedrock are not well known

FIGURE 2   (A) Sensitivity of bedrock, and soils derived from bedrock, to acid precipitation in southeastern Canada. (B) Sensitivity of glacial and postglacial sediments to acid precipitation.

After W. W. Shilts, "Sensitivity of Bedrock to Acid Precipitation: Modification by Glacial Processes" in Paper 81–14, 191, Geological Survey of Canada, Natural Resources Canada. Reproduced with the permission of the Minister of Supply and Services Canada.

Cement and concrete also contain calcite and are therefore also susceptible to damage from acid rain (figure 5.4). Calcite dissolution is gradually destroying delicate sculptural features and eating away at the very fabric of many buildings in urban areas, where acid rain is particularly common.

A ctually, all rain is naturally somewhat acidic, though there is much talk now of the "acid rain" caused by human activities (see box 5.1). The acidity of a solution is proportional to the concentration of hydrogen (H^+) ions in it. In air, some carbon dioxide (CO_2) reacts with water vapor (H_2O) to make carbonic acid, a weak acid (H_2CO_3). The carbonic acid dissociates in solution, breaking up into H^+ ions and HCO_3^- (*bicarbonate*) ions, thereby making rain acidic. The acidity is increased by addition of pollutant gases containing nitrogen and sulfur, which leads to formation of nitric and sulfuric acids in the atmosphere.

Silicates tend to be somewhat less susceptible to chemical weathering and to leave other minerals behind when they are attacked. Feldspars principally weather into clay minerals. Because feldspars are the most common minerals in the crust, clays are particularly common in sediments and soils. Ferromagnesian silicates leave behind insoluble iron oxides and hydroxides and sometimes clays, depending on the specific ferromagnesian mineral in question, while other chemical components are dissolved away. Those residual iron compounds are responsible for the red or yellow colors of many soils. In most climates, quartz is extremely resistant to chemical weathering, dissolving only slightly. Representative chemical weathering reactions are shown in table 5.1.

FIGURE 5.4 Sculptural details on this grotesque are being eaten away by acidic rainfall. (Base is much newer than original figure; white support at front is segment of pipe, more resistant to weathering.)

TABLE 5.1 *Some Chemical Weathering Reactions**

Solution of calcite (no solid residue):

$CaCO_3 + 2 H^+ = Ca^{2+} + H_2O + CO_2$ (gas)

Breakdown of ferromagnesian minerals (possible mineral residues include iron compounds, clays):

$FeMgSiO_4$ (olivine) $+ 2 H^+ = Mg^{2+} + Fe(OH)_2 + SiO_2$

$2 KMg_2FeAlSi_3O_{10}(OH)_2$ (biotite) $+ 10 H^+ + 1/2 O_2$ (gas) $= 2 Fe(OH)_3 + Al_2Si_2O_5(OH)_4$ (kaolinite, a clay) $+ 4 SiO_2 + 2 K^+ + 4 Mg^{2+} + 2 H_2O$

Breakdown of feldspar (clay is the common residue):

$2 NaAlSi_3O_8$ (sodium feldspar) $+ 2 H^+ + H_2O = Al_2Si_2O_5(OH)_4 + 4 SiO_2 + 2 Na^+$

Solution of pyrite (making dissolved sulfuric acid, H_2SO_4):

$2 FeS_2 + 5 H_2O + 15/2 O_2$ (gas) $= 4 H_2SO_4 + Fe_2O_3 + H_2O$

Note: Hundreds of possible reactions could be written; the above are only examples of the kinds of processes involved.

*All ions (charged species) are dissolved in solution, as silica (SiO_2) commonly is also; all other substances, except water, are solid unless specified otherwise.

A lthough in the context of weathering we often speak generically of "clay," different clay minerals may have distinct properties and uses. The aluminum-rich *kaolinite* is a white clay used in manufacturing ceramics. The weathering of volcanic glass and tuff commonly yields abundant *montmorillonite,* an "expansive clay" that can absorb water and expand or dry out and shrink, thereby contributing to slope instability (see chapter 16). *Illite* has characteristics of both clay and mica; it is common in shales and is named for the state of Illinois, where much of the early study of clay minerals occurred.

The susceptibility of many silicates to chemical weathering can be inferred from the conditions under which they originally formed. Given several silicates that have crystallized from the same magma, those that formed at the highest temperatures tend to be the least stable, or most easily weathered, at the low temperatures of soil formation at the earth's surface, and vice versa. Also, many of the higher-temperature silicates, like olivine and pyroxene, have simpler structures than do such low-temperature silicates as quartz and mica, and this, too, contributes to their comparative susceptibility to breakdown. The relative ease of chemical weathering of many common silicates can be estimated by inverting Bowen's Reaction Series (see figure 5.5).

A rock's tendency to weather chemically is determined, in turn, by its mineralogical composition. For example, a gabbro, formed at high temperatures and rich in ferromagnesian minerals and calcic plagioclase, generally weathers more readily than does a granite rich in quartz and low-temperature feldspars. In outcrop, differential resistance to weathering is reflected in differential relief between adjacent rock units (figure 5.6).

Climate plays a major role in the intensity of chemical weathering. Most of the relevant chemical reactions involve water, or at least dissolved chemicals. All else being equal, the more water, the more chemical weathering. Also, most chemical reactions proceed more rapidly at high temperatures than at low ones. Therefore, warmer climates are conducive to more rapid chemical weathering. Plants, animals, and microorganisms develop more abundantly and in greater variety in warm, wet climates, too. Many organisms produce compounds that react with and dissolve or break down minerals. Such chemical weathering involving biological activity is thus more intense in warm, moist climates, just as inorganic chemical weathering generally is.

The rates of chemical and mechanical weathering are interrelated. Chemical weathering may accelerate the mechanical breakup of rocks if the minerals being dissolved are holding the rock together by cementing the mineral grains, as in

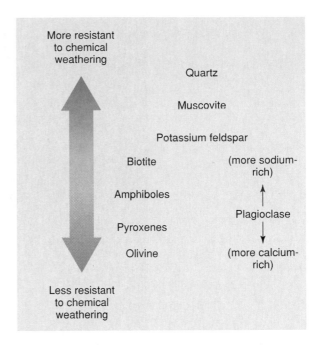

FIGURE 5.5 Bowen's Reaction Series and susceptibility to chemical weathering.

some sedimentary rocks. Increased mechanical weathering may, in turn, accelerate chemical weathering through the increase in exposed surface area, since it is only at grain surfaces that minerals, air, and water interact. The higher the ratio of surface area to volume—that is, the smaller the particles—the more rapid the chemical weathering.

The relationship of surface area to weathering rate can also be seen in the tendency of angular chunks of rock to become more rounded and spheroidal with progressive weathering (figure 5.7). The corners of a block are under attack on several surfaces; edges are being attacked from two sides; the faces of the block are being weathered from only one direction. Therefore, the corners are rounded most rapidly. Once the rock mass has assumed a rather spheroidal shape, further weathering may proceed in curved layers inward from the surface in a process known as **spheroidal weathering** (figure 5.8). This process may be promoted by the weathering of feldspars into clays, with an attendant increase in volume, the expansion contributing to the flaking-off of successive layers of weathered rock.

SOIL

The result of mechanical and chemical weathering, together with the accumulation of decaying remains from organisms living on the land, is the formation of a layer of soil between

A

B

FIGURE 5.6 Differential susceptibility of rock units to weathering and erosion results in differential relief. (A) These quartz veins are more resistant than the sedimentary rock they crosscut. (B) In Bryce Canyon, Utah, the more resistant sedimentary layers are prominent as shelves in the weathered landscape.

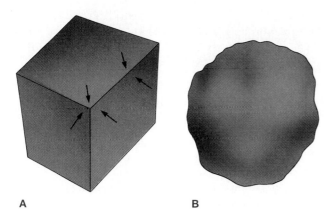

FIGURE 5.7 Angular fragments are rounded by weathering. (A) Corners are attacked from three sides, edges from two, surfaces from only one direction. (B) In time, fragments become rounded as a result.

bedrock and atmosphere. A cross section of this soil blanket usually reveals a series of layers of different colors, compositions, and physical properties. The number of layers and their thicknesses vary.

The Soil Profile

A basic, generalized soil profile as developed over bedrock (*residual soil*) is shown in figure 5.9. At the top is the **A horizon.** It consists of the most intensively weathered rock material, being the zone most exposed to surface processes. It also contains most of the organic remains. Unless the local water table is exceptionally high, precipitation infiltrates down through the A horizon. In so doing, the water may dissolve soluble minerals and carry them away. This process is known as **leaching,** and the A horizon is also known as the *zone of leaching.* Fine-grained minerals, like clays, may also be carried physically downward with percolating water.

FIGURE 5.8 Spheroidal weathering of basalt in Hawaii.

Many of the minerals leached from the A horizon accumulate in the layer below, the **B horizon,** also known as the *zone of accumulation* or *zone of deposition*. Soil in the B horizon is often coarser-grained than the A-horizon soil because it has been somewhat protected from surface processes. In other cases, accumulation of clays in the B horizon contributes to a layer of fine, compact soil there. Organic matter from the surface is also less well mixed into the B horizon.

Below the B horizon is a zone consisting principally of very coarsely broken-up bedrock and little else. This is the **C horizon,** which does not resemble our usual idea of soil at all. Sometimes, the bedrock or parent material itself is designated as a fourth horizon, the D horizon or (in more current usage) R layer.

The boundaries between adjacent soil horizons may be sharp or indistinct. In some instances, one horizon may be divided into several recognizable subhorizons—for example, the top of the A horizon might consist of a layer of topsoil so rich in organic matter that a distinct "O horizon" can be designated. Subhorizons may also exist that are gradational between A and B or B and C horizons. One or more horizons may be absent from the soil profile. Soil developed on sediment (*transported soil,* meaning one developed on transported material) may show a different characteristic profile than soil developed on bedrock.

Variations in soil characteristics arise from the different mix of soil-forming processes and starting materials found from place to place. In general, soil formation is a slow process; many centuries may be required to produce a thick soil with well-developed horizons. The overall total thickness of soil above the bedrock is partly a function of the local rate of soil formation and partly a function of the rate of soil erosion. The latter reflects the work of wind and water, the topography, and the extent and kinds of human activities.

Color, Texture, and Structure of Soils

The physical properties of the soil are affected by its mineralogy, the texture of the mineral grains (coarse or fine, well or poorly sorted by grain size, rounded or angular, and so on), and any organic matter present.

Soil *color* tends to reflect compositional characteristics. Soils rich in organic matter tend to be black or brown, while those poor in organic matter are paler in color, often white or gray. When iron is present and has been oxidized by reaction with oxygen in air or water, it adds a yellow or red color. (Rust on iron or steel is also produced by oxidation of iron.)

Soil *texture* is related to the sizes of fragments in the soil. The U.S. Department of Agriculture recognizes three size components: sand (grain diameters 2 to 0.05 millimeters), silt (0.05 to 0.002 mm), and clay (less than .002 mm). Soils are named on the basis of the dominant grain size(s) present (figure 5.10). An additional term, *loam,* describes a soil that is a mixture of all three particle sizes in similar proportions (10 to 30% clay, the balance nearly equal amounts of sand and silt). Soils of intermediate particle-size distribution have mixed names: for example, "silty clay" would be a soil consisting of about half silt-sized particles, half clay-sized particles. The significance of soil texture is primarily the way it influences drainage. Sandy soils, with high permeability, drain quickly.

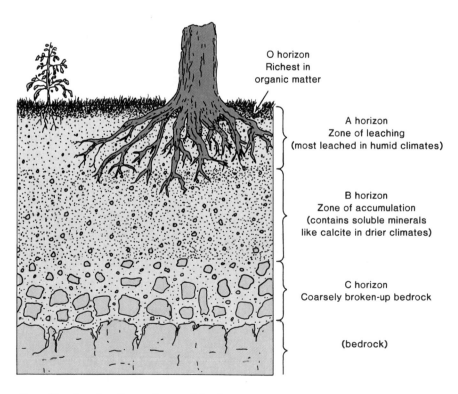

O horizon
Richest in organic matter

A horizon
Zone of leaching
(most leached in humid climates)

B horizon
Zone of accumulation
(contains soluble minerals
like calcite in drier climates)

C horizon
Coarsely broken-up bedrock

(bedrock)

FIGURE 5.9 A representative soil profile, showing the various soil horizons.

In an agricultural setting, this may mean a need for frequent watering or irrigation. In a flood-control context, it means relatively rapid infiltration. Clay-rich soils, by contrast, may hold a great deal of water but be relatively slow to drain, by virtue of their much lower permeability, a characteristic shared by their sedimentary-rock counterpart, shale.

Soil *structure* relates to the soil's tendency to form lumps or clods of soil particles. These clumps are technically called *peds,* their name deriving from the Latin root *pedo-,* meaning "soil." A soil that clumps readily may be more resistant to erosion. On the other hand, soil consisting of very large peds with large cracks between them may be a poor growing medium for small plants with fine roots. Abundant organic matter may promote the aggregation of soil particles into crumblike peds especially conducive to good plant growth; mechanical weathering can break up larger clumps into smaller, just as it breaks up rock fragments.

Soil Composition and Classification

Mechanical weathering merely breaks up rock without changing its composition. In the absence of pollution, wind and rainwater rarely add many chemicals, and runoff water may carry away some leached chemicals in solution. Thus, chemical weathering tends to involve a net subtraction of elements from rock or soil, and a primary control on a soil's composition is the composition of the bedrock (or sediment) on which it is formed. If the bedrock is low in certain critical plant nutrients, the soil produced from it will also be low in those nutrients, and chemical fertilizers may be needed to grow particular crops on that soil even if the soil has never been farmed before.

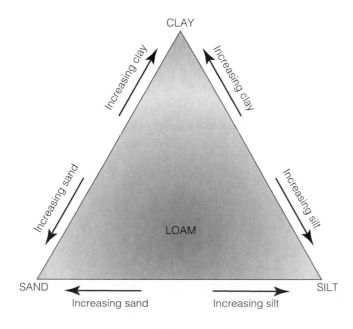

Figure 5.10 Soil-texture terminology reflects particle size, not mineralogy. This diagram shows simplified U.S. Department of Agriculture classification scheme.

The extent and balance of weathering processes involved in soil formation determine the extent of further depletion of the soil in various elements. Mechanical weathering usually dominates only in cooler, drier climates; under other conditions, chemical weathering, often accelerated by biological activity, is the dominant process. The weathering processes also influence the mineralogy of the soil, the compounds in which the soil's elements occur. The physical properties of the soil are affected by its mineralogy, the texture of the mineral grains (coarse or fine, rounded or angular, and so on), and any organic matter present.

Early attempts at soil classification emphasized compositional differences among soils and thus principally reflected the effects of chemical weathering. Two broad categories of soil were recognized. The **pedalfer** soils were seen as characteristic of more humid regions. Where the climate is wetter, there is naturally more extensive leaching of the soil, and what remains is enriched in the less-soluble oxides and hydroxides of aluminum and iron and, especially in the B horizon, clay. The term *pedalfer* comes from the Latin prefix *pedo-* ("soil") and the Latin words for aluminum (*alumium*) and iron (*ferrum*). In North America, pedalfer-type soils are found in higher-rainfall areas like the eastern United States and most of Canada. Pedalfer soils are typically acidic.

Where the climate is drier, such as in the plains states and the western and especially the southwestern areas of the United States, leaching is much less extensive. Even quite soluble compounds like calcite remain in the soil, especially in the B horizon, where they may have been redeposited after leaching from near-surface layers. From this observation came the term **pedocal** for the soil of a dry climate. The presence of calcium carbonate makes pedocal soils more alkaline.

The pedocals are thus not highly leached by infiltrating water from above. Moreover, in many of these dry areas, net water flow is *upward*. Where evaporation near the surface is rapid, soil moisture may be drawn upward along the grain surfaces by **capillary action.** The effect is similar to the way in which kerosene is drawn up the wick in a kerosene lamp as fuel is burned at the top of the wick. In the case of pedocal soil, the water rises, carrying with it any dissolved salts, and then, as it evaporates, it deposits those salts. Some pedocal soils become so salty that they are unsuitable for agriculture. Capillary action contributes to formation of a crusty, calcite-rich layer called **caliche,** or *hardpan,* in a pedocal soil.

One problem with the simple pedalfer/pedocal classification scheme is that, for it to be strictly applied, the soils it describes must have formed over suitable bedrock. For example, if the bedrock and early soils in a region contained little calcium, then the later soils would not be likely to contain a great deal of calcite either, regardless of how little leaching had occurred. Conversely, a rock poor in iron and aluminum, such as a limestone, will not leave an iron- and aluminum-rich residue, no matter how extensively it is leached. Still, the terms *pedalfer* and *pedocal* can be used generally to indicate, respectively, more and less extensively leached soils.

TABLE 5.2 *Soil Orders of the Seventh Approximation, with General Descriptions of Terms*

Order	Description
1. Entisols	Soils without layering, except perhaps a plowed layer
2. Vertisols	Soils with upper layers mixed or inverted because they contain expandable clays (clays that expand when wet, contract and crack when dry)
3. Inceptisols	Very young soils with weakly developed soil layers and not much leaching or mineral alteration
4. Aridosols	Soils of deserts and semiarid regions, and related saline or alkaline soils
5. Mollisols	Grassland soils, mostly rich in calcium; also, forest soils developed on calcium-rich parent materials; characterized by a thick surface layer rich in organic material
6. Spodosols	Soils with a light, ashy-gray A horizon and a B horizon containing organic matter and clay leached from the A horizon
7. Alfisols	Includes most other acid soils with clay-enriched subsoils
8. Ultisols	Similar to alfisols but with weathering more advanced; includes some lateritic soils
9. Oxisols	Still more weathered than the ultisols; includes most laterites
10. Histosols	Bog-type soils

Source: Data from C. B. Hunt, *Geology of Soils,* Copyright © 1972 W. H. Freeman and Company.

FIGURE 5.11 Lateritic soil in northern Australia, showing the bright red color of iron oxide minerals.

PROBLEMS RELATING TO SOIL

The presence of certain types of soil can present problems because of the soil's particular physical or agricultural properties. The absence of soil, the result of erosion, is increasingly problematic as well.

Lateritic Soil

One widely recognized soil type is sufficiently distinctive and important, especially to many less-developed nations, to justify special consideration. This is **lateritic soil.** A *laterite* may be regarded as an extreme kind of pedalfer. Lateritic soils develop in tropical climates with high temperatures and heavy rainfall, so they are extensively leached. Even quartz may have been dissolved out of the soil under these conditions. Lateritic soil (figure 5.11) often contains very little besides insoluble aluminum and iron oxide compounds.

Soils of lush tropical rain forests are commonly lateritic, which seems to suggest that lateritic soils have great potential as farmland. Surprisingly, however, the opposite is true, for two reasons.

The highly leached character of lateritic soils is one reason. Even where the vegetation is dense, the soil itself has few soluble nutrients left. The forest holds a huge reserve of nutrients, but there is no corresponding reserve in the soil. Further growth is supported by the decay of earlier vegetation. As one plant dies and decomposes, the nutrients it contained are quickly either taken up by other plants or leached away. If the rain forest is cleared to plant crops, most of the nutrients are cleared away with it, leaving little in the soil to nourish the crops. Many natives of tropical climates practice

Modern soil classification has become considerably more sophisticated and complex, taking into account characteristics of present soil composition and texture, the type of bedrock on which the soil was formed, the present climate of the region, the degree of chemical "maturity" of the soil, and the extent of development of the different soil horizons. Different countries have adopted different soil classification schemes. One United Nations Educational, Scientific, and Cultural Organization (UNESCO) world map uses 110 different soil map units. This is, in fact, a small number compared to the number of distinctions made under certain other classification schemes.

The U.S. comprehensive soil classification, known as the Seventh Approximation, has ten major categories (orders) that are subdivided through five more levels of classification into a total of some 12,000 soil series. A short summary of the ten orders is presented for reference in table 5.2. Certain of the orders are characterized by a particular environment of formation and the distinctive soil properties resulting from it—for example, the histosols, which are bog soils. Others are characterized principally by physical properties—for example, the entisols, which lack horizon zonation, and the vertisols, in which the upper layers are mixed because these soils contain expansive clays. Most of the oxisols and some ultisols are soils of a type that has serious implications for agriculture, especially in much of the Third World: lateritic soils.

"slash-and-burn" agriculture, cutting and burning the jungle to clear the land. Some of the nutrients in the burned vegetation do settle into the topsoil temporarily, but relentless leaching by the warm rains makes the soil nutrient-poor and infertile within a few growing seasons.

A second problem with lateritic soil is suggested by the term *laterite* itself, which is derived from the Latin for "brick." A lush rain forest shields the soil from the drying and baking effects of the sun, while vigorous root action helps to keep the soil well broken up. When its vegetative cover is cleared and it is exposed to the baking tropical sun, lateritic soil can quickly harden into a solid, bricklike consistency that resists infiltration by water or penetration by crops' roots. What crops can be grown provide little protection for the soil and do not slow the hardening process very much. Within five years or less, a freshly cleared field may become completely uncultivatable. The tendency of laterite to bake into brick has been used to advantage in some tropical regions, where blocks of hardened laterite have served as building materials.

Still, the problem of how to maintain crops remains. Often, the only recourse is to abandon each field after a few years and clear a replacement. This results, over time, in the destruction of vast tracts of rain forest where only a moderate expanse of farmland is needed. Also, once the soil has hardened and efforts to farm it have been abandoned, it may revegetate only slowly, if at all. This can result in the loss of arable land, just as can desertification, described in chapter 17.

In Indochina, in what is now Cambodian jungle, are the remains of the Khmer civilization that flourished from the ninth to the sixteenth centuries. There is no clear evidence of why that civilization vanished. A major reason may have been the difficulty of agriculture in the region's lateritic soil. Perhaps the Mayas, too, moved north into Mexico to escape the problems of lateritic soils. Historically, some countries in Africa and Asia with latcritic soils have achieved agricultural success only because frequent flooding of major rivers has deposited fresh, nutrient-rich sediment over the depleted laterite. Recent moves to control the flooding and the attendant damage and loss of life could, ironically, create an agricultural disaster for such regions.

Soil Erosion and Its Control

Weathering is the breakdown of rock or mineral material in place, while *erosion* involves physical removal of material from one place to another. Weathering may accelerate erosion. Later chapters examine the nature of erosion by the action of water, wind, and ice in more detail. This section focuses specifically on erosion as it affects soil.

Rain striking the ground helps to break soil particles loose. Surface runoff and wind together carry loosened soil away. Surface runoff washing down over a slope surface is **sheet wash.** Where the water begins to erode small, preferred channels, **rill erosion** occurs. Should the rills enlarge to produce channels so deep that normal cultivation will not

FIGURE 5.12 Gullying on unprotected farmland.
Photograph courtesy of USDA Soil Conservation Service.

erase them, the result is **gullying** (figure 5.12). The faster that wind and water travel, the larger the particles and the greater the load they can move. Therefore, high winds cause more erosion than do calmer ones, and fast-flowing surface runoff moves more soil than does slow runoff. This in turn suggests that, all else being equal, steep and unobstructed slopes are more susceptible to erosion by water, for surface runoff flows more rapidly over them. Flat land lacking vegetation is correspondingly more vulnerable to wind erosion. The physical properties of the soil—for example, how cohesive it is—also influence its vulnerability to erosion.

Rates of soil erosion can be estimated in a variety of ways. Over a large area, erosion resulting from surface runoff may be judged by estimating the sediment load of streams draining the area. On small plots, runoff can be collected and its sediment load measured. Wind is harder to trap or monitor comprehensively, especially over a range of altitudes, so the extent of wind erosion is more difficult to estimate. Generally, it is much less significant than erosion through surface-water runoff, except under drought conditions. Controlled laboratory experiments can be used to simulate wind and water erosion and to measure their impact.

Estimates of the total amount of soil erosion in the United States vary widely, but U.S. Soil Conservation Service estimates put the figure at over 4 billion tons per year. The present rates of erosion have been accelerated by human activities, construction, and especially farming (recall figure 1.10). Some 412 million acres of U.S. cropland suffer soil-erosion losses averaging an estimated 4.8 tons per acre per year from water alone. In very round numbers, that is a thickness of about 0.04 centimeter of soil removed each year, on average.

It is difficult to generalize how this compares with the rate of soil *formation* because the rate of soil formation is so sensitive to climate, the nature of the parent rock material, and other factors. Some upper limits on soil-formation rates

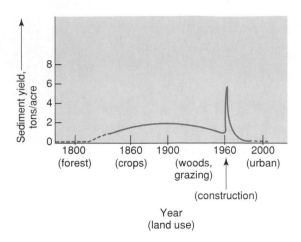

Figure 5.13 Rates of erosion vary dramatically as a function of land use.

From M. G. Wolman, "A Cycle of Sedimentation and Erosion in Urban River Channels" in Geografiska Annaler, *Vol. 49, Series A. Copyright © Swedish Society of Anthropology and Geography. Reprinted by permission.*

can be estimated by looking at soils in areas of the northern United States last scraped clean by the glaciers tens of thousands of years ago. Where the parent material was glacier-deposited sediment, which should weather more easily than solid rock, about 1 meter of soil has formed in 15,000 years in the temperate, fairly humid climate of the upper Midwest. The corresponding average rate of 0.006 centimeter of soil formed per year is less than one-sixth the average rate of cropland erosion by water alone. Furthermore, soil formation in many areas, especially over more resistant bedrock, is slower still. In some places in the Midwest, virtually no soil has formed on glaciated bedrock; in the drier Southwest, soil formation is also likely to be much slower. Soil erosion, too, can locally be much more rapid than the average: over 90 tons per acre per year may be lost from cultivated hillsides where no soil-conservation measures are practiced.

Figure 5.13 shows that erosion during active urbanization (highway and building construction and the like) is considerably more severe than erosion on any sort of undeveloped land. However, the total area undergoing active development is only about 1.5 million acres each year. Also, this disturbance is of relatively short duration. Once the land is covered by buildings, pavement, or established landscaping, erosion rates typically drop below even natural predevelopment levels. The problems of erosion during construction are thus very intensive but typically localized and short-lived. The remaining discussion in this section, therefore, focuses on the volumetrically larger problem of erosion from agricultural land.

Though not always viewed as such, soil is a resource, like ores or fuels. The slowness of its production relative to observed erosion rates in many places suggests that, locally at least, it is an exhaustible resource. The organic-matter-rich topsoil, with its higher content of nutrients and better moisture retention, is especially fertile and suitable for agriculture. But it is also the topsoil that is naturally lost first as the soil erodes. Topsoil erosion from cropland then leads to reduced productivity and crop quality and thus reduced agricultural income.

Also, the soil eroded from one place is deposited, sooner or later, somewhere else. If a large quantity is moved and dumped on other farmland, the crops on which it is deposited may be stunted or destroyed (although *small* additions of fresh topsoil may enrich cropland and make it more productive). A subtle consequence of soil erosion in some places has been increased persistence of toxic residues of herbicides and pesticides in the soil: loss of nutrients and organic matter through topsoil erosion may decrease the activity of soil microorganisms that normally speed the breakdown of toxic agricultural chemicals.

Another major soil-erosion problem is sediment pollution. In the United States, about 750 million tons of eroded sediment end up in lakes and streams each year. This decreases the water quality and may harm wildlife. The problem is still more acute when those sediments contain toxic chemical residues. A secondary consequence of this sediment load is the filling in of stream channels and reservoirs, restricting navigation and decreasing the volume of reservoirs and thus their usefulness for their intended purposes, whether for water supply, hydroelectric power generation, or flood control.

The wide variety of approaches for reducing erosion on farmland basically all involve either reducing the velocity of an eroding agent or protecting the soil from the effects of that eroding agent. Under the latter heading come such practices as leaving stubble in the fields after a crop has been harvested and planting cover crops in the off season between cash crops. In either case, the plants' roots help to hold the soil in place, and the foliage itself, to some extent, shields the soil from wind and rain.

The lower the wind or runoff-water velocity, the less material carried. Wind can be slowed down by the use of *windbreaks* along the borders of fields. These can be trees or hedges, planted in rows perpendicular to the dominant wind direction, or low fences similarly arrayed. This does not altogether stop the soil movement, as evidenced by the ridges of soil that pile up along windbreaks, but it does reduce the distance over which soil is transported, and some of the soil caught along the windbreaks can be redistributed over the fields. *Strip cropping*—planting strips of crops of different heights, to disrupt wind flow across the land surface (figure 5.14)—is another strategy.

Surface runoff can be slowed on moderate slopes by *contour plowing* (figure 5.15A). Plowing in rows that run around the hill, parallel to the contours of the hill and thus perpendicular to the direction of water flow, creates a ridged land surface down which water does not rush as readily. Other slopes may require *terracing*, which involves breaking up a single slope into a series of shallower slopes, or even into steps that slant backward into the hillside (figure 5.15B). Again, surface runoff making its way down a terraced slope

FIGURE 5.14 Strip cropping breaks up the surface topography, disrupting sweeping winds.
Photograph courtesy of USDA Soil Conservation Service.

A

B

FIGURE 5.15 (A) Contour plowing to slow sheet wash and channelized water runoff. (B) Terracing also reduces the velocity of surface runoff. Modern terraces in Midwestern farmland.
Photographs courtesy of USDA Soil Conservation Service.

does so more slowly, if at all, and therefore carries far less sediment with it. Terracing has been practiced since ancient times.

Other strategies can minimize erosion in nonfarm areas. With urban construction projects, one reason for the severity of the resulting erosion is that it has been common practice to clear a whole area, such as an entire housing-development site, at the beginning of the work, even though only a small portion of the site will be worked actively at any given time. Clearing land in stages as needed, and thus minimizing the length of time the soil is exposed in any one place, is a way to reduce urban erosion. Stricter mining regulations requiring reclamation of strip-mined land (see chapter on mineral and energy resources) are already significantly reducing soil erosion and related problems in mined areas.

SUMMARY

Soil forms from the breakdown of rock materials through chemical and mechanical weathering. Weathering rates are closely related to climate. The character of the resultant soil reflects the nature of the parent material and the kinds and intensities of weathering processes through which the soil formed. The wetter the climate, the more leached the soil; the hotter the climate, the more vigorous the chemical weathering, provided that water is present. The pedalfer soils are the somewhat leached soils of temperate climates with moderate rainfall. Pedocal soils, formed in drier climates, retain many more soluble minerals. The lateritic soils of tropical climates are particularly unsuitable for agriculture: not only are they highly leached of nutrients, but when exposed by clearing of the forest cover, they harden to bricklike solidity. The texture of soil is a major determinant of its drainage characteristics; soil structure is related to its suitability for agriculture.

Soil erosion by wind and water is a natural part of the rock cycle. Where accelerated by human activity, however, it can also be a serious problem, especially on farmland or, locally, in areas subject to construction or strip mining. Erosion rates far exceed inferred rates of soil formation in many places. Loss of soil fertility or productivity is a particular concern on farmland. A secondary problem associated with soil erosion is the resultant sediment pollution of lakes and streams. Strategies to reduce soil erosion focus on reducing wind and water velocity—through windbreaks, terracing, or contour plowing—or providing vegetative cover to protect and anchor soil in place.

TERMS TO REMEMBER

acid rain 72
A horizon 77
B horizon 78
caliche 79
capillary action 79
chemical weathering 70
C horizon 78

exfoliation 70
frost wedging 70
gullying 81
lateritic soil 80
leaching 77
mechanical weathering 70
pedalfer 79
pedocal 79

regolith 70
rill erosion 81
sediment 70
sheet wash 81
soil 70
spheroidal weathering 75
weathering 70

QUESTIONS FOR REVIEW

1. What is the distinction between soil and regolith? Between soil and sediment?
2. Define *mechanical weathering,* and give two examples.
3. What is *exfoliation,* and why is it believed to occur?
4. Chemical weathering is more rapid in warmer and wetter climates than in cooler and drier ones. Why?
5. Rank the following minerals in terms of their expected resistance to chemical weathering (most resistant first): biotite, calcite, quartz, calcium-rich plagioclase, amphibole.
6. Sketch a generalized soil profile. Indicate the A, B, and C horizons, zone of accumulation, and zone of leaching. Explain the latter two terms.
7. Distinguish between pedalfer and pedocal types of soil. In what kind of climate is each more common?
8. What is customarily meant by the phrase "acid rain"? Is the impact of acid rainfall the same everywhere in North America? Explain.
9. Describe how the tendency of a soil to form clumps, or peds, can influence its susceptibility to erosion and its ability to support plant growth.
10. What is a *laterite?* Where are laterites common? Are they fertile soils for agriculture? Explain.
11. Cite and briefly explain any three strategies for reducing soil erosion.

FOR FURTHER THOUGHT

1. Given the reactions in table 5.1, suggest why underground caverns form in limestone but not in granite, even if the granite is weathered.

2. Rates of weathering, and the relative resistance of different rock types to weathering, can often be evaluated with more accuracy using stone buildings, monuments, or even gravestones in cemeteries than by examining natural rocks in the field. Why? Suggest some possible limitations on the use of such artificial structures to estimate weathering rates.

3. Inspect a local construction site or farmland, and look for evidence of soil erosion. Consider what might be done to reduce any problems you identify.

SUGGESTIONS FOR FURTHER READING

Birkeland, P. W. 1984. *Soils and geomorphology*. New York: Oxford University Press.

Blaxter, Sirk. 1986. *People, food, and resources*. Cambridge, England: Cambridge University Press.

Bridges, E. M. 1978. *World soils*. 2d ed. Cambridge, England: Cambridge University Press.

Buol, S. W., Hole, F. D., and McCracken, R. J. 1980. *Soil genesis and classification*. 2d ed. Ames, Iowa: Iowa State University Press.

Crosson, P. R., and Rosenberg, N. J. 1989. Strategies for agriculture. *Scientific American* 261 (September): 128–35.

Gerard, A. J. 1981. *Soils and landforms*. Boston: Allen and Unwin.

Harlin, J. M., and Berardi, G. M. 1987. *Agricultural soil loss: Processes, policies, and prospects*. Boulder, Colo.: Westview Press.

Hunt, C. B. 1972. *Geology of soils*. San Francisco: W. H. Freeman.

Lal, R. 1987. Managing the soils of sub-Saharan Africa. *Science* 236: 1069–76.

Lindsay, W. L. 1979. *Chemical equilibria in soils*. New York: John Wiley and Sons.

Morgan, R. P. C. 1979. *Soil erosion*. London: Longman Group.

Paton, T. R. 1978. *The formation of soil material*. Boston: Allen and Unwin.

Shilts, W. W. 1981. *Sensitivity of bedrock to acid precipitation: Modification by glacial processes*. Geological Survey of Canada Paper 81–14.

Sposito, G. 1989. *The chemistry of soils*. New York: Oxford University Press.

Thompson, L. M., and Troeh, F. R. 1978. *Soils and soil fertility*. 4th ed. New York: McGraw-Hill.

Troeh, F. R., Hobbs, J. A., and Donahue, R. L. 1980. *Soil and water conservation*. Englewood Cliffs, N.J.: Prentice-Hall.

6

SEDIMENT AND SEDIMENTARY ROCKS

*Layers enriched in iron oxides help define the windswept crossbeds
formed as the Navajo Sandstone was deposited. Sample near Kanab, UT.
© Doug Sherman/Geofile*

INTRODUCTION

As noted in the last chapter, sediments are the products of weathering and erosion, transported and then deposited by the action of wind, water, or ice. When sediments become consolidated into a cohesive mass, they become sedimentary rock. The composition, texture, and other features of a sedimentary rock can provide clues about its origin and source materials and the setting in which the sediment was deposited. The igneous rocks discussed in chapter 3 represent one end of the temperature spectrum. Sedimentary rocks represent the other end. The latter part of this chapter is devoted to the formation, characterization, and classification of sedimentary rocks.

TYPES OF SEDIMENTS

Sediments are classified, first, on the basis of the way in which they are formed, and second, on the basis of composition or particle size. The principal division is into *clastic* versus *chemical* sediments. Each of these types of sediment can include a biological contribution.

Clastic Sediments

Clastic sediments, which take their name from the Greek word *klastos,* meaning "broken," are composed of broken-up fragments of preexisting rocks and minerals. Individual fragments in a clastic sediment may be single mineral grains or bits of rock comprising several minerals.

Clastic sediments, and the rocks formed from them, are named on the basis of fragment sizes (table 6.1). Gravels and boulders are the most coarse-grained clastic sediments. When they are transformed into rock, the corresponding rock is called either **conglomerate** or **breccia,** depending on whether the fragments are, respectively, well rounded or angular (figure 6.1). The next finer sediments are sands, which,

when consolidated, make **sandstone** (figure 6.2). As particle size decreases further, individual grains become indistinct to the naked eye, and the sediments feel less gritty to the touch. These very fine sediments are the silts and clays. Properly, the corresponding rocks are *siltstones* and *claystones.* However, since the grains cannot be seen with the naked eye, siltstones and claystones cannot readily be distinguished in handsample. They are often lumped together under the general heading of **mudstone,** any rock made from fine, muddy sediment. Mudstones commonly show a tendency to break roughly

A

B

FIGURE 6.1 Coarse-grained clastic sedimentary rocks.
(A) Conglomerate—rounded fragments (note quarter for scale).
(B) Breccia—angular fragments.
(B) © Times Mirror Higher Education Group, Inc./Doug Sherman, photographer.

TABLE 6.1 *Simplified Classification Scheme for Clastic Sediments and Rocks*

Particle Size Range	Sediment	Rock
Over 256 mm (10 in)	Boulder	Conglomerate (rounded fragments) or breccia (angular fragments)
2 to 256 mm (0.08 to 10 in)	Gravel	Conglomerate or breccia
1/16 to 2 mm (0.0025 to 0.08 in)	Sand	Sandstone
1/256 to 1/16 mm (0.00015 to 0.0025 in)	Silt	Siltstone*
Less than 1/256 mm (less than 0.00015 in)	Clay	Claystone*

*Both *siltstone* and *claystone* are also known as *mudstone,* commonly called *shale* if the rock shows a tendency to split on parallel planes.

FIGURE 6.2 Example of sandstone, a clastic rock of intermediate grain size.
© *Times Mirror Higher Education Group, Inc./Bob Coyle, photographer.*

FIGURE 6.3 Shale, a variety of mudstone. Note that individual grains are too fine to be seen with the naked eye.
© *Times Mirror Higher Education Group, Inc./Bob Coyle, photographer.*

along planes corresponding to depositional layers, in which case they are given the name **shale** (figure 6.3). Shale is the most abundant of the sedimentary rock types.

None of the previous terms carries any implications about the composition of the fragments or even about whether the fragments are mineral grains or chunks of rock. Even the terms *clay* and *claystone* do not necessarily mean that the sediments consist of clay minerals, although clay minerals usually form only very fine crystals, and many mudstones and shales do indeed consist predominantly of clay minerals. However, clay-sized fragments of quartz, feldspar, obsidian, or other materials can also occur.

As already mentioned, the most coarse-grained clastic sediments—conglomerate and breccia—are distinguished on the basis of roundness or angularity of fragments. *Roundness* and *sphericity* are distinct properties, as illustrated in figure 6.4. Angular fragments have sharp corners; the more well-rounded the fragments become, the smoother their corners. Fragments need not be spheroidal, like little balls, to be considered well-rounded; they must simply have smooth, rounded corners and edges. Rounding may occur mechanically, by abrasion, or chemically, by preferential solution of projecting points and corners, as described earlier. With finer-grained sediments, distinctions are not made on the basis of roundness. In the very finest sediments, roundness cannot, after all, be determined with the naked eye because individual grains cannot be seen. Also, by the time rock and mineral fragments have been broken down into very fine pieces, they are usually somewhat rounded by abrasion anyway.

Another textural characteristic of clastic sediments and rocks is the degree of **sorting** they show. Sorting is a measure of the range of grain sizes present. A well-sorted sediment is one in which the grains are all similar in size (figure 6.5); a poorly sorted sediment shows wide variation in grain size (figure 6.6). The finer material filling in the pores between coarser grains in a poorly sorted sediment is termed the **matrix.** Logically, more potential for poor sorting exists among

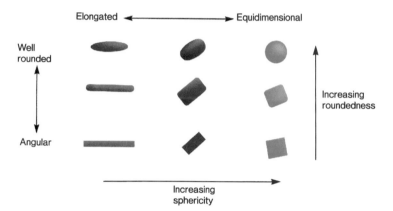

FIGURE 6.4 Roundness versus grain shape. A grain can be elongated but well-rounded or blocky (equidimensional) but angular.

the coarser sediments: grain-size variation is limited among clay-sized fragments when all must, by definition, be smaller than 1/256 millimeter in diameter! Sorting, in turn, influences other properties of sediment, such as the volume of open pore space between grains in the sediment. Sorting also provides clues to the mode of sediment transport and environment of deposition. For example, sediments deposited directly by melting ice are typically poorly sorted; those deposited by wind or flowing water are often better sorted. The reasons for this are explored more fully in this and later chapters.

Chemical Sediments

Chemical sediments are those precipitated from solution. The rocks formed from them are named, for the most part, on the basis of composition, not grain size. Several major types are summarized in table 6.2.

Volumetrically, the most important chemical sediment is **limestone,** a rock composed of carbonate minerals

Figure 6.5 Well-sorted sediment: sample of Tapeats sandstone (close-up view).
© *Times Mirror Higher Education Group, Inc. /Doug Sherman, photographer.*

Figure 6.6 Sedimentary rock with poorly sorted fragments.

TABLE 6.2 *Some Chemical Sedimentary Rocks and Constituent Minerals*

Rock Type	Composed of
Carbonate	
Limestone	Calcite, aragonite (both polymorphs of $CaCO_3$)
Dolomite	Dolomite ($CaMg(CO_3)_2$)
Evaporite	Variable—possibilities include halite ($NaCl$), gypsum (hydrated calcium sulfate, $CaSO_4 \cdot 2H_2O$), anhydrite ($CaSO_4$), borax ($Na_2B_4O_7 \cdot 10H_2O$)
Chert	Silica (SiO_2)
Banded ironstone	Hematite, magnetite (iron oxides, Fe_2O_3, Fe_3O_4), usually with quartz or calcite

Figure 6.7 Limestone, a chemical sedimentary rock. Note chalky appearance and tendency to break along bedding planes.
Photograph by I. J. Witkind, USGS Photo Library, Denver, CO.

(figure 6.7). (Any rock consisting mostly of calcium and magnesium carbonate can properly be called limestone in the broad sense. However, the term is sometimes restricted to rocks consisting of calcite, calcium carbonate, while a rock made of the calcium-magnesium carbonate mineral **dolomite** may itself be called dolomite, or dolostone.) Carbonates may be precipitated from either fresh or salt waters. The oceans are enormous reservoirs of dissolved carbonate, and marine limestones hundreds of meters thick are known. Observations of modern sedimentation indicate that the majority of carbonate sediments are precipitated as calcium carbonate, usually through biological means. Most dolomite is believed to form from later chemical changes brought about through the addition of magnesium dissolved in pore waters circulating

through the sediment, reacting with calcite to form dolomite. Many limestones include large quantities of the carbonate shells of marine organisms, large and small.

When salt water is trapped in a shallow sea and dries up, dissolved minerals precipitate out as an **evaporite** deposit. Evaporites can contain a variety of minerals. Halite (sodium chloride) is the principal one, since the oceans contain considerable dissolved salt. Other important minerals in evaporites are the calcium sulfates gypsum and anhydrite, and borax.

Under other conditions, dissolved silica (SiO_2) may precipitate out of solution as a noncrystalline solid or even a watery gel. When solidified, such a deposit forms the rock *chert*. In some ancient sedimentary rocks, layers of iron oxides

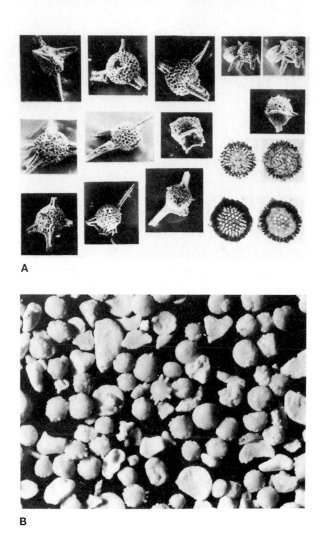

FIGURE 6.9 A limestone (coquina) rich in shell fragments.
© *Times Mirror Higher Education Group, Inc. /Doug Sherman, photographer.*

FIGURE 6.8 Microorganisms that make up many marine sediments. (A) Radiolaria, which have silica skeletons. (B) Foraminifera, with carbonate skeletons.

alternate with layers of chert or carbonate. These *banded ironstones* are important iron ores today (see chapter on mineral and energy resources). Clays may sometimes precipitate from solution; some mudstones, therefore, are chemical sediments also.

Biological Contributions

Biological sediments do not really constitute a distinct class of sediments, except insofar as organisms are involved in their formation. Most are simply chemical sediments, organically precipitated. A particular mineral is precipitated by an organism living in water, generally to build a shell or a skeleton; when the organism dies, and the soft body parts decay away, the mineral material remains. Coral reefs form carbonate rocks. In other cases, rocks are formed from sediments that consist of the remains of countless microorganisms that have accumulated on the sea floor. Cherts can be formed from the silica skeletons of sponges, diatoms, and radiolarians (figure 6.8A). Skeletons of calcareous (calcite-secreting) microorganisms can accumulate into a sediment from which limestone forms (figure 6.8B). In the modern oceans, such biogenic

(life-formed) carbonate sediments may be more common than inorganically precipitated ones. Less commonly, coarser biological debris, such as carbonate shells, is transported and accumulated by wave action or currents in a deposit of carbonate sediment that produces a clastic limestone (figure 6.9).

Other biological sedimentary materials, which are discussed in more detail in the chapter on mineral and energy resources, are **coal** and the related substances *peat* and *lignite.* All of these are formed from the remains of land plants, which have been converted through heat and pressure and accompanying chemical reactions to a carbon-rich solid that can be burned as fuel.

FROM SEDIMENT TO ROCK: DIAGENESIS AND LITHIFICATION

To become rock, a sediment must be **lithified** (from the Greek *lithos,* which means "stone"). The set of processes by which this is accomplished is collectively termed **diagenesis.**

Diagenesis occurs at rather low temperatures and pressures. Higher temperatures and pressures would lead to metamorphism, as described in chapter 7, and the resultant rock would be a metamorphic rock. The distinction between diagenesis and weak metamorphism is not sharp, but by convention, diagenetic processes are regarded as those occurring in the upper few kilometers of the crust, at temperatures below about 200°C (approximately 400°F). Also, diagenesis of sediments may occur without lithification; the sediments may or may not become a cohesive solid (rock). Examples of diagenetic processes follow.

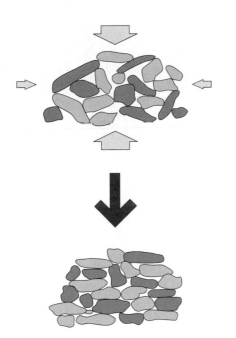

FIGURE 6.10 Compaction (schematic). Grains are packed more tightly together, often during sediment burial.

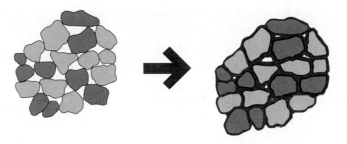

FIGURE 6.11 Cementation (schematic). Precipitation of calcite between grains helps to stick them together, promoting lithification.

As sediments pile atop other sediments, the earliest ones become ever more deeply buried. The weight of sediments above puts pressure on those below. Sediment under pressure tends to undergo **compaction** (figure 6.10). It is squeezed more tightly together, the volume of pore space between grains decreases, pore water is squeezed out, and the grains may begin to stick together. Platy grains may align themselves perpendicular to the direction of principal stress (horizontally, if the pressure is mainly from the weight of rocks above). Such alignment is common in shales and accounts for their tendency to split preferentially in one direction, parallel to and between the flat mineral grains. Compaction alone is generally insufficient to make the sediment cohesive enough to qualify as rock, though it is more effective on fine-grained sediments than on coarser ones. Cementation is almost always involved as well.

Cementation, as the name suggests, is the sticking together of mineral grains by additional material between the grains. Most sediments are deposited in water and have fluid in the pore spaces between grains. That fluid, in turn, contains dissolved minerals. As pore fluids flow through buried sediment, silica, calcite, or other dissolved mineral materials may begin to precipitate on grain surfaces (figure 6.11). In time, enough of this intergranular material may precipitate to effectively glue the sediment together into a rigid, cohesive mass: a sedimentary rock. Like compaction, cementation also tends to decrease pore space, as cracks and pores are filled in by precipitating minerals. Compaction and cementation often occur simultaneously. However, any cementing material must be introduced into the rock before compaction has proceeded so far that the flow of the cement-bearing fluid through the rock is severely restricted.

Burial subjects sediment not only to added pressure but also to increased temperature, due to the increase in temperature with depth in the earth. The combination of increased pressure and temperature may bring about recrystallization of minerals that are stable only at very low temperatures. Pore waters flowing through sediment can dissolve soluble materials, in addition to adding minerals as cement. Solution tends to increase the volume of available pore space, creating more room in the rock to be occupied by water, oil, or other fluids. This process thus has practical, as well as theoretical, significance. Chemicals dissolved in circulating pore waters can also react with minerals in the sediment to form new minerals. *Dolomitization*—the conversion of calcium carbonate to calcium-magnesium carbonate, as mentioned earlier—is an example.

Diagenesis thus comprises a variety of processes. Many of them change the texture or composition of the material and thus make it more difficult to draw inferences about the original sediment from the rock eventually produced.

STRUCTURES AND OTHER FEATURES OF SEDIMENTARY ROCKS

Sedimentary rocks can contain various kinds of structures. Some are useful in identifying the setting in which the sediments were deposited; some indicate the direction of flow of the wind or water transporting the sediment; some help in determining which side was originally up, in cases where the sedimentary rocks have been deformed or displaced after lithification. We survey several important examples.

Structures Related to Bedding

Virtually all sediments, clastic or chemical, are deposited in layers. For many, this layering, or **bedding,** is their most prominent characteristic. A *bedding plane* is a distinct surface upon which sediments have been deposited. Bedding that is very fine or thin is often termed **lamination** (figure 6.12A). On the other hand, a single bed may be many meters thick (figure 6.12B). When sediments are deposited in water, the bedding is almost always horizontal or nearly so. Each visually distinct layer that corresponds to a particular period of deposition is a single **stratum** (plural, *strata*).

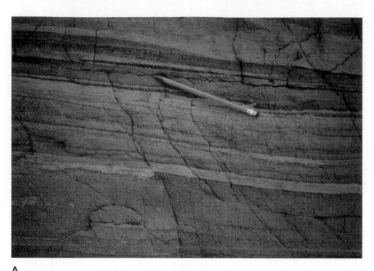

A

B

FIGURE 6.12 (A) Lamination, very fine bedding. (B) Bedding on a grand scale: the Grand Canyon.

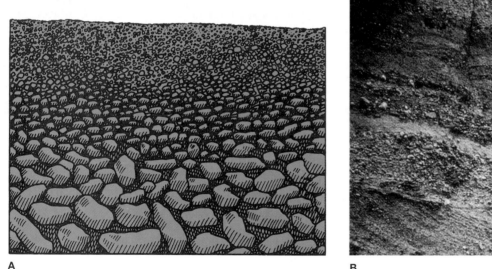

A

B

FIGURE 6.13 Graded bedding. (A) Graded bedding becoming finer near the top (schematic). (B) Example of graded beds.

Bedding does not always take the form of a stack of distinct layers, each of which is internally homogeneous. Particularly in clastic sediments, **graded bedding** may be observed. A graded bed is one in which grain size grades from coarse to fine, or vice versa, vertically within the bed (figure 6.13). Most commonly, the grading goes from coarse at the bottom of the bed to fine at the top. Graded bedding may develop if a mass of poorly sorted sediment is suspended in water suddenly, as, for example, by a submarine landslide churning up gravel, sand, and silt. The coarser particles tend to settle fastest, while very fine material may remain suspended for some time and settle out only slowly. (The process can be demonstrated by stirring a handful of poorly sorted sediment or soil into a large jar of water and then letting it settle.) Sediment transport by wind and water is also related to velocity of flow, in that

the faster these agents flow, the larger (coarser) the sediments they can move. A sediment-laden current of water that gradually slows down drops the coarsest sediments first, then successively finer ones.

Bedding is not always locally horizontal. Wind or water flowing across sloping surfaces during sediment deposition drapes layers of sediment over the existing topography. Successive layers are deposited farther and farther downstream or downwind. The resultant inclined bedding within a thicker sedimentary layer that is broadly horizontal is known as **cross-bedding** (figure 6.14). Should the direction of flow change, so will the orientation of subsequent crossbeds. In lithified desert sands in which many windblown dunes were superimposed one atop another, or in stream sediments subjected to many changes of flow pattern through time, the cross-bedding can become very complex (figure 6.14C).

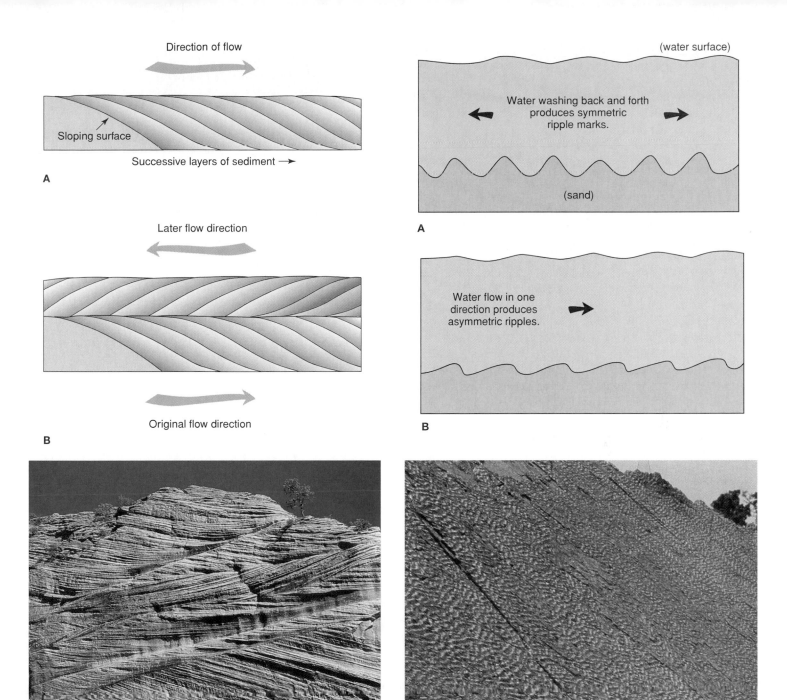

Figure 6.14 Cross-bedding. (A) Formation of crossbeds (schematic). (B) Change in current direction results in different orientation of successive crossbeds. (C) Example of cross-bedded sandstone: the Navajo sandstone of the southwestern United States.
(C) © Times Mirror Higher Education Group, Inc. /Doug Sherman, photographer.

Figure 6.15 Ripple marks. (A) Symmetric ripple marks, formed by water washing back and forth. (B) Asymmetric ripple marks, formed by currents flowing consistently in one direction. (C) Example of ripple marks in shale near Parrsboro, Nova Scotia, Canada.
(C) © David Laing.

Surface Features

The rhythmic motion of waves or flow of currents produces surface topography on fine- to medium-grained sediments, in the form of **ripple marks** (figure 6.15). The geometry of ripple marks depends on the mode of formation. Ripples that are *symmetric* in cross section are produced by the alternating back-and-forth motion of water near shore (figure 6.15A). They are subjected to an equal flow of water from each side, hence the symmetry. Sediments on a lake bottom near shore might show symmetric ripple marks. Ripples produced by consistent flow of wind or water in one direction are *asymmetric* in cross section, flattened on the side from which the current flows, steeper on the down-current side (figure 6.15B). Thus, asymmetric ripples indicate not only deposition in the presence of flowing currents but also the direction of flow. They are the usual form of ripple found in stream beds. The reasons for their shape are explored further in the discussions of wind in chapter 17.

When fine-grained, water-laid sediment dries out, it often shrinks and forms polygonal **mud cracks** on the surface (figure 6.16). These, too, may be preserved during lithification. They are formed particularly in finer-grained sediments for several reasons. For one, fine sediments often pack much less tightly during deposition than do sands and gravels (a mud may be 50 to 60% water), so volume decreases dramatically as the water is removed. Clay minerals are common in fine-grained sediments, too. Many clays have the capacity to absorb water when wet and release it when dried, expanding and contracting in the process. If such clays are deposited and then baked dry by exposure to air and sun, they shrink, and the drying mud forms cracks as they do so. Mud cracks, then, indicate two things about the conditions of sedimentation: that the sediments were initially deposited in water, and that they were subsequently dried out by exposure to air before lithification. Suitable conditions for formation of mud cracks exist in several environments, including tidal flats and on the bottoms of lakes that dry up as a result of changing climatic conditions.

Additional Features of Sedimentary Rocks

From sedimentary rocks comes nearly all our knowledge of earlier life-forms, through **fossils.** Fossils are the remains or evidence of ancient life. The fossil record is an incomplete one, for not all organisms, or all parts of each organism, are equally well preserved.

Much less common but quite distinctive features of some carbonate sediments are **oolites** (figure 6.17). These are spheroidal objects, typically the size of coarse sand grains, formed by precipitation of layer after layer of calcium carbonate around a core of mineral or shell fragment. Their formation is somewhat analogous to that of pearls, but oolites are completely inorganic. The spheroidal shape is a result of the setting in which they form—on shallow-water carbonate platforms (discussed in the next section). There, they are rolled around by the water and constantly stirred by waves and currents, becoming evenly coated on all sides. Where present, they are valuable indicators of sedimentary environment.

SEDIMENTARY ENVIRONMENTS AND SEDIMENT CHARACTER

The characteristics of a sedimentary rock are plainly determined in large measure by the depositional setting of the sediments. For clastic rocks, the nature of and distance from the source of the sediment are also factors. In this section, we survey briefly some of the common sedimentary environments and principal sediment characteristics of each.

FIGURE 6.16 Modern mud cracks near Cainville, Utah.
© *Times Mirror Higher Education Group, Inc. /Doug Sherman, photographer.*

Clastic Sedimentary Environments

Deposition of clastic sediment in the absence of significant water occurs principally in deserts. Here, the sediment is found in windblown dunes, usually of sand-sized particles. These dunes may show internal crossbeds. Finer materials are commonly carried away, to be deposited by other means in different settings.

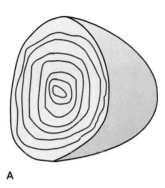

A

B

FIGURE 6.17 Oolites. (A) Cross section (schematic). (B) Oolitic limestone.

(B) © Times Mirror Higher Education Group, Inc./Bob Coyle, photographer.

Rivers provide other depositional environments on the continents. Stream-deposited sediments commonly consist of the coarser clastics—sand and gravel—with relatively little mud; the finer sediments often stay in suspension in the flowing water. (When the stream floods, it may blanket the flooded land with that fine sediment.) Deposits of stream sediments are also characteristically elongated in shape and restricted in areal extent, which helps to distinguish them from similar materials deposited in large basins. Asymmetric ripple marks can indicate the direction of streamflow. Characteristics of stream-deposited sediments are explored in greater detail in chapter 14.

Where land and sea meet, several kinds of sedimentation are possible. Streams deposit at their mouths large, fan-shaped *deltas* of continentally derived sands and muds that interfinger, or mix, with marine sediments. The high-energy *beach* environment, with active waves and tides, is characterized by well-sorted and generally well-rounded sands and gravels: the vigorous water action abrades and rounds the sediment grains while washing away the finest size fractions. The more placid setting of a *tidal flat* may provide a place for

fine silt and clay to settle out in muddy deposits, which may develop mud cracks and show evidence of biological activity (animal burrows and tracks, for instance).

Farther offshore from the beach and tidal flat, but still close to the continent, the character of sediment is rather like that of a stream delta—a mix of fine- to medium-grained continental clastics with marine sediment, perhaps including carbonate or shells of marine organisms—but it is spread by currents flowing along the coast into a more sheetlike deposit. Only the finest clastics are carried far offshore, either in the water by currents or from the continents by winds. Clastic sediments of the deep-sea floor are finely laminated clays that accumulate very slowly. Marine sedimentation is discussed further in chapter 12.

Chemical Sedimentary Environments

Chemical sedimentation inevitably involves water, so all depositional environments of chemical sediments are under, or close to, water.

Shallow-water environments in warm waters account for most carbonate sedimentation. Unlike most chemicals, calcite dissolves more readily in cold water than in warm, or in other words, it is more readily precipitated from warmer waters. Off the coasts of some continents are broad, shallowly submerged shelves; in near-equatorial regions, the waters over these shelves are quite warm. In such areas, *carbonate platforms* may be built up by direct precipitation of carbonate muds or by the accumulation of shell or skeletal fragments of carbonate-secreting organisms that thrive under the same conditions. Reefs, likewise made of calcium carbonate, are also common in this environment. Areas of active carbonate precipitation include the Bahamas and the Florida Keys. It should also be noted that carbonates dominate sediment only when clastic input is relatively low.

Restricted warm basins into which seawater can flow and from which it then evaporates provide the appropriate setting for deposition of evaporites. Such basins once existed over much of North America. Drilling into sedimentary rocks under the Mediterranean Sea likewise suggests that, at one time, water flow into and out of it was much more limited than it is now, so that seawater would flow in, be trapped, warmed, and evaporated, and leave a salt deposit behind. The thickness of many evaporite deposits, including those of the Mediterranean basin, is far greater than could be accounted for by taking a single basinful of seawater, even one as deep as the modern ocean, and drying it up. Apparently, thick evaporite beds form through the cumulative effect of repeated cycles of basin filling and evaporation.

Deep, cold basins of the sea floor may be sites of predominantly chemical sedimentation, if they are far enough removed from clastic input from the continents. Deep bottom waters are too cold for the preservation of calcium carbonate, but they are suitable for the preservation of silica. Biochemical silica-rich muds accumulate as the siliceous skeletons of diatoms and radiolarians settle to the sea floor.

OTHER FACTORS THAT INFLUENCE SEDIMENT CHARACTER

As was noted earlier, the underlying control on the composition of a clastic sediment is the composition of the material from which it is derived. If the source region consists of granitic rock, the resultant sediment will not contain olivine grains or fragments of limestone. The bulk of that sediment will consist—depending on the climate—of varying proportions of quartz, feldspar, clays, and iron and aluminum oxides. If the source region is basaltic, the sediment will not be rich in quartz grains.

Distance from the sediment source is particularly relevant to water-transported clastic sediments; it influences both the sediment's texture and its composition. In terms of grain size, all else being equal, the farther from the source the rock and mineral fragments have traveled, the finer the fragments. The longer the distance traveled, the more opportunity for grains to be partially dissolved, broken, or abraded, and therefore the smaller they get.

Distance from source also influences the **chemical** or **mineralogical maturity** of the sediment. Not all minerals dissolve or undergo chemical reactions equally readily. The longer the distance traveled, the greater the proportion of the easily dissolved or easily altered material that is eliminated. The residue becomes more and more enriched in so-called resistant minerals. For example, most ferromagnesians weather easily; quartz, in most climates, is fairly resistant to both chemical and physical breakdown. A quartz-bearing sediment transported a long way in a stream channel is likely to become more and more quartz-rich along the way.

These principles are far less valid for clastics transported by wind or by ice. Wind generally moves only very fine material (rarely as large as sand-sized fragments), so further breakup in transit is not readily detectable; also, wind does not attack minerals chemically. Materials frozen into ice are subject to far less abrasion or contact with other rock fragments than they would be traveling in water, so further physical breakup is minimized; in addition, ice tends not to react chemically with other minerals.

As noted earlier, both temperature and moisture influence the extent of chemical weathering. Most such reactions involve water; therefore, the more rainfall, the more chemical breakdown. Most chemical reactions also proceed more rapidly at high temperatures than at low, so warm climates favor rapid chemical weathering. Sediments in warm, moist climates approach maturity—in which the sediment is rich in resistant minerals like quartz—more rapidly than sediments under cold, dry conditions. In the latter case, even easily weathered minerals like olivine and calcic plagioclase may persist for some time.

Topography principally affects the relative rates of sediment transport and chemical breakdown. That is, the steeper or more rugged the terrain, the more gravity will work with wind, water, or ice to move sediment quickly, allowing less time for chemical breakdown. The transportation itself may also be more vigorous, increasing mechanical breakup of rock and mineral fragments. In very flat terrain, sediment transport is likely to be much slower, unless there are unusually strong winds or fast-flowing streams. The sediment is thus exposed to prolonged chemical breakdown. Sediments in rugged terrain, then, commonly are mineralogically immature, preserving even readily weathered minerals, but are more likely to be rapidly broken up mechanically. Sediments in gently sloping terrain may suffer only slow breakup but considerable chemical attack and tend to be more mature mineralogically.

THE FACIES CONCEPT: VARIABILITY WITHIN ROCK UNITS

Implicit in the foregoing discussion is the idea that a single rock unit shows a single set of physical and chemical characteristics. This is not necessarily true, however, especially for units of large areal extent. A single mass of sedimentary material deposited during one time interval may be spatially continuous over a broad area but vary greatly in character (composition, texture) locally.

Sedimentary Facies

Consider the example shown in figure 6.18, which illustrates deposition along a shoreline to which sediment is supplied by streams. Near the beach, the sediment is sandy. Finer, muddy sediments are carried farther offshore before they settle to the bottom. The quantity of continentally derived sediment generally decreases with increasing distance from shore until it becomes insignificant; sedimentation then takes the form of calcite precipitation from seawater, with or without biological action.

After diagenesis and lithification, these sediments would form several quite distinct rock types. Yet they are all, in a sense, part of the same sedimentary package. The original sediments were deposited simultaneously in a continuum of sedimentary environments. There will be no sharp demarcation between the limestone, mudstone, and sandstone formed—each will grade into the next. For instance, if these rocks were later uplifted and exposed, and one walked from the mudstone to the limestone, sampling along the way, the nature of the rocks would change gradually from a mudstone with a little carbonate, to mixed rocks with similar proportions of each, to limestone with a little silt or clay in it, to, perhaps, nearly pure limestone.

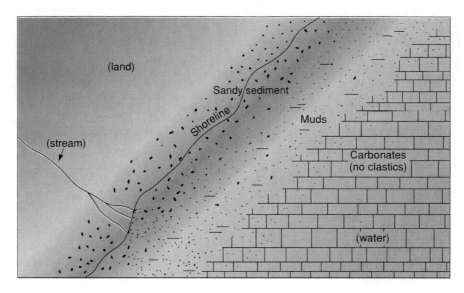

FIGURE 6.18 Facies changes at a shoreline (map view).

The different rock types mentioned in the previous paragraph represent different *facies* of the same basic unit. The term **facies** is used to describe collectively a set of conditions that lead to a particular type of rock (or sediment) or to distinguish that portion of the rock itself that represents a distinct set of conditions. (For clarity, the term **lithofacies,** literally "rock facies," is sometimes used to designate the rocks corresponding to a particular facies, or set of conditions.) The hypothetical sedimentary rock unit described could be said to have a sandstone facies, a shale or mudstone facies, and a limestone facies. The sediments that went into the rocks of the shale facies would be those deposited in a setting that favors deposition of muds—here, one with some clastic input but far enough from source and shore to exclude sandy or coarser grains, with water currents gentle enough not to sweep the muds away, and an abundant contribution of clastics relative to precipitated carbonate.

The facies phenomenon makes interpretation of the rock record more difficult. Where rocks are very well exposed at the surface and one can walk through the facies changes, it is fairly easy to recognize what is, or was, going on. Where rocks are poorly exposed, so that one finds here an outcrop of shale, there another of limestone, it is more difficult to know whether the two are different facies of the same unit, equivalent in age and deposited as part of the same overall depositional package, or are wholly unrelated rocks. Correlation using fossils (chapter 8) can be helpful in such cases.

Facies and Shifting Shorelines

Most shoreline sedimentation is characterized by some sequence of sedimentary facies that changes in the direction perpendicular to the shoreline. Therefore, if the position of the shoreline changes, later deposition may lay down sediments of different facies atop earlier sediments in a given spot. Various events, including worldwide rise and fall of sea level or localized rise and fall of the edge of a continent, can cause the shoreline to shift landward or seaward.

For example, if worldwide sea level rises (as is happening now as polar ice melts), the effect is that shorelines move landward. The impact on the distribution of sedimentary facies in map view and in cross section is shown in figure 6.19. As the sea encroaches on the land, each facies effectively moves inland, too: beach sands are deposited at the new shore, which previously was dry land; muds overlie the earlier beach; limestone is deposited over older muds. The process of the ocean advancing over what was dry land is termed a **transgression** of the sea.

Now imagine this whole region buried by later events and the sediments lithified. If one were to drill a hole through the resultant rock at point ▲ in figure 6.19A, the sequence of rocks would appear as in figure 6.19C: sandstone on the bottom, grading upward into mud and then limestone. This collection of rock types is a typical *transgressive sequence,* reflecting deepening water and changing sedimentation patterns at ▲ as the shoreline migrates inland. The reverse process, **regression,** is reflected in a correspondingly inverted sequence of rock types.

PALEOGEOGRAPHIC RECONSTRUCTION

It is often possible to reconstruct the distribution of land and sea across a continent at various times in the past. Such **paleogeographic maps,** maps of ancient geography, are constructed predominantly through the interpretation of sedimentary

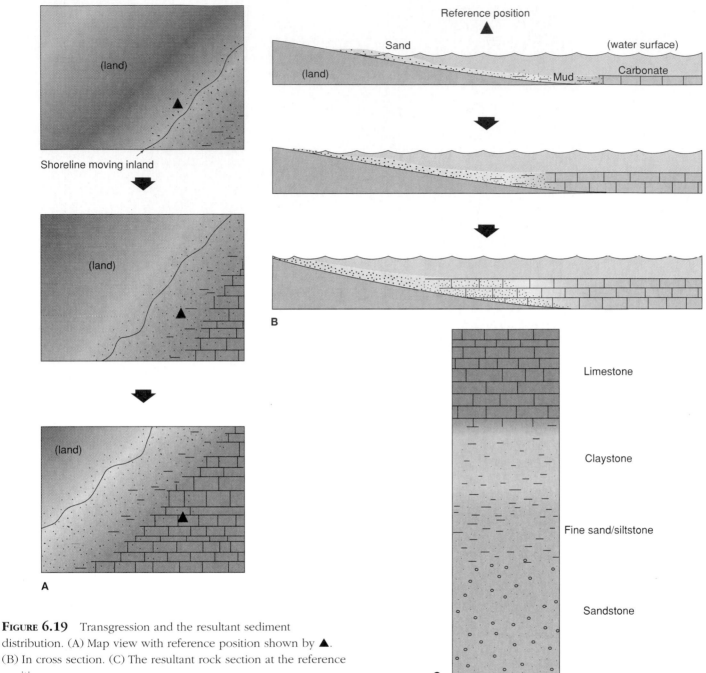

Figure 6.19 Transgression and the resultant sediment distribution. (A) Map view with reference position shown by ▲. (B) In cross section. (C) The resultant rock section at the reference position.

rocks because it is the sedimentary rocks, formed near the surface from materials exposed at the surface, that supply the most relevant information. Frequently, the paleogeography of a region bears little resemblance to its present geography: sandstone made from desert sands may be found in the rock record of a region now temperate and well vegetated; remains of ancient tropical soils may be found at now-cold latitudes; deep-water marine sediments may be found high in mountain ranges. Interpreting a single rock unit and

variations in its character over a large area can provide considerable information about sources and depositional settlings of sediment.

We have already noted that the grain size of clastic sediments tends to become finer with increasing distance from the source of the sediments. Another commonly used interpretive tool is the orientation of *paleocurrents* (ancient currents that prevailed at the time the sediments were deposited). Paleocurrent directions are deduced from sedimentary structures,

Box 6.1

What's in a Rock Pile?

NAMES AND SIGNIFICANCE OF SOME SEDIMENTARY ROCKS IN THE GRAND CANYON

Especially when the geology of a region is dominated by layer upon layer of sedimentary rock (as is the Grand Canyon, for example—recall figure 6.12B), it is convenient to represent the physical sequence of layers diagrammatically in a geologic **column,** or **section.** A portion of the Grand Canyon section is shown in figure 1. By convention, different rock types are represented by different symbols: horizontal dashes for shale, dots for sandstone, bricks for limestone, and so on. These distinctive, recognizable units are then named. The physical sequence also implies a chronological sequence, as is discussed in chapter 8.

The basic unit is the **formation,** a rock unit representing deposition under a uniform set of conditions and at one time, producing a recognizable, mappable unit. A formation need not all be one rock type, but it should reflect a predominance of some particular depositional conditions. Boundaries between formations are drawn where definite changes in depositional conditions can be identified. Formation names as presently assigned have two parts: a locality name, corresponding to a place where that formation is well exposed, and a rock type—for example, "Temple Butte limestone." (In the past, a prominent characteristic of the rock might have been substituted for a locality—for example, "Redwall limestone," named for its color and tendency to form vertical cliffs—but this is not correct modern practice.) If the rock types are so varied within a formation that no one type dominates in the section, the word "formation" may be substituted, as in "Supai formation."

Within a single formation, varying rock types that can themselves be recognized and mapped may be distinguished. These rock units are termed **members.** Several members are identified in the Redwall limestone: the Whitmore Wash

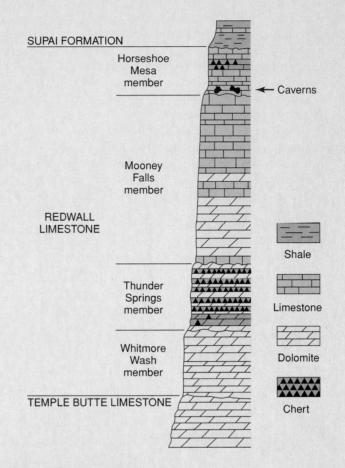

FIGURE 1 The Grand Canyon section.

Source: After E. D. McKee, "Paleozoic Rocks of the Grand Canyon," in Geology of the Grand Canyon, *edited by W. J. Breed and E. Roat, pp. 42–64, 1978, Classic Printers, Prescott, AZ.*

member (dolomite), the Thunder Springs member (interlayered chert and dolomite), the Mooney Falls member (limestone and dolomite), and the Horseshoe Mesa member (limestone with chert). Note the use of localities in the naming of members. The boundary between the Redwall limestone and the

Supai formation is put at the change in rock type from carbonate to mostly shale, which reflects an increase in clastic input.

Sometimes, but not always, formations genetically related in some way are described together as a **group,** but many formations do not belong to any higher-order unit.

including asymmetric ripple marks and crossbeds. Since local eddies in wind or water can cause anomalous orientations in single samples, many measurements of apparent paleocurrent directions are made in a single rock unit, and the results often are interpreted with the aid of statistics.

We have also noted that the nature of the rocks of the source region is a fundamental control on the composition of clastic sediments. The principle can be turned around: the composition of clastic sediments may be used to infer the source rocks, within limits, or to differentiate among several possible source regions. If a sedimentary basin is adjacent to several quite different possible source regions—for example, one consisting mainly of granite, one of basalt, and one of limestone and shale—it may be possible to decide which one(s) contributed to the sediments. Abundant coarse quartz and feldspar grains would suggest the granite as source; many limestone fragments in the sediments would suggest the sedimentary source area; and so forth. If the possible source areas are quite similar compositionally—for instance, all basically composed of granitic rock—then it would be necessary to look at more subtle details of the chemistry and mineralogy of the sediments and possible sources, perhaps in combination with other evidence, such as paleocurrent indicators.

The *paleoclimate,* or ancient climate, can frequently be deduced, at least in a general way, from sediment character. A chemically mature clastic sediment in which fragments are still quite angular (which have therefore not been transported far) suggests the rapid chemical weathering of a warm, wet climate. Conversely, a very fine-grained sediment with well-rounded fragments that still contain easily weathered minerals indicates a colder, drier climate. Carbonate platform sediments suggest warm, shallow seas. Latitude is a primary control on earth surface temperatures, so climatic inferences can be used to deduce the approximate latitudes at which the sediments were originally deposited, and they also aid in the recognition of global warming and cooling trends. The presence and types of fossils, if any, may also indicate climate and, in the case of water-laid sediments, even water depth. Paleogeographic reconstruction is an integral component of interpretation of earth's history, as will be seen in chapters later in the text.

SUMMARY

Sedimentary rocks are those formed at low temperatures, near the earth's surface, from sediments. Sediments may be either clastic (formed from mechanical breakup of preexisting rocks) or chemical (precipitated from solution). Water-dwelling organisms that secrete shells or skeletons made from dissolved chemicals may also contribute to sediment. Clastic sedimentary rocks are classified principally on the basis of grain size. Chemical sedimentary rocks are subdivided according to composition. Sediments are lithified through various diagenetic processes, including compaction, cementation, and recrystallization. Clastic sedimentary rocks, in particular, may show structures that reflect the conditions of sedimentation—graded bedding, cross-bedding, ripple marks, mud cracks, and the like. Sediments of all kinds may contain the fossilized remains of ancient creatures.

Different sedimentary environments are conducive to deposition of sediments of different composition, texture, and structure. Individual sedimentary environments are not necessarily sharply bounded but often grade laterally into one another. A single rock unit deposited at one time but over a large area may thus show several different facies, with different characteristics indicative of the varying depositional settings from place to place.

Reconstruction of the paleogeography of a region may be based on inferences about the depositional environment(s), composition of the sources of clastic sediments, distance from sources as indicated by grain size and rounding, paleocurrents, and paleoclimate.

TERMS TO REMEMBER

bedding 91
biological sediments 90
breccia 87
cementation 91
chemical maturity 96
chemical sediment 88
clastic 87
coal 90
column 99
compaction 91
conglomerate 87
cross-bedding 92
diagenesis 90

dolomite 89
evaporite 89
facies 97
formation 99
fossils 94
graded bedding 92
group 99
lamination 91
limestone 88
lithification 90
lithofacies 97
matrix 88
member 99

mineralogical maturity 96
mud cracks 94
mudstone 87
oolites 94
paleogeographic maps 97
regression 97
ripple marks 93
sandstone 87
section 99
shale 88
sorting 88
stratum 91
transgression 97

QUESTIONS FOR REVIEW

1. On what basis are clastic sediments subdivided and named?
2. What is meant by saying that a sediment is well sorted? Well rounded?
3. Name and describe two kinds of chemical sediments.
4. Discuss how compaction and cementation contribute to lithification of sediments.
5. Explain how a sediment's volume of available pore space, and the ease with which fluids might flow through the sediment, are modified as the sediment is lithified.
6. Describe *graded bedding* and *cross-bedding,* and briefly explain one way in which each can arise.
7. Mud cracks form most readily in finer-grained sediments. Why?
8. Compare the kinds of sediment you would expect in a beach environment and in a tidal flat.
9. How are sediments modified as they are transported farther and farther from their source region?
10. What is a *facies?* How does the existence of sedimentary facies complicate the identification or mapping of individual rock units?
11. What is a *transgression?* Describe how this might be recognized in the rock record.
12. Cite and briefly describe two possible paleocurrent indicators.

FOR FURTHER THOUGHT

1. Many organisms that we now know only as fossils lived over a limited span of geologic time. Consider how this fact might be used to recognize different facies of the same rock unit in an area in which the rocks are generally not well or continuously exposed.

2. In which of the following sediments might you hope to find dinosaur footprints: (a) a coarse sandstone with pronounced crossbeds; (b) a mudstone with symmetric ripple marks; (c) an oolitic limestone? Explain.

3. Much of North America is covered by sedimentary rock, miles thick. Where might such an immense volume of sediment have come from?

SUGGESTIONS FOR FURTHER READING

Blatt, H., Middleton, G., and Murray, R. 1980. *Origin of sedimentary rocks.* 2d ed. Englewood Cliffs, N.J.: Prentice-Hall.

Boggs, S. 1992. *Petrology of sedimentary rocks.* New York: Macmillan.

Collinson, J. D. 1982. *Sedimentary structures.* Boston: Allen and Unwin.

Folk, R. L. 1974. *Petrology of sedimentary rocks.* Austin, Tex.: Hemphill.

Freeman, T. J. 1971. *Field guide to layered rocks.* Boston: Houghton-Mifflin.

Garrels, R. M., and Mackenzie, F. T. 1971. *Evolution of sedimentary rocks.* New York: W. W. Norton.

Laporte, L. F. 1968. *Ancient environments.* Englewood Cliffs, N.J.: Prentice-Hall.

McKee, E. D. 1978. Paleozoic rocks of the Grand Canyon. In *Geology of the Grand Canyon,* edited by W. J. Breed and E. Roat, 42–64. Prescott, Ariz.: Classic Printers.

Reading, H. G., ed. 1986. *Sedimentary environments and facies.* 2d ed. Boston: Blackwell Scientific.

Reineck, H. E., and Singh, I. B. 1975. *Depositional sedimentary environments.* New York: Springer-Verlag.

Selley, R. C. 1982. *An introduction to sedimentology.* New York: Academic Press.

Shilts, W. W. 1981. *Sensitivity of bedrock to acid precipitation: Modification by glacial processes.* Geological Survey of Canada Paper 81-14.

Stewart, J. H., McMenamin, M. A. S., and Morales-Ramirez, J. M. 1984. *Upper Proterozoic and Cambrian rocks in the Cabora Region, Sonora, Mexico.* U.S. Geological Survey Professional Paper 1309.

7

METAMORPHISM AND METAMORPHIC ROCKS

Stretched pebbles and boulders attest to deformation of conglomerate under the stress of metamorphism. Outcrop near Wawa, Ontario, Canada. © Doug Sherman/Geofile

The term **metamorphism** is derived from the Greek for "change of form." It has parallels in other sciences. For example, biologists use the term *metamorphosis* to describe collectively the complex changes by which larvae become butterflies and moths, or tadpoles become frogs. Geologic metamorphism is likewise a process of change—but in rocks: change in the composition, mineralogy, texture, or structure of a rock by which it is transformed into a distinctly different rock. In this chapter, we survey some of the causes and consequences of metamorphism, describe situations that lead to metamorphism, and define some common metamorphic rock types.

FACTORS IN METAMORPHISM

Rocks are changed when they are subjected to physical or chemical conditions different from those under which they formed. The two most common agents of metamorphism are heat and pressure. Each mineral is stable only within certain limits of pressure and temperature. Held for long enough under conditions in which it is unstable, a mineral changes to a more stable form. How long is long enough? That depends on the particular mineral and physical conditions.

M any metamorphic minerals are not truly stable at surface conditions. Nor are many other minerals, such as diamonds, which are formed at the still-higher pressures and temperatures of the mantle. Carbon's stable mineralogical form at low pressure is graphite. When rocks containing such minerals are exposed at the surface, therefore, why don't the now-unstable minerals all break down or change into low-temperature, low-pressure minerals?

The main reason is lack of time. Both chemical reactions and structural changes like recrystallization proceed in solid materials quickly at high temperatures and much more slowly at low temperatures. Even over millions of years, some technically unstable minerals are preserved as a result of slow reaction rates. So diamonds are not forever. But neither will the diamonds in our watches and jewelry turn to graphite for many more human generations.

Temperature

Metamorphic temperatures are limited at the low end by diagenesis, at the high end by melting. As already noted in chapter 6, there is no sharp division between diagenesis and metamorphism—geologists disagree on how to define the boundary—but it generally corresponds to temperatures in the range of 100 to 200°C (200 to 400°F). The variability at the other end of the metamorphic temperature range is much greater because the temperature at which a rock melts depends on many factors, including its composition, the prevailing pressure, and the presence or absence of fluids. Typical maximum metamorphic temperatures that most rocks can sustain without melting are in the range of 700 to 800°C (1300 to 1500°F), but dry mafic rocks rich in high-melting-temperature ferromagnesians can be heated to more than 1000°C (1850°F) without melting. Overall, the prevailing temperature influences not only the stability of various minerals but also the rates of chemical reactions, which generally proceed faster at higher temperatures.

There are two principal sources of heat for metamorphism. One is the normal increase in temperature with depth already described in chapter 3, the **geothermal gradient.** This is a moderate effect in most places. Typical geothermal gradients are about 30°C per kilometer, or close to 100°F per mile of depth. Local geothermal gradients can be increased by plutonism, emplacement of hot magma into the crust. Also, rocks very close to a pluton can be raised to near-magmatic temperatures directly. Magmatic activity, then, is a second source of heat for metamorphism, one commonly associated with plate boundaries, as noted in chapter 4.

Pressure

Pressure takes two forms. Rocks in the earth are all subject to a **confining pressure,** an ambient pressure equal in all directions, imposed by the surrounding rocks, all of which are under pressure from other rocks above. Confining pressure generally increases with depth. In addition, rocks may or may not be subject to **directed stress,** which, as the name implies, is not uniform in all directions.

The distinction between confining pressure and directed stress can be seen by analogy with a person standing on the earth's surface. The body experiences a confining pressure from the surrounding atmosphere (about 1.04 kilograms per square centimeter, or 14.7 pounds per square inch, of body surface at sea level), and atmospheric pressure acts equally all around the body. (If it did not, the body would be compressed in the directions from which higher pressures were exerted, stretched in others.) In addition, the feet are subjected to a vertical compressive stress applied by the downward pull of gravity on the mass of the body, and typically, the feet are somewhat flattened in response.

Directed stress in rocks frequently arises in connection with mountain-building processes, when rocks are squeezed, crumpled, and stretched. For example, compression is common at plate boundaries where two plates converge or collide. These tectonic processes are explored further in chapters 9 and 11. Directed stress also can develop on a smaller scale, as, for example, when magma is forcibly injected into cracks in rocks during igneous intrusion. Fault-

FIGURE 7.1 The folds in this rock are the result of stress during metamorphism.

ing is also associated with directed stress, as is discussed in chapters 10 and 11. The common result of directed stress is deformation (figure 7.1).

Chemistry and Metamorphism

In addition to temperature and pressure, a third factor that enters into some metamorphic situations is a change in chemical conditions in a rock. If rocks are fairly permeable, fluids flow readily through them, and either the fluids themselves, or one or more elements dissolved in them, may react with existing rock to produce new minerals. This can occur, for example, when warm fluid associated with magma migrates away from the silicate melt, carrying dissolved gases and other elements. **Hydrothermal** activity, involving the action of hot water seeping through fractures, is common in the country rock around igneous intrusions. Gases may pass through or escape from permeable rocks. Gain or loss of fluids and the elements they contain changes the composition of the resultant rocks.

The Role of Fluids

Even in the absence of fluids, chemical elements can migrate through rocks, but they do so slowly. The presence of fluids facilitates this migration and thus makes possible changes in bulk rock chemistry and mineralogy. For the same reason, fluids also facilitate mineralogical changes even if the rock as a whole behaves as a closed system, with no gain or loss of elements. Elements can migrate, or diffuse, through rock more readily in the presence of fluids than without them, so

atoms can more readily rearrange themselves into new minerals in the presence of fluids. The hydroxide minerals and hydrous silicates, those containing (OH^-) groups, require the presence of some water (H_2O) even to form or to remain stable. Finally, as noted in chapter 3, wet rocks melt at lower temperatures than dry rocks do, so the presence of fluid has some bearing on how high temperatures can get before the line between metamorphism and melting is crossed.

The fluids involved in metamorphism may be derived from several sources, including the hydrothermal fluids already mentioned, pore fluids squeezed out of rocks undergoing deep burial or tectonic compression, and water and other volatiles released through mineralogical breakdown during metamorphism, as described in the next section.

EFFECTS OF METAMORPHISM

The details of metamorphic changes depend both on the chemical composition and mineralogy of the **parent rock,** or starting material, and on the specific pressure, temperature, and other conditions of metamorphism, such as whether fluids or dissolved chemicals are entering or leaving the system. However, some generalizations are possible.

Effects of Increased Temperature

As temperatures are increased, minerals that are stable at low temperatures react or break down to form high-temperature minerals. A common kind of reaction is one in which a hydrous mineral breaks down. Recall from Bowen's Reaction Series that the common hydrous silicates—amphiboles and micas—form at moderate to low temperatures. At higher temperatures, these minerals become unstable. For example, amphibole breaks down to form pyroxene and release water; quartz is also produced, balancing the chemical equation:

$$(Mg,Fe)_7Si_8O_{22}(OH)_2 = 7\,(Mg,Fe)SiO_3 + SiO_2 + H_2O$$
amphibole   pyroxene   quartz water

At even lower temperatures, the hydrous clay minerals in sedimentary rocks break down to form more stable micas. If temperatures are increased further, the micas, in turn, break down (to feldspar and other minerals, plus water). Certain nonsilicate mineral groups may break down in analogous fashion. For instance, calcite, a carbonate, may break down with release of carbon dioxide gas (CO_2). Examples of some possible metamorphic reactions are shown in table 7.1.

Effects of Increased Pressure

The effects of increased pressure depend, in part, on whether or not the stress is directed. In general, increased confining pressure favors denser minerals, those that pack

TABLE 7.1 Common Types of Metamorphic Reactions

Reactions in Silicate Rocks

$(Mg,Fe)_7Si_8O_{22}(OH)_2$ amphibole	$=$	$7\ (Mg,Fe)SiO_3$ pyroxene	$+$	SiO_2 quartz	$+$	H_2O water		
$(Mg,Fe)_3Si_2O_5(OH)_4$ chlorite	$+$	$KAlSi_3O_8$ potassium feldspar	$=$	$K(Mg,Fe)_3AlSi_3O_{10}(OH)_2$ biotite	$+$	$2\ SiO_2$ quartz	$+$	H_2O water
$3\ (Mg,Fe)_3(Al,Si)_2O_5(OH)_4$ chlorite $+$	Fe_3O_4 magnetite $+$	$6\ SiO_2$ quartz	$=$	$3\ (Mg,Fe)_3(Fe,Al)_2Si_3O_{12}$ garnet	$+$	$6\ H_2O$ water $+$		$\tfrac{1}{2}\ O_2$ oxygen
$KFe_3AlSi_3O_{10}(OH)_2$ iron-rich biotite	$+$	$\tfrac{1}{2}\ O_2$ oxygen	$=$	$KAlSi_3O_8$ potassium feldspar	$+$	Fe_3O_4 magnetite	$+$	H_2O water

Reactions in Carbonate Rocks

$CaCO_3$ calcite	$+$	SiO_2 quartz	$=$	$CaSiO_3$ wollastonite*	$+$	CO_2 carbon dioxide	
$CaMg(CO_3)_2$ dolomite	$+$	$2\ SiO_2$ quartz	$=$	$CaMgSi_2O_6$ pyroxene	$+$	$2\ CO_2$ carbon dioxide	

*Silicate similar to pyroxenes

Note: In all of the above reactions, the water involved is water vapor, carbon dioxide and oxygen are gases, and all other species are solids.

Many more reactions are possible, including various polymorphic transitions in which a low-temperature or low-pressure polymorph is converted to a higher-pressure or higher-temperature form. Potassium feldspar, for example, has several polymorphs—microcline, orthoclase, and sanidine—each stable under somewhat different pressure-temperature conditions. Microcline is the stable form at surface conditions, while sanidine is the high-temperature form often found in volcanic rocks.

The above reactions should be regarded only as examples; actual reactions may be complicated still more by the effects of solid solution.

FIGURE 7.2 Garnet-bearing metamorphic rock (gneiss). See also figure 7.14.

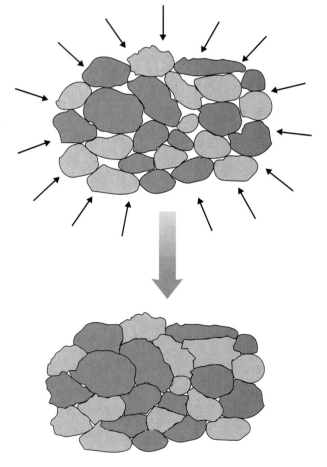

FIGURE 7.3 Effects of recrystallization during metamorphism (schematic). The grains become tightly interlocking, and porosity decreases. Arrows in top diagram show confining pressure.

more mass into less space. Garnet, for example, is a very common metamorphic mineral in moderate- to high-pressure situations (figure 7.2). Iron-magnesium garnets are dense even by comparison with other ferromagnesian silicates: while the specific gravities of various pyroxenes and amphiboles are 3.2 to 3.9 and 2.0 to 3.4, respectively, ferromagnesian garnets range from 3.8 to 4.3. Another effect of increased confining pressure that is particularly evident in metamorphosed sedimentary rocks is a tendency for the rock's texture to become much more compact. The minerals recrystallize and in so doing decrease the rock's porosity and increase its density (figure 7.3).

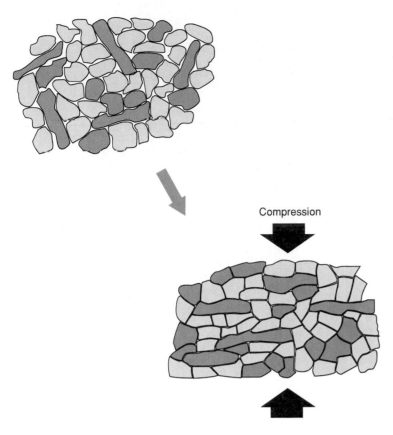

Compression

FIGURE 7.4 Development of foliation in the presence of directed stress. Elongated or platy minerals are reoriented, becoming aligned in parallel planes. Tensile stress would act parallel to elongation, perpendicular to the direction shown for compressive stress.

FIGURE 7.5 Slate, showing characteristic slaty cleavage. Compared to shale, slate is typically harder, denser, more compact, and more cohesive.
© *Times Mirror Higher Education Group, Inc. /Bob Coyle, photographer.*

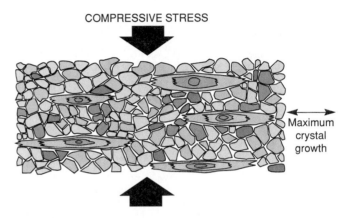

COMPRESSIVE STRESS

Maximum crystal growth

FIGURE 7.6 Growing micas grow most readily in the plane perpendicular to the applied compressive stress, leading to development of schistosity.

When directed stress is applied, the result is typically a rock with some compositional or textural banding. This can take several forms. If there are already flat, platy minerals, such as clays and micas, in the rock, they tend to become aligned parallel to each other. *Tensional* stress tends to stretch the object to which it is applied; *compressive* stress tends to compress or squeeze the object. Platy minerals in a rock will tend to be drawn out parallel to a tensional stress, positioned perpendicular to a compressive stress (figure 7.4). Even when the grains are too small to see with the naked eye, the rock often shows a tendency to split parallel to the aligned plates (rather than through or across the plates). This behavior is **rock cleavage.** It is analogous to mineral cleavage, but rock cleavage is a tendency to break between planar mineral grains rather than between planes of atoms in a crystal. Rock cleavage is demonstrated by *slate,* which splits readily into flat slabs from which tiles and flagstones can be made (figure 7.5). In fact, rock cleavage developed in such a fine-grained rock is often called **slaty cleavage** after this common example of the phenomenon.

As metamorphism intensifies at higher temperatures, crystals of various metamorphic minerals may grow progressively coarser (larger). Many metamorphic minerals have platelike or elongated, needle-shaped crystals (for ex-

ample, micas and amphiboles, respectively). In the presence of directed stress, these crystals grow in similar preferred orientation. For example, flakes of mica tend to grow sideways to, rather than directly opposed to, a compressive stress (figure 7.6). The resultant rock texture, in which coarse-grained platy minerals show preferred orientation in parallel planes, is described as **schistosity** (figure 7.7). Slaty cleavage in fine-grained rocks and schistosity in coarser-grained ones are both examples of **foliation,** from the Latin for "leaf" (as in the parallel leaves, or pages, of a closed book). The analogous term for parallel alignment of elongated needlelike or linear crystals, such as those of many amphiboles, is **lineation.**

Another texture sometimes regarded as a form of foliation is compositional, rather than textural, layering. In some metamorphic rocks, especially those subjected to strong metamorphism, recrystallizing minerals in the rock segregate into bands of differing composition or texture. Often, the result is a rock of striped appearance, with light-colored bands rich in quartz and feldspar alternating with dark-colored

FIGURE 7.7 Schistosity promotes rock cleavage on planes parallel to the foliation. Beaver Tail Lighthouse, Rhode Island.

FIGURE 7.8 Gneiss, a compositionally banded metamorphic rock. See also figure 7.1.

© Times Mirror Higher Education Group, Inc./Bob Coyle, photographer.

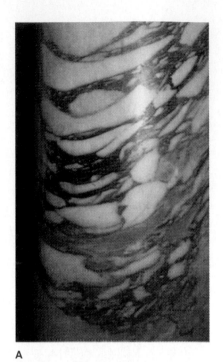

A

Stretched fossil shells

Undeformed fossil shape

B

FIGURE 7.9 Examples of rocks deformed by stress. (A) A column of metamorphosed marble breccia, showing stretched fragments. (B) Deformation of fossils as host rock is deformed (schematic).

bands rich in ferromagnesian minerals (figure 7.8). This mineralogical or compositional banding is *gneissic* (pronounced "nice-ic") texture. Although gneissic texture most commonly reflects compositional differences between layers, the term can also be applied to banded rocks with alternating schistose and equigranular layers. Most gneisses are of granitic composition, but the term need not have compositional implications.

On a megascopic scale, even rock fragments and fossils can be deformed or reoriented by directed stress. If a tennis ball is squeezed, it is flattened in the direction of compression and broadened in the plane perpendicular to it. A stretched piece of foam rubber is elongated in the direction of tension and thinned perpendicular to it. Some sedimentary rocks contain objects whose original shape is known, such as oolites, symmetric fossils, or spheroidal

quartz pebbles in conglomerate. Metamorphism may deform these objects in such a way that the orientation of the deforming forces can be determined (figure 7.9).

Effects of Chemical Changes

It is more difficult to generalize about the consequences of chemical changes during metamorphism. Briefly, addition of a particular element tends to stabilize minerals rich in that element and may cause more such minerals to form. For ex-

Box 7.1

Stress and Strain Under the Microscope

Strain, or deformation, on a gross scale is often visible as folds or other physical changes in outcrop (see figure 7.1). Even when a rock does not show such features obviously, the effects of stress can often be seen in thin section. Compare figures 1 and 2. Figure 1 is a photomicrograph of a thin section of granite, made up mostly of large, clear crystals of quartz (white to dark gray, featureless), coarse feldspar (flecked and striped), and biotite mica (variegated blocks and laths). Figure 2 is a metamorphic rock (schist) of similar mineralogy, but the once-platy biotite crystals are bent, and even the quartz grains have a mottled appearance due to small, stress-induced irregularities in their crystal structures.

FIGURE 1 Thin section of granite. The darker, blocky grains are biotite mica; the gray and white grains are quartz and potassium feldspar.

FIGURE 2 Thin section of biotite schist. Note the kinking or bending of biotite and the mottled appearance of quartz grains due to stress.

ample, if migrating fluids carry dissolved potassium ions into a silicic rock, more potassium feldspars and micas might form. Such introduction of ions in fluids, called **metasomatism,** is not uncommon around plutons, from which fluids may escape into the surrounding rocks.

Conversely, loss of a specific chemical constituent has a destabilizing effect. For instance, if water is driven out of a rock during metamorphism, any remaining hydrous minerals (clays, micas, amphiboles, and so forth) tend to break down into nonhydrous phases (feldspars and pyroxenes, for example) more readily.

TYPES OF METAMORPHIC ROCKS

Metamorphic rocks are subdivided into *foliated* and *nonfoliated* rocks on the basis of the presence or absence of preferred orientation of elongated or platy minerals. Further subdivisions within the foliated rocks typically are based on texture. Most unfoliated rocks are named on the basis of composition.

Foliated Rocks

Many of the common foliated rock types have names directly related to their characteristic textures. As noted previously, for example, **slate** (recall figure 7.5) is a metamorphic rock exhibiting slaty cleavage. Typically, the individual mineral grains in slate are too fine to be seen with the naked eye. With progressive metamorphism of slate, the growing mica crystals may become coarse enough to begin to reflect light strongly, and the cleavage planes in the rock take on a shiny appearance in consequence. Such a rock is called **phyllite** (figure 7.10). The individual mica crystals in a phyllite may just be distinguishable with the unaided eye, but the rock is still basically fine-grained.

FIGURE 7.10 Phyllite, showing satiny or glossy surface due to parallel mica flakes.
© *Times Mirror Higher Education Group, Inc. /Bob Coyle, photographer.*

FIGURE 7.12 Quartzite, metamorphosed quartz sandstone.
© *Times Mirror Higher Education Group, Inc. /Bob Coyle, photographer.*

FIGURE 7.11 Mica schist, metamorphosed shale.

Further progressive metamorphism, with the growth of coarser mica crystals in the presence of directed stress, leads to the development of obvious schistosity, and the corresponding rock is termed a **schist** (figure 7.11). A compositionally layered **gneiss**—a rock with gneissic texture, as shown in figure 7.8—may form either through still-further metamorphism of a schist, with breakdown of some of the micas at higher pressures and temperatures, or during strong metamorphism of other rock types, such as granite.

The rock names of the foliated metamorphic rocks have few compositional implications. Parallel alignment of mica flakes creates slaty cleavage and schistosity, so slates and schists necessarily contain micas. Even so, there are many different compositions of micas, and other minerals may be present as well. The term *gneiss* indicates the existence of compositional layering without specifying anything at all about the minerals in the rock. Compositional terms can be added to the rock name to provide more information. For example, a rock might be described as a "garnet-biotite schist" if those two minerals are prominent in it; a "granitic gneiss" would be a metamorphic rock showing gneissic texture but having a mineralogic composition like that of granite (rich in quartz and feldspar).

Nonfoliated Rocks

Rocks that consist predominantly of equidimensional grains do not show foliation. If they consist mainly of a single mineral, they may be named on the basis of composition. When a quartz-rich sandstone is metamorphosed, for example, the quartz grains are recrystallized and the rock compacted into a denser product with tightly interlocking grains. The resulting metamorphic rock is termed **quartzite** (figure 7.12). In hand specimen, its appearance typically differs from that of unmetamorphosed sandstone in that the recrystallized quartz grains in quartzite have a shinier or more glittery appearance than that of the original abraded grains in the sandstone. Sometimes the quartzite has what is described as a sugary appearance. Quartzite also is usually denser than sandstone as a result of the compaction under pressure. A metamorphosed limestone similarly recrystallized is **marble** (figure 7.13). Usually, a marble is assumed to consist predominantly of calcite. If in fact it consists mainly of the mineral dolomite, it may be called dolomitic marble for clarity.

FIGURE 7.13 Marble, metamorphosed limestone.

Q uartzite and marble do not typically show foliation. If, however, they contain impurities of elongated or platy minerals, such as mica, these minor minerals may show a preferred orientation. Also, because they have formed from sedimentary rocks that may have been bedded or compositionally layered, some quartzites and marbles do show banding unrelated to the presence of directed stress.

An **amphibolite** (figure 7.14) is a metamorphic rock rich in amphiboles. Strictly speaking, amphibolites are not necessarily unfoliated rocks, for amphiboles commonly form in elongated or needlelike crystals that may take on a preferred orientation in the presence of directed stress. A textural term may then be inserted to indicate this: a foliated amphibolite, for example, is one in which the amphibole crystals lie in parallel planes, while a lineated amphibolite is one in which the crystals are all aligned in the same direction (figure 7.15).

Sometimes a metamorphic rock neither shows a distinctive texture nor has a sufficiently simple composition to justify applying one of the previous terms. In such a case, if the nature of the parent rock can be recognized, the prefix *meta-* is simply added to the parent rock name to describe the present rock—for example, *metaconglomerate* or *metabasalt* to describe, more concisely, a metamorphosed conglomerate or basalt.

FIGURE 7.14 Garnet-bearing amphibolite from Gore Mountain, New York.
© Doug Sherman/Geofile.

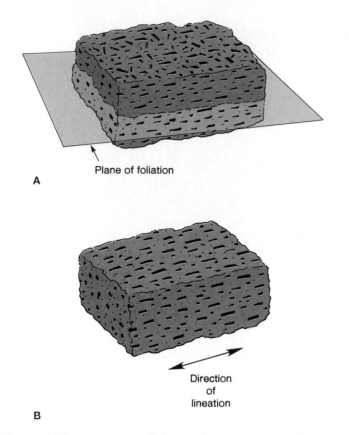

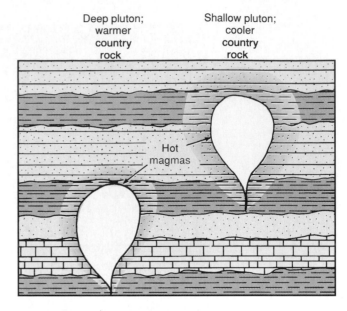

FIGURE 7.16 Contact aureoles around plutons emplaced at deep and shallow crustal levels (schematic).

FIGURE 7.15 Orientation of elongated amphibole crystals in amphibolite. (A) Foliated amphibolite: crystals lie in parallel planes. (B) Lineated amphibolite: crystals all line up in the same direction.

ENVIRONMENTS OF METAMORPHISM

A variety of geologic settings and events lead to metamorphism. Most metamorphic processes can be subdivided into regional, contact, and fault-zone metamorphism.

Regional Metamorphism

Regional metamorphism is, as its name implies, metamorphism on a grand scale, a regional event. Regional metamorphism is commonly associated with mountain-building events, when large areas are uplifted, downwarped, or stressed severely and deformed, as during plate collisions. Rocks pushed to greater depths are subjected to increased pressures and temperatures and are metamorphosed. Collision adds directed stress, producing abundant lineated and foliated rocks. Regional metamorphism involving changes in both pressure and temperature is also called **dynamothermal metamorphism.** The emplacement of very large batholith complexes, which may or may not be associated with mountain building, raises crustal temperatures over broad areas as heat is released from crystallizing magmas and cooling plutons, so batholith formation may likewise result in metamorphism on a regional scale. An individual, smaller pluton, however, will have much more localized ef-

fects, confined to the rocks immediately surrounding the pluton and clearly associated directly with it; this is contact metamorphism.

Contact Metamorphism

Contact metamorphism is named for the setting in which it occurs, near the contact of a pluton. The pluton, emplaced from greater depths, is hotter than the country rock, and if it is significantly hotter, the adjacent country rock is metamorphosed. The pluton then is surrounded by a zone of metamorphic rock, also known as a contact **aureole,** or halo. (The term *aureole* comes from the Latin for "golden," as a crown or halo might be.) Higher-temperature metamorphic minerals are found close to the contact, lower-temperature minerals farther away. Metasomatism may or may not occur in the contact aureole. When it does, it may result in the formation of hydrothermal ore deposits, as magmatic fluids rich in dissolved metals seep into the country rock, cool, and deposit metallic ore minerals.

Contact metamorphism, by definition, is a more localized phenomenon than regional metamorphism, since it is confined to the immediate environs of the responsible pluton. Contact-metamorphic effects also tend to be most marked around plutons emplaced at shallow depths in the crust. Figure 7.16 and the understanding that temperature generally increases with depth in the crust should make this clearer. If a mass of magma at some specific temperature crystallizes deep in the crust, the country rock around it is already at somewhat elevated temperatures and should already contain moderately high-temperature minerals. The modest increase in temperature around the pluton may not be enough to destabilize these minerals. Close to the earth's

A

B

FIGURE 7.17 (A) Typical Martian impact crater, with raised rim; note lobes of debris (right) ejected from the crater by impact. (B) Satellite photographs reveal possible impact structures on earth: Clearwater Lakes, Quebec.

(A) © NSSDC/Goddard Space Flight Center. (B) © NASA.

surface, however, the country rock is much cooler, perhaps only 100 to 200°C. Emplacement of a magma at 700 to 800°C or above changes the temperature of the adjacent country rock more dramatically and is more likely to cause the breakdown of minerals stable only at low temperatures and the crystallization of higher-temperature ones. Similarly, emplacement of a mafic magma at a given depth tends to result in a more pronounced, and generally larger, contact aureole than does emplacement of a silicic magma at the same depth, because of the greater temperature contrast between the country rock and the hotter mafic magma.

Fault-Zone Metamorphism

Fault-zone metamorphism is another localized effect, of relatively minor importance volumetrically. A *fault* is a planar break in rocks, along which the rocks on one side have moved relative to the other. Right along the fault zone, as movement occurs, the rocks may be severely stressed, crushed, and fragmented. There is some increase in temperature due to frictional heating, but except in the case of very large faults, the temperature increase is typically much less than that associated either with deep burial or with magma emplacement. The principal effects result from the stress applied. A rock formed from the angular fragments of rocks broken up in the fault zone is a **fault breccia.** It may contain minerals that crystallize either only at high pressures, or at low temperatures but only in the presence of strong directed stress.

An even more specialized metamorphic environment is that of a meteorite-impact crater (figure 7.17A). When a meteorite first strikes the earth's surface, shock waves dissipate the energy of the meteorite through the ground. The ground is rapidly and severely compressed, commonly fractured; some may be melted or vaporized. After passage of the shock wave, the surface rock is rapidly decompressed, and material is flung violently out from the impact site to form a crater. A shock wave and subsequent decompression affect the meteorite, too, and it commonly shatters and becomes part of the debris. The rocks are subjected to a brief but very intense, ultra-high-pressure pulse of metamorphism. Even a small meteorite impact involves a great deal of energy: an iron meteorite only 20 meters in diameter striking the surface at a velocity of 10 kilometers per second would produce a crater nearly 600 kilometers across.

The passage of the shock wave induces **shock metamorphism,** effects of which can include production of distinctive fracture patterns in the rocks; disruption of crystal structures (detectable microscopically or by X rays); and production of very-high-pressure polymorphs, not normally found in the crust (conveniently, the very common mineral quartz has two such polymorphs—*coesite* and *stishovite*). The distinctive characteristics of shock-metamorphosed rocks have, in fact, been helpful in identifying old impact craters on the earth, where vegetation and millions of years of weathering may have tended to erase them (figure 7.17B).

METAMORPHISM AND THE FACIES CONCEPT

The concept of **facies** has already been introduced in the context of sedimentary rocks (chapter 6). An analogous concept can be applied to metamorphic rocks. A metamorphic facies is a set of physical conditions that gives rise to characteristic mineral assemblages. That is, rocks of a particular metamorphic facies contain one or more minerals indicative of a particular, restricted range of pressures and temperatures. Figure 7.18 shows many of the principal metamorphic facies arranged in the appropriate positions on a pressure-temperature diagram. The pressures are equated to depths on the assumption that these pressures result from the weight of overlying crustal rock of average density. Where additional stress is present—as during plate collision, for instance—there

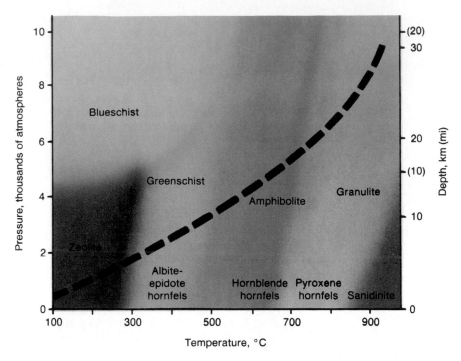

FIGURE 7.18 Metamorphic facies as functions of temperature and pressure (depth). Dashed curve is average continental geothermal gradient.

may be higher pressures at shallower depths. Similarly, it has already been noted that the geothermal gradient is not the only cause of elevated temperature. The facies reflected in a metamorphic rock is useful as a guide to the pressure-temperature conditions under which it formed and to possible unusual pressure-temperature conditions, but some facies span a broad range of conditions. The boundaries between facies are not sharp, because the stabilities of minerals depend, in part, upon factors other than pressure and temperature (such as rock composition and fluids present).

Contact-metamorphic facies are the facies of low pressure and a range of temperatures. The **sanidinite facies** is characterized by the presence of sanidine, a very-high-temperature polymorph of potassium feldspar. Sanidinite-facies conditions are not commonly achieved except very close to extremely high-temperature plutons, such as plutons of ultramafic rock. The term **hornfels** is applied to a variety of fine-grained contact-metamorphic rocks formed at intermediate temperatures. They can be subdivided into several facies of more restricted temperature range. For example, the **pyroxene-hornfels facies** represents temperatures just below those of the sanidinite facies; the **hornblende-hornfels facies** is a lower-temperature facies, since the hydrous hornblende (an amphibole) is not stable at the higher temperatures at which most pyroxenes form. The zeolites are a group of hydrous silicates, stable only at low pressure and temperature, that gives its name to the **zeolite facies.** A given contact aureole may span several facies, grading outward from a high-temperature facies, or mineral assemblage, closest to the pluton, through one or more lower-temperature facies, to unmetamorphosed country rock farther away.

Regional-metamorphic facies are characterized by elevated pressure and temperature. The **greenschist facies** is so named because greenschist-facies rocks commonly contain one or both of the greenish silicates chlorite (a micalike sheet silicate) and epidote. Many, but not all, greenschist-facies rocks are also texturally true schists. Many amphiboles are stable over the range of conditions represented by the **amphibolite facies,** and rocks of this facies are generally rich in amphiboles. Garnets are also common under these conditions. Rocks of the **granulite facies,** which are often gneissic in texture, are characterized by various high-temperature, high-pressure mineral assemblages. The hydrous minerals have commonly broken down by the time granulite-facies conditions are reached.

The dashed curve in figure 7.18 represents a typical continental geothermal gradient. The curve indicates what happens to a rock subjected to progressive metamorphism resulting from deeper and deeper burial in the crust—for example, during regional deformation. The rock passes progressively through conditions of the greenschist, amphibolite, and ultimately, granulite facies, with appropriate changes in mineralogy occurring as low-temperature, low-pressure minerals break down and are replaced by minerals stable at higher pressures and temperatures. Under extreme conditions of metamorphism, at the upper limit of the granulite facies, the minerals with the lowest melting temperatures (generally quartz and potassium feldspar) indeed begin to melt as the boundary between metamorphism and magmatism is reached. If this partly melted rock is then cooled and fully solidified again, the resulting **migmatite,** or "mixed rock," has a mix of igneous and metamorphic characteristics. The rock in figure 7.1 is a migmatite, the quartz and feldspar having melted in part.

The contact-metamorphic facies are characterized by lower pressures for a given temperature than those of the corresponding regional-metamorphic facies. Attainment of

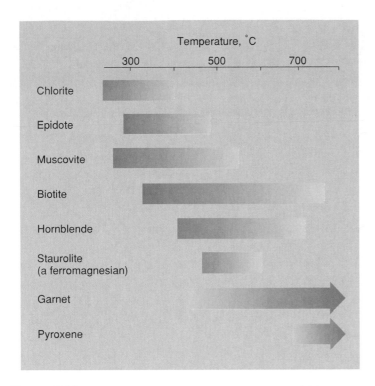

Temperature, °C

FIGURE 7.19 Approximate temperature ranges over which some representative index minerals are stable.

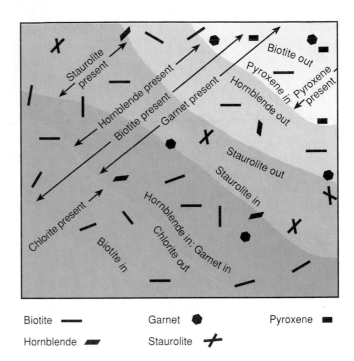

Biotite — Garnet ⬣ Pyroxene ▬

Hornblende ▰ Staurolite ✛

FIGURE 7.20 Isograds on a map show regional trends in metamorphic grade.

high temperatures without concurrent increases in pressure is achieved by emplacement of hot magma at shallow depths, rather than by burial.

The **blueschist facies** is characterized by high pressure but low temperature. Its name derives from the several bluish silicates—especially kyanite (an aluminum silicate) and glaucophane (an amphibole)—that form under these conditions. Again, blueschist-facies conditions are not encountered in normal continental crust; unusual pressures or stresses are required. The precise geologic setting in which blueschist-facies rocks form is not well understood. One setting in which the necessary conditions might occur is examined below and in chapter 9.

In general, plate-tectonic activity and associated magmatism account for a great deal of metamorphism. Plutons produced by subduction can individually cause contact metamorphism. Large-scale heating from batholiths, together with the compressive stress of plate collision, leads to regional metamorphism. Rocks transmit heat relatively poorly, so a subducted slab may stay relatively cold for a long time, while subjected to considerable compressive stress; blueschists might be produced in such a situation.

INDEX MINERALS AND METAMORPHIC GRADE

The **metamorphic grade** of a rock is a general description of the overall intensity of metamorphism to which that rock was subjected. That is, a high-grade metamorphic rock is one that shows textural or mineralogic evidence of having been subjected to high temperatures and/or pressures; a

low-grade metamorphic rock, the reverse. Blueschists and rocks of the granulite facies are examples of high-grade rocks, while zeolite-facies rocks are low-grade rocks. In a contact-metamorphic aureole, metamorphic grade increases with proximity to pluton contact. In a regional context, metamorphic grade increases with deeper burial.

Where regional-metamorphic rocks spanning a range of metamorphic grades are exposed at the surface, the direction of increasing or decreasing grade may be determined using a sequence of **index minerals,** each of which is stable over a restricted temperature range (figure 7.19). The presence of each of the index minerals can be mapped and lines drawn where a particular mineral appears or disappears (figure 7.20). The directional order of these lines, or **isograds** (lines joining points of equal metamorphic grade), shows the regional trend in metamorphic grade. In the example in figure 7.20, the metamorphic grade increases to the northeast. The event causing this regional metamorphism, then, may have been more intense to the northeast. Alternatively, perhaps the rocks now found exposed in the northeast were originally the more deeply buried, and have been uplifted and eroded more since metamorphism, so that higher-grade rocks are now exposed there. Such information is useful in reconstructing past geologic events.

Not all minerals are equally useful as index minerals. Quartz and potassium feldspar, for example, are stable over the whole range of typical regional-metamorphic conditions. Their presence does not restrict the possible pressures or temperatures under which a particular rock formed. Most of the ferromagnesian silicates, on the other hand, are stable over a narrower range of conditions, and many of these are useful index minerals.

SUMMARY

Metamorphism is change in rocks (short of complete melting) brought about by changes in temperature, pressure, or chemical conditions. With progressive metamorphism, existing minerals are commonly recrystallized, the crystals often growing larger in the process, and minerals stable at low temperatures and pressures may break down, to be replaced by other minerals stable at higher-grade conditions. Directed stress may also lead to formation of foliated rocks, in which elongated or platy minerals assume a preferred orientation.

Foliated rocks are named on the basis of their particular texture. Nonfoliated rocks are most often named on the basis of mineralogic composition. Most metamorphism is either contact metamorphism, which occurs in country rock close to the contacts of invading plutons and is characterized by relatively low-pressure mineral assemblages, or regional metamorphism, which is caused principally by plate-tectonic activity and/or emplacement of large batholiths,

and is characterized by elevated pressure as well as elevated temperature. Metamorphic rocks are assigned to a particular facies, corresponding to a specific range of pressures and temperatures, on the basis of the mineral assemblages they contain. Index minerals are useful in assessing the general metamorphic grade of a rock and in determining regional trends in metamorphic grade.

TERMS TO REMEMBER

amphibolite 111
amphibolite facies 114
aureole 112
blueschist facies 115
cleavage (rock) 107
confining pressure 104
contact metamorphism 112
directed stress 104
dynamothermal metamorphism 112
facies 113
fault breccia 113
foliation 107
geothermal gradient 104

gneiss 110
granulite facies 114
greenschist facies 114
hornblende-hornfels facies 114
hornfels 114
hydrothermal 105
index mineral 115
isograd 115
lineation 107
marble 110
metamorphic grade 115
metamorphism 104
metasomatism 109

migmatite 114
parent rock 105
phyllite 109
pyroxene-hornfels facies 114
quartzite 110
regional metamorphism 112
sanidinite facies 114
schist 110
schistosity 107
shock metamorphism 113
slate 109
slaty cleavage 107
zeolite facies 114

QUESTIONS FOR REVIEW

1. What is *metamorphism?* What limits the maximum temperatures possible in metamorphism?
2. Describe two sources of heat that cause metamorphism.
3. Explain the distinction between confining pressure and directed stress. Which leads to such deformation as folding or development of foliation?
4. Describe two ways in which the presence of fluids can influence metamorphism.

5. How do rock cleavage and schistosity develop?
6. What is *metasomatism?* Describe one environment in which it might commonly occur.
7. Define the following terms: *phyllite, schist, gneiss, amphibolite.*
8. Explain the difference between contact and regional metamorphism.

9. Outline the progressive metamorphism of rocks buried deeper and deeper in the crust. Describe the facies through which they might pass and some of the changes to be expected.
10. What is a *migmatite?*
11. Explain what characteristics make a mineral useful as a metamorphic index mineral.

FOR FURTHER THOUGHT

1. Many reactions are so slow at near-surface temperatures that minerals unstable at those temperatures can nevertheless be preserved. Why does this allow geologists to gather much more information than they could if all mineral assemblages constantly adjusted to new pressure, temperature, or other conditions?

2. Sometimes, during cooling from peak metamorphic temperatures, rocks undergo *retrograde* reactions, in which lower-grade minerals form again from higher-grade ones. Considering some of the metamorphic reactions in table 7.1, how important would you expect fluids to be in retrograde metamorphism? Explain.

SUGGESTIONS FOR FURTHER READING

Best, M. G. 1982. *Igneous and metamorphic petrology.* San Francisco: W. H. Freeman.

Hyndman, D. W. 1985. *Petrology of igneous and metamorphic rocks.* 2d ed. New York: McGraw-Hill.

Mason, R. 1978. *Petrology of the metamorphic rocks.* Boston: Allen and Unwin.

Turner, F. J. 1981. *Metamorphic petrology: Mineralogical, field, and tectonic aspects.* 2d ed. New York: McGraw-Hill.

Winkler, H. G. F. 1979. *Petrogenesis of metamorphic rocks.* 5th ed. New York: Springer-Verlag.

8

GEOLOGIC TIME

Though dinosaurs predated humans by tens of millions to hundreds of millions of years, even the dinosaurs appeared relatively late in earth's history. Tools for dating geologic materials have helped unravel more than four billion years of that history.

A student once wrote on an examination: "Geologic time is much slower and made up of very large units of time." This answer, while not quite the expected one, contains a large element of truth. Time passes no more slowly for geologists than for anyone else, but they do necessarily deal with the very long span of the earth's history. Consequently, they often work with very large units of time, millions or even billions of years long.

Much of the understanding of geologic processes, including the rates at which they occur and, therefore, the kinds of impacts they may have on human activities, has been made possible through the development of various means of measuring ages and time spans in geologic systems. In this chapter, we explore some of these methods. A final section examines several applications of geologic age determinations to the study of process rates.

Relative Dating

Before any techniques for establishing numerical ages of rocks or geologic events were known, it was sometimes at least possible to place a sequence of events in the proper order. This is known as **relative dating.**

Arranging Events in Order

Among the earliest efforts to arrange geologic events in sequence were those of Nicolaus Steno. In 1669, he set forth several basic principles that could be applied to sedimentary rocks. The first, the **Law of Superposition,** pointed out that, in an undisturbed pile of sediments (or sedimentary rocks), unaffected by later folding, faulting, or other physical disruption, those on the bottom had been deposited first, followed in succession by the layers above them, ending with the youngest on top (figure 8.1). Today, this idea may seem

rather obvious, but at the time, it represented a real step forward in thinking logically about rocks. Steno's second concept, the **Law of Original Horizontality,** was based on the observation that, especially on a large scale, sediments are commonly deposited in approximately horizontal, flat-lying layers (figure 8.2). Therefore, if one comes upon sedimentary rocks in which the major strata are folded or dipping steeply, the rocks must have been displaced or deformed after deposition and lithification (figure 8.3). The third, the **Law of Lateral Continuity,** noted that at the time it is deposited, a water-deposited layer of sediment will be continuous in all directions until it reaches the edge of the sedimentary basin (shore, stream bank, etc.) or thins out to the point of no deposition.

Time-breaks in the sedimentary record also exist; they may or may not be easily recognized. Any such break in the rock record is a *hiatus.* An **unconformity** is a surface within a sedimentary sequence on which there was a lack of sediment deposition, or where active erosion may even have occurred, for some period of time. When the rocks are later examined, that time span will not be represented in the record, and if the unconformity is erosional, some of the record once present will have been lost. An unconformity, then, can represent a long interval of time and one or more significant geologic events. The most difficult kind of unconformity to recognize is a **disconformity,** an unconformity at which the sedimentary rock layers above and below it are parallel

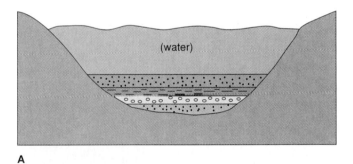

A

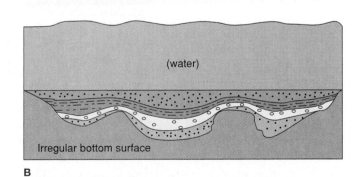

B

FIGURE 8.2 The Law of Original Horizontality. (A) Sediments tend to be deposited in horizontal layers. (B) Even where the sediments are draped over an irregular surface, they tend toward the horizontal.

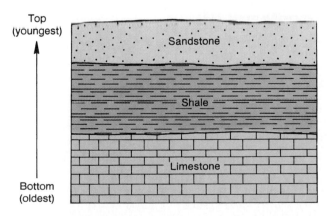

FIGURE 8.1 The Law of Superposition. In an undisturbed sedimentary sequence, the rocks on the bottom were deposited first, and the depositional ages become younger higher in the pile.

FIGURE 8.3 Tilted sedimentary rocks, deformed after deposition. (Compare to figure 8.2B.)
Photograph by M. R. Mudge, USGS Photo Library, Denver, CO.

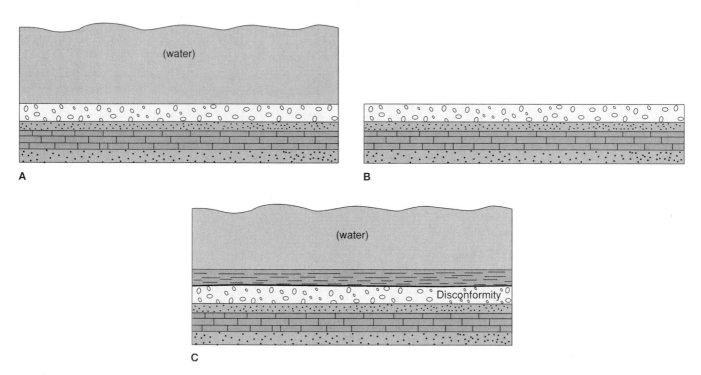

FIGURE 8.4 Development of a disconformity: sediment deposition is interrupted for a period of time. (A) Sediment deposition in water. (B) Water now absent, deposition ceases. (C) Deposition resumes after some lapse of time; disconformity is established.

(figure 8.4). More obvious is the **angular unconformity,** at which the bedding planes in rock layers above and below the unconformity are not parallel (figure 8.5). The presence of an angular unconformity usually implies a significant period of erosion, as well as uplift. The new depositional surface for the younger rocks may have been the land surface for a considerable time after uplift and tilting, especially if it has then been planed flat by wind and water. An angular unconformity thus suggests a great lapse of time, or in other words, the loss of a significant chunk of the geologic record from the rocks.

In later centuries, reasoning about the relative ages of rocks was extended to include igneous rocks in rock sequences. If a dike or other pluton cuts across layers of sedimentary rocks, the sedimentary rocks must have been there

120   *Chapter 8*

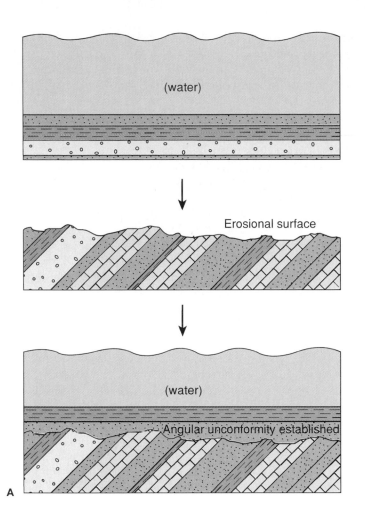

B

FIGURE 8.5 Development of an angular unconformity requires some deformation and erosion before sedimentation is resumed. (A) Sequence of events (top to bottom): deposition; rocks tilted, eroded; subsequent deposition. (B) Angular unconformity in the Grand Canyon.

(B) © Times Mirror Higher Education Group, Inc./Doug Sherman, photographer.

first, the igneous rock introduced later (figure 8.6). This is sometimes called the *Principle* (or *Law*) *of Crosscutting Relationships*. Often, there is a further clue to the correct sequence: The hotter igneous rock may have "baked," or metamorphosed, the country rock immediately adjacent to it, so again the igneous rock must have come second. (This is a particular help when the pluton is concordant.) If a pluton contains xenoliths, the rock from which the xenoliths came must predate the intrusion (*Principle of Inclusion*).

Such geologic common sense can be applied to many rock associations. If a strongly metamorphosed sedimentary rock is overlain by a completely unmetamorphosed one, for instance, the metamorphism must have occurred after the first sedimentary rock formed but before the sediments of the second were deposited. Quite complex sequences of geologic events can be unraveled by taking into account such simple underlying principles. See figure 8.7 for an example.

Correlation

Fossils also play a role in the determination of relative ages. The concept that fossils could be the remains of older life-forms dates back at least to the ancient Greeks, but for some time, the idea fell out of favor. It was seriously revived in the eighteenth century, and around the year 1800, the **Law of Faunal Succession** was advanced. Its basic concept is that life-forms change through time; old ones disappear from and

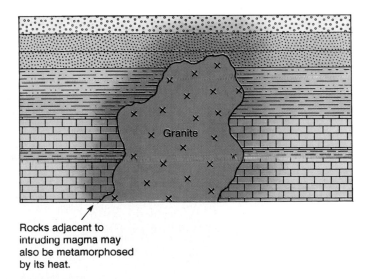

Rocks adjacent to intruding magma may also be metamorphosed by its heat.

FIGURE 8.6 A pluton cutting across older rocks.

new ones appear in the fossil record, but the same form is never exactly duplicated independently at two different times in history, and one form succeeds another in a regular, consistent fashion. (A theoretical basis for the Law of Faunal Succession was later provided by Charles Darwin's theories of evolution and natural selection.) According to the Law of Faunal Succession, then, once the basic sequence of fossil

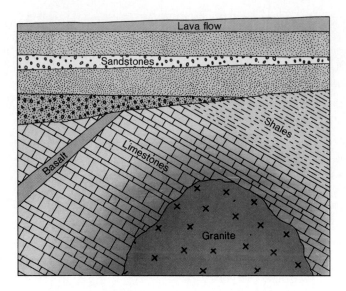

FIGURE 8.7 Deciphering a complex rock sequence: The limestones must be the oldest (Law of Superposition), followed by the shales. The granite pluton and basalt must both be younger than the limestone they crosscut. Note the metamorphosed zone around the granite also. It is not possible to determine whether the igneous rocks predate or postdate the shales or whether the sedimentary rocks were tilted before or after the igneous rocks were emplaced. After the limestones and shales were tilted, they were eroded, and then the sandstones were deposited on top. Finally, a lava flow covered the entire sequence.

forms in the rock record is determined, rocks can be placed in their correct relative chronological position on the basis of the fossils contained in them. The same principle, in turn, implies that when one finds exactly the same type of fossil organism preserved in two rocks, even if the rocks are quite different compositionally, and geographically widely separated, those rocks should be the same age. This conviction is strengthened if the same set of several different types of fossils—the same *faunal assemblage*—is found in both rocks. This allows age **correlation** between rock units exposed in different places (figure 8.8).

Correlation need not be based only on fossils. Where rock exposures are good and the bedding regular, visual correlation is possible, supported by the Law of Lateral Continuity (figure 8.9). Units can also be correlated on the basis of distinctive mineralogy or chemistry, or a particular depositional sequence, as will be explored in later chapters. Igneous and metamorphic rocks that can be dated directly may be correlated on the basis of a combination of age and compositional data. But for sedimentary rocks, precise direct dating may not be possible, and fossils can be especially helpful.

Correlation of sedimentary rocks on a global scale is facilitated by the use of **index fossils.** Good index fossils are those derived from particular plant or animal species that were distributed widely over the earth but existed for a very limited period of time. A rock in which a given index fossil is found, then, has a rather narrowly constrained age and must correspond closely in age with other rocks containing the

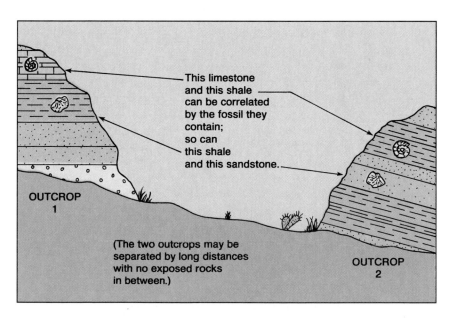

FIGURE 8.8 Similarity of fossils suggests similarity of ages, even in different rocks widely separated in space.

Figure 8.9 Where rock outcrops are discontinuous, visual correlation may nevertheless be possible, on the basis of characteristics such as color, susceptibility to weathering, and relative thickness of strata. Monument Valley, Arizona.

same fossil. Index fossils also need to be easily identified/recognized, and readily preserved.

The usefulness of fossil correlation and of the concept of faunal succession is limited, however, because these principles can be applied only to rocks in which fossils are well preserved, which are almost exclusively sedimentary rocks. Correlation also can be based on the occurrence of unusual rock types, distinctive rock sequences, or other geologic similarities.

The foregoing ideas are all useful in clarifying relative age relationships among rock units. They do not, however, help to answer such questions as, "How old is this granite?" "How long did it take to deposit this limestone?" "How recently, and over how long a period, did this apparently extinct volcano erupt?" "Has this fault been active in modern times?" Questions requiring numerical replies went unanswered until the twentieth century.

How Old Is the Earth?—Early Answers

How old is the earth? Geologists and nongeologists alike have been fascinated for centuries with this very basic question in geologic time. Many have attempted to answer it, but with little success until the last few decades.

One of the earliest widely publicized estimates of the earth's age was the seventeenth-century work of Archbishop Ussher of Ireland. He painstakingly counted up the generations in the Bible and arrived at the conclusion that the earth was created in 4004 B.C. Later, an English theologian named Lightfoot declared that the precise time was 9 A.M. on Tuesday, October 26, in that year. This very young age was hard for many geologists to accept. The complex geology of the modern earth seemed to require far longer to develop.

Even less satisfactory from that point of view was the estimate of the philosopher Immanuel Kant. Kant tried to find a maximum possible age for the earth by assuming that the sun had always shone down on earth and that the sun's tremendous energy output is due to the burning of some sort of conventional fuel. But a mass of fuel the size of the sun would burn up in only 1000 years, given the rate at which the energy is being released, plainly an impossible result in view of several millennia of recorded human history. Kant, of course, knew nothing of the nuclear-fusion processes by which the sun actually generates its energy.

About 1750, Georges L. L. de Buffon attacked the question from another angle. He assumed that the earth was initially molten, modeled it as a ball of iron, and calculated how long it would take this quantity of iron to cool to present surface temperatures. The result was about 75,000 years. To most geologists, this still was not nearly long enough.

Around 1850, physicist Hermann L. F. von Helmholtz approached the problem by supposing that the sun's luminosity was due to infall of matter into its center, converting gravitational potential energy to heat and light. This gave a maximum age for the sun (and presumably for the earth) of 20 to 40 million years. Again, his assumptions were wrong, so his answer was also incorrect.

In the late 1800s, Lord William T. Kelvin reworked Buffon's calculations, modeling the earth more realistically in terms of rock properties rather than those of metallic iron. Interestingly, he, too, arrived at an age of 20 to 40 million years for the earth. What he did not take into account, because natural radioactivity was then unknown, was that some heat is continually being *added* to the earth's interior through radioactive decay, so it has taken correspondingly longer to cool down to its present temperature regime.

The calculations went on, and there was something wrong with each. In 1893, U.S. Geological Survey geologist Charles D. Walcott tried to compute the total thickness of the sedimentary rock record throughout geologic history and, dividing by typical modern sedimentation rates, to estimate how long that pile of sediments would have taken to accumulate. His answer was 75 million years. Walcott was hampered in his efforts by several factors, including the gaps in many sedimentary rock sequences (the unconformities previously described), periods during which no sedimentation occurred and periods during which some sediments were eroded away. (In fact, in most places on earth, we have very little of the vast history of the earth preserved as rock record.) Walcott further assumed sedimentation rates comparable to present ones but actually had no proof that ancient sedimentation rates could not have been quite different. He also had no real idea of the total

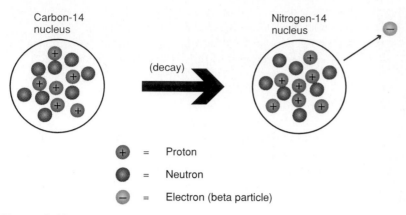

+ = Proton

● = Neutron

⊖ = Electron (beta particle)

FIGURE 8.10 The decay of a carbon-14 nucleus to a nitrogen-14 nucleus (schematic). (Since electrons are not involved, they have been omitted for simplicity.)

thickness of sediments deposited in the time before organisms capable of preservation as fossils became widespread, which turns out to be most of earth's history.

In 1899, physicist John Joly published calculations based on the salinity of the oceans. Taking the total load of dissolved salts delivered to the seas by rivers, and assuming that the ocean was initially pure water, he determined that it would take about 100 million years for the present concentrations of salts to be reached. (This method, incidentally, had been proposed two centuries before by Sir Edmund Halley, of Halley's Comet fame.) Aside from Joly's unjustified assumption that rates of weathering and salt input into the oceans throughout earth's history were constant, he did not consider that the buildup of salts is slowed by the removal of some of the dissolved material—for example, as chemical sediments.

So the debate continued, frequently heated, until the discovery of radioactivity around the year 1900 provided a much more powerful and accurate tool with which to undertake a solution.

RADIOMETRIC DATING

Henri Becquerel did not set out to solve any geologic problems or even to discover radioactivity. He was interested in a curious property of some uranium salts: if exposed to light, they continued to emit light (phosphoresce) for awhile afterward, even in a dark room. On one occasion, he had stored a vial of uranium salts in a drawer on top of some photographic plates well wrapped in black paper. To his surprise, when he next examined the photographic plates, they were fogged with a faint image of the vial, as if they had somehow been exposed to light right through the paper. The uranium was emitting something that could pass through light-opaque materials. We now know that uranium isotopes are among several dozen natural isotopes that are radioactive, that undergo spontaneous decay. Once the phenomenon of radioactive decay was reasonably well understood, physicists and geoscientists began to realize that it could be a very useful tool for investigating the earth's history.

Radioactive Decay and Dating

A **radioactive** isotope is one with an unstable nuclear configuration. That is, the particular combination of protons and neutrons is not truly stable, and therefore, in time, the nucleus decays, changing into a different nucleus. Up to the instant at which decay occurs, the nucleus of a radioactive isotope behaves no differently from a stable nucleus. The decay, when it does occur, involves emission of radiation, typically consisting of one or more particles plus energy. Principal kinds of radiation emitted are *alpha particles* (helium nuclei, consisting of two protons and two neutrons), *beta particles* (electrons or their positively charged antiparticles, positrons), or *gamma rays* (electromagnetic radiation, analogous to X rays but more penetrating). The decaying nucleus is, by convention, called the **parent** nucleus, while the nucleus into which it decays is called the **daughter** nucleus.

By way of example, consider the radioactive isotope carbon-14. Its nucleus contains eight neutrons and the six protons characteristic of carbon. When the parent carbon-14 decays, a neutron is converted to a proton plus an electron (figure 8.10). The proton remains in the nucleus, while the electron (beta particle) is ejected. Energy is also released during the decay. The resulting daughter nucleus is nitrogen-14 (having seven protons and seven neutrons).

If decay of unstable nuclei were instantaneous or occurred at altogether random times, the phenomenon would have no application to dating. Indeed, the exact moment at which any single nucleus of a radioisotope will decay cannot be predicted. However, given a large number of atoms of the same radioisotope, it can be predicted that a certain fraction of them will decay over a given period of time. Radioactive decay is thus a statistical phenomenon, one that obeys the laws of probability. In that sense, it is like flipping a coin. One cannot know whether any single flip of an (unbiased) coin will come up heads or tails, but statistically, over a large number of flips, half should come up heads, half tails.

Each radioisotope can be characterized by a parameter called its **half-life,** defined as the length of time required for half of a given initial number of atoms of that isotope to decay. If the half-life of a particular radioisotope is 10 million years, and we begin with 1000 atoms of it, then after 10 million years, half of it will have decayed away, and 500 atoms will be left. After another 10 million years, half of the remaining 500 atoms will have decayed, and 250 will be left; and so on.

Not only is the half-life of each radioisotope characteristic and measurable in the laboratory; it is also constant. That is, a given radioisotope will decay at the same rate regardless of its chemical or physical state, the compound in which it may occur, or the temperature or pressure to which it is subjected. The half-life of carbon-14, for example, is 5730 years. If we start with a billion atoms of carbon-14, then after 5730 years, we will have half a billion atoms of carbon-14 left, whether we have the carbon in the form of diamond, coal, organic compounds in human tissues, carbon dioxide gas, or whatever.

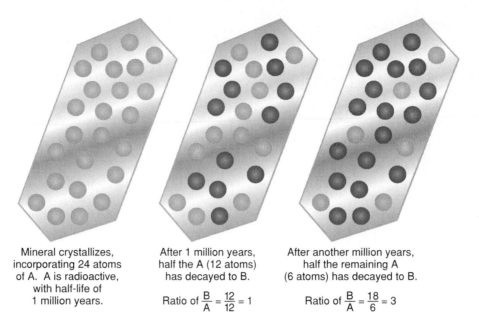

Mineral crystallizes, incorporating 24 atoms of A. A is radioactive, with half-life of 1 million years.

After 1 million years, half the A (12 atoms) has decayed to B.

Ratio of $\frac{B}{A} = \frac{12}{12} = 1$

After another million years, half the remaining A (6 atoms) has decayed to B.

Ratio of $\frac{B}{A} = \frac{18}{6} = 3$

FIGURE 8.11 With radioactive decay, half of the parent isotope has decayed with each passing half-life. Ratio of daughter isotope to parent increases over time as a function of half-life.

H uman activities do not affect the rate of decay of radioactive materials. An unstable nucleus cannot be made more or less stable; we cannot speed up, slow down, or stop the decay of a radioisotope. This is part of the problem with radioactive waste materials. They cannot be treated to make them nonradioactive. Instead, if they pose a radiation hazard, they must be isolated for as long as it takes them to decay away to the point that the quantity left is harmless.

Given the constancy of radioactive decay rates, one can then, in principle, use the relative amounts of a parent isotope and the product daughter isotope into which it decays to find the age of a sample. Figure 8.11 illustrates the concept. Suppose that parent isotope A decays to daughter B with a half-life of 1 million years. In a rock that contained some A and no B when it formed, A will gradually decay and B will accumulate. After 1 million years (one half-life of A), half of the A initially present will have decayed to yield an equal number of atoms of B. After another half-life, half of the remaining half will have decayed, so three-fourths of the initial amount of A will have been converted to B and one fourth will remain. The ratio of B to A will be 3:1. After another million years, there will be seven atoms of B for every atom of A, and so on. If a sample contains 24,000 atoms of B and 8000 atoms of A, then, assuming no B present when the rock formed and noting the 3:1 ratio of B to A, geologists would conclude that the sample was two half-lives of A (in this case, 2 million years) old. The changes in amounts of A and B, and thus in their relative abundances, as a function of time, are shown graphically in figure 8.12A.

To take a more realistic geologic example, suppose A is uranium-235 and B is lead-207; the half-life of uranium-235 is 704 million years. In a sample of uranium ore—uranium-rich,

essentially lead-free—it would take 704 million years for the number of accumulated lead-207 atoms formed by uranium decay to equal the number of uranium-235 atoms: half the uranium-235 will have decayed after one half-life. If the ratio of lead-207 to uranium-235 were 3:1, then as shown in figure 8.12B, the age of the ore sample would be three half-lives of uranium-235, or 1408 million years. In natural systems, where samples are rarely a whole number of half-lives old, the calculations may be somewhat more complex, but the underlying principles are the same. For example, if the numbers of uranium and lead atoms present are such that we determine that 90% of the original uranium-235 had decayed to lead-207, leaving 10% of the uranium-235, we can estimate the sample's age graphically from the decay curve, noting that 10% of the parent will be left after 3.32 half-lives, or 2340 million (2.34 billion) years.

The foregoing is a considerable simplification of the dating of natural geologic samples. Many samples contain some of the daughter isotope, as well as the parent, at the time of formation, and correction for this must be made. Various methods exist for making these corrections. Other samples have not remained chemically closed throughout their histories and have gained or lost atoms of the parent or daughter isotope of interest, which will make the calculated date incorrect. The trained specialist can recognize rock or mineral samples in which this has occurred.

"Absolute" Ages

Dates calculated on the basis of decay of radioisotopes are sometimes inaccurately termed **absolute ages.** What is really meant by the term is a numerical age, as distinguished from the relative ages described earlier in the chapter. There are various kinds of so-called absolute ages—for example, dating

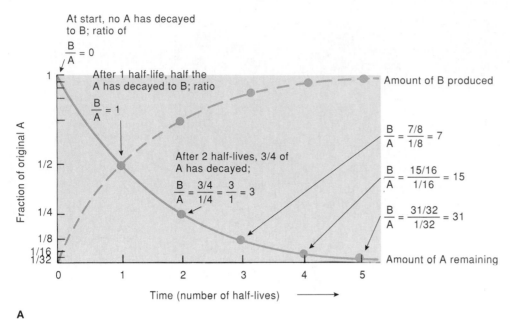

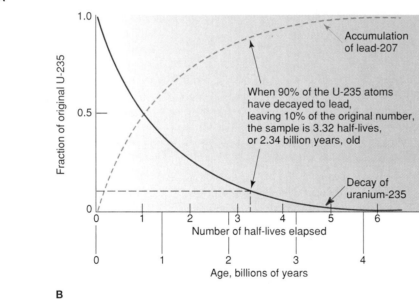

Figure 8.12 (A) Radioactive decay of parent isotope A proceeds at a regular rate, with corresponding accumulation of daughter isotope B. (B) Using the decay of uranium-235 to lead-207 as an example, we see how age can be determined from the decay curve by considering how much of the original parent remains.

be measurable. Either the daughter isotope must not normally be incorporated into the sample initially, or there must be a means of correcting for the amount initially present. The half-life of the parent must be appropriate to the age of the event being dated: long enough that some parent atoms are still present but short enough that some appreciable decay and accumulation of daughter atoms has occurred. A radioisotope with a half-life of ten days would be of no use in dating a million-year-old rock; the parent isotope would have decayed away completely long ago, and there would be no way to tell *how* long ago. Conversely, a radioisotope with a half-life of 10 billion years would be useless for dating a fresh lava flow because the atoms of the daughter isotope that would have accumulated in the rock would be too few to be measurable.

Fewer than a dozen isotopes are widely used in dating geologic samples. Several of the most important are listed in table 8.1, along with the materials in which the parent isotope in each case is readily concentrated.

What Is Being Dated?

Radiometric dates are not the ages of the elements in the rock or the age of the earth. Where valid dates can be obtained, they pertain to some aspect of the history of that particular rock. What kinds of events are dateable can be explored by taking the potassium-argon scheme as an example.

Potassium-40 decays to argon-40. Argon is one of the inert gases and therefore does not readily bond chemically with other atoms or ions in a crystal. When a potassium-bearing mineral is crystallizing, whether from a melt or a solution, argon is not ordinarily incorporated into the growing crystal. Once the crystal is formed, however, any argon-40 produced by decay of potassium-40 is trapped within the rigid crystal structure. After a period of time, a quantity of argon-40 will have accumulated that is proportional to the amount of potassium-40 in the crystal and to the length of time since it formed. From measuring the quantities of potassium-40 and argon-40, and knowing the half-life of potassium-40, geologists can determine the time since the mineral crystallized. The date of crystallization of a potassium-rich igneous mineral (or rock) or chemical sedimentary mineral (rock) can thereby be determined.

a tree by counting the tree rings (annual growth rings) is such a method. The more correct modern term for a numerical age measured using the phenomenon of radioactive decay is **radiometric age.** The geologist who specializes in radiometric dating is a *geochronologist*. The analytical procedure is explored in box 8.1.

Choice of Isotopic System

Several conditions must be satisfied by an isotopic system to be used for dating geologic materials. The parent isotope chosen must be abundant enough in the sample for its quantity to

Box 8.1

Tools of the Trade

MASS SPECTROMETRY

The development of radiometric dating required not only recognition of the phenomenon of radioactive decay but also the construction of sophisticated analytical equipment capable of counting or measuring the quantities of different isotopes of an element: the **mass spectrometer** (figure 1).

Prior to analysis, the sample is dissolved in strong acid and chemically processed to separate individual elements of interest. A drop of solution containing such an element is loaded onto a metal filament, which is heated in the spectrometer under vacuum. Ions of the element are vaporized off the filament and moved along a tube through a magnetic field to a collector/counter. The magnetic field deflects the stream of ions, isotopes of different masses being deflected slightly differently, allowing each to be counted

FIGURE 1 A mass spectrometer.
Photograph by J. N. Lytwyn.

separately. The relative numbers of different isotopes are indicated by the counter. On a modern spectrometer, the data are processed by a computer, which also operates the instrument wholly or partially.

Refinements in analytical techniques have gradually reduced the amount of material needed for analysis. Less than a tenth of a gram (0.0035 ounce) of powdered rock may be enough for rubidium-strontium analysis. Some methods for uranium-lead analysis of zircons can be applied to a single zircon crystal the size of a small sand grain.

TABLE 8.1 *Radiometric Decay Schemes Commonly Used in Geology*

Parent Isotope	Half-Life (Years)	Daughter Isotope	Parent Abundant in
Potassium-40	1.3 billion	Argon-40	Potassium-rich minerals (including feldspar, micas)
Rubidium-87	48.8 billion	Strontium-87	Potassium-rich minerals
Thorium-232	14 billion	Lead-208	Zircon and other minor minerals
Uranium-235	704 million	Lead-207	Uranium ores; zircon and other minor minerals
Uranium-238	4.5 billion	Lead-206	(Same as uranium-235)
Carbon-14*	5730	Nitrogen-14	Organic matter; atmospheric CO_2; dissolved carbonate

*This method works somewhat differently from the other methods listed. Also, with its very short half-life, carbon-14 cannot be used to date samples more than 50,000 to 100,000 years old. It is used principally for dating wood, cloth, paper, bones, and so on from archaeological sites.

Metamorphism tends to blur the results. When a crystal is heated and the atoms in the structure begin to vibrate more energetically, trapped argon can more readily escape. In general, the higher the temperature, the more easily the argon is lost. Losing some of the argon from a sample makes the sample appear too young: when it is later analyzed, it will have too little argon for its true original age and potassium content. The rock's apparent age will thus have no exact significance in the history of that rock. The date will fall somewhere between the rock's original age and the time of metamorphism. On the other hand, in the case of intense, high-grade metamorphism, *all* of the accumulated argon can be lost, especially if the minerals are recrystallized during metamorphism. The isotopic clock is then effectively *reset* to the time of metamorphism. If the rock remains undisturbed thereafter,

the date it yields when analyzed will be the correct date of the metamorphism.

Physical breakup typically is not accompanied by extensive loss of accumulated daughter isotopes. The formation of clastic sediments is therefore not an event that tends to reset the isotopic systems of the constituent minerals. Clastic sediments are said to "inherit" a portion of their daughter isotopes from their source rocks. When analyzed long after deposition, they tend to yield apparent ages older than the time of deposition, for they will contain not only the daughter isotopes accumulated as a result of radioactive decay following deposition but also some inherited daughter isotopes. If the grains lost none of the daughter isotopes during breakup, transport, and deposition, the sediment may even preserve the age of the source rocks, although this is rare. In any case, isotopic methods do not generally yield the time of deposition of a clastic sediment, and therefore, radiometric methods are not generally applied to such materials.

Each isotopic system behaves a bit differently because of differences in the chemical behavior of the parent and daughter isotopes in each case. Some systems are more easily reset than others. Because different elements are concentrated in different minerals and rock types, not all isotopic systems work equally well on a given rock. As a very general rule, however, the time of crystallization of an igneous rock, the date of a high-grade metamorphic event, or the time of precipitation of a chemical sediment from solution can often be determined quite reliably using an appropriate isotopic system. Where weak metamorphism has been superimposed on a preexisting rock, radiometric ages obtained are often geologically meaningless, in that they fall somewhere between the primary rock age and the date of metamorphism. Sometimes, the age of the source rocks of clastic sediments can be measured, but the time of sediment deposition cannot be.

The foregoing discussion has noted some of the limitations of radiometric dates. Recognition of those limitations allows geochronologists to know when a date obtained is credible, as well as whether a given material can usefully be dated. Several decades of research into radiometric dating of geologic materials and of laboratory experiments on the behavior of geologic and chemical systems have increased understanding to the point that the geochronologist can readily determine which dates are indeed reliable. Some criteria are geologic, relating to the nature of the rock and its thermal history, as deduced from rock type and mineralogy. Other criteria are chronologic. For example, it was noted previously that different isotopic systems behave differently in nature. When several chemically different methods (rubidium-strontium, potassium-argon, and uranium-lead, for example) give the same date on a sample, that date is particularly well constrained, for it would be virtually impossible for such agreement to arise by coincidence.

In other words, there are good and bad radiometric dates, but it is possible to distinguish which is which. Geochronologists avoid dating inappropriate materials and apply stringent tests to those dates obtained to assess their value. The vast majority of radiometric dates obtained today are highly reliable, and the availability of these powerful methods has added immensely to the understanding of the earth and of the solar system.

Radiometric and Relative Ages Combined

Because accurate radiometric dates cannot be determined for many rocks and fossils, radiometric and relative dating methods are often used in conjunction to constrain the ages of units not easily dated. Undateable sedimentary rocks, crosscut and perhaps metamorphosed by a dike or pluton, must be older than the invading igneous rock, which may be dateable. This at least puts a lower (younger) limit on the age of the sedimentary rock. A fossil found in a rock sandwiched between two dateable lava flows has its age bracketed by the ages of the volcanic rocks. (If strata containing a recognized index fossil can be dated with some precision, its usefulness in fossil correlation is enhanced.) An unmetamorphosed sedimentary rock directly in contact with a high-grade metamorphic rock must postdate the metamorphism, which puts an upper limit on the time of sedimentation. Following such reasoning, geologists are often able to work out a complex sequence of events from a single exposure of rock and to put age limits, if not exact ages, on the various rock units (see figure 8.13).

THE GEOLOGIC TIME SCALE

Particularly in the days before radiometric dating, it was necessary to establish some subdivisions of earth history by which particular periods of time could be indicated. This initially was done using those rocks with abundant fossil remains. Originally, the resultant timescale was strictly a scheme of relative ages; it was subsequently refined and made quantitative with the aid of radiometric dates.

The Phanerozoic Eon

The Law of Faunal Succession and the appearance and disappearance of particular fossils in the sedimentary record were first used to mark the boundaries of the time units (table 8.2). The **Phanerozoic** eon encompassed all that span of time from which fossil remains were then known. The principal divisions were the **eras—Paleozoic, Mesozoic,** and **Cenozoic,** meaning, respectively, "ancient life," "intermediate life," and "recent life." The eras were subdivided into periods and the periods into epochs. The names of these smaller time units were assigned in various ways. Often, they were named for the location of the *type section,* the place where rocks of that age were well exposed and that time unit was first defined. For example, rocks of the Cambrian period are well exposed in Wales, and the type section is there. *Cambria* is a Latin form of the native Welsh name for Wales. The Jurassic period is named for the Jura Mountains of France. Other periods are named on the basis of some characteristic of the type rocks: The Cretaceous period derives its name from the Latin *creta* ("chalk"), for the chalky strata of that age in southern England and northern Europe. Rocks of Carboniferous age commonly include coal beds. All of the relatively unfossiliferous rocks that seemed to be older (lower in the sedimentary sequence) than the Cambrian period were lumped together as pre-Cambrian, later formally named **Precambrian.**

With the advent of radiometric dating, numbers could be attached to the units and boundaries of the timescale. It became apparent that, indeed, geologic history spanned long periods of time, and that the most detailed part of the scale—the Phanerozoic (Cambrian and later)—was by far the shortest, comprising less than 15% of earth history.

Radiometric Dating and the Timescale

Assigning very precise dates to the fine subdivisions of the Phanerozoic has proved difficult due to the nature of those subdivisions and of radiometric dating. As is apparent from

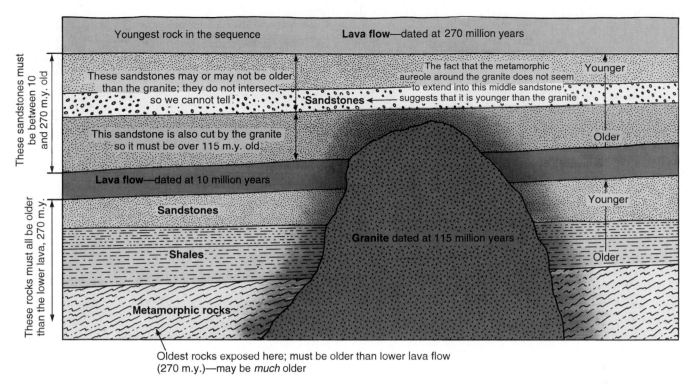

Figure 8.13 The ability to date some units in an outcrop allows the ages of units not directly dateable to be constrained.

Table 8.2 *The Geologic Time Scale (Circa 1900)*

Era	Period	Epoch	Distinctive Life-Forms
Cenozoic	Quaternary	Recent	Modern humans
		Pleistocene	Stone-Age humans
		Pliocene	
		Miocene	Flowering plants common
	Tertiary	Oligocene	Ancestral pigs, apes
		Eocene	Ancestral horses, cattle
		Paleocene	
Mesozoic	Cretaceous		Dinosaurs become extinct; flowering plants appear
	Jurassic		Birds, mammals appear
	Triassic		Dinosaurs, first modern corals appear
Paleozoic	Permian		Rise of reptiles, amphibians
	Carboniferous		Coal forests; first reptiles, winged insects
	Devonian		First amphibians, trees
	Silurian		First land plants, coral reefs
	Ordovician		First fishlike vertebrates
	Cambrian		First widespread fossils
Precambrian			Scant invertebrate fossils

previous discussion, the time of deposition of most sedimentary rocks cannot readily be dated. Very few fossils can be dated directly either, and in any case, fossilization may occur long after the organism is dead and buried. Yet the eras, periods, and epochs of the Phanerozoic were defined almost entirely on the basis of the sedimentary/fossil record. In practice, then, it was often necessary to approach the ages of the units by indirect dating and determination of age limits for sedimentary rocks, as described earlier, and then making correlations with rocks in the type sections.

A further complication with respect to the finest subdivisions is that, like any measurements, most radiometric dates have some inherent uncertainty associated with them, arising out of geologic disturbance of the samples, laboratory and statistical uncertainties, and so on. In the best cases, these uncertainties are much less than 1% of the date determined; in unfavorable cases, they may amount to 10% or more of the age. That is, geologists cannot date a rock at exactly 174,692,361 years; the age might instead be reported as 174 ± 3 million years, meaning that the rock has been determined to be between 171 million and 177 million years old. On a rock billions of years old, the uncertainty may be tens of millions of years; on a million-year-old sample, it might be only tens of thousands of years. Then, too, analytical methods and laboratory instrumentation have been constantly improving, and radiometric dates have been refined. Redating of units dated previously has yielded new, more precise ages that may differ slightly from the less precise ages determined earlier.

The approximate time framework of the Phanerozoic was readily established, but the various technical limitations have caused persistent, small uncertainties in the exact dates in question. The student should not be surprised, therefore, to find that texts published over the last few decades may differ somewhat in the exact ages shown. A recent revision of the Phanerozoic timescale, with radiometric dates, is shown in table 8.3.

The Precambrian

The Precambrian has continued to pose something of a problem. A unit that spans 4 billion years of time seems to demand subdivision—but, in the virtual absence of fossils, on what basis? The first division into the *Archean* ("ancient") and *Proterozoic* ("pre-life") eons split the Precambrian into two nearly equal halves, but there was considerable disagreement as to how to define the boundary geologically. Although the time around 2.5 billion years ago was a time of widespread igneous activity, metamorphism, and mountain building, no single event of global impact could be identified to mark the boundary. There was still less agreement on how the Archean and Proterozoic should be further subdivided. Eventually, an international commission convened to address these and other related problems and recommended adoption of the scheme shown in table 8.4. The dates of the boundaries are arbitrary in the sense that they do not represent well-defined events, such as the extinction of the dinosaurs. Still, the existence of these divisions allows events to be placed in a time framework by referring to them as "middle Proterozoic," "late Archean," or whatever, without resorting to a more cumbersome phrase, such as "falling between 2100 million years and 1500 million years in age."

How Old Is the Earth?—A Better Answer

Because the earth is geologically still very active, no rocks have been preserved unchanged since its formation. The age of the earth, therefore, cannot be determined directly, even by isotopic methods. However, evidence indicates that the earth, moon, and meteorites all formed at the same time, during the formation of the solar system. Fortunately, some of these other materials have had quite different subsequent histories. The moon, for instance, is much smaller than the earth, cooled more rapidly after accretion, and has been geologically inactive for billions of years. The oldest of the samples returned by the Apollo lunar missions are approximately 4.6 billion years old.

TABLE 8.3 *Modern Geologic Time Scale for the Phanerozoic, with Radiometric Ages*

Era	Period	Epoch*	Date of Boundary (Millions of Years Ago)†
Cenozoic	Quaternary	Holocene	0.1
		Pleistocene	2
	Tertiary	Pliocene	5
		Miocene	24
		Oligocene	37
		Eocene	58
		Paleocene	66
Mesozoic	Cretaceous		144
	Jurassic		208
	Triassic		245
Paleozoic	Permian		286
	Carboniferous‡		
	Pennsylvanian		320
	Mississippian		360
	Devonian		408
	Silurian		438
	Ordovician		505
	Cambrian		570

*Pre-Cenozoic periods are also subdivided into epochs, but there is little uniformity of nomenclature for these worldwide.

†Dates used from compilation of Geological Society of America for Decade of North American Geology.

‡Not generally subdivided outside the United States.

Geologists also have numerous samples of *meteorites,* fragments of rock or metal of extraterrestrial origin that have fallen to earth. Some of these have been disturbed after formation by collisions in space or by other processes, but the majority of meteorites yield ages in the range of 4.5 to 4.6 billion years. Strong chemical similarities between the earth and meteorites support the idea that the earth and meteorites formed from the same materials (solar nebula), presumably at the same time.

On the basis of the foregoing and other evidence, the earth is inferred to have formed at about 4.55 billion years ago, and this date is typically taken as the beginning of Precambrian time. The oldest terrestrial rocks that have actually been accurately dated directly are close to 4 billion years old. Rocks close to this age are found on nearly every continent. The oldest rocks on each continent are generally between 3.6 and 3.9 billion years old. A very few isolated samples as old as 4.2 billion years old have now been analyzed.

Geologists have no rocks that date back to the time of the earth's formation, probably for two reasons. First, the early earth was very hot—the surface may even have been molten for some time—and the radiometric "clocks" would not start until rocks formed and cooled to temperatures more nearly approaching modern earth surface temperatures. Sec-

TABLE 8.4 *Subdivisions of the Precambrian Proposed by the North American Commission on Stratigraphic Nomenclature*

Eon	Era	Date at Boundary (millions of years)
(Phanerozoic)		570
Proterozoic	Late	900
	Middle	1600
	Early	2500
Archean	Late	3000
	Middle	3400
	Early	3900 (?—age of oldest terrestrial rocks)

Note: Periods and finer subdivisions have not been established.

ond, the earth has undergone so much geologic change, with formation and destruction and alteration of rocks, that survival of unaltered samples of the earliest rocks to have formed is highly unlikely.

DATING AND GEOLOGIC PROCESS RATES

The rates at which geologic processes occur can be estimated in a variety of ways, many of which rely on radiometric dating techniques.

Examples of Rate Determination

As will be seen in the next chapter, the continents and sea floor move over the earth's surface. A few approximately fixed reference points (the magnetic poles, for example) allow the determination of how far the continents and sea floor have moved. If one can determine the radiometric ages of rocks formed at one time, when a continent was at point *A,* and those formed at another time, when the continent was at point *B,* and measure the distance between *A* and *B,* the average rate of movement during that period can be calculated. Suppose that a continent is shown to have moved 400 kilometers (about 250 miles) in 10 million years. The average rate of movement is then

$$\frac{400 \text{ km}}{10,000,000 \text{ year}} \times \frac{1000 \text{ m}}{1 \text{ km}} \times \frac{100 \text{ cm}}{1 \text{ m}} = \frac{4 \text{ cm}}{\text{year}}$$

or 4 centimeters/year. This is, of course, only an average rate for the whole 10 million years. The continent may have moved 10 cm/year or more at some times, 1 cm/year or less at others.

By dating the volcanic rocks from one or a group of volcanoes, geologists can judge how long volcanic areas remain active and how frequently volcanoes erupt. Such information can be very useful in assessing volcanic hazards. A typical small volcano may erupt intermittently over a period of 100,000 years; a major volcanic center may stay active for 1 to 10 million years. The building of a major mountain range, with all the attendant igneous and metamorphic activity, may take 100 million years.

Some rates can only be broadly constrained. The minimum rate of uplift of rocks in a mountain range might be estimated from the age of marine sedimentary rocks in the mountains, which were once deposited underwater and are now found high above sea level. Such uplift rates are typically 1 centimeter/year or less, often much less. Beaches formed on Scandinavian coastlines during the last Ice Age have been rising since the ice sheets melted and their enormous mass was removed from the land. From the beach deposits' ages and present elevation above sea level, uplift rates of this postglacial rebound have been approximated at 1 centimeter/year.

Finally, the development of isotopic dating methods has made possible a much better understanding of the rates of organic evolution and the whole history of life on earth.

The Dangers of Extrapolation

One must be somewhat cautious about extrapolating present (modern) process rates too far into the past or future, and one must not assume that rates measured over a short time are representative of a long period. This point is admirably illustrated by the following passage from Mark Twain's *Life on the Mississippi.* As described in chapter 14, rivers that have developed contorted, twisted channels full of bends (meanders) may change course during times of high water flow. The meanders are cut off, or bypassed, as the flowing water seeks a straighter, shorter, more direct path. Here, Twain speculates on the implications of the shortening of the Mississippi River by this process:

> In the space of 176 years, the lower Mississippi has shortened itself 242 miles. This is an average of a trifle over one mile and a third per year. Therefore, any calm person, who is not blind or idiotic, can see that, in the Old Oolitic Silurian Period just a million years ago next November, the Lower Mississippi was upwards of 1,300,000 miles long and stuck out over the Gulf of Mexico like a fishing rod. And by the same token, any person can see that 742 years from now the Lower Mississippi will be only a mile and three-quarters long, and Cairo and New Orleans will have joined their streets together, and be plodding along comfortably under a single mayor and a mutual board of aldermen. There is something fascinating about science. One gets such wholesale returns of conjecture out of such a trifling investment of fact.

Still, it is evident from the data we have accumulated on the ages of rocks and the typical rates of common geologic processes—plate motions, uplift and erosion of mountains, deposition of sediments, and many more—that by and large, the earth has been shaped by seemingly slow processes acting over great spans of time. Continents moving apart at a few centimeters a year leave an ocean basin the size of the Atlantic between them in only a couple of hundred million years—and the whole of earth's history is more than twenty times as long as that. In contemplating human impacts on the geological environment, as well as in trying to understand natural geologic processes, it is critical to keep the vastness of geologic time in mind.

SUMMARY

Before the discovery of natural radioactivity, only relative age determinations for rocks and geologic events were possible. Field relationships and fossils were the principal tools used for this purpose. Rocks could be placed in relative age sequence or, with the aid of fossils, correlated from place to place.

Radiometric dating has made quantitative age measurements possible, although not all geologic materials can be so dated. Most of the commonly used isotopic dating methods work best for

igneous and high-grade metamorphic rocks. It is rarely possible to determine the time of deposition of sedimentary rocks. This has led to some difficulties in the assignment of very precise ages to the units of the Phanerozoic timescale, which is subdivided largely using fossils and sedimentary rocks. The Phanerozoic, with its abundant life-forms, represents only a relatively small part of the earth's history. The age of the earth is now estimated radiometrically at approximately 4.55 billion years.

Radiometric ages can be used to explore the rates at which different kinds of geologic processes occur and have considerably advanced understanding of the earth's development. However, while observations of present geologic processes can be used to understand past geologic history, it cannot be assumed that those processes have always proceeded at rates comparable to those presently observed.

TERMS TO REMEMBER

absolute age 125
angular unconformity 120
Cenozoic 128
correlation 122
daughter (nucleus) 124
disconformity 119
era 128
half-life 124

index fossil 122
Law of Faunal Succession 121
Law of Lateral Continuity 119
Law of Original Horizontality 119
Law of Superposition 119
mass spectrometer 127
Mesozoic 128
Paleozoic 128

parent (nucleus) 124
Phanerozoic 128
Precambrian 128
radioactive 124
radiometric age 126
relative dating 119
unconformity 119

QUESTIONS FOR REVIEW

1. Describe the significance of the Law of Superposition and the Law of Original Horizontality to relative dating of sedimentary sequences.
2. What is the distinction between a disconformity and an angular unconformity? What do they have in common?
3. Explain two ways in which you might determine the relative ages of a pluton and surrounding sedimentary rocks.

4. How is the correlation of rock units made easier by the Law of Faunal Succession? What is a limitation on its use?
5. Why is it important to radiometric dating that radioactive isotopes have constant half-lives?
6. Describe any three requirements that must be satisfied in order for a radiometric decay scheme to be useful in dating geological materials.
7. It has proven somewhat difficult to establish radiometric dates for the units of the Phanerozoic

timescale, because the subdivisions were defined using sedimentary rocks. Explain. How do geologists address this problem?
8. When the geologic time scale was first established, the Precambrian was not subdivided. Why?
9. Why is it not possible to determine the age of the earth directly by radiometric methods? On what basis is its age estimated?

FOR FURTHER THOUGHT

1. As noted in the chapter, the best index fossils are those that are found widely distributed over the earth and derived from organisms that existed only for geologically short periods of time. Consider why these two criteria are important. What sorts of organisms might satisfy the first criterion especially well?

2. Each of the various radioactive isotopes has a distinct and unique half-life. What would be the impact on radiometric dating if all radioisotopes had the same half-life? If there were only one naturally occurring radioisotope?

3. The equation for decay of a radioactive isotope is
$$N = N_0 e^{-\lambda t}$$
where N_0 is the initial number of parent atoms, N the number remaining after time t, and λ the *decay constant* characteristic of that parent isotope (proportional to the rate at which it decays). If each parent atom decays to one daughter atom D, and there is no D present in the sample initially, then after time t, the number of atoms of daughter D present is given by
$$D = N_0 - N = N(e^{\lambda t} - 1).$$
Suppose that you have analyzed a zircon sample in which the ratio of lead-206 atoms to (parent) uranium-238 atoms is 4094:5280. The decay constant λ for uranium-238 is 1.55×10^{-10} per year. What is the age of the zircon?

SUGGESTIONS FOR FURTHER READING

Cloud, P. 1988. *Oasis in space: Earth history from the beginning*. New York: W. W. Norton.

Dott, R. H., Jr., and Batten, R. L. 1981. *Evolution of the earth*. 3d ed. New York: McGraw-Hill.

Eicher, D. L. 1968. *Geologic time*. 2d ed. Englewood Cliffs, N.J.: Prentice-Hall.

Faure, G. 1986. *Principles of isotope geology*. 2d ed. New York: John Wiley and Sons.

Harland, W. B. 1978. *Geochronologic scales*. American Association of Petroleum Geologists Studies in Geology 6, pp. 9–32.

Hurley, P. M. 1959. *How old is the earth?* New York: Doubleday. (This classic, written in the early days of radiometric dating of geological materials, is particularly aimed at the nonspecialist reader.)

Johns, R. B., ed. 1986. *Biological markers in the sedimentary record*. New York: Elsevier.

McLaren, D. J. 1978. *Dating and correlation, a review*. American Association of Petroleum Geologists Studies in Geology 6, pp. 1–7.

Moorbath, S. 1977. The oldest rocks and the growth of continents. *Scientific American* 236 (March): 92–104.

Newson, H. E., and Jones, J. H. 1990. *Origin of the Earth*. New York: Oxford University Press.

Nisbet, E. G. 1987. *The young earth: An introduction to Archean geology*. Boston: Allen and Unwin.

Windley, B. F. 1984. *The evolving continents*. 2d ed. New York: John Wiley and Sons.

9

PLATE TECTONICS

Water accumulates in topographic lows; here, a so-called sag pond has formed in the depression marking the San Andreas fault in the Carizzo Plain, CA. View is northward along the fault, which marks the boundary between the Pacific plate and the North American plate. © Doug Sherman/Geofile

INTRODUCTION

More than three centuries ago, observers looking at world maps noticed the similarity in outline between the eastern coast of South America and the western coast of Africa (figure 9.1). Francis Bacon remarked upon the resemblance in the early seventeenth century. In 1855, Antonio Snider went so far as to publish a sketch showing how the two continents fit together, jigsaw-puzzle fashion. Such reconstruction gave rise to the bold suggestion that perhaps these continents had once been part of the same landmass, which had later broken up.

This concept of **continental drift**—the idea that individual continents could shift position on the globe—had an especially vocal champion in Alfred Wegener, a German meteorologist. He began to publish his ideas on the subject of continental drift in 1912 and continued to do so for nearly two decades. Several other prominent scientists found the idea plausible. However, most people, scientists and nonscientists alike, had difficulty visualizing how something as massive as a continent could possibly "drift" around on a solid earth, or why it should do so. The majority of reputable scientists scoffed at the idea or, at best, politely ignored it.

As it happens, most of the supporting evidence was simply undiscovered or unrecognized at the time. Beginning in the 1960s, however, data of many different kinds began to accumulate that indicated that the continents have indeed moved. Continental drift turns out to be just one aspect of a broader theory known as *plate tectonics* that has evolved over the last two decades. **Tectonics** is the study of large-scale movement and deformation of the earth's outer layers. **Plate tectonics** relates such deformation to the existence and movement of rigid "plates" of rock over a weak or plastic layer in the upper mantle.

THE EARLY CONCEPT OF CONTINENTAL DRIFT

Alfred Wegener had based his proposition of continental drift on several lines of evidence, of which the fit of the edges of the continents was only one. Another involved the similarities of rock types and structures on opposing shores of the Atlantic. Fossil correlation provided further evidence.

Fossil Evidence

Wegener had become aware that paleontologists studying fossil remains had found some plant and animal forms that occurred in the rock record over limited areas of several continents now widely separated in space. Leaves of a distinctive fossil plant, *Glossopteris,* were found in southern Africa, Australia, South America, India, and even Antarctica. Certain dinosaur and other vertebrate fossils were likewise found only over limited areas of several different continents. How did a given life-form develop, identically and simultaneously, on different continents now widely distributed around the globe?

Some paleontologists postulated long-distance transport of seeds and spores by wind or water over the intervening

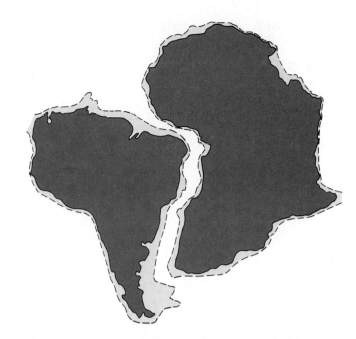

FIGURE 9.1 The jigsaw-puzzle fit of South America and Africa suggests that they might once have been joined together and were subsequently separated by continental drift.

oceans. For the larger animals, they proposed the past existence of land bridges, now vanished, between continents. Wegener suggested instead that, at the time these organisms flourished, the continents had been united, and that the continents had split and drifted apart after the plant and animal remains had been entombed in the rocks.

Climatic Evidence

Climatic considerations also seemed to provide support for the theory of continental drift. Many factors determine a region's climate, but a dominant one is latitude. In general, equatorial regions tend to be warmest, polar regions coldest, with more moderate temperatures in between. Sedimentary rocks, formed at the earth's surface, often preserve evidence of the climatic conditions under which they formed. Fossil remains of plants known to thrive in moist heat imply a tropical climate; sandstones in which windblown desert dunes are preserved suggest dry conditions; some sediments can be identified as having been deposited by glaciers.

In many cases, the ancient climate of a given region appears to have differed drastically from the same region's present climate. For example, there is evidence of widespread past glaciation over India, in places where ice is unknown today. Moreover, such discrepancies cannot be accounted for simply in terms of *global* climatic changes, for the rock record does not show the same warming or cooling trends simultaneously on all continents. Wegener's interpretation was that rocks indicating cold climatic conditions were deposited when a continent was near a pole, and vice versa, and that apparent changes in temperature regime were the result of the continents' drifting from one latitude to another (figure 9.2).

Limitations of the Early Model

Wegener envisioned the continental blocks as plowing through the ocean floor, piling up mountains at their leading edges as they went. But there was no topographic evidence of disruption of the sea floor such as would be expected from this process. Nor could Wegener identify a sufficiently powerful driving force that could move continents without seriously violating basic physical principles. He died decades before his basic vision was validated.

PLATES AND DRIFTING CONTINENTS

One of the many obstacles to accepting the idea of continental drift was the difficulty of imagining solid continents moving over solid earth. However, as can be demonstrated by geophysical methods (see chapter 10), the earth is not completely solid from the surface to the center of the core. In fact, a plastic, locally partly molten zone lies relatively close to the surface, so that in a sense, a thin, solid, rigid shell of rock floats on a weaker, semisolid layer below (figure 9.3).

Lithosphere and Asthenosphere

The earth's outermost, solid layer is the **lithosphere,** named from the Greek word *lithos,* meaning "rock." The name reflects the rigid quality of lithosphere. The lithosphere varies in thickness from place to place on the earth. It is thinnest underneath the oceans, where it extends to a depth of about 50 kilometers (about 30 miles). Lithosphere under the continents is thicker, extending in places to depths of over 100 kilometers (60 miles).

The layer below the lithosphere, the **asthenosphere,** derives its name from the Greek word *asthenes,* meaning "without strength." This term reflects the more plastic quality of the asthenosphere, which can flow slowly, as contrasted with the rigid, more brittle lithosphere. The asthenosphere, which lies entirely within the upper mantle, extends to an average depth of about 500 kilometers (300 miles). Its lack of strength or rigidity may result, in part, from melting in the upper asthenosphere. This zone is certainly not all molten; at most, only a small percentage of magma exists in otherwise solid rock. Even where melting has not actually occurred, temperatures in the asthenosphere are very close to melting temperatures. Under such high-temperature, high-pressure

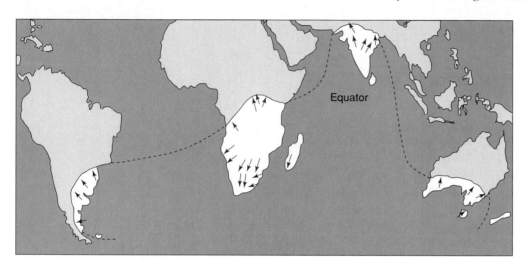

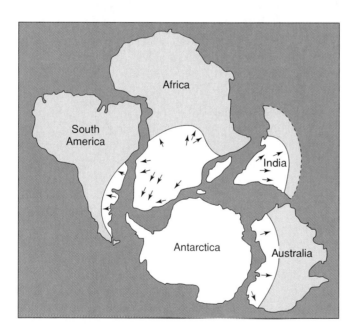

FIGURE 9.2 Glacial deposits across the southern continents suggest past juxtaposition. Arrows indicate direction of ice flow. Top diagram reflects modern distribution of continents; bottom diagram shows how they might have been assembled during the glaciation.
Source: Data from Arthur Holmes, Principles of Physical Geology, *2d ed., 1965, Ronald Press, New York, NY.*

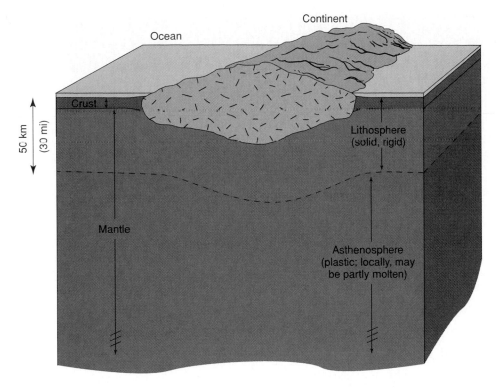

FIGURE 9.3 The outer zones of the earth (not to scale). The terms *crust* and *mantle* have compositional implications; *lithosphere* and *asthenosphere* describe physical properties. The lithosphere includes the crust and uppermost mantle. The asthenosphere lies entirely within the upper mantle.

conditions (confining pressures in the asthenosphere are tens of thousands of times atmospheric pressure), the rocks behave plastically and can flow slowly under stress, even without quite starting to melt. (If it is difficult to picture a solid that can nevertheless flow, consider the behavior of asphalt on a hot day, or of modeling clay.)

The presence of the asthenosphere makes the concept of continental drift more plausible. The continents need not drag across or plow through solid rock. Instead, they can be pictured as sliding over a softened or very viscous fluid layer underneath.

Locating Plate Boundaries

The distribution of earthquakes and volcanic eruptions on a map indicates that these phenomena are, for the most part, concentrated in belts or linear chains (figure 9.4). This pattern suggests that the rigid shell of lithosphere is cracked in places, broken up into pieces or plates. The volcanoes and earthquakes are concentrated at the boundaries of these

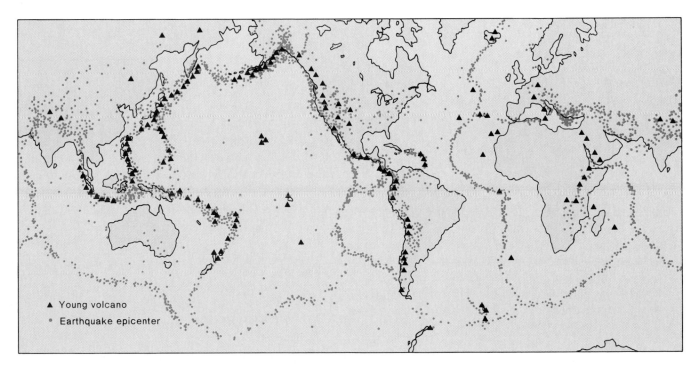

FIGURE 9.4 Locations of modern volcanoes and earthquakes around the world.
Source: Map plotted by the Environmental Data and Information Service of the National Oceanic and Atmospheric Administration; earthquakes from U.S. Coast and Geodetic Survey.

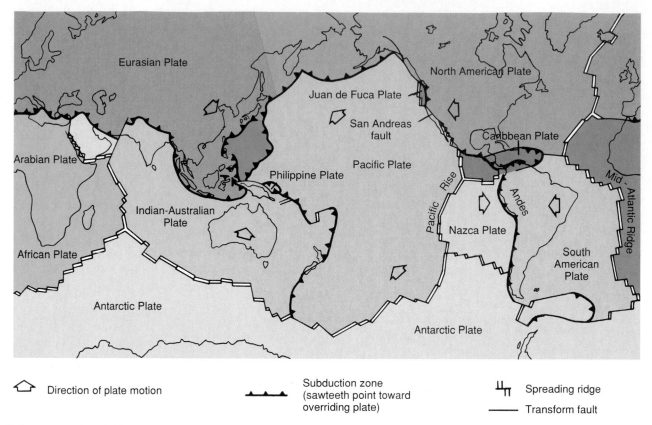

 Direction of plate motion

Subduction zone (sawteeth point toward overriding plate)

⊔⊓ Spreading ridge

— Transform fault

FIGURE 9.5 Principal world lithospheric plates, inferred from information such as that shown in figure 9.4 and other data. *Source: After W. Hamilton, U.S. Geological Survey.*

lithospheric plates, where plates jostle or scrape against one another. (The effect is somewhat like ice floes on an arctic sea: most of the grinding and crushing of ice and the spurting up of water from below occurs at the edges of the blocks of ice, while the floes' thick, solid, central portions are relatively undisturbed.) About half a dozen very large lithospheric plates and many smaller ones have now been identified (figure 9.5).

Recognition of the existence of the asthenosphere made plate motions more plausible, but it did not *prove* that they had actually occurred. Likewise, the apparent existence of discrete plates of rigid lithosphere near the earth's surface did not prove that those plates had ever moved. Many additional observations had to be gathered, and suggested explanations tested, before most scientists would accept the concept of plate tectonics. In time, however, it even became possible to document the rates and directions of plate movements.

PLATE MOVEMENTS— ACCUMULATING EVIDENCE

If South America and Africa really have moved apart, one might expect to see some evidence of the continents' passage on the sea floor between them. A topographic map of the floor of the Atlantic Ocean shows an obvious ridge running north–south about halfway between those continents (figure 9.6). This midocean ridge could be the seam from which the two continents have moved apart. Similar ridges are found on the floors of other oceans. (Seafloor ridges are explored in detail in chapter 12.) It remained for scientists studying the ages and magnetic properties of seafloor rocks to recognize the significance of the ocean ridges to plate tectonics.

FIGURE 9.6 The mid-Atlantic ridge, one prominent seafloor spreading ridge.

Source: "World Ocean Floor Panorama" by Bruce C. Heezen and Marie Tharp. Copyright © 1977 Marie Tharp.

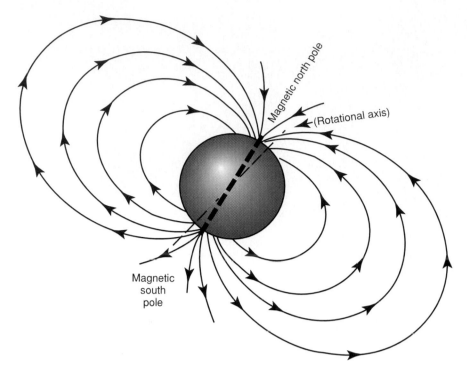

FIGURE 9.7 Lines of force of the earth's magnetic field with "normal" polarity (as it is today).

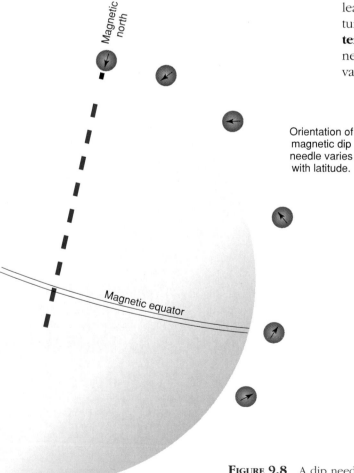

FIGURE 9.8 A dip needle varies in orientation with latitude, lying horizontally at the magnetic equator, pointing vertically at the poles.

Magnetism in Rocks

The earth possesses a magnetic field that can, to a first approximation, be described as similar to what would be expected from a huge bar magnet buried at the earth's center (figure 9.7). Magnetic lines of force run around the earth from the south magnetic pole to the north; the magnetic poles lie close to the geographic (rotational) poles, though they do not coincide. A compass needle aligns itself parallel to the lines of the magnetic field, north–south, and points to magnetic north. A dip needle (a magnetic needle suspended in such a way that it can pivot vertically, not just horizontally) likewise dips along the trend of the magnetic field lines at that latitude (figure 9.8). Magnetic dip, also termed magnetic inclination, varies with latitude: the dip needle is horizontal near the equator, vertical at either pole, and at some intermediate angle at latitudes in between. The magnetic dip thus indicates distance from the magnetic pole.

Most iron-bearing minerals are at least weakly magnetic at surface temperatures. Each magnetic mineral has a **Curie temperature** above which it loses its magnetic properties. The Curie temperature varies from mineral to mineral, but it is always below a mineral's melting temperature. A hot magma is therefore not magnetic, but as it cools and solidifies, and ferromagnesian silicates and other iron-bearing minerals crystallize in it, those magnetic minerals tend to line up in the same direction. Like tiny compass or dip needles, but free to move in three dimensions, they align themselves parallel to the lines of force of the earth's magnetic field. They point to magnetic north and indicate a magnetic dip consistent with their magnetic latitude. They retain their internal magnetic orientation unless they are heated again. This is the basis for the study of **paleomagnetism,** ancient magnetism in rocks. Paleomagnetism can also be detected in some iron-rich clastic sediments, in which the minerals have settled during deposition into alignment parallel to the earth's field, and in some metamorphic rocks, in which iron-rich minerals have recrystallized at elevated temperatures to take on a similar magnetic orientation.

Magnetic north has not always coincided with its present position. In the early 1900s, scientists investigating the direction of magnetization of a sequence of young volcanic rocks in France discovered that some of the earlier flows appeared to be magnetized in the opposite direction from the rest, their magnetic minerals pointing south instead of north. Confirmation of this discovery in many places around the world led to the suggestion, in the late 1920s, that the earth's magnetic field at some past time had "flipped," or reversed polarity, with north and south magnetic poles switching places.

Although geologists still do not know just why or how this occurs, the phenomenon of magnetic reversals has now been well documented with the aid of radiometric dating. Rocks crystallizing at times when the earth's field has been in its present orientation are said to be *normally magnetized;* rocks crystallizing when the field was oriented the opposite way are described as *reversely magnetized.* Over the earth's history, the magnetic field has reversed many times, but rocks of a given age show a consistent polarity. Through magnetic measurements and radiometric dating of the same rocks, geologists have been able to reconstruct the reversal history of the earth's magnetic field in detail.

Paleomagnetism and Seafloor Spreading

The ocean floor is made up largely of basalt, rich in ferromagnesian minerals. During the 1950s, the first large-scale surveys of the magnetic properties of the sea floor produced an entirely unexpected result. The floor of the ocean was found to consist of alternating "stripes" or bands of normally and reversely magnetized rocks, symmetrically arranged around the ocean ridges. This seemed so incredible that, at first, the instruments or measurements were assumed to be faulty. Other studies, however, consistently obtained the same results. For several years, geoscientists were baffled.

Then, in 1963, an elegant explanation was proposed by the team of F. J. Vine and D. H. Matthews and, independently, by L. W. Morley. A few years previously, geophysicist Professor Harry Hess of Princeton University had suggested the possibility of **seafloor spreading,** the idea being that the sea floor had split and spread away from the ridges. Seafloor spreading could account very simply for the magnetic stripes on the sea floor.

If the oceanic lithosphere splits and plates move apart, a 50-kilometer-deep rift in the lithosphere begins to open up. But the rift does not stay open because, as it begins to form, it provides a path for the escape of some of the magma derived from the asthenosphere. The magma rises, cools, and solidifies to form new basaltic rock, which becomes magnetized in

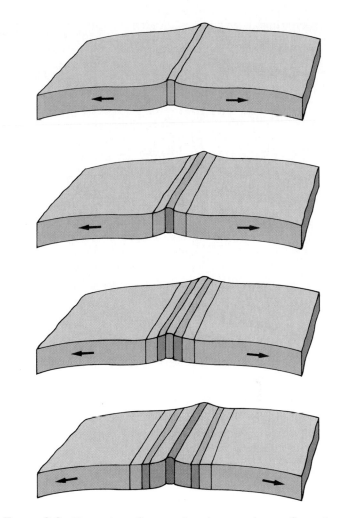

FIGURE 9.9   Formation of magnetic stripes on the sea floor. As each new piece of sea floor forms at the ridge, it becomes magnetized in a direction dependent on the orientation of the earth's field at that time. The magnetism is "frozen in" or preserved in the rock thereafter. Past reversals of the field are reflected in alternating bands of normally and reversely magnetized rocks.

the prevailing direction of the earth's magnetic field. As the plates continue to move apart, the new rock also splits and parts, making way for more magma to form still younger rock, and so on.

If, during the course of seafloor spreading, the polarity of the earth's magnetic field reverses, the rocks formed after the reversal are polarized oppositely from those formed before it. The ocean floor is a continuous sequence of basalts formed over many tens of millions of years, during which time there have been dozens of polarity reversals. The basalts of the sea floor have acted as a sort of magnetic tape recorder throughout that time, preserving a record of polarity reversals in the alternating bands of normally and reversely magnetized rocks (figure 9.9).

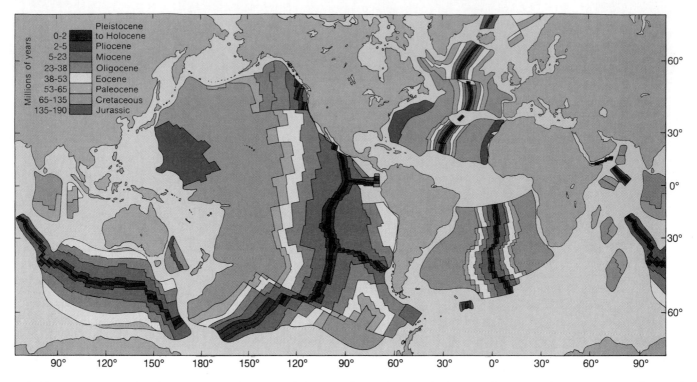

FIGURE 9.10 The pattern of seafloor ages on either side of the spreading ridges results from seafloor spreading. Younger rocks are closer to the ridge, and vice versa.

Source: Adapted from W. C. Pitman III, R. L. Larson, and E. M. Herron, Geological Society of America, 1974.

Age of the Sea Floor

If any geologic model is correct, it should be possible to use it to make predictions about other kinds of data before the corresponding measurements are made (the geologic version of the scientific method). The seafloor-spreading model, in particular, implies that rocks of the sea floor should be younger close to the spreading ridges and progressively older farther away.

Specially designed research ships can sample sediment from the deep-sea floor and drill into the basalt beneath. The time at which an igneous rock like basalt crystallized from its magma can be determined radiometrically, as described in chapter 8. When this is done for many samples of seafloor basalt, the predicted pattern emerges. The rocks of the sea floor, and the bottommost sediments deposited on them, are youngest close to the ocean ridges and become progressively older the farther away they are from the ridges on either side (see, for example, figure 9.10). Like the magnetic striping, the age pattern is symmetric across each ridge, again as one would predict. This is powerful confirmation of the seafloor-spreading hypothesis. As spreading progresses, previously formed rocks are continually spread apart and moved farther from the ridge, while fresh magma rises from the astheno-sphere to form new, younger lithosphere at the ridge. The oldest rocks recovered from the sea floor, well away from the ridges, are about 200 million years old.

Polar-Wander Curves

Evidence for plate movements does not come only from the sea floor. For reasons outlined later in this chapter and in chapter 11, much older rocks are preserved on the continents than beneath the oceans, so longer periods of earth history can be investigated through continental rocks.

Studies of paleomagnetism in continental rocks can span many hundreds of millions of years and yield quite complex data. Magnetized rocks of widely different ages on a single continent may point to very different apparent magnetic-pole positions. The apparent ancient magnetic north and south poles may not simply be reversed but may be rotated or displaced in latitude from the present magnetic north–south. When the directions of magnetization of many rocks of various ages from a single continent are determined and plotted on a map, it appears that the magnetic poles have meandered far over the surface of the earth, if the position of the continent is assumed to have been fixed on the earth throughout time. The resulting curve, showing the apparent movement of the magnetic poles relative to one continent or region as a function of time, is the **polar-wander curve** for that landmass (see figure 9.11).

The discovery and construction of polar-wander curves was initially troublesome, because there are good geophysical reasons to believe that the earth's magnetic poles should

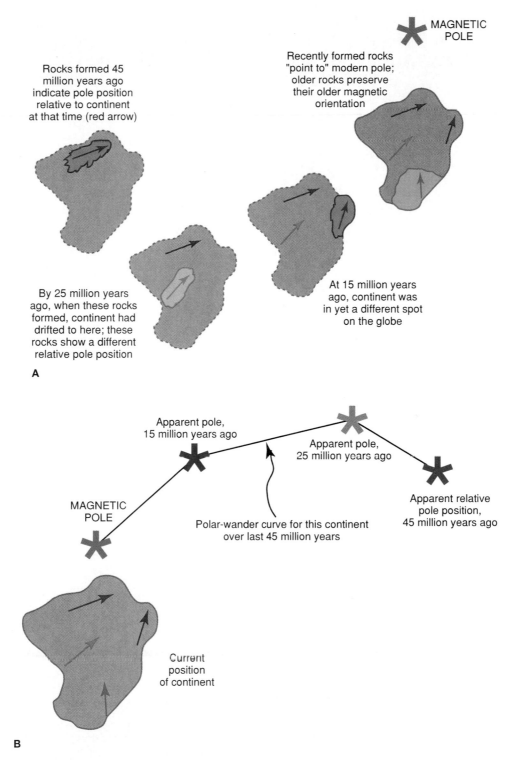

Figure 9.11 Polar-wander curves actually reflect wandering continents. (A) As rocks crystallize, their magnetic minerals align with the contemporary magnetic field. But continental drift changes the relative position of continent and magnetic pole over time. (B) Assuming a stationary continent, the shifting relative pole positions suggest "polar wander."

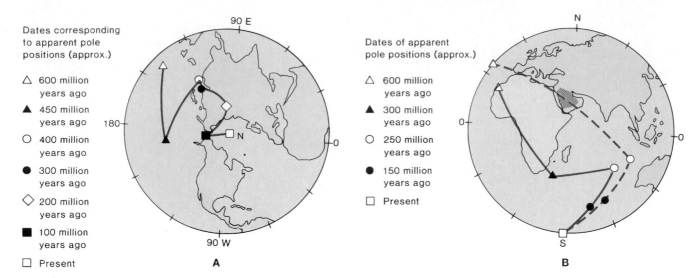

FIGURE 9.12 Examples of polar-wander curves. The apparent position of the magnetic pole relative to the continent is plotted for rocks of different ages, and these data points are connected to form the curve. The present positions of the continents are shown for reference. (A) Polar-wander curve for North America for the last 600 million years, as viewed from the North Pole. (B) Polar-wander curves for Africa (solid line) and Saudi Arabia (dashed line) suggest that these landmasses moved quite independently up until about 250 million years ago, when the polar-wander curves converged.

After M. W. McElhinny, Paleomagnetism and Plate Tectonics. *Copyright 1973 Cambridge University Press, New York, NY. Reprinted by permission.*

remain close to the geographic (rotational) poles, as they now are. Furthermore, polar-wander curves for different continents do not match. Rocks of exactly the same age from two different continents may seem to point to two entirely different sets of magnetic poles! (See figure 9.12.)

This confusion can be eliminated, however, if it is assumed that the magnetic poles *have* always remained close to the geographic poles but that the *continents* have moved and rotated. The polar-wander curves then provide a way to map the directions in which the continents have moved through time, relative to the approximately stationary magnetic poles and relative to each other. In other words, polarity reversals aside, the north/south magnetic poles and the north/south geographic (rotational) poles have approximately coincided over time. As continents have moved and rotated over the surface of the globe, igneous rocks have formed at various times, and those rocks preserve "snapshots" of the relative positions of continent and pole at the times the magmas crystallized. And those snapshots confirm the motion of the plates relative to the poles.

The Jigsaw Puzzle Refined

As mentioned earlier, the apparent similarity of the coastlines of Africa and South America triggered early speculation about the possibility of continental drift. Actually, the pieces of this jigsaw puzzle fit even better if we look not at coastlines but at the true edges of the continents—the outer edges of the continental shelves (dashed lines in figure 9.1), beyond which the depth of the ocean increases rapidly to typical ocean-basin

depths. The continental shelves represent the edges of the rifts that formed when a larger continental mass was split apart by plate motions. Computers have been used to help determine the best physical fit of the puzzle pieces. The results are imperfect, perhaps because not every bit of continent has been perfectly preserved in the tens or hundreds of millions of years since continental breakup. Overall, however, the fit is remarkably good.

Continental reconstructions can be refined using details of continental geology—rock types, rock ages, evidence of glaciation, fossils, ore deposits, mountain ranges, and so on. If two now-separate continents were once part of the same landmass, then the geologic features presently found at the margin of one should have counterparts at the corresponding edge of the other. In short, the geology should match when the puzzle pieces are reassembled, as in the example shown in figure 9.13.

Pangaea

Efforts to reconstruct the ancient locations and arrangements of the continents have been relatively successful, at least for the not-too-distant geologic past. It has been shown, for example, that a little more than 200 million years ago, there was a single, great supercontinent, which has been named **Pangaea** (from the Greek for "all lands"; see figure 9.15). In the early stages of its fragmentation, Pangaea broke up into two smaller landmasses: *Laurasia,* which comprised the northern continents of North America, Europe, and Asia, and *Gondwanaland,* from which the southern continents were

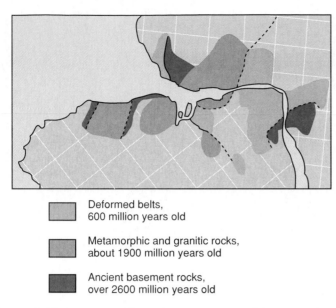

■ Deformed belts,
600 million years old

■ Metamorphic and granitic rocks,
about 1900 million years old

■ Ancient basement rocks,
over 2600 million years old

FIGURE 9.13 Recognizable geologic provinces of distinct rock types and different ages can be correlated between western Africa and eastern South America.

From P. M. Hurley and J. R. Rand, in The Ocean Basins and Margins, *Volume I. Copyright © 1973 Plenum Publishing Company, New York, NY. Reprinted by permission of the publisher and authors.*

derived. The present seafloor-spreading ridges are the lithospheric scars of the breakup of Pangaea. However, they are not the only kind of boundary found between plates, as will be seen later in the chapter.

HOW FAR, HOW FAST, HOW LONG, HOW COME?

Rates and directions of plate movement can be determined in a variety of ways. Radiometric dating also contributes to answering the question of how long plate-tectonic processes have been active.

Past Motions, Present Velocities

As previously discussed, polar-wander curves from continental rocks can be used to determine the directions in which the continents have drifted. Seafloor spreading provides another way of determining plate movement. The direction of seafloor spreading is usually obvious: away from the ridge. Rates of seafloor spreading can be found very simply by dating rocks at different distances from the spreading ridge and dividing the distance moved by the rock's age (the time it has taken to move that distance from the ridge at which it formed).

Still another way to monitor rates and directions of plate movement is by using mantle hot spots. These are isolated areas of volcanic activity, usually not associated with plate boundaries. Possible causes of hot spots were explored in chapter 4. If mantle hot spots are assumed to remain fixed in

position while the lithospheric plates move over them, the result should be a string of volcanoes of differing ages, with the youngest closest to the hot spot.

A good example can be seen in the north Pacific Ocean (see figure 9.14). A topographic map shows an L-shaped chain of volcanic islands and submerged volcanoes. When rocks from these volcanoes are dated radiometrically, they show a progression of ages, from about 75 million years at the northwestern end of the chain, to about 40 million years at the bend, through progressively younger islands to the still-active volcanoes of the island of Hawaii, at the eastern end of the Hawaiian Island group. The latter now sits over the hot spot responsible for the whole chain. The age progression is reflected not only in the radiometric dates but in topography and surface features: the farther west one goes in the chain, in general, the more extensively weathered and eroded the islands and the lower their relief above the surrounding ocean.

E ven as the volcanoes on the island of Hawaii continue to build that island above sea level, the newest volcano in the chain rises up from the sea floor to the east. Named Loihi, the submarine volcano will be the next Hawaiian Island.

From the distances and age differences between pairs of points in the chain, the rate of plate motion can be determined. For instance, Midway Island and Hawaii are about 2700 kilometers apart. The volcanoes of Midway were active about 25 million years ago. Over the last 25 million years, then, the Pacific Plate has moved over the mantle hot spot at an average rate of (2700 kilometers)/(25 million years), or about 11 centimeters per year. The orientation of the volcanic chain shows the direction of plate movement—most recently west–northwest. The kink in the chain at about 40 million years ago indicates that the Pacific Plate changed direction at that time.

Hot spots occur under continents as well as beneath ocean basins. They are, however, somewhat easier to detect in oceanic regions, perhaps because the associated magmas can more readily work their way up through the thinner oceanic lithosphere.

Looking at many determinations of plate movement from all over the world, geologists find that average rates of plate motion are 2 to 3 centimeters (about 1 inch) per year. In a few places, movement at rates of up to about 18 centimeters per year is observed; elsewhere, rates may be slower, but a few centimeters per year is typical. This seemingly trivial amount of motion adds up through geologic time. Movement of 2 centimeters per year for 100 million years means a shift of 2000 kilometers, or about 1250 miles! When the motions are extrapolated back into the past, the breakup of Pangaea can be reconstructed (figure 9.15).

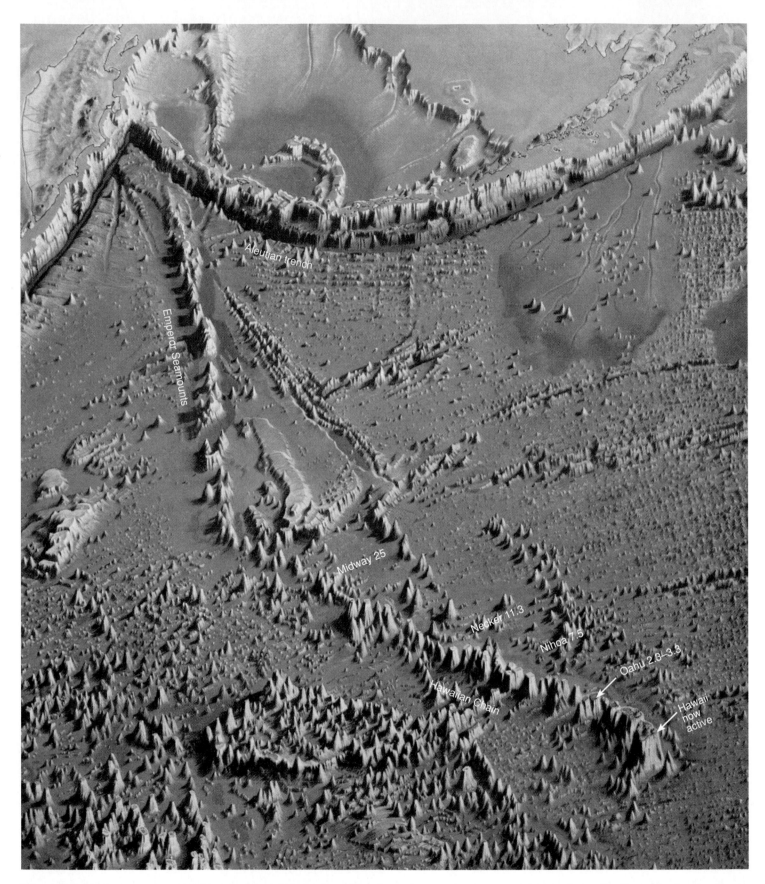

FIGURE 9.14 The Hawaiian Islands and other volcanoes in a chain formed over a hot spot. Numbers indicate the dates of volcanic eruptions, in millions of years. Movement of the Pacific Plate has carried the older volcanoes far from the hot spot, now located under the active volcanic island of Hawaii.

From "World Ocean Floor Panorama" by Bruce C. Heezen and Marie Tharp. Copyright © 1977 Marie Tharp.

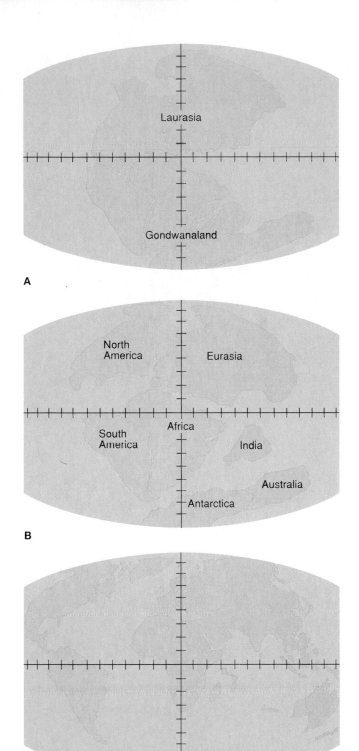

FIGURE 9.15 Reconstructed plate movements during the last 200 million years: the breakup of Pangaea. Although these changes occur very slowly on the timescale of a human lifetime, they illustrate the magnitude of the natural forces to which we must adjust. (A) 200 million years ago. (B) 100 million years ago. (C) Today.

From R. S. Dietz and J. C. Holden, Journal of Geophysical Research, *Volume 75, (4): 939–56. 1970, copyright © by the American Geophysical Union.*

What preceded Pangaea? A different configuration of continents and oceans. The buoyant continents have drifted around the globe, sometimes being split, sometimes having pieces added to their margins, probably as long as continental landmasses have been preserved on earth. We simply have less detailed information on their sizes, shapes, and locations earlier in earth's history, given the impermanence of rocks over time.

Why Do Plates Move?

A driving force for plate tectonics has not been definitely established. Several possibilites are consistent with existing data. The most widely accepted explanation over the past two decades, first developed extensively by Hess in connection with his seafloor-spreading model, is that the plastic asthenosphere is slowly churning in large **convection cells** (figure 9.16). According to this scenario, hot mantle material rises at the spreading ridges; some escapes as magma to fortablished. Several possibilities are consistent with existing data. The most widely accepted explanation over the past two decades, first developed extensively by Hess in connection with hism new lithosphere, but most does not. The rest spreads out sideways beneath the lithosphere, slowly cooling in the process. As it flows outward, it drags the overlying lithosphere outward with it, thus continuing to open the ridges. When it cools, the flowing material becomes dense enough to sink back deeper into the mantle. This may be happening under subduction zones.

Debate continues about the vertical dimensions of the convection cells. They need not necessarily be confined to the asthenosphere. Geophysical evidence indicates that oceanic lithosphere can be subducted to depths of about 700 kilometers (400 miles); perhaps convection cells operate down to those depths. Some researchers have proposed convection spanning virtually the whole mantle, from the base of the lithosphere to the core-mantle boundary, 2900 kilometers down. There is no definite evidence supporting either of these possibilities over the other.

An alternative explanation for plate motions that is gaining support from recent research attributes horizontal plate motion, indirectly, to gravity. The weight of the dense, cold, down-going slab of lithosphere in a subduction zone may simply pull the rest of the trailing plate along with it. This, in turn, would open up the spreading ridges so magma could ooze upward. In this case, friction between the plate and the asthenosphere below could help to drive convection as the plates move, rather than the reverse.

A variant on a gravity-driven model considers the elevation of spreading ridges by hot material rising from the asthenosphere at the ridges. Pushing up the spreading edges of the plates creates force that will tend to make the plates push downward and outward to restore equilibrium. This could drive seafloor spreading and thus plate motions.

The full answer may be a combination of these mechanisms and of other mechanisms not yet considered.

How long plate-tectonic processes have been active is not entirely clear. The magnetic stripes characteristic

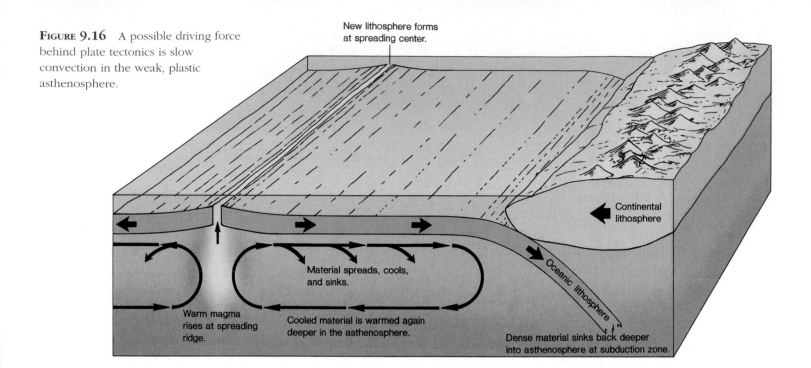

FIGURE 9.16 A possible driving force behind plate tectonics is slow convection in the weak, plastic asthenosphere.

New lithosphere forms at spreading center.

Continental lithosphere

Oceanic lithosphere

Material spreads, cools, and sinks.

Warm magma rises at spreading ridge.

Cooled material is warmed again deeper in the asthenosphere.

Dense material sinks back deeper into asthenosphere at subduction zone.

of seafloor spreading are apparent over even the oldest, 200-million-year-old ocean floor. From continental rocks, apparent polar-wander curves going back more than a billion years can be reconstructed, although the relative scarcity of undisturbed ancient rocks makes such efforts more difficult for the earth's earliest history. It seems clear that the continents have been shifting in position over the earth's surface for billions of years, though not necessarily at the same rates or with exactly the same results as at present. For instance, if mantle convection is responsible for the motion, it may well have been more rapid in the past, when the earth's interior was hotter. Plate-tectonic processes of some sort have certainly long played a major role in shaping the earth. They are likely to continue doing so for the foreseeable future.

TYPES OF PLATE BOUNDARIES

Different things happen at the boundaries between lithospheric plates, depending on the relative motions of the plates and on whether continental or oceanic lithosphere is at the edge of each plate.

Divergent Plate Boundaries

At a **divergent plate boundary,** such as a midocean spreading ridge, lithospheric plates move apart, magma wells up from the asthenosphere, and new lithosphere is created (figure 9.17). A great deal of volcanic activity thus occurs at spreading ridges. Iceland, sitting squarely on the mid-Atlantic ridge, gives geologists a special opportunity to observe a divergent boundary above sea level. The pulling-apart of the plates of lithosphere also results in earthquakes along these ridge plate boundaries (see also chapter 10).

Continents can be rifted apart, too (figure 9.18). This is less common, perhaps because continental lithosphere is so

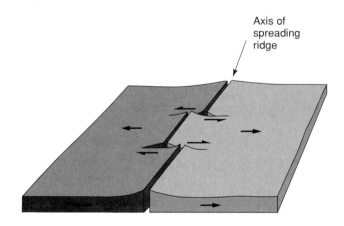

Axis of spreading ridge

FIGURE 9.17 A spreading ridge, a divergent plate boundary. Arrows indicate direction of plate motions. Note transform boundary between offset ridge segments (see figure 9.19 for detail).

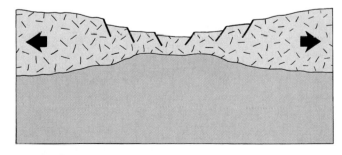

FIGURE 9.18 Continental rifting. The continental crust is thinned and fractured, and in time, a new ocean basin is formed.

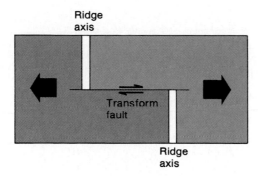

Figure 9.19 Transform fault between offset segments of a spreading ridge (map view). Arrows indicate the direction of plate movement. Along the transform fault between ridge segments, plates move in opposite directions.

From Charles C. Plummer and David McGeary, Physical Geology, 5th ed. Copyright © Times Mirror Higher Education Group, Inc., Dubuque, Iowa. All Rights Reserved. Reprinted by permission.

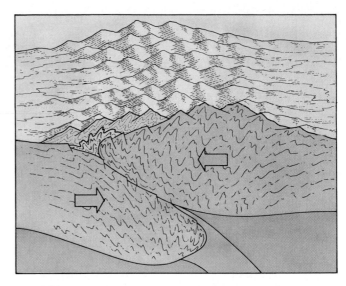

Figure 9.20 Continent–continent collision at a convergent boundary. Rocks are deformed and some lithospheric thickening occurs, but neither plate is subducted to any great extent.

much thicker than oceanic lithosphere. In the early stages of continental rifting, the rising of warm asthenospheric material and thermal expansion of overlying lithosphere cause doming of the continent. Volcanoes may erupt along the rift, as in East Africa today, or great flows of basaltic lava may pour out through fissures in the continent. As the continental crust is stretched and thinned, a shallow inland sea may inundate the rift zone. If the rifting continues, a new ocean basin floored with basalt eventually forms between the torn-apart pieces of continent. The Red Sea formed from such a rift in continental lithosphere. Continued rifting is slowly ripping apart East Africa: the easternmost strip of the continent is being rifted apart from the rest, and another ocean may one day separate the resulting pieces. In the same way, the Atlantic Ocean formed from the breakup of Pangaea.

Transform Boundaries

The actual structure of a seafloor-spreading ridge is more complex than a single, straight crack. A close look at a mid-ocean spreading ridge reveals that it is not a continuous break thousands of kilometers long (see figure 9.6). Rather, ridges consist of many short segments slightly offset from one another. The offsets are a special kind of fault, or break in the lithosphere, known as a **transform fault** (figure 9.19). The opposite sides of a transform fault belong to two different plates, which are moving apart in opposite directions. As the plates scrape past each other, earthquakes occur along the transform fault between ridge segments. Beyond the spreading ridges to either side, both sides of the fracture belong to the same plate and move in the same direction, which accounts for the lack of earthquake activity in these regions.

Transform faults are a geometric necessity on a spheroidal earth. On a flat earth, it would be quite possible to have long, unbroken spreading ridges. But the lithosphere wraps around a curved earth, the curvature of which varies with latitude. The transforms allow adjustment of the displaced plates to the underlying curved earth over which they move.

The famous San Andreas fault in California is an example of a transform fault that slices a continent sitting along a spreading ridge. The East Pacific Rise, a seafloor-spreading ridge off the western coast of North America, disappears under the edge of the continent, to reappear farther south in the Gulf of California (see figure 9.4). The San Andreas is the transform fault between these segments of spreading ridge. Most of North America is part of the North American Plate. The thin strip of California on the west side of the San Andreas fault, however, is moving northwest with the Pacific Plate.

Convergent Plate Boundaries

At a **convergent plate boundary,** as the name indicates, plates are moving toward each other. Just what happens depends on what sort of lithosphere is at the leading edge of each plate. Continental lithosphere is relatively low in density and buoyant, so it tends to float on the asthenosphere. Oceanic lithosphere is closer in density to the underlying asthenosphere, and its buoyancy is therefore less, so it is more easily forced down into the asthenosphere as the plates move together. (The relative buoyancy of crust and mantle is explored further in chapter 11.)

In a continent–continent collision, the two landmasses come together, crumple, and deform (figure 9.20). One may

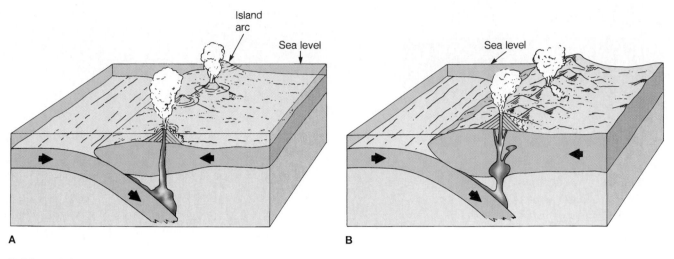

FIGURE 9.21 Subduction zones are formed at a convergent boundary when a slab of oceanic lithosphere is forced into the mantle. (A) Ocean–ocean convergence: An island arc is formed by volcanic activity above the subducted slab. (B) Ocean–continent convergence: The magmatic activity accompanying subduction produces volcanic mountains on the continent.

partially override the other, but neither sinks into the mantle, and a very great thickness of continent may result. Earthquakes are frequent during active collision, as a result of the large stresses involved in the process. The extreme height of the Himalaya Mountains is attributed to just this sort of continent–continent collision. India was not always a part of the Asian continent. Paleomagnetic evidence indicates that it drifted northward over tens of millions of years until it ran into Asia. The Himalayas were built up in this collision.

More commonly, oceanic lithosphere is at the leading edge of one or both of the converging plates. One plate of oceanic lithosphere may be pushed under the other plate and descend into the asthenosphere. This type of plate boundary, where one plate is carried down below (*subducted* beneath) another, is called a **subduction zone** (figure 9.21). Topographically, there is commonly a depression (trench) marking the boundary on the sea floor.

The subduction zones of the world balance the seafloor equation. Because oceanic lithosphere is constantly being created at spreading ridges, an equal amount of sea floor must be destroyed somewhere or the earth would simply keep getting bigger. This excess sea floor is consumed in subduction zones. The subducted plate is heated by the hot asthenosphere and, in time, becomes hot enough to melt. Some of the melt rises to form volcanoes on the overriding plate. Some of the melt may eventually migrate to and rise again at a spreading ridge, to make new sea floor. In a sense, the oceanic lithosphere is constantly being recycled through this process; entire plates may be consumed by subduction. This explains why no very ancient seafloor rocks are known. The buoyant continents are not so easily reworked; therefore, very old rocks may be preserved on the continents.

Subduction zones are, geologically, very active places. Sediments eroded from the continents may accumulate in the trench formed by the down-going plate. Some of these sediments are caught in the fractured oceanic crust and carried down into the asthenosphere to be melted along with the

sinking lithosphere. Volcanoes form where the melted material rises up through the overlying plate to the surface. Where the overriding plate also consists of oceanic lithosphere, the volcanoes may form a string of islands known as an **island arc** (figure 9.21A). The Aleutians are island-arc volcanoes, bordered on the south by a trench (as seen in figure 9.14). The Pacific Plate is being subducted under southern Alaska. Where continent overrides ocean, the volcanoes are built up as mountains on the continent (figure 9.21B). The Cascades and the Andes are examples. The great stresses involved in convergence and subduction give rise to numerous earthquakes. Parts of the world near or above subduction zones, and therefore prone to both volcanic and earthquake activity, include the Andes region of South America, western Central America, parts of the northwestern United States and Canada, the Aleutian Islands, China, Japan, and much of the rim of the Pacific Ocean basin (see figure 9.4). Earthquake distribution at subduction zones and the nature of volcanism there are described in more detail in chapters 4 and 10, respectively. Subduction zones have also been proposed as possible waste-disposal sites; see box 9.1.

PLATE TECTONICS AND THE ROCK CYCLE

In chapter 2, we noted that all rocks may be considered related by the concept of the rock cycle. We can also look at the rock cycle in a plate-tectonic context, as illustrated in simplified form in figure 9.22.

New igneous rocks form from magmas rising out of the asthenosphere at spreading ridges or over subduction zones. The heat released by the cooling magmas can cause metamorphism in the continental crust, with recrystallization at an elevated temperature changing the texture and/or the mineralogy of the surrounding rocks. Some of these surrounding rocks may themselves melt to form new igneous rocks. The

Box 9.1

Plate Tectonics and Radioactive Waste: New Solutions or Wishful Thinking?

When plate-tectonic theory was being developed and the existence and nature of subduction zones discovered, some people began to see subduction zones as the ultimate in waste disposal. In particular, subduction zones were suggested as possible disposal sites for canisters of radioactive wastes.

The appeal of the idea is the image of these potentially hazardous materials being carried deep into the earth, there to vanish, effectively, forever (at least from a human perspective). The principal drawbacks are twofold. First, given the extremely slow rates of plate motion, complete subduction of wastes would take thousands of years at least. Unless the waste canisters were somehow emplaced well into the subducted plate, they would sit exposed to interaction with seawater in the meantime. Seawater, especially seawater warmed by decaying radioactive wastes, is highly corrosive and might breach the canisters, allowing leakage of the wastes. Second, a plate undergoing subduction does not slip quietly under the plate above like a spatula under a pancake. Subduction is accompanied by earthquakes and deformation of the plates and the sediments carried on them. Some of the overlying sediments are not, in fact, subducted but are scraped off onto the overriding plate. It is possible that some waste canisters could be ruptured or caught up in the scraped-off sedimentary pile.

As understanding of plate-tectonic processes has increased, enthusiasm for the subduction-zone disposal scheme has diminished. However, now that the significance of plate tectonics to the formation and destruction of the sea floor has been recognized, there *is* interest in using the stable interiors of large oceanic plates as possible radioactive-waste-disposal sites. An international committee is presently studying this idea.

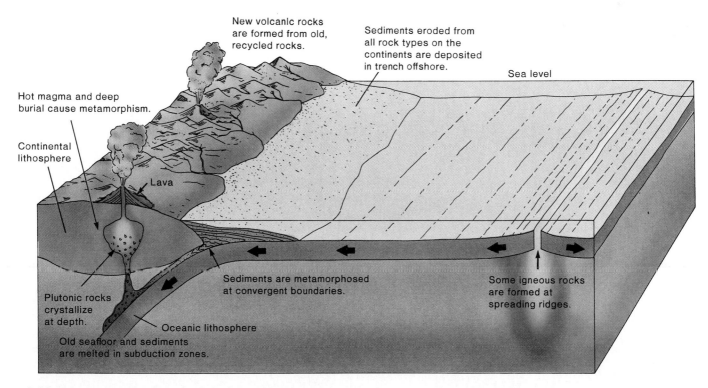

FIGURE 9.22 The rock cycle, interpreted in plate-tectonic terms.

forces of plate collision at convergent margins also contribute to metamorphism by increasing the pressures and directed stresses acting on the rocks.

Weathering and erosion on the continents wear down preexisting rocks of all kinds into sediment. Much of this sediment is eventually transported to the edges of the continents, where it is deposited in deep basins and trenches. Through burial under more layers of sediment, it may become lithified into sedimentary rock. Sedimentary rocks, in turn, may be metamorphosed or even melted by the stresses and igneous activity at the plate margins. Some of these sedimentary or metamorphic materials may also be carried down with subducted oceanic lithosphere, to be melted and eventually recycled as igneous rock. Plate-tectonic activity thus plays a large role in the formation of new rocks from old that is constantly underway on the earth.

SUMMARY

The solid outermost layer of the earth is the 50- to 100-kilometer-thick lithosphere, which is broken up into a series of rigid plates. The lithosphere is underlain by a plastic, partly molten layer of the mantle, the asthenosphere, over which the plates can move. This plate motion gives rise to earthquakes and volcanic activity at the plate boundaries. At seafloor-spreading ridges, which are divergent plate boundaries, new sea floor is created from magma rising from the asthenosphere. The sea floor moves in conveyor-belt fashion, ultimately to be destroyed in subduction zones, a type of convergent plate boundary, where it is carried down into the asthenosphere and eventually remelted. Convergence of continents forms high mountain ranges.

Evidence for seafloor spreading includes the age distribution of seafloor rocks and the magnetic stripes on the ocean floor. Continental drift can be demonstrated by such means as polar-wander curves and evidence of ancient climates as revealed in the rock record. Past "supercontinents" can be reconstructed by fitting together modern continental margins and matching similar geologic features and fossil deposits from continent to continent.

Present rates of plate movement average a few centimeters a year. A mechanism for moving the plates has not been proven definitively. Possible driving forces are slow convection in the asthenosphere (and perhaps in the deeper mantle), and lateral motion due to the downward pull of gravity on downgoing lithosphere in subduction zones. Although plate motions are less readily determined in ancient rocks, plate-tectonic processes have probably been more or less active for much of the earth's history. They play an integral part in the rock cycle and can be related to earthquake occurrence, volcanic activity, and formation of mineral deposits.

TERMS TO REMEMBER

asthenosphere 136
continental drift 135
convection cells 147
convergent plate boundary 149
Curie temperature 140
divergent plate boundary 148

island arc 150
lithosphere 136
paleomagnetism 140
Pangaea 144
plate tectonics 135
polar-wander curve 142

seafloor spreading 141
subduction zone 150
tectonics 135
transform fault 149

QUESTIONS FOR REVIEW

1. Are the terms *lithosphere* and *asthenosphere* equivalent, respectively, to *crust* and *mantle?* Explain.
2. What property of the asthenosphere gives it its name? Why does it behave in this way?
3. How are major plate boundaries identified?
4. What is a *Curie temperature,* and what does it have to do with paleomagnetic studies?
5. Describe the origin of the magnetic stripes on the sea floor.

6. The ages of seafloor rocks and sediments show a regular pattern around a spreading ridge. Describe the pattern.
7. What is a *polar-wander curve?* Is the name an accurate one?
8. What is a *transform fault,* and where are such faults found?
9. Contrast what occurs at a convergent plate boundary when the advancing edge of each plate is continental lithosphere with what occurs when one plate is oceanic lithosphere. How does this help to account for the relative youth of the sea floor?

10. What is an *island arc,* and where and how would one form?
11. Describe two means of determining rates of plate motion.
12. Convection cells in the asthenosphere may drive plate motions. Explain.
13. Describe an alternative possible driving force, other than convection cells, for plate tectonics.
14. Briefly explain the rock cycle in the context of plate tectonics.

FOR FURTHER THOUGHT

1. It has been proposed (not by geologists) that the phenomenon of apparent polar wander can be explained by a flipping or rotation of the earth's whole lithosphere as a single unit, rather than by plate tectonics and independent movements of different continents. Can you suggest any evidence that is inconsistent with that proposal?

2. The moon, though smaller than the earth, has a much thicker lithosphere: the moon's radius is only 1740 kilometers, yet its lithosphere is about 1000 kilometers thick. Would you expect plate-tectonic activity and subduction to occur on the moon as they do on earth? Why or why not?

SUGGESTIONS FOR FURTHER READING

Anderson, D. L., Tanimoto, T., and Zhang, Y. 1992. Plate tectonics and hotspots: The third dimension. *Science* 256: 1645–51.

Bird, J. M., ed. 1980. *Plate tectonics*. 2d ed. Washington, D.C.: American Geophysical Union.

Burke, K., and Wilson, J. T. 1976. Hot spots on the earth's surface. *Scientific American* 235 (August): 46–57.

Condie, K. 1982. *Plate tectonics and crustal evolution*. 2d ed. New York: Pergamon Press.

Dewey, J. F. 1972. Plate tectonics. *Scientific American* 226 (May): 56–68.

Dietz, R. S., and Holden, J. C. 1970. The breakup of Pangaea. *Scientific American* 223 (April): 30–41.

Hoffman, K. A. 1988. Ancient magnetic reversals: Clues to the geodynamo. *Scientific American* 258 (May): 76–83.

Hurley, P. M. 1968. The confirmation of continental drift. *Scientific American* 218 (April): 52–64.

Marvin, U. B. 1973. *Continental drift: The evolution of a concept*. Washington, D.C.: Smithsonian Institution Press.

McElhinny, M. W. 1973. *Paleomagnetism and plate tectonics*. Cambridge, England: Cambridge University Press.

Menard, H. W. 1986. *The oceans of truth: A personal history of global tectonics*. Princeton, N.J.: Princeton University Press.

Molnar, P., and Tapponier, P. 1977. The collision between India and Eurasia. *Scientific American* 236 (April): 30–41.

Mutter, J. C. 1986. Seismic images of plate boundaries. *Scientific American* 254 (February): 66–75.

Nance, R. D., Worsley, T. R., and Moody, J. B. 1988. The supercontinent cycle. *Scientific American* 259 (July): 72–79.

Paton, T. R. 1986. *Perspectives on a dynamic earth*. Boston: Allen and Unwin.

Wegener, A. 1924. *The origin of continents and oceans*. London: Methuen.

Weyman, D. 1981. *Tectonic processes*. London: Allen and Unwin.

10

EARTHQUAKES, SEISMIC WAVES, AND THE EARTH'S INTERIOR

About 90% of the nearly 5400 casualties in the January 1995 earthquake in Kobe, Japan, occurred when houses like this one, traditional Japanese woodframe structures, collapsed into piles of debris. Commercial and municipal structures designed for earthquake resistance generally fared better. Photograph by C. S. Prentice, U.S. Geological Survey; courtesy T. L. Holzer and Geological Society of America.

San Francisco, 8 April 1906

The first thing I was aware of was being wakened sharply to see my bureau lunging solemnly at me across the width of the room. . . . Then I remember standing in the doorway to see the great barred leaves of the entrance on the second floor part quietly as under an unseen hand . . . and suddenly an eruption of nightgowned figures crying out that it was only an earthquake. . . . Almost before the dust of ruined walls had ceased rising, smoke began to go up against the sun. . . . South of Market, in the district known as the Mission, there were cheap man-traps folded in like pasteboard, and from these, before the rip of the flames blotted out the sound, arose the thin, long scream of mortal agony. . . . In the park were the refugees huddled on the damp sod with insufficient bedding and less food and no water. . . . Hot, stifling smoke billowed down upon them, cinders pattered like hail. . . . I came out . . . and saw a man I knew hurrying down toward the gutted district. . . . "Bob," I said, "It looks like the day of judgment." He cast back at me over his shoulder unveiled disgust at the inadequacy of my terms. "Aw!" he said, "It looks like hell!" (Austin 1981)

Terrifying as that San Francisco earthquake was, it caused only an estimated 700 deaths and $4 million in property damage. At that, most of the damage was due to uncontrolled fires following the quake. History records other earthquake disasters claiming over a million lives or $100 billion in damages (table 10.1 on page 156).

Averaged over millennia, plate motions are slow, but they are not always so on a shorter timescale. When the strength of the lithosphere fails and it snaps or shifts suddenly in response to built-up stress, the result is the phenomenon called an earthquake. In this chapter, we survey the nature of earthquakes, their causes, and their effects. We also look at some approaches to minimizing the potential damage from earthquakes and examine the current status of earthquake-prediction efforts.

The seismic waves associated with earthquakes do have a useful side. In several previous chapters, reference has been made to various aspects of the physical or chemical properties of the earth's interior. But how do geoscientists know so much about the interior? Seismic data account for much of their information. The latter part of this chapter assembles the various kinds of evidence used to determine the physical and chemical makeup of the earth's interior.

EARTHQUAKES—BASIC PRINCIPLES

Devastating earthquakes demonstrate dramatically that the earth is a dynamic, changing system. They occur along **faults,** planar breaks in rock along which there is displacement (movement) of rocks on one side relative to the other, and they happen when that displacement occurs suddenly in response to stress.

Stress, Strain, and the Strength of Geological Materials

An object is under **stress** when force is being applied to it. The stress may be **compressive,** tending to squeeze the object, or it may be **tensile,** tending to pull the object apart. A **shearing stress** is one that tends to cause different parts of the object to move in different directions across a plane, or to slide past one another, as when a deck of cards is spread out on a tabletop by a sideways sweep of the hand.

Strain is deformation resulting from stress. It may be either temporary or permanent, depending on the amount and type of stress and the material's ability to resist it. If the deformation is **elastic,** the amount of deformation is proportional to the stress applied, and the material returns to its original size and shape when the stress is removed. A gently stretched rubber band shows elastic behavior. Rocks, too, may behave elastically, although much greater stress is needed to produce detectable strain. Once the **elastic limit** of a material is reached, it may go through a phase of **plastic deformation** with increasing stress. During this stage, relatively small added stresses yield large corresponding strains, and the changes of shape are permanent: the material does not return to its original dimensions after removal of the stress. A glassblower, an artist shaping clay, a carpenter fitting caulk into cracks around a window, and a blacksmith shaping a bar of hot iron into a horseshoe are all making use of plastic behavior of materials.

If stress is increased further, solids eventually break, or **rupture,** as a rubber band will do when stretched too far. In **brittle** materials, rupture may occur before there is any plastic deformation. Brittle behavior is characteristic of most rocks at near-surface conditions and is illustrated by line *A* in figure 10.1. At greater depths, where temperatures are higher and rocks are confined (in a sense, supported by surrounding

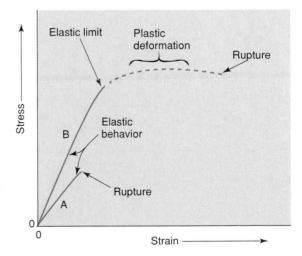

FIGURE 10.1 Stress-strain relationships for brittle materials and those capable of plastic deformation.

TABLE 10.1 Selected Major Historic Earthquakes

Year	Location	Magnitude*	Deaths	Damages†
70 b.c.	China	(IX)	6,000	Unknown
a.d. 342	Turkey		40,000	Unknown
365	Crete: Knossos		50,000	Unknown
1201	Egypt, Syria		1,100,000	Over $25 million
1290	China	6.75	100,000	Unknown
1456	Italy	(XI)	40,000+	Over $25 million
1556	China	(XI)	830,000	Over $25 million
1667	USSR: Shemakh	6.9	80,000	Over $25 million
1693	Italy, Sicily		100,000	$5–25 million
1703	Japan	8.2	200,000	$5–25 million
1737	India: Calcutta		300,000	$1–5 million
1755	Portugal: Lisbon	(XI)	62,000	Over $25 million
1780	Iran: Tabriz		100,000	Unknown
1850	China	(X)	20,600	Over $25 million
1854	Japan	8.4	3,200	Over $25 million
1857	Italy: Campania	6.5	12,000	Unknown
1868	Peru, Chile	8.5	52,000	$300 million
1883	Java		100,000	Over $25 million
1906	United States: San Francisco	8.3	700+	Over $25 million
1906	Ecuador	8.9	1,000	Unknown
1908	Italy: Calabria	7.5	58,000	Over $25 million
1920	China: Kansu	8.5	200,000	Over $25 million
1923	Japan: Tokyo	8.3	143,000	$2.8 billion
1933	Japan	8.9	3,000+	Over $25 million
1939	Chile	8.3	28,000	$100 million
1939	Turkey	7.9	30,000	Over $25 million
1948	USSR: Ashkhabad	7.3	19,800	Over $25 million
1964	Alaska	8.4	115	$540 million
1967	Venezuela	7.5	1,100	Over $140 million
1970	N. Peru	7.8	66,800	$250 million
1971	United States: San Fernando	6.4	60	$500 million
1972	Nicaragua	6.2	5,000	$800 million
1976	Guatemala	7.5	23,000	$1.1 billion
1976	NE Italy	6.5	1,000	$8 billion
1976	China: T'ang Shan	8.0	655,000	Over $25 million
1978	Iran	7.4	20,000	Unknown
1980	S. Italy	6.9	3,100	$5–25 million
1983	Japan: Honshu	7.8	104	$416 million
1983	Turkey	6.9	2,700	Over $25 million
1985	Chile	7.8	177	$1.8 billion
1985	SE Mexico	8.1	5,600+	$5–25 million
1988	USSR, Afghanistan	6.8	25,000	Unknown
1989	United States: Loma Prieta, California	7.1	63	$5.6–7.1 billion
1990	Iran	6.4	40,000+	Over $7 million
1992	United States: Landers, California	7.4	1	Over $100 million
1994	United States: Northridge, California	6.7	57	$13–15 billion
1995	Japan: Kobe	7.2	5,200+	$95–140 billion

Note: This table is intended not to be a complete or systematic compilation but to give examples of major damaging earthquakes throughout history

*Where actual magnitude is not available, maximum Mercalli intensity is sometimes given in parentheses (see table 10.3).

†Estimated in 1979 dollars. Amounts "over $25 million" may be considerably over this amount.

Source (except 1988 and later entries): *Catalog of Significant Earthquakes 2000 b.c.–1979* (and supplement to 1985), World Data Center A, Report SE–27.

FIGURE 10.2 Sidewalk warped by creep along the Hayward Fault, Hollister, California.
Photograph courtesy of USGS Photo Library, Denver, CO.

rocks at high pressure), rocks may behave plastically (dashed section of curve *B*). The effect of temperature can be seen in the behavior of warm and cold glass. A rod of cold glass is brittle and snaps under stress before appreciable plastic deformation occurs, while a sufficiently warmed glass rod may be bent and twisted without breaking. Frozen butter is brittle and can be broken; room-temperature butter flows. In rocks, plastic deformation is often reflected in the production of folds, as described in chapter 11.

As already indicated, the physical behavior of a rock is affected by external factors, such as temperature or confining pressure, as well as by the intrinsic characteristics of the rock itself. Also, rocks respond differently to different types of stress. Most are far stronger under compression than under tension: a given rock may rupture under a tensile stress only one-tenth as large as the compressive stress required to break it at the same pressure and temperature. Consequently, the term *strength* has no single, simple meaning when applied to rocks unless all of these variables are specified.

When movement along existing faults occurs gradually and smoothly, it is termed **creep.** Creep causes broken curbstones, offset fences, and the like (figure 10.2). In buildings, dams, or other structures built directly across a fault, creep can stress and deform walls to the point of failure over a period of time. Damage is very localized, however, and lives are rarely lost as a consequence of slow fault creep.

By contrast, when friction between rocks on either side of a fault is such as to prevent the rocks from slipping easily, or when the rock under stress is not already fractured, some elastic deformation occurs before failure (figure 10.3). When the stress at last exceeds the rupture strength of the rock (or the friction between rocks along an existing fault), sudden movement occurs along the fault: an **earthquake.** The stressed rocks, released by the rupture, snap back elastically to their previous dimensions, a phenomenon known as **elastic rebound.** The occurrence of movement and stress release is reflected in the relative displacement of the rocks on either

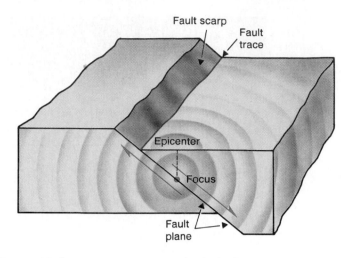

FIGURE 10.4 Simplified diagram of a fault, illustrating component parts and associated earthquake terminology.

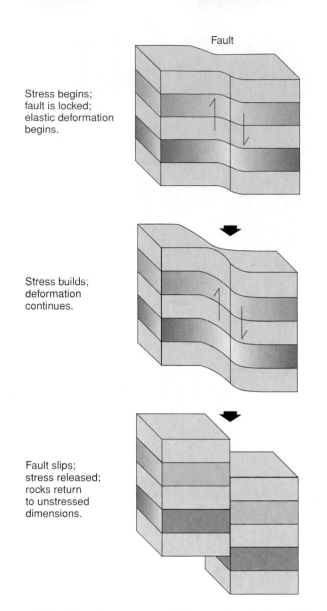

FIGURE 10.3 The phenomenon of elastic rebound along fault zones. The rocks deform elastically under stress until failure and then snap back to their original, undeformed condition after the earthquake.

side of the fault following the earthquake. A section of an existing fault along which friction prevents the gradual slippage of creep can be described as **locked.**

Faults and fractures come in all sizes, from microscopically small to hundreds of kilometers long. Likewise, earthquakes come in all sizes, from tremors so small that even sensitive instruments can barely detect them to massive shocks that can level cities.

Earthquake Terminology

The point on a fault at which the first movement or break occurs during an earthquake is called the earthquake's **focus,** or *hypocenter* (figure 10.4). In the case of a large earthquake, a section of fault many kilometers long may slip, but there is always a point at which the first movement occurs, and this is the focus. The point on the earth's surface directly above the focus is the **epicenter.** News accounts that tell where an earthquake happened report the location of the epicenter. The line along which the fault plane intersects the earth's surface is the fault **trace.** If there is vertical movement along the fault, the cliff formed is called a fault **scarp.** Note that subsequent erosion may cause the scarp to be eroded back from its original position along the fault trace proper.

Earthquake Locations

A map of the locations of major earthquake epicenters over nearly a decade shows that most are concentrated in linear belts corresponding to plate boundaries (figure 10.5A). Not all earthquakes occur at plate boundaries, but most do. These areas are where plates jostle, collide with, or slide past each other, where plate movements may build up very large stresses, where major faults or breaks already exist along which further movement can occur.

A map showing only deeper-focus earthquakes (earthquakes with focal depths of over 100 kilometers) looks somewhat different: the spreading ridges have disappeared, while subduction zones are still shown by their frequent earthquakes (figure 10.5B). The explanation is that earthquakes occur in the lithosphere, where rocks are more rigid and brittle and are therefore capable of breaking or slipping suddenly. In the warmer, plastic asthenosphere, material flows, rather than snaps, under stress. Therefore, the deep-focus earthquakes are concentrated in subduction zones, the only place where cold, brittle lithosphere is pushed deep into the mantle. Note also that these earthquakes cannot be the result of friction between plates, for they occur within the single subducted plate, at depths below the thickness of the overriding lithospheric plate, as stresses are abruptly released.

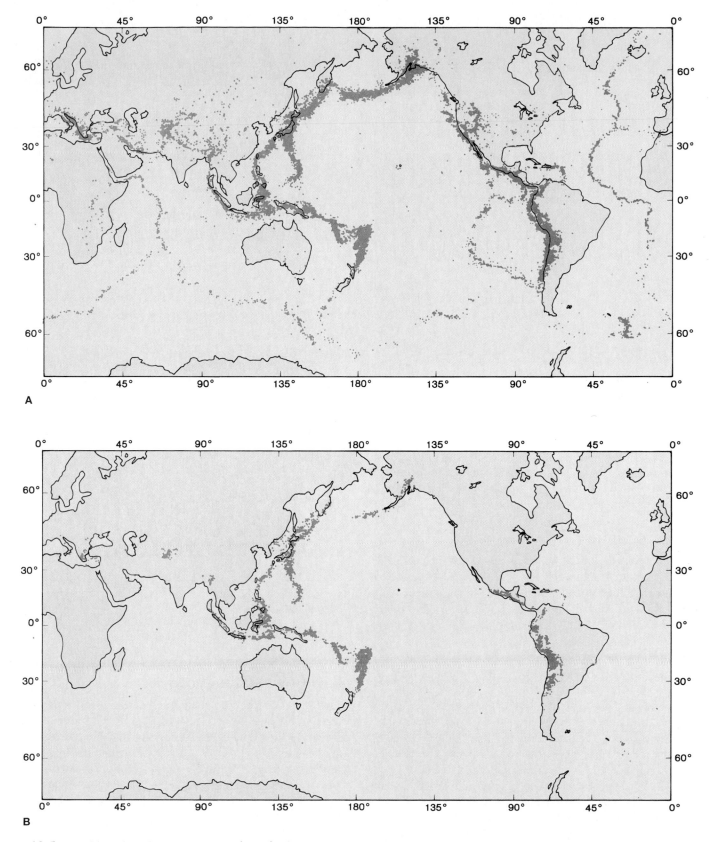

FIGURE 10.5 World earthquake epicenters, 1961–1967, from U.S. Coast and Geodetic Survey. (A) All earthquakes. (B) Earthquakes with focal depths greater than 100 kilometers.

From Barazangi and Dorman, "World Seismicity Maps, 1961–1967," Bulletin of the Seismological Society of America, *vol. 59, no. 1, pp. 360–380. Used by permission.*

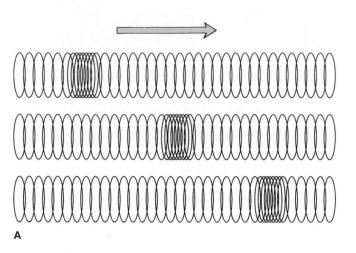
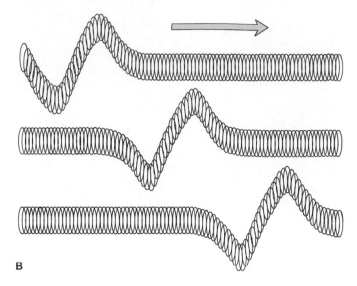

FIGURE 10.6 Seismic body waves, illustrated using a Slinky™ toy (schematic). In both cases, the time sequence is from top of figure to bottom; arrows indicate the direction in which the wave is traveling. (A) P wave (compressional). (B) S wave (shear).

SEISMIC WAVES AND EPICENTER LOCATION

When an earthquake occurs, the stored-up energy is released in the form of **seismic waves** that travel away from the focus. There are several types of seismic waves. **Body waves** (*P waves* and *S waves*) travel through the interior of the earth. **Surface waves,** as their name suggests, travel along the surface. The use of body waves to explore the earth's internal structure is described later in this chapter.

Types of Seismic Waves

P waves are compressional waves. As P waves travel through matter, the matter is alternately compressed and expanded. P waves travel through the earth, then, much as sound waves travel through air. A compressional type of wave can be illustrated with a Slinky™ toy by stretching the coil to a length of several feet along a smooth surface, holding one end in place, pushing in the other end suddenly, and then holding that end still also. A pulse of compressed coil travels away from the end moved (figure 10.6A).

    S waves are shear waves, involving a side-to-side sliding motion of material. Shear-type waves can also be demonstrated with a Slinky™ by stretching the coil out as before, but this time twitching one end of the coil sideways (perpendicular to its length). As the wave moves along the length of the Slinky™, the loops of coil move sideways relative to each other, not closer together and farther apart as with the compressional wave (figure 10.6B).

    Surface seismic waves are somewhat analogous to surface waves on water, which are described in chapter 13. That is, they cause rocks and soil to be displaced in such a way that the ground surface ripples or undulates. *Rayleigh waves* cause vertical ground motions; *Love waves* cause horizontal motions. Most earthquake damage is caused by surface waves, particularly those with shearing motion.

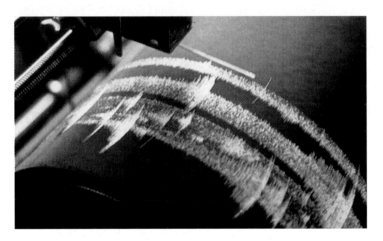

FIGURE 10.7 A seismograph, used for detecting and recording ground motions; the larger the ground motion, the greater the oscillation of the pen.
Photo courtesy of USGS Photo Library, Denver, CO.

Locating the Epicenter

Earthquake epicenters can be located using seismic body waves. P waves travel faster through rocks than do S waves. Therefore, at points some distance from the site of an earthquake, the first P waves arrive somewhat before the first S waves. (Indeed, "P wave" derives from "primary wave;" "S wave," from "secondary wave.") Both types of body waves cause ground motions detectable using a **seismograph** (figure 10.7), and a trained seismologist can distinguish these first arrivals of P and S waves from other ground displacements (including surface waves and background noise such as passing traffic). The difference in arrival times of the first P and S waves is a function of distance to the earthquake epicenter.

    The effect can be illustrated by considering a pedestrian and a bicyclist traveling the same route, starting at the same time. If the bicyclist can travel faster, he or she will arrive at

FIGURE 10.8 Use of seismic waves in locating earthquakes. (A) Difference in times of first arrivals of P waves and S waves is a function of distance from the focus. (B) Triangulation using data from several seismograph stations allows location of the earthquake epicenter.

the destination first. The longer the route to be traveled, the greater the difference in time between the arrival of the cyclist and the later arrival of the pedestrian. Likewise, the farther the receiving seismograph is from the earthquake epicenter, the greater the time lag between the first arrivals of P waves and S waves. The principle is illustrated graphically in figure 10.8A.

Once several recording stations have determined their distances from the epicenter in this way, the epicenter can be located on a map (figure 10.8B). If the epicenter is 1350 kilometers from San Francisco, it is located somewhere on a circle with a 1350-kilometer radius around that city. If it is also found to be 1900 kilometers from Dallas, the epicenter must fall at either of the two points that are both 1350 kilometers from San Francisco and 1900 kilometers from Dallas. If the epicenter is also determined to be 2400 kilometers from Chicago, the position of the epicenter is uniquely identified: northeastern Utah, near Salt Lake City.

In practice, precise location of epicenters typically requires data on distances from more than three seismograph stations because geologic variations in the earth (differences in rock type or density, for example) cause local distortions in the arrival times at individual stations. Computers are often used to refine the location of the epicenter further.

TABLE 10.2 *Frequency and Energy Release of Earthquakes of Various Magnitudes*

Description	Magnitude	Number per Year	Approximate Energy Released (ergs)
Great earthquake	Over 8.0	1 to 2	Over 5.8×10^{23}
Major earthquake	7.0 to 7.9	18	2–42×10^{22}
Destructive earthquake	6.0 to 6.9	120	8–150×10^{20}
Damaging earthquake	5.0 to 5.9	800	3–55×10^{19}
Minor earthquake	4.0 to 4.9	6,200	1–20×10^{18}
Smallest usually felt by humans	3.0 to 3.9	49,000	4–72×10^{16}
Detected but not felt	2.0 to 2.9	300,000	1–26×10^{15}

Note: For every unit increase in Richter magnitude, ground displacement increases by a factor of 10, while energy release increases by a factor of 30. Therefore, most of the energy released by earthquakes each year is released not by the hundreds of thousands of small tremors, but by the handful of earthquakes of magnitude 7 or larger, the so-called major or great earthquakes. For comparison, an earthquake of magnitude 6 releases about as much energy as a 1-megaton atomic bomb.

Source: Frequency data from B. Gutenberg and C. F. Richter, *Seismicity of the Earth and Associated Phenomena.* Copyright © 1954, Princeton University Press, Princeton, NJ.

SIZES OF EARTHQUAKES

All seismic waves represent means of energy release and transmission; they cause the ground shaking that people associate with earthquakes. There are various ways of describing the size of an earthquake. The two parameters most commonly used are *magnitude* and *intensity.*

Magnitude

The amount of ground shaking (amount of vertical motion) is related to the **magnitude** of the earthquake. Earthquake magnitude is most often reported using the *Richter magnitude scale,* named after geophysicist Professor Charles Richter, who developed it. A magnitude number is assigned to an earthquake on the basis of the amount of ground displacement or shaking that the earthquake produces, as measured by a seismograph. The reading at a given station is adjusted for the distance of the instrument from the earthquake epicenter because ground motion naturally decreases with increasing distance from the site of the earthquake as the energy is dissipated. Thus, measuring stations in different places arrive at approximately the same magnitude value. The Richter scale is a logarithmic one, meaning that an earthquake of magnitude 4 causes ten times as much ground movement as one of magnitude 3, one hundred times as much as one of magnitude 2, and so on. The amount of energy released rises even faster with increasing magnitude—by about a factor of thirty for each unit of magnitude: an earthquake of magnitude 4 releases approximately thirty times as much energy as one of magnitude 3, and nine hundred times as much as one of magnitude 2.

Technically, there is no upper limit to the Richter scale. The largest recorded earthquakes have had magnitudes of about 8.9. Earthquakes of this size have occurred in Japan and in Ecuador. Although we only hear of the very severe, damaging earthquakes, hundreds of thousands of earthquakes of all sizes occur each year. Table 10.2 summarizes the frequency and effects of earthquakes in different magnitude ranges. An important point to notice is that more energy is released by a single great earthquake (magnitude over 8) than in all the other lower-magnitude earthquakes put together. This has implications for the possibility of earthquake control.

Intensity

An alternative way of describing the size of an earthquake is by the earthquake's **intensity.** Intensity is a measure of the earthquake's effects on humans and on surface features. It is not a unique, precisely defined characteristic. The surface effects produced by an earthquake of given magnitude vary considerably as a function of such factors as local geologic conditions, quality of construction, and distance from the epicenter. A single earthquake, then, can produce effects of many different intensities in different places, though it will have only one magnitude assigned to it (figure 10.9A).

Also, the extent of the area experiencing a given intensity of damage varies with local geology, even for earthquakes of similar magnitude (figure 10.9B). Seismic waves may be transmitted very far and efficiently through the igneous and metamorphic bedrock of central North America, for example; these rocks are compact, relatively homogeneous, and elastic. Seismic waves travel less effectively through either unconsolidated coastal sediments or compositionally varied, complexly faulted and broken-up rocks, such as are common near the San Andreas fault.

Intensity is a somewhat subjective measure in that it is based on direct observation by individuals, rather than on instrumental measurements. Different observers in the same spot may assign different intensity values to a single earthquake. On the other hand, intensity is a more direct indication of the human impact of a particular seismic event in a

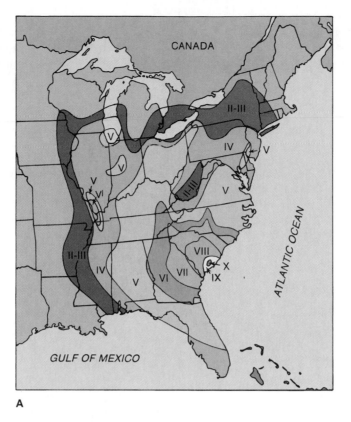

A

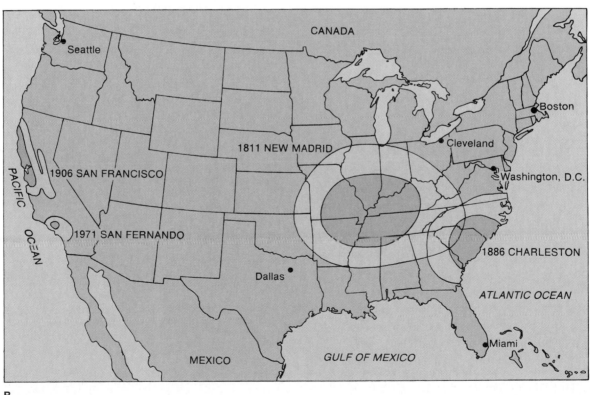

B

FIGURE 10.9 Regional variations in earthquake intensity. (A) Zones of different intensity (as reported using the Modified Mercalli scale—see table 10.3) for a single earthquake, the Charleston earthquake of 1886. (B) Areas of equal-intensity damage for several major earthquakes. The outer zone in each case experiences Mercalli intensity VI, the inner zone intensity VI or above.
(A) Source: Data from U.S. Geological Survey. (B) Source: After U.S. Geological Survey Professional Paper 1240-B.

TABLE 10.3 *Modified Mercalli Intensity Scale (abridged)*

Intensity	Description
I	Not felt.
II	Felt by persons at rest, on upper floors.
III	Felt indoors. Hanging objects swing. Vibration like passing of light trucks.
IV	Vibration like passing of heavy trucks. Standing automobiles rock. Windows, dishes, doors rattle; wooden walls or frame may creak.
V	Felt outdoors. Sleepers wakened. Liquids disturbed, some spilled. Small objects may be moved or upset. Doors swing; shutters and pictures move.
VI	Felt by all; many frightened. People walk unsteadily. Windows, dishes broken. Objects knocked off shelves, pictures off walls. Furniture moved or overturned. Weak plaster cracked. Small bells ring. Trees, bushes shaken.
VII	Difficult to stand. Furniture broken. Damage to weak materials, such as adobe; some cracking of ordinary masonry. Fall of plaster, loose bricks, tile. Waves on ponds; water muddy; small slides along sand or gravel banks. Large bells ring.
VIII	Steering of automobiles affected. Damage to, partial collapse of ordinary masonry. Fall of chimneys, towers. Frame houses moved on foundations if not bolted down. Changes in flow of springs and wells.
IX	General panic. Frame structures shifted off foundations if not bolted down; frames cracked. Serious damage even to partially reinforced masonry. Underground pipes broken; reservoirs damaged. Conspicuous cracks in ground.
X	Most masonry and frame structures destroyed with their foundations. Serious damage to dams, dikes. Large landslides. Rails bent slightly.
XI	Rails bent greatly. Underground pipelines out of service.
XII	Damage nearly total. Large rock masses shifted; objects thrown into the air.

given place than is magnitude. Several dozen intensity scales are in use worldwide. The most widely applied intensity scale in the United States is the Modified Mercalli scale, a modern version of which is summarized in table 10.3.

SEISMIC WAVES AT THE SURFACE: EARTHQUAKE-RELATED HAZARDS

Earthquakes can have a variety of harmful effects, some obvious, some more subtle. Earthquakes of the same magnitude occurring in two different places can cause very different amounts of damage, depending on such variables as the nature of the local geology, whether the affected area is near the coast, and whether the terrain is steep or flat. Some of the hazards can be minimized; some can only be avoided by staying clear of the dangerous area.

Types of Hazards

Ground shaking and *displacement along the fault* are obvious hazards. Ground shaking by seismic waves causes damage to and sometimes complete failure of buildings (figure 10.10). Sudden shifts of even a few tens of centimeters can be devastating, especially to weak materials, such as adobe. Offset between rocks on opposite sides of the fault can also break power lines, pipelines, buildings, roads, bridges, dams, and other structures that cross the fault. In the 1906 San Francisco earthquake, maximum relative horizontal displacement across the San Andreas fault was more than 6 meters (nearly 20 feet). In the Alaskan earthquake of 1964, maximum displacement was on the order of 10 meters. Such effects of fault displacement, of course, are most severe on or

FIGURE 10.10 Building failure from 1985 earthquake, Mexico City: pancake-style collapse of fifteen-story reinforced-concrete structure.
Photograph by M. Celebi, USGS Photo Library, Denver, CO.

very close to the fault, so the simplest strategy would be not to build near fault zones. However, many cities have already developed near major faults.

Short of moving whole towns, can anything be done in such places? Power lines and pipelines can be built with extra slack where they cross a fault zone, or they can be designed

FIGURE 10.11 Bends designed into the Trans-Alaska Pipeline allow "give" in case of thermal expansion/contraction as well as stress due to earthquakes.

FIGURE 10.12 Damage resulting from the 1971 earthquake in San Fernando, California: remains of freeway interchange, Los Angeles County.
Photograph courtesy of USGS Photo Library, Denver, CO.

with other features to allow some "give" as the fault slips and stretches them. Such considerations had to be taken into account when the Trans-Alaska Pipeline was built, for it crosses several major known faults along its route (figure 10.11).

Designing so-called earthquake-resistant buildings is a greater challenge and a relatively new idea developed mainly in the last few decades. Engineers have studied how different types of buildings have fared, and often failed, in real earthquakes. Scientists can conduct laboratory experiments on scale models of skyscrapers and other buildings, subjecting them to small-scale shaking designed to simulate the kinds of ground movement to be expected during an earthquake. On the basis of their findings, special building codes for earthquake-prone regions can be developed.

This approach, however, has many limitations. For one thing, there are very few reliable records of just how the ground actually moves in a severe earthquake. To obtain such records, sensitive instruments must be in place near the fault zone beforehand, and those instruments must survive the earthquake. Even with good records from an actual earthquake, there are no guarantees that the laboratory experiments accurately simulate a real earthquake or that model buildings will respond to model earthquakes in the laboratory in the same ways that real skyscrapers or other structures respond to real earthquakes. Such uncertainties raise major concerns about the safety of dams and nuclear power plants near active faults.

A further complication is that the same building codes cannot be applied everywhere. Not all earthquakes of given magnitude produce the same patterns of ground motion. The 1994 Northridge earthquake underscored the dependence of reliable design on accurately anticipating ground motion. The epicenter was close to that of the 1971 San Fernando earthquake that destroyed many freeways (figure 10.12). They were rebuilt to be earthquake-resistant, assuming the predom-

inantly horizontal displacement and ground motion that is characteristic of the San Andreas. The Northridge quake, however, had a strong vertical-displacement component as well—and many of the rebuilt freeways collapsed again. It is also important to consider not only how structures are built but what they are built *on*. Buildings built on solid bedrock seem to suffer far less damage than those built on deep soil. Mexico City is underlain by thick layers of weak volcanic ash and clay; most smaller and older buildings lack the deep foundations needed to reach more stable sand layers at depth. This is one reason why damage from the 1985 Mexican earthquakes was so extensive in Mexico City and why so many buildings completely collapsed. By contrast, Acapulco, which was much closer to the epicenter, suffered far less damage because it stands firmly on bedrock. In the 1906 San Francisco and 1989 Loma Prieta earthquakes along the San Andreas fault, structures built on filled land rimming San Francisco Bay were far more severely damaged than nearby structures on firmer ground. Similar damage patterns were seen in the 1995 Kobe earthquake.

The characteristics of the earthquakes in a particular region also must be taken into account. For example, severe earthquakes are generally followed by many **aftershocks,** earthquakes that are weaker than the principal tremor. The main shock usually causes the most damage, but when aftershocks are many and are nearly as strong as the main shock, they may also cause serious destruction. One day after the magnitude-8.1 main shock of the 1985 Mexican earthquake, another major shock, with magnitude close to 7.3, leveled more buildings and deepened the rubble, hampering rescue efforts. The duration of an earthquake also affects how well a building survives it. In reinforced concrete, ground shaking leads to the formation of hairline cracks, which then widen and lengthen as the shaking continues. A concrete building that can withstand a 1-minute main shock might collapse in an earthquake in which

FIGURE 10.13 Landslide in Turnagain Heights area, Anchorage, Alaska, 1964. Close inspection reveals a number of houses amid the jumble of downdropped blocks in the foreground.
Photograph courtesy of USGS Photo Library, Denver, CO.

FIGURE 10.14 Effects of soil liquefaction during an earthquake. The buildings themselves were designed to be earthquake-resistant, and they toppled over intact. Niigata, Japan, 1964.
Photograph courtesy of National Geophysical Data Center.

the main shock lasts 3 minutes. Many of the California building codes, used as models around the world, require that buildings be designed to withstand a 25-second main shock, but earthquakes can last ten times that long.

Finally, a major problem is that even the best building codes are typically applied only to new construction. Where a large city is located near a fault zone, thousands of vulnerable older buildings may already have been built in high-risk areas. The costs to redesign, rebuild, or even modify all of those buildings would be staggering. Most legislative bodies are reluctant to require such efforts—indeed, many do nothing even about municipal buildings built in fault zones.

A secondary hazard of earthquakes is *fire,* which may be more devastating than ground movement. In the 1906 San Francisco earthquake, 70% of the damage was due to fire, not to simple building failure. As it was, the flames were confined to a 10-square-kilometer area only by dynamiting rows of buildings around the burning section. Fires start as fuel lines and tanks and power lines are broken, touching off flames and fueling them. At the same time, water lines are broken, leaving no way to fight the fires effectively. Putting numerous valves in all water and fuel pipeline systems helps to combat these problems because breaks in pipes can then be isolated before too much pressure or liquid is lost.

Landslides can be a serious earthquake-related hazard in hilly areas, since earthquakes are one of the possible triggering mechanisms of sliding of unstable slopes (figure 10.13). The best solution is not to build in such areas. Even if a whole region is hilly, detailed engineering studies of rock and soil properties and slope stability may make it possible to avoid the most dangerous sites. Visible evidence of past landslides is another indication of especially dangerous areas.

Ground shaking may cause a further problem in areas where the ground is very wet—in filled land, near the coast, or in places with a high water table. This problem is **liquefaction.** When wet soil is shaken by an earthquake, the soil particles may be jarred apart, allowing water to seep in between them. This greatly reduces the friction between soil particles that gives the soil strength, and it causes the ground to become somewhat like quicksand. When this happens, buildings can just topple over or partially sink into the liquefied soil—the soil has no strength to support them. The effects of liquefaction were dramatically illustrated after a major earthquake in Niigata, Japan, in 1964. One multistory apartment building tipped over to settle at an angle of 30 degrees to the ground—while the structure remained intact! (See figure 10.14.)

In some areas prone to liquefaction, improved underground drainage systems may be installed to try to keep the soil drier, but little else can be done about this hazard, beyond avoiding areas at risk. Not all areas with wet soils are subject to liquefaction. The nature of the soil or fill plays a large role in the extent of the danger.

Coastal areas, especially those around the Pacific Ocean basin where so many large earthquakes occur, may also be vulnerable to **tsunamis.** These are seismic sea waves, sometimes improperly called "tidal waves," although they have nothing to do with tides. When an undersea or near-shore earthquake occurs, sudden movement of the sea floor may set up waves traveling away from that spot, like ripples on a pond caused by a dropped pebble. Tsunamis can also be triggered by major submarine landslides or by violent explosion of volcanoes in ocean basins, such as Krakatoa (see chapter 4).

Contrary to modern movie fiction, a tsunami is not seen as a huge breaker in the open ocean that topples ocean liners in one sweep. In the open sea, the tsunami is only an unusually broad swell or ripple on the water surface. Like all waves, tsunamis only develop into breakers as they approach shore

FIGURE 10.15 Boats washed into the heart of Kodiak, Alaska, by tsunami of 1964.
Photograph courtesy of National Geophysical Data Center.

and the undulating waters touch bottom (see chapter 13). The breakers associated with tsunamis, however, can easily be over 15 meters (45 feet) high and may reach up to 65 meters (close to 200 feet) in the case of larger earthquakes. Their effects can be correspondingly dramatic (figure 10.15). Several such breakers may crash over the coast in succession; between waves, the water may be pulled swiftly seaward, emptying a harbor or bay, and perhaps pulling unwary onlookers along. Tsunamis can travel very quickly—speeds of 1000 kilometers per hour (600 miles per hour) are not uncommon—and a tsunami set off on one side of the Pacific may still cause noticeable effects on the other side of the ocean. A tsunami caused by a 1960 earthquake in Chile was still vigorous enough to make 7-meter-high breakers when it reached Hawaii some fifteen hours later, and twenty-five hours after the earthquake, the tsunami was detected in Japan. The tsunami from the 1964 Alaskan earthquake caused eleven deaths in California, 2500 kilometers to the south.

Given the speeds at which tsunamis travel, little can be done to warn those near the earthquake epicenter, but people living some distance away can be warned in time to evacuate, saving lives, if not property. In 1948, two years after a devastating tsunami hit Hawaii, the U.S. Coast and Geodetic Survey established a Tsunami Early Warning System, based in Hawaii (where more lives have been lost to tsunamis than anywhere else). Whenever a major earthquake occurs in the Pacific region, tidal (sea-level) data are collected from a series of monitoring stations around the Pacific. If a tsunami is detected, data on its source, speed, and estimated time of arrival can be relayed to areas in danger, and people can be evacuated as necessary.

Earthquake Control?

Since earthquakes are ultimately caused by forces strong enough to move continents, it is clear that human efforts will not stop them from occurring. However, moderation of some of their most severe effects may be possible.

As noted earlier, friction between rocks along existing faults may prevent movement until sufficient stress has built up to overcome the friction. If friction is relatively small, slippage occurs when little stress has accumulated, and creep or rather small earthquakes result. The more stress built up before failure occurs, the stronger the eventual earthquake.

Many of the worst earthquakes have happened along sections of major faults that had temporarily become locked, or immobilized. Maps of the locations of earthquake epicenters along major faults show that there are sections with little or no seismic activity, while small or moderate earthquakes continue along other sections of the same fault zone. Such quiescent sections of otherwise-active fault zones are termed **seismic gaps.** These areas may be sites of future serious earthquakes. On either side of a locked section, stresses are being released by creep and/or earthquakes. In the seismically quiet locked sections, friction is apparently sufficient to prevent the fault from slipping, so the stresses are simply building up. The fear, of course, is that the stresses will build up so far that, when that locked section of fault finally does slip again, a very large earthquake will result.

At one time, it was actually suggested that carefully placed nuclear explosions could be used to unstick locked faults. This idea is not being considered very seriously any more, partly because of concern about radiation release and partly because the sudden, poorly controlled jolt of a nuclear explosion in a locked section of fault with a great deal of built-up stress could itself cause the feared large earthquake. What is needed is a way to release locked faults gently, in a more controlled way.

In the mid-1960s, the city of Denver began to experience small earthquakes. The earthquakes were not particularly damaging, but they were puzzling, since Denver had not previously been earthquake-prone. In time, a connection with an Army liquid-waste-disposal well at the nearby Rocky Mountain Arsenal was suggested. The Army denied any possible link, but the frequency of the earthquakes was correlated with the quantities of liquid pumped into the well at different times (figure 10.16). The earthquake foci were also concentrated near the arsenal well. It seemed likely that the liquid, by increasing the fluid pressures in rocks along old faults, was decreasing the resistance to shearing and allowing the rocks to slip. Experiments later conducted at an abandoned oil field near Rangely, Colorado, suggested similar possibilities. When the fluid pressure of liquids pumped into the ground exceeded a certain level, old faults were reactivated. Existing stresses in the rocks were greater than the fluid-reduced shear strength, and earthquakes occurred as the faulted rocks slipped. Other observations and experiments around the world have supported the concept that fluids in fault zones may facilitate movement along a fault.

Such observations have prompted speculation by many scientists that **fluid injection** might be used along locked sections of major faults to allow the release of built-up stress. Unfortunately, geologists are presently far from sure of the results to be expected from injecting fluid (probably water) along large, locked faults. There is no guarantee that only small earthquakes would be produced. Indeed, injecting fluid in an area where a fault had been locked for a long time could lead to release of all the stress at once, in a major, damaging earthquake, just as might happen if a nuclear explosion jarred it loose.

Possible casualties and damage are a tremendous concern, as are the legal and political consequences of a serious, human-induced earthquake. Whether it is even possible

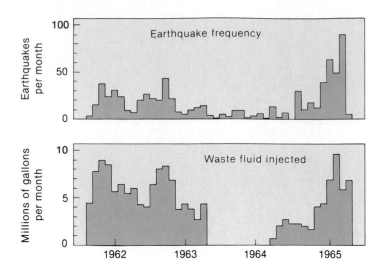

FIGURE 10.16 Correlation between waste disposal at the Rocky Mountain Arsenal (lower diagram) and frequency of earthquakes in the Denver area.

From David M. Evans, "Man-made Earthquakes in Denver" in Geotimes, *vol. 10, (9):11–18, May/June 1966. Used by permission.*

to release large amounts of energy through small earthquakes only, considering how little energy is released in low-magnitude earthquakes (recall table 10.2), is also not known.

The fluid-injection technique might more safely be used in major fault zones that have not been seismically quiet for long, where less stress has built up, to keep the fault slipping and to prevent the stress from continuing to build. Also, the first attempts are likely to be made in areas of low population density, to minimize the potential for inadvertently causing serious harm. Certainly, much more research is needed before this method can be applied safely and reliably on a large scale. It is, however, an intriguing possibility for the future.

Earthquake Prediction?

Earthquake prediction is based on the study of earthquake **precursor phenomena,** things that happen or rock properties that change prior to an earthquake. It may be possible to identify warning signs of an earthquake before it occurs. Many different, possibly useful precursor phenomena are being studied. For example, the ground surface may be uplifted and tilted prior to an earthquake. P-wave velocities in rocks near the fault (measured using artificially caused shocks, such as those from small explosions) drop and then rise before an earthquake. The same pattern is seen in the ratio of P-wave to S-wave velocity. Electrical resistivity (the resistance of rocks to the flow of electric current through them) increases, then decreases, before an earthquake. The amount of radon—a chemically inert, radioactive gas produced naturally in rocks from decay of uranium—seems to increase in well waters prior to an earthquake. The Chinese have had some success using anomalous animal behavior in prediction; it was a significant factor in the Haicheng prediction described on the next page. (Interestingly, they observed that animals that spend part or all of their time underground—snakes, for example—exhibited unusual behavior days before most surface-dwelling creatures, a not-unexpected observation.)

The hope, with the study of precursor phenomena, is that one can identify patterns of precursory changes that can be used with confidence to issue earthquake predictions that are precise enough as to time to allow precautionary evacuations or other preparations. Unfortunately, the precursors have not proven to be very reliable. Not only does the length of time over which precursory changes are seen before an earthquake vary, but the pattern of precursors—which parameters show changes, and of what sorts—also varies; and, most problematic, some earthquakes seem quite unheralded by recognizable precursors at all. Loma Prieta was one of these. So was the Northridge earthquake. And only months after the devastating Kobe earthquake did scientists document increases in groundwater radon and other chemical changes in water that developed days to weeks before the earthquake.

I n February 1975, after months of smaller earthquakes, radon anomalies, days of unusual animal behavior, and increases in ground tilt followed by a rapid increase in both tilt and microearthquake frequency, Chinese scientists predicted an imminent earthquake near Haicheng in northeastern China. The government ordered several million people out of their homes, into the open. Nine and one-half hours later, a major earthquake struck, and many lives were saved because people were not crushed by collapsing buildings.

Over the next two years, the earthquake scientists successfully predicted four large earthquakes in the Hebei district. They also concluded that a major earthquake could be expected near T'ang Shan, about 150 kilometers southeast of Beijing. In the latter case, however, they could say that the event was likely to occur sometime during the following two months. When the earthquake—magnitude 8.0 with aftershocks up to magnitude 7.9—did occur, there was no immediate warning, no sudden change in the precursor phenomena, and over 650,000 people died.

In the near term, it seems that the more feasible approach is earthquake *forecasting*, identifying levels of earthquake probability in fault zones within relatively broad time windows, as described below. This at least allows for long-term preparations, such as structural improvements. Ultimately, short-term, precise predictions that can save more lives are still likely to require recognition and understanding of precursory changes.

Studies of the dates of large historic and prehistoric earthquakes along major fault zones have suggested that they may be broadly periodic, occurring at more-or-less regular intervals (figure 10.17). This is interpreted in terms of an **earthquake cycle** that would occur along a (non-creeping) fault: a period of stress buildup, sudden fault rupture in a major earthquake, followed by a brief interval of aftershocks reflecting minor lithospheric adjustments; then another extended period of stress buildup (figure 10.18). The periodicity can be understood in terms of two considerations: First, assuming that the stress buildup is primarily associated with the slow,

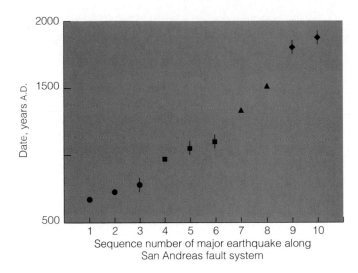

FIGURE 10.17 Periodicity of earthquakes assists in prediction/forecasting efforts. Long-term records of major earthquakes on the San Andreas show both broad periodicity and some tendency toward clustering of two to three earthquakes in each active period. Prehistoric dates are based on carbon-14 dating of faulted peat deposits.
Source: Data from U.S. Geological Survey Circular 1079.

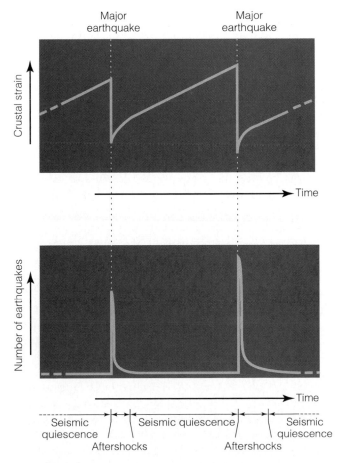

FIGURE 10.18 The "earthquake cycle" concept.
Source: After U.S. Geological Survey Circular 1079, fig. 6, p. 15.

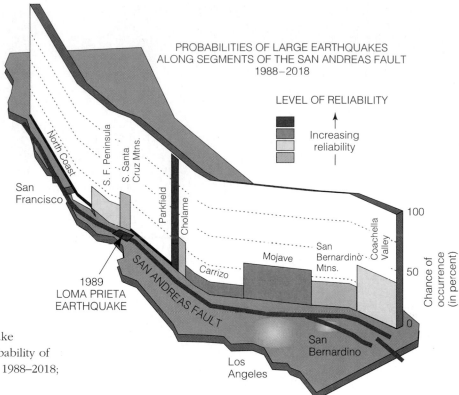

FIGURE 10.19 Earthquake forecast map issued in 1988 by the Working Group on California Earthquake Probabilities. Height of bar indicates estimated probability of a magnitude 6.5 to 7 earthquake during the period 1988–2018; color of bar indicates confidence in forecast.

Source: After U.S. Geological Survey Circular 1079, fig. 15, p. 29.

ponderous, inexorable movements of lithospheric plates, which at least over decades or centuries will move at roughly constant rates, one might reasonably expect an approximately constant rate of buildup of stress (or accumulated strain). Second, the rocks along a given fault zone will have particular physical properties, which allow them to accumulate a certain amount of energy before failure, or fault rupture, and that amount would be approximately constant from earthquake to earthquake. Those two factors together suggest that periodicity is reasonable. Once the pattern for a particular fault zone is established, one may use that pattern together with measurements of strain accumulation in rocks along fault zones to project the time window during which the next major earthquake is to be expected along that fault zone, and to estimate the likelihood of the earthquake's occurrence in any particular time period. Also, if a given fault zone or segment can only store a certain amount of accumulated energy before failure, the maximum size of earthquake to be expected can be estimated.

This is, of course, an oversimplification of a very complex reality. For example, though we speak of the San Andreas fault, it is not a single simple, continuous slice through the lithosphere, nor are the rocks it divides either identical or homogeneous on either side. Both stress buildup and displacement are distributed across a number of faults and small blocks of lithosphere. Earthquake forecasting is correspondingly more difficult. Nevertheless, the U.S. Geological Survey's Working Group on California Earthquake Probabilities published, in 1988, a set of forecasts for different segments of the San Andreas (figure 10.19). Note that the time window in question was the 30-year period 1988–2018. Such projections,

longer-term analogues of the meteorologist's forecasting of probabilities of precipitation in coming days, are revised as small (and large) earthquakes occur and more data on slip in creeping sections are collected.

Only four nations—Japan, the former Soviet Union, the People's Republic of China, and the United States—have developed government-sponsored earthquake-prediction programs. Such programs typically involve intensified monitoring of active fault zones to expand the observational data base, coupled with laboratory experiments designed to increase understanding of precursor phenomena and the behavior of rocks under stress. The notable lack of success of prediction efforts even when such active research programs exist attests to the difficulty of the task.

In the United States, earthquake predictions have not been regarded as reliable or precise enough to justify such actions as large-scale evacuations. It may someday be possible to predict the timing and size of major earthquakes accurately enough that populated areas can be evacuated in an orderly way when serious earthquakes are imminent, thereby saving many lives. Electricity travels faster than seismic waves, and even a few minutes' warning could allow people to seek safer shelter, shut off pipeline valves, and so on. Property damage is inevitable as long as people persist in living and building in earthquake-prone areas. In any case, consistently reliable earthquake predictions are probably still more than a decade in the future. As the predictions become more commonplace, different problems may arise; see box 10.1. In the interim, it is at least possible to identify proper responses when an earthquake has struck (table 10.4).

Box 10.1

Seismic Risks and Public Response

Assuming that routine earthquake prediction becomes reality, some planners have begun to look beyond the scientific questions to possible social or legal complications of such predictions. For example, individuals have speculated that, if a major earthquake were predicted several years ahead for a particular area, property values would plummet because no one would want to live there. A counterargument to that view, perhaps, is San Francisco: despite the widespread acceptance of the idea that another large earthquake like the 1906 earthquake will strike, the city continues to thrive.

The logistics of evacuating a large urban area on short notice if a near-term earthquake prediction is made is another concern. Many urban areas have hopelessly snarled traffic every rush hour; what happens if everyone in the city wants to leave at once? Some critics say that a rapid evacuation might involve more casualties than the eventual earthquake. And what if a city is evacuated, people are inconvenienced or hurt, property is perhaps damaged by vandals and looters, and then the predicted earthquake never comes? Will the issuers of the warning be sued? Will future warnings be ignored?

In the People's Republic of China, vigorous public education programs (and several recent major earthquakes) have made earthquakes a well-recognized hazard. Strong government support and the involvement of large numbers of citizens in earthquake-prediction efforts have resulted in high visibility and widespread community support for those efforts. "Earthquake drills," which stress orderly response to an earthquake warning, are held in Japan on the anniversary of the 1923 Tokyo earthquake, in which more than 140,000 people died.

By contrast, in the United States, surveys have repeatedly shown that even people living in such high-risk areas as along the San Andreas fault are often unaware of the earthquake hazard. Many of those who *are* aware believe that the risks are not very great and indicate that they would not take any special action even if a specific earthquake warning were issued. Past mixed responses to tsunami warnings underscore doubts about public response to earthquake predictions.

Yet another type of concern was raised in 1990, when a biologist-turned-consultant predicted a major earthquake for the New Madrid fault zone of the midwestern United States to occur about 3 December 1990. The basis for the prediction was variation in tidal stresses acting on the earth, which has been investigated by the geological community and found not to be correlated with major earthquakes; seismologists officially declared the "prediction" without merit. But uncritical media coverage and lack of public understanding fueled considerable local panic, as schools were closed, stores stripped of supplies, and so on . . . and the earthquake never came.

Since 1976, the director of the U.S. Geological Survey has had the authority to issue warnings of impending earthquakes and other potentially hazardous geologic events (volcanic eruptions, landslides, and so forth). An Earthquake Prediction Panel reviews scientific evidence that might indicate an earthquake threat and makes recommendations to the director regarding the issuance of appropriate public statements. These statements could range in detail and immediacy from a general notice to residents in a fault zone of the existence and nature of earthquake hazards there, to a specific warning of the anticipated time, location, and severity of an imminent earthquake.

There have been no very specific predictions of larger earthquakes, such as the 1989 Loma Prieta or 1994 Northridge quake, since the Earthquake Prediction Panel was formed. Public response to a formal, official prediction of such a serious event has yet to be tested. On the other hand, the fact that the unsanctioned New Madrid "prediction" was followed by no earthquake has caused some fear that when an official, scientific prediction of a major earthquake *is* someday made, it may simply be ignored. In principle, earthquake prediction seems desirable. In practice, the challenge of predicting earthquakes may be matched by the challenge of predicting the response, and channeling it appropriately.

TABLE 10.4 *Earthquake Safety Rules*

During the Shaking:

1. *Do not panic.*
2. If you are indoors, stay there. Seek protection under a table or desk, or in a doorway. Stay away from glass. Do not use matches, candles, or any open flame; douse all fires.
3. If you are outside, move away from buildings and power lines, and stay in the open. Do not run through or near buildings.
4. If you are in a moving car, bring it to a stop as quickly as possible, but stay in it. The car's springs will absorb some of the shaking; the car will offer you protection.

After the Shaking:

1. Check, but do *not* turn on, utilities. If you smell gas, open windows, shut off the main valve, and leave the building. Report the leak to the utility, and do not reenter the building until it has been checked out. If water mains are damaged, shut off the main valve. If electrical wiring is shorting, close the switch at the main meter box.
2. Turn on radio or television (if possible) for emergency bulletins.
3. Stay off the telephone except to report an emergency.
4. Stay out of severely damaged buildings that could collapse in aftershocks.
5. Do not go sightseeing; you will only hamper the efforts of emergency personnel and repair crews.

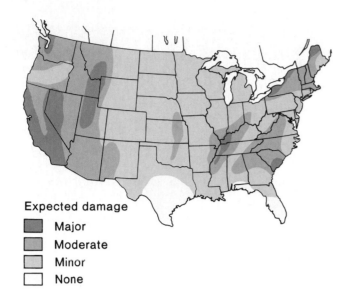

Expected damage
- Major
- Moderate
- Minor
- None

FIGURE 10.20 Seismic-risk map for the contiguous United States. *Source: National Oceanic and Atmospheric Administration.*

FIGURE 10.21 Consequences of the 1989 Loma Prieta earthquake included the collapse of the upper deck of I-880 onto the lower deck, with failure of supporting columns. Damage was particularly severe because the freeway was founded in soft sediment. Alameda County, California. *Photograph from USGS Open-File Report 889–687; by M. Rymer.*

Current and Future Seismic-Hazard Areas in the United States

Figure 10.20 is a seismic-risk map for the United States. Several of the shaded high-risk areas are not normally associated with earthquake activity—the ones near the center of the country, for instance, or those in the southeastern coastal region. This map is based partly on the distribution of known faults, partly on historical records of earthquakes. It reflects not just the frequency of past earthquakes but also their severity, which is taken to indicate the possible severity of future earthquakes in terms of ground motion.

When the possibility of another large earthquake in San Francisco is raised, most geologists debate not "whether" but "when?". At present, that section of the San Andreas fault has been locked for some time. The last major earthquake there was in 1906. At that time, movement occurred along at least 300 kilometers, and perhaps 450 kilometers, of the fault. The 1971 San Fernando earthquake was only of magnitude 6.4, yet it cost sixty-four lives and an estimated $1 billion in property damage. The death toll would have been far higher had it not occurred early in the morning.

The 1989 Loma Prieta earthquake, at magnitude 7.1, was the most severe quake along the San Andreas fault near San Francisco since 1906. Because its epicenter was about 100 kilometers south of San Francisco, in a relatively sparsely populated region of the southern Santa Cruz mountains—and because the freeways that would normally have been choked with commuters had been emptied of the many baseball fans who had gone home early to watch the World Series—the confirmed death toll was only sixty-three. However, the property damage in San Francisco, Oakland, Santa Cruz, and elsewhere totaled more than $5 billion (figure 10.21). And this was not "the big one" for San Francisco. The peninsular segment of the San Andreas is still locked (figure 10.22); in-

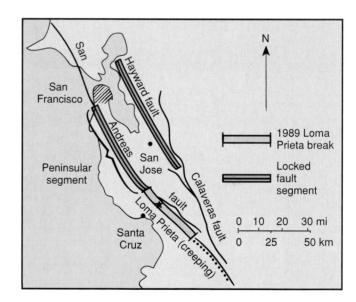

FIGURE 10.22 Creeping and locked sections of the San Andreas fault are separated by the section ruptured in the 1989 Loma Prieta quake.

deed, the rupture of the previously locked Loma Prieta segment, which shifted about 2 meters in the 1989 earthquake, effectively puts *more* stress on the peninsular segment, increasing the earthquake potential there. The estimated probability of an earthquake of magnitude 7 or greater occurring within the next thirty years either along the peninsular segment of the San Andreas or on the nearby Hayward fault (quiescent since 1868) has been increased from 50% to 75% since Loma Prieta.

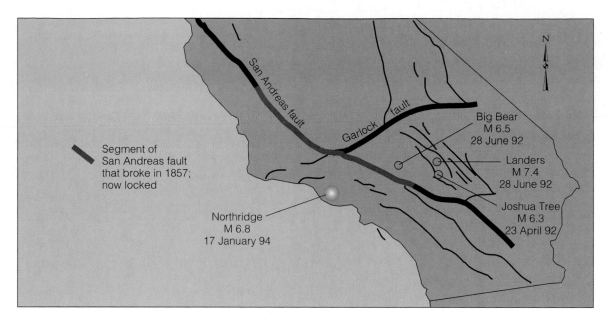

FIGURE 10.23 The 1994 Northridge earthquake is one of a series clustered around the southern locked segment of the San Andreas fault.

Events of the past few years have suggested that there may, in fact, be a greater near-term risk along the southern San Andreas, near Los Angeles, than previously thought. In the spring and summer of 1992, a set of three significant earthquakes, including the Landers quake, occurred east of the southern San Andreas. The 1994 Northridge earthquake (figure 10.23) then occurred on a previously unrecognized fault. That quake was unexpected in several respects—lack of precursors, occurrence on a "new" fault, and the type of fault displacement. It seems to have been a thrust fault, along which rocks on one side of the fault are thrust up and over those below, so there is vertical displacement and net shortening of the crust, not just horizontal slip. The southern San Andreas in this region, which last broke in 1857, has remained locked through all of this, so the accumulated strain has not been released by this cluster of earthquakes. Some believe that their net effect has been to increase the stress and the likelihood of failure along this section of the San Andreas. Moreover, in the late 1800s, there was a significant increase in numbers of moderate earthquakes around the northern San Andreas, followed by the 1906 San Francisco earthquake. The 1992–1994 activity may represent a similar pattern of increased activity building up to a major earthquake along the southern San Andreas. Time will tell.

On Good Friday, 27 March 1964, an earthquake with magnitude estimated at approximately 8.5 struck southern Alaska. The main shock, which lasted three to four minutes, was felt more than 1200 kilometers (750 miles) from the epicenter. About 12,000 aftershocks, many over magnitude 6, occurred during the next two months, and more continued intermittently through the following year. Structures were damaged over more than 100,000 square kilometers; ice on rivers and lakes cracked over an area of 250,000 square kilometers. Many areas suffered permanent uplift or depression. The uplifts—up to 12 meters on land and over 15 meters in places on the sea

floor—left some harbors high and dry and destroyed the habitats of marine organisms. Downwarping of 2 meters or more flooded some coastal communities. Tsunamis following the main shock destroyed four villages and seriously damaged many coastal cities; they were responsible for about 90% of the 115 deaths, including 16 deaths in Oregon and California. Some of these waves traveled as far as Antarctica. Landslides—both submarine and aboveground—were widespread and accounted for most of the remaining casualties. Levels of water in wells shifted abruptly as far away as South Africa. Total damage was estimated at $300 million. Southern Alaska sits above a subduction zone. The area can certainly expect more earthquakes, some of which could be severe. The coastal regions there also continue to be vulnerable to tsunamis.

Perhaps an even more dangerous situation than that of a known area of earthquake risk is that in which people are *not* accustomed to regard their area as earthquake-prone—for example, in the central United States. Earthquakes occur most frequently along plate boundaries, but severe earthquakes can occur within plates too. Though most Americans think "California" when they hear "earthquake," the largest earthquakes ever in the United States occurred in the vicinity of New Madrid, Missouri, during 1811–1812. The three strongest shocks are estimated to have had Richter magnitudes of 8.6, 8.4, and 8.7. They were spaced out over almost two months, from 16 December 1811 to 7 February 1812. Between the December 16 earthquake and the following March 15, 1,874 shocks in all were recorded by one resident engineer *without* the aid of sensitive modern seismographs. Some towns were leveled, others drowned by flooding rivers. The most vigorous tremors were felt from Quebec to New Orleans and along the East Coast from New England to Georgia. Lakes were uplifted, "blown up," and emptied, while elsewhere the ground sank to make new ones. Boats were flung out of rivers; an estimated 150,000 acres of timberland

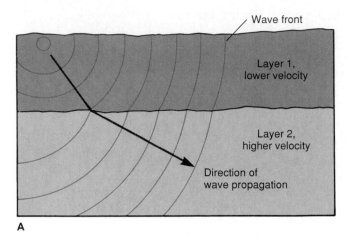

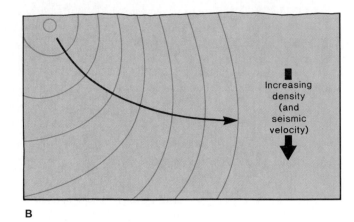

FIGURE 10.24 (A) Refraction of seismic wave at a boundary between two layers of different density/seismic velocity. (B) When density increases gradually with depth, continuous refraction in small increments produces a curved seismic-wave path.

were destroyed. Aftershocks continued for more than *ten years.* Total damage and casualties have never been accurately estimated.

The potential damage from even one earthquake as severe as the worst of the New Madrid shocks is enormous: twelve million people now live in the immediate area. Very few of those twelve million are probably aware of the risk. The big earthquakes there were a long time ago, beyond the memory of anyone now living. Minor earthquakes along the New Madrid fault zone in the summer of 1987 only briefly brought the danger to public attention. The 1990 "prediction" noted in box 10.1 was more widely publicized, but interest faded quickly when nothing came of it. Unfortunately, the danger of more such serious earthquakes is real. Beneath the midcontinent is a failed rift zone, where the continent began to rift apart, then stopped. Long and deep faults in the litho-sphere there remain a zone of weakness in the continent and a probable site of more major tremors. It is only a matter of time. It is important to bear in mind, too, that humans have been observing and recording earthquakes systematically for only a few hundred years; there may well be buried faults capable of major earthquakes of which even seismologists are unaware, simply because those faults have not slipped since the observations began.

SEISMIC WAVES AND THE STRUCTURE OF THE EARTH'S INTERIOR

Seismic body waves provide a unique means of investigating the nature of the earth's interior below the depths from which samples can be obtained. Body waves from large earthquakes can be detected all over the earth, and their paths and travel times through the earth allow geoscientists to deduce such properties as the density and physical state of the materials at depth.

In a general way, body-wave velocities are proportional to the density of the material through which the body waves propagate: the denser the medium, the faster a given type of body wave travels. If body waves cross a boundary between two materials of distinctly different densities, they are **refracted,** or deflected in a different direction (figure 10.24A). In a compositionally uniform material that increases continually in density with depth as a result of pressure, a body wave is continuously refracted, and the resulting path is curved (figure 10.24B).

A ll kinds of waves can be refracted. Light is similarly deflected on passing from air into either water or a transparent mineral. A person spearfishing from a boat must allow for this effect because the fish is not really where it appears to be from above the water surface. The effects of the refraction of light waves can also be observed by putting a hand into an aquarium or a pencil into a partially filled glass of water.

The Moho

One of the first internal features to be detected seismically was the crust/mantle boundary. It is marked by an increase in seismic-wave velocities in the mantle relative to the crust, which corresponds to a change in composition and a consequent increase in density (confirmed by xenolith evidence). This boundary is the **Mohorovičić discontinuity,** often known as the **Moho** for short, named after the Yugoslavian seismologist who first reported the seismic break by which it is identified.

The Moho is most easily recognized beneath the continents, where the greatest compositional contrasts between lower crust and upper mantle exist. Even under the continents, there is considerable lateral variation in seismic-wave velocities in crust and mantle, related to composi-

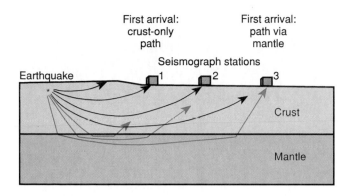

FIGURE 10.25 The thickness of the earth's crust can be determined from seismic velocities in the crust and upper mantle and from the distance from an earthquake source at which waves traveling through the mantle reach a detecting seismograph ahead of waves traveling a shorter path entirely through the crust.

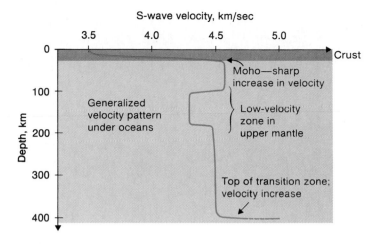

FIGURE 10.26 The low velocity zone as detected using S waves.

tional variations. Commonly, the Moho in any spot is defined as the depth at which P-wave velocities first reach 7.8 kilometers/second or higher.

The higher seismic velocities in the upper mantle and the phenomenon of wave refraction together make possible the measurement of the thickness of the earth's crust. The principle is illustrated in figure 10.25. When an earthquake occurs, the seismic body waves traveling only through the crust reach stations near the site ahead of those following a longer, deeper path to the mantle and back up to the surface. But at some distance away, the higher seismic velocities of the mantle compensate for the greater distance traveled by waves following the deeper routes, and those waves traveling part of the way in the mantle arrive ahead of waves traveling entirely in the crust.

The effect can be visualized by analogy. Suppose that a person has some distance to travel and can either walk directly to the destination or take a (faster) bus, which requires walking a bit out of the way at either end of the trip. For very short trips, the direct walk is the faster way to go, since it avoids the detour to the bus. But for long trips, the time spent getting to and from the bus is more than compensated by the time saved along most of the length of the route.

How long the trip must be before taking the bus becomes time-effective depends partly on how much faster the bus travels than the pedestrian does, and partly on how far out of the way the bus route is. Likewise, how far apart earthquake and seismograph must be for the body waves going down to the mantle to arrive ahead of those traveling wholly through the crust depends on the relative seismic velocities in the crust and mantle and on the thickness of the crust. Since the seismic velocities of samples of crust and upper-mantle rocks can be determined experimentally in the laboratory, the thickness of the crust can be deduced. In the

actual case, the problem is further complicated by seismic-velocity variations in the mantle, and particularly in the heterogeneous crust, as well as by local variations in crustal thickness. Still, the seismic approach is vastly simpler and less costly than is deep drilling.

Fine Structure of the Upper Mantle

The weak asthenosphere is important to the plausibility of plate tectonics, for it provides a means of explaining plate movements. The existence of the asthenosphere was deduced from seismic-wave studies. Body waves passing through this zone showed unusually long travel times, from which a decrease in seismic velocities in this zone was deduced (figure 10.26). This seismic **low-velocity zone** extends from the base of the lithosphere (by definition) to, typically, depths of 175 to 250 kilometers (about 110 to 155 miles). It does not encircle the whole earth uniformly; nor does it everywhere include partly molten material. Rocks of the low-velocity layer, as previously noted, are all quite close to their melting temperatures, however, and are therefore relatively less elastic and more plastic in their behavior. Plastic materials, which tend to flow under prolonged stress, transmit both compression and shearing deformation less efficiently than do elastic materials, if at all. Under the short-term stress of a passing seismic wave, the rocks of the low-velocity layer may be sufficiently elastic to transmit the waves, though more slowly than in fully elastic rocks. Under sustained stress, these same rocks flow and deform permanently. Similarly, a ball of wax or putty may bounce when dropped quickly against a hard surface and retain its original shape, but the same material may flow plastically and become permanently deformed when a weight is applied over a long period of time. As can be seen in figure 10.26, the seismic velocities in the low-velocity zone

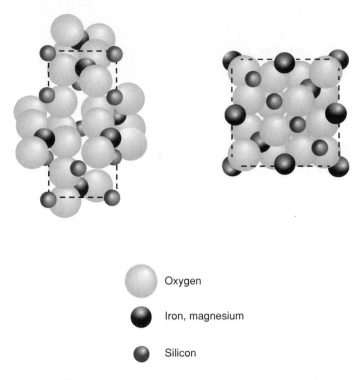

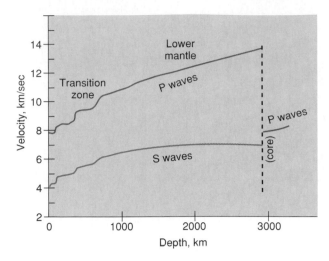

FIGURE 10.28 Seismic-wave velocity variations with depth in the mantle. Below the transition zone, velocities increase smoothly with increasing depth/pressure.

Oxygen

Iron, magnesium

Silicon

FIGURE 10.27 One of the increases in seismic velocity in the transition zone is attributed to a change in the crystal structure of olivine, from its low-pressure form (left) to a denser, more compact form (right) at depth.
Source: George D. Garland. Introduction to Geophysics, *2d ed., Holt, Rinehart and Winston, 1979.*

are still substantially higher than crustal velocities, but they are markedly lower than velocities in the more rigid mantle immediately above and below that layer.

The Moho represents a compositional change, the boundary between the crust and the more mafic mantle. Both the crust and the mantle are solid at this depth. The lithosphere/asthenosphere boundary, still deeper, represents a change in physical state—from a rigid solid to a more plastic or even partially molten material—with no significant change in composition.

Still deeper in the upper mantle, beneath the low-velocity zone, seismic velocities increase again with depth and density. For the most part, the increases are gradual, as if caused simply by the increasing density to be expected as a compositionally uniform mantle becomes more and more compressed with increasing depth. There are, however, several small jumps in velocity within the **transition zone,** between depths of approximately 400 and 700 kilometers (250 to 450 miles). These are attributed to **phase changes,** changes in mineralogy or crystal structure without, necessarily, any change in composition. Several specific phase changes have been proposed to explain the velocity increases. The jump at about 400 kilometers, for example, has been attributed to the collapse of the magnesian olivine abundant in upper-mantle rocks to a more compact, denser structure, like that of the mineral spinel (figure 10.27). Labo-

ratory experiments on olivine under pressure have suggested that the change occurs at the appropriate depths/pressures. Deeper still, the various silicates of the mantle might break down into simple oxides (MgO, FeO, SiO_2, and so on) capable of achieving very dense crystal structures in response to the extremely high pressures.

Below the transition zone, from about 700 kilometers to the base of the mantle, seismic velocities resume their gradual increase with depth (figure 10.28). There seem to be no sharp compositional or phase changes, then, in the lower mantle. (This may be an argument against breakdown of silicates into oxides at depth.) Continuous changes—for example, a change in the ratio of iron to magnesium with depth—are still possible within the limits of the data.

The Shadow Zone

The strength of seismic waves from a major earthquake is such that the body waves could, in principle, be expected to reach the opposite side of the earth. P waves indeed do so. S waves do not, and the pattern of their absence led to recognition of another major zone of the interior and the determination of its size.

Compressional waves can pass through solids or liquids. Sound waves, which are compressional, can travel through water (in addition to solids and air), and P waves can travel through rock, magma, and other fluids, although their velocities naturally differ in the different media. Liquids do not, however, support and propagate a shear wave. A liquid subjected to shear stress basically flows in response, and there is no tensile elasticity to pull it back again. S waves, therefore, do not travel through a liquid.

When a major earthquake occurs, no direct S waves are detected at distances greater than about 103 degrees of arc from the source (figure 10.29). A simple explanation for this

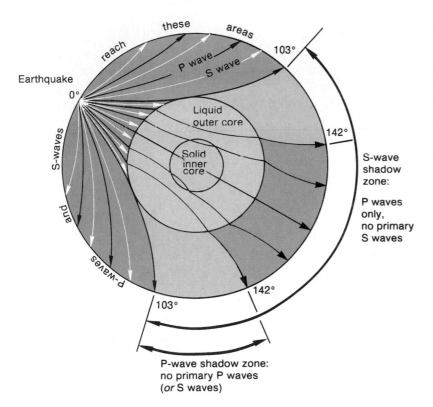

FIGURE 10.29 Seismic shadow zones for S waves and P waves caused by the liquid outer core. Note curvature of the wave paths due to refraction in the mantle and core.

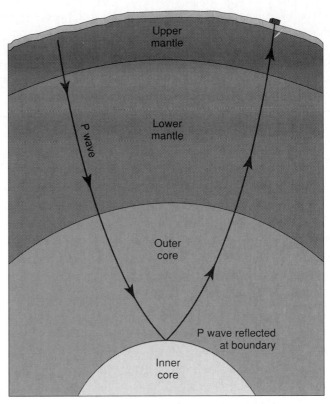

FIGURE 10.30 Travel times of P waves that bounce off the inner/outer core boundary allow the depth to that boundary to be determined.

S-wave **seismic shadow** is that a large fluid mass is blocking S-wave transmission. This is the earth's liquid outer core. Its physical state (liquid) is indicated by its ability to block S waves; its size is apparent from the size of the seismic shadow it casts. The outer core also casts a conical or ringlike P-wave shadow, as seen in figure 10.29. This much smaller shadow zone, extending in a band from 103 degrees to 142 degrees of arc away from the earthquake source, is a consequence of the differences in elastic properties between solids and liquids, and the correspondingly anomalous refraction that P waves undergo at the core-mantle boundary.

An emerging area of research—**seismic tomography,** a technique that involves mapping of seismic-velocity variations in three dimensions—has begun to reveal the topography of the core-mantle boundary. Somewhat unexpectedly, that boundary seems to have considerable relief on it. Why a liquid outer core should have an irregular "surface" (interface with the mantle), or what the irregularities mean, is not yet clear. See also "Temperature" section following.

The Inner Core

Seismic waves can also bounce off, or be reflected from, the boundaries between layers of different seismic properties. It is in this way that the presence of a further zone within the core was detected (figure 10.30). Some P waves are reflected from the boundary between the inner and outer core. P-wave velocities in the outer core can be approximated from the travel times of those P waves passing only through the outer core (and shallower zones). Once velocities in mantle and outer core are known, the travel times of P waves reflected off the inner core can be used to deduce the depth to the inner/outer-core boundary.

Travel times of P waves traveling straight through the earth, inner core and all, indicate significantly higher velocities in the inner core than in the outer. From this, it is inferred that the inner core is again solid, rather than liquid. Temperatures in the deepest interior are not believed to increase very much with depth, but pressures, especially in the very dense core, certainly do. Therefore, even if the inner and outer core are compositionally identical, the inner core, at much higher pressures, could be solid while the outer core is molten.

Seismic Data and the Composition of the Core

The earth has a sizeable magnetic field, which could, so far as its orientation is concerned, be represented approximately by a giant bar magnet aligned nearly through the rotational poles. It might be tempting to postulate a large lump of solid iron, or perhaps iron-nickel alloy such as is found in some meteorites, in the middle of the earth. Iron is a relatively abundant element, it is strongly magnetic, and it is dense. Unfortunately, a mass of solid iron in the interior does not account for the magnetic field as neatly as it might at first appear to do. Recall from chapter 9 that each magnetic material has a *Curie temperature,* above which it loses its magnetic properties. For iron, the Curie temperature is less than 800°C. Yet magmas from the upper mantle frequently have temperatures of 1000°C or more; temperatures in the deep interior must be hotter still. Any solid iron deep within the earth, then, is well above its Curie temperature. The explanation for the magnetic field must be more complex.

Comparison of the results of laboratory measurements of seismic velocities and densities of various materials under very high pressures with whole-earth seismic data shows that silicate materials are unlikely candidates for the core. Metallic iron gives a fairly good fit to actual data, but it is somewhat too dense, especially for the outer core. Iron meteorites are typically made of an iron-nickel alloy, so it has been suggested that nickel is present in the earth's core with iron. But nickel is denser than iron; if iron is a little too dense for the seismic data, an iron-nickel alloy would be a worse fit. Therefore, it is generally accepted that there must be at least one other element present in moderate quantity (perhaps 10%) to "lighten" the core, lowering its density to conform to actual values estimated from seismic data. Several elements that are both cosmically abundant and miscible with molten iron have been proposed as minor constituents of the outer core, including metallic silicon, oxygen, and sulfur.

The demonstration of the presence of a liquid outer core and the inference that it consists mostly of iron together permit an explanation of the earth's magnetic field after all. A magnetic field can be generated by flow in an electrically conducting fluid, much as a dynamo generates electricity, even at temperatures above the Curie point of the corresponding solid. Molten metallic iron seems a suitable fluid. Direct proof of this model for the magnetic field is lacking, largely because geoscientists cannot make appropriate direct measurements of motions in or electrical conductivity of the outer core. However, the model is consistent with available data and with the observation that the magnetic poles are close to the rotational poles: intuitively, it makes sense that fluid motions in the outer core might be influenced, controlled, or even generated by the earth's rotation. Further research suggests that flow in the outer core may be related to much slower plastic flow in the solid lower mantle, to which the cooling core transfers its heat.

The phenomenon of magnetic reversals, well documented by paleomagnetic studies, has yet to be satisfactorily explained by this or any other model of the earth's magnetic field.

CHEMISTRY OF THE INTERIOR

Direct Sampling and Chemical Analysis

Analysis of surface samples and drill cores reveals the compositions of a variety of crustal rocks of all types—igneous, metamorphic, and sedimentary. Where tectonic deformation

FIGURE 10.31 Xenoliths of ultramafic mantle material (olivine-rich dunite) in basalt.

has brought deep plutonic and high-grade metamorphic rocks closer to the surface, geoscientists have obtained some samples even of lower-crustal rocks, though these exposures are few. Crustal rocks are extremely varied in composition, but the sampling is sufficiently thorough that there is general agreement on the crust's average composition.

Deeper samples, from the upper mantle beneath the continents, are conveniently brought to the surface by volcanic rocks. Rising magmas formed by partial melting in the mantle may carry along with them fragments of still-unmelted mantle, or they may pick up pieces of mantle or crustal rocks above the depth of melting as they ascend. These rock inclusions—the *xenoliths* first described in chapter 3—provide valuable clues to the composition of the deep crust and upper mantle (figure 10.31). The depths from which xenoliths come can frequently be deduced on the basis of the mineral assemblages in them. In any case, they cannot have come from depths below the source region of the magma containing them. When the same volcanic rock contains several types of xenoliths, the xenoliths clearly are samples of several different subsurface rock units present below that point, although the vertical sequence of those units may be indeterminable. Even the sampling of xenoliths only extends a short distance into the earth, and these deepest xenoliths are few. They do, at least, include some upper-mantle material. The deepest-source magmas for which there is good evidence of the depth of formation are the diamond-bearing ultramafic rocks known as **kimberlites,** which originate at depths of approximately 200 kilometers and occur as volcanic pipes on the continents.

A few samples of the uppermost mantle beneath the oceans are obtained from *ophiolites,* slices of oceanic lithosphere occasionally caught up in plate collisions and thus emplaced onto the continents (see chapter 12). However, as samples of the upper mantle, rocks from ophiolites are com-

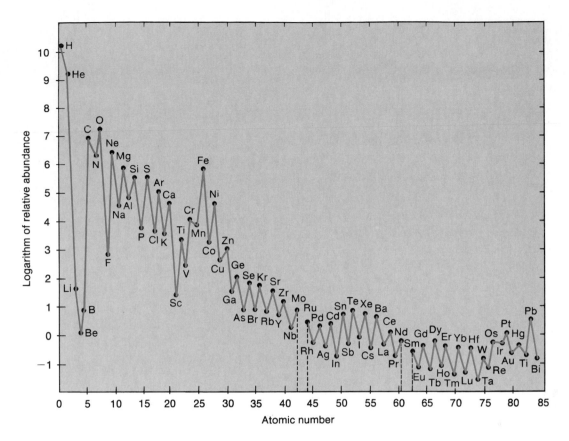

FIGURE 10.32
So-called cosmic abundance curve of the elements. Note that the scale on the vertical axis is logarithmic. Abundances are reported per million (10^6) atoms of silicon.

monly of rather poor quality, for they are ultramafic and therefore very easily weathered once exposed at the surface. Their original mineralogy can be determined, but chemical weathering will have altered their chemical composition.

Further inferences about the composition of the upper mantle can be drawn from the compositions of the magmas produced from it, combined with laboratory experiments on melting and magmatic crystallization. The nature of a magma's parent mantle rocks can frequently be deduced from the composition of the partial melt. However, geologists have no examples of volcanic rocks formed from melts that can be shown to have originated below about 200 kilometers, and most magmas are of much shallower origin.

Cosmic Abundances of the Elements

The compositions of stars can be determined from analyses of their spectra; different wavelengths of light correspond to different elements. Most stars are quite similar in composition, at least in terms of the principal elements detected; 90% of them are compositionally similar to our sun. The sun's light spectrum can thus be analyzed to estimate the composition of most stars, which, in turn, encompass most of the mass of the universe. One limitation of this approach is that the presence and amounts of rarer elements are difficult to detect.

Fortunately, the major-element composition of the sun (excluding gases) is remarkably similar to that of primitive meteorites. This suggests that the relative amounts of the rarer elements in the same meteorites could be used to estimate the abundances of those elements in the rest of the solar system, and perhaps the universe.

In such ways, the so-called **cosmic abundance curve** of the elements can be assembled (figure 10.32). Relative abundances of major elements in the earth's crust are also very similar to those shown. The principal discrepancies are in such elements as the inert gases and some other volatile elements (like hydrogen, oxygen, and carbon), and these may simply have failed to condense in the primitive earth as it formed from the solar nebula.

This, in turn, suggests that the initial composition of the solar nebula can be estimated from the cosmic abundance curve, and, if the major nonvolatile elements in the earth's crust have relative abundances generally similar to cosmic abundances, all of the nonvolatile elements should have condensed in the primitive earth. The estimated bulk composition of the earth thus determined is shown in table 10.5. If the crust is relatively depleted in iron, nickel, and magnesium by comparison, then these elements can be inferred to be concentrated deeper in the earth. The abundance of ferromagnesians in upper-mantle samples is consistent with such

TABLE 10.5 *Approximate Average Composition of the Whole Earth*

Element	Percentage in Earth
Iron (Fe)	33.3
Oxygen (O)	29.8
Silicon (Si)	15.6
Magnesium (Mg)	13.9
Nickel (Ni)	2.0
Calcium (Ca)	1.8
Aluminum (Al)	1.5
Sodium (Na)	0.2

Note: All concentrations in weight percent; averages of several independent estimates.

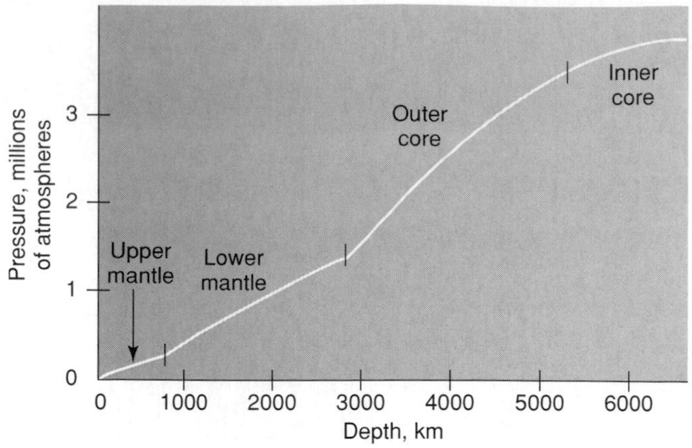

FIGURE 10.33 Pressure as a function of depth in the earth.

reasoning. Conversely, one cannot realistically postulate that the earth's deep interior is made up of something that is, cosmically speaking, relatively rare, such as lead or gold.

CONSTRAINING PRESSURE AND TEMPERATURE

Just as geoscientists cannot sample the earth's deep interior directly to determine its composition, so they must also infer the pressures and temperatures in the deep mantle and core.

Pressure

Internal pressures are comparatively easy to estimate. Below the lithosphere, in which additional lateral stresses arise from plate motions, pressures are effectively determined by the thicknesses and densities of overlying rocks at any point. The densities of various layers can be obtained from seismic body-wave data, as noted earlier. This makes possible the calculation of pressure as a function of depth in the earth, shown graphically in figure 10.33. Maximum pressures, in the inner core, are over 3.5 million times atmospheric (surface) pressure.

Temperature

Internal temperatures are somewhat more difficult to deduce than internal pressures because temperatures do not increase in any predictable way with the thickness of overlying rock. If the problem is approached by simply extrapolating crustal geothermal gradients downward, projected temperatures quickly become impossibly high. Extrapolating at a temperature increase of 30°C per kilometer, for example, yields expected temperatures of about 21,000°C (37,000°F) at the base of the upper mantle and 86,700°C (156,000°F) at the core-mantle boundary. Rocks could not be solid, or even liquid, under such pressure-temperature conditions, so actual temperatures in the mantle and core must be substantially lower.

Fortunately, knowledge of the composition of the interior, coupled with the demonstrated variations in physical state of the interior, permits geoscientists to put upper or lower limits on the temperatures of various zones. For example, the fact that there is some partial melting in the low-velocity zone indicates that temperatures there must be at or above the minimum melting temperature of the ultramafic rock of the upper mantle. On the other hand, most of the mantle is solid. So, assuming no major compositional changes deeper in the mantle (and there is no seismic evidence for any such changes), temperatures deeper in the mantle must be *below* the melting curve for such material at appropriate pressures.

Similar reasoning can be applied to the core. The core is at least predominantly iron, if not pure iron, so geoscientists can approximate core melting properties by those of iron. The outer core is molten, so temperatures there must be above the melting curve for iron at those pressures. The inner core is solid, so temperatures there must be below the melting point of iron. (Small amounts of one or more additional elements in the core, as indicated by seismic data, tend to re-

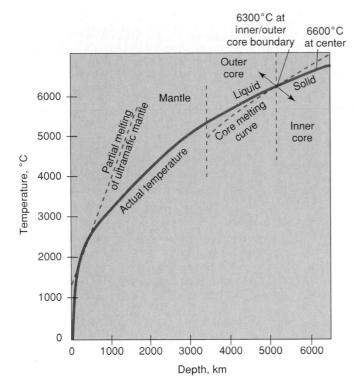

6300°C at inner/outer core boundary   6600°C at center

FIGURE 10.34 Temperature as a function of depth in the earth. Major constraints are the melting curves of ultramafic rock (mantle) and iron (core).

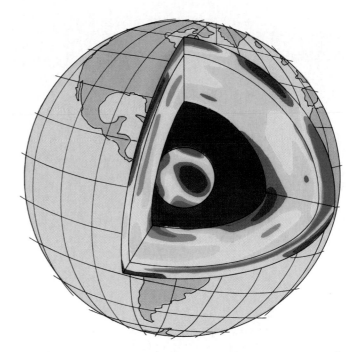

FIGURE 10.35 Cutaway view of seismic-velocity variations in the earth's interior. Blue (colder?) regions have above-average velocities, red (warmer?) have below-average. Velocity variations in the upper mantle are for S waves; total variation may be up to 10%. In deeper regions, variations are for P-wave velocities, which vary laterally by less than 2%.

Redrawn with permission from A. M. Dziewonski, "Global Images of the Earth's Interior," Science 236:37, April 13, 1987. Copyright 1987 by the American Association for the Advancement of Science.

duce the melting temperature of iron somewhat at a given pressure, but the effect is probably not sufficient to invalidate temperature estimates based on the properties of pure iron.) Recent experiments using special apparatus capable of sustaining ultra-high pressures like those in the core suggest that the maximum temperature at the earth's center is close to 5000°C, or about 9000°F, and that at the inner/outer-core boundary, the temperature is approximately 4500°C (8100°F). Seismic studies of the mantle also suggest a jump in temperature as the mantle-core boundary is crossed.

Temperatures estimated in this way for the mantle and core allow geoscientists to project the geothermal gradient at depth, as shown in figure 10.34. These temperature constraints, however, are limits only. That is, geoscientists may be able to say that temperatures at a given depth in the core must be above or below the melting temperature of iron, but they cannot say *how much* above or below. In particular, the temperature of the inner core may be considerably higher than shown in figure 10.34, given the great increases in pressure with depth there.

The seismic tomography mentioned earlier can be used to detect lateral variations in seismic velocity within the various layers of the interior, which are interpreted as reflecting temperature variations (figure 10.35). It has confirmed upwelling of hot material at spreading ridges and sinking of cold material under subduction zones. Variations at greater depth are more subtle, but it is hoped that in time, seismic tomography can be used to help answer such questions as whether mantle convection is confined to the asthenosphere and whether there is a single convecting layer or more than one.

Rocks subjected to stress may behave elastically or plastically. Earthquakes result from sudden rupture of brittle or elastic rocks or sudden movement along fault zones in response to stress. Most earthquakes occur at plate boundaries and are related to plate-tectonic processes. Earthquakes are confined to the cold, rigid lithosphere; the necessary sudden failure cannot occur in the warm, plastic asthenosphere. The pent-up energy of an earthquake is released through seismic waves, which include both compressional and shear body waves, and surface waves, which cause the most structural damage. Earthquake hazards include ground rupture and shaking, fire, liquefaction, landslides, and tsunamis.

While earthquakes cannot be stopped, their negative effects can be limited by (1) seeking ways to cause locked faults to slip gradually and harmlessly, perhaps by using fluid injection to reduce frictional resistance to shear; (2) designing structures in active fault zones to be more resistant to earthquake damage; (3) identifying and, wherever possible, avoiding development in areas at particular risk from earthquake-related hazards; (4) increasing public awareness of and preparedness for earthquakes in threatened areas; and (5) learning enough about earthquake cycles and precursor phenomena to make accurate and timely predictions of earthquakes.

Detailed studies of the propagation of seismic body waves have provided most of the information on the structure of the earth's interior and the densities and physical states of the zones within it. Seismic data have made possible the identification of the silicate mantle, including a low-velocity layer underlain by several phase transitions in the upper mantle, and the much denser, iron-rich core. The S-wave shadow zone demonstrates that the outer core is liquid and indicates its size. High-pressure laboratory experiments and seismic data suggest that one or more elements besides iron and nickel are present at least in the outer core, as its density is somewhat too low for pure iron or iron-nickel alloy.

In addition to seismic data, information of several other kinds contributes to the determination of the physical state and chemical composition of the earth's interior, shown in figure 10.36. Chemical and mineralogical data for the crust and uppermost mantle are obtained by direct sampling and chemical analysis, coupled with laboratory studies of magmas and melting. The bulk composition of the earth can be constrained using compositional data from meteorites and stars. Pressures in the interior are estimated from the densities of the various zones in the earth. Internal temperature variation with depth is constrained using the melting curves of ultramafic rocks (for the mantle) and of iron (for the core) for calculated pressures corresponding to those prevailing in the interior. Lateral temperature variations are detected using seismic tomography; they may provide clues to the nature of mantle convection and details of internal structure.

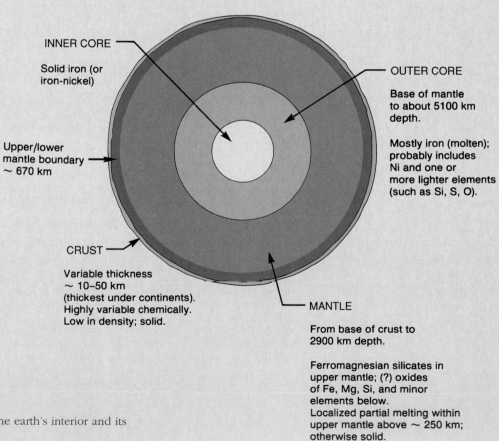

INNER CORE

Solid iron (or iron-nickel)

Upper/lower mantle boundary ~ 670 km

CRUST

Variable thickness ~ 10–50 km (thickest under continents). Highly variable chemically. Low in density; solid.

OUTER CORE

Base of mantle to about 5100 km depth.

Mostly iron (molten); probably includes Ni and one or more lighter elements (such as Si, S, O).

MANTLE

From base of crust to 2900 km depth.

Ferromagnesian silicates in upper mantle; (?) oxides of Fe, Mg, Si, and minor elements below. Localized partial melting within upper mantle above ~ 250 km; otherwise solid.

FIGURE 10.36 Composite picture of the earth's interior and its various zones.

aftershocks 165
body waves 160
brittle 155
compressive stress 155
cosmic abundance curve 179
creep 157
earthquake 157
earthquake cycle 169
elastic deformation 155
elastic limit 155
elastic rebound 157
epicenter 158
fault 155
fluid injection 168
focus 158

intensity 162
kimberlites 178
liquefaction 166
locked fault 158
low velocity zone 175
magnitude 162
Mohorovičić discontinuity (Moho) 174
phase changes 176
plastic deformation 155
precursor phenomena 168
P waves 160
refraction 174
rupture 155
scarp 158
seismic gap 168

seismic shadow 177
seismic tomography 177
seismic waves 160
seismograph 160
shearing stress 155
strain 155
stress 155
surface waves 160
S waves 160
tensile stress 155
trace 158
transition zone 176
tsunami 166

QUESTIONS FOR REVIEW

1. Compare and contrast elastic and plastic behavior in rocks. What two factors may cause rocks to behave more plastically?
2. What determines whether creep or large earthquakes will occur along an existing fault?
3. What is the distinction between an earthquake's epicenter and its focus?
4. Deep-focus earthquakes are confined to subduction zones. Why?
5. Name and briefly describe the two kinds of seismic body waves.
6. A given earthquake has one magnitude but a range of intensities. Explain.
7. Cite and discuss any three factors that limit the extent to which building codes can be designed for earthquake-resistant structures.
8. Describe the phenomenon of liquefaction and what happens to structures when it occurs.
9. What is a *tsunami?* Are tsunamis hazardous in the open ocean, near shore, or both?

10. Earthquakes in Denver indirectly suggested a possible future means of reducing the danger of major earthquakes along locked faults. Explain the concept involved.
11. Cite any three earthquake precursor phenomena and describe how they change prior to an earthquake.
12. Describe the assumptions behind the earthquake-cycle concept of earthquake forecasting. How do earthquake *prediction* and earthquake *forecasting* differ?
13. A seismic-risk map of the United States shows an area of high risk in the middle of the country. Why?
14. What is the *Moho,* and how is the depth to the Moho determined?
15. What causes the decrease in seismic velocities associated with the low-velocity layer?
16. Geoscientists do not believe that there are significant compositional changes in the lower mantle. Why not?

17. Describe the origin of the S-wave shadow zone. What does it tell geoscientists about the earth's interior?
18. How is the depth to the inner/outer-core boundary determined?
19. Can the earth's magnetic field be due to a lump of solid iron in the core? Why or why not? How else can the field be explained?
20. The core is believed to be neither pure iron nor iron-nickel alloy. Why?
21. By what means do geoscientists obtain samples of the deep crust and uppermost mantle, below depths to which they can drill?
22. From what information is the cosmic abundance curve of the elements derived, and how is the cosmic abundance curve used to constrain the composition of the earth's interior?
23. Can temperatures in the earth's interior be determined by extrapolation from geothermal gradients in the crust? How else are internal temperatures constrained? Discuss briefly.

FOR FURTHER THOUGHT

1. Investigate the history of any modern seismic activity in your area. What is the geologic explanation for this activity, and how probable is significant future activity? (You might consult the U.S. Geological Survey or state geological surveys for information.) If a significant hazard is perceived, what plans are in place to respond to it?

2. Research one or more past earthquake predictions made in the United States or elsewhere. How specific were the predictions? How accurate? What was the public response, if any?

3. If you live near a major urban area, collect data on the population in the city on a workday and the rate at which commuters can be moved into or out of the city during rush hour. Speculate on the implications of your findings for the impact of short term earthquake predictions.

4. The relatively high cosmic abundance of iron suggests that iron is the dense element that forms the earth's core. In the absence of that constraint, consider whether one could demonstrate in some other way that the core is not made of another metal, such as gold or lead, and if so, how.

SUGGESTIONS FOR FURTHER READING

Asada, T., ed. 1982. *Earthquake prediction techniques.* Tokyo: University of Tokyo Press.

Austin, M. 1981. No more wooden towers for San Francisco, 1906. In *Language of the earth,* edited by F. H. T. Rhodes and R. O. Stone, 64–67. New York: Pergamon Press.

Bercovici, D., Schubert, G., and Glatzmaier, G. A. 1989. Three-dimensional spherical models of convection in the earth's mantle. *Science* 244: 950–55.

Gass, I. G., Smith, P. J., and Wilson, R. C. L., eds. 1971. *Understanding the earth.* Cambridge, Mass.: MIT Press. (See especially "Mohole: Geopolitical fiasco" by D. S. Greenberg, 342–48; "The composition of the earth" by P. Harris, 52–69; and "The earth's heat and internal temperatures" by J. H. Sass, 80–87.)

Gere, J. M., and Shuh, H. C. 1984. *Terra non firma.* New York: W. H. Freeman.

Hanks, T. C. 1985. *National Earthquake Hazard Reduction Program—Scientific status.* U.S. Geological Survey Bulletin 1659.

Hays, W. W., ed. 1981. *Facing geologic and hydrologic hazards.* U.S. Geological Survey Professional Paper 1240-B. (See especially section Z, "Hazards from Earthquakes," 6–38.)

Henderson, P. 1982. *Inorganic geochemistry.* New York: Pergamon Press.

Hoffman, K. A. 1988. Ancient magnetic reversals: Clues to the geodynamo. *Scientific American* 258 (May): 76–83.

Johnston, A. C., and Kantner, L. R. 1990. Earthquakes in stable continental crust. *Scientific American* 262 (March): 68–75.

Kerr, R. A. 1993. Parkfield quakes skip a beat. *Science* 259 (19 February): 1120–22.

Kobe Earthquake: An urban disaster. 1995. *EOS* 76 (7 February) 49–51.

Lay, T., *et al.* 1990. Studies of the Earth's deep interior: Goals and trends. *Physics Today* 43 (October): 44–52.

Liu, L., and Bassett, W. A. 1986. *Elements, oxides, and silicates: High-pressure phases with implications for the earth's interior.* New York: Oxford University Press.

Page, R. A., Boore, D. M., Buckman, R. C., and Thatcher, W. R. 1992. *Goals, opportunities, and priorities for the USGS Earthquake Hazards Reduction Program.* U.S. Geological Survey Circular 1079.

Penick, J. L., Jr. 1981. *The New Madrid earthquakes.* 2d ed. Columbia, Mo.: University of Missouri Press.

Reid, T. R. 1995. Kobe wakes to a nightmare. *National Geographic* 188 (July): 112–136.

Scientific American 249 (September 1983). (Contains a series of articles on the earth and its various zones.)

Spence, W., Herrmann, R. B., Johnston, A. C., and Reagor, G. 1993. *Responses to Iben Browning's prediction of a 1990 New Madrid, Missouri, earthquake.* U.S. Geological Survey Circular 1083.

Tank, R., ed. 1983. *Environmental geology.* 3d ed. New York: Oxford University Press. (This collection of readings includes six chapters relating to earthquakes.)

United Nations Educational, Scientific, and Cultural Organization. 1984. *Earthquake prediction.* Paris: UNESCO Press.

U.S. Geological Survey Staff. 1990. The Loma Prieta, California earthquake: An anticipated event. *Science* 247 (January): 286–93.

Volcanoes and the earth's interior. 1983. San Francisco: W. H. Freeman. (A selection of readings from *Scientific American,* 1975–1982.)

Wesson, R. L., and Wallace, R. E. 1985. Predicting the next earthquake in California. *Scientific American* 252 (February). 35–43.

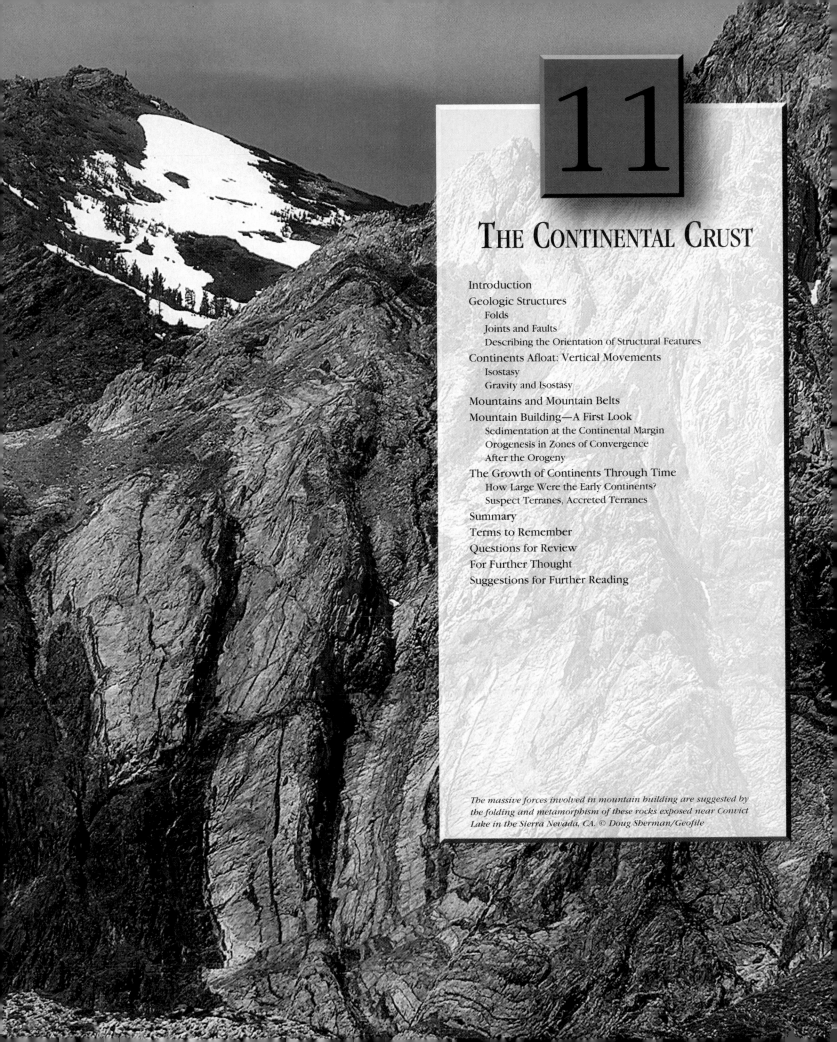

11

THE CONTINENTAL CRUST

The massive forces involved in mountain building are suggested by the folding and metamorphism of these rocks exposed near Convict Lake in the Sierra Nevada, CA. © Doug Sherman/Geofile

The surfaces of the continents are the part of the earth most accessible for observation, measurement, and sample collection. In favorable cases, geologists can extrapolate features recognized at the surface some distance into the crust. The first portion of this chapter describes common kinds of geologic structures and other features formed in the continental crust and how they are identified. (This is not to imply that folds and faults are confined to the continental crust/lithosphere, but they are particularly well developed and occur in greater variety here.) Next follows a brief survey of some geophysical techniques—including seismic methods discussed in chapter 10—that allow probing of the deeper crust. A discussion of the nature of mountain building and the different kinds of mountains is next. The chapter concludes with a look at the continents through time and what can be said about the longevity and growth of continents.

GEOLOGIC STRUCTURES

We have already looked at a variety of rock textures produced during the formation of a single rock unit—for example, the foliation of schists or the porphyritic or vesicular textures of some volcanic rocks. Additional structure can be imposed on rocks after formation, through the application of stress. The type of structure produced depends on the nature of the stress and on whether the rock behavior is more brittle or more plastic. Brittle behavior favors the formation of fractures; plastic behavior, folds. (For a review of these terms and related rock properties, see chapter 10.) It follows that folds are particularly likely to form in the deep crust, where temperatures and confining pressures are higher. Cold, shallow crust fractures and faults more readily. The orientation of folds and faults, in turn, can provide information about the directions from which the stresses came, which may be related to ancient plate motions.

Folds

Plastic deformation under tensional stress stretches or thins the rocks. When plastic deformation results from compression or shear stress, the resultant crumpling of the rocks commonly produces folds. Folds, like faults, come in all sizes— from microscopic, through handsample size, to the scale of mountains (figure 11.1). Each fold has several components (figure 11.2). The **hinge** is the most sharply curved part of the fold. The **axis** is the direction around which the fold is curved. It lies parallel to the hinge. The **axial surface** divides the two **limbs,** or sides, of the fold.

Large-scale folds (of regional scale) are subdivided on the basis of whether they arch upward or are bowed down into trough shapes. If the rocks folded were originally flat lying sedimentary rocks, the resulting structures differ not only in the orientation of the beds but also in the relative ages of the rocks in various parts of the fold. An arching fold is an

FIGURE 11.1 Mountain-scale folding. Scapegoat Mountain, Montana.
Photograph by M. R. Mudge, USGS Photo Library, Denver, CO.

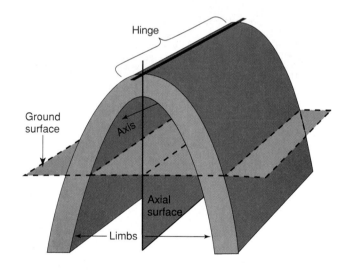

FIGURE 11.2 Nomenclature of the component parts of a fold.

antiform. The limbs of the antiform dip away from the hinge. If the units folded were originally undeformed sedimentary rocks, the rocks at the core of the antiform will be the oldest, and the structure is then termed an **anticline** (figure 11.3). A trough-shaped fold, a **synform,** shows the reverse pattern. The limbs of the fold dip toward the center of it, and if a sedimentary sequence is folded, the youngest units are folded into the center of the synform, which is then more precisely termed a **syncline** (figure 11.4).

Folds need not have either vertical axial surfaces or horizontal axes. A fold with a dipping axis is a **plunging fold**

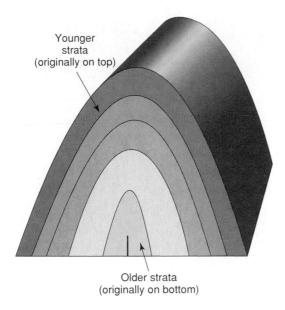

Younger
strata
(originally on top)

Older strata
(originally on bottom)

FIGURE 11.3 An anticline, or arching fold, in layered sediments. Note that the oldest strata are at the center.

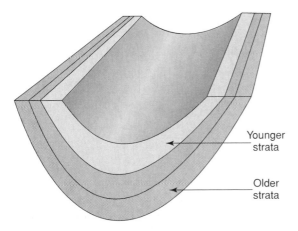

Younger
strata

Older
strata

FIGURE 11.4 A syncline, showing the reverse age pattern from that of an anticline.

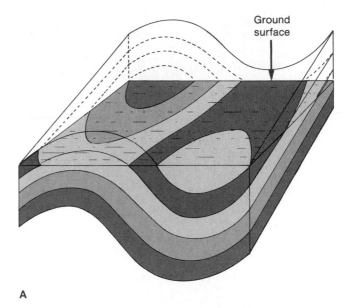

Ground
surface

A

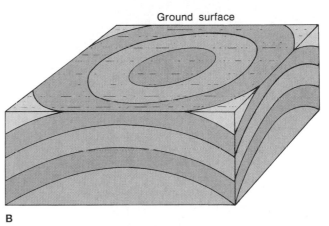

Ground surface

B

C

FIGURE 11.5 Plunging folds. (A) A simply plunging anticline and syncline. (B) A dome. (C) An eroded dome. Sinclair, Carbon County, Wyoming.

(C) Photograph by J. R. Balsley, USGS Photo Library, Denver, CO.

(figure 11.5A). The plunge of the fold is described by the orientation of the axis. A special case of plunging fold is that in which the fold effectively plunges radially in all directions. The upwarped fold in this case is a **dome** (figure 11.5B); the corresponding downwarped fold is a **basin.**

When the amount of compression is great, the axial surface of a fold may be tilted close to the horizontal as the rocks are doubled back upon themselves. The result is a **recumbent fold** (figure 11.6). If the axial surface tilts past the horizontal, the fold is described as **overturned** (figure 11.6B). In such complex cases, erosion can produce unusual patterns of repeated rock units and relative ages. For example, figure 11.6C shows that erosion of this overturned anticline in sedimentary rocks produces a synform, but the rocks at the center of the trough are the oldest in the sequence, as would be appropriate for an anticline.

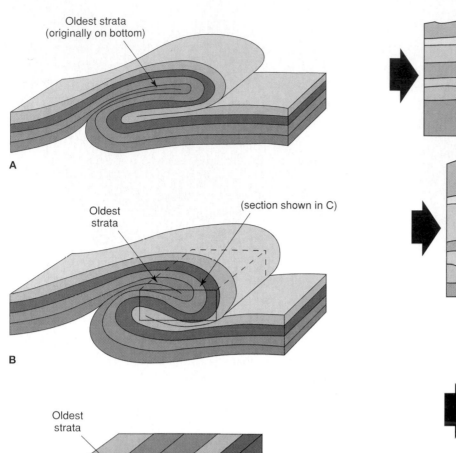

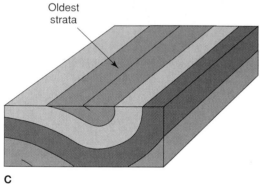

FIGURE 11.6 Extreme crustal shortening produces recumbent or overturned folds (see also figure 11.7). (A) A recumbent anticline. (B) An overturned anticline. (C) Erosion of the overturned anticline reveals a synformal structure, but the relative ages of the rocks are inconsistent with a syncline.

Any kind of rock can be folded, just as rocks of all kinds can be faulted. The emphasis here has been on layered sedimentary rocks mainly because fold structures are particularly easy to see and describe in such rocks. Likewise, the age relationships described for anticlines and synclines are observed only for rocks spanning an age progression in an orderly way, such as layered sediments or a sequence of lava flows. In principle, a massive rock unit such as granite pluton can be folded or otherwise deformed, but the rocks in it are all the same age, and the nature of the folding may be much harder to see.

Folding due to compression produces a net horizontal shortening of the crust, as well as some thickening (figure 11.7).

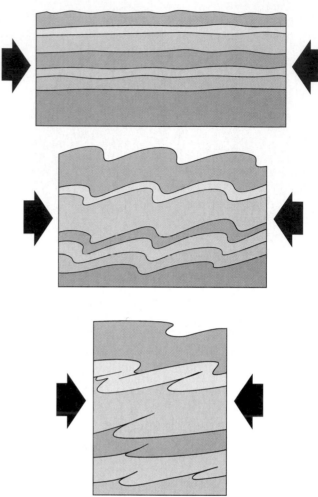

FIGURE 11.7 Folding as a mechanism of crustal thickening.

Plate collision is a source of immense compressive stresses, so it is no accident that major fold belts are associated with collisions at continental margins. Folded mountains are a common result, as will be explored further in the section on mountain building, later in this chapter.

Joints and Faults

As noted in chapter 10, brittle rocks fracture in response to stress. If the rocks on one side of the fracture do not move relative to the rocks on the other side, the planar fractures are described as **joints.** The columnar jointing of a basalt flow is one example: the tensional stress created by contraction during cooling causes fracturing of the flow in place but not displacement or rearrangement of the columns. Jointing can also result from compression. Often, the imposition of stress on a regional scale results in multiple fractures with the same orientation, a **joint set** (figure 11.8).

When there is relative movement between rocks on either side of the fracture, the fracture is then a **fault,** as discussed in chapter 10. The nature of the movement, or type of faulting, may be more important in interpreting the geology and tectonics than is the mere existence of the fault. Faults

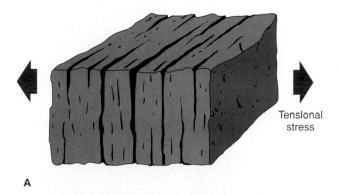

Tensional
stress

A

B

FIGURE 11.8 A joint set is made up of parallel planar fractures. (A) A joint set (schematic). (B) Joint sets in shale. Bedding dips toward right of photograph; joint sets are nearly perpendicular to bedding planes. Shoshone County, Idaho.
(B) Photograph by F. C. Calkins, USGS Photo Library, Denver, CO.

may be described further in terms of the steepness of the fault plane and the direction of relative movement. The relative movement may be horizontal, vertical, or a combination of the two. The orientation and sense of displacement along the fault, in turn, indicate the nature of the stresses involved.

Tension may produce very steeply sloping faults, as brittle rocks fracture. With further tension, vertical movement along the faults can occur. This is common in rift zones. If the crust is bulged up and stretched—for example, by a subcontinental hot spot—the tension associated with crustal stretching over the bulge often causes high-angle (steeply dipping) normal or reverse faults to form. Some blocks may settle into the space created by the stretching; others may be pushed up by the buoyancy of rising magma. The role of such *block faults* in mountain building is discussed later in the chapter.

FIGURE 11.9 The San Andreas fault, a transform fault. Note the offset of stream channels across the fault, which runs nearly horizontally across the center of the picture.
Photograph by R. E. Wallace, USGS Photo Library, Denver, CO.

A transform fault, like the San Andreas (figure 11.9), reflects the action of shear stress in the horizontal plane: one side of the fault is sliding past the other, with essentially no vertical movement. Again, we can relate the fault displacement to the underlying stresses, which here result from relative movements of plates over the earth's surface.

Compression acting on brittle rocks produces shallowly sloping faults, rather than folds. Such a fault is a **thrust fault,** in which the rocks on one side of the fault are thrust up and over those on the other (figure 11.10). Thrust faulting on a regional scale may have the effect of thickening the crust by piling slices of rock atop one another. The direction of fault movement is a direct reflection of the orientation of the compression responsible for the faulting. When plates collide, we might expect thrust faulting to contribute to crustal shortening and thickening. The same plate collision might cause thrust faulting in shallower, colder, more brittle rocks and folding (with fold axes perpendicular to the compressive stress) in deeper, warmer, more plastic rocks.

Describing the Orientation of Structural Features

As noted already, what is important is frequently not only the kind of geologic structure present but also its orientation. This provides information about the directions from which the deforming stresses have come, which may, in turn, aid in tectonic interpretation. The orientation of planes can be described in terms of two parameters: *strike* and *dip*. **Strike** is a compass direction, measured parallel to the earth's surface. **Dip,** as the name suggests, is displacement downward from the horizontal, measured in degrees.

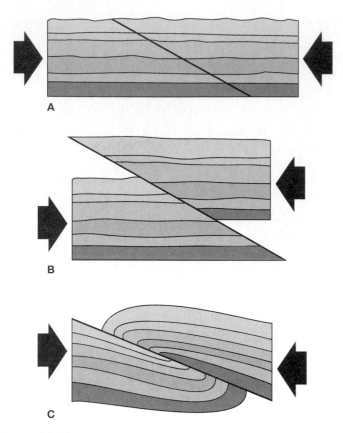

FIGURE 11.10 Thrust faulting, perhaps accompanied by some folding, also thickens and shortens the crust when compressive stress is applied.

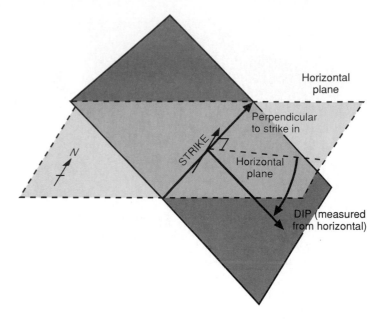

FIGURE 11.11 Strike and dip of a plane.

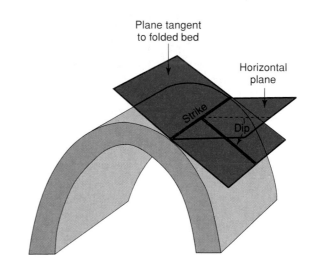

FIGURE 11.12 Strike and dip of a plane tangent to a folded bed.

Figure 11.11 illustrates the strike and dip of a plane. The strike of a plane is the compass orientation of the line of intersection of that plane with a horizontal plane. It is conventionally measured as degrees east or west from north. Values of strike, then, range from 0 to 90 degrees east or west. For example, the plane in figure 11.11 would be described as striking N 20°E. Dip, by definition, is measured perpendicular to strike. It is the vertical angle between a line perpendicular to strike in the horizontal plane and in the plane of interest. Dip angles therefore range from 0 degrees (horizontal) to 90 degrees (vertical). For any dip between these limits, the direction of dip is also specified. For the example shown in figure 11.11, the dip would be described as 45° SE. Note that for any (nonhorizontal) plane, one could measure a displacement from the horizontal in many different directions, and the measured angle would differ depending on the direction chosen. The angle measured perpendicular to the strike is the *maximum* of these values. Specifying that dip is to be measured perpendicular to the strike of the plane ensures consistency in reported data and also provides information on the maximum slope of the plane.

The concepts of strike and dip of a plane can be explored by immersing a sheet of stiff cardboard or a thin board part way into a basin of water. The strike of the board is the orientation of the line formed where the board meets the water surface; dip is measured at right angles to that direction. In an outcrop, the direction of dip of a sloping surface can be determined most readily by pouring water on the outcrop: the water trickles down the steepest slope, which is also in the direction of dip. The strike is the trend of the horizontal line perpendicular to the dip.

Many geologic features are planar: for example, sedimentary beds, igneous dikes and sills, metamorphic foliation. However, these planes may be deformed into nonplanar shapes that do not maintain the same orientation over broad areas. If a limestone bed is folded, for example, the bed will be curved, and its orientation becomes more complex to describe. Strike and dip can still be measured at any single point by considering the orientation of a hypothetical plane tangent to (just touching) the curved surface at that point (figure 11.12). To characterize the geometry of the structure

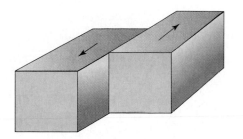

FIGURE 11.13 Strike-slip faulting. The San Andreas (figure 11.9) is a right-lateral strike-slip fault.

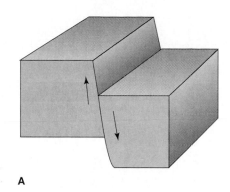

A

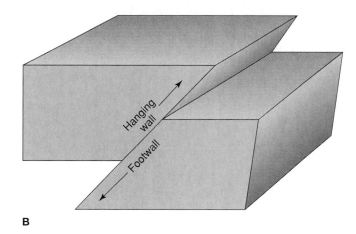

B

FIGURE 11.14 Dip-slip faulting: (A) normal and (B) reverse. With dip-slip faults, a hanging wall and a footwall can be identified.

fully, however, it is necessary to take strike and dip measurements at many points, not just one or two. Faults, too, are planar features to which strikes and dips can be assigned; so are the axial surfaces of folds.

Faults are sometimes described on the basis of the relationship between fault displacement and the orientation of the fault plane. For example, movement only in the horizontal direction corresponds to movement along the strike of the fault, which is then called a **strike-slip fault** (figure 11.13). Transform faults are one kind of strike-slip fault. Along simple strike-slip faults, features that were originally directly opposite one another across the fault are separated horizontally by the displacement. If the displacement is such that an observer facing the fault can find the matching features across the fault displaced to the right, the fault is described as *right-lateral*. A *left-lateral* fault shows the opposite movement. If the fault has cut and displaced one or more features that can be recognized as having once been continuous across the fault plane, the amount of displacement can easily be determined by the horizontal distance now separating matching features along the fault. The San Andreas is also a right-lateral strike-slip fault.

If the relative movement along the fault is entirely vertical, either up or down in the direction of the dip of the fault, the fault is a **dip-slip fault** (figure 11.14). Unless the fault plane is perfectly vertical, a *hanging wall* and a *footwall* can be distinguished (figure 11.14B). These can be identified by imagining oneself standing within the fault between the rock blocks: the footwall would be underfoot, below the inclined fault plane, with the hanging wall hanging overhead. If the hanging wall moves downward relative to the footwall, the fault is a **normal fault;** if the hanging wall moves upward relative to the footwall; it is a **reverse fault.** A thrust fault is also a reverse fault with a shallowly dipping (low-angle) fault plane. There are also faults along which the relative movement has both a horizontal and a vertical component. Such faults are termed **oblique-slip faults.**

CONTINENTS AFLOAT: VERTICAL MOVEMENTS

The lithosphere is less dense than the asthenosphere, and both continental and oceanic lithosphere therefore float buoyantly on the asthenosphere, as an iceberg floats on water. The overall density contrast is greater for continental lithosphere. The mantle portion of all lithosphere is very similar in density to the underlying asthenosphere: both are about 3.3 grams per cubic centimeter (g/cc). The iron-rich basaltic oceanic crust has a somewhat lower density of about 3 g/cc. The granitic continental crust is still lower in density, about 2.7 g/cc, so continental lithosphere is relatively more buoyant.

The effect of density on buoyancy can be demonstrated by comparing the behavior in water of an ice cube, a block of dry wood, a cork, and a Ping-Pong ball. The ice is only slightly less dense than water. Therefore, although ice floats, most of its volume is submerged below the surface. Cork and most woods are less dense than ice, so a higher proportion of each of these objects projects above the water surface. The density of the Ping-Pong ball is not far above the density of the air it encloses, and most of its volume bobs above the water. While much of the difference in elevation between the continents' surfaces and the sea floor can be ascribed to the much-greater thickness of the continental lithosphere, the continents also possess additional buoyancy resulting from the lower density of continental crust. Figure 11.15 illustrates the effects of thickness and density differences for a set of floating blocks.

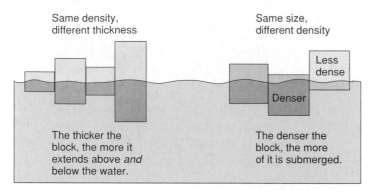

FIGURE 11.15 Blocks of similar density but different thickness float at different heights in water. Density differences affect the proportion of each block that projects above the water.

Isostasy

The tendency of lithospheric masses to float at elevations consistent with their relative densities is **isostasy,** named from the Greek for "same standing." The phenomenon of isostasy implies that, if the load on a region changes, the lithosphere shifts vertically in adjustment (figure 11.16). When the lithosphere and underlying asthenosphere are in **isostatic equilibrium,** the mass of the overlying column of rock above a certain depth in the mantle is the same everywhere. For example, as a mountain erodes, the thickness of the continental crust is decreased, and the total mass of the column of lithosphere beneath the mountain's surface is reduced. The lithosphere there will warp upward in compensation, allowing additional dense asthenosphere to flow in beneath the thinned crust. For a crude illustration of the phenomenon, float a stack of two or three flat slabs of wood in water; note the elevation of the bottom slab relative to the water surface; then remove the top slab and look again.

If lithospheric blocks over the asthenosphere behaved exactly like blocks of wood bobbing in water, isostatic equilibrium would readily be maintained. However, two factors slow the restoration of equilibrium in the lithosphere/asthenosphere system when it has been disturbed. One is the viscosity of the asthenosphere, its resistance to flow. Just as a pile of bricks does not immediately sink into warm asphalt, so the lithosphere does not immediately sink into the asthenosphere as it is loaded. Conversely, as erosion or the melting of an ice sheet unloads the lithosphere, the lithosphere does not immediately bob up in response. The second factor is the limited flexibility of the lithosphere. The thin, pliable rubber surface of an inner tube or water bed deforms easily when stressed and bounces back quickly when the stress is removed. Lithosphere tens of kilometers thick is far more rigid; the relatively cold crust is particularly stiff. Some time may elapse between loading or unloading and the corresponding deformation.

The slowness of isostatic adjustment is dramatically illustrated by the modern rebound of the Scandinavian region. That region was weighted down by thick ice sheets (about 2.5 kilometers, or 1.5 miles, thick) during the Ice Age. The ice sheets

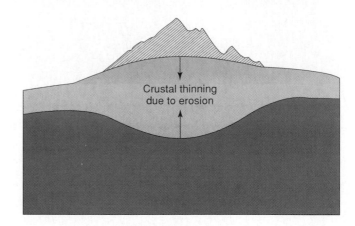

Crustal thinning due to erosion

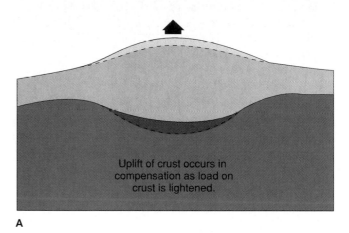

Uplift of crust occurs in compensation as load on crust is lightened.

A

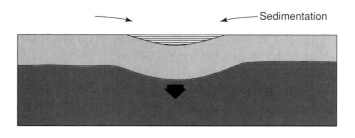

Sedimentation

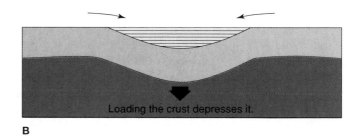

Loading the crust depresses it.

B

FIGURE 11.16 Isostatic compensation. (A) As a mountain erodes and the continent thins, the decrease in elevation is compensated by uplift of the whole lithosphere, decreasing the depth of the root below. (B) As a basin is filled, the thickened crust sinks.

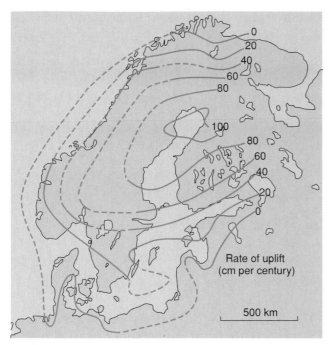

FIGURE 11.17 The rebound of Fennoscandia illustrates the slowness of isostatic adjustment.
From B. Gutenberg, Physics of the Earth's Interior. *Copyright © 1959 Academic Press, Orlando, FL. Reprinted by permission of Stephanie Sugar and Arthur W. Gutenberg.*

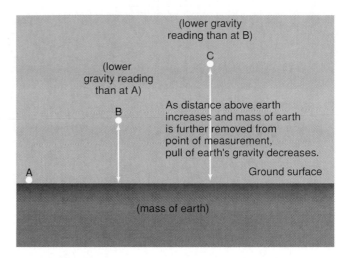

FIGURE 11.18 The effect of elevation on gravity.

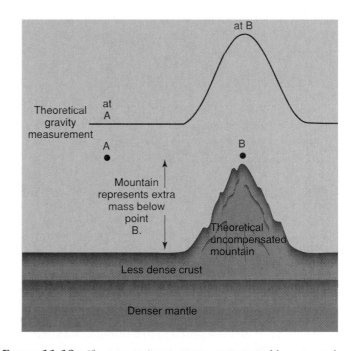

FIGURE 11.19 If a mountain were uncompensated by a crustal root, the gravitational pull atop a mountain should be greater than the pull at the same altitude where no mountain exists.

had completely melted by about 10,000 years ago. Yet the region is still rising, or rebounding, from the removal of that ice load (figure 11.17). Isostatic adjustment is continuing, but slowly. Present uplift rates are 1 centimeter per year or less.

Gravity and Isostasy

The force of gravity attracts all matter on earth toward the earth's center. The strength of the gravitational attraction between two objects is a function of the mass of each and of the distance between them. If the earth were perfectly homogeneous and spherical, the gravitational force acting on an object at its surface would be the same at any point on the surface. The same would be true if the earth consisted of perfectly spherical, concentric shells of materials of different densities, provided that each shell were homogeneous. In either case, the same amount of mass, distributed the same way, would be pulling inward from any point on the surface. In practice, because the earth is not so uniform, there are small, lateral variations in the measured force of gravity from place to place, which are termed **gravity anomalies.**

One source of variation is the elevation at which the measurement is made (figure 11.18). The gravitational attraction between two objects decreases as the distance between them increases. As an object at the surface is raised or lowered, the total gravitational pull on it decreases or increases, respectively, as the distance from object to earth is increased or decreased. A gravity measurement made from an airplane

over some point on the earth's surface gives a lower value than does the corresponding measurement made at the surface. All gravity data are routinely corrected to their equivalent values as if measured at mean sea level, before any possible anomalies due to geology are considered.

Another possible source of local gravity variation is the additional mass represented by topographic variations (figure 11.19). That is, if the continents are assumed to be uniform in composition below a certain elevation—sea level, for instance—but the surface is not uniform in elevation,

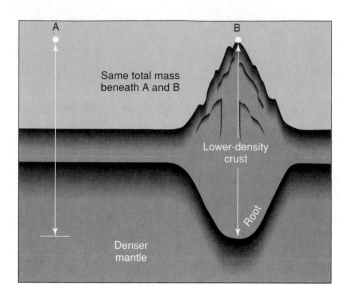

FIGURE 11.20 Gravity measurements corrected only for topography and for the mass of the mountains above sea level show an apparent mass defect, caused by the crustal root displacing denser mantle.

then the total mass in a vertical column beneath the surface should vary from place to place. Where there is more total mass, gravity should be a little stronger, and vice versa.

If the objective is to detect deep-crustal features, this kind of effect should be subtracted out. From sampling the surface or near-surface rock types, geologists can choose reasonable densities for the rocks of, for example, a mountain range. Then they subtract the gravity increment due to those rocks, correcting the data over a region to the values that would be measured at sea level if all rocks above sea level were stripped away.

The results provide strong evidence that, in most cases, the topography of the continental surface is mirrored by the profile of the base of the continental crust (Moho), that mountains are isostatically compensated by "roots" of crustal rock (figure 11.20). Mountain ranges typically show negative gravity anomalies when the raw data are corrected only for topography and for the rock masses above sea level. In other words, the pull of gravity is less than would be expected if the column of rock below sea level under the mountains corresponded in densities and thicknesses of layers to the rock column below low-elevation areas. But if the less-dense granitic rocks of the continental crust extend to greater-than-average depths beneath the mountains, displacing denser mantle rocks, the apparent mass deficiency is explained.

Given the densities of crustal and mantle rocks, geologists can calculate the size of the crustal root needed to compensate for a mountain range of particular size, or for any particular topography. They can then assume the appropriate compensation and further correct the gravity measurements on that basis.

Any anomalies that still remain can be explained in two principal ways. One is by the presence of localized, anomalous masses within the crust. A dense mafic pluton within the crust, for example, may cause a local positive gravity anomaly above it. A second kind of anomaly, of regional scale, may arise as a consequence of the time lag in isostatic readjustment. For example, in the case of the ice-depressed Fennoscandian region of northern Europe, the ice mass that caused the original downwarping is gone, but the lithosphere has not yet fully rebounded or adjusted isostatically (recall figure 11.17). This results in a negative gravity anomaly, an apparent mass defect, over the region. When isostatic adjustment is complete—when the crust has risen again to an elevation consistent with the new lighter load, and denser mantle material has moved in underneath the rising lithosphere—this residual gravity anomaly will be eliminated. Regional gravity anomalies, then, can be used to detect areas that are not in complete isostatic equilibrium.

MOUNTAINS AND MOUNTAIN BELTS

A *mountain,* by convention, is a distinct topographic high that rises 300 meters or more above the surrounding land and has a limited summit area (to distinguish it, for example, from a plateau). The topographic prominence of mountains tends to inspire investigation of and speculation about the nature of mountain ranges and how they form. One immediate observation is that mountains tend to occur in ranges or belts rather than in isolation. Forces and processes of sufficient scale to cause the amount of crustal thickening represented by a mountain a thousand or more meters high usually also have a regional impact, and they create not one mountain but many. The principal exception is a volcano built up over an isolated hot spot, and even in that case, as the overlying lithospheric plate moves, a trail of additional volcanoes will be formed over the hot spot.

The current topography of a mountain range may or may not closely reflect either the nature of the processes by which the mountains formed or the structures in the rocks in the mountain range. The sculpturing effects of water and ice not only tend to reduce the overall elevation of the land through time, they also shape and reshape the eroded rocks. The height and form of a mountain range are likely to be as much a function of the age of the mountain-building or crust-thickening events (old mountains are typically more severely eroded and of lower overall relief), and of the readiness with which the rocks erode, as of the nature of the mountain-building process.

Mountains form in a variety of ways, some by a combination of processes or with a variety of structures. Often, a single process or structure predominates throughout a range, simply because the rock types and the stresses to which the rocks were subjected, or the tectonic setting of the region, were the same throughout the range.

The relationship between volcanic activity and formation of mountains is easily visualized and is explored in some detail in chapter 4. Lava or pyroclastics, or both, pile up

FIGURE 11.21 Volcanic mountains: Mount St. Helens (foreground); Mount Rainier (rear).
Photograph by D. R. Mullineux, USGS Photo Library, Denver, CO.

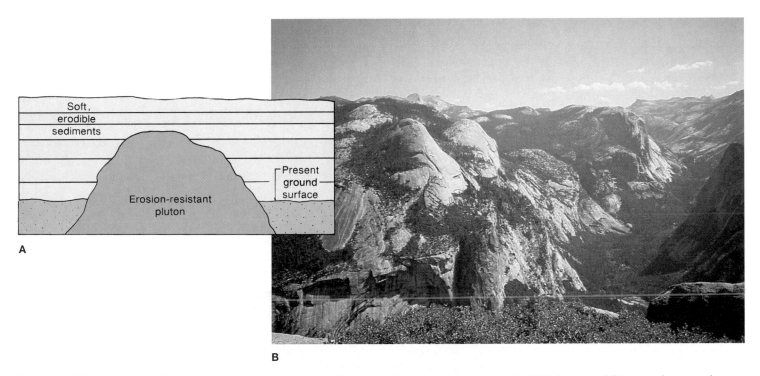

FIGURE 11.22 Formation of a mountain by subtraction of erodible rocks from above and around it. (A) Schematic. (B) Exposed granite domes in Yosemite National Park formed in this way.

around a volcanic vent, creating a topographic high at that spot (figure 11.21). The chains of volcanoes that form above subduction zones contribute to the formation of mountains on continents along convergent plate boundaries, although other processes contribute to crustal thickening there as well, as we will see later in the chapter.

Some mountains form, in effect, by subtraction (figure 11.22A). Magma invades a sequence of relatively easily weathered rocks and solidifies into one or more plutons. If the region is elevated so that it is subjected to erosion, that erosion removes the softer country rocks preferentially, and what is left is the durable, resistant, igneous core of pluton.

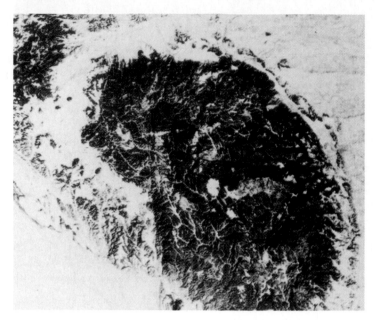

FIGURE 11.23 The Black Hills of South Dakota, formed by doming followed by erosion of sedimentary rocks to reveal a granite core (Landsat satellite photograph). © *NASA*.

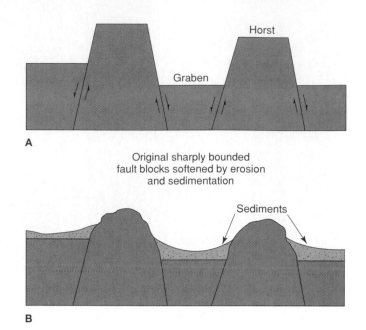

FIGURE 11.24 Fault-block mountains. (A) Formation of horsts and grabens creates fault-block mountains. (B) The topography is later smoothed by combined erosion and deposition.

The granite peaks of Yosemite National Park (figure 11.22B) and the White Mountains of New Hampshire are examples of residual mountains formed in this way. Relatively erosion-resistant sedimentary or metamorphic rocks can also form residual mountains; the Catskill Mountains of New York are an example.

Broad arching or doming on a regional scale may produce a topographic high that ultimately forms a mountain range. The Black Hills of South Dakota are mountains of this kind (figure 11.23). They are cored by ancient igneous and metamorphic rocks, uplifted at the center of the dome which resisted the erosion that stripped away the overlying blanket of sedimentary rocks.

As noted earlier, tension or compression on a regional scale can lead to the formation of large-scale joint sets and parallel faults. With continued stress and movement along the faults, crustal blocks are shifted up or down relative to one another (figure 11.24). A block dropped down relative to the adjacent blocks is a **graben;** a block uplifted relative to those on either side is a **horst.** If the relative vertical movement associated with the block faulting is significant, fault-block mountains are formed. A common characteristic of fault-block mountains, especially young ones, is that they rise sharply from the surrounding terrain, since they are bounded by steeply dipping fault scarps. The Teton Mountains are one of several ranges of fault-block mountains in the western United States (figure 11.25).

The great vertical relief associated with major mountain ranges—the Appalachians, the Andes, and many others—requires a degree of crustal thickening that is virtually impossible to achieve without some folding on a large scale. Fold-

ing is thus an important feature of all major mountain belts, although magmatic activity and faulting are invariably present also, to some extent. Differential erosion of different rock units within the folds occasionally causes the fold structures to stand out starkly (figure 11.26). Folding and thrust faulting are often interrelated, too: if rocks are not plastic enough, evolving folds may break and thrust faults may then be developed in the deformed rocks during compression.

MOUNTAIN BUILDING—A FIRST LOOK

The term **orogenesis** (from the Greek *oros,* meaning "mountain," and *genesis,* meaning "birth") is used to describe collectively the set of processes by which mountains are formed. It thus comprises a variety of folding, faulting, volcanic, plutonic, and metamorphic activities. An **orogeny** is the set of events occurring within a specific time period that has led to the formation of a particular mountain system.

All continents have regions that have remained geologically quiet or stable for a long time. ("Long," in this context, would typically mean hundreds of millions of years, or more.) Such a region is called a **craton.** Within many cratons are **shield** areas consisting of exposed Precambrian igneous and metamorphic rocks, which are flanked by younger stable areas. In North America, the continental shield is exposed principally in central Canada. Much of the central United States is also part of the North American craton, but the upper-crustal rocks are younger and more often sedimentary than igneous or metamorphic. Orogenesis, by definition, typically is confined to the margins of a craton, which often

FIGURE 11.25 The Tetons rise abruptly from the surrounding plains.

correspond to the continental margin. At the margin, accumulation of sediments, which later become incorporated into the developing mountain belt, commonly precedes orogenesis.

Sedimentation at the Continental Margin

Continental margins are classified as **active** or **passive,** depending on whether there is an active plate boundary at or near the margin. An active margin is one at which two plates meet; relative motion between the two plates creates seismic and perhaps volcanic activity. No such plate boundary exists at a passive margin, which is correspondingly tectonically quiet. (Active and passive margins will be discussed further in chapter 12.)

At a passive, tectonically quiet margin, typical sediments are a mix of relatively mature sands and shales eroded from the stable continental interior, and marine limestones precipitated from seawater. Thick sequences of sediments with well-developed, regular bedding can accumulate at passive margins. As sediments deepen, particularly on the oceanic crust, isostatic adjustment may depress the sedimentary basins further, allowing still greater thicknesses of sediment to accumulate. The Gulf Coast and Atlantic margins of North America are passive margins.

At an active margin—most often associated with ocean–continent convergence, as along the west coast of

FIGURE 11.26 The folded structure of the Appalachian Mountains near Harrisburg, Pennsylvania, is obvious in this satellite image. *© NASA.*

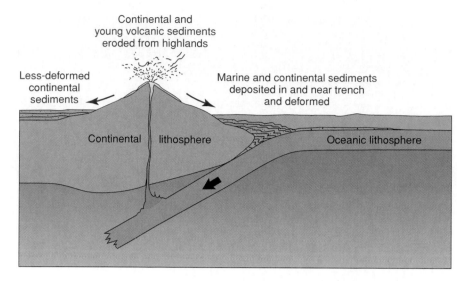

FIGURE 11.27 Sedimentation at a convergent margin.

South America or the northwestern margin of North America—the pattern of sedimentation is more complex, involving a greater diversity of sediment (figure 11.27). Preexisting continental rocks are eroded and the resulting sediments deposited in adjacent sedimentary basins. Magmatism above the subduction zone adds an additional supply of volcanic and plutonic material to the continent and may create a topographic high, from which sediments are shed both landward toward the craton and seaward toward the trench. Volcanic activity adds a volcanogenic component to the sediment. Erosion may also be more rapid in the higher-relief region of the magmatic belt, and therefore the sediments produced may be less mature chemically, less extensively weathered, and more voluminous than the continental sediments on a passive margin. The higher proportion of continental clastics reduces the relative importance of marine limestones in the sequence. The sedimentary basin nearer the continent will be less subject to deformation, while sediments to the seaward side and especially in the trench may be considerably deformed.

Orogenesis in Zones of Convergence

A passive margin does not undergo orogenesis until conditions change so as to transform it into an active margin. For example, plate movements may change in such a way that a subduction zone develops along the margin. The situation then becomes similar to that of the active margin previously described.

As convergence progresses, stress may crumple and fracture the rocks of the preexisting continent and the magmatic additions associated with subduction. Strong horizontal compression may cause tight folding and development of thrust faults in the continental crust, contributing to its thickening (figure 11.28). The continent may also be extended, as some of the sediments in the trench are wedged against the margin of the overriding plate. Here, too, compression results in reverse faulting, while metamorphism is associated both

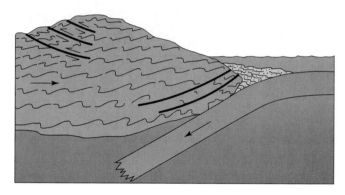

FIGURE 11.28 While oceanic lithosphere is subducted at a convergent margin, folding and faulting on the continent cause crustal thickening.

with the stress of convergence and with the heat of magma rising into the crust.

Where the margin is passive for some time prior to the initiation of subduction, a considerable period of sedimentation precedes orogenesis. Once subduction has begun, sedimentation and orogenesis proceed simultaneously thereafter for the duration of subduction.

The situation becomes more complex when the subducting plate carries with it a continent near its leading edge. As sea floor is consumed by subduction, the continent advances toward the subduction zone. The continent's buoyancy, however, prevents it from being subducted to any significant extent. Typically, after a period of compressive deformation, the two plates become **sutured** together, consolidated into a single unit that moves and acts as one plate, and subduction ceases.

If a continent lies at the advancing edge of the overriding plate, then, as noted in chapter 9, the result is continent–continent collision and the production of a belt of greatly thickened continental crust at the suture zone (figure 11.29). Isosta-

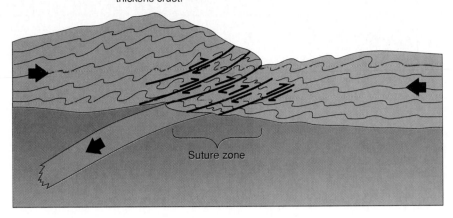

Continent-continent collision, with tight folding and thrust faulting, thickens crust.

Suture zone

FIGURE 11.29 A continent–continent collision builds unusually thick lithosphere.

tic considerations dictate that the thicker crust also rides higher above the asthenosphere, and particularly high mountains are thus formed. It is no coincidence that the world's highest peak (Mount Everest) is found in a mountain range formed in this way (the Himalayas). Continent–continent collision is believed to be the cause of major mountain ranges, such as the Urals, now found within cratonic regions. The Appalachian Mountains were formed when Europe and Africa collided with North America, prior to the opening of the Atlantic Ocean.

If the overriding plate is of oceanic lithosphere, with an island arc upon it fed by magma from a subducting plate, having oceanic lithosphere at its leading edge but a continent close behind, a similar suturing process can eventually occur. As the subducting oceanic plate is consumed, arc and continent converge, and the island arc then becomes sutured onto the edge of the continent. In this case, however, the thinner oceanic lithosphere of the old overriding plate may break out beyond the island arc, under continued compressive stress. That plate, being the less buoyant, may then start to be subducted in turn (figure 11.30). The same arc may then be supplied with magma derived from the plate of which it was once a part. The net effect of this process is to increase the size of the continent.

After the Orogeny

The rate of collision, orogenic deformation, and crustal thickening may outstrip isostatic adjustment to the new vertical distribution of more- and less-dense material. A common sequel to orogeny, and the accompanying cessation of compression, is slow uplift (and erosion). The uplift may contribute to the formation of fault-block mountains: the upper crust is stretched as uplift proceeds; it may fracture, and continued upward pressure may then push up horsts, from which erosion will carve complex mountain topography. Material eroded from the horsts is often deposited in the adjacent grabens, which contributes to the leveling of the surface through time.

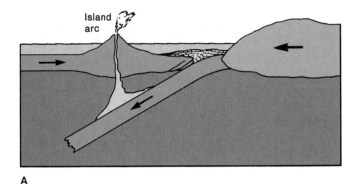

A

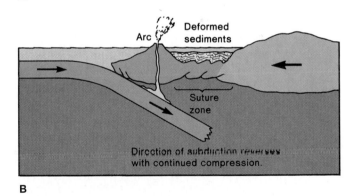

B

FIGURE 11.30 The direction of subduction reverses when a continent on the subducting plate arrives at a subduction zone. (A) Subduction of plate carrying continent brings arc and continent into collision. (B) Arc becomes accreted onto advancing edge of continent; plate behind arc is detached and begins to be subducted as convergence continues.

When the surface has been smoothed by erosion and isostatic equilibrium has been restored, the mountain belt has become *cratonized,* incorporated into the stable continental block. By the time cratonization is complete, the mountain belt formed by orogeny has largely been erased topographically. Only the deformed metamorphic and plutonic rocks

Box 11.1

New Theories for Old

PLATE TECTONICS AND GEOSYNCLINES

Before the advent of plate-tectonic theory, the prevailing theory used to explain mountain building was the geosynclinal theory. A **geosyncline** is a syncline of regional scale—hundreds or thousands of kilometers long. Geosynclines, acting as sedimentary basins at the margin of a continent, were believed to be the first phase of a mountain-building episode. Later elaboration of the model resulted in subdivision of the geosyncline into a *miogeosyncline,* on the edge of the continent proper, in which "clean" limestones, sandstones, and some shales were deposited, and a deeper *eugeosyncline* to the seaward side, characterized by the accumulation of a much more diverse or "dirty" sedimentary package, including more poorly weathered rock fragments and volcanic material (figure 1). With progressive deepening of the basin, a point was presumably reached at which melting of the most deeply buried sediments began in the eugeosyncline. The magmas thus formed rose to invade the rocks above; the geosyncline became unstable and began to deform. Ultimately, the sediments of the eugeosyncline were transformed into an igneous/metamorphic complex. Mountains were built, and the crust then stabilized, with the newly formed mountains representing a marginal extension of the continent.

Theories must be modified or abandoned in response to new or improved data. When the geosynclinal theory was developed, it represented the best, most coherent, most logical explanation of the data then available. At the time, radiometric dating was in its infancy; many of the geophysical

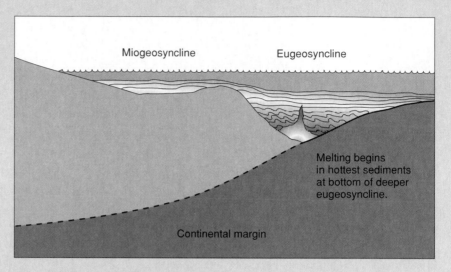

FIGURE 1   Eugeosyncline and miogeosyncline, part of a pre-plate-tectonics theory of mountain building.

techniques now used routinely were undeveloped. As these more powerful investigative tools were employed to gather more data, new theories had to be developed to accommodate the newer information along with the old. Negative gravity anomalies in trenches could not be explained by simple basin filling and sediment loading. Magnetic stripes on the sea floor needed explaining. Magnetic anomalies and radiometric dating together yielded polar-wander curves, tracers of continental drift. Fragments of weathered sea floor were found on continents (see next chapter). Eventually, the geosynclinal theory was essentially discarded, to be replaced by a more sophis-

ticated plate-tectonic model of orogenesis at convergent margins.

The ultimate orogenic model may remain to be developed. The discovery of suspect terranes (discussed later in the chapter) has suggested that further refinements are necessary. This is the nature of the science. It deals not with a static body of facts and truths but with an ever-growing body of knowledge. Where additional data are consistent with an established theory, confidence in the theory is increased. Inconsistencies suggest a need for modification. Irreconcilable inconsistencies may require that the old theory be discarded altogether, to be replaced by one that *can* account for all available data.

that formed the core of the belt, now exposed at the surface, testify to the orogenic activity of the past.

The advent of plate-tectonic theory has modified understanding of orogeny considerably. See box 11.1 for an earlier model of the process.

THE GROWTH OF CONTINENTS THROUGH TIME

A subject of considerable interest (and heated debate) is the extent and nature of the continental crust through time. Were the continental masses essentially created completely

as a result of the initial melting and differentiation of the primitive earth, and have they always existed in approximately their present forms and extent? With the development of paleomagnetic techniques, it has become apparent that continents have *moved* through time, but that does not address the question of the total mass of continental material present at any given time.

How Large Were the Early Continents?

The oldest reliably dated continental material is 3.8 to 4.2 billion years old, and some rocks of approximately that age exist on virtually every continent. Early geochronologic stud-

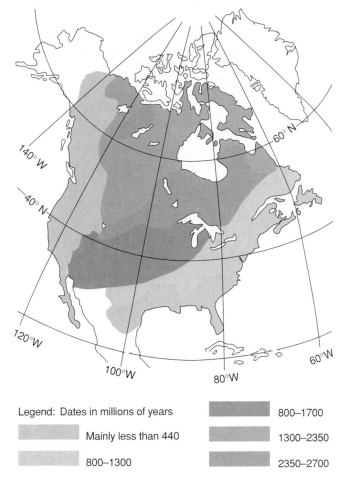

FIGURE 11.31 Pattern of younger ages toward the margins of North America.

Redrawn with permission. From P. M. Hurley and J. R. Rand, "Pre-drift continental nuclei" in Science, 164: 1231, June 1969. Copyright © 1969 by the American Association for the Advancement of Science, Washington, DC. Reprinted by permission of the publisher and authors.

Legend: Dates in millions of years

- Mainly less than 440
- 800–1300
- 800–1700
- 1300–2350
- 2350–2700

ies suggested that continents might show a pattern of an older shield region flanked by successively younger regions (figure 11.31). The chemistry of the younger rocks was also consistent with their derivation, in large measure, from mantle rather than from crustal material. These observations were interpreted to indicate that continents had grown continuously through time. Such an interpretation is also consistent with the observation that many orogenic processes have the net effect of adding material at the continental margins.

Other geochemical evidence suggests that the extent of depletion of the mantle in those elements that are concentrated in the crust has *not* increased comparably and continually through time. That would be an argument for a nearly constant volume of continental material over billions of years.

These lines of evidence can be reconciled through plate tectonics, considering the crustal recycling for which plate-tectonic activity is responsible. While some sediment is scraped off the down-going slab at a convergent margin, some continentally derived sediments are also subducted along with oceanic lithosphere at an ocean–continent convergence. That material is then remixed with the much larger volume of

upper-mantle material. Elements concentrated in the crust are returned to the mantle and at the same time diluted in it. Fresh magma extracted from that mantle shows predominantly mantle-like chemistry. Also, the rocks crystallized from such magma yield an age that reflects the time of last extraction of the magma from the mantle, even though some of the elements in those rocks previously spent some time in the crust.

The majority view at present, therefore, is that the most extensive creation of continental crust from the mantle occurred in the latter part of the Archean, between $3\frac{1}{2}$ and $2\frac{1}{2}$ billion years ago. The total volume of continental crust seems to have increased more slowly in the last $2\frac{1}{2}$ billion years. This more gradual continental growth has been accompanied by repeated recycling of continental material through the mantle, especially through plate-tectonic processes.

Suspect Terranes, Accreted Terranes

An interesting new twist on models of continental growth has been provided by the discovery, within the last two decades, of continental regions whose geology does not appear to be consistent or continuous with that of surrounding areas. Geologists use the term **terrane** to describe a region or a group of rock units of a common age, type, or deformational style. Recent studies of the geology of western North America have revealed an apparent jumble of juxtaposed, diverse terranes. Their rocks and histories appear unrelated except insofar as the rocks are now together in one place. The question arose whether some of these terranes might have formed far apart and then been added to the continent later. This would account for marked geologic differences between adjacent terranes.

Regions for which this was proposed were named **suspect terranes.** A suspect terrane that is confirmed to be a crustal fragment transported and then accreted onto the continent is more precisely described as an **accreted terrane.** Accreted terranes are often recognized, in part, because they are bounded by major faults. If the amount of transport or movement of the accreted terrane with respect to the continent of which it is now a part is shown to be large, the accreted terrane may be termed an *exotic terrane.* However, it is not always possible to determine accurately just how far an accreted terrane has traveled. North–south movement, across latitude lines, can be detected by paleomagnetism (recall that magnetic dip varies systematically with latitude), but east–west movement cannot be similarly recognized or quantified.

Paleomagnetic and fossil evidence has since confirmed that some of these accreted terranes have indeed traveled long distances to reach their present positions. The Yakutat block of northwestern North America has moved thousands of kilometers over the last 45 million years (figure 11.32) to collide with the continent in the Gulf of Alaska region. Another example is the terrane called Wrangellia (figure 11.33). The data suggest that, before it was added to the western margin of North America 100 million years ago, Wrangellia traveled nearly 6000 kilometers, probably from somewhere south of the equator. Since its accretion onto the continent,

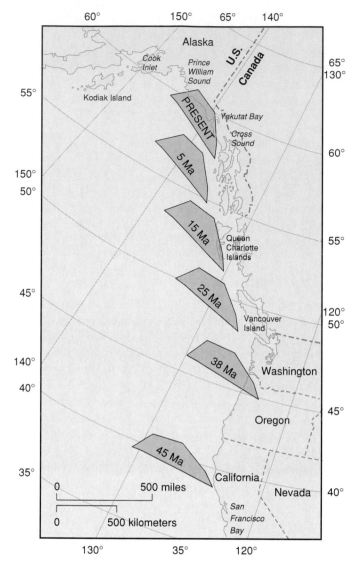

FIGURE 11.32 Path of the Yakutat block over the past 45 million years.

After T. R. Bruns, "Model for the Origin of the Yakutat Block . . ." in Geology *11, December 1983, figure 3, page 720, Geological Society of America.*

FIGURE 11.33 Wrangellia.

After U.S. Geological Survey 1982 Annual Report.

Wrangellia has been further broken up and the fragments displaced by younger strike-slip faulting. As shown in figure 11.33, further data suggest that a broad band of rocks along the western margin of the continent consists of accreted terranes, formed separately from the continent and later attached to it. The full interpretation of the geology of the region is going to be considerably more complex than was formerly believed. Similar mosaics of accreted terranes are now being identified in mountain belts worldwide.

The origin of the small landmasses making up accreted terranes is not yet known. There may be several sources. Some accreted terranes could have originated as large island-arc systems. Others may have been small blocks of continental crust separated by continental rifting or strike-slip faulting from a larger continent, much as a slice of North America west of the San Andreas fault is being propelled northward and East Africa is being separated from the rest of Africa. Oceanic plateaus, described in chapter 12, some underlain by crustal roots up to 40 kilometers thick, are clearly fragments of continental crust. These may have formed during the breakup of Pangaea. As plate movements continue, many of these plateaus are destined to collide eventually with larger continental landmasses, perhaps to become future accreted terranes.

SUMMARY

Deformation of the continental crust takes the form of folds (plastic deformation) and faults (brittle deformation). The nature and orientation of these structures provide information on the kinds and orientations of the stresses that formed them. Both folds and thrust faults can result in both horizontal crustal shortening and crustal thickening.

The lower density and greater thickness of the continental crust relative to oceanic crust give continents greater buoyancy and account for their relatively high relief. When a continental block is in isostatic equilibrium, the extra relief above sea level is compensated by a low-density root extending into the mantle. When erosion, orogenesis, glaciation, or other processes change the distribution of crustal mass, isostatic adjustment occurs, with the lithosphere rising or sinking as appropriate until equilibrium is restored. Gravity measurements, suitably corrected, can aid in assessing the present extent of isostatic adjustment and in recognizing anomalous mass distribution within the crust.

Large mountain ranges are formed through orogenesis, which comprises deformation, metamorphism, and magmatic activity. The present topography of individual mountains is, in most cases, due to erosion by water or ice rather than to the nature of the crustal structures involved in the mountains' formation. Orogenesis most commonly occurs at convergent plate boundaries, where new material is accreted or sutured onto the margin of a craton. Some of this added material, now found in the so-called accreted terranes, has been transported over considerable distances. That material is added to continental margins by successive orogenic events does not necessarily mean that there is a large net increase in the total mass or extent of the continents through time, for continental material is also cycled back into the mantle through subduction.

TERMS TO REMEMBER

accreted terrane 201
active margin 197
anticline 186
antiform 186
axial surface 186
axis 186
basin 187
craton 196
dip 189
dip-slip fault 191
dome 187
fault 188
geosyncline 200
graben 196

gravity anomalies 193
hinge 186
horst 196
isostasy 192
isostatic equilibrium 192
joint 188
joint set 188
limbs 186
normal fault 191
oblique-slip fault 191
orogenesis 196
orogeny 196
overturned fold 187
passive margin 197

plunging fold 186
recumbent fold 187
reverse fault 191
shield 196
strike 189
strike-slip fault 191
suspect terrane 201
suture 198
syncline 186
synform 186
terrane 201
thrust fault 189

QUESTIONS FOR REVIEW

1. Define the *strike* and *dip* of a plane.
2. What is the distinction between *joints* and *faults?* Between a *thrust fault* and a *normal fault?*
3. Describe two processes by which continental crust may be thickened.
4. How are *anticlines* and *synclines* distinguished when they occur in sedimentary rock sequences?
5. Under what circumstances do overturned folds develop?
6. Explain the concept of *isostatic adjustment,* giving an example. Why is such adjustment not instantaneous?
7. How are the crustal roots of continents detected?
8. Name and briefly describe three kinds of mountains.
9. Define the following terms: *craton, shield, active margin, passive margin.*
10. Summarize the principal orogenic processes that occur at a continental margin with an adjacent subduction zone.
11. What is a *terrane?* A *suspect terrane?* An *exotic terrane?*

For Further Thought

1. The identification of suspect terranes is difficult, as demonstrated by their relatively recent recognition. Suppose that you have found a batholith adjacent to a sequence of interlayered sedimentary and volcanic rocks. What features or properties might you look for to decide whether or not these are two distinct terranes, formed as different continental blocks, that have become juxtaposed?

2. In general, would you expect shield regions to be more or less likely to be in isostatic equilibrium than continental margins? Why? Would you expect differences in the extent of isostatic equilibrium between active and passive margins? Explain.

Suggestions for Further Reading

Bruns, T. R. 1983. Model for the origin of the Yakutat block, an accreting terrane in the northern Gulf of Alaska. *Geology* 11 (December): 718–21.

Condie, K. 1989. *Plate tectonics and crustal evolution.* 3d ed. New York: Pergamon Press.

Davis, G. H. 1984. *Structural geology of rocks and regions.* New York: John Wiley and Sons.

Hobbs, B. E., Means, W. D., and Williams, P. F. 1976. *An outline of structural geology.* New York: John Wiley and Sons.

Howell, D. G. 1989. *Tectonics of suspect terranes.* New York: Chapman and Hall.

Hsu, K. J. 1982. *Mountain building processes.* New York: Academic Press.

Lisle, R. J. 1988. *Geological structures and maps.* New York: Pergamon Press.

Lyttleton, R. A. 1982. *The earth and its mountains.* New York: John Wiley and Sons.

Meissner, R. 1986. *The continental crust: A geophysical approach.* Orlando, Fla.: Academic Press.

Miyashiro, A., Aki, K., and Sengor, A. M. C. 1982. *Orogeny.* New York: John Wiley and Sons.

Raymo, C. 1982. *The crust of our earth: An armchair traveler's guide to the new geology.* Englewood Cliffs, N.J.: Prentice-Hall.

Sharma, P. V. 1976. *Geophysical methods in geology.* New York: Elsevier.

Tarling, D. H., ed. 1978. *Evolution of the earth's crust.* New York: Academic Press.

Weyman, D. 1981. *Tectonic processes.* London: Allen and Unwin.

Windley, B. 1984. *The evolving continents.* 2d ed. New York: John Wiley and Sons.

12

THE OCEAN BASINS

A scuba diver can explore the shallow ocean basins, as in this underwater limestone cave in Australia; specially designed submersible research vessels can go far deeper, allowing investigation of midocean ridges. © Darryl Torckler/Tony Stone Images

INTRODUCTION

For many centuries, the deep ocean basins were virtually unknown territory—dark, cold, inaccessible. That changed, however, with the development of echo-sounding and seismic methods for exploring the topography and structure of the sea floor. Later tools included sampling devices capable of collecting cores of sediment from the sea floor and dredging rocks off the bottom and cameras to photograph the features of the deep. Still more recently, submersible research vessels, like miniature submarines, have allowed scientists to observe, sample, and photograph such features as ridge systems directly. This chapter begins with a brief discussion of the principal structural features of the ocean basins and their relation to plate-tectonic processes. It then proceeds to shallower depths to consider the patterns of sediment distribution on the sea floor and principal aspects of ocean circulation.

TOPOGRAPHIC REGIONS OF THE SEA FLOOR

If the oceans were drained of water, a complex topography would be revealed (figure 12.1). Far from being featureless, as they once were thought to be, the oceans contain mountains and valleys that rival in scale similar features on the continents. The sea floor can be subdivided into several distinct physiographic, or topographic, regions: the abyssal regions (broadly defined, the "deep ocean," where the bottom lies below 1000 fathoms—6000 feet, or nearly 2 kilometers), the continental margins, and the oceanic ridge systems. Figure 12.1 reveals additional smaller topographic features—including mountains, canyons, and plateaus—which are described separately later in the chapter. The characteristics of the continental-marginal zones differ, depending on whether those margins are active or passive.

Abyssal Plains

The **abyssal plains** are essentially flat areas occupying what is also called the "deep ocean basin," away from continental margins and ridges. Except for the trenches, the abyssal plains constitute the deepest regions of the sea floor. Although the plains themselves are quite level, they are not featureless (figure 12.1). Often, they are dotted with either **abyssal hills,** low hills rising several hundred meters above the surrounding plains, or the higher *seamounts*. These features, of volcanic origin, are described further later in the chapter, as they are not unique to the abyssal plains. Some abyssal plains are also cut by channels resembling shallow stream channels. The channels appear to be related to, or extensions of, the submarine canyons of many continental margins (also described

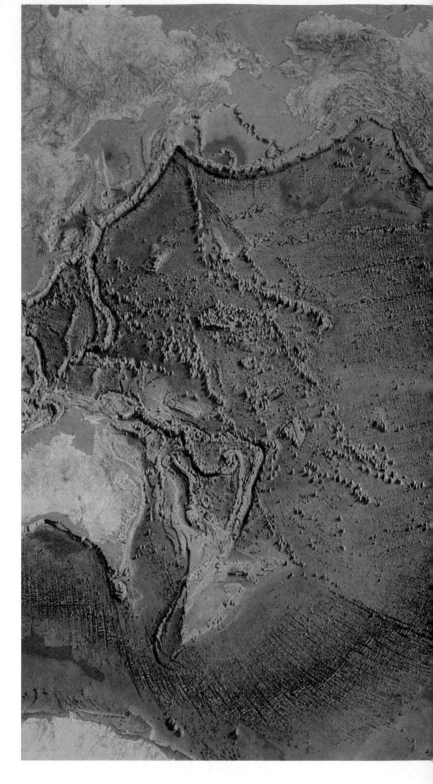

FIGURE 12.1 Physiographic map of the world ocean floor.
The whole-world map from "World Ocean Floor Panorama" by Bruce C. Heezen and Marie Tharp. Copyright © 1977 Marie Tharp.

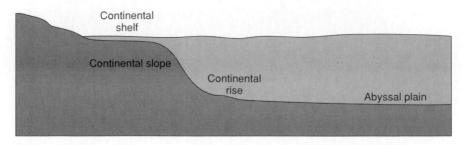

FIGURE 12.2 Cross section of the topography of a passive continental margin, showing continental shelf, slope, and rise.

later in the chapter). If so, they are probably carved by the same density currents, although it is difficult to understand how sea-bottom currents could continue to flow across nearly flat plains with sufficient velocity to cause much erosion.

Passive Continental Margins

A *passive margin,* as noted in chapter 11, is seismically and volcanically quiet, not associated with an active plate boundary. There is no relative motion between plates here, no collision, no subduction; continent and sea floor belong to the same plate at a passive margin. A generalized topographic profile across a passive continental margin shows that the marginal region consists of several zones distinguished topographically on the basis of slope (figure 12.2).

Immediately offshore from the continent is the **continental shelf,** a nearly flat, shallowly submerged feature. The average slope of the world's continental shelves is about one-tenth of a degree, or 10 feet of drop per mile of distance offshore. This accounts, in part, for the generally shallow waters above continental shelves, the majority of which lie no deeper than 100 fathoms (600 feet) at their outer limits. The average width of all continental shelves is about 70 kilometers (40 miles); the shelves at passive margins are systematically broader than those at active margins, with a few passive-margin shelves extending nearly 1000 kilometers (600 miles) offshore from the adjacent continent. Onshore from a passive margin one commonly finds wide beaches, broad coastal plains, and relatively eroded topography: mountain building and volcanism are well in the past. The Atlantic margin of North America is an example of a passive margin.

The continental shelves are a natural site for the accumulation of sediment derived from the continent, and many shelves have quite flat surfaces as a consequence of the deposition of layer upon layer of sediment over the underlying bedrock. This is not true of all shelves. During the last ice age, worldwide sea levels were lowered to such an extent that much of the now-submerged continental shelves was dry land. Rivers now terminating at the present shoreline flowed out tens or hundreds of kilometers across the continental shelves, carving valleys and depositing sediments just as they do now on the exposed continents. In regions directly subjected to glaciation, the ice sheets themselves carved up the

shelves and left behind deposits of sand, gravel, and boulders that now give the shelves added topographic relief.

The outer limit of the continental shelf is defined by the abrupt break in slope to the steeper **continental slope.** The angle of the continental slope varies widely, from nearly vertical to only a few degrees. The continental slope is the principal transition zone from near-continental elevations to deep-ocean depths. At the bottom (outside) edge of the continental slope, the topography again flattens out, the slope decreasing once more to less than 1 degree. The depth at which this occurs is highly variable, ranging from 1.5 to 3 kilometers below sea level. The region beyond this second break in slope is the **continental rise.** The sea bottom continues to deepen across the rise until the deepest, flat abyssal plains are reached. (The continental rise in some areas is not well developed, and in these cases, the continental slope terminates in the deep-ocean plains.)

The nature and formation of passive continental margins can be understood in plate-tectonic terms. Passive margins are believed to have begun as continental rift systems, torn apart by mantle plumes and/or convection systems welling up from underneath (figure 12.3A). The continental lithosphere is stretched and thinned during this process. It ultimately fractures, along long and deep faults perpendicular to the direction of stretching, and with continued tension, grabens form along the rift, making a long rift valley. This is now occurring in East Africa, which, if rifting continues, will eventually be split off from the rest of the continent. Already, the floors of some of the graben-formed valleys of the East African rift system lie below sea level. They remain dry only because mountains presently block the ingress of the sea.

Eventually, rifting proceeds to the point that seawater floods the rift valleys (figure 12.3B). While the new seas remain shallow, rapid evaporation may cause deposition of salt in these basins. This is the probable origin of the thick salt beds now buried along many continental margins.

With further tension, the rifting of the continent is completed (figure 12.3C). The rift valley is now floored by oceanic crust; an oceanic spreading ridge divides the fragments of continent; a narrow strip of true ocean basin lies between the separated continental masses. The Red Sea is in this stage of development. With continued spreading, the ocean widens, as the North Atlantic has following the breakup of Pangaea.

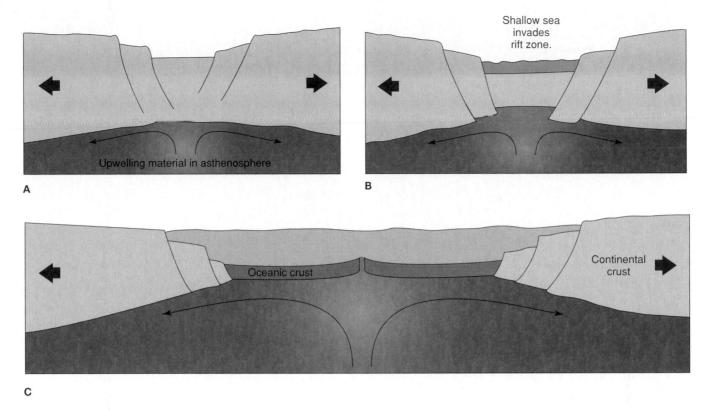

FIGURE 12.3 Formation of a passive margin. (A) The continent is stretched and fractured. (B) As rifting progresses, a shallow sea forms. Evaporation may leave salt deposits in the rift zone. (C) When rifting is complete, oceanic crust is formed between the fragments of continent.

Throughout this process, the newly formed continental margins are passive in the sense that there is no subduction and therefore none of the associated seismic and volcanic activity. Limited volcanism does occur as magma works its way up through the fractured continent, and earthquakes are associated with the down-dropping of the grabens as well as with the spreading ridge when it has evolved. However, once the spreading ridge is formed, it becomes the active plate boundary in the system. Where the oceanic and continental lithosphere meet at each margin is not a plate boundary; on a given side of the ridge, ocean and continent belong to the same plate. The passive margin is created by rifting, but once the intervening ocean is formed, further rifting is restricted to the seafloor-spreading ridge; hence the subsequent tectonic quiescence of the continental margin so formed.

The lithospheric thinning and block-faulting together account for the thinning of continental crust and the deepening of the ocean seaward from the continent. The exact geometry of the faulting on a given margin exercises a fundamental control on the margin's topography. For example, the steep continental slopes at some passive margins are probably fault scarps formed during the development of the margin. The topography is then modified by sedimentation. The extent to which sedimentary blankets soften the contours of the margin is principally a function of the quantity of sediment supplied from the continent. If the accumulated

sedimentary sequence is thick, its weight may continue to depress the margin and deepen the sedimentary basin, as may be occurring in the Gulf of Mexico.

Active Continental Margins

Active margins are those at or near the subduction zones; they are "active" in a tectonic sense. Like passive margins, active margins have continental shelves, but the shelves adjacent to active margins are narrower than those at passive margins. Onshore from an active margin, young mountain ranges and volcanoes are common, topography may be more rugged than is typical at a passive margin, and earthquakes rattle the land. The continental slope of an active margin is characteristically steeper than that of a passive margin, increasing in slope with depth. The slope of an active margin terminates not in a rise but in a **trench,** a long, deep, steep-walled valley trending approximately parallel to the continental margin. The locations of principal oceanic trenches are shown in figure 12.4 (see also figure 12.1). Note that trenches occur both at active continental margins and at ocean–ocean convergence zones, parallel to island arcs.

The trenches represent the greatest ocean depths. The deepest trenches plunge to more than 10½ kilometers (6½ miles) below sea level, and their depths below the sea floor rival the elevations of the highest mountains on the

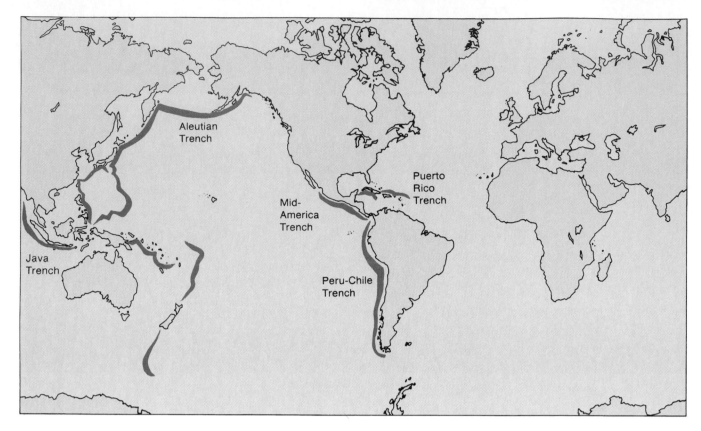

FIGURE 12.4 Locations of principal trenches on the sea floor.

continents. Trenches mark a plate boundary at which subduction is occurring and serve as sites of sediment accumulation if there is adjacent land to act as a sediment source. Because most subduction zones occur in the Pacific Ocean, the trenches are concentrated there also. The convergence and subduction signified by those trenches are slowly shrinking the Pacific Ocean basin. The Atlantic basin, expanding with the spreading of the mid-Atlantic ridge, correspondingly lacks such trenches.

Before the development of plate-tectonic theory, the trenches were a major puzzle to marine geologists. No erosional mechanism for forming them could be imagined. Trenches are characterized by negative gravity anomalies, which indicates that they are significantly out of isostatic equilibrium. However, it was difficult to picture a large volume of low-density lithosphere below the trench, and certainly there was no evidence for this. (In fact, the lithosphere is apparently depressed or held down by the stresses attendant on plate convergence, and perhaps by the weight of the cold down-going slab.) Trenches also show abnormally low heat flow, which is now attributed to the presence of the cold descending slab of lithosphere. The existence of that slab is further supported by the pattern of seismic activity at trenches: earthquake epicenters are located not only at the trench proper but along a plane dipping away from the trench into the mantle, termed a **Benioff zone** (figure 12.5). Earthquakes

occur there because brittle lithosphere is being subducted under another plate. The existence and orientation of the Benioff zone is further evidence of the subduction that accounts for the presence of the trench.

Certain specialized active margins arise in unusual tectonic settings. Subduction along the western edge of North America has caused a seafloor ridge system to be partially overridden by the continent. A portion of this margin, along the California coast, is now a transform fault—the San Andreas—separating offset segments of a spreading ridge off the coast of Oregon and Washington from other spreading segments in the Gulf of California.

Oceanic Ridge Systems

Although the zones of active spreading are conventionally drawn as lines on a map, an oceanic ridge system is a broad feature (recall figure 12.1). The rift and its associated structures span wide areas of the sea floor. Two principal topographic features may be observed: the sea floor slopes away from the ridge crest symmetrically on either side and there is a valley at the very center of the ridge.

The valley is a **rift valley,** similar to a continental rift valley and formed in much the same way. The tensional stresses of rifting (spreading) cause deep, high-angle faulting, and with spreading, grabens bounded by these faults drop

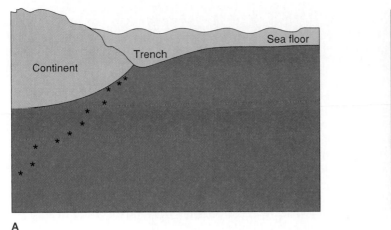
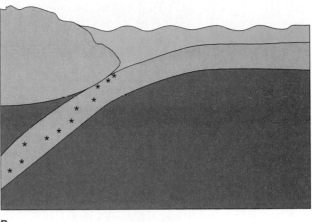

A **B**

FIGURE 12.5 The Benioff zone of progressively deeper earthquakes dipping away from a trench along the subducted slab. (A) The data. (B) The interpretation in plate-tectonic terms.

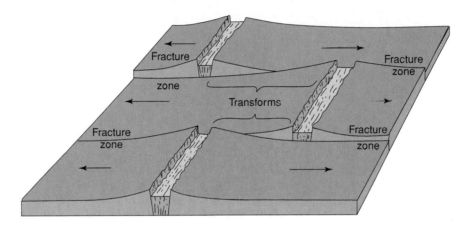

FIGURE 12.6 Fracture zones are the seismically quiet extensions of transform faults.

into the gap. The decreasing elevation of the sea floor away from the ridge was once thought to be due to relative uplift of the central portion of the ridge by upwelling mantle material. However, calculations have shown that the profile can be explained by thermal contraction. Newly formed sea floor at the ridge, though solid, is warm. As it cools, it contracts. The cooling and contraction of the oceanic lithosphere as it moves away from the ridge at which it formed account for its declining elevation. That the sea floor is somewhat elevated even tens or hundreds of kilometers away from the rift is an indication of the slowness of the cooling, a function of rocks' poor heat-conduction properties. Recall also from chapter 9 that the production and rifting of new sea floor at the ridge leads to symmetric patterns of magnetic anomalies across the ridge.

Chapter 9 already described the distribution of seismicity along a rift system and noted that earthquakes are concentrated along the ridge proper and the transform faults between offset ridge segments. The fractures formed

between offset segments do not disappear when plate movement has carried them beyond the spreading ridge (figure 12.6). They become seismically quiet where both plates move in the same direction, but the fractures remain, and there may be large scarps across the fractures. The fracture systems may be detectable outward from the rift over most of the region of elevated topography.

Along the ridge crest, seawater seeps into the young, hot lithosphere through fractures in the crust. The water is warmed as it circulates through the warmed crust and reacts with the fresh seafloor rocks. It may escape again at **hydrothermal** (hot-water) **vents** along the ridge, rising up into the overlying colder seawater (figure 12.7). Some hydrothermal vents are crusted with minerals, once dissolved in the circulating waters and later deposited as the temperature and chemistry of the hydrothermal waters changed upon reaction with uncirculated seawater. Some of these minerals are possible future mineral resources (see chapter on mineral and energy resources).

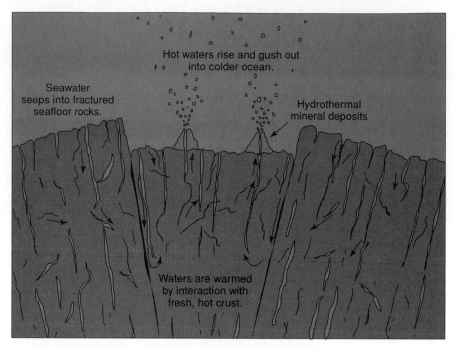

FIGURE 12.7 Seawater circulating through new sea floor at a spreading ridge produces hydrothermal vents.

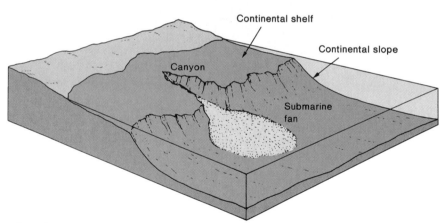

FIGURE 12.8 Topography of a submarine canyon.

he crews of submersible vehicles exploring hydrothermal vents made a particularly surprising discovery: life. In some areas, complex communities, including such higher organisms as giant clams and tube worms (of species previously unknown), were found. How could they live? Similar organisms in near-surface waters derive their energy ultimately from plants. But no sunlight to stimulate plant growth reaches the ridge vents, and little organic matter rains down from above. Bacteria seem to be a key part of the answer. The warm waters around the vents teem with bacteria, which derive their energy from chemical reactions involving compounds released at the vents (for example, hydrogen sulfide, H_2S). Some of the higher organisms feed directly on the bacteria; others contain bacterial colonies that produce the organic carbon the animals need to grow.

OTHER STRUCTURES OF THE OCEAN BASINS

A variety of structural features—submarine canyons (distinct from the trenches), hills of various sizes, and shallowly submerged platforms—are found in the ocean basins.

Submarine Canyons

A notable feature of many continental slopes, at both active and passive margins, is the presence of deep, steep-walled **submarine canyons** that cut across the shelf/slope system (figure 12.8). These canyons are usually V-shaped, sometimes winding; sometimes, they have tributaries on the continental shelves. Typically, they enlarge and deepen seaward and extend to the base of the continental slope, where they often terminate in a fanlike deposit of sediment resembling an alluvial fan (see chapter 14). They are cut into hard rock as well as into soft sediment.

The origin of submarine canyons has long been debated. A few are continuous with, and appear to be extensions of, major stream systems on land. Logically, it might be supposed that they were cut by those stream systems at a time when the continental shelves were exposed by a lowering of sea level. However, the canyons dissect not only the continental shelf but, in many cases, the whole length of the continental slope as well. Streams stop downcutting where they flow into the sea, and it seems highly unlikely that sea level would ever have been lowered to the base of the continental slope. It is therefore impossible to attribute these canyon systems entirely to rivers. Moreover, many submarine canyons have no corresponding stream systems on land; they seem to have formed independently of stream action. Another hypothesis attributed the canyons to submarine landslides, but this idea was not borne out by subsequent experiments and measurements; nor would it easily explain canyons in solid rock.

A possible mechanism of canyon carving was suggested, in part, by observations following an earthquake off the Grand Banks of Newfoundland in 1929. Several transcontinental telephone cables that had been run along the continental slope were broken. Furthermore, many of the breaks occurred hours after the earthquake, and in apparent progressive sequence with increasing distance from the epicenter (figure 12.9). The overall time span was too great for the breaks to be attributed directly to seismic-wave action. The most plausible explanation is that the breakage was caused by a fast-moving **turbidity current,** a density current of suspended sediment and water rushing down the continental slope. Turbidity currents are analogous to pyroclastic flows: just as the latter are denser than air ash clouds that flow along the ground as a consequence of their density, so the turbidity current is a denser-than-water mass that flows along the ocean bottom.

Fast-flowing turbidity currents could be forceful enough not only to break underwater cables but also to carve canyons in rock or sediment. With the shallowing of slope at the continental rise, the currents would slow down and begin to deposit their sediment load. Individual beds of sediment at the bases of canyons often show graded bedding fining upward. This would be consistent with the sediments having been deposited by gradual settling out of suspension, beginning with the coarsest material. Multiple graded beds, one atop another, could reflect successive turbidity flows. A turbidity current could be formed as sediment is stirred up by an earthquake, or perhaps in conjunction with an undersea landslide. Since sediment is an essential component of turbidity currents, they are most common where there is an abundant sediment supply on the continental shelf.

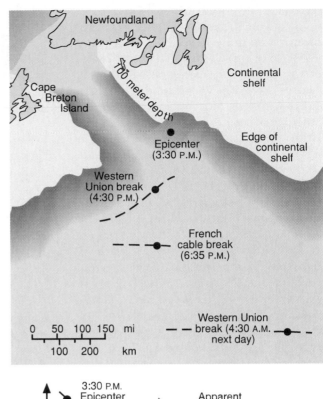

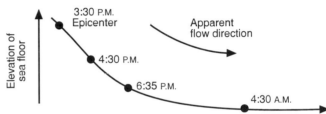

FIGURE 12.9 Cable breaks following the Grand Banks earthquake provided some of the strongest indirect evidence for turbidity currents. *From Francis P. Shepard, The Earth Beneath the Sea, p. 25. Copyright 1959, © 1967 Johns Hopkins University Press, Baltimore, MD. Reprinted by permission.*

One remaining puzzling feature is the rapid velocities—up to 60 kilometers per hour, or over 40 miles per hour—inferred for some turbidity currents. Such velocities exceed those measured for any undersea currents, though the measurements are admittedly limited in number. (Measuring turbidity-current velocities directly would require anticipating the time and place of such an event so as to have instruments in place.)

Turbidity currents may not account for all submarine canyons, but they appear to be the most plausible mechanism for forming the majority of the canyons. Indeed, those canyons not associated with continental river systems are difficult to account for in any other way.

Seamounts, Guyots, and Coral Reefs

The ridge systems and deep ocean basins are dotted with hills, some tall enough to break the surface and form islands, some not so high. These hills are generally of volcanic origin, whether associated with hot spots or with volcanism near the spreading ridges. **Seamounts** are those larger hills rising more than 500 fathoms (3000 feet, or about 1 kilometer) above the surrounding ocean floor but not extending above the sea surface. Seamounts may or may not occur in linear chains (see figure 12.1). When a row of seamounts exists, age data may indicate formation in succession as the oceanic plate moved over a stationary hot spot, as described in chapter 9.

A flat-topped seamount is called a **guyot** (pronounced "ghee-oh," with a hard *g*). The simplest explanation for the flat top is that a conical hill was planed off at sea level through wave action, but then how does the guyot come to be submerged? (Many lie below the lowest sea levels expected during past glacial episodes.) Two possible mechanisms, operating separately or in conjunction, may account for the submergence of guyots. The first, which is most relevant to guyots found near spreading ridges, is related to the thermal contraction of sea floor as it moves away from the ridge (figure 12.10). As the sea floor cools and contracts, it sinks deeper and carries the guyot with it. A second mechanism is isostatic compensation: the mass of the guyot loads the lithosphere, which should sink or warp downward locally in response. Hot-spot volcanoes on the sea floor, becoming inactive as they move away from the hot spot, may evolve into guyots by this mechanism. Refer back to figure 9.14; note the guyots in the Emperor Seamount chain, and imagine the evolution of these seamounts and the Hawaii–Midway Island chain as the Pacific Plate slides away from the Hawaiian hot spot and slowly sinks as it moves toward the Aleutian trench.

The formation of ringlike **coral atolls** is also linked to the existence of oceanic islands (figure 12.11). The corals require a solid foundation on which to build a reef. They also must be near the surface, in the lighted (*photic*) zone of the water, where the microscopic plants on which they feed, and with which they coexist, are abundant. Thus reef construction begins around the rim of an island. Weathering erodes the island, which may also be sinking, for reasons already described, or becoming submerged because sea level is rising worldwide. Reef growth continues upward, to keep the live colonies in the photic zone; meanwhile, the island's surface moves downward. Eventually, a large ring of reef may be all that projects above the water surface. Not all oceanic islands are hospitable to corals. Corals also require water warmer than 20°C (68°F), which further restricts their occurrence. Corals are therefore useful paleoenvironmental indicators.

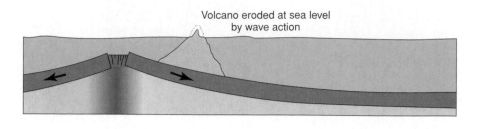

Volcano eroded at sea level by wave action

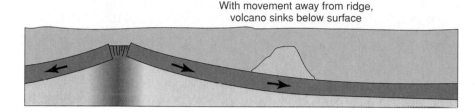

With movement away from ridge, volcano sinks below surface

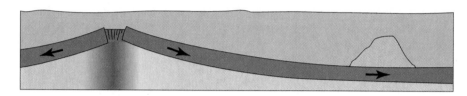

FIGURE 12.10 Possible mode of formation of a guyot: A volcano formed near a ridge sinks as the lithosphere is carried away from the ridge. Meanwhile, the volcano is planed off by waves. A similar model would apply to guyots formed initially as hot-spot volcanoes.

S till another requirement for corals to thrive is relatively clear (sediment-free) water of particular salinity. Major streams draining from continents into the ocean usually both lower the salinity of and dump suspended sediment into near-shore waters. Therefore, few coral reefs form in such settings. The extensive development of Australia's Great Barrier Reef may be due not only to appropriate water conditions and temperatures in that part of the world, but also to a lack of rivers flowing into that particular coastal area.

Plateaus

In addition to ridge systems and seamount/island chains, the ocean basins contain broader **plateaus**, shallowly submerged or projecting partially above sea level. Some of the larger examples are mapped in figure 12.12. Seismic and gravity data have shown that a number of these plateaus consist of continental, not oceanic, crust, and are compensated isostatically by roots of continental crust, just as the continents themselves are (figure 12.13). The origin of oceanic plateaus of continental affinity is unclear. The most likely explanation is that they are splinters off larger continents, perhaps split

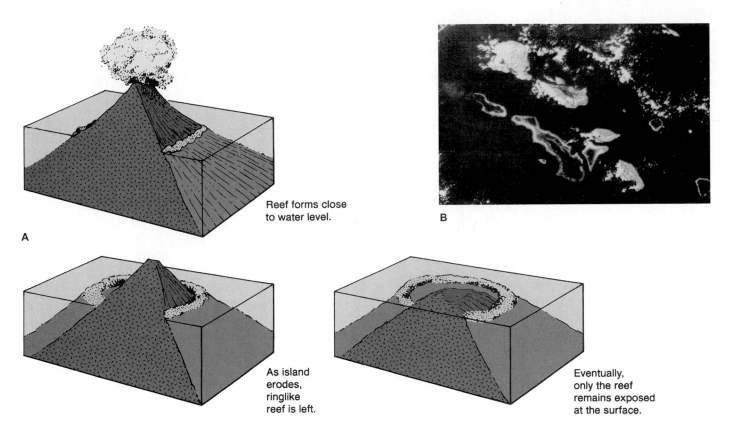

FIGURE 12.11 Coral atolls are closely related to marine volcanoes. (A) Formation of an atoll (schematic). A reef forms around a volcanic island, which then erodes and/or sinks, leaving a circular atoll. (B) Coral atolls in the Pacific Ocean (satellite photograph). *(B) © NASA.*

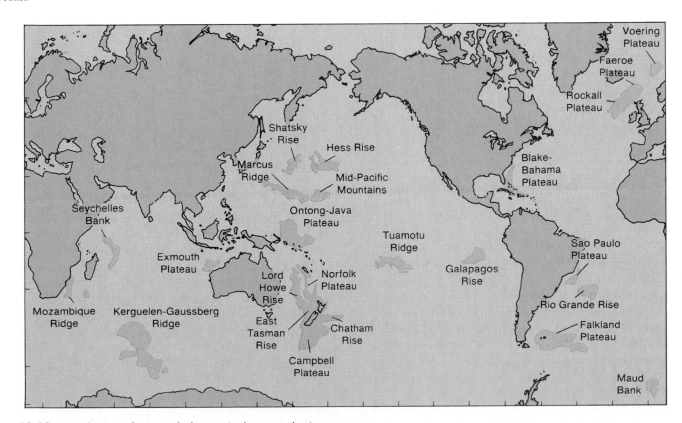

FIGURE 12.12 Distribution of principal plateaus in the ocean basins.

Redrawn with permission. From Zvi Ben-Avraham, et al., "Continental accretion: From oceanic plateaus to allochthonous terranes" in Science, *213:47–54, 1981. Copyright © 1981 by the American Association for the Advancement of Science, Washington, DC. Reprinted by permission of the publisher and authors.*

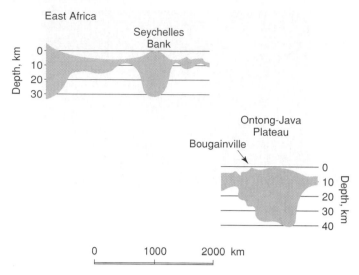

East Africa

Seychelles
Bank

Ontong-Java
Plateau

Bougainville

0   1000   2000 km

FIGURE 12.13   Some plateaus have a crustal structure that closely resembles that of continental crust.

From Zvi Ben-Avraham, "The movement of continents" in American Scientist, *69:285–99, 1981. Reprinted by permission of the author.*

away by continental rifting during the formation of ocean basins. Erosion and subsequent sinking as the lithosphere moved away from the ridge/rift zone would account for why they are now submerged, like seamounts.

Given the longevity of the sea floor (or lack of it) over geologic time, it seems that the plateaus will, sooner or later, be transported via plate movements to a subduction margin at which sea floor is being consumed. The greater buoyancy of continental lithosphere will prevent the plateaus' subduction. Instead, they might be pasted onto the edges of overriding continents. Perhaps the plateaus represent one possible origin of the suspect terranes found in the present continents.

The Structure of the Sea Floor; Ophiolites

Seismic evidence puts some constraints on the structure of the oceanic lithosphere (figure 12.14A). The oceanic crust is subdivided into three layers. Layer 1, which averages about 450 meters (1500 feet) thick, consists of unconsolidated or un-lithified sediment. This is known both from the low measured seismic velocities and from direct core sampling. Layer 2, an average of 1.75 kilometers (1.1 miles) thick, has also been sampled by deep coring and shown to consist of basalt. Layer 2 basalts often show a pillowed structure characteristic of submarine extrusion. This observation is entirely reasonable considering that Layer 2 would be at the top of new oceanic lithosphere being created at a spreading ridge. The thickest layer of the oceanic crust, Layer 3, is about 4.7 kilometers (3 miles) thick. It has not been sampled directly. However, its seismic velocity is also consistent with a composition of basalt, or basalt's coarse-grained equivalent, gabbro. Samples of gabbro that may have come from Layer 3 have been

collected by dredging along the scarps in fracture zones. The base of Layer 3 is the Moho. Below that, seismic velocities jump to values typical of the ultramafic rocks of the mantle (probably peridotite, rich in olivine with some pyroxene).

Deep coring through the oceanic crust would be both technologically difficult and extremely expensive. However, samples of oceanic crust may be accessible on land, in the form of **ophiolites** (figure 12.14B). An ophiolite is a particular mafic/ultramafic rock sequence that is widely believed, on the basis of similarities between the types and arrangement of rocks in the ophiolites and analogous layers of the sea floor, to be a slice of old oceanic crust. Ophiolites contain pillow basalts like those of Layer 2. These are overlain by beds of *chert,* a microcrystalline rock formed from silica-rich sediment; some areas of the sea floor today are sites of accumulation of siliceous sediment. Below an ophiolite's pillow basalt is a complex of *sheeted dikes* (closely spaced, parallel, vertical dikes of basaltic composition) and gabbro. The gabbro would be the coarse plutonic rock crystallized at depth during the formation of new sea floor, with the sheeted dikes being the feeder fissures of the pillow-basalt eruptions; abundant high-angle fractures would certainly be expected at a spreading ridge. Below the gabbros, ophiolites contain ultramafic rock, plausible upper-mantle material. This ultramafic rock is typically highly altered by weathering or low-grade metamorphism.

The layers in ophiolite sequences correspond closely with what is known of the sea floor, but the correspondence is not perfect. Not all marine sediments are siliceous, yet chert is the principal component of the sedimentary part of the classic ophiolite sequence. The exact thicknesses of the individual layers in the ophiolite sequence are difficult to know, for ophiolites are commonly highly fractured and deformed, but the reconstructed thicknesses seem approximately right for oceanic crust. Also, geologists cannot be sure that oceanic Layer 3 consists of sheeted dikes and gabbro; however, the seismic velocities are about right, and the sheeted dikes make sense tectonically. On balance, it seems highly likely that the ophiolites are indeed samples of oceanic crust and a bit of upper mantle. It then remains to explain how denser oceanic crust comes to be perched on the continents.

Ophiolites are exposed in mountain ranges. Mountains are commonly formed by convergence. It may be that, during convergence, especially when an ocean is slowly disappearing between two approaching continents, a few bits of ocean floor can be caught between converging plates and forced up onto a continent. This would be consistent both with where ophiolites are found and with their typically highly deformed condition. The process by which a slab of oceanic lithosphere is slipped up onto a continent has sometimes been termed **obduction,** to distinguish it from *sub*duction, in which the oceanic plate goes below the continent. On the basis of rock density, one would expect obduction to be relatively rare, and ophiolites are rather rare. That oceanic

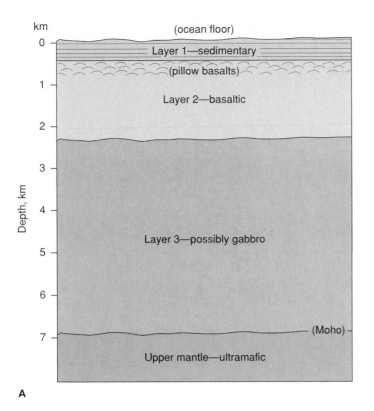

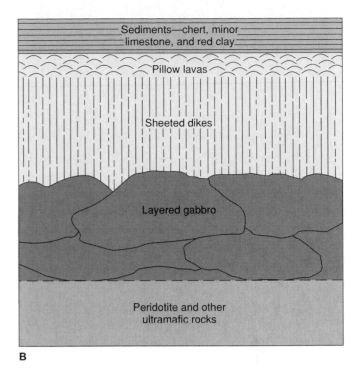

FIGURE 12.14 Ophiolites may be samples of the oceanic crust and upper mantle. (A) Typical cross section of the oceanic lithosphere. (B) Typical ophiolite sequence as reconstructed from outcrops on the continents.

lithosphere should ordinarily be subducted rather than obducted may, in fact, imply that ophiolites are in some way not altogether typical of "normal" oceanic lithosphere.

SEDIMENTATION IN OCEAN BASINS

Several sources supply sediment to the ocean basins. The principal sources are (1) the continents, from which the products of weathering and erosion are transported to the sea by wind, water, or ice; (2) seawater, with its content of dissolved chemicals; and (3) the organisms that live in the sea, whose shells and skeletons provide a biogenic component to oceanic sediment. There is even a small but detectable contribution from the micrometeorites that rain unnoticed through the atmosphere. Figure 12.15 is a generalized map of the distribution of seafloor sediment types.

Terrigenous Sediments

Continental-margin sediments are generally dominated by terrigenous material. **Terrigenous**—literally, "earth-derived"—**sediment** is clastic sediment derived from the continents. As already noted in chapter 5, the coarser sediments tend to be deposited first, closest to shore, with progressively finer materials transported farther seaward before they settle out of the water.

Sediments at the margins are not static, once deposited. They are redistributed by currents flowing along the continental shelf and slope, and by turbidity currents and submarine landslides that carry them down the continental slope and beyond. The presence of a trench or rise near the margin may limit the extent to which terrigenous sediments, especially coarser ones, are transported away from the continent.

The chemical and mineralogical makeup of the terrigenous sediments are as varied as the weathering products of the diverse rock types of the continents. The finest size fraction is typically dominated by clay minerals.

Pelagic Sediments

This finest size fraction of sediments stays in suspension most readily and may be transported beyond topographic barriers on the sea floor to settle into abyssal plains. It is one component of **pelagic sediment.** Pelagic sediments (named from the Greek *pelagos,* meaning "sea") are the sediments of the open ocean (away from continental margins), regardless of origin. As can be seen from figure 12.15, however, the pelagic sediments of much of the sea floor are dominated by material not of terrigenous origin.

Many of the larger marine creatures with shells or skeletons live in the shallower waters of near-shore regions. They contribute relatively little to pelagic sediment. Volumetrically,

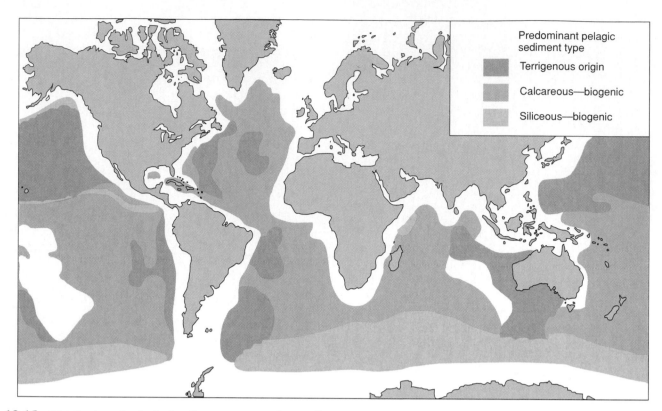

FIGURE 12.15 Distribution of principal sediment types on the sea floor.

From G. Arrhenius, "Pelagic Sediments" in M. N. Hill, The Sea, *Volume 3. Copyright © 1963 by John Wiley & Sons, Inc. New York, NY. Reprinted by permission of John Wiley & Sons, Inc.*

the most important biological contributors to pelagic sediment are microorganisms: the *foraminiferans* and *radiolarians* illustrated in chapter 5, and diatoms. The majority of species of foraminiferans have calcareous ($CaCO_3$-rich) hard parts; the radiolarians and diatoms, siliceous (SiO_2-rich) ones. All these microorganisms live in the near-surface waters; when they die, they settle toward the bottom, the soft organic matter decaying away in the process. Although each shell or skeleton is tiny, the immense numbers of organisms involved collectively produce a significant accumulation of sediment. The result is a fine-grained, water-saturated, calcareous or siliceous **ooze** on the bottom. The distribution of the biogenic oozes on the sea floor is only partially a reflection of the distribution of the corresponding organisms in near-surface waters. Both calcium carbonate and silica are also subject to dissolution in deep seawater, which selectively removes them from the sediments into solution.

Other Oceanic Sediments

Another locally important biogenic component of marine sedimentary rock is the calcareous reef, which may be built either by corals or by calcareous algae. As already noted, reefs must be built on a solid base, but within the photic zone. This requirement, plus the effect of temperature on calcium carbonate solubility, dictates that most reefs are found in warm, shallow waters, on continental shelves or shallowly submerged platforms.

In relatively warm waters, such as are found on platforms, especially in tropical latitudes, carbonates may be precipitated directly out of seawater to form limestone beds. Some clays are precipitated directly from seawater also, although the bulk of marine clay is probably of terrigenous origin. And over much of the deep-sea floor, where overall sedimentation rates are slow, precipitation of manganese oxides and hydroxides forms the lumpy **manganese nodules** that may become an important mineral resource in the future (figure 12.16).

alcite dissolves more readily in cold water than in warm; CO_2 and other gases are more soluble in colder water, making the water more acidic (recall the formation of acid rain) and thus better able to dissolve the calcite. It also dissolves more readily at higher pressure. The ocean-bottom waters are very cold and under high pressure from the overlying water column above. Calcite crystallized in warm, shallow waters and then carried to the bottom begins to dissolve under the new pressure/temperature conditions. In an ocean, the *carbonate compensation depth* is that depth below which calcium carbonate (especially calcite) tends to dissolve. It is between 4 and 5 kilometers deep in the modern Pacific Ocean, shallower in the Atlantic.

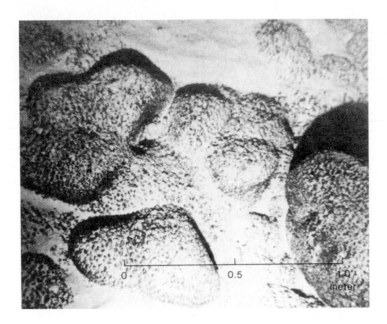

FIGURE 12.16 Where terrigenous sedimentation is limited, manganese nodules may litter the sea floor.
Photograph by K. O. Emery, USGS Photo Library, Denver, CO.

OCEAN CIRCULATION— AN INTRODUCTION

The details of oceanic circulation patterns are beyond the scope of this chapter, but some general comments and observations about ocean currents and controls on their movement can be made.

The principal driving force behind the motion of oceanic currents, particularly near surface currents, is atmospheric circulation. On a superficial level, friction between low-altitude winds and the sea surface generates ripples and waves on the water. Surface-water circulation patterns are established largely by atmospheric pressure differences from place to place. Atmospheric circulation patterns can be described in terms of belts of easterly or westerly winds (chapter 17). These provide the principal drive behind near-surface water transport also. The interaction of the momentum imparted by surface winds and the rotation of the earth results in net water movement oriented somewhat differently from the directions of wind flow. The dominant flow directions of shallow oceanic currents are shown in figure 12.17.

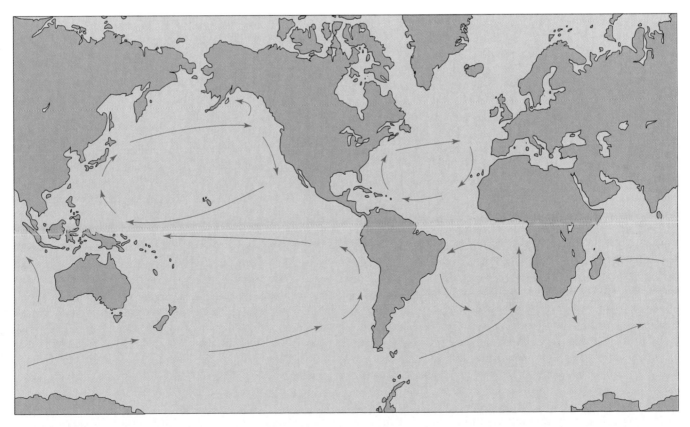

FIGURE 12.17 Principal near-surface circulation patterns of the world's oceans.

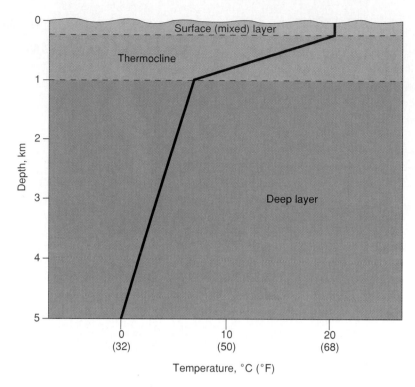

FIGURE 12.18 Vertical temperature distribution in the oceans.

Figure 12.17 shows that there are several belts or zones of water **convergence,** where opposing currents bring waters together to form different directions. The result is a pileup of water, a measurable mound on the sea surface, up to several meters in height. The principal convergences are at about 30 degrees north and south latitude. The additional water height and mass associated with these zones create a pressure gradient with respect to surrounding waters, which further modifies current flow. Water flows down the pressure gradient away from such a high; a rotational component is added by the earth's rotation. The result is a **gyre,** a nearly closed circulation pattern. See, for example, the North Atlantic circulation in figure 12.17.

Circulation patterns are further modified by the presence of continents disrupting the free flow of currents. Many of the surface currents flow predominantly east–west in the open ocean, which keeps them in approximately the same climatic zone along their flow paths. When they are deflected by a continent, these currents begin to move more nearly north–south, across climatic zones. This is the origin of currents markedly warmer or colder than the general regional surface temperature of the surroundings. The Gulf Stream is one example. Formed by the northward deflection of a warm, low-latitude current, the Gulf Stream carries that warmer water northward, moderating temperatures along the Atlantic coast of North America. Temperatures in the Gulf Stream may be more than 5°C higher than in the surrounding water masses on either side.

With the exception of turbidity currents, most of the vigorous circulation of the oceans is confined to the near-surface waters. These waters are also distinct from most of the volume of the oceans in terms of temperature (figure 12.18). Only the shallowest waters, within 100 to 200 meters of the surface, are well mixed, by waves, currents, and winds. The mixing produces a thin layer of fairly uniform temperature and salinity. The photic zone of water lies within this surface layer, too, so plant growth is confined to this zone, as are many organisms that require marine plants for food. The average temperature of the surface layer is about 15°C.

Below the surface layer is the **thermocline,** a zone of rapidly decreasing temperature that extends 500 to 1000 meters below the surface. Temperatures at the base of the thermocline are typically about 5°C. The thermocline is the interface between the warm mixed layer and the so-called **deep layer** of cold, slow-moving, rather isolated water. Temperatures in the deep layer continue to decrease with depth but more slowly than in the thermocline. The temperature of the bottommost water is close to freezing and may even be slightly below freezing (the water prevented from freezing solid by its dissolved salt content). This cold, deep layer originates largely in the polar regions. Dense, cold polar water masses sink and then slowly begin to migrate toward the warmer climate of the equatorial region. The deep waters do move, but slowly; they may take centuries to travel from poles to equator.

When winds blow parallel or nearly parallel to a coastline, the resultant currents may cause warm surface

Box 12.1
Upwelling and El Niño

As described in the text, upwelling contributes to fertility of near-shore surface waters by providing fresh supplies of nutrient-rich water to organisms in the photic zone. From time to time, however, the upwelling is suppressed for a period of weeks or longer (figure 1). The reasons are not precisely known, and probably several factors are involved. Abatement of coastal winds would reduce the pressure gradient driving the upwelling. Other wind changes in the western Pacific can cause anomalous west-to-east flows of warm surface water, depressing the thermocline and thickening the warm surface layer in the eastern Pacific. The reduction in upwelling of the fertile cold waters has a catastrophic effect on the Peruvian anchoveta industry. Such an event is called *El Niño* ("the (Christ) Child") by the fishers, for it commonly occurs in winter, near the Christmas season. The intensities of and intervals between El Niño events, however, are quite variable.

Significant El Niño conditions occurred in 1957–1958, 1965, 1972–1973, 1982–1983, and 1991–1993. During each of these events, various other meteorological problems arose worldwide—droughts in some places, torrential rains elsewhere. The coincidence in time, together with the high probability that meteorological factors do influence El Niño in some measure, has prompted various investigators to blame El Niño for the droughts and floods also. The 1991 eruptions of Mount Pinatubo likewise seemed to be correlated with El Niño conditions. A great deal more research is required to clarify any causative links among meteorological phenomena and El Niño episodes. Given the negative economic impacts of El Niño episodes on the fishing industry, this is a subject of very active research.

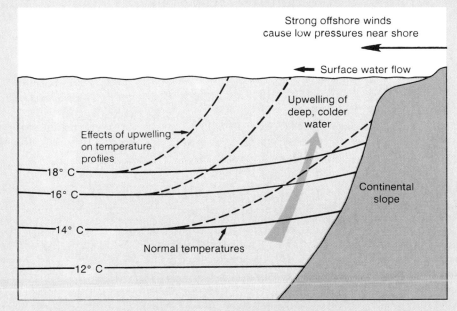

FIGURE 1   Upwelling results when strong offshore winds cause a low-pressure zone near shore, and deep waters rise in response.

waters to be blown offshore. This, in turn, creates a region of low pressure and may result in an **upwelling** of deep waters to replace the displaced surface waters. The deeper waters are relatively cold and also relatively enriched in dissolved nutrients, in part because few organisms live in the cold, dark depths to consume those nutrients. When the nutrient-laden waters rise into the photic zone, they can support abundant plant life and, in turn, animal life that feeds on the plants. Many rich fishing grounds are located in zones of coastal upwelling. Given the distribution of continents, winds, and currents, conditions tend to be particularly favorable to the development of upwelling off the western coasts of continents. The west coasts of North and South America and of Africa are subject to especially frequent upwelling events, and the productivity of anchoveta fisheries off the coast of Peru is noteworthy. When the upwelling is disrupted, the corresponding economic impact can be great; see box 12.1.

SUMMARY

The sea floor can be subdivided into continental-margin regions, spreading-ridge systems, and abyssal plains. The continental margins are divided, on the basis of tectonics, into active and passive margins. A passive continental margin, formed during continental rifting and the creation of a new ocean, typically has a broad continental shelf; the angle of the continental slope is variable. An active, convergent margin has a narrow shelf, usually a steep continental slope, and a trench formed by subduction at the foot of the slope. The shelves and slopes of continental margins are often dissected by canyons, most likely cut by turbidity currents. Oceanic ridge systems possess a central rift valley; the topography slopes away from the ridge crest, and fracture zones (extensions of transform faults) stretch for tens or hundreds of kilometers from the ridge crest. Hydrothermal activity occurs as a result of circulation of water through the hot rocks along the ridge. The abyssal plains are the deepest parts of the oceans. They may be dotted with hills, seamounts, and guyots.

The relatively thin oceanic crust is divided into Layer 1 (sedimentary), Layer 2 (basaltic), and an unsampled Layer 3 that may be gabbroic. Ophiolites may be slices of ancient, obducted oceanic crust. The crust underlying some of the large plateaus in the ocean basins, by contrast, has a density and thickness comparable to that of continental crust; these plateaus may be future accreted terranes. Terrigenous sedimentation in the oceans is largely confined to the continental margins. Pelagic sediment consists predominantly of fine clays (partially terrigenous) and calcareous and siliceous oozes.

The most active oceanic circulation is confined to the well-mixed surface layer, the uppermost 100 to 200 meters, and is predominantly wind-driven. The deep layer of water below the thermocline is much colder and slower-moving. In favorable settings, these deep, nutrient-rich waters rise in upwelling events along continental coastlines, greatly increasing the biological productivity in the near-surface waters of the affected areas.

TERMS TO REMEMBER

abyssal hills 206
abyssal plains 206
Benioff zone 210
continental rise 208
continental shelf 208
continental slope 208
convergence 220
coral atoll 214
deep layer 220

guyot 214
gyre 220
hydrothermal vents 211
manganese nodules 218
obduction 216
ooze 218
ophiolite 216
pelagic sediment 217
plateau (oceanic) 214

rift valley 210
seamount 214
submarine canyon 212
terrigenous sediment 217
thermocline 220
trench 209
turbidity current 213
upwelling 221

QUESTIONS FOR REVIEW

1. What are the principal topographic features of a passive continental margin?
2. Describe the process of ocean-basin creation that leads to formation of a passive margin.
3. Where and how are trenches formed? What evidence supports the plate-tectonic explanation of trenches?
4. Sketch a topographic cross section of a spreading ridge, and account for the central valley and the general slope of the topography.
5. What are the *abyssal plains?* How flat and featureless are they?
6. Describe the nature and probable origin of a turbidity current. What features of the sea floor might be explained by turbidity currents?
7. What is a *guyot?* Explain one mechanism by which guyots might be formed.
8. How do coral atolls develop?
9. The oceanic plateaus may be the suspect terranes of the future. Explain briefly.
10. What is an *ophiolite?* Why are ophiolites believed to be samples of oceanic lithosphere, and how might they come to be found on the continents?
11. Cite and briefly explain two major components of pelagic sediment.
12. Describe how seawater temperature changes with depth. How well mixed are the oceans?
13. Briefly describe the phenomenon of coastal upwelling.

FOR FURTHER THOUGHT

1. In what ways might you be able to distinguish a sediment sample from the abyssal plain from a sediment sample collected at the base of the continental slope of a passive margin?

2. Suppose that, while drilling in the sediments on the continental shelf off New England, you find buried fossil coral reefs. The water there today is too cold for corals. Suggest at least two possible explanations for the presence of the fossil corals.

SUGGESTIONS FOR FURTHER READING

Beer, T. 1983. *Environmental oceanography*. New York: Pergamon Press.

Bishop, J. M. 1984. *Applied oceanography*. New York: John Wiley & Sons.

El Niño. 1984. *Oceanus* 27 (2, summer).

Fanning, K. A., and Manheim, F. T., eds. 1982. *The dynamic environment of the ocean floor*. Lexington, Mass.: D. C. Heath.

Heezen, B. C., and Hollister, C. D. 1971. *The face of the deep*. New York: Oxford University Press.

Hill, M. N., ed. 1962–1983. *The sea*. New York: Interscience Publishers. (The eight-volume set, though somewhat dated in parts, still assembles in one place a wealth of data and basic information on the oceans and ocean basins.)

Kennett, J. P. 1982. *Marine geology*. Englewood Cliffs, N.J.: Prentice-Hall.

Malpas, J. 1993. Deep drilling of the oceanic crust and upper mantle. *GSA Today* 3 (March): 53–57.

Pethick, J. 1984. *An introduction to coastal geomorphology*. London: Edward Arnold.

Pickard, G. L. 1979. *Descriptive physical oceanography*. 3d ed. New York: Pergamon Press.

Scrutton, R. A., ed. 1982. *Dynamics of passive margins*. Washington, D.C.: American Geophysical Union.

Scrutton, R. A., and Talwani, M., eds. 1982. *The ocean floor*. New York: Wiley-Interscience.

Seibold, E., and Berger, W. H. 1982. *The sea floor*. New York: Springer-Verlag.

Shepard, F. P. 1977. *Geological oceanography*. 3d ed. New York: Crane, Russak.

Walsh, J. J., ed. 1988. *On the nature of continental shelves*. New York: Academic Press.

Woods Hole Oceanographic Institution. 1989. The oceans and global warming. *Oceanus* 32 (2).

13

COASTAL ZONES AND PROCESSES

Breakers crashing on the shore near Coos Bay, OR, are visible evidence of the energy of the sea working to reshape the shoreline. © Doug Sherman/Geofile

Coastal areas vary greatly in character and in the kinds and intensities of geologic processes that occur along them. They may be dynamic and rapidly changing under the interaction of land and water, or they may be comparatively stable. What happens in coastal zones is of direct concern to a large fraction of the people in the United States: thirty of the fifty states abut a major body of water (Atlantic or Pacific Ocean, Gulf of Mexico, one of the Great Lakes), and those thirty states are home to approximately 85% of the nation's population. In this chapter, we will briefly review some of the processes that occur at shorelines, look at several different types of shorelines, and consider the impacts that various human activities have on them (and vice versa).

WAVES AND TIDES

Waves and associated currents are the principal forces behind change along coasts. Waves, in turn, are produced by the flow of wind across the water surface. Small, local differences in air pressure create undulations in the water surface. The alternating rise and fall of the water surface is a **wave.** If the wind continues to blow, the undulations may begin to move laterally across the water surface and even to enlarge.

Waves and Breakers

The shape and apparent motion of a wave reflect the changing form of the water surface; the actual motion of the water molecules is different (figure 13.1). In open, deep water, the water moves in circular orbits, relatively large near the surface and decreasing in diameter with increasing depth. The water surface takes on a rippled form. The peak or top of each ripple is a wave **crest;** the bottom of each intervening low is a **trough.** The **height** of the waves is simply the difference in elevation between crest and trough, and the **wavelength** is the horizontal distance between adjacent wave crests (or troughs). The **period** of a set of waves is the time interval between the passage of successive wave crests

(or troughs) past a fixed point. The **wave base** is the depth at which water motion and associated sediment transport are negligible. This depth is approximately half the wavelength; commonly, wave base occurs at a depth of about 10 meters.

Only when a wave begins to interact with the bottom does it begin to develop into a *breaker.* This first occurs when the water depth has decreased to about the wave base. Friction with the bottom distorts and ultimately breaks up the orbits, and the water arches over in a breaking wave (figure 13.2). Bottom topography thus plays a major role in the nature of waves reaching the shore. If the coast consists

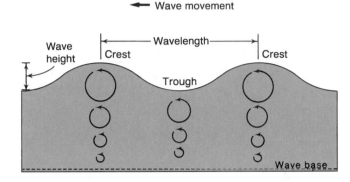

FIGURE 13.1 Waves and wave terminology. Note the motion of water beneath the surface.

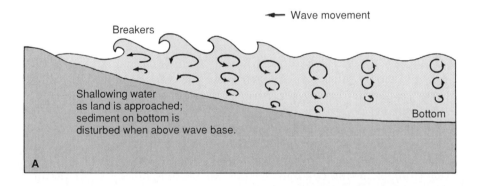

FIGURE 13.2 Breakers form as undulating waters approaching shore "touch bottom." (A) Note distortion of water orbits as water shallows (schematic). (B) Evenly spaced breakers form as successive waves touch bottom along a gently sloping shore.

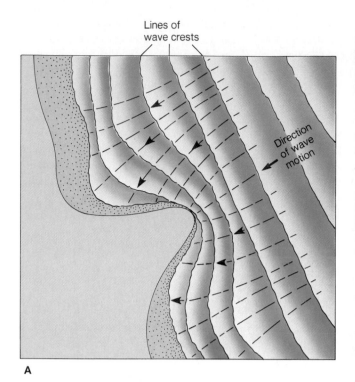

Lines of
wave crests

Direction
of wave
motion

A

FIGURE 13.3 Wave refraction. (A) The energy of waves approaching a jutting point of land is focused on it by refraction. (B) Wave refraction caused by nearshore coral reefs, Oahu, Hawaii.

B

of cliffs and the water deepens sharply offshore, breaker development may be minimal. The **surf zone** extends outward from the beach to the outermost limit of the occurrence of breakers.

The influence of interaction with the ocean bottom is also seen in the phenomenon of **wave refraction** (figure 13.3). As waves approach a coast, they are slowed first where they first touch bottom, while continuing to move more rapidly elsewhere. As a result, the line of crests is bent, or refracted. If the slope of the bottom is similar all along the coast, the waves are deflected toward projecting points of land, and their energy is concentrated at those points. Refracted waves tend to approach the shoreline more squarely, with wave crests more nearly parallel to the shore as they move toward it.

Waves, Tides, and Surges

As noted previously, waves are localized water-level oscillations set up by wind. **Tides** are broader, regional changes in water level caused primarily by the gravitational pull of the sun and moon on the envelope of water surrounding the earth.

If one takes a bucketful of water and swings it quickly in a circle at arm's length, the water stays in the bucket because forces associated with the circular motion push the water outward from the center of the circle (and thus into the bucket). Similarly, as the earth rotates, matter would be flung outward from it were it not for the restraining influence of gravity. Even so, the earth and the watery envelope of oceans surrounding it show the effects of these rotational forces. The velocity of the moving surface of the earth is greatest near the equator and decreases toward the poles, so earth and water deform to create an equatorial bulge.

Superimposed on this effect of the earth's rotation are the effects of the gravitational pull of the sun and especially the moon. The closer one is to an object, the stronger its gravitational attraction. The moon therefore pulls most strongly on matter on the side of the earth facing it, least strongly on the opposite side. The combined effects of gravity and rotation cause two bulges in the water envelope: one facing the moon, where the moon's gravitational pull on the water is greatest, and one on the opposite side of the earth, where the gravitational pull is weakest and rotational forces dominate (figure 13.4).

As the rotating earth spins through these two bulges of water each day, overall water level at a given point on the surface rises and falls twice daily. This is the phenomenon recognized as tides. Tidal extremes are greatest when sun,

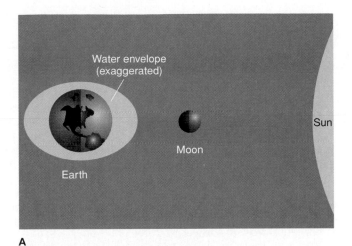

A

B

Figure 13.4 Tides. (A) Spring tides: sun, moon, and earth are all aligned (near times of full and new moons). (B) Neap tides: when moon and sun are at right angles, tidal extremes are reduced. (Note: though the diagram is shown in the plane of the page, the moon does not orbit the earth in the plane in which the earth circles the sun.)

moon, and earth are all aligned, and the sun and moon are thus pulling together. The resultant tides are **spring tides** (figure 13.4A). (They have nothing in particular to do with the spring season of the year, but rather occur when the moon is new or full.) When the sun and moon are pulling at right angles to each other, the difference between high and low tides is minimized. These are **neap tides** (figure 13.4B), associated with the first-quarter and third-quarter phases of the moon. The magnitude of water-level fluctuations in any one spot is also controlled, in part, by the underwater topography. The oscillations of waves are superimposed on the tidal regime.

Waves rise and fall in seconds, tides over several hours. A water-level rise of intermediate duration, usually associated with storms, is known as a **surge.** It results from some combination of a significant drop in air pressure over an area (which causes a local bulge or rise in water elevation beneath it) and strong winds blowing from the sea toward the land. Surges from severe storms can easily be several meters high and are most serious when they coincide with the already high spring-tide water levels at times of full or new moons. On top of that, of course, strong storms can also cause unusually large waves, which mean not only very high breakers, but also ocean-bottom disturbance at greater depths as the wave base deepens. The combination of surges and high waves can make coastal storms especially devastating, as was illustrated by Hurricane Gilbert in the fall of 1988. The scale of surges possible can be better appreciated by noting the relation of hurricane size (category) to surge in table 13.1.

BEACHES AND RELATED FEATURES

A **beach** is a gently sloping shore covered by silt, sand, or gravel that is washed by waves and tides. Most beaches are sandy, and the discussion that follows focuses on this type of beach. Beaches vary in detail, but certain features are commonly observed.

TABLE 13.1 *Saffir-Simpson Hurricane Scale*

Category	Wind Speed (in miles per hour)	Storm Surge (feet above normal tides)	Evacuation
No. 1	74–94	4.5	No.
No. 2	96–110	6–8	Some shoreline residences and low-lying areas, evacuation required.
No. 3	111–130	9–12	Low-lying residences within several blocks of shoreline, evacuation possibly required.
No. 4	131–155	13–18	Massive evacuation of all residences on low ground within 2 miles of shore.
No. 5	155	18	Massive evacuation of residential areas on low ground within 5–10 miles of shore possibly required.

Source: National Weather Service, NOAA.

The Beach Profile

A representative profile or cross section of a beach is shown in figure 13.5A. The **beach face** is that portion of the beach exposed to direct overwash of the surf. Its slope varies with the grain size of the beach sediment and the energy of the waves. To the seaward side of the beach face is typically a more shallowly sloping zone, below the low-tide level. Within this zone, sediment may be moved by currents, but it is not actually under attack by waves. Landward of the beach face, there may be a flat or landward-sloping terrace, or **berm,** backing up the beach. Some beaches have no berms; others have more than one. The landward limit of the beach can be defined in several ways, depending on the particular situation—for example, by the presence of a rocky cliff or by a zone of permanent vegetation.

Beach profiles are not static. Even a single beach may not always be characterized by the same slope or other features throughout the year. The surges and increased wave action of storms bring about changes that are both sudden and dramatic (see figure 13.6). As storms vary seasonally in intensity and frequency, so beach geometry may vary seasonally also. Indeed, most major alterations of beach geometry are associated with storms, not with the ordinary actions of everyday surf and tides.

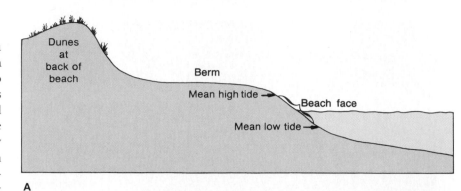

A

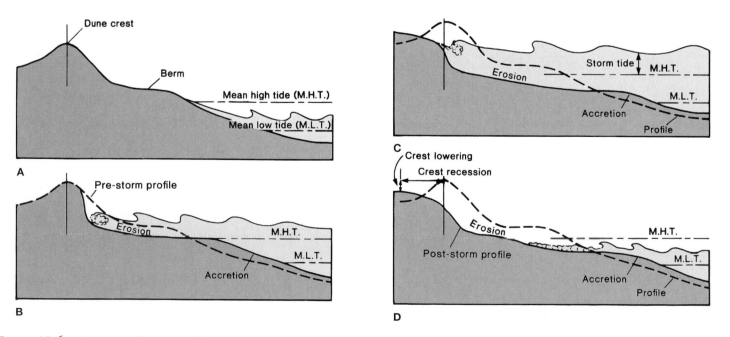

B

FIGURE 13.5 (A) A simplified beach profile. (B) The rush of surf up the beach face, more energetic than the return flow, produces asymmetric ripples, flattened toward the sea. Heron Island, Great Barrier Reef, Australia.

FIGURE 13.6 Alteration of beach profile due to accelerated erosion by unusually high storm tides. (A) Normal wave action. (B) Initial attack of storm waves. (C) Storm wave attack on foredune. (D) After storm wave attack, normal wave action. Beach profile has been permanently changed.

Source: "Shore Protection Guidelines," U.S. Army Corps of Engineers.

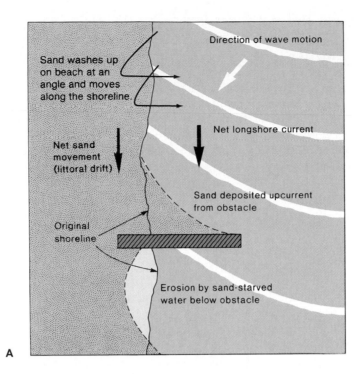

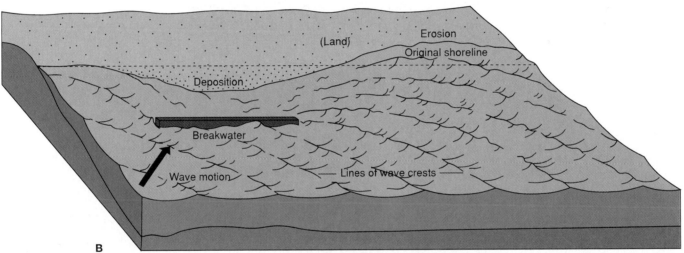

Figure 13.7 Longshore currents and their effect on sand movement. (A) Dashed profile is shoreline after modification. (B) A breakwater causes sand deposition.

Sediment Transport at Shorelines

As with streams, coastal currents move material more efficiently the faster they flow, and sediment-laden waters deposit their load as they are slowed. One consequence of wave refraction along an irregular coastline is that wave energy is dissipated in recessed bays between projecting points of land. Therefore, beaches are most often found in the bays; the concentration of energy at headlands tends to wash sand away.

On any beach, the rush of water up the beach face after waves have broken **(swash)** tends to push sediment upslope toward the land. Gravity and the retreating **backwash** together carry it down again. (It is the backwash that is sometimes felt by swimmers as undertow.) A beach profile

is described as being in equilibrium when its slope is such that the net upslope and downslope transport of sediment are equal.

If waves approach the beach at an oblique angle, sediment may be moved along the length of the beach, rather than just up and down the beach face perpendicular to the shoreline. Wave refraction tends to deflect waves so that the wave crests approaching the beach are more nearly parallel to the shoreline, but they may still strike the beach face at a small angle. As the water washes up onto and down off the beach, then, it also moves laterally along the coast. This is a **longshore current** (figure 13.7). Likewise, any sand caught up by and moved along with the flowing water is not carried straight up the beach but is transported at an angle to the shoreline. The net result is **littoral drift,** sand movement along the beach in the same general direction as the motion of the longshore current. Currents tend to move consistently in certain preferred directions on any given beach, which means that, over time, there is continual transport of sand from one end of the beach to the other. On many natural beaches where this occurs, the continued existence of the beach is assured by a fresh supply of sediment produced locally by wave erosion or delivered by streams from farther inland.

Littoral drift is a common and natural beach process. However, beachfront-property owners, concerned that their beaches (and perhaps their homes or businesses) will wash away, may erect structures to try to "stabilize" the beach, which generally further alter the beach's geometry (figure 13.7). One common method is the construction of one or more groins or jetties—long, narrow obstacles set more or less perpendicular to the coastline—or breakwaters parallel to the coastline (figure 13.8). By disrupting the usual flow and velocity patterns of the currents, these structures change the coastline. Coastal currents slowed by such a barrier tend to drop their load of sand up-current from it. Below (down-current from) the barrier, the water picks up more sediment to replace the lost load, and the beach is eroded. The common result is that a formerly stable, straight shoreline develops an unnatural scalloped shape. The beach is built out in the area up-current of the groins and eroded landward below. Beachfront properties in the eroded zone may be more severely threatened after construction of the "stabilization" structures than they were before.

Any interference with sediment-laden waters can cause redistribution of sand along beachfronts, just as natural bottom and coastline irregularities influencing current flow velocity can cause deposition of sand bars parallel to the shore or spits of sand projecting out from it. A marina may cause some deposition of sand around it, and perhaps within the protected harbor. Farther along the beach, the now-unburdened waters can more readily take on a new sediment load of beach sand. Even modifications far from the coast can affect the beach. One notable example is the practice, increasingly common over the last century or so, of damming large rivers for flood control, power generation, or other purposes. One consequence of the construction of artificial reservoirs is the trapping behind dams of the sediment load carried by each stream. The cutoff of sediment supply to coastal beaches near the mouth of the stream can lead to erosion of the sand-starved beaches. It may be a difficult problem to solve, too, because no readily available alternate supply of sediment might exist near the problem beach areas.

Figure 13.8 Construction of groins and jetties, obstructing longshore currents, causes deposition of sand: groins on Rockaway Beach, New York, here produced a scalloped shoreline.
© Bruce F. Molnia/TERRAPHOTOGRAPHICS/BPS.

Flood-control dams constructed on the Missouri River are believed to be the main reason why the sediment load delivered to the Gulf of Mexico has dropped by more than half over the last thirty-five years. This may, in turn, explain the recently observed coastal erosion in parts of the Mississippi Delta. Impounding of sediment by the Aswan Dam is also leading to degradation of the Nile Delta.

When beach erosion is rapid and development (especially tourism) is widespread, efforts have sometimes been made to import replacement sand to maintain wide beaches. The initial cost of such an effort can be millions of dollars for every kilometer of beach so restored. Moreover, if no steps are taken to reduce the erosion rates, or if no clear-cut steps *can* be taken to slow or halt the erosion in that particular area, the sand will have to be replenished over and over, at ever-rising cost. Sometimes, too, when it has not been possible (or thought necessary) to duplicate the mineralogy or grain size of the sand originally lost, the result has been further environmental deterioration. When coarse sands are replaced by finer ones, softer and muddier than the original sand, the finer material more readily stays suspended in the water, clouding it. This increased water *turbidity* is not only unsightly but also can be deadly to organisms. Off Waikiki Beach in Hawaii and off Miami Beach, delicate coral-reef communities have been damaged or killed by such increased turbidity resulting from beach "nourishment" programs.

EMERGENT AND SUBMERGENT COASTLINES

Water levels vary relative to coastal land, over the long term, as a result of tectonic processes or from changes in either land elevation or water level related to glaciation. When the coastal land rises or sea level falls, the coastline is described as *emergent*. A *submergent* coastline is found when the land is sinking relative to sea level, or sea level is rising.

Causes of Long-Term Sea-Level Change

As plates move, crumple, and shift, continental margins may be uplifted or dropped down (see, for example, figure 13.9). Such movements, when associated with earthquakes, can shift the land by several meters in a matter of seconds to minutes. The results are abrupt, permanent changes in the geometry of the land/water interface and the patterns of erosion and deposition.

FIGURE 13.9 Coastal flooding at Portage, Alaska, due to tectonic subsidence resulting from 1964 earthquake.
Photograph courtesy of USGS Photo Library, Denver, CO.

In regions overlain and weighted down by massive ice sheets in the last ice age, the lithosphere was downwarped by the load. Tens of thousands of years later, the lithosphere is still slowly rebounding isostatically to its pre-ice elevation. Where thick ice extended to the sea, a consequence is that the coastline is slowly rising relative to the sea.

Ice caps, as noted in chapter 18, represent an immense reserve of water. As this ice melts, sea levels rise worldwide. Such simultaneous, global changes in sea level are termed **eustatic** changes. In terms of the relative elevation of land and sea, postglacial isostatic rebound locally counters the eustatic sea-level rise resulting from the melting ice.

All of these changes in the relative elevation of land and water may produce distinctive coastal features, which can sometimes be used to identify the kind of vertical movement that is occurring. Descriptions of several examples follow. Formation of the barrier islands and estuaries described later in the chapter also may involve relative changes in sea level.

Wave-Cut Benches and Terraces

Given sufficient time, wave action tends to erode the land down to the level of the water surface, creating a wave-cut platform at sea level. If the land rises in a series of tectonic shifts and stays at each new elevation for some time before the next movement, each rise results in the erosion of a portion of the coastal land down to the new water level. The eventual product is a series of steplike surfaces, called **wave-cut benches** (figure 13.10). These develop most readily on rocky coasts, rather than along coasts made of soft, unconsolidated material. The surface of each such step—each bench—represents an old water-level marker on the continent's edge. Wave-cut benches can be formed when the continent is rising relative to the sea, or when sea level is falling with respect to the land. In the Caribbean, notably Barbados, coral terraces have formed when tectonic uplift of the land has coincided with eustatic sea-level rise, so the zone of reef building (just below sea level) has stayed at the same elevation relative to the land for an extended period, and a great mass of coral has built up. Continued land uplift and subsequent drop of sea level have exposed the terraces.

Drowned Valleys and Estuaries

When, on the other hand, sea level rises or the land drops, one result is that streams that once flowed out to sea now have the sea rising partway up the stream valley from the mouth. A portion of the floodplain may be filled by encroaching seawater, forming a **drowned valley** (figure 13.11).

A variation on the drowned valley is produced by the advance and retreat of coastal glaciers. Glaciers flowing off land and into water do not just keep eroding deeper valleys under water until they melt. Ice floats, and glacial ice, made of fresh water, floats especially readily on denser salt water. Therefore, the carving of glacial valleys essentially stops at

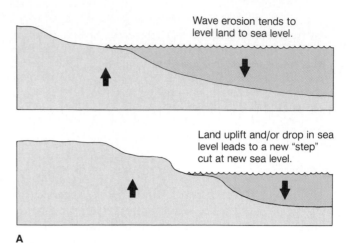

B

Figure 13.10 Wave-cut benches. (A) Wave-cut benches form when land is elevated or sea level falls (schematic). (B) Wave-cut benches of Mikhail Point, Alaska.
(B) Photograph by J. P. Schafer, courtesy of R. E. Wilcox, USGS Photo Library, Denver, CO.

the shore. During the last ice age, when sea level worldwide was as much as 100 meters lower than it now is, glaciers at the edges of continental landmasses cut valleys into what are now submerged continental-shelf areas. With the retreat of the glaciers and the concurrent rise in sea level, these old glacial valleys were emptied of ice and partially filled with seawater. This is the origin of the steep-walled *fjords* so common in Scandinavian countries, British Columbia, and parts of New Zealand.

An **estuary** is a body of water along a coastline, open to the sea, in which the tide rises and falls, and which contains a mix of fresh and salt water (*brackish* water). San Francisco Bay, Chesapeake Bay, Puget Sound, and Long Island Sound are examples. Many estuaries form near the mouths of streams, especially in drowned valleys.

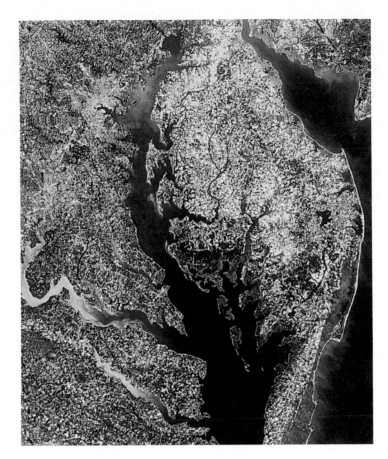

FIGURE 13.11 A drowned valley. Chesapeake Bay, Maryland (satellite image).
Courtesy Chesapeake Bay Foundation.

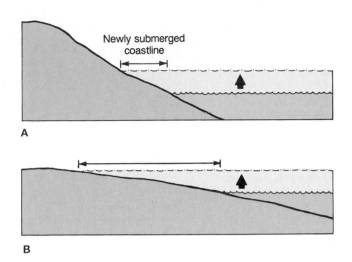

FIGURE 13.12 Effects of small rises in sea level. (A) On steeply sloping shoreline, relatively little new land is inundated. (B) On gently sloping coastline, a little increase in water depth may mean a large area flooded. Dashed lines indicate new, higher sea level.

O ver time, the complex communities of organisms in each estuary have adjusted to the salinity of that particular water. The salinity itself reflects the balance between freshwater input (usually river flow) and salt water. Any modifications that alter this balance change the salinity and can have a catastrophic impact on the organisms. Also, water circulation in estuaries is often very limited. This makes them especially vulnerable to pollution; because they are not freely flushed out by vigorous water flow, pollutants can accumulate. In this sense, it may be unfortunate that many of the world's large coastal cities are located beside estuaries.

Present and Future Sea-Level Trends

Much of the coastal erosion presently plaguing the United States is the result of gradual but sustained eustatic sea-level rise, probably from the melting of the remaining polar ice. The rise is estimated at about ⅓ meter (1 foot) per century. While this does not sound particularly threatening, two additional factors should be considered. First, many coastal areas slope very gently or are nearly flat, so that a small rise in sea level translates into a far larger inland retreat of the shoreline (figure 13.12). (This is why a few meters of storm surge may require evacuation of a miles-wide strip of coastal land, as shown in table 13.1.) Rates of shoreline retreat (landward movement of the shoreline) due to rising sea level have, in fact, been measured at several meters per year in some low-lying coastal areas. A ⅓-meter rise in sea level could translate into 30 to 60 meters of inland beach erosion. Second, there is concern that global warming trends will begin to melt remaining ice caps more rapidly, accelerating the sea-level rise (see the discussion of the *greenhouse effect* in chapter 18). At the same time, global warming would cause thermal expansion of ocean waters, which would exacerbate the sea-level rise.

A lthough the terms *coastline* and *shoreline* are sometimes used interchangeably, the **shoreline** is technically the line made by the water's edge on the land, a small, local feature that is constantly shifting and changing shape. The term **coastline** is properly used to describe the overall geometry of the margin of the land, which encompasses a much larger area of the coastal region and is a somewhat more permanent feature. The coast divides land from water, extending inland to the first significant change in types of landforms.

Consistently rising sea levels are a major reason that efforts to stabilize coastlines repeatedly fail. The problem is not only that the high-energy coastal environment presents difficult engineering challenges. The problems themselves are intensifying, as shoreline retreat brings water higher and farther inland, pressing ever closer to and more forcefully against more and more structures developed along the coast.

COASTAL EROSION AND COASTAL HAZARDS

Littoral drift along beaches is not the only cause of coastal erosion. Even rocky cliffs are vulnerable, attacked either by the direct pounding of waves or by the grinding effect of sand, pebbles, and cobbles propelled by waves, which is termed **milling.** Solution and chemical weathering of the rock may cause still more rapid erosion. Because wave action is concentrated at the waterline, above wave base, cliff erosion is often most vigorous there (figure 13.13). Also, wave refraction accelerates erosion of projecting points of land.

Sea-Cliff Erosion

Sediments are obviously much more readily eroded than are solid rocks, and erosion of sandy cliffs may be especially rapid. Removal of material at and below the waterline undercuts the cliff, leading, in turn, to slumping and sliding of sandy sediments and the swift landward retreat of the shoreline. Many who build on coastal cliffs are unpleasantly surprised to discover how quickly the coastline can change. Unanticipated and rapid cliff erosion can be an especially dramatic threat. Sandy cliffs can be cut back by several meters a year.

Various measures, often as unsuccessful as they are expensive, may be tried to halt the erosion. A common practice is to place some type of barrier at the base of the cliff to break the force of wave impact (figure 13.14). The

FIGURE 13.13 Sea arch formed by wave erosion at the waterline on a lava coast, Hawaii. As erosion undercuts the rock at sea level, chunks of fractured basalt above collapse, creating the arch.

A

B

FIGURE 13.14 Examples of cliff-protection structures. (A) Seawall along the Cliff Walk, Newport, Rhode Island. (B) Riprap at base of cliff, El Granada, California. The structure was built in 1973; it may have to be abandoned within decades as erosion continues on either side of the riprap.
(B) Photograph courtesy of USGS Photo Library, Denver, CO.

Figure 13.15 Barrier islands along North Carolina's Outer Banks. *Photograph by R. Dolan, USGS Photo Library, Denver, CO.*

protection may take the form of a solid wall (*seawall*) of concrete or other material, or a pile of large boulders or other blocky debris (*riprap*). If the obstruction is placed only along a short length of cliff directly below a threatened structure, the water is likely to wash in beneath, around, and behind the obstruction, rendering it largely ineffective. Erosion continues unabated on either side of the barrier; loss of the protected structure will be delayed but probably not prevented. Wave energy may also bounce off, or be reflected from, a short length of smooth seawall, to attack a nearby unprotected cliff with greater force.

An alternative approach to direct shoreline protection is to erect breakwaters farther away from and parallel to the shore, to reduce the energy of the pounding waves reaching shore. This may slow the erosion but, again, is unlikely to stop it, especially if the cliffs are sandy or made of other weak or unconsolidated materials. Moreover, a breakwater also tends to cause altered patterns of sediment deposition offshore, which may prove inconvenient or may damage marine life. See also box 13.1.

Barrier Islands

Barrier islands are long, low, narrow islands paralleling a coastline (figure 13.15). Exactly how or why they form is not known. Some theories suggest that they form through the action of longshore currents on delta sands deposited at the mouths of streams. Their formation may also require relative changes in sea level. However they have formed, they now provide important protection for the water and shore inland from them because they constitute the first line of defense against the fury of high surf from storms at sea. These islands often are so low-lying that water several meters deep may wash completely over them during unusually high storm tides, such as occur during hurricanes, especially along the Gulf and Atlantic coasts of North America (figure 13.16).

Because they are usually subject to higher-energy waters on their seaward sides than on their landward sides, most barrier islands retreat landward with time. Typical rates of retreat on the Atlantic coast of the United States are ½ to 2 meters per year, but rates in excess of 20 meters per year have been noted—for example, along some barrier islands in Virginia. Clearly, such settings represent particularly

Box 13.1

Trouble on the Texas Coast

The Texas coast, like much of the Atlantic margin and Gulf Coast of the United States, is rimmed with barrier islands and barrier peninsulas. The area is particularly vulnerable to alteration by natural forces because it is frequently subject to severe storms, which are accompanied by elevated tides (storm surges, commonly several meters or more above normal high tides), strong winds, and high waves. Over the last century, an average of one tropical storm a year has made landfall somewhere along the Texas coast, and a hurricane has struck more than once every five years. A particularly fierce storm hit Galveston, Texas, in 1900, washing away two-thirds of the city's buildings and causing 6000 deaths. Modern meteorological monitoring and improved communications have greatly reduced the number of hurricane deaths in recent decades, but the shifting shoreline poses a continuing challenge to structures.

A period of quiet weather in the 1950s was accompanied by a rush of development along the Texas coast; another period of accelerated building occurred in the 1970s. The number of structures at risk continues to grow. Meanwhile, the landward retreat of the beaches also continues, at an average rate of 2 to 7 meters (6 to 23 feet) per year. In some especially unstable areas, shoreline changes of more than 20 meters per year have been recorded.

As in other dynamic environments, stabilization efforts have had mixed results. In 1902, in response to the devastation of the 1900 hurricane, the Galveston seawall (figure 1) was built. The nearly 16-kilometer-long, 6-meter-high structure, built at a cost of $15.5 million (including the cost of raising houses by as much as 4 meters behind the seawall), has offered valuable protection to structures on land, especially during subsequent storms. However, it has also demonstrated some of the permanent changes that can result from seawall construction.

As noted earlier, a portion of wave energy is reflected back from smooth-faced seawalls, so the sand in front of them is more actively eroded than it might be in the absence of the wall. Longshore currents are commonly strengthened along a seawall, and the seawall cuts off the supply of sand from any dunes at the back of the beach to the area in front of the seawall. The result, typically, is gradual loss of the sandy beach in

FIGURE 1   The Galveston seawall. Note absence of beach.
Photograph by W. T. Lee, USGS Photo Library, Denver, CO.

front of the seawall within fifty years of its construction. The beach in front of the Galveston seawall, once up to several hundred meters wide, is virtually gone now. Moreover, the unprotected area at either end of the seawall is being eroded very rapidly. Changes caused by the seawall have made it likely that the seawall will eventually have to be replaced by another, larger, more expensive structure.

Beach replenishment has not often been practiced on the Texas coast, partly for lack of funds and partly for lack of suitable replacement sand. Such an effort was undertaken at Corpus Christi, however, using sand from a nearby river. It has succeeded in the sense that a sandy beach has been preserved there. On the other hand, the river sand is much coarser-grained than the original beach sand. It has stabilized at a steeper

slope angle, both above and below the waterline, than characterized the original beach, and the beach has consequently become less suitable for use by small children. Also, of course, the replenishment efforts must continue if the new beach is not to be eroded away in its turn. The vulnerability of the area is illustrated by the breaching of sandy barrier islands by sand washout during storms (figure 2).

Beach-replenishment efforts in barrier-island regions elsewhere on the east coast of the United States have been rather less successful. If "beach life span" is defined as the length of time required for 50% of the new sand to be lost, that life span has been estimated at less than two years in 40% of cases, two to five years in nearly 50% of cases, and more

than five years in only 12% of cases. Such observations have led one investigator to conclude that "(1) the parameters used to design beaches don't work, (2) predictions of beach durability are *always* wrong, and (3) nobody in the coastal engineering community evaluates past projects, so no progress has been made in understanding beach replenishment. In addition the public is unaware of the uncertainties of beach replenishment and, consequently, the taxpayers take it on the chin." (Pilkey 1989, p. 308) Whether or not this is overly pessimistic, experience certainly supports the general truth of two principles: first, that shoreline engineering permanently alters the shoreline, and second, that once shoreline engineering is begun, it must be continued unless structures or beaches are ultimately to be abandoned.

FIGURE 2 Storm overwash from Hurricane Beulah (1967) reopened old inlets and seriously eroded these barrier islands near Corpus Christi, Texas.
Photograph courtesy of USGS Photo Library, Denver, CO.

FIGURE 13.16 Storm damage on low-lying barrier islands, North Carolina's Outer Banks: results of overwash of low ground by storm surge. *Photograph by R. Dolan, USGS Photo Library, Denver, CO.*

unstable locations in which to build, yet the aesthetic appeal of long beaches has led to extensive development on privately owned sections of barrier islands. About 1.4 million acres of barrier islands occur in the United States, and approximately 20% of this total area has been developed. Thousands of structures are at risk, including many in large coastal cities like Miami, Florida, and Galveston, Texas.

On barrier islands, shoreline-stabilization efforts—building groins and breakwaters and replenishing sand—tend to be especially expensive and, frequently, futile. At best, the benefits are temporary. Construction of artificial stabilization structures may easily cost tens of millions of dollars. At the same time, the structures may destroy the natural character of the shore and even the beach that was the principal attraction for developers in the first place. An equally costly alternative is to keep moving buildings or rebuilding roads ever farther landward as the beach before them erodes. Expense aside, this is clearly a short-term "solution" only, and in many cases, there is no place left to which to retreat. Given the inexorable eustatic rise of sea level at present, more barrier-island land is being submerged more frequently, so the problems will only intensify.

It is often possible to identify the most unstable or threatened areas along a coast, so as to avoid building on them. The best setting for building near a beach or on an island, for instance, is at a relatively high elevation (5 meters or more above normal high tide, to be above the reach of storm surges) and in a spot protected by many high dunes between the proposed building site and the water. Thick vegetation, if present, will help to stabilize the beach sand. Information about what has happened in major storms in the past also is very useful. Was the site flooded? Did the overwash cover the whole island? Were protective dunes destroyed? On either beach or cliff sites, one very important factor is the rate of erosion (if any). Information might be obtained from people who have lived in the area for some time, or more reliably, by comparison of old maps or photographs with the modern coastline geometry. (It should be kept in mind that shoreline retreat in the future may be more rapid than it has been in the past, as a consequence of more rapidly rising sea levels.) On cliff sites, too, there is landslide potential to consider, and in a seismically active area, the dangers are greatly magnified. It is also advisable to find out what shoreline modifications are in place or are planned, not only close to the site of interest but elsewhere along the coast.

SUMMARY

Coastal areas are, by nature, dynamic geologic environments, and many are undergoing rapid change. Erosion is a major factor, with waves and currents the principal agents. Waves are generated by the action of wind on the water surface. They develop into breakers as they approach shore, where they may also be deflected by wave refraction in shallowing water. Storms intensify wave erosion, both by increasing wave heights and by causing storm surges that raise overall water levels. In addition to net erosion or deposition of sediment, many beaches are subject to lateral transport of sand by longshore currents. Interference with this process of littoral drift—for example, through construction of piers or jetties—results in redistribution of sediment and altered patterns of sediment erosion and deposition. The same is often true of other shore-protection structures.

An overall rise or fall of the land relative to the sea may result, respectively, in the formation of wave-cut benches, or drowned valleys and fjords. At present, eustatic sea-level rise, probably caused by the ongoing melting of ice caps, is intensifying many coastal problems, especially in low-lying, vulnerable areas, such as barrier islands.

Even changes far from the shore may affect it, as when trapping of stream sediments by flood-control dams starves coastal beaches of sand and accelerates their disappearance. Prospective residents of coastal areas can investigate the long-term stability of the coastline through old maps or photographs, in addition to observing present processes active along the coast.

TERMS TO REMEMBER

backwash 229
barrier island 235
beach 227
beach face 228
berm 228
coastline 233
crest 225
drowned valley 232
estuary 232

eustatic 232
height (wave) 225
littoral drift 230
longshore current 229
milling 234
neap tides 227
period 225
shoreline 233
spring tides 227
surf zone 226

surge 227
swash 229
tides 226
trough 225
wave 225
wave base 225
wave-cut bench 232
wavelength 225
wave refraction 226

QUESTIONS FOR REVIEW

1. Sketch a cross section of several waves, and indicate the following: wave crests, wavelength, how water is moving beneath the surface.
2. Do breakers typically form in the open ocean? Explain.
3. Briefly describe the origin of tides.
4. Under what circumstances does littoral drift occur? Why does it not necessarily result in complete removal of a beach's sand over time?

5. Describe how a shoreline is altered when groins or breakwaters are erected to slow littoral drift.
6. Cite two possible problems associated with artificial beach-sand replenishment.
7. What is an *emergent coastline?* Does it necessarily reflect uplift of the coastal land? Cite one erosional feature characteristic of some emergent coastlines.

8. Describe one cause of eustatic sea-level changes.
9. What is a *barrier island?* Describe how such islands commonly migrate through time, and comment on the impact of rising world sea levels.
10. Name and describe any three factors that should be considered before undertaking development on a coastal site.

FOR FURTHER THOUGHT

1. Choose any major coastal city, and investigate the extent of flooding that could be expected in the case of a eustatic sea-level rise of (a) 1 meter and (b) 5 meters. For an example of possible problems and responses to them, investigate what is occurring in the city of Venice, Italy, where a combination of tectonic sinking, surface subsidence, and sea-level rise is causing increasing flooding problems.

2. If you live in a near-coastal area, calculate how soon water would reach your home at a rate of sea-level rise of 1 meter per century.

SUGGESTIONS FOR FURTHER READING

American Geological Institute. 1981. Old solutions fail to solve beach problems. *Geotimes,* December, 18–22.

Barnes, R. S. K., ed. 1977. *The coastline.* New York: John Wiley and Sons.

Bird, E. C. F. 1969. *Coasts.* Cambridge, Mass.: M.I.T. Press.

Carter, W. 1988. *Coastal environments.* New York: Academic Press.

Environmental Protection Agency. 1989. Can our coasts survive more growth? *EPA Journal* 15 (September/October).

Fisher, J. S., and Dolan, R. eds. 1977. *Beach processes and coastal hydrodynamics.* Stroudsburg, Penn.: Dowden, Hutchinson, and Ross.

Heikoff, J. M. 1980. *Marine shoreland resources management.* Ann Arbor, Mich.: Ann Arbor Science.

Horton, T. 1993. Hanging in the balance: Chesapeake Bay. *National Geographic* 183 (June): 3–35.

MacLeish, W. H., ed. 1981. The coast. *Oceanus* 23 (4).

Morton, R. A., et al. 1983. *Living with the Texas shore.* Durham, N.C.: Duke University Press.

Pethick, J. 1984. *An introduction to coastal geomorphology.* London: Edward Arnold.

Pilkey, O. H. 1989. The engineering of sand. *Journal of Geological Education* 37 (November): 308–11.

U.S. Army Corps of Engineers. 1971. *Shore protection guidelines.* Washington, D.C.: U.S. Government Printing Office.

Walden, D. 1990. Raising Galveston. *American Heritage of Invention and Technology* (Winter): 8–18.

Woods Hole Oceanographic Institution. 1989. The oceans and global warming. *Oceanus* 32 (2).

14

STREAMS

Streams flow placidly when their channels slope gently and water volume is low; here, sandstone cliffs in Zion National Park, UT, are reflected in the Virgin River. © Doug Sherman/Geofile

When rain falls on the earth's surface, or accumulated snow melts, the water either sinks into the ground (**infiltrates**) or remains at the surface. Subsurface waters are considered in detail in chapter 15.

Surface water tends to flow from higher to lower elevations. It may wash in broad sheets over the ground surface, but commonly, the flowing water is collected into a channel, forming a **stream.** The term *stream* is used to describe any body of flowing water confined within a channel, regardless of its size. In this chapter, we begin by considering streams from the perspective of their place in the global water cycle. Discussions of stream characteristics, and of streams as agents of erosion, sediment transport, and deposition, follow. Later in the chapter, we consider the phenomenon of stream flooding, and the interplay of flooding and human activities.

STREAMS AND THE HYDROLOGIC CYCLE

The **hydrosphere** includes all the water at and near the surface of the earth. Recall from chapter 1 that most of this water is believed to have been outgassed from the earth's interior early in its history, when earth's temperature was higher. Now, except for occasional minor contributions from volca-

noes bringing additional water up from the mantle, the quantity of water in the hydrosphere remains essentially constant.

All of the water in the hydrosphere is caught up in the **hydrologic cycle,** illustrated in figure 14.1. The main processes in the cycle are evaporation into and precipitation out of the atmosphere. Solar energy drives the evaporation, as well as the winds that distribute the moisture. Precipitation onto land can either reevaporate (directly from the ground surface or indirectly through plants by evapotranspiration), infiltrate into the ground, or run off over the ground surface. Water that sinks into the ground can also flow (see chapter 15). Both surface and subsurface runoff act to return water to streams and, ultimately, to the oceans in most cases. With their vast exposed surface area, the oceans are the principal source of evaporated water.

The total amount of water moving through the global hydrologic cycle is large, more than 100 million billion gallons per year. A portion of this water is diverted for human use, but it eventually makes its way back into the global water cycle by a variety of routes, including release of municipal sewage, evaporation from irrigated fields, or discharge of industrial wastewater into streams. Water in the hydrosphere many spend extended periods of time—even tens of thousands of years—in one or another of the water reservoirs, such as a glacier or a groundwater system, but from the longer perspective of geologic history, it

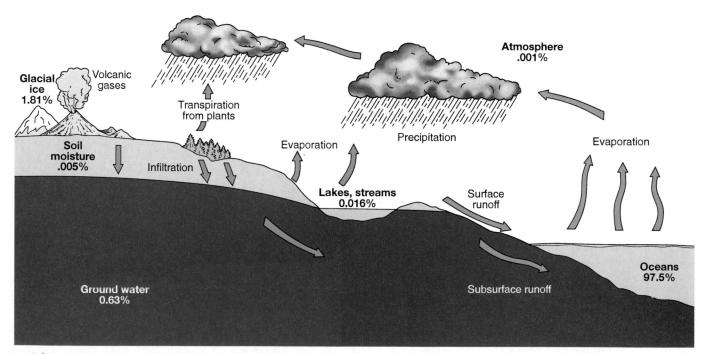

FIGURE 14.1 Principal processes and reservoirs of the hydrologic cycle.

is still regarded as moving continually through the hydrologic cycle. In the process, it may contribute to other geologic cycles and processes as well; for example, streams eroding rock and moving sediment, and subsurface waters dissolving rock and transporting dissolved chemicals, are also contributing to the rock cycle.

STREAMS AND SURFACE DRAINAGE

The geographic area from which a stream system draws water is its **drainage basin,** sometimes also called its *watershed* (figure 14.2). The volume of water in a stream at any spot is related, in part, to the size of the area drained, as well as to such other factors as the amount of precipitation in the area and how readily the water infiltrates into the ground. The boundary between adjacent drainage basins is a topographic high known as a **divide.**

O n a grand scale, **continental divides** can be identified. In North America, the continental divide runs through the Rocky Mountains. East of it, streams drain to the Atlantic Ocean or Gulf of Mexico; west of it, to the Pacific Ocean.

Typically, only very small drainage basins are drained by a single channel. As small streams flow into larger ones, drainage patterns become more complex. One way to describe the complexity of drainage is in terms of stream *order*. A **tributary stream** is one that flows into another stream. A *first-order stream* is one into which no tributaries flow. A *second-order stream* is one that has only first-order streams as tributaries, and so on. In general, the higher the order of the stream, the larger the area drained by the stream and its tributaries. Deciding when a trickle of surface water in a channel properly constitutes a stream, and thereby determining the order of streams farther down in a given drainage system, is, in part, a matter of judgment and the scale of map being used. The U.S. Geological Survey has identified between 1.5 and 2 million streams in the United States. The more than 1.5 million first-order streams have an average drainage area of 2.5 square kilometers each; the 4200 fifth-order streams average about 1300 square kilometers in drainage area; the Mississippi River, the single tenth-order stream, drains 320 million square kilometers.

The majority of stream systems exhibit a branching drainage pattern described as **dendritic,** from the Greek word *dendros,* meaning "tree." In a dendritic drainage pattern, stream channels are irregular, and tributaries join larger streams at a variety of angles. The systems in figures 14.2 and 14.3A are examples.

Local geology may cause the drainage network to assume a more symmetric geometry. Where sets of parallel fractures cause zones of weakness and/or topographic lows, streams may establish a rectilinear **trellis drainage** pattern in which tributaries join the main stream at right angles. Trellis drainage may also develop where the topography is dominated by parallel ridges of resistant rock, perhaps created as a consequence of folding or tilting of strata, and

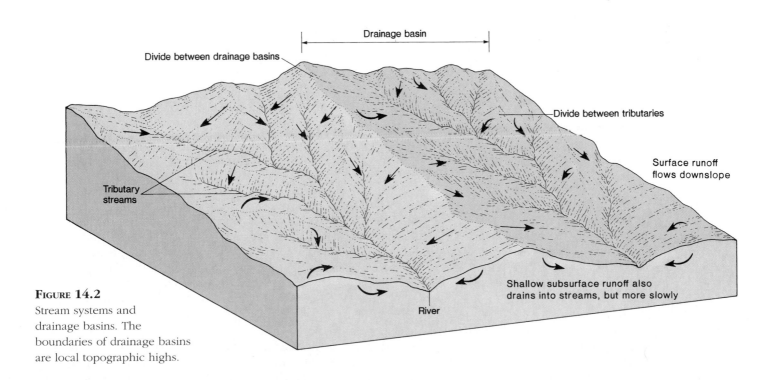

FIGURE 14.2
Stream systems and drainage basins. The boundaries of drainage basins are local topographic highs.

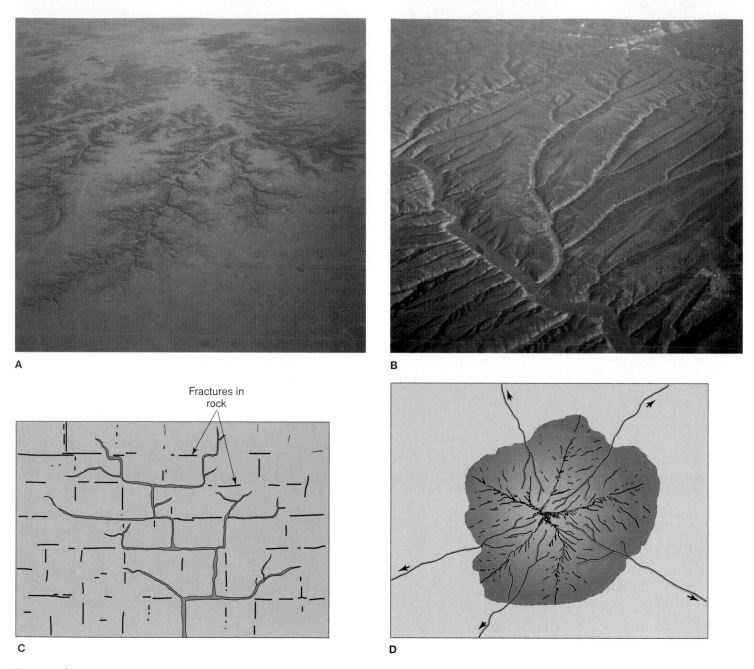

FIGURE 14.3 Specialized geometry in drainage networks arises from special geologic conditions. (A) Dendritic drainage on midwestern farmland. (B) Fracture-controlled trellis drainage. (C) Rectangular drainage controlled by intersecting joint sets. (D) Radial drainage around a conical hill.

streams tend to flow between those ridges (figure 14.3B). Intersecting joint sets may create **rectangular drainage,** with right-angle bends in all streams, and stream segments of relatively equal length between bends (figure 14.3C). Where there exists a conical topographic high, such as a volcano, **radial drainage** carries water away from the center of the high in all directions (figure 14.3D).

The size of a stream at any point can be defined in terms of its **discharge,** the volume of water flowing past a given point (or more precisely, through a given cross section) in a specified period of time. Discharge is calculated as the product of the cross-sectional area of the stream times the flow velocity (figure 14.4). Conventionally, discharge is measured in cubic feet per second or cubic meters per second. As a practical matter, discharge can be difficult to measure, for the velocity of the stream is not everywhere the same, nor, for that matter, are natural stream channels constant in shape. Flow is generally slower along the channel bed and sides, where there is friction between water and channel, and faster near the center of the channel and surface of the stream. Where a stream channel is relatively straight, the cross section tends to be more symmetric and

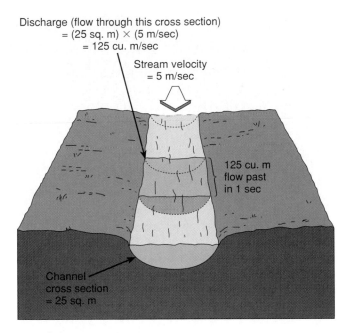

Discharge (flow through this cross section)
= (25 sq. m) × (5 m/sec)
= 125 cu. m/sec

Stream velocity
= 5 m/sec

125 cu. m
flow past
in 1 sec

Channel
cross section
= 25 sq. m

FIGURE 14.4 Discharge = area × velocity.

FIGURE 14.5 V-shaped valley of a rapidly downcutting stream: the Grand Canyon of the Yellowstone River.

shallower relative to its width. Where the channel curves, flow is faster on the outside of the bend, and the resultant increased erosion tends to deepen the channel there, producing a more triangular cross section. Discharge is a function of many factors, including climate (precipitation, availability of water), flow velocity, characteristics of the surrounding soil that influence water infiltration, and the overall size of the drainage basin from which the streamflow is drawn. A stream's discharge, in turn, has a bearing on the stream's effectiveness in erosion and sediment transport.

STREAMS AS AGENTS OF EROSION, TRANSPORT, AND DEPOSITION

Moving water is the principal erosive agent at the earth's surface. Although they are most effective in eroding unconsolidated sediments, water and waterborne sediment can even scour solid rock. Stream erosion widens, deepens, and modifies the channel proper and, over time, alters the geometry of the valley surrounding the channel as well. An important contributing factor is gravity, which helps to deliver sediment to the stream that then sweeps it away.

A stream eroding downward into its bed is engaged in **downcutting.** All else being equal, the higher the velocity of streamflow, the more rapid the erosion. How rapidly downcutting occurs is partly a function of the nature of the streambed material: solid rock is generally much more resistant to erosion than is sediment. If the stream itself is transporting sediment, that sediment can abrade the channel bottom and sides, hastening erosion. The water can also attack soluble rocks chemically; a limestone, for example, is especially susceptible to dissolution. Where a stream is engaged in relatively rapid downcutting, it characteristically carves a narrow, steep-walled, V-shaped valley (figure 14.5).

Meanders

Where downcutting is less rapid and stream velocity slower, the straight channel begins to develop lateral displacements. Small irregularities in the channel cause local variations in flow velocity, which, in turn, contribute to variations in erosion and sediment deposition from point to point. Bends, or **meanders,** begin to develop along the channel.

Once meanders begin to form, they tend to move both laterally and along the length of the stream. This can be understood in terms of flow velocity within the channel (figure 14.6). The water strikes with greatest force the outer, downstream bank of the meander, which causes increased erosion in that area, sometimes described as the **cut bank.** Flow is slower on the inside and upstream bank of a meander, so

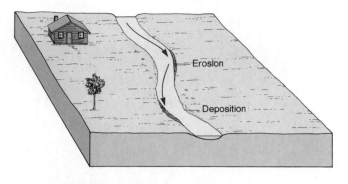

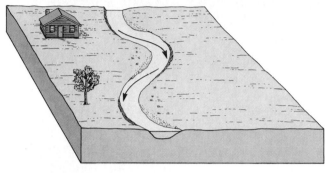

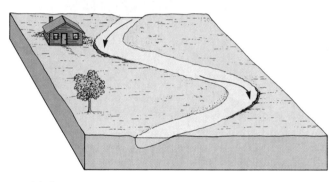

FIGURE 14.6 Map view of development of meanders. Channel erosion is greatest on the outside of curves and on the downstream side, where flow velocity is higher; deposition occurs in the sheltered area on the inside of curves where the stream's flow is slower. Over time, meanders migrate both laterally and downstream.

sediment deposition tends to occur there, for reasons described later in the chapter. Over time, then, meanders can enlarge sideways, migrate laterally, and shift downstream by this combined erosion and sedimentation.

Meanders do not enlarge indefinitely. Large, looping meanders along a stream represent sizeable detours for the flowing water. Eventually, the stream may make a shortcut, cutting off the meanders and abandoning the longer, irregular channel for a shorter, more direct route (figure 14.7). The cutoff meanders are termed **oxbows.** If they remain filled with water, they are *oxbow lakes.* Meander cutoff most commonly occurs during flood events, when the larger volume of water flowing faster in the channel may have enough additional momentum to carve the shortcut, bypassing the meander. The stream gradient is then increased—the water

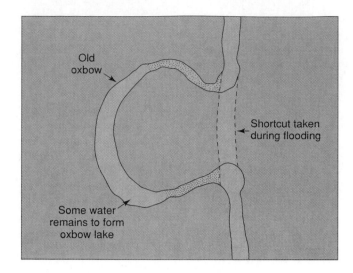

FIGURE 14.7 Map view of formation of oxbow by meander cutoff. See also figure 14.9.

travels a shorter path in dropping the same vertical distance—so the stream's velocity, and consequently its discharge, also increases after meander cutoff.

O n an intuitive level, meandering makes sense. It is more difficult to explain why many streams develop regular, sinuous meanders of similar size and shape over a considerable length of the stream. No fully satisfactory theoretical model has been developed to account for this. Nor is it as easy to understand how meanders become established in bedrock as it is to visualize them forming in loose, unconsolidated sediment or soil.

The Floodplain

Over time, the processes of lateral erosion associated with meandering, sediment deposition behind migrating meanders, and additional sedimentation during flood events when the stream has overflowed its banks collectively work to create a **floodplain** (figure 14.8). A floodplain is a flat or gently sloping region around a stream channel into which the stream flows during flood events (figure 14.9). At any given time, the meandering stream commonly occupies only a portion of the breadth of the floodplain. The overall width of the floodplain is a function of many factors, including the size of the stream, the relative rates of meander migration and of downcutting, the strength of the valley walls, and time.

While climatic and tectonic conditions remain fairly constant, slow lateral channel migrations over time tend to maintain a single floodplain with a fairly level surface. If these external conditions are suddenly changed significantly, relative rates of meandering and downcutting may also be changed in such a way as to alter the consequent floodplain geometry. If a stream has established a broad floodplain and

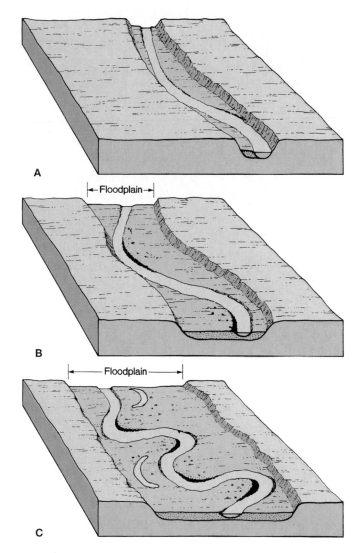

FIGURE 14.8 A floodplain is broadened by meandering. (A) Initially, the stream channel is relatively straight. (B) Small bends in the channel enlarge and migrate over time; meanders broaden. (C) Ultimately, a broad, flat floodplain is developed around the stream channel, with the aid of sediment deposition as well as meandering.

FIGURE 14.9 Floodplain carved by a meandering stream: the Sweetwater River, Wyoming.
Photograph by W. R. Hansen, USGS Photo Library, Denver, CO.

FIGURE 14.10 Terraces within a valley, above the present floodplain. Terraces of the Snake River, western Wyoming.
Photograph by Arthur H. Doerr.

there is then regional uplift, for instance, downcutting may be accelerated and the old floodplain cut up by development of a new, narrower, deeper one within the broader valley of the old. The uplift, by increasing stream gradient, could cause a shift from a situation in which erosion and deposition are in balance to one dominated by erosion. This is one mechanism that has been proposed for the formation of **terraces,** steplike plateaus of unconsolidated material at some higher elevation above the present floodplain (figure 14.10). Multiple levels of terraces may be found within a single broad valley. It is possible, however, that none of the old terrace surfaces represents a former floodplain level of the present stream.

Another, somewhat puzzling erosional process is the formation of **incised meanders,** sometimes also known as *entrenched meanders* (figure 14.11). As their name suggests, these are meanders deeply cut into rock. The stream is not surrounded by a floodplain at nearly the same elevation but is entrenched within a deep, winding, steep-sided valley

FIGURE 14.11 Incised meanders in rock, with no floodplain surrounding the channel. Goosenecks of the San Juan River, Utah.

closely matching the meandering course of the stream. Several modes of formation have been proposed, and it is not always possible to determine the one responsible in each instance. Some streams, especially slowly flowing ones on gently sloping surfaces, exhibit meanders from their inception, and if downcutting is relatively rapid and the valley wall rocks strong enough to maintain steep cliffs, these early meanders may be dug deeper and deeper. Alternatively, meanders may be developed initially in soft or unconsolidated materials and subsequently fixed in form as the stream cuts down below the softer rock into rocks less prone to lateral erosion. The meander pattern in this instance is imposed on the underlying rock after formation in softer layers above.

Stream Piracy

Headward erosion at the head of a stream valley may lengthen the valley from its upper end. In time, headward erosion may cut through a divide. When this occurs, some part of one stream system is diverted into the other, in a process known as **stream piracy** (figure 14.12). The redistribution of water that results from stream piracy can be expected to cause changes in erosion, sediment transport and deposition, and channel geometry in both stream systems.

Sediment Transport

Streams serve as conveyors for sediment and dissolved chemicals, carrying these materials from elevated areas to depositional basins. Often, the sediment is modified in the process—rounded, sorted, abraded, and dissolved. The stream may also impose a distinctive structure on the sediment as the sediment is deposited. Streams can move material in several ways. The total quantity of material that a stream transports is called its **load.** The load is subdivided on the basis of how the material is transported.

The heaviest debris may be rolled or pushed along the stream bed. This material is described as the **bed load** of the stream. The stream's **competence** at any point is the largest size of particle it can move in the bed load, which is largely a function of the stream's velocity. Stream **capacity** is the total quantity of material that a stream can potentially move, which depends on discharge (that is, on both velocity and quantity of water present). The **suspended load** consists of material that is light or fine enough that it can be moved along suspended in the stream, supported by the flowing water. The maximum suspended load that can be transported is also dependent on discharge. Suspended sediment clouds a stream and gives the water a muddy appearance.

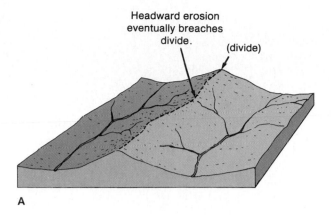

A

B

FIGURE 14.12 Stream piracy (schematic). (A) Headward erosion of one stream valley penetrates the divide between drainage basins. (B) This results in diversion of some of the drainage.

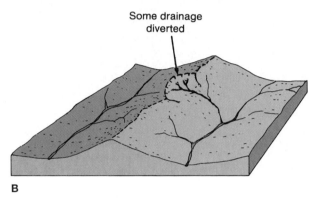

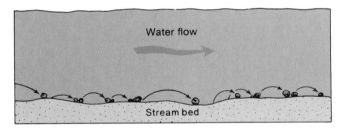

FIGURE 14.13 Saltation (schematic). Heavier particles move in a series of short hops, jostled up into the flowing water and then sinking again to the bottom.

Material of intermediate size may be carried in short hops along the stream bed by a process called **saltation** (figure 14.13). Finally, some substances can be completely dissolved in the water (**dissolved load**).

How much of a load is actually transported depends on the availability of sediments or soluble material: a stream flowing over solid, insoluble bedrock is not able to dislodge much material, while a similar stream flowing through sand may move a considerable load.

Variations in a stream's velocity along its length are reflected in the sediments deposited at different points. The

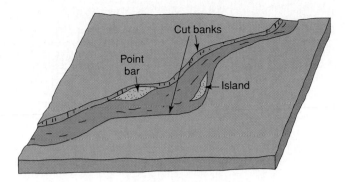

FIGURE 14.14 Point bars and islands form in a sediment-laden stream where the water flow is slowed.

sediments found motionless in a stream bed at any point are those too big or heavy for that stream to move at that point. Where the stream flows most quickly, it can carry gravel and even boulders along with the finer sediments. If velocity controls the maximum particle size that can be moved, it follows that as the stream slows down, it selectively drops the heaviest, largest particles first, continuing to move along the lighter, finer materials. As velocity continues to decrease, successively smaller particles are dropped. In a very slowly flowing stream, only the finest sediments and dissolved materials are still being carried. This link between the velocity of water flow and the size of particles moved accounts for one common characteristic of alluvial sediments in channels: they are often well sorted by size or density, with materials deposited at a given point tending to be similar in size or weight. It also accounts for the increased erosion on the outside banks of meanders (where water flows faster) and deposition inside the curves.

When the stream eventually deposits its load, the resulting deposit is called **alluvium,** a term derived from the French and Latin for "to wash over." *Alluvium* is a general term for any stream-deposited sediment. Certain alluvial deposits take distinctive forms that are given more specific names.

Depositional Features

As a meander migrates laterally, eroding its outer (cut) bank, there is generally corresponding deposition along the inner bank, where the water flow is slower. The sedimentary feature thus built up is a **point bar** (figure 14.14). Where water slows along the channel bed, through friction, it may blanket the bottom with sediment. If the slowdown is a localized one—caused, for example, by obstacles such as isolated boulders on the bottom—the sediment deposition may be correspondingly localized into an island within the channel. Once such a feature has formed, it may further impede stream flow and may therefore tend to enlarge as water continues to be slowed and more sediments are deposited.

FIGURE 14.15 Braided streams, with many channels dividing and rejoining, in an Alaskan glacial-outwash plain.

Where the sediment load carried is large in relation to capacity, channel islands may grow until they reach the water surface, where they may be further stabilized by vegetation. The stream channel then has become divided into two channels, a process called *braiding*. A **braided stream** is one with several (perhaps many) channels that divide and then rejoin downstream (figure 14.15). Patterns of braided channels may become very complex. Where many shallow, braided channels cross a broad area of easily eroded sediment, the channels may shift constantly and the patterns change daily or even hourly.

A floodplain is itself a depositional feature, in part. As channels shift, point bars build land where stream waters flowed. Additional deposition in the floodplain can occur outside the channel, during flood events. The whole floodplain can be blanketed in sediment as the channel overflows and the floodwaters are slowed by vegetation or other obstacles. Such sediments are descriptively termed **overbank deposits.** Much of the coarser sediment is deposited close to the channel, where the velocity of the flooding stream first slows abruptly as the stream overflows its banks. The overbank deposits may then assume a distinctive form, making ridges along the edge of the channel. These are natural **levees** (figure 14.16). Development of levees is generally most pronounced where the suspended sediment load is relatively coarse. Streams carrying primarily fine, silty sediments may not form obvious levees.

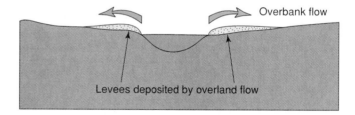

FIGURE 14.16 Natural levees form along the margins of a stream channel as a result of sediment deposition during flooding. (Levees are vertically exaggerated; natural levees are rarely this obvious.)

As previously noted, a decrease in velocity is typically accompanied by deposition, beginning with the coarsest sediments. An abrupt drop in velocity may result in a large deposit. This is not uncommon at the **mouth,** or end, of a stream, where the stream flows into another body of water or terminates in a dry valley or plain.

If a stream flows into a body of standing water, such as a lake or ocean, the flow velocity drops to zero, and all but the dissolved portion of the load is deposited. The resulting fan-shaped sediment wedge is called a **delta** (figure 14.17). Over time, deposition of delta sediments may actually build the land outward into the water, with a series of beds draped one atop another, somewhat as crossbeds form.

A similar fan- or wedge-shaped deposit may form wherever a sediment-laden stream slows suddenly: where a

FIGURE 14.17 The Mississippi River Delta (satellite image). Suspended sediment clouds the water beyond.
© *NASA*.

FIGURE 14.18 An alluvial fan deposited where a fast-moving, sediment-bearing mountain stream (in this case, unleashed by dam failure) flowed into a plain and slowed. Rocky Mountain National Park, Colorado.

fast-moving mountain stream flows out into a plain or flatter valley, or where a tributary flows into a more slowly moving stream. Such a deposit is an **alluvial fan** (figure 14.18).

STREAM CHANNELS AND EQUILIBRIUM

For stream waters to flow, there must be a difference in elevation between the **source,** where the first perceptible channel indicates the stream's existence, and the mouth. The **gradient** of the stream is a measure of the steepness of the channel's slope. It is the difference in elevation between two points along the stream divided by the length of the channel between them. The higher the gradient, the steeper the channel and, all else being equal, the higher the water velocity.

The water level at the mouth of a stream, where it generally flows into an ocean, lake, or other stream, is the stream's **base level.** Base level represents the lowest level to which a stream can cut down. Once elevation upstream reaches base level, the gradient is zero, velocity should become zero, and no further erosion is possible.

Streams are dynamic systems, their forms changing in response to changes in environmental factors. If, over time,

regional precipitation and stream discharge increase, increased erosion will scour a larger channel to accommodate the larger volume of water. An influx of sediment may result in increased sedimentation. If land in the drainage basin is uplifted, the increasing stream gradient and the consequent increases in velocity and rate of downcutting will hasten the stream's reapproach to its base level.

Equilibrium Profile

Over time, provided that environmental factors remain approximately constant, streams and their channels tend to adjust to the prevailing conditions and approach a condition of

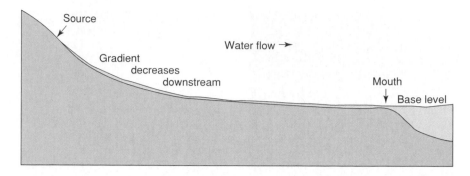

FIGURE 14.19 Longitudinal profile of a graded stream, concave upward, with steeper gradient near the source.

dynamic equilibrium. In terms of erosion and deposition, this means that sediment transport is just sufficient to balance the input of sediment from the drainage basin, and erosion of and deposition in the stream channel are equal. A stream in which the channel and its gradient have been thus adjusted, so that input and outflow of sediment are just equal, is sometimes described as a **graded stream.**

A graded stream tends to show a characteristic pattern of decreasing gradient from source to mouth, illustrated in the **longitudinal profile** of figure 14.19. Generally, the gradient is steepest near the source, flattening toward the mouth. In natural streams, however, velocity does not necessarily decrease downstream, despite the decreasing gradient. The explanation lies in the increase in water volume and channel cross section downstream, associated with the addition of tributaries to the stream and increasing drainage area. The added weight of water being pulled down toward base level by gravity partially compensates for the decrease in gradient, and as cross section increases, a relatively smaller fraction of the water is slowed by friction with the channel base and sides. Overall discharge does increase downstream.

Geology can cause deviations from the idealized stream profile. Along the stream channel, there may be steep drops in elevation, known as **knickpoints** (figure 14.20A). Knickpoints can arise from streamflow across different rock types of different erodibility, or by faulting. Also, above the junction with a tributary, the channel may be cut less deeply than below, where an abruptly greater volume of water is present. Major knickpoints may be marked by waterfalls (figure 14.20B). Once established, such scarps tend to erode headward, or upstream, as continued bed erosion wears them back or as material is carried downstream from the base of the scarp and the undercut scarp slumps and fails. The upstream migration of Niagara Falls is a classic example of such **headward erosion** of a knickpoint.

The concept of a stream in perfect equilibrium is an abstraction. Over the long term, gravity tends to level the land, and ultimately, in the absence of tectonic uplift, volcanic eruption, or other mountain-building activities, a net downcutting should be expected until the land is flattened to sea level. However, the approach to equilibrium can be so close that, on a human timescale, the net change is negligible.

Effects of Changes in Base Level

Tectonic uplift can raise the upper portion of a drainage basin relative to the base level of the stream system (an adjacent ocean, for example). The net effect is to increase the overall gradient of the system and the rate of downcutting toward base level. Also, changes in sea level, such as are caused by glaciation (see chapter 18), move base level up or down and decrease or increase average gradients accordingly.

When human activities alter base levels, more complex modifications of the stream system are possible. For example, if a dam with a reservoir behind it is built along a previously graded stream, the water level in the reservoir becomes a new local base level for the stream above the dam (figure 14.21). Its average gradient is decreased, and any downcutting should likewise decrease. Moreover, when the stream reaches the still reservoir, and flow ceases, the stream drops its (nondissolved) sediment load. This begins to fill in the reservoir, reducing the volume available for water storage, whether for irrigation, water supply, flood control, or whatever. Below the dam, the water released is free of suspended sediment or bed load and is therefore capable of increased erosion through uptake of sediment. The net result of all of these changes is reflected in a modified longitudinal profile, as shown in figure 14.21.

FLOODING AND FLOOD HAZARDS

In most (moderately humid) climates, the size of a stream channel adjusts to accommodate the average maximum annual discharge. Much of the year, the water level (stream **stage**) may be well below the stream bank height. From time to time, the stream will reach **bankfull stage.** Should the water rise above bankfull stage, overflowing the channel, the stream is at **flood stage.** In other words, floods occur when the rate at which water reaches some point in a stream exceeds the bankfull discharge there. Floods are not unnatural or particularly unusual events. Instead, flooding is a perfectly normal and, to some extent, predictable phenomenon.

Factors Influencing Flood Severity

Many factors together determine whether a flood will occur. The quantity of water involved and the rate at which it is

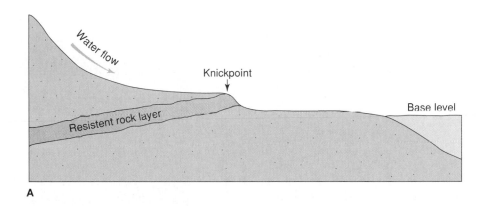

A

B

FIGURE 14.20 Knickpoints in the profile. (A) Slower erosion of a resistant rock layer produces a cliff (schematic). (B) Niagara Falls formed where a nearly flat-lying ledge of Lockport dolomite has produced a knickpoint.

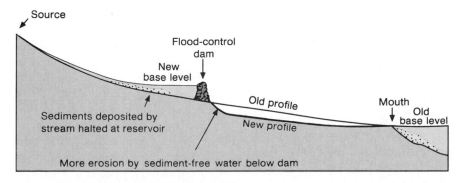

FIGURE 14.21 Effects of a dam and reservoir on a stream profile. The new base level of the reservoir causes deposition both in the reservoir and upstream from it. Erosion increases below the dam.

put into the stream system are among the major factors. Worldwide, the most intense rainfall events occur in Southeast Asia, where storms have dumped up to 200 centimeters (80 inches) of rain in less than three days. (To put such numbers in perspective, that amount of rain is more than double the average *annual* rainfall over the United States!) In the United States, heavy rainfall events occur in the southeastern states, which are vulnerable to storms from the Gulf of Mexico; the western coastal states, which are subject to prolonged storms from the Pacific Ocean; and the midcontinent states, where hot, moist air from the Gulf of Mexico collides with cold air sweeping down from Canada. Streams that drain the Rocky Mountains are particularly likely to flood during snowmelt events, especially when rapid spring thawing follows a winter of unusually heavy snow.

The risk of flooding may be moderated by infiltration, the rate of which, in turn, is controlled by the soil type and how much soil is exposed. Soils, like rocks, vary both in proportion of pore space and in the rate at which water seeps through them (*permeability*). A very porous and permeable soil can allow a great deal of water to sink in relatively rapidly. If the soil is less permeable, the proportion of water that runs off over the surface increases. Also, once even permeable soil is saturated with water and all the pore space is filled (as, for instance, by previous storms), any additional moisture is necessarily forced to become part of the surface runoff. Topography also influences the extent of surface runoff: the steeper the terrain, the more readily water runs off over the surface and the less it tends to sink into the soil.

Water that infiltrates into the soil tends to flow downhill, like surface runoff, and may, in time, also reach the stream. However, the underground runoff water, flowing through soil or rock, generally moves much more slowly than the surface runoff. The more gradually the water reaches the stream, the better the chances that the stream discharge will be adequate to carry the water away without flooding. Therefore, the relative amounts of surface and subsurface runoff, which are strongly influenced by the near-surface geology of the drainage basin, are fundamental factors affecting the severity of stream flooding.

Vegetation may reduce flood hazards in several ways: by providing a physical barrier to surface runoff, slowing it down; through plants' root action, which keeps the soil looser and more permeable, thereby increasing infiltration and decreasing surface runoff; and by soaking up some of the water and later releasing it by evapotranspiration from foliage. Vegetation can also be critical to preventing soil erosion. When the vegetation is removed and erosion increased, much more soil may be washed into streams. There it can begin to fill in the channel, decreasing channel volume and thus reducing the stream's ability to carry water away quickly.

Parameters for Describing Flooding

During a flood, stream stage, velocity, and discharge all increase as a greater mass of water is pulled downstream by gravity. The magnitude of the flood can be described either by the maximum discharge measured or by the maximum stage.

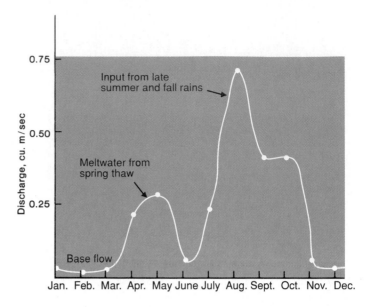

FIGURE 14.22 Sample hydrograph for Horse Creek near Sugar City, Colorado, spanning a one-year period.
Source: U.S. Geological Survey Open-File Report 79-681.

The stream is said to **crest** when the maximum stage is reached. This may occur within minutes of the influx of water, as with flooding just below a failed dam. However, in places far downstream from the water input, or when surface runoff has been delayed, the stream may not crest until several days after the flooding episode begins. In other words, the worst is not necessarily over just because the rain has stopped.

Fluctuations in stream stage or discharge through time can be plotted on a *hydrograph* (figure 14.22). Hydrographs spanning long periods of time are very useful in constructing a picture of the normal behavior of a stream and of that stream's response to flood-causing events. A flood shows as a peak on the hydrograph. The height and width of that peak and its position in time relative to the water-input event(s) depend, in part, on where the measurements are being taken, relative to where the excess water is entering the system. Upstream, where the drainage basin is smaller and the water need not travel so far to reach the stream, the peak is likely to be sharper—higher crest, more rapid rise and fall of the water level—and to occur sooner after the influx of water. Downstream, in a larger drainage basin, where some of the water must travel a considerable distance to the stream, the arrival of the water spans a longer time. The peak is spread out, so that the hydrograph shows both a later and a broader, gentler peak (see figure 14.23). Similarly, in the case of a localized flood high in a drainage system, caused by an event such as dam failure or a local cloudburst, measurements made near the point where the excess water is entering the system show an earlier, sharper peak. By the time that water pulse has moved downstream to a lower point in the drainage basin, it has dispersed somewhat, so that the peak on the hydrograph is again later, slower, and of longer duration. A short event like a severe cloudburst tends to produce a sharper peak than a more prolonged event like several days of steady rain or snowmelt, even if the same amount of water is involved.

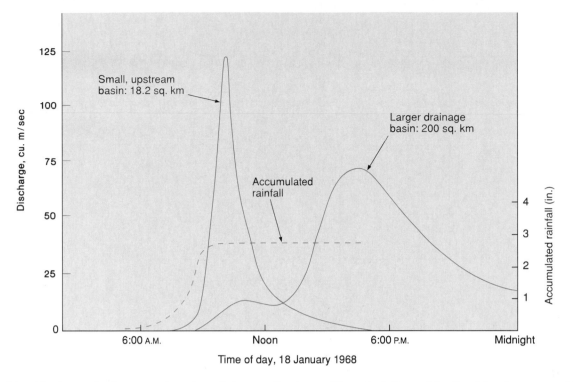

Figure 14.23 Flood hydrographs for two points along Calaveras Creek near Elmendorf, Texas. The flood has been caused by heavy rainfall. Upstream, flooding quickly follows rain. Downstream in the basin, response to water input is more sluggish.
Source: U.S. Geological Survey Water Resources Division.

Flood-Frequency Curves

Another way of looking at flooding is in terms of the frequency or probability of flood events of differing severity. The availability of long-term records makes it possible to construct a curve showing discharge as a function of recurrence interval, or probability of occurrence, for a particular stream or section of one. The sample flood-frequency curve shown in figure 14.24 indicates that, on average, an event producing a discharge of 675 cubic feet/second occurs once every ten years, an event with discharge of 350 cubic feet/second occurs once every two years, and so on. A given flood event can thus be described by its **recurrence interval** (how frequently a flood of that severity occurs, on average, for that stream). For the stream of figure 14.24, for example, a flood with discharge of 675 cubic feet/second would be called a *10-year flood,* meaning that a flood of that size occurs about once every ten years; a discharge of 900 cubic feet/second corresponds to a *40-year flood;* and so on. An alternative way of looking at flooding is in terms of probability, which is inversely related to recurrence interval. That is, a 10-year flood has a ¹⁄₁₀ (or 10%) probability of occurrence in any one year; a 2-year flood has a ½ (50%) probability in any one year; and so on.

Flood-frequency curves can be extremely useful in assessing regional flood hazards. In principle, even if a 100- or 200-year flood event has not occurred within memory, planners can, with the aid of topographic maps, project how much of a region would be flooded in such events. This, in turn, is useful in siting new construction projects so as to minimize the risk of flood damage or in making property owners in threatened areas aware of dangers.

I t is important to remember that the recurrence intervals assigned to floods are *averages.* Over many centuries, a 50-year flood should occur an average of once every fifty years; or, in other words, there is one chance in fifty that that discharge will be exceeded in any one year. However, that does not mean that two 50-year floods could not occur in successive years, or even in the same year. Therefore, it would be foolish to assume that, just because a very severe flood has recently occurred in an area, the area is in any sense "safe" for awhile. Statistically, another such severe flood probably will not occur again for some time, but there is no guarantee. The probability terminology may be less misleading to the nonspecialist, for it carries no implication of a long time interval between major floods. However, the recurrence-interval terminology is more commonly used by many public agencies dealing with flood hazards, such as the U.S. Army Corps of Engineers.

Unfortunately, the best-constrained part of the curve is, by definition, that for the lower-discharge, more-frequent, less-serious floods. Much of the United States has been settled for a century or less. Reliable records of stream stages, discharges, and the extent of past floods typically extend back only a few decades. Many areas, then, may never have recorded a 50- or 100-year flood. Moreover, when the odd severe flood does occur, how does one know whether it is

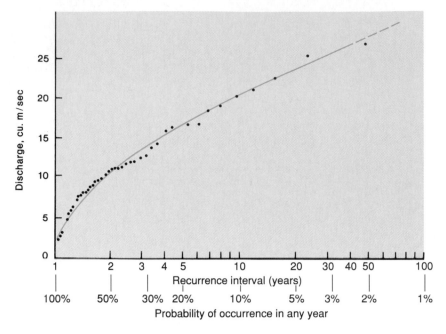

FIGURE 14.24 Sample flood-frequency curve for the Eagle River at Red Cliff, Colorado. Records span 46 years, so fairly accurate assessments can be made of the likelihood of moderate flood events.
Source: U.S. Geological Survey Open-File Report 79-1060.

a 60-year flood, a 110-year flood, or whatever? The high-discharge events are rare, and even when they happen, their recurrence interval can often only be estimated. Consider-able uncertainty can exist, then, about just how serious a particular stream's 100- or 200-year flood might be. This problem is explored further in box 14.1.

Another complication is that streams in heavily populated areas are being affected by human activities. The way a stream responded to 10 centimeters of rain from a thunderstorm 100 years ago may be quite different from the way it responds today. The flood-frequency curves are therefore changing with time. Except for measures specifically designed for flood con-trol, most human activities have tended to aggravate flood haz-ards. What may have been a 100-year flood two centuries ago might be a 50-year flood now, or even a more frequent one.

Flooding and Human Activities

Obviously, the more people settle and build in floodplains, the more damage flooding will do. What people often fail to realize is that floodplain development can actually increase the likelihood or severity of flooding.

Two factors affecting flood severity are the proportion and rate of surface runoff. The materials extensively used to cover the ground when cities are built, such as asphalt and concrete, are relatively impermeable and greatly reduce infil-tration. Therefore, when a considerable area is covered by these materials, surface runoff tends to be much more concen-trated and rapid than before, increasing the risk of flooding. Another problem is that buildings in a floodplain also can in-crease flood heights: the buildings occupy volume that water

formerly could fill, and a given discharge then corresponds to a higher stage. Filling in floodplain land for construction simi-larly decreases the water storage volume available to stream water and further aggravates the situation (figure 14.25). City storm sewers are installed to keep water from swamping streets during heavy rains, and often the storm water is chan-neled straight into a nearby stream. This works fine if the total flow is moderate enough, but by decreasing the time normally taken by the water to reach the stream channel, such measures increase the probability of flooding.

The **peak lag time** of a flood event can be defined as the length of time between precipitation and maximum flood discharge (or stage). Comparison of stream response before and after urbanization shows that urbanization typi-cally decreases peak lag time (figure 14.26). Peak discharge (or stage) also increases.

A basic strategy for reducing the risk of flood damage is to identify, as accurately as possible, the area at risk. Care-ful mapping, coupled with accurate stream discharge data, allows identification of those areas threatened by floods of different recurrence intervals. Satellite images (figure 14.27) can also be helpful in identifying vulnerable areas. Land that could be inundated often—by the 25-year floods, perhaps—might best be restricted to land uses not involving much building, such as for livestock grazing pasture or for parks or other recreational purposes. Where land is at a premium and there is pressure to build even in the floodplain (as in many urban areas), buildings can be raised on stilts, so that the lowest floor is above the expected 200-year flood stage, for example. This at least minimizes both the danger of damage to the building and its contents and the interference

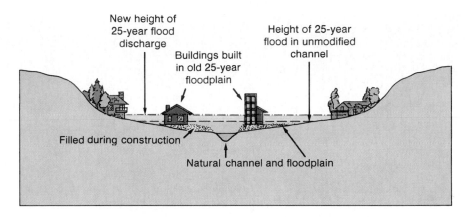

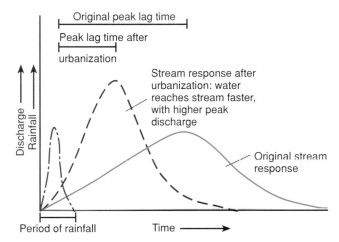

FIGURE 14.25 The filling-in of floodplain land and the building of structures can result in higher flood stages for the same discharge in later flood events.

FIGURE 14.26 Hydrograph reflects modifications of stream response to precipitation following urbanization: peak discharge increases, lag time to peak decreases.

FIGURE 14.27 Mississippi River basin flooding, 19 July 1993. The Missouri River (entering from top of image) and Kansas River (left) have spread broadly across their floodplains. Within the flooded areas, the rivers' normal channels are outlined by vegetation (red in image). The Kansas City, Missouri, municipal airport is just east of the junction of the rivers. Such images are useful in assessing flood hazards and damage. *Photograph © NASA.*

Box 14.1

How Big Is the 100-Year Flood?

The difficulty of knowing the true recurrence intervals of floods of various magnitudes can be illustrated by an example that demonstrates the need for long-term records.

The usual way of estimating the recurrence interval of a flood of given size is as follows: Suppose that records of maximum discharge (or stage) reached by a particular stream each year have been kept for N years. Each of these yearly maxima can be given a rank M, ranging from 1 to N, 1 being the largest, N the smallest. Then the recurrence interval R of a given annual maximum is defined as

$$R = (N + 1)/M.$$

For example, table 1 shows the maximum one-day mean discharges of the Big Thompson River, as measured near Estes Park, Colorado, for twenty-five consecutive years, 1951–1975. If these values are ranked, 1 to 25, the 1971 maximum of 1030 cubic feet/second is the seventh largest and, therefore, has an estimated recurrence interval of $(25 + 1)/7$ = 3.71 years.

Suppose, however, that only ten years of records are available, for 1966–1975. The same 1971 maximum discharge happens to be the largest in that period of record. On the basis of the shorter record, its estimated recurrence interval is $(10 + 1)/1 = 11$ years. Alternatively, if we look at only the first ten years of record, 1951–1960, the recurrence interval for the 1958 maximum discharge of 1040 cubic feet/second (a discharge of nearly the same size as the 1971 maximum) can be estimated at 2.2 years.

Which estimate is right? Perhaps none of them, but their differences illustrate the need for long-term records to

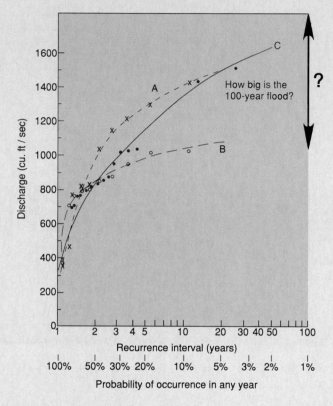

FIGURE 1   Short-term records (curves A and B) can be misleading about the true recurrence intervals or probabilities of floods of different magnitudes. Compare with curve C, based on a 25-year record.

smooth out short-term anomalies in streamflow patterns.

The point is further illustrated in figure 1. It is rare to have 100 years or more of records for a given stream, so the magnitude of 50-year, 100-year, or larger floods is commonly estimated from a flood-frequency curve. Curves A and B in figure 1 are based, respectively, on the first and last ten years of data for the Big

Thompson River (last two columns of table 1). These two data sets give estimates for the size of the 100-year flood that differ by more than 50%. These results, in turn, both differ from an estimate based on the full twenty-five years of data (curve C). Figure 1 graphically illustrates how important long-term records can be in the projection of recurrence intervals of larger flood events.

TABLE 1 *Calculated Recurrence Intervals for Discharges of Big Thompson River at Estes Park, Colorado*

Year	Maximum Mean One-Day Discharge (cu. ft/sec)	For Twenty-Five-Year Record		For Ten-Year Record	
		M (rank)	R (years)	M (rank)	R (years)
1951	1220	4	6.50	3	3.67
1952	1310	3	8.67	2	5.50
1953	1150	5	5.20	4	2.75
1954	346	25	1.04	10	1.10
1955	470	23	1.13	9	1.22
1956	830	13	2.00	6	1.83
1957	1440	2	13.0	1	11.0
1958	1040	6	4.33	5	2.20
1959	816	14	1.86	7	1.57
1960	769	17	1.53	8	1.38
1961	836	12	2.17		
1962	709	19	1.37		
1963	692	21	1.23		
1964	481	22	1.18		
1965	1520	1	26.0		
1966	368	24	1.08	10	1.10
1967	698	20	1.30	9	1.22
1968	764	18	1.44	8	1.38
1969	878	10	2.60	4	2.75
1970	950	9	2.89	3	3.67
1971	1030	7	3.71	1	11.0
1972	857	11	2.36	5	2.20
1973	1020	8	3.25	2	5.50
1974	796	15	1.73	6	1.83
1975	793	16	1.62	7	1.57

Source: *Data from U.S. Geological Survey Open-File Report 79–681.*

by the structure with flood flow. Another possibility is to provide compensating capacity elsewhere for every loss of water capacity caused by filling in floodplain land. A major limitation of any floodplain zoning plan, whether it involves prescribing design features for buildings or banning them entirely, is that it almost always applies only to new construction, not to scores of older structures already at risk.

If open land is available, flood hazards along a stream may be reduced greatly by the use of **retention ponds.** These are large basins that trap some of the surface runoff, keeping it from flowing immediately into the stream. They may be elaborate artificial structures, old abandoned quarries, or, in the simplest cases, fields dammed by dikes of piled-up soil. Often the land can still be used for other purposes, such as farming, except on those rare occasions of heavy runoff when it is needed as a retention pond. Retention ponds are frequently a relatively inexpensive option and have the advantage that they do not attempt to alter the character of the stream itself.

Channelization is a general term for various modifications of the stream channel itself that are usually intended to increase the velocity of water flow, the volume of the channel, or both. These modifications, in turn, increase the discharge of the stream, and hence the rate at which surplus water is carried away. The channel can be widened or deepened, especially where soil erosion and subsequent sediment deposition in the stream have partially filled in the channel. Alternatively, a stream channel might be rerouted—for example, by deliberately cutting off meanders to provide a more direct path for the water flow. Such measures do indeed tend to decrease the flood hazard upstream from where they are carried out. But a meandering stream often tends to keep meandering or to revert to old meanders. Channelization is not a one-time effort. Constant maintenance is required to limit channel bank erosion and to keep the river in the redirected channel. Also, by causing more water to flow downstream faster, channelization often increases the likelihood of flooding downstream from the alterations.

This last complication is also often true with the building of artificial levees that raise the height of the stream banks so that the water can rise higher without flooding the surrounding country. This ancient technique was practiced thousands of years ago on the Nile by the Egyptian pharaohs. Confining the water to the channel rather than allowing it to flow out into the floodplain effectively shunts the water downstream faster during high-discharge events, however, which may increase flood risks downstream. It artificially raises the stage of the stream for a given discharge, which can increase the risks upstream, too. Another problem is that levees may make people feel so safe about living in the floodplain that development will be far more extensive than if the stream were allowed to flood naturally from time to time. If the levees have not, in fact, been built high enough, and a very severe flood overtops them, or if they simply fail and are breached during a high-discharge event, far more lives and property may be lost as a result. The catastrophic flooding in the Mississippi River basin in the summer of 1993 (figure 14.28), which caused fifty deaths and over $10 billion in property damage, provided a dramatic demonstration of this problem. Also, if the levees are overtopped, they then trap water *outside* the stream channel, where it may stand for some time after the stream stage subsides, until infiltration returns it to the stream (figure 14.29).

Yet another approach to moderating streamflow to prevent or minimize flooding is the construction of flood-control dams at one or more points along the stream. Excess water is held in the reservoir formed upstream from such a dam and may then be released at a controlled rate so as not to overwhelm the capacity of the channel beyond. As previously mentioned, however, such an artificial change in local stream base level alters channel geometry. If the stream normally carries a high sediment load, silting-up of the reservoir forces repeated dredging, which can be expensive and presents the problem of where to dump the dredged sediment. Sometimes, the reduction in sediment carried to the stream mouth is also a problem there. So many dams along the Mississippi River system have impounded sediment that, overall, the river delta is no longer building up and is eroding in many places. Some large reservoirs, such as Lake Mead behind Hoover Dam, have even been found to cause earthquakes. The water in the reservoir represents an added load on the rocks, increasing the stresses on them, while infiltration of water into the ground under the reservoir increases pore fluid pressures in rocks along old faults. Since Hoover Dam was built, at least 10,000 small earthquakes have occurred in the vicinity. Earthquakes caused in this way are usually of low magnitude, but their foci, naturally, are close to the dam. This raises concerns about the possibility of catastrophic dam failure caused, in effect, by the presence of the dam itself.

A   **B**

FIGURE 14.28   Consequences of Mississippi River flooding, 1993. (A) Flooding along the riverfront at Davenport, Iowa; note submerged baseball stadium in foreground. (B) The flooding turned rural areas into lakes dotted with trees and scattered houses.
Sources: Photograph (A) © Doug Sherman/Geofile. Photograph (B) © Jim Shaffer.

FIGURE 14.29   When insufficiently high levees are overtopped, water is trapped behind them. Several days after this flood on the Kishwaukee River in DeKalb, Illinois, the stage of the river (right) had dropped considerably. Water level behind the levee (left) remained much higher, prolonging flooding and increasing flood damage.

SUMMARY

All water at and near the earth's surface moves through the hydrologic cycle, the principal component processes of which are evaporation, precipitation, and surface and groundwater runoff. Streams are a major agent of surface runoff. A stream is any body of flowing water usually confined within a channel. The size of a stream, as measured by its discharge, is a function of climate, near-surface geology, and the size of the drainage basin from which the water is drawn. The force of the flowing water may cause both downcutting and lateral erosion of banks, including meandering. Meander formation and migration, coupled with deposition of alluvial sediments in and around the stream, contribute to the creation of floodplains. Material is moved by streams in solution, in suspension, or along the stream bed. The faster the water flow, the coarser the particles moved, and the greater the capacity of the stream to transport sediment. Alluvium deposited where water velocity decreases may take the form of bars or islands in the channel, overbank deposits (including natural levees) in the floodplain, or alluvial fans and deltas at stream mouths. Over time, most streams tend toward an equilibrium longitudinal profile, with gradient decreasing from source to mouth. When sediment influx and outflow are in balance, the result is a graded stream.

Floods are the stream's response to an input of water too large and/or too rapid for the discharge to be accommodated within the channel. The risk of flooding is a function of climate, soil character, presence or absence of vegetation, and topography. Human activities may unintentionally increase flood hazards by filling in the channel or floodplain during construction, covering soil with impermeable materials, or draining runoff to streams quickly via sewers. Use of flood-frequency records may help to identify the areas most at risk, but the changes wrought recently by human activities limit the usefulness of historic records. Strategies used to reduce flood hazards include restrictive zoning; the building of retention ponds, artificial levees, or flood-control dams; and various kinds of channelization. These practices, however, are not without some disadvantages or even risks.

TERMS TO REMEMBER

alluvial fan 257
alluvium 249
bankfull stage 252
base level 251
bed load 248
braided stream 250
capacity 248
channelization 260
competence 248
continental divide 243
crest (flood) 254
cut bank 245
delta 250
dendritic drainage 243
discharge 244
dissolved load 249
divide 243

downcutting 245
drainage basin 243
dynamic equilibrium 252
floodplain 246
flood stage 252
graded stream 252
gradient 251
headward erosion 252
hydrologic cycle 242
hydrosphere 242
incised meanders 247
infiltrates 242
knickpoint 252
levees 250
load 248
longitudinal profile 252
meanders 245
mouth 250

overbank deposit 250
oxbows 246
peak lag time 256
point bar 249
radial drainage 244
rectangular drainage 244
recurrence interval 255
retention pond 260
saltation 249
source 251
stage 252
stream 242
stream piracy 248
suspended load 248
terraces 247
trellis drainage 243
tributary stream 243

QUESTIONS FOR REVIEW

1. Briefly summarize the principal processes and reservoirs of the hydrologic cycle. (You may find a sketch helpful.)
2. What is a *stream?* A *drainage basin?*
3. What determines the order of a stream?
4. Describe the kind of geologic situation in which you might expect each of the following to develop: (a) rectangular drainage, (b) radial drainage.
5. What is a stream's *discharge,* and how is it related to stream capacity and load?

6. Briefly describe the development and migration of meanders. How are meanders and oxbow lakes related?
7. Under what circumstances does stream piracy occur?
8. What is a *braided stream,* and how does braiding develop?
9. Why does a stream commonly deposit a delta at its mouth?
10. Sketch the equilibrium longitudinal profile of a graded stream, indicating base level. How is the profile changed by construction of a dam along the stream?

11. What is a *flood?* Outline briefly how flood hazards are affected by each of the following: (a) intensity of precipitation, (b) soil type, (c) presence of vegetation, (d) construction that adds extensive impermeable cover.
12. What is a *flood-frequency curve?* Is the term accurate?
13. Cite and briefly explain three strategies for reducing flood hazards, noting the strengths and weaknesses of each.

FOR FURTHER THOUGHT

1. Make a point of visiting a nearby stream. Walk along the channel, noting meandering, areas where the bank is eroding or point bars are being deposited, and any evidence of human modification of the channel.

2. Investigate the availability of flood-hazard maps for a nearby stream. How current are the maps and the data upon which they are based? Have any efforts been made to reduce the flood hazards? If so, what have they been, what have they cost, and what negative effects, if any, have they had?

3. For a river that has undergone channel modification, investigate the history of flooding before and after that modification, and compare the costs of flood damage to costs of flood-hazard-reduction strategies. (The Army Corps of Engineers engages in many such projects and may be able to supply the information.)

SUGGESTIONS FOR FURTHER READING

Bolt, B. A., et al. 1975. *Geological hazards.* New York: Springer-Verlag. (Chapter 7 surveys causes of floods and discusses flood-hazard mitigation.)

Chin, E. H., Skelton, J., and Guy, H. P. 1975. *The 1975 Mississippi River basin flood.* U.S. Geological Survey Professional Paper 937.

Illinois Department of Transportation. 1982. *Protect your home from flood damage.* Springfield, Ill: Illinois Dept. of Transportation.

Knighton, D. 1984. *Fluvial forms and processes.* London: Edward Arnold.

Leopold, L. B., Wolman, M. G., and Miller, J. P. 1964. *Fluvial processes in geomorphology.* San Francisco: W. H. Freeman.

Mairson, A. 1994. The great flood of '93. *National Geographic* 185 (January): 44–87.

Morisawa, M. 1985. *Rivers.* New York: Longman.

Petts, G. E., and Foster, G. P. 1985. *Rivers and landscape.* Baltimore, MD: Edward Arnold.

Richards, K. (ed.) 1987. *River channels: Environment and process.* New York: Blackwell.

Ritter, D. F. 1986. *Process geomorphology.* 2d ed. Dubuque, Iowa: Wm. C. Brown Publishers.

Tank, R. W. 1983. *Environmental geology, text and readings.* 3d ed. New York: Oxford University Press. (The section on floods, pp. 218–77, includes several readings on responses to flood hazards.)

Whipple, W., Grigg, N. S., Grizzard, T., Randall, C. W., Shubinski, R. P., and Tucker, C. S. 1983. *Stormwater management in urbanizing areas.* Englewood Cliffs, N.J.: Prentice-Hall.

15

GROUND WATER AND WATER RESOURCES

Solution of limestone by subsurface water hollows out caverns; calcite deposited by dripping, evaporating water then forms stalactites and stalagmites. Carlsbad Caverns, NM. © Doug Sherman/Geofile

Streams and oceans are water in its most visible form. An increasing proportion of the water that we actually *use,* however, comes from below the surface, from groundwater supplies. This chapter begins with an examination of the relationship between surface and subsurface waters and the global water budget. A detailed discussion of the nature of ground water and the consequences of its presence and consumption follows. The problem of groundwater pollution is briefly outlined. The chapter concludes with a survey of the water-supply situation in the United States, including some options for extending freshwater supplies.

Table 15.1 shows how the water in the hydrosphere is distributed; recall also figure 14.1. An important point that emerges immediately from these data is that there is, relatively speaking, very little fresh liquid water on the earth. By far the largest reservoir is the very salty ocean. Most of the fresh water is locked up as ice, mainly in the large polar ice caps. Even the ground water beneath the surface of the continents is not all fresh, although it is so classified in the table for lack of analytical data on all ground water. These facts underscore the need for restraint in the use of fresh water.

SUBSURFACE WATER

If soil on which precipitation falls is sufficiently permeable, infiltration occurs. Gravity draws the water downward until an impermeable rock or soil layer is reached. The water then begins to accumulate above that layer. Immediately above the impermeable material is a zone of rock or soil that is saturated (in which water fills all the accessible pore space), called the **phreatic zone,** or *zone of saturation.* Above the phreatic zone is rock or soil in which the pore spaces are filled partly with water, partly with air (the **vadose zone** or *zone of aeration*). All of the water occupying pore space below the ground surface is, logically, termed **subsurface water.** True **ground water,** however, is the water in the zone of saturation, or phreatic zone, only. It is distinguished from **soil moisture,** which is water held in small pores or on grain surfaces in unsaturated soil (in the vadose zone). The **water table** is defined as the top of the zone of saturation, where the saturated zone is not confined by overlying impermeable rocks (see the discussion of aquifer geometry later in this chapter). These relationships are illustrated in figure 15.1. In nature, then, precipitation is the ultimate source of subsurface water.

Porosity and Permeability

The abundance, availability, and movement of subsurface water are directly related to the *porosity* and *permeability* of geologic materials. **Porosity** is the proportion of void space (holes or cracks) in the material. Porosity includes empty spaces between mineral grains in the rock (*intergranular*

TABLE 15.1 *The Water in the Hydrosphere*

Reservoir	Percentage of Total Water*	Percentage of Fresh Water†	Percentage of Unfrozen Fresh Water†
Oceans	97.54	—	—
Ice	1.81	73.9	—
Ground water	0.63	25.7	98.4
Lakes and streams			
Salt	0.007		
Fresh	0.009	0.36	1.4
Atmosphere	0.001	0.04	0.2

*These figures account for over 99.9% of the water. Some water is also held in organisms (the biosphere) and some as soil moisture.

†This assumes that all ground water is more or less fresh water, since it is not all readily accessible to be tested and classified.

Source: Data from J. R. Mather, *Water Resources,* 1984, John Wiley & Sons, Inc., New York, NY.

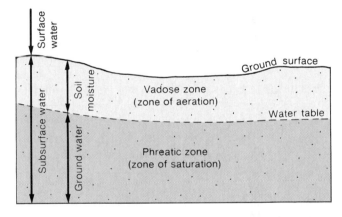

FIGURE 15.1 Nomenclature for surface and subsurface waters. Ground water is the water in the zone of saturation, below the water table.

porosity), and unfilled fractures. It may be expressed either as a percentage (for example, 1.5%) or as an equivalent decimal fraction (0.015). The pore spaces may be occupied by gas, liquid, or a combination of the two. **Permeability** is a measure of how readily fluids pass through the material. It is related to the extent to which pores or cracks are interconnected.

The porosity and permeability of geologic materials are both influenced by the shapes of the mineral grains or rock fragments in the material, the range of grain sizes present, and the ways in which the grains fit together (figure 15.2). Rocks that consist of tightly interlocking crystals (such as igneous rocks) usually have both little porosity and low permeability, unless they have been broken up by fracturing or weathering. By contrast, materials consisting of well-rounded,

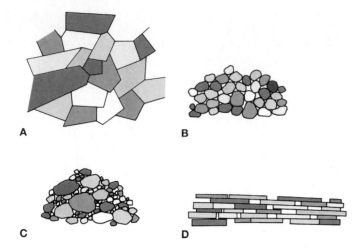

FIGURE 15.2 Porosity and permeability vary with grain shapes and the ways in which grains fit together. (A) Low porosity in an igneous rock with tightly interlocking crystals. (B) A sandstone with well-rounded, well-sorted grains has more pore spaces. (C) Poorly sorted sediment: fine grains fill pores between coarser ones. (D) Packing of plates of clay in a shale may result in high porosity but low permeability. (Note: clay grains are much smaller than sand; they are enlarged here for clarity.)

TABLE 15.2 *Representative Porosities and Permeabilities of Geologic Materials*

Material	Porosity (%)	Permeability (m/day)
Unconsolidated		
Clay	45–55	Less than 0.01
Fine sand	30–52	0.01–10
Gravel	25–40	1000–10,000
Glacial till	25–45	0.001–10
Consolidated (rock)		
Sandstone and conglomerate	5–30	0.3–3
Limestone (crystalline, unfractured)	1–10	0.00003–0.1
Granite (unweathered)	Less than 1–5	0.0003–0.003
Lava	1–30; mostly less than 10	0.0003–3, depending on presence or absence of fractures or interconnected gas bubbles

Source: Data from T. Dunne and L. B. Leopold. *Water in Environmental Planning,* 1978, W. H. Freeman and Company, New York, NY.

equidimensional grains of similar size—a well-sorted sandstone, for instance—may have quite high intergranular porosity and permeability. (This can be illustrated by pouring water through a pile of marbles.) In poorly sorted materials, finer grains can fill the gaps between coarser ones, reducing porosity, though permeability may remain high. Conversely, a rock in which mineral grains are platy or slab-shaped may have those grains arranged in such a way that porosity is high but the pores are poorly connected, so permeability is low, especially perpendicular to the grains' flat surfaces. (Imagine a stack of closed, empty, flat boxes.) This situation is characteristic of many shales. Porosities and permeabilities of some representative geologic materials are shown in table 15.2.

Ground Water

The water table is not always below the ground surface. Wherever surface water persists, as in a lake or stream, the water table is locally above the ground surface, and the water's surface is the water table. Whether the ground water feeds the stream or the stream replenishes the ground water depends on geology, climate, and season. With any perennial stream (one in which some water is always flowing) can be associated a certain level of **base flow,** supported by ground water from adjacent rock and soil and distinguished from the direct surface runoff associated with precipitation or melting events (see figure 15.3). When the flow is base flow only, the stream bed must be below the water table; the stream in this case receives water from adjacent units. This situation is more common between intense runoff events in moderately humid climates. When excess surface runoff drains into the stream, it

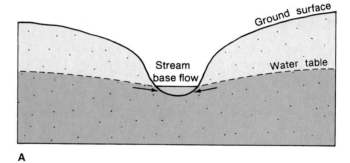

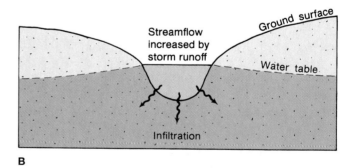

FIGURE 15.3 The amount of base flow in a stream is dependent on the height of the regional water table. Base flow is distinguished from storm runoff. (A) In dry conditions, flow is entirely base flow supported by ground water. (If the water table is too low, the stream dries up entirely.) This is an *effluent stream.* (B) During storms or melting episodes, surface runoff contributes to streamflow. From this *influent stream,* water infiltrates to recharge ground water.

FIGURE 15.4 The water gushing from the rock wall above this stream is spring-fed by water percolating through the limestone; it flows whether or not rain has recently fallen, confirming that it is fed by ground water.

supplements the base flow and increases stream discharge. The stream stage is raised above the adjacent water table, and the stream contributes water to the adjacent rock and soil. This occurs during the rare precipitation runoff events in arid climates and during high-precipitation events in moister areas. **Springs,** where water flows out spontaneously onto the earth's surface, are also places where, locally, the water table intersects the surface (figure 15.4).

The water table should not be pictured as flat like a table-top; it may undulate with the surface topography and with the changing distribution of permeable and impermeable rocks underground. The depth to the water table also varies. The water table is highest when the ratio of input water to water removed is greatest, typically in the spring, when rain is heavy or snow and ice accumulations melt. In dry seasons, or when human use of ground water is heavy, the water table drops, and the amount of available ground water remaining decreases. Ground water does not move only vertically. It can also flow laterally through permeable soil and rock, from higher elevations to lower, from areas of abundant infiltration to drier ones, or from areas of little groundwater use toward areas of heavy use. The direction of flow can be predicted, and just as there are surface-water divides, there are groundwater divides, from which ground water flows in different directions. The processes of infiltration and migration through which ground water is replaced are collectively termed **recharge.**

Aquifers

Rocks and soils vary greatly in porosity and permeability. If ground water drawn from a well is to be used as a source of water supply, the porosity and permeability of the surrounding rocks are critical. The porosity controls the total amount of water available. In most places, there are not vast pools of open water underground to be tapped, just the water in the rocks' pore spaces. This pore water at most amounts to a few percent of the rocks' volume, and usually much less. The permeability, in turn, governs both the rate at which water can be withdrawn and the rate at which recharge can occur. All the pore water in the world would be of no practical use if it were so tightly locked in the rocks that it could not be pumped out. Even in relatively permeable rocks, the rate of groundwater flow is typically slow (recall table 15.2).

A rock that holds and transmits enough water to be useful as a source of water is an **aquifer.** Many of the best aquifers are sandstones, but any other type of rock may serve if it is sufficiently porous and permeable. An **aquitard** is a rock in which permeability is low and water flow is very much slower, so that it is not useful as a water source. (Its extreme, sometimes called an **aquiclude,** would be a rock that is effectively impermeable on a human timescale and that thus acts as a barrier to water flow, though in fact there are few, if any, truly impermeable rocks.) Shales are common aquitards.

When an aquifer is overlain only by permeable rocks and soil, it is described as **unconfined** (figure 15.5). In an unconfined aquifer, the water is not under unusual pressure. In a well drilled into an unconfined aquifer, the water will rise to the same height as the water table in the adjacent aquifer rocks. The water must be actively pumped up to the ground surface. An unconfined aquifer may be recharged by infiltration over the whole area underlain by that aquifer because there is nothing to stop the downward flow of water from surface to aquifer.

A **confined aquifer** is one overlain by an aquitard or aquiclude. Because the vertical movement of water is restricted, water in a confined aquifer may be under considerable pressure from overlying rocks or as a consequence of lateral changes in aquifer elevation. The fluid pressure is also known as **hydrostatic pressure.** The water level in the satu rated zone may be confined well below the height to which the hydrostatic pressure would push the water if it were unconfined (figure 15.6).

If a well is drilled into a confined aquifer, the extra fluid pressure will indeed cause ground water to rise in the well above its level in the aquifer. This is an **artesian** system, one in which ground water rises above its (confined) aquifer under its own pressure. The water in an artesian system may or may not rise all the way to the ground surface; some pumping may still be necessary to bring it to the surface for use. Advertising claims notwithstanding, this is all that "artesian water" means—it is no different chemically, no purer, no better tasting or more wholesome than any other ground water; it is just under natural pressure.

 rtesian behavior can be demonstrated very simply with a water-filled length of rubber tubing or hose held at a steep angle. The water is confined by the hose, and the water in the low end is under pressure from the weight of the water above it. If a small hole is poked in the top side of the lower end of the hose, a stream of water will shoot up to the level of the water at the upper end of the hose. In a natural aquifer system, some of the pressure is dissipated through friction between the water and the aquifer rocks through which the water flows; the rate of water flow is slower than in the previous hose analogy, but the principle is similar.

The Potentiometric Surface

In any aquifer, one can refer to the **potentiometric surface,** which at any point represents the height to which water would rise in a well or pipe drilled into that aquifer at that point. It is a reflection of local hydrostatic pressure. In an unconfined aquifer, the potentiometric surface is the water table. In a confined aquifer, the added hydrostatic pressure means that the potentiometric surface will be somewhat higher than the top of the aquifer, where the rocks are saturated, and it may be above the ground surface (see again figure 15.6). In an artesian well, the water rises to the elevation of the potentiometric surface.

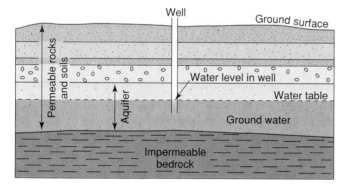

FIGURE 15.5 An unconfined aquifer system.

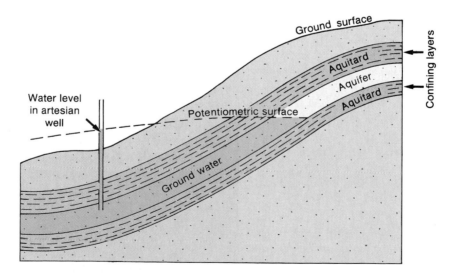

FIGURE 15.6 Water in a confined aquifer, sandwiched between aquitards, builds up extra hydrostatic pressure and may exhibit artesian conditions.

FIGURE 15.7 A geyser gushes as hot subsurface water is changed to steam. Yellowstone National Park.

EFFECTS OF THE PRESENCE OF SUBSURFACE WATER

Subsurface water warmed by magmatic heat may feed hot springs and geysers (figure 15.7). The intermittent gushing of the latter occurs when some of the water is heated enough to be converted to steam, which expands and forces water to the surface with a rush. These are relatively uncommon features, however. Much more common are features produced by the simple process of solution.

An abundance of surface and near-surface water may indicate an ample water supply, but it also means more water available to dissolve rocks. Most rocks are not very soluble, so this is not a concern in all areas. A few rock types, however, are extremely soluble. The most common of these is limestone; others include gypsum and halite. Limestone is particularly soluble in acidic water. All rainfall is naturally somewhat acidic because carbon dioxide in the air reacts with water vapor to form carbonic acid (H_2CO_3), so limestone is readily dissolved by infiltrating rainwater.

Over long periods of time, underground water may dissolve large volumes of limestone, creating and slowly enlarging underground caves and eroding support for the land above (figure 15.8). There may be no obvious evidence at the surface of what is occurring until the ground collapses abruptly into the void, producing a **sinkhole** (figure 15.9).

The collapse of a sinkhole may be triggered by a drop in the water table as a result of drought or water use that leaves unsupported rocks and soil that were previously buoyed up by water pressure. Failure may also be caused by rapid input of large quantities of water, as from heavy rains that wash overlying soil down into the cavern, or by an increase in subsurface-water flow rates.

Sinkholes come in many sizes. The larger ones are quite capable of swallowing up many houses at a time; they may be over 50 meters deep and cover several tens of acres. A single sinkhole in a developed area can cause millions of dollars

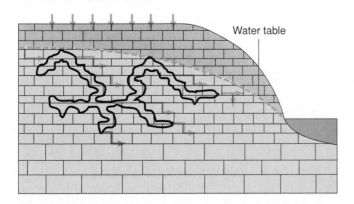

FIGURE 15.8 A cave is formed by solution of limestone or other soluble rock.

FIGURE 15.9 Homes are lost to a sinkhole in Winter Park, Florida. Note people and buildings on upper rim for scale.
Photograph courtesy of USGS Photo Library, Denver, CO.

in property damage. Sudden sinkhole collapses beneath bridges, roads, and railways have caused accidents and even (rarely) deaths.

Sinkholes are rarely isolated phenomena. Limestone most often is formed from chemical sediments deposited in shallow seas, and limestone beds commonly cover broad areas. Therefore, where there is one sinkhole in a region underlain by limestone, there are likely to be others. Thousands have formed in Alabama alone since the year 1900. An abundance of sinkholes in an area is a strong indication that more can be expected. Areas with many sinkholes commonly are characterized by the presence of soluble rocks close to the surface and a high water table or abundant precipitation, leading to extensive solution of those rocks. The high concentration of subsurface cavities means that streams in such regions commonly drain into the subsurface, rather than carrying surface runoff out of the region.

The situation may sometimes be easily recognized through aerial or satellite photography, which can show a region to be pockmarked with the circular lakes or holes characteristic of the solution-dominated **karst topography** (figure 15.10). Karst is common in the Gulf Coast states of North America. Clearly, in such an area, the subsurface situation should be investigated before buying or building a home or business. Circular patterns of cracks on the ground or conical depressions in the ground surface may be early signs of trouble developing below. In karst regions, groundwater flow can be unusually rapid, too, as a result of the greatly increased porosity and permeability due to solution.

FIGURE 15.10 Telltale karst topography of many round lakes formed in sinkholes (satellite image of central Florida).
© NASA.

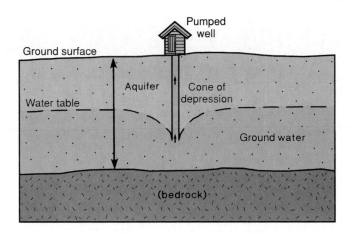

Figure 15.11 Cone of depression formed around a pumped well in an unconfined aquifer.

Consequences of Groundwater Withdrawal

When ground water must be pumped from an aquifer for use, the rate at which water flows in from surrounding rock to replace that which is extracted is generally slower than the rate at which water is taken out. In an unconfined aquifer, the result is a conical lowering of the water table immediately around the well, which is termed a **cone of depression** (figure 15.11). (A geometrically similar feature is produced on the surface of a liquid in a drinking glass when the liquid is sipped hard through a straw.) If many wells are closely spaced, the cones of depression of adjacent wells may overlap, further lowering the water table between wells. If, over a period of time, groundwater-withdrawal rates consistently exceed recharge rates, the regional water table may drop. A clue that this is happening is the need for wells throughout a region to be drilled deeper periodically to keep the water flowing.

The process of deepening wells to reach water cannot continue indefinitely. Ground water does not exist even through the whole thickness of the crust. In most places, ground water can be found, at most, a few kilometers into the crust. (In the deep crust and below, pressures on rocks are so great that compression and plastic flow close up any pores that ground water might fill.) In many areas, impermeable rocks are reached at much shallower depths. Therefore, there is a bottom to the groundwater supply, though the depth to that bottom usually cannot be determined without drilling. When the water table is lowered to the base of the aquifer system, groundwater use in that area is terminated for some time. As noted earlier, groundwater flow rates are commonly very slow—millimeters to meters per year—so recharge of significant amounts of ground water can require decades or centuries.

Regional decreases in water pressure due to water use can produce features analogous to cones of depression in potentiometric surfaces, and those surfaces can be lowered, too. Again, this reflects groundwater withdrawal at a rate exceeding the recharge rate, and gradual depletion of local ground water. Box 15.1 provides an example.

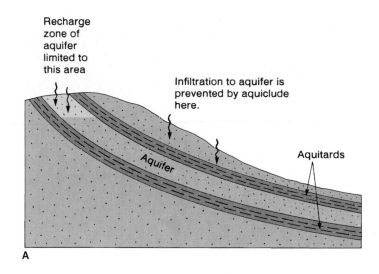

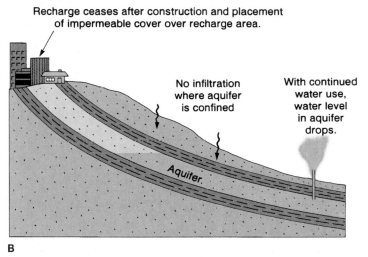

Figure 15.12 Recharge of a confined aquifer. (A) The recharge area of this confined aquifer is limited to the area where permeable rocks intersect the surface. (B) Construction over the recharge area reduces recharge, leading to depletion of water in the aquifer with continued use.

Where ground water is being depleted by too much withdrawal too fast, one can speak of "mining" ground water. The idea is not necessarily that the water will never be recharged, but that the rate is so slow on the timescale of human affairs as to be insignificant. From the human point of view, we may indeed use up ground water in some heavy-use areas. Also, human activities in some places have reduced or halted natural recharge, so ground water consumed may not be replaced, even slowly.

Impermeable cover over one part of a broad area underlain by an unconfined aquifer has relatively little impact on that aquifer's recharge. Infiltration will continue over most of the region. In the case of a confined aquifer, however, the available recharge area may be very limited, since the overlying confining layer prevents direct downward infiltration in most places. If impermeable cover is built over the restricted recharge area, recharge can be considerably reduced, thus aggravating the water-supply situation (figure 15.12).

Box 15.1

The Great Depression—in Ground Water

The Ogallala formation, a sedimentary aquifer in which the most productive units are sands and gravels, underlies most of Nebraska and sizeable portions of Colorado, Kansas, and the Texas and Oklahoma panhandles (figure 1). The area is one of the largest and most important agricultural regions in the United States. It accounts for about 25% of U.S. feed-grain exports and 40% of wheat, flour, and cotton exports. More than 14 million acres of land are irrigated with water pumped from the Ogallala. Yields on irrigated land may be triple the yields on similar land cultivated by "dry farming" (no irrigation).

The Ogallala's water was, for the most part, stored during the retreat of the Pleistocene continental ice sheets, described in chapter 18. Present recharge is negligible, except near the Platte River and in a few isolated locations where current water use is not very intense (see figure 1). The original groundwater reserve in the Ogallala is estimated to have been approximately 2 billion acre-feet. (One acre-foot is the amount of water needed to cover an area of 1 acre to a depth of 1 foot; it is about 326,000 gallons.) But each year, farmers draw from the Ogallala more water than the entire flow of the Colorado River. According to estimates by the U.S. Geological Survey, the average thickness of the saturated zone of the Ogallala formation in 1940 was nearly 200 feet. By 1993, it had declined by about 6%, to under 188 feet. Over the past decade or so, the average drop in the water table has been about 2 inches (5 centimeters) per year. And while that is significant in itself, it is important to keep in mind the great regional variations in drawdown, as shown in figure 1; depletion in some areas is acute.

The effects of excessive pumping can be seen even in areas widely regarded as receiving adequate moisture. Consider, as an example, northern Illinois. Sandstones of the Cambro-Ordovician aquifer system have long been used as the principal or sole source of municipal water in many parts of northern Illinois. When the height of the potentiometric surface for the con-

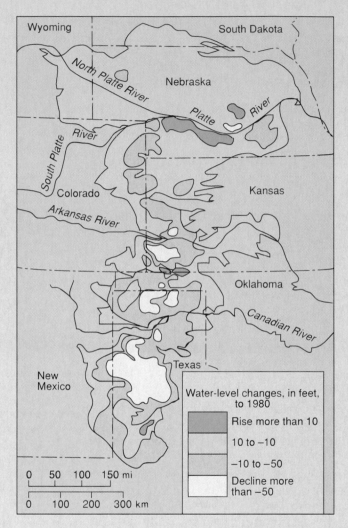

Figure 1 Map extent of the High Plains aquifer system.
Source: U.S. Geological Survey 1982 Annual Report.

fined aquifer is mapped, as in figure 2, a dramatic drop is seen in the Chicago metropolitan area, from over 750 feet (250 meters) above sea level to, locally, more than 100 feet (30+ meters) below sea level. What the contours really represent is a pressure drain resembling a giant cone of depression in the potentiometric surface. Just over the last decade, water levels in some wells have dropped more than 30 meters. Close to Lake Michigan, the situation can be alleviated by using proportionately more lake water. However, communities and rural homesteads lacking ready access to the lake waters are finding it increasingly difficult and expensive to keep the ground water flowing. Even if all groundwater withdrawal ceased for a period of time—an unlikely possibility—centuries of recharge at least would be required to restore the water levels.

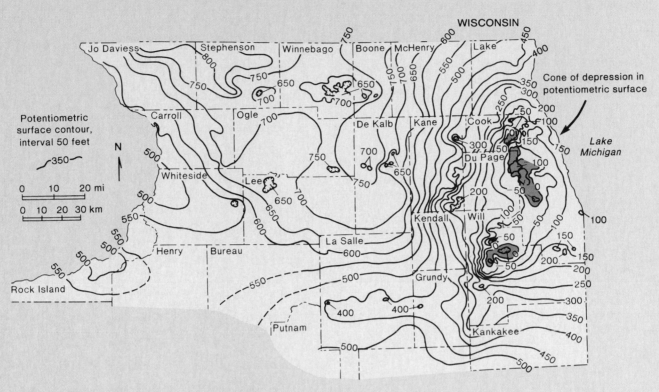

FIGURE 2 Potentiometric surface of the principal aquifer system of northern Illinois. Contours in feet above sea level. Shaded area denotes where potentiometric surface is below sea level. The city of Chicago is located in Cook County.

After R. T. Sasman et al., Illinois State Water Survey. Circular 113, 1973. Used by permission.

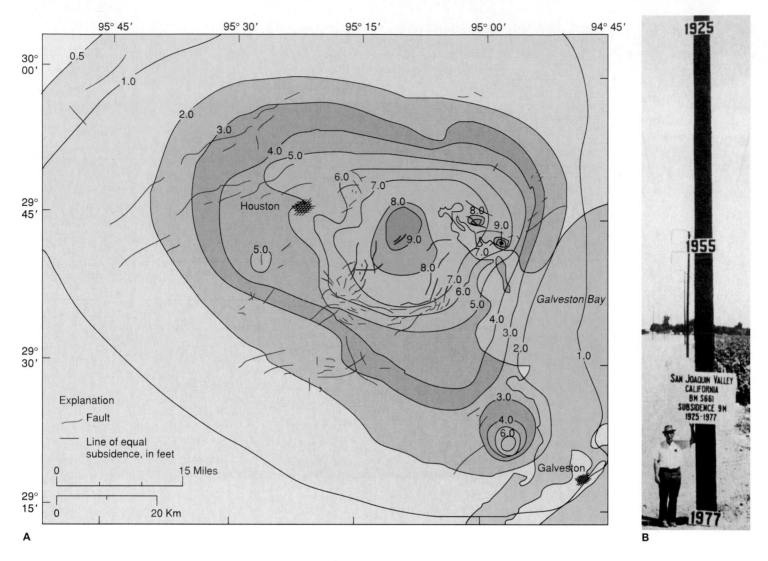

Figure 15.13 Subsidence due to groundwater withdrawal. (A) Surface subsidence near Galveston Bay, Texas, since 1906. (B) Subsidence of as much as 9 meters occurred between 1925 and 1977 as a result of groundwater withdrawal in the San Joaquin Valley, California. Signs indicate former ground surface elevation.

Sources: (A) U.S. Geological Survey 1982 Annual Report. *(B) Courtesy U.S. Geological Survey.*

Filling in wetlands is a common way to provide more land for construction. This, too, can interfere with recharge. Especially if surface runoff is rapid elsewhere in the area, a stagnant swamp, holding water for long periods, can be a major source of infiltration and recharge. Filling in the swamp so that water no longer accumulates there, and then topping the fill with impermeable cover, again may greatly reduce local recharge of ground water.

Lowering of the water table can contribute to sinkhole formation, as previously noted. Another possible result is that the aquifer, no longer saturated with water, may become compacted from the weight of overlying rocks. This decreases the rocks' porosity, permanently reducing their water-holding capacity, and it may also decrease their permeability. At the same time, as the rocks below compact and settle, the ground sur-

face itself may subside. Where water use is heavy, the surface subsidence may be several meters, as it has been, for example, in the Houston/Galveston area of Texas (figure 15.13A) or the San Joaquin Valley of California (figure 15.13B).

At high elevations, or in inland areas, this subsidence causes only structural problems, as building foundations are disrupted. In low-elevation coastal regions, the subsidence may lead to extensive flooding, as well as to increased rates of coastal erosion. The city of Venice, Italy, is in one such slowly sinking coastal area. Many of its historic architectural and artistic treasures are threatened by the combined effects of the gradual rise in worldwide sea levels, the tectonic sinking of the Adriatic coast, and surface subsidence from extensive groundwater withdrawal. Drastic and expensive engineering efforts will be needed to save them. In such areas,

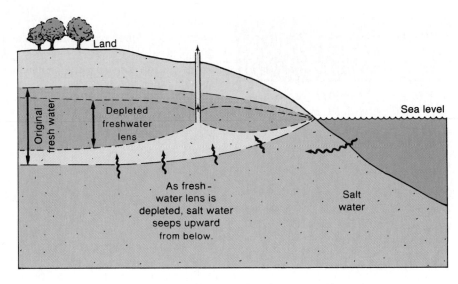

Figure 15.14 Saltwater intrusion and upconing as freshwater lens is depleted in a coastal zone.

simple solutions (like pumping water back underground) are unlikely to work. The rocks may have been permanently compacted. Also, what water is to be used? If groundwater consumption is heavy, it is probably because there simply is no great supply of fresh surface water. Pumping in salt water, in a coastal area will, in time, make the rest of the water in the aquifer salty too. And presumably, a supply of fresh water is still needed for local use, so groundwater withdrawal cannot easily or conveniently be curtailed.

A further problem arising from groundwater use in coastal regions, aside from the possibility of surface subsidence, is **saltwater intrusion** (figure 15.14). When rain falls into the ocean, the fresh water promptly becomes mixed with the salt water. However, fresh water falling on land does not mix so readily with saline ground water at depth because water in the pore spaces in rock or soil is not vigorously churned by currents or wave action. Fresh water is also less dense than salt water. So the fresh water accumulates in a lens that floats above the denser salt water.

If water use approximately equals the rate of recharge, the freshwater lens stays about the same thickness. However, if consumption of fresh ground water is heavier, the freshwater lens thins, and the denser saline ground water moves up to fill in pores emptied by removal of fresh water. **Upconing** of salt water below cones of depression in the freshwater lens may also occur. Wells that had been tapping the freshwater lens may begin pumping unwanted salt water instead, as the limited freshwater supply gradually decreases. Saltwater intrusion destroyed useful aquifers beneath Brooklyn, New York, in the 1930s, was a problem on Long Island in the 1960s, and is an ongoing concern in many coastal areas of the southeastern and Gulf coastal states and in some densely populated parts of California.

WATER QUALITY

As noted earlier, most of the water in the hydrosphere is in the very salty oceans, and almost all of the remainder occurs as ice. That leaves relatively little surface or subsurface water as potential freshwater sources. Moreover, much of the water on and in the continents is not strictly fresh. Even rainwater, long the standard for "pure" water, contains dissolved chemicals of various kinds, especially in industrialized areas with substantial air pollution. Once precipitation reaches the ground, it reacts with soil, rock, and organic debris, dissolving still more chemicals naturally, aside from any pollution generated by human activities. Water quality thus must be a consideration when evaluating water supplies.

Measurement of Water Quality

Water quality can be described in a variety of ways. A common approach is to express the amount of a dissolved chemical substance present as a concentration in parts per million (ppm) or, for very dilute substances, parts per billion (ppb). These units are analogous to percentages (which are really "parts per hundred") but are used for lower (more dilute) concentrations. That is, if water contains 1% salt, it contains 1 gram of salt per 100 grams of water, or 1 ton of salt per 100 tons of water, or whatever unit one wants to use. Likewise, if the water contains 1 ppm salt, it contains 1 gram of salt per 1 million grams of water (about 250 gallons), and so on.

Another way to express overall water quality is in terms of *total dissolved solids* (*TDS*), the sum of the concentrations of all dissolved solid chemicals in the water. How low a level of TDS is required or acceptable varies with the application. Standards might specify a maximum of 500 or 1000 ppm TDS for drinking water; 2000 ppm TDS might be acceptable for

TABLE 15.3 *Concentrations of Some Dissolved Constituents in Average Rainwater, River Water, and Seawater*

Constituent	Average Concentration (in ppm)		
	Rain-water*	River Water†	Seawater
Silica (SiO_2)	—	13	6.4
Calcium (Ca)	1.41	15	400
Sodium (Na)	0.42	6.3	10,500
Potassium (K)	—	2.3	380
Magnesium (Mg)	—	4.1	1350
Chloride (Cl)	0.22	7.8	19,000
Fluoride (F)	—	—	1.3
Sulfate (SO_4)	2.14	11	2700
Bicarbonate (HCO_3)	—	58	142
Nitrate (NO_3)	—	1	0.5

*One-year average, inland United States.
†World average (estimated).
Source: Data from J. D. Hem, *Study and Interpretation of the Chemical Characteristics of Natural Water*, 2d ed., U.S. Geological Survey Water-Supply Paper 1473, 1970.

watering livestock; in industrial applications where water chemistry is important, the water might need to be even purer than normal drinking water. At least as important as the quantities of impurities present, however, is *what* those impurities are. If the main dissolved component is calcite from a limestone aquifer, the water may taste fine and be perfectly wholesome with well over 1000 ppm TDS in it. Conversely, if iron or sulfur is the dissolved substance, even a few parts per million may be enough to make the water taste bad. Many synthetic chemicals that have leaked into water through improper waste disposal are toxic even at concentrations of 1 ppb or less.

Overall, groundwater quality is highly variable. It may be nearly as pure as rainwater or saltier than the oceans. Some representative analyses of different waters in the hydrosphere are shown in table 15.3 for reference.

Hard Water

Aside from the issue of health, water quality may be of concern because of the particular ways certain dissolved substances alter water properties. In areas where water supplies have passed through soluble carbonate rocks, like limestone, the water may be described as "hard." **Hard water** simply contains substantial amounts of dissolved calcium and magnesium. There is no single cutoff figure for calcium and magnesium concentrations that divides "hard" water from "soft," although hardness becomes objectionable in the range of 80 to 100 ppm.

Perhaps the most troublesome routine problem with hard water is the way it reacts with soap, preventing the soap from lathering properly, causing bathtubs to develop rings and laundered clothes to retain a grey soap scum. Hard water or water otherwise high in dissolved minerals may also leave mineral

deposits in plumbing and in such appliances as coffeepots and steam irons. Primarily for these reasons, many people in hard-water areas use water softeners, which remove calcium, magnesium, and certain other ions from water in exchange for sodium ions. The sodium ions are replenished from the salt (sodium chloride) supply in the water softener. (While softened water containing sodium ions in moderate concentration is unobjectionable in taste or for household use, it may be of concern to those on diets requiring restricted sodium intake.)

H ard water actually may be more healthful than soft. Studies of regional patterns of disease occurrence have shown an inverse correlation between heart-disease rates and the hardness of public water supplies. This correlation has been found in many such studies worldwide. A clear medical reason for the observed relationship, however, has not yet been identified.

Groundwater Pollution

Beyond the natural addition of soluble minerals to water, various kinds of pollution arise through human activities. Any soluble material discharged into the air, left exposed on the ground surface, or buried unsealed underground has the potential to pollute groundwater supplies (figure 15.15). Surface-water pollution, too, may lead to groundwater pollution—for example, where a polluted stream contributes to groundwater recharge, or when acid runoff from a surface mine seeps into the soil. Underground mines may also be sources of acid drainage that pollutes ground water; see the chapter on mineral and energy resources.

Air pollutants can react with or be dissolved in rainwater. When that rain falls and infiltrates into the soil, the pollutants may be carried along. The acid rain formed when sulfurous exhaust gases (such as from the burning of coal) react with oxygen and water vapor to make sulfuric acid can acidify both surface and groundwater supplies. Volatile metals, such as lead and mercury, are a particular problem in the air near smelters and other metal-processing plants; these metals, too, may be scavenged out of the air by rain and eventually added to groundwater supplies.

Sewage is a major waste-disposal problem. In less densely populated areas, sewage disposal by underground septic tanks on individual home sites is common. A normally functioning septic tank releases sewage slowly into surrounding permeable soil where, ideally, it is decomposed through reaction with oxygen in the soil pore spaces and by the action of soil microorganisms. In an improperly designed septic system, or where the water table is too high, too close to the septic system, sewage may mingle with ground water before it has been fully decomposed, contaminating the ground water.

Dumps and landfill sites, legal or illegal, also are potential sources of groundwater pollution. If they are underlain by permeable rock and soil, liquid wastes can seep out and down toward groundwater supplies below. Rainwater infiltrating from above and dissolving additional soluble chemicals

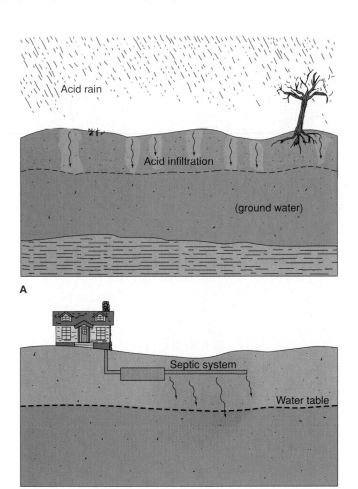

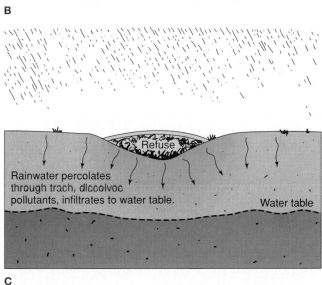

FIGURE 15.15 Some sources of groundwater pollution. (A) Acid rain. (B) Septic system too close to water table. (C) Leakage from landfill site.

from the wastes compounds the problem. Disposal sites for industrial wastes are not the only hazard in this regard, though they may represent especially concentrated sources of toxic chemicals. A variety of dangerous household chemicals—cleansers, paints and solvents, pesticides, and many others—are likely to be found in municipal landfills.

In principle, toxic wastes in disposal sites can be isolated by impermeable materials (clay layers, plastic liners) above and below. In practice, however, many carefully designed sites have subsequently leaked, and there is now some question as to whether a landfill site can ever be perfectly sealed, although leakage can be minimized by thoughtful design.

There are many other sources of groundwater pollution: Road salt applied in winter washes off, dissolves in rain or meltwater, and seeps into the soil. Herbicides and pesticides from farmland dissolve and infiltrate, perhaps down to the water table. Old underground gasoline tanks leak.

Once the ground water is polluted, the pollution may be difficult to correct or even to detect. Smoke belching from an exhaust stack is very visible. Oddly colored water discharged into a stream likewise attracts attention. Groundwater pollution, on the other hand, is hidden, insidious, often unnoticed until long after the fact. Pollutants escaping from an old dump site may first be detected in well water some distance away, long after the contamination has begun. By the time groundwater pollution is detected, it may be very widespread, and the exact extent of the problem may not be readily determined without the drilling of many monitoring wells across the affected area. Even the source can be hard to identify unless the chemistry of the contamination is distinctive.

Treatment at that point in time may simply not be possible, at least while the water remains underground. Natural filtration by passage through an aquifer can remove particles but not dissolved chemicals. Ground water moves slowly and mixes sluggishly. A polluted lake may be treatable by mixing in chemicals to remove or neutralize the pollutants. An equivalent treatment of ground water is impractical: how can the added chemicals be dispersed quickly throughout the aquifer as needed? The usual approach to treating groundwater pollution is to halt the pollution at the source—if that source can be identified—and to treat the water only after withdrawal, as it is used.

WATER USE, WATER SUPPLY

Biologically, humans require about a gallon of water per day per person, or about 250 million gallons per day for the United States. Yet Americans divert some 450 *billion* gallons of water for use each day—for cooking, washing, or other household use, for industrial processes, for livestock and irrigation. Even that figure does not include the several trillion gallons of water that flow through U.S. hydroelectric plants every day. Of the total withdrawn, over 100 billion gallons per day are consumed, meaning that the water is not returned directly as wastewater. Most of the consumed water is lost to evaporation; some is lost in transport. The overall U.S. water budget is shown in figure 15.16.

Surface Water Versus Ground Water as Supply

Surface waters are much more accessible than subsurface waters. Why, then, use ground water at all? One basic reason is that, in many dry areas, there is little or no surface

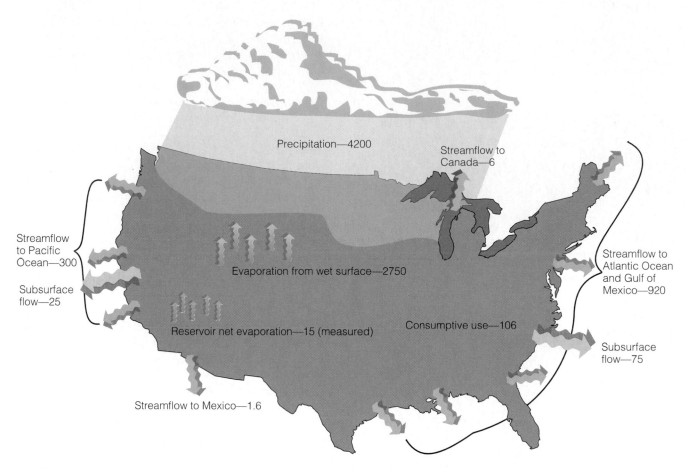

FIGURE 15.16 In terms of gross water flow, the U.S. water budget seems ample. Figures are in billions of gallons per day.
Source: Data from The Nation's Water Resources 1975–2000, *Vol. 1, U.S. Water Resources Council.*

water available, while there may be a large supply of water deep underground. Tapping the latter supply allows people to live and farm in areas that would otherwise be quite uninhabitable. Then, too, streamflow varies seasonally. During dry seasons, the water supply may be inadequate. Building dams and reservoirs allows a reserve of water to be accumulated during the wet seasons to be drawn on in dry times. However, some of the negative consequences of reservoir construction were discussed in chapter 14. Furthermore, if a region is so dry at some times that a reservoir is necessary, the rate of evaporation of water from the broad, still surface of the reservoir may itself represent a considerable water loss and aggravate the water-supply problem.

Precipitation (the primary source of abundant surface runoff) varies widely geographically (figure 15.17); so does population density. In many areas, the concentration of people may far exceed what can be supported by available local surface water, even during the wettest season. Also, streams and large lakes have historically been used as disposal sites for wastewater and sewage, often unpurified. This makes the surface water decidedly less appealing as drinking water. Finally, ground water is by far the largest reservoir of unfrozen fresh water (recall table 15.1).

For a variety of reasons, then, underground water may be preferred as a supplementary or even sole water source. When subsurface waters are used, ground water is drawn from the saturated zone. (Trying to pump water out of the unsaturated zone is a bit like trying to drink root beer by sipping at the foam in the top of the glass; it is not possible to draw up much liquid that way.) Water that has passed through the rock of an aquifer has also been naturally filtered to remove some impurities—soil or sediment particles, and even the larger bacteria—although it can still contain many dissolved chemicals. Sandy rocks and sediments can be particularly effective filters.

Regional Variations in Water Use

Water withdrawal varies regionally (figure 15.18), as does water consumption. Aside from hydropower generation, four principal categories of use can be identified: municipal supplies (home use and some industrial use in urban and suburban areas), rural use (supplying domestic needs for rural homes and watering livestock), irrigation, and self-supplied industrial use (by industries for which water supplies are separate from municipal sources). The quantities withdrawn for and consumed by each of these categories of users are summarized in figure 15.19.

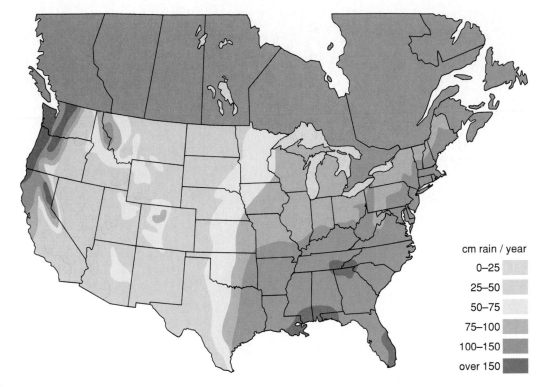

FIGURE 15.17 Distribution of average annual precipitation over the United States.
Source: U.S. Water Resources Council.

cm rain / year

0–25	
25–50	
50–75	
75–100	
100–150	
over 150	

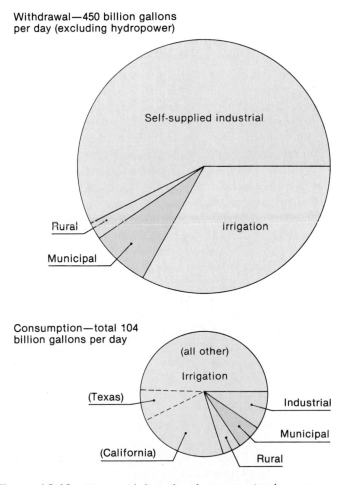

Withdrawal—450 billion gallons per day (excluding hydropower)

Self-supplied industrial

Irrigation

Rural

Municipal

Consumption—total 104 billion gallons per day

(all other)

Irrigation

(Texas)

Industrial

Municipal

(California)

Rural

FIGURE 15.19 Water withdrawal and consumption by sector.

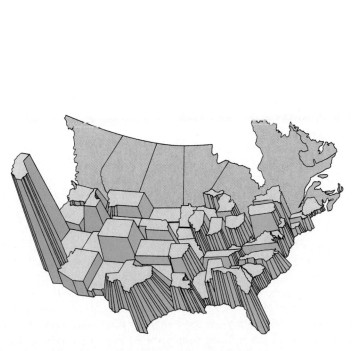

FIGURE 15.18 Regional variations in total water use in the United States.
Source: U.S. Geological Survey Water Supply Circular 1001.

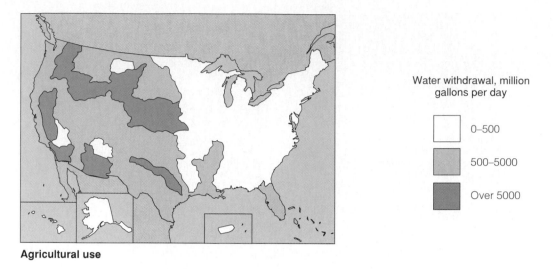

Water withdrawal, million gallons per day

☐ 0–500

▨ 500–5000

▩ Over 5000

Agricultural use

FIGURE 15.20 Regional water use for agriculture.

Source: Data from The Nation's Water Resources 1975–2000, *Vol. 1, U.S. Water Resources Council.*

R egional variations in population distribution do not coincide with variations in water supply. Indeed, some of the larger population centers are very poorly supplied with water. This leads to interstate or interregional political disputes over water rights, as we will see later in the chapter.

Industrial Versus Agricultural Use

A point that quickly becomes apparent from figure 15.19 is why industry may be called the major water *user,* while agriculture is the big water *consumer.* Self-supplied industrial users account for more than half the water withdrawn, but nearly all their wastewater is returned, as liquid water, at or near the point in the hydrologic cycle from which it was taken. Most of these users are diverting surface waters and dumping wastewaters back into the same lake or stream. All told, industrial users *consume* only about 10 billion gallons per day, or approximately 10% of total consumption.

Irrigation water—83 billion gallons per day—is nearly *all* consumed: lost to evaporation, lost through transpiration from plants, or lost because of leakage from ditches and pipes. Moreover, 40% of the water used for irrigation is ground water. Most of the water lost to evaporation drifts with air currents out of the area, to come down as rain or snow somewhere far removed from the irrigation site. It then does not contribute to recharge of the aquifers or to runoff to the streams from which the water was drawn. Where irrigation use of water is heavy, water tables have dropped by tens of meters and streams have been drained nearly dry, while we have become increasingly dependent on the crops. The case of the Ogallala formation (box 15.1) is just one example. Perhaps not surprisingly, those regions where agricultural use of water is heaviest (figure 15.20) are generally those with less precipitation (recall figure 15.17).

EXTENDING THE WATER SUPPLY

The most basic approach to improving the water-supply situation is conservation. In particular, the heavy drain by irrigation must be moderated if the rate at which water supplies are being depleted is to be reduced appreciably. For example, either crops should be grown where natural rainfall is adequate to support them, or irrigation methods must be made more efficient—by replacing open ditches or sprinkler systems with water distribution through perforated pipes at or below ground level, for instance—to reduce evaporation losses.

Part of the supply problem is purely local. That is, people persist in settling and farming in areas that may not be especially well supplied with fresh water, while other areas with abundant water go undeveloped. If people cannot be persuaded to move to areas with abundant water, perhaps the water can be redirected. This is the idea behind interbasin transfers, moving surface waters from one drainage basin to another, where demand is higher. California pioneered the idea with the Los Angeles Aqueduct. Completed in 1913, it carried nearly 150 million gallons of water per day from the eastern slopes of the Sierra Nevada to Los Angeles, a distance of some 200 miles (over 300 kilometers). New York City draws on several reservoirs in upstate New York. Dozens of interbasin transfers of surface water have been proposed, including transfers of water from little-developed areas of Canada to high-demand areas of the United States and Mexico. Such proposals, which could involve transporting water over thousands of kilometers, are expensive, as well as technically demanding. (One such scheme, the North American Water and Power Alliance, had a projected cost of $100 billion; it was not implemented.) International water-transport plans presume a continued willingness on the part of other nations to share their water. Even within the United States, serious political objections to interbasin water transfer may be raised. For example,

FIGURE 15.21 This ball field on Long Island doubles as an artificial recharge basin in wet weather.
Photograph courtesy of USGS Photo Library, Denver, CO.

during the mid-1980s, officials in states bordering the Great Lakes took a firm public stand against the suggestion that some of the lakes' water be diverted to dry states in the southern and southwestern United States.

In steeply sloping areas or those with low-permeability soils, well-planned construction that includes artificial recharge basins can aid in increasing groundwater recharge (figure 15.21). The basin acts similarly to a flood-control retention pond, in that it is designed to catch some of the surface runoff during high-runoff events (heavy rain or snowmelt). Trapping the water allows more time for slow infiltration, and thus more recharge, in an area from which the fresh water might otherwise be quickly lost to streams and carried away. Recharge basins are a partial solution to the problem in areas where groundwater use exceeds natural recharge rates, but, of course, they are only effective where there is surface runoff to catch. Where natural infiltration is slow, or where the aquifer to be recharged is confined, artificial recharge may require the use of injection wells to pump water back into the aquifer.

Another alternative for extending the water supply is to improve the quality of waters not now used, purifying them sufficiently to make them usable. Desalination of seawater, in particular, would allow coastal regions to tap the vast ocean reservoirs. Also, some ground water is not presently used for water supplies because it contains excessive concentrations of dissolved materials. The two basic methods used to purify water of dissolved minerals are filtration and distillation.

In a filtration system, the water is passed through fine filters or membranes that screen out dissolved impurities. The advantage of filtration is that it can operate rapidly on a great quantity of water. A disadvantage is that the method works best on water that does not contain very high levels of dissolved minerals. Anything as salty as seawater quickly clogs the filters. This method, then, is most useful for cleaning up only moderately saline ground water or lake or stream waters.

Distillation involves heating or boiling water full of dissolved minerals. The water vapor driven off is pure water, while the minerals stay behind in what remains of the liquid. This is true regardless of how concentrated the dissolved minerals are, so the method works as well on seawater as on less saline waters. A difficulty with this method, however, is the nature of the necessary heat source. Any conventional fuel is costly in large quantity. Some solar distillation facilities now exist, but their efficiency is limited because solar heat is low-intensity heat. If a large quantity of desalinated water is required rapidly, the water to be heated must be spread out shallowly over a large area. A major city might need a solar facility covering thousands of square kilometers to provide adequate water, and construction on such a scale would be prohibitively expensive even if sufficient space were available.

SUMMARY

Collectively, the oceans are, by far, the largest single water reservoir in the hydrologic cycle. Most of the fresh water is stored in ice sheets. Ground water is the largest reservoir of unfrozen fresh water.

A rock that is sufficiently porous and permeable to be useful as a source of ground water is an aquifer; relatively impermeable rocks are aquitards. The top of the saturated, or phreatic, zone is the water table, which locally may be above the ground surface. In an unconfined aquifer, water in an unpumped well will rise to the height of the water table. In a confined aquifer, additional pressure acting on the water may create artesian conditions, in which the water will rise in an unpumped well to the level of the potentiometric surface, which may be far

above the aquifer and even above the ground surface. The presence of ground water in soluble rocks contributes to the formation of caverns and sinkholes. Groundwater withdrawal exceeding recharge can lead to surface subsidence, compaction of aquifer rocks, and saltwater intrusion, as well as to lowering of the water table and general groundwater depletion. Urbanization that modifies runoff patterns or covers recharge areas may ultimately decrease groundwater supplies.

Groundwater quality varies widely. Natural water quality is diminished by the solution of soluble rocks and minerals. Hard water is common where limestones are abundant. Additional toxic chemicals are sometimes added by human activities, especially through

improper waste disposal or runoff from highways and farmland. Once established, groundwater pollution can be slow to appear, difficult to trace or monitor, and impossible to treat while the water remains in its aquifer.

Industry accounts for the largest share of water withdrawn for use in the United States, but most of that water is not consumed during use. By contrast, most of the water withdrawn for irrigation *is* consumed, especially through evaporation. About 40% of irrigation water is ground water, from aquifers that may or may not currently be undergoing recharge. Possible strategies for extending water supplies in areas of shortage include conservation, interbasin water transfer, and desalination of either seawater or saline ground water.

TERMS TO REMEMBER

aquiclude 267
aquifer 267
aquitard 267
artesian 268
base flow 266
cone of depression 271
confined aquifer 268
ground water 265

hard water 276
hydrostatic pressure 268
karst topography 270
permeability 265
phreatic zone 265
porosity 265
potentiometric surface 268
recharge 267
saltwater intrusion 275

sinkhole 269
soil moisture 265
spring 267
subsurface water 265
unconfined aquifer 268
upconing 275
vadose zone 265
water table 265

QUESTIONS FOR REVIEW

1. How is ground water distinguished from subsurface water?
2. Define and compare the properties *porosity* and *permeability*.
3. What is an *aquifer?* A *confined aquifer?*
4. Explain the nature of artesian conditions and how they arise.
5. What is a *water table,* and how is it related to a *potentiometric surface?* What is a *cone of depression?* How does it develop?

6. What is *karst topography?* Under what conditions does it tend to develop?
7. Groundwater withdrawal that exceeds recharge may lead to surface subsidence. Explain.
8. What is *saltwater intrusion,* and in what circumstances does it occur?
9. How might urbanization reduce recharge of a confined aquifer? Describe one strategy for increasing recharge to an aquifer (confined or unconfined).
10. What is *hard water?* Is it harmful?

11. Cite and explain any two ways in which ground water can become polluted. Is it practical to treat groundwater pollution in place? Why or why not?
12. Why is ground water frequently preferred to surface water as a source of water supply?
13. Industry is the principal water *user;* agriculture is the major water *consumer.* Explain.
14. Compare and contrast distillation and filtration as means of desalinating water, noting the advantages and disadvantages of each.

FOR FURTHER THOUGHT

1. Where does your water come from? What is the quality at the source? Is the water treated before consumption? If the source is ground water, is there evidence that the supply is being depleted? If so, what plans exist for averting future shortages?

2. If there is a landfill or other waste-disposal site near you, investigate how possible groundwater pollution is monitored. What tests are made and how often? If appreciable pollution has been detected in the past, what steps have been taken to reduce or eliminate it?

3. Some activities are commonly cited as water-wasters. One is letting the tap run while brushing one's teeth. Try plugging the drain, then doing this; then measure the volume of water accumulated during a single tooth-brushing (a measuring cup and a bucket may be helpful). Consider the implications for water use, bearing in mind that the U.S. population is about 265 million, and our strictly biological need of about 1 gallon per person per day.

SUGGESTIONS FOR FURTHER READING

Cherry, J. A., et al. 1972. *Hydrogeology of the Rocky Mountains and interior plains*. Montreal, Canada: XXIV International Geological Congress, Guidebook.

Dolan, R., and Goodell, H. G. 1986. Sinking cities. *American Scientist* 74 (January–February): 38–47.

Dunne, T., and Leopold, L. B. 1978. *Water in environmental planning*. San Francisco: W. H. Freeman.

Ford, D. C., and Williams, P. W. 1989. *Karst geomorphology and hydrology*. Winchester, Mass: Unwin Hyman Academic.

Francko, D. A., and Wetzel, R. G. 1983. *To quench our thirst*. Ann Arbor, Mich.: University of Michigan Press.

Goodman, A. S. 1984. Principles of water resources planning. Englewood Cliffs, N.J.: Prentice-Hall.

Heppenheimer, T. A. 1991. The man who made Los Angeles possible. *Invention and Technology* (Spring/Summer): 11–18.

Howe, C. W., and Easter, K. W. 1971. *Interbasin transfers of water, economic issues and impacts*. Baltimore, Md.: Johns Hopkins Press.

Jermar, M. K. 1986. *Water resource management*. New York: Elsevier.

Mather, J. R. 1984. *Water resources*. New York: John Wiley and Sons.

National water summary 1983. 1984. U.S. Geological Survey Water-Supply Paper 2250. Washington, D.C.: U.S. Government Printing Office.

Sasman, R. T., et al. 1973. *Water level decline and pumpage in deep wells in the Chicago region, 1966–1971*. Illinois State Water Survey Circular #113.

Smith, Z. A. 1988. *Groundwater in the West*. New York: Academic Press.

Solley, W. B., Chase, E. B., and Man, W. B. IV. 1983. *Estimated use of water in the United States in 1980*. U.S. Geological Survey Circular 1001.

Sun, R. J. 1985. Investigations of the regional aquifer systems in the United States, 1978–1984. U.S. Geological Survey, *Annual Report 1985*, 48–51.

U.S. Water Resources Council. 1978. *The nation's water resources 1975–2000*. Washington, D.C.: U.S. Government Printing Office.

Zwingle, E. 1993. Ogallala aquifer: wellspring of the High Plains. *National Geographic* 184 (March): 80–109.

16

LANDSLIDES AND MASS WASTING

Large blocks of quartzite have tumbled down in a landslide in the Endless Chain Range, Jasper National Park, Alberta, Canada.
© Doug Sherman/Geofile

While the internal heat of the earth drives mountain-building processes, just as inevitably, the force of gravity acts to wear the mountains down again. Gravity is the great leveler. It constantly pulls downward on every mass of material everywhere on earth, causing a variety of phenomena collectively termed **mass wasting,** whereby geological materials are moved downslope from one place to another, without a transporting agent such as wind, water, or ice. Erosion is one form of mass wasting. The movement associated with mass wasting can be slow, subtle, and almost undetectable on a day-to-day basis, but if continued over days or years, the cumulative effects can be very large. Alternatively, that movement can be sudden, as in a rockslide or an avalanche, and the resultant changes swift, devastating, and obvious.

Landslide is a general term for rapid mass movement. In the United States alone, landslides and related mass movements cause over $1 billion in property damage each year, though losses of human life are relatively small. Many landslides occur quite independently of human activities. In some areas, active steps have been taken to control downslope movement or to limit the resultant damage. On the other hand, certain human activities aggravate local landslide dangers, usually as a result of failure to take those hazards into account. Large areas of the United States—and not necessarily only mountainous regions—are potentially at risk from landslides (figure 16.1).

CAUSES AND CONSEQUENCES OF MASS MOVEMENTS

Basically, mass movements occur whenever the downward pull of gravity overcomes the forces resisting sliding or flow. The downslope pull tending to cause mass movement, called the **shearing stress,** is related to the mass of material and to slope angle (figure 16.2). Counteracting the shearing stress is *friction* or, in the case of a cohesive solid, **shear strength.** When shearing stress exceeds friction or shear strength, sliding occurs. Sudden movements may be set off by a triggering mechanism, such as an earthquake. Mass movements, in turn, may cause secondary problems, such as flooding, as described later in the chapter.

Another way to consider the factors contributing to landslides is by grouping them into categories; most can be classified as either *static* factors or *dynamic* factors. The static factors are those that are inherent in the system, that don't change (at least on a human timescale)—for example, the nature of the material or geologic structures present. The dynamic factors are those that can be changed by natural processes or human activities: slope, load, vegetation, moisture content, ground movement. Both types of factors contribute to susceptibility to sliding; typically, a change in one or more of the dynamic factors sets it off.

Effects of Slope

Steepness of slope is one major factor contributing to the probability of a landslide: all else being equal, the steeper the slope, the greater the shearing stress.

For unconsolidated material, the **angle of repose** is the maximum slope angle at which the material is stable (figure 16.3). This angle varies with the material. Smooth, rounded particles tend to support only very low-angle slopes (imagine

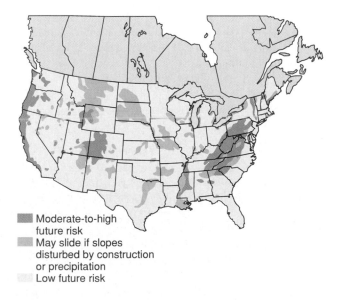

Moderate-to-high future risk
May slide if slopes disturbed by construction or precipitation
Low future risk

FIGURE 16.1 Landslide hazards in the United States.
Source: After U.S. Geological Survey Professional Paper 950, 1978.

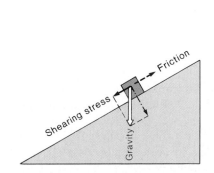
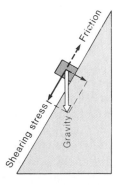

FIGURE 16.2 Effects of slope geometry on slide potential. The mass of the block, and thus the total downward pull of gravity, is the same in both cases, but the steeper the slope, the greater the shearing stress component.

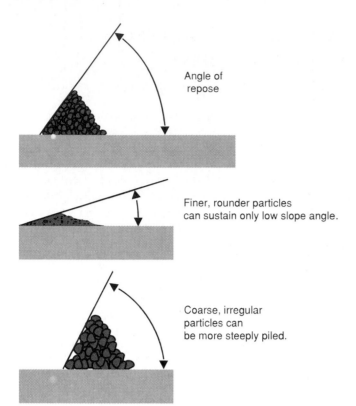

FIGURE 16.3 Angle of repose, an indicator of an unconsolidated material's resistance to sliding. Coarser and rougher particles can maintain steeper stable slope angles.

making a heap of marbles or ball bearings), while rough, sticky, or irregular particles can be piled more steeply without becoming unstable.

Solid rock can be perfectly stable even at a vertical slope but may lose its strength if it is broken up by weathering or fracturing. Also, in layered sedimentary rocks, there may be weakness along bedding planes, where different rock units are imperfectly held together; or some units may themselves be weak or even slippery (clay-rich layers, for example). Such planes of weakness are potential slide planes.

Slopes may be steepened to unstable angles by natural erosion by water or ice. Erosion can also undercut rock or soil, removing the support beneath a mass of material and thus leaving it susceptible to falling or sliding. This is a common contributing factor to landslides in coastal areas (figure 16.4). Over long periods of time, slow tectonic deformation can also alter the angles of slopes and bedding planes, making them steeper or gentler. This is most likely to be a significant factor in young, active mountain ranges, such as the Alps, or the coast ranges of California. Steepening of slopes by tectonic movements was suggested as the cause of a large rockslide in Switzerland in 1806 that buried the township of Goldan under a block of rock nearly 2 kilometers long, 300 meters wide, and 60 to 100 meters thick. Human activities, too, can steepen slopes; examples include quarrying and the excavation of roadcuts.

FIGURE 16.4 This coastal cliff in California is not truly stable; here, part of it has collapsed, the sediment assuming its customary angle of repose.
Photograph by J. T. McGill, USGS Photo Library, Denver, CO.

Effects of Fluid

Aside from its role in erosion, water can greatly increase the likelihood of mass movements in other ways. For example, it can seep along bedding planes in layered rock, reducing friction and making sliding more likely. As noted in connection with earthquakes and faulting, an increase in pore water pressure in saturated rocks decreases the rocks' resistance to shearing stress, which can also facilitate sliding.

In unconsolidated materials, the role of water is variable. A little moisture can add some cohesion (it takes damp sand to build a sand castle). However, substantial increases in water content both increase pore pressure and reduce the friction between particles that provides strength. The very mass of water in saturated soil may add enough extra weight, enough additional downward pull, to set off a landslide on a slope that was stable when dry (see figure 16.5). This is one

FIGURE 16.5 Landslide following storms in Brazil, triggered by heavy rainfall.

Photograph by F. O. Jones, USGS Photo Library, Denver, CO.

FIGURE 16.6 Slumping in expansive clay seriously damaged this road near Boulder, Colorado.

Photograph courtesy of USGS Photo Library, Denver, CO.

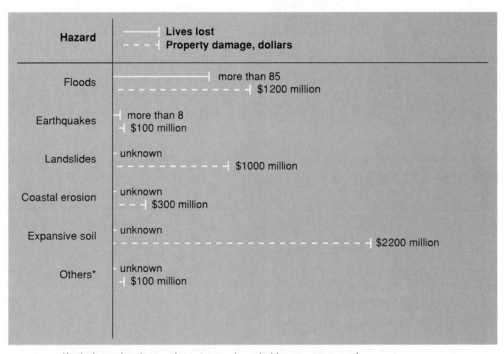

*Includes volcanic eruptions, tsunamis, subsidence, creep, and other phenomena.

FIGURE 16.7 Relative magnitude of annual loss of life and property damage in the United States from various geologic hazards.

Source: U.S. Geological Survey Professional Paper 950.

reason landslides may be especially frequent in tropical climates or, in any climate, during times of heavy rainfall or rapid snowmelt. Frost wedging and the associated fracturing can also weaken rocks.

Some soils rich in clays absorb water readily; one type of clay, *montmorillonite,* may absorb twenty times its weight in water and form a weak gel. Such material fails easily under stress. Other clays expand when wet, contract when dry, and can destabilize a slope in the process (figure 16.6). In terms of property damage, these **expansive clays** are, in fact, the most costly geologic hazard in the United States (see figure 16.7).

Effects of Vegetation

Vegetation tends to stabilize slopes. Plant roots, especially those of larger shrubs and trees, can provide a strong interlocking network to hold unconsolidated materials together

FIGURE 16.8 An old, 3-million-cubic-meter slide near Thistle, Utah, was reactivated in the wet spring of 1983. It blocked Spanish Fork Canyon and cut off highway and rail routes. A number of local residents were driven from their homes as the lake behind the landslide grew.
Photograph courtesy of U.S. Geological Survey.

FIGURE 16.9 Examination of this Alaskan roadcut shows surface soil held in place by grass roots, with erosion and slope failure below.

ust as landslides can be a result of earthquakes and heavy precipitation, other events, notably floods, can be produced by landslides. A stream in the process of cutting a valley may create unstable slopes. Subsequent landslides into the valley can dam up the stream, creating a natural reservoir. The filling of the reservoir makes the area behind the earth dam uninhabitable, though this usually happens slowly enough that lives are not lost. (See figure 16.8.) A further danger is that the unplanned dam formed by a landslide may later fail. In 1925, an enormous rockslide a kilometer wide and over 3 kilometers long in the valley of the Gros Ventre in Wyoming blocked that valley, and the resulting lake stretched to 9 kilometers long. After spring rains and snowmelt, the "dam" failed. Floodwaters swept down the valley below. Six people died, and the toll would have been far higher in a more populous area. Conversely, flooding may promote landslides as soils are saturated and slopes are undercut by fast-flowing water.

and prevent flow (figure 16.9). (This benefit may continue for several years after trees are cut down, before the roots decompose.) Actively growing vegetation also takes up moisture from the upper layers of soil and can thus reduce the overall moisture content of the mass.

Earthquakes as Triggers

Landslides are a common consequence of earthquakes in hilly terrain, as noted in chapter 10. Seismic waves passing through rock stress and fracture it. The added stress may be as much as half that already present due to gravity. Ground shaking also jars apart soil particles and rock masses, reducing the friction that holds them in place.

One of the most lethal earthquake-induced landslides occurred in Peru in 1970. One slide had already occurred below the steep, snowy slopes of Nevados Huascarán, the highest peak in the Peruvian Andes, in 1962. This slide, which was not triggered by an earthquake, killed approximately 3500 people. In 1970, a magnitude-7.7 earthquake centered 130 kilometers to the west shook loose a much larger debris avalanche that buried most of the towns of Yungay and Ranrachirca, and more than 18,000 people with them (figure 16.10). Some of the debris was estimated to

A

B

FIGURE 16.10 Nevados Huascarán debris avalanche. (A) Aerial view. (B) Aerial view of lower part of avalanche.
Photographs by G. Plafker, USGS Photo Library, Denver, CO.

FIGURE 16.11 Quick clay. Turnagain Heights, Anchorage, Alaska.

TABLE 16.1 *Types of Mass Movements (Excluding Creep) and Typical Rates of Movement*

	Material moves as coherent unit	Chaotic, incoherent movement
Free-falling motion	Rockfall Speed: Rapid (meters/second)	Soil fall (rare) Speed: Rapid (meters/second)
In contact with surface below	Slides: Rockslide Slump (soil) Speed: Highly variable (meters/year to meters/second)	Flows: Snow avalanche Debris avalanche Pyroclastic flow Solifluction Earthflow Mudflow Lahar Speed: Rapid (meters/day to meters/second)

sediment is later uplifted above sea level by tectonic movements, it contains salty pore water. The sodium chloride in the pore water acts as a "glue," holding the clay particles together. Fresh water subsequently infiltrating the clay washes out the salts, leaving a delicate, honeycomb-like structure of particles. Vibration from seismic waves breaks the structure apart, reducing the quick clay to a fraction of its original strength and creating a finer-grained equivalent of quicksand that is highly prone to sliding.

So-called sensitive clays are similar in behavior to quick clays but differently formed. Weathering of volcanic ash, for example, can produce a sensitive clay sediment, *bentonite*. Such deposits are locally common in the western United States, where there has been much relatively recent volcanism.

TYPES OF MASS MOVEMENTS

Mass movements may be subdivided on the basis of the type of material moved and the nature of the movement, including the rate of movement. The material moved can be either unconsolidated, fairly fine material (for example, soil or snow) or large, solid blocks or sheets of rock. It can be quite uniform in size or a mix of fine sediment and boulders. A few examples may clarify the different types of mass movements, key characteristics of which are summarized in table 16.1.

A **fall** is a free-falling action in which the moving material is not always in contact with the ground below. Falls are most often *rockfalls* (figure 16.12A), though soil falls also occur (recall figure 16.4). Falls frequently occur on very steep slopes or cliffs, when rocks high on the slope, weakened and broken up by weathering, lose support as materials under them are eroded away. The undercutting agent can be a stream, glacier, ocean, or lake, or, in rare cases, wind-borne

have moved at 1000 kilometers per hour. In the 1964 Alaskan earthquake, the Turnagain Heights section of Anchorage was heavily damaged by landslides. California contains not only many fault zones but also many sea cliffs and hillsides prone to landslides when earthquakes occur. Recall from chapter 4 that, in the 1980 explosion of Mount St. Helens, the volcanic activity caused an earthquake of magnitude over 5, which shook loose a landslide from the bulging north slope of the volcano, which, in turn, removed enough mass from the mountainside that the pressurized gases inside could blast out.

Quick Clays

A geologic factor that contributed to the 1964 Anchorage landslides and that continues to add to the landslide hazards in Alaska, California, parts of northern Europe, and elsewhere is a material known as **quick clay** or **sensitive clay.** True *quick clays* (figure 16.11) are most common in northern latitudes. They are formed from glacial rock flour deposited in a marine environment. When this extremely fine

A

B

FIGURE 16.12 Rockfall. (A) Schematic. (B) Rockfalls at Devil's Postpile National Monument leave large talus slope.

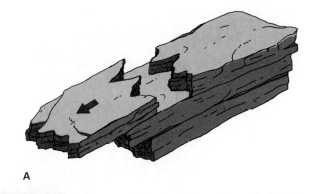

A

B

FIGURE 16.13 Rockslides. (A) A coherent mass of rock moves as a unit when inadequately supported (schematic). (B) Rockslide in tilted sandstone beds. Garfield County, Colorado. Note man at lower right for scale.
(B) Photograph by D. J. Varnes, USGS Photo Library, Denver, CO.

sediment. Repeated rockfalls in one place over a period of time can result in the accumulation of piles of blocky debris, termed **talus** (figure 16.12B).

In a **slide,** a fairly coherent unit of material slips downward along a clearly defined surface or plane. Rockslides most often involve either movement along a bedding plane between successive layers of sedimentary rocks or slippage where differences in the original composition of the sedimentary layers result in a weakened layer or a surface with little cohesion (figure 16.13).

If the rock or soil mass moves over a short distance, the slide may be termed a **slump.** Slumps in soil, rather than rock, are often characterized by a rotational movement, on a slip plane curved concave upward, that accompanies the downslope movement of the soil mass. The surface at the top of the slide may be relatively undisturbed. The clifflike scar at the head of the slump is a type of **scarp.** The lower part of the slump block may end in an earthflow (figure 16.14).

Landslides involving unconsolidated, noncohesive material are extremely common. These are **flows,** in which material

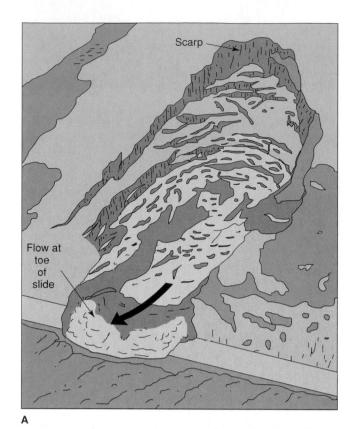

A

B

Figure 16.14 Slumps in soil. (A) Most of the soil mass moves coherently, and the surface is disturbed relatively little (schematic). (B) An example of a soil slump, ending in a flow across the road below. Note the scarp at the head of the slump.
(B) Photograph by J. T. McGill, USGS Photo Library, Denver, CO.

Figure 16.15 Beginning of a snow avalanche, set off by the slight disturbance of a skier passing over an unstable patch of snow.
Photograph by Ludwig, courtesy of U.S. Forest Service/USGS Photo Library, Denver, CO.

Figure 16.16 Recent earthflows in grassland. Los Banos, California.
© Times Mirror Higher Education Group, Inc./Doug Sherman, photographer.

moves in a chaotic, disorganized fashion, with mixing of particles within the flowing mass, as a fluid moves. Flows need not involve only soils. Snow avalanches are one kind of flow (figure 16.15); pyroclastic flows are another, associated only with volcanic activity (see chapter 4). Where soil is the flowing material, flow phenomena may be described as *earth-* *flows* (fairly dry soil; see figure 16.16) or *mudflows* (if saturated with water), which include the volcanic lahars. Mudflows can occur even in desert regions, during the rare periods of heavy rainfall. A kind of wet-soil flow especially characteristic of alpine regions is a result of near-surface melting of ice in frozen soil, while deeper soil remains frozen

FIGURE 16.17 Debris avalanche at Puget Peak following 1964 Alaskan earthquake.
Photograph by G. Plafker, USGS Photo Library, Denver, CO.

FIGURE 16.18 This broken wall on a California slope is the result of creep.
Photograph by J. T. McGill, USGS Photo Library, Denver, CO.

and impermeable. The resultant movement of sodden material over solid, impermeable ground is **solifluction.** When a wide variety of materials—soil, rocks, trees, and so on—is incorporated together in a single flow, the result is called a **debris avalanche** (figure 16.17; see also figure 16.10). Regardless of the nature of the materials moved, all flows have in common the chaotic movement of the particles or objects in the flow.

Mass movements can occur on a variety of scales, as well as at a variety of rates. They may involve a few cubic meters of material or more than a billion cubic meters. Total displacement may be over a distance of only a few centimeters or even millimeters, or the falling, sliding, or flowing material may travel several kilometers.

In the most rapid mass movements, which include most rockfalls, avalanches, and mudflows, materials can travel at speeds up to several tens of meters per second. (For reference, 30 meters per second is over 100 miles per hour.) Because there is little time for anyone to react once these events start, such events are associated with the greatest proportion of mass-movement casualties. Slides generally move at more moderate rates, in the range of a few meters per week to meters per day, and slow slides, slumps, and earthflows may move as slowly as a millimeter per year or less.

Extremely slow movement, so slow as to be undetectable to the naked eye, is described as **creep.** Its ultimate cause, as with all mass movements, is gravity. Creep can be promoted, in temperate climates, by the expansion and contraction associated with freezing/thawing cycles of water, which help to dislodge individual particles. Soil creep is more common than rock creep. The principal impact of the slower kinds of mass wasting is property damage (figure 16.18).

LANDSLIDES AND HUMAN ACTIVITIES

Human activities can increase the risk of landslides in many ways. One is by the clearing of stabilizing vegetation. In some instances, where clear-cutting logging operations have exposed sloping soil, for example, earthflows and mudflows have become far more frequent or been more severe than before. Dam construction and the subsequent saturation of reservoir rocks has also been a factor in some cases; see box 16.1.

Recognizing Past Mass Movements

How can past mass movements be recognized? Rockfalls tend to be quite obvious, especially in a generally vegetated area. Talus slopes are inhospitable to most vegetation, so rockfalls tend to remain barren of trees and plants, as in figure 16.12B. Lack of mature vegetation may also mark the paths of past debris avalanches or other soil flows or slides (figure 16.19). These scars on the landscape point plainly to slope instability.

Landslides are not the only kinds of mass movements that recur in the same places. Snow avalanches disappear when the snow melts, but historical records of past avalanche occurrences can pinpoint particularly dangerous areas. Records of the character of past volcanic activity and an examination of the typical products of a particular volcano can similarly be used to assess that volcano's tendency to produce pyroclastic flows.

Very large slumps and slides may be less obvious, especially when viewed from ground level. The coherent nature of rock and soil movement in most slides means that vegetation growing atop the slide may not be greatly disturbed by the movement. Aerial photography or high-quality

Box 16.1

A Landslide of Human Making

THE VAIONT DAM DISASTER

The Vaiont River flows through an old glacial valley in the Italian Alps. The valley is underlain by a thick sequence of sedimentary rocks, principally limestones with some clay-rich layers, that were folded and fractured during the building of the Alps. The beds on either side of the valley dip down toward the valley floor (figure 1). The rocks themselves are relatively weak and are particularly prone to sliding along the clay-rich layers. Evidence of old rockslides can be seen in the valley and was noted in drill cores taken in the early 1960s, shortly after construction of the Vaiont Dam was completed. Extensive solution of the carbonates by ground water further weakens the rocks, producing sinkholes and underground caverns and channels.

The Vaiont Dam, built for power generation, is the highest "thin-arch" dam in the world, over 265 meters (875 feet) high at its highest point. Modern engineering methods were used to stabilize the rocks in which it is based. The capacity of the reservoir behind the dam was 150 million cubic meters—initially.

As the water level in the reservoir behind the dam rose following completion of the dam in 1960, pore pressures of ground water in the rocks of the reservoir walls rose also. This tended to buoy up the rocks and swell the clays of the clay-rich layers, further decreasing their strength and making sliding easier. In 1960, a block of 700,000 cubic meters of rock slid from the slopes of Mount Toc, on the south wall, into the reservoir. Creep was noted over a still greater area. A set of monitoring stations was established on the slopes of Mount Toc to track any further movement. In 1960–1962, measured creep rates occasionally reached 25 to 30 centimeters (nearly 1 foot) per week.

Late summer and fall of 1963 were times of heavy rainfall in the Vaiont Valley. The saturated rocks represented more mass pushing downward on zones of weakness. Groundwater flow increased, while the water table and reservoir level rose by over 20 meters. Measured creep rates increased. What was still not realized was that the rocks were slipping not as many small blocks but as a single, coherent unit. On October 1, animals that had been grazing on the slopes of Mount Toc moved off the hillside and would not return.

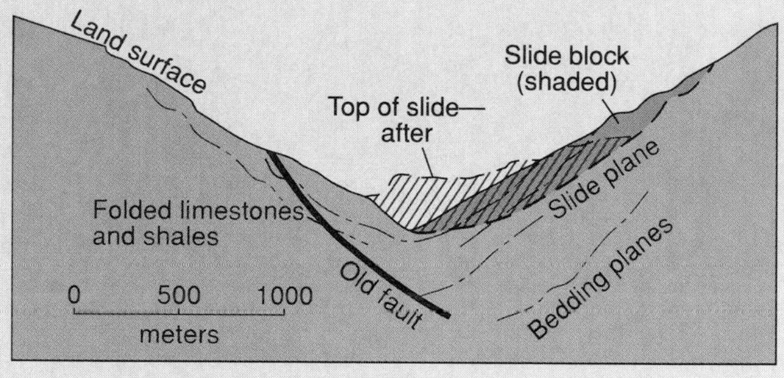

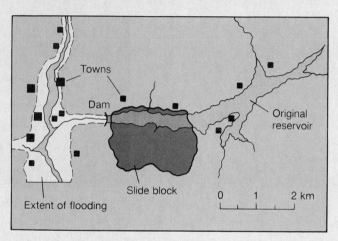

FIGURE 1 Cross section of the Vaiont River valley.

From G. A. Kiersch, "The Vaiont River disaster," Mineral Information Service 18, no. 7, pp. 129–38, 1965. State of California, Division of Mines and Geology, Sacramento, CA. Reprinted by permission.

FIGURE 2 The Vaiont Reservoir slide block. Note areas and towns inundated by resultant flooding.

From G. A. Kiersch, "The Vaiont River disaster," Mineral Information Service 18, no. 7, pp. 129–38, 1965. State of California, Division of Mines and Geology, Sacramento, CA. Reprinted by permission.

The rains continued, and so did the creep, rates of which increased to 20 to 30 centimeters per *day*. Finally, on October 8, the engineers realized that all the creep-monitoring stations were moving together. They also discovered that a far larger area of hillside was moving than they had thought. They tried to reduce the water level in the reservoir by opening the gates of two outlet tunnels. But the water level continued to rise: the silently creeping mass had begun to encroach on the reservoir, reducing the reservoir's volume. By October 9, creep rates were as high as 80 centimeters per day.

At about 10:40 on the night of October 9, in the midst of another downpour, disaster struck. A resident of Casso later reported that, at first, there was a sound of rolling rocks, ever louder. Then a gust of wind hit his house, smashing windows and raising the roof so that the rain poured in. Just as abruptly, the wind ceased and the roof collapsed. He was hurt and shaken, but he, at least, survived. Others downstream were not so fortunate.

A 240-million-cubic-meter chunk of hillside had slid into the reservoir

(figure 2). The shock of the slide was detected on seismometers in Rome, Brussels, and elsewhere in Europe. The sudden movement set off the corresponding shock wave of wind that rattled Casso and drew the water 240 meters upslope out of the reservoir after it, on the far bank. The displaced water crashed over the dam, in a wall 100 meters above the dam crest, and rushed down the valley below. The water wave was still over 70 meters high some 1.5 kilometers downstream, where the Vaiont River flows into the Piave River. The energy of the rushing water was such that some of it flowed *upstream* in the Piave for more than 2 kilometers. Within about five minutes, nearly 3000 people were drowned and entire towns obliterated.

It is a great tribute to the designer of the Vaiont Dam that the dam itself held through all of this, resisting forces far beyond its design specifications. However, those who chose the site had every reason to realize the landslide risks since there was ample evidence of persistent slope instability. In fact, the dam builders in this instance were later found guilty of gross negligence and imprisoned.

FIGURE 16.19 Areas prone to landslides may be recognized by the failure of vegetation to establish itself on unstable slopes. East side of Rock Creek valley, Montana.

topographic maps can be helpful in such a case. In a regional overview, the mass movement often shows very clearly, revealed by a scarp at the head of the slump or an area of hummocky, disrupted topography relative to surrounding, more stable areas.

With creep, individual movements are only over short distances, and the whole process is slow, so vegetation may continue to grow in spite of the slippage. More detailed observation, however, can reveal the movement. For example, trees are biochemically "programmed" to grow vertically upward. If the soil in which they are growing begins to creep downslope, the tree trunks may be tilted, and this indicates the soil movement. Further tree growth will continue to be vertical. If slow creep is prolonged over a considerable period of time, during which growth proceeds, curved tree trunks may result (figure 16.20). Inanimate objects can reflect soil creep also. Slanted utility poles and

fences and the tilting of once-vertical gravestones or other monuments likewise indicate that the soil is moving (figure 16.21). The ground surface itself may show cracks parallel to (across) the slope.

Even where slides are not necessarily evident, one may anticipate relative risks by considering slope angles and how slip-prone the materials making up the slopes are; shaly rocks dipping downslope, or clay-rich soils, are much more slide-prone than granite bedrock. The state geological survey can be a further source of information on slide risk.

Effects of Construction

Many types of construction lead to oversteepening of slopes. Highway roadcuts, quarrying or open-pit mining operations, and construction of stepped home-building sites on hillsides are among the activities that can cause problems

FIGURE 16.20 Tilted tree trunks are a consequence of creep. These curved tree trunks have developed on an imperfectly stabilized slope above a roadcut. Lassen Park, California.

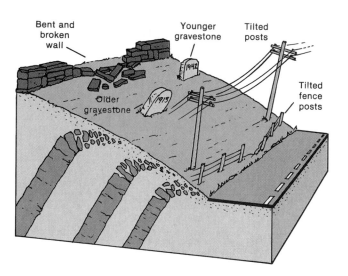

FIGURE 16.21 Other signs of creep: tilted monuments, fences, utility poles.

(figure 16.22). Where dipping layers of rock are present, removal of material at the lower end of the layers may leave large masses of rock unsupported, held in place only by friction between the layers. Slopes cut in unconsolidated material at angles higher than the angle of repose of that material are by nature unstable, especially if no stabilizing vegetation is planted (figure 16.23).

Building a house above a naturally unstable or artificially steepened slope adds weight to the slope, thereby increasing the shearing stress acting on the slope. Indeed, con-struction of the homesites itself may have involved adding masses of fill to level sites, in addition to steepening slopes as shown in figure 16.22. Other activities connected with the presence of housing developments on hillsides can increase the risk of landslides in more subtle ways. Watering the lawn, using a septic tank for sewage disposal, and even installing an in-ground swimming pool from which water can seep slowly are all activities that increase the soil's moisture content and can render the slope more susceptible to sliding. Irrigation and the use of septic systems both increase the flushing of fresh water through soils and sediments. In areas underlain by quick or sensitive clays, these practices may hasten the washing-out of salty pore waters and the destabilization of the clays. On the other hand, the planting associated with deliberate landscaping can reduce the risk of sliding.

A prospective home buyer should look for indications of unstable land underneath. Ground slippage may have caused cracks in driveways, garage floors, or brick or concrete walls; doors and windows may jam because the frame has warped due to differential movement in the soil and foundation. If there has already been enough movement to cause obvious structural damage, it may not be possible to stabilize the slope adequately, except perhaps at great expense. It may be especially important to investigate the site stability if the site has a slope of more than 15%, or if much steeper slopes lie above or below it. Note, however, that steep slopes are not required for sliding to occur.

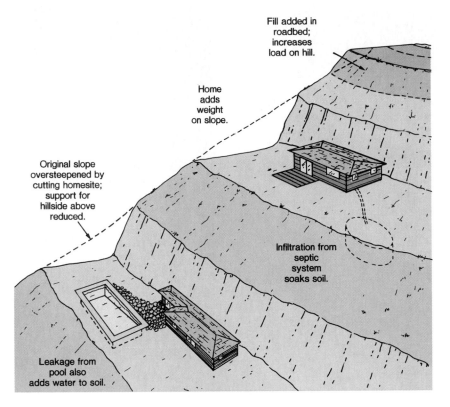

FIGURE 16.22 Effects of construction and human habitation on slope stability.

FIGURE 16.23 Failure of steep, unvegetated roadcut now threatens this cabin.

A

B

FIGURE 16.24 Avalanche-protection structure built over railroad track or road in snow-avalanche area deflects the flow. (A) Schematic. (B) Example in Canada's Glacier National Park.

Landslide Prevention

In places where the structures to be protected are few or small, and the slide zone is narrow, it may be economically feasible to bridge structures and simply let the slides flow over them. For example, this might be done to protect a railway line or road running along a valley below a particularly steep slope from avalanches—snow or debris (figure 16.24). This solution would be far too expensive on a large scale, however, and no use at all if the base on which the structure was built were also sliding. Avoiding the most landslide-prone areas altogether would greatly limit damage, but as is true with fault zones, floodplains, and other hazardous settings, developments may already be in place in areas at risk, and economic pressure for more development can be strong. Certain steps to reduce the potential for mass movements may then be undertaken.

If a slope is too steep to be stable under the load it carries, then one can either (1) reduce the slope angle, (2) place additional supporting material at the foot of the slope to prevent a slide or flow at the base of the slope, or (3) reduce the load (weight) on the slope by removing some of the rock or soil (or artificial structures) high on the slope. See, for example, figures 16.25 and 16.26. These measures may be used in combination but, depending on the instability of the slope, may need to be executed cautiously. If earth-moving equipment is being used to remove soil at the top of a slope, for instance, the added weight of the equipment itself, or the vibrations from it, may trigger a landslide. Where retaining walls are placed below a slope, the greatest success has generally been achieved with low, thick walls placed at the toe of a relatively coherent slope; high, thin walls have been less successful, given the distribution of stresses on an unstable slope.

To stabilize exposed, near-surface soil, ground covers or other vegetation may be planted. The preferred varieties are fast-growing plants with sturdy and extensive root systems, such as the pink-flowered crown vetch commonly used along highways. Where plants are an impractical solution, surface application of thin sheets of concrete or other material may be an alternative, although this is not always effective (figure 16.27).

The other principal strategy for reducing landslide hazards is to reduce the water content, or pore pressure, of the rock or soil, thereby increasing frictional resistance to sliding. This might be done by covering the surface completely with an impermeable material and diverting surface runoff above the slope. Alternatively, subsurface drainage might be undertaken. As a preliminary step, old water wells in the area can be vigorously pumped. Additional networks of underground boreholes can then be drilled to increase drainage, and pipelines can be installed to carry the water out of the slide area. All such moisture-reducing techniques naturally

A

B

FIGURE 16.25 Slope stabilization. Lake Tahoe. (A) A naturalistic appearance is achieved by using irregular rocks formed into building blocks with wire mesh. (B) Side view of resultant retaining wall.

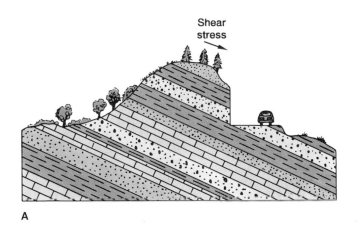

A

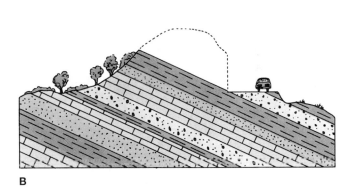

B

FIGURE 16.26 Removal of some of the material on a hillside reduces load and slope angle, increasing stability. (A) Before: roadcut leaves steep, unsupported slope. (B) After: material removed to reduce slope angle and load.

have the greatest impact where rocks or soil are relatively permeable. Where the rock or soil drains only slowly, hot air may be blown through the boreholes to dry out the ground more rapidly.

Anchoring an unstable mass is yet another approach. Vertical piles driven into the foot of a shallow slide may hold the sliding block in place, but only where the slide is comparatively solid (loose soils simply flow between piles), thin (so that piles can be driven deep into stable material below), and on a low-angle slope (otherwise, the shearing stresses may just snap the piles). This strategy has been tried on several European slides, generally not very effectively.

Greater success has been achieved with the use of *rock bolts* to stabilize rocky slopes and, occasionally, rockslides.

Rock bolts have long been used in tunneling and mining to stabilize rock walls. It is sometimes also possible to anchor a rockslide with giant steel bolts driven into stable rocks below the slip plane (figure 16.28). Again, this works best on thin slide blocks of very coherent rocks on low-angle slopes.

Additional procedures occasionally used on unconsolidated materials include hardening unstable soil by drying and baking it with heat (this can work well with clay-rich soils) or treating with portland cement. By far the most common strategies, however, are modification of slope geometry and load, or dewatering, or a combination of these two. The more ambitious engineering efforts are correspondingly more expensive and usually reserved for large construction projects, rather than individual homesites.

FIGURE 16.27 Broken retaining wall with freshly repaired road below is evidence of ongoing slope instability.

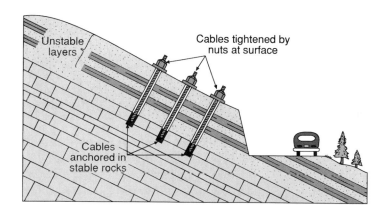

Unstable layers

Cables tightened by nuts at surface

Cables anchored in stable rocks

FIGURE 16.28 Use of rock bolts to stabilize a rockslide.

etermining the absolute limits of slope stability is an imprecise science. This was illustrated in Japan in 1969, when a test hillside being deliberately saturated with water slid prematurely, killing several of the researchers. The same point is made less dramatically every time a slope-stabilization effort fails.

SUMMARY

Mass movements occur when the shearing stress acting on rocks or soil exceeds the shear strength of the material to resist it. Gravity provides the main component of the shearing stress. The basic susceptibility of a slope to failure is determined by a combination of factors, including geology (nature and strength of materials), geometry (slope angle), and moisture content. Sudden failure usually also involves a triggering mechanism, such as vibration from an earthquake, addition of moisture, steepening of slopes (naturally or through human activities), or removal of stabilizing vegetation. One possible secondary consequence of mass movements is flooding.

The slowest mass wasting is creep. More rapid movements, collectively termed landslides when they involve rock or soil material, are subdivided on the basis of the nature of the movement and the materials moved. Falls occur when unsupported materials tumble down in free-falling style, making little contact with the underlying slope or cliff. Slumps and slides involve movement of coherent rock or soil masses. Flows, which include snow avalanches and pyroclastic flows as well as debris avalanches and soil flows, occur only in unconsolidated materials and are characterized by more chaotic movement of particles within the flowing mass.

Recognized landslide hazards may be reduced by modifying the slope geometry, reducing the weight acting on the slope, planting vegetation, or dewatering the rocks or soil.

TERMS TO REMEMBER

angle of repose 285
creep 293
debris avalanche 293
expansive clay 287
fall 290

flow 291
landslide 285
mass wasting 285
quick clay 290
scarp 291
sensitive clay 290

shearing stress 285
shear strength 285
slide 291
slump 291
solifluction 293
talus 291

QUESTIONS FOR REVIEW

1. What is the driving force behind mass movements? What properties of a material resist the shearing-stress component?
2. Explain the concept of angle of repose and its relationship to slope stability. To what kinds of material does this concept apply?
3. Briefly describe the role of fluid in mass movements.
4. Name two possible triggers for sudden slope failure.
5. What is a *quick clay?*
6. What is the distinction between a fall and a slide? between a slide and a flow?
7. Name four kinds of flow, noting the differences among them.
8. Rank the following in terms of velocity, from slowest to fastest: soil slump, creep, debris avalanche.
9. Describe any four ways in which human activities can destabilize slopes.
10. Suggest one possible approach to stabilizing (a) a steep roadcut in severely weathered granite; (b) a sandy soil slope; (c) a shallow but unstable slope in thinly bedded shale, where the bedding dips parallel to the slope.

1. What kind of geologic features might you look at to obtain an estimate of the angle of repose of (a) sand and (b) volcanic ash?

What factors might affect the results of your observations or limit their applicability?

2. Spring is landslide season in mountainous areas of many temperate climates. Suggest at least three reasons for this.

SUGGESTIONS FOR FURTHER READING

Aune, R. A. 1983. Quick clays and California's clays: No quick solutions. In *Environmental geology, text and readings,* 3d ed., edited by R. W. Tank. New York: Oxford University Press.

Bolt, B. A., et al. 1975. Hazards from landslides. Chap. 4 in *Geological Hazards.* New York: Springer-Verlag.

Crozier, M. J. 1986. *Landslides: Causes, consequences, and environment.* Dover, N.H.: Croom Helm.

Fleming, R. W., and Taylor, F. A. 1980. *Estimating the cost of landslide damage in the United States.* U.S. Geological Survey Circular 832.

Kiersch, G. A. 1965. The Vaiont Reservoir disaster. *Mineral Information Service* 18 (7): 129–38.

Radbrich-Hall, D. H., et al. 1976. Preliminary landslide overview map of the conterminous United States. U.S. Geological Survey Miscellaneous Field Studies Map MF-771.

Schultz, A. P., and Jibson, R. W., eds. 1989. *Landslide processes of the eastern U.S. and Puerto Rico.* Geological Society of America Special Paper 236.

Schuster, R. L., and Krizek, R. J., eds. 1978. *Landslides, analysis and control.* Washington, D.C.: National Research Council, Transportation Research Board Special Report 176.

Schuster, R. L., Varnes, D. J., and Fleming, R. W. 1981, Landslides. In *Facing geologic and hydrologic hazards,* edited by W. W. Hays. U.S. Geological Survey Professional Paper 1240-B.

Small, R. J., and Clark, M. J. 1982. *Slopes and weathering.* New York: Cambridge University Press.

Utgard, R. O., McKenzie, G. D., and Foley, D. 1978. *Geology in the urban environment.* Minneapolis: Burgess.

Varnes, D. J. 1984. *Landslide hazard zonation: A review of principles and practice.* Paris: UNESCO Press.

Voight, B., ed. 1978. *Rockslides and avalanches.* New York: Elsevier.

Zaruba, Q., and Mencl, V. 1969. *Landslides and their control.* New York: Elsevier.

17

WIND AND DESERTS

Windblown sand grains help to shape the surface features of arid regions like Arches Park, UT.

Wind is another agent of change at the earth's surface, eroding, transporting, and depositing material, but it is considerably less efficient in that role than is ice or water. On average, worldwide, winds move only a small percentage as much material as do streams; collectively, mass transport by winds is comparable to mass transport by glaciers, though the latter are really far more restricted. Wind lacks the ability to attack rocks chemically by solution or by wedging during freezing/thawing cycles. Even in many deserts, more sediment is moved during the brief periods of intense surface runoff following occasional rainstorms than by wind during the prolonged dry periods. However, it is in deserts and semiarid regions that the effects of wind action are usually most pronounced and most easily observed, so in that sense, the subjects of wind-related processes and deserts are linked. Wind erosion can be significant in glacial areas and at high altitudes also. In this chapter we examine the origin of wind, principal atmospheric circulation patterns, and wind's erosional and depositional characteristics.

The extent of both deserts and glaciers is related to climate and may be influenced by human activities. In the short term, desertification accelerated by too-intensive human use of dry lands may enlarge desert regions. Over the longer term, global warming may also play a role in the changing extent and distribution of deserts, as patterns of evaporation and precipitation shift. Global warming will be explored more fully in the next chapter.

THE ORIGIN OF WIND: ATMOSPHERIC CIRCULATION

The flow of streams is driven by the downward pull of gravity. Gravity likewise pulls downward on the atmosphere, but this is not the principal driving force behind the horizontal flow of air at the earth's surface. Air moves over the surface primarily in response to differences in pressure, which commonly correspond to differences in temperature. Local conditions then modify the details of flow direction and speed.

A basic factor in surface temperature variation is latitude. The sun's rays fall most directly, most intensely, on the earth's surface near the equator. Solar radiation is more dispersed near the poles. On a nonrotating earth with a uniform surface, the result would be very simple air-circulation patterns. The surface would be heated more strongly near the equator, less strongly near the poles; correspondingly, air over the equator would be warmer than polar air. Warmer, less dense (lower-pressure) air would rise at the equator, while cooler, denser (higher-pressure) air from the poles would move in toward the lower-pressure region. The rising warm air would spread out laterally, cool, and eventually sink. Large circulating air cells would develop, moving somewhat like mantle convection cells, cycling air from equator to poles and back.

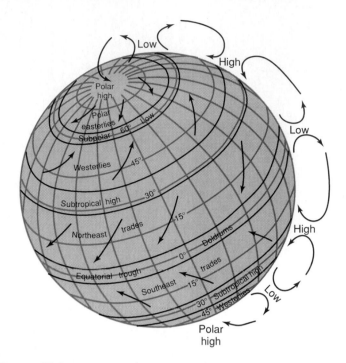

FIGURE 17.1 Principal present atmospheric circulation patterns.
From Arthur Getis, et al., Introduction to Geography, *3d. ed. Copyright © 1991 Times Mirror Higher Education Group, Inc., Dubuque, Iowa. All rights reserved. Reprinted by permission.*

The actual situation is considerably more complicated. For one thing, land and water are heated differentially by sunlight. Moreover, surface temperatures over the continents generally fluctuate much more on a daily basis than temperatures over adjacent oceans. Thus, the irregular distribution of land and water influences the distribution of high- and low-pressure regions, thereby modifying air flow. Another factor is that the earth rotates on its axis, which adds an east/west component to air movement as viewed from the surface. Also, the earth's surface is not flat; topographic relief introduces further irregularities in air circulation. Friction between moving air masses and the surface alters both wind direction and wind speed. For example, the presence of tall, dense vegetation can reduce near-surface wind speeds by 30 to 40% over the vegetated areas.

A generalized view of large-scale global air-circulation patterns is shown in figure 17.1. Different latitude belts are characterized by different prevailing wind directions. Most of the United States is in a zone of westerlies, in which winds generally blow from west/southwest to east/northeast. Local weather conditions and geography, of course, produce local deviations from this pattern on a day-to-day basis.

The global air-circulation pattern, in turn, influences the availability of precipitation. Warm air holds more moisture than cold. Similarly, a mass of air on which pressure is increased also can hold more moisture. Air spreading outward from the equator at high altitudes is chilled and at low pressure, since air pressure and temperature decrease with increasing altitude. Thus, the air holds little moisture. When that air circulates downward, at about 30 degrees north and

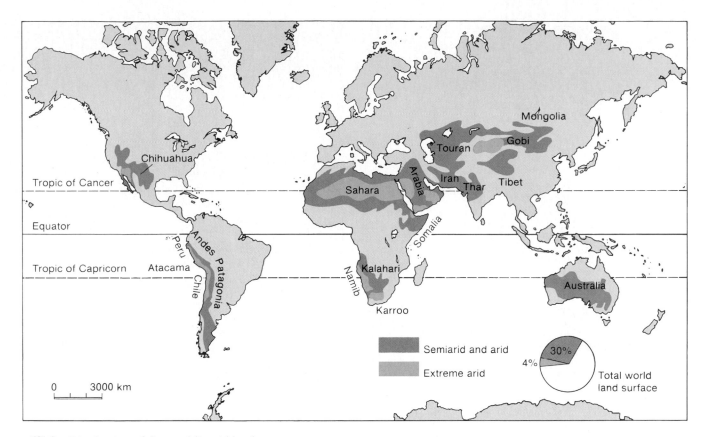

FIGURE 17.2 Distribution of the world's arid lands.

Source: From A. Goudie and J. Wilkinson, The Warm Desert Environment. *Copyright © 1977 Cambridge University Press, New York, NY. Reprinted by permission.*

south latitudes, it is warmed as it approaches the surface and also subjected to increasing pressure from the deepening column of air above it. It can then hold considerably more water, so when it reaches the earth's surface, it causes rapid evaporation. Note in figure 17.2 that many of the world's major deserts fall in belts close to these zones of sinking air at 30 degrees north and south of the equator.

WIND EROSION
AND SEDIMENT TRANSPORT

Flowing air and flowing water have much in common as agents of sediment transport. Both can move particles by rolling them along the surface, by saltation, or in suspension. Both also transport coarser and denser material the faster they move. However, the low density of air relative to sediment makes wind a less effective agent of sediment transport than water or ice. At a given velocity, wind cannot transport as coarse material by a given method as can water. Wind rarely moves coarse gravel or cobbles at all. Sand grains are moved by rolling or by saltation but are rarely lifted very high off the ground. Only the finest dust stays suspended in the air for long.

Like water, wind erodes sediment more readily than solid rock. In fact, wind alone has little impact on rock. However, wind-transported *sediment* can wear away at rock by the process of **abrasion.** Wind abrasion is a sort of natural sandblasting, very similar to milling by sand-laden waves or to glacial abrasion by ice-borne sediment. Abrasion during dust storms is capable of stripping paint and frosting and pitting the glass of car windshields, as well as eroding rocks. Because wind-borne sand is seldom lifted far off the ground, for reasons already noted, wind abrasion is generally a near-surface process only.

If winds blow consistently from one or a few directions, exposed cobbles and boulders may be planed off where they face the wind, in time taking on a faceted shape (figure 17.3). Grooves may also be cut in the direction of wind flow. Tall rocks may show undercutting close to ground level (figure 17.4). Rocks sculptured by abrasion are called **ventifacts**—literally, "wind-made" rocks (from the Latin).

Wholesale removal of unconsolidated sediment by wind action is **deflation.** Deflation is naturally most active where winds are unobstructed and the sediment exposed, unprotected by vegetative or artificial cover: deserts, beaches, unplanted farmland. In some dry areas, the selective removal

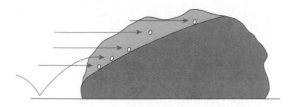

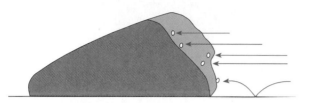

A

B

FIGURE 17.3 Ventifacts. (A) Ventifact formation by abrasion from one or several directions. (B) Example of a ventifact: rock polished and faceted by windblown sand at San Gorgonio Pass near Palm Springs, California.

(B) Photograph © Doug Sherman/Geofile.

FIGURE 17.4 Undercutting by near-surface wind abrasion. Garden of the Gods, Colorado.

of finer sediments by wind, often assisted by seasonal surface runoff, produces a surface sediment of residual coarser material. This **desert pavement** effectively protects underlying finer sediments from further erosion (figure 17.5). Desert pavement is sometimes also known as *deflation armor*. A desert pavement surface, once established, can be very stable. If the protective layer of coarser gravel and boulders is disturbed, however, the newly exposed fine sediment may be subject to rapid wind erosion.

The key role of vegetation in retarding wind erosion of sediment and soil was demonstrated especially dramatically in the United States in the early twentieth century. After the Civil War, there was a major westward migration of farmers to the plains states. They found flat or gently rolling land, much of it covered by prairie grasses and wildflowers rather than thick forests, so it was easy to clear and adapt for farming. Over much of the area, native vegetation was removed and the land plowed and planted to seasonal crops. Elsewhere, grazing livestock cropped the prairie. In the 1930s, several years of drought killed the crops, leaving the soil bare and vulnerable to erosion by the unusually strong winds that followed. This was the Dust Bowl era. Hundreds of millions of tons of soil were removed, transported, and then dumped as the winds abated, burying homes and farms. The native prairie vegetation had been adapted to the climate, while many of the crops were not. Once the crops died,

FIGURE 17.5 Desert pavement on surface of old gravel. Death Valley National Monument, California.
Photograph by C. B. Hunt, USGS Photo Library, Denver, CO.

there was nothing left to hold down the soil and protect it from the west winds sweeping unobstructed across the plains. See also box 17.1 later in the chapter.

EOLIAN DEPOSITS

Wind-related processes and products are also termed **eolian** (or, in older literature, *aeolian,* named for Aeolus, Roman god of the winds). The relatively few kinds of eolian sedimentary deposits are commonly well sorted. As with water-laid sediments, this sorting results because the maximum particle size moved depends on flow velocity. Winds selectively move finer, lighter materials; as they slow, they drop coarser, denser grains first.

Dunes

The principal eolian depositional landform is the **dune,** a low mound or ridge of sediment. When they hear the word *dune,* most people think "sand dune." However, dunes can be formed of sediment of different sizes, and even of snow. Dunes begin to form when sediment-bearing winds en-

counter an obstacle that slows them down. With reduced velocity, the wind begins to drop the coarsest, heaviest fraction of its load. The deposition, in turn, creates a larger obstacle that constitutes more of a windbreak, causing more deposition, in a self-reinforcing cycle. Once started, a dune can grow very large. Typical dunes range from 3 to 100 meters in height, but dunes 200 meters or more high are occasionally found. What ultimately limits a dune's size is not known. One consideration is that a dune is not a permanent, static object. Once formed, dunes tend to migrate, particularly if winds continue to blow predominantly from a single direction.

Dunes may be particularly prominent in desert climates, but they can be formed anywhere that sediment is exposed to wind action, especially in the absence of much vegetation. Thus they are also found on beaches, unplanted farmland, or alluvial or glacial deposits.

Dune Migration

A dune assumes a characteristic profile in cross section (cut parallel to the direction of wind flow): gently sloping on the

A

B

Figure 17.6 (A) Sand slides down slip face of sand dune. Oregon Dunes National Recreation Area. (B) Deposition of multiple sloping sand layers produces eolian crossbeds. Oregon Dunes National Recreation Area. See also figure 6.14C.

windward side (the side facing the wind, sometimes called the transport slope), steeper on the downwind side. With continued wind flow, particles are rolled or moved by saltation up the shallower upwind slope and tumble down the steeper face, or **slip face** (figure 17.6A). The slope of the slip face is characteristic of the sediment of the dune, as it is the angle of repose of the dune sediment; for sand, it is commonly 30 to 35 degrees from the horizontal. Layering often develops on the slip face. If winds and depositional patterns shift, eolian crossbeds result (figure 17.6B). When such crossbeds are found in sedimentary rocks (as shown in figure 6.14C), they reveal the desert origins of the sedimentary deposit.

The net effect of the many individual grain movements under sustained winds from one direction is that the dune itself is moved slowly downward (figure 17.7A). In a desert setting away from civilization, dune migration is largely of academic interest. However, migrating dunes—especially large ones—can be a real menace where they march across roads, through forests, and even over buildings (figure 17.7B). The costs to clear and maintain roads along sandy beaches can be high, for dunes can move several meters or more in a year. The usual approach to dune stabilization is to plant vegetation. However, many dunes exist where they do, in part, because the climate is too dry to support vegetation. Aside from any water limitations, young plants may be difficult to establish in shifting dune sands because their tiny roots cannot secure a hold.

Dune Forms

There are several types of dune forms. In a given area, one type usually predominates, depending on the particular balance among sediment supply, wind characteristics, and abundance of vegetation.

Transverse dunes are elongated perpendicular to the prevailing wind direction (figure 17.8). Many of these have a crescent shape, with arms or "horns" pointing downwind (in the direction of the slip face); these are **barchan dunes.** Very long, narrow barchan dunes grade into continuous *transverse ridges,* with the same cross-sectional profile but not appearing as discrete mounds of sand. Transverse dunes are common where sediment supply is abundant but may also occur where it is scarce, provided that vegetation is also very limited.

Longitudinal dunes occur where sediment supply is limited and winds are relatively strong. These dunes are elongated parallel to the direction of wind flow. They may form as the limited quantity of sediment is strung out gradually by the wind.

Where vegetation is more abundant, though not so plentiful as to prevent dune formation altogether, **parabolic dunes** tend to form (figure 17.9). At first glance, they appear similar in shape to the barchan dunes, but the arms of parabolic dunes point *upwind,* while the slip face is, as always, downwind. This appears to result from vegetation anchoring the arms of the dune. Wind velocity slows around the vege-

FIGURE 17.7 (A) Dune migration occurs as a result of many individual grain movements. (B) Marching sand dune encroaching on trees. Oregon Dunes National Recreation Area.

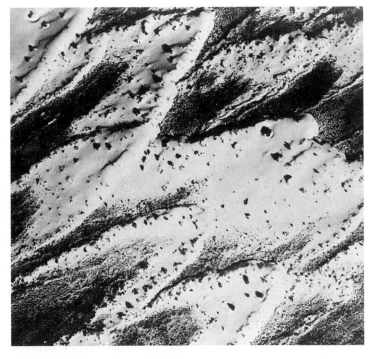

FIGURE 17.8 Transverse dunes: crescent-shaped barchan dunes. Sherman County, Oregon.
Photograph by E. D. McKee, USGS Photo Library, Denver, CO.

FIGURE 17.9 Parabolic dunes, the arms of the dunes anchored by vegetation.
Photograph by E. D. McKee, USGS Photo Library, Denver, CO.

tation, allowing the vegetation to hold sediment in place, while the bulk of the dune marches on. Parabolic dunes may be common in lightly vegetated beach areas.

Loess

Rarely is the wind strong enough to move sand-sized or larger particles very far or very rapidly. Fine dust, on the other hand, is more easily suspended in the wind and can be carried many kilometers before it is dropped. A deposit of windblown silt is known as **loess.** The rock and mineral fragments in loess are in the range of 0.01 to 0.06 millimeters (0.0004 to 0.0024 inches) in diameter.

The principal loess deposits in the United States are in the central part of the country and in southeastern Washington State in an area known as the Palouse. Their spatial

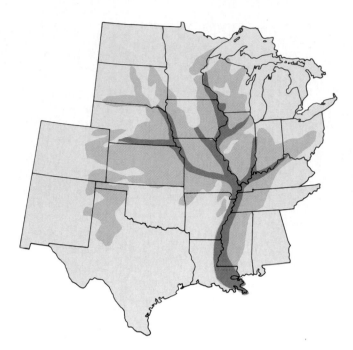

FIGURE 17.10 Distribution of loess deposits (shaded) in the central United States and location of major glacier-fed stream valleys.

After J. Thorp and H. T. U. Smith, "Pleistocene Eolian Deposits of the United States, Alaska, and Parts of Canada," Geological Society of America map, 1952.

distribution provides a clue to their source. The loess of the Palouse was deposited after the development of the Cascade Mountains created a rain shadow (see the discussion of rain shadows later in the chapter) that dried up lakes and marshes in central Washington. The deposits of the central United States are concentrated around the Mississippi River drainage basin, particularly on the east sides of major rivers of that basin (figure 17.10). Those same rivers drained away much of the meltwater from retreating ice sheets in the last ice age. Glacially produced sediment was washed down and deposited along the river valleys, and the lightest material, much of it rock flour, very finely ground sediment, was blown farther eastward by the wind.

Because dry glacial erosion does not involve as much chemical breakdown as does stream erosion, as noted earlier in this chapter, many soluble minerals are preserved in glacial rock flour. These minerals provide some valuable plant nutrients to the farmland soils now developed on the loess. Newly deposited loess is also quite porous and open in structure, so it has good moisture-holding capacity. These two characteristics together make the farmlands developed on midwestern loess particularly productive. The Palouse is also noted for its agricultural productivity.

Not all loess deposits are ultimately of glacial origin. Some in the southwestern United States may have formed from dust blown off the nearby deserts. Loess derived from the Gobi Desert covers large areas of China, and additional loess deposits are found downwind of other major deserts. Loess may also be derived from the finest fractions of volcanic ash deposits.

Loess does have drawbacks with respect to applications other than farming. While its light, open structure is reasonably strong when dry and not heavily loaded, it may not make suitable foundation material. Loess is subject to *hydrocompaction,* meaning that, when wetted, it tends to settle, crack, and become denser and more consolidated, to the detriment of any structures built on top of it. The very weight of a large building can also cause settling and collapse of loess.

DESERTS AND DESERTIFICATION

A **desert** is a region with so little vegetation that no significant population can be supported on that land (figure 17.11). It need not be hot or even, technically, dry. Polar ice caps are a kind of desert. In more temperate climates, deserts are characterized by very little precipitation—25 centimeters (about 10 inches) per year or less—but they may be consistently hot, cold, or variable in temperature with the season or time of day. The distribution of the arid regions of the world (exclusive of polar deserts) was shown in figure 17.2.

Causes of Natural Deserts

A variety of factors contribute to formation of a desert. One is moderately high surface temperatures. Most vegetation, under such conditions, requires abundant rainfall and/or slow evaporation of what precipitation does fall. This accounts for the abundance of deserts in the belts of rapid evaporation at 30 degrees north latitude (where, for example, the Sahara and Arabian deserts are found) and 30 degrees south latitude (for example, the Australian deserts) in figure 17.2.

Topography also plays a role in controlling the distribution of precipitation. A high mountain range along the path of principal air currents between the ocean and a desert area may be the cause of the latter's dryness. As moisture-laden air from over the ocean moves inland across the mountains, it is forced to higher altitudes, where the temperatures are colder and the air thinner (lower pressure). Under these conditions, much of the moisture originally in the air mass is forced out as precipitation, and the air is much drier when it moves farther inland and down out of the mountains. In effect, the mountains cast a **rain shadow** on the land beyond (figure 17.12). Rain shadows cast by the Sierra Nevada of California, by the Cascades in Washington and Oregon, and by the southern Rockies contribute to the dryness of the western United States.

Because the oceans are the major source of moisture in the air, simple distance from the ocean (in the direction of air movement) can be a factor contributing to the formation of a desert. The longer an air mass is in transit over dry land, the greater chance it has of losing some of its moisture through precipitation. This accounts, in part, for the existence of the Gobi Desert. On the other hand, under special circumstances, even coastal areas can have deserts. If the land is hot and the adjacent ocean cooled by cold currents, the moist air coming off the ocean will also be cool and carry less moisture than would warmer air over an ocean. As that cooler air warms

FIGURE 17.11 Sparse vegetation of desert in Monument Valley, Arizona, shows lack of capacity to support life.

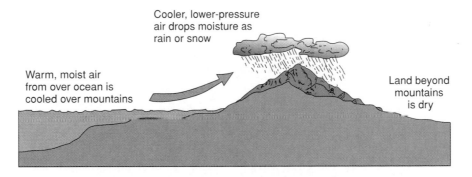

Cooler, lower-pressure air drops moisture as rain or snow

Warm, moist air from over ocean is cooled over mountains

Land beyond mountains is dry

FIGURE 17.12 Rain shadow cast by mountains lying in the path of winds from the ocean.

over the land and becomes capable of holding still more moisture, it causes rapid evaporation from the land, rather than precipitation. This phenomenon is observed along portions of the western coasts of Africa and South America.

Desert Landforms and Phenomena

As already discussed, dunes constitute the principal landform of many deserts. However, several other features result from the climatic conditions of arid deserts, especially the limited precipitation and sporadic streamflow.

Streams in deserts, which may flow only during a short rainy season or even just for a matter of hours after a cloudburst, are termed **ephemeral streams.** Their channels lie so far above the water table that there is no base flow from ground water to keep them flowing between precipitation events. These storm-fed streams may be characterized by **flash flooding,** a very sharp rise and fall of water level during and immediately after the cloudburst. As the stream rushes down its channel, with minimal interaction between the rapidly flowing water and the channel sides, and little

FIGURE 17.13 Alluvial fans formed where mountain streams flow into a dry plain near Albuquerque, New Mexico.

meandering or lateral channel movement, there is usually rapid infiltration into the parched ground, which may ultimately consume all the water. While streamflow persists, the rapid flow can move a large bed load.

Should the stream flow out of the mountains into a flat plain, especially if it is constantly losing water to infiltration, the combination of decreased velocity and decreased water flow may cause the sediment load to be deposited in an **alluvial fan** at the edge of the plain (figure 17.13). Multiple fans may overlap as several streams flow out of the same mountain range, forming a *bajada* (from the Spanish for "slope").

Slow erosion of bedrock at the mountain front may also lead to formation of a bedrock platform, sloping gently away from the foot of the mountains down into the desert. This is a **pediment,** its name derived from the Latin for "foot." From a distance, a pediment may appear very similar to a series of alluvial fans, but pediments are erosional features, made of bedrock with little or no sediment cover, while alluvial fans are deposits of sediment.

Many deserts are characterized by **internal drainage,** in which streams terminate in enclosed basins rather than flowing out of the region. The base level of these streams is a plain, not a sea. Into the basin the streams carry fine sediments, which are deposited as the waters evaporate or infiltrate. To these clastic sediments are added salts crystallized out of solution from the stream waters as they evaporate and the dissolved salts become more concentrated. The resultant "dry lakes" are termed **playas** in North America (nomenclature varies worldwide). The fine-grained sediments of playas are especially subject to shrinkage as they dry out, so mud cracks are common surface features.

Weathering in deserts is typically very slow, because of the scarcity of water to dissolve minerals and promote chemical reactions. Desert soils may retain high concentrations of soluble salts and calcium carbonate (recall the conditions that promote formation of pedocal soils). Indeed, where deserts or semiarid lands have been irrigated, the result has sometimes been to make the soils saltier still: water evaporates rapidly near the surface, capillary action wicks deeper, salt-laden water up from below, and continued evaporation deposits more salts in upper soil levels.

Desertification: Causes and Impacts

Climatic zones, like many other surface features, shift over time. In addition, topography changes, global temperatures change, and plate motions move landmasses to different latitudes. Amid these changes, new deserts develop in areas that previously had more extensive vegetative cover. The term **desertification,** however, is generally restricted to apply only to the relatively rapid development of deserts caused by the impact of human activities.

Exact definition of the lands at risk is difficult. *Arid* and *semiarid* lands are commonly defined as those with annual rainfall of less than 60 centimeters (24 inches), though the extent to which vegetation will thrive in low-precipitation areas also depends on such additional factors as temperature and local evaporation and infiltration rates. Many of the arid lands border true desert regions. Desertification does not involve the advance or expansion of desert regions as a result of forces originating within the desert. Rather, desertification is a patchy conversion of dry-but-habitable land to uninhabitable desert as a consequence of land-use practices (perhaps accelerated by such natural factors as drought).

Vegetation in dry lands is, by nature, limited. At the same time, it is a precious resource, which may in various cases provide food for people or for livestock, wood for shelter or energy, and protection for the soil from erosion. Desertification typically involves severe disturbance of that already sparse and often fragile vegetation. The environment is not a resilient one, and its deterioration, once begun, may be irreversible and even self-accelerating.

On land used for farming, native vegetation is routinely cleared to make way for crops. While the crops thrive, all may be well. If the crops fail, or if the land is left unplanted for a time, several consequences follow. One, as in the Dust Bowl, is erosion. A second, linked to the first, is loss of soil fertility. The topmost soil layer, rich in organic matter, is the most nutrient-rich and also is the first lost to erosion. A third result may be loss of soil structural quality. Under the baking sun typical of many dry lands, and with no plant roots to break it up, the soil may crust over, becoming less permeable. This increases surface runoff, correspondingly decreasing infiltration by what precipitation does fall and thus decreasing reserves of soil moisture and ground water on which future crops may depend. All of these changes together make it that much harder for future crops to succeed, and the problems intensify.

Similar results follow from the raising of numerous livestock on the dry lands. In drier periods, vegetation may be reduced or stunted. Yet it is precisely during those periods that livestock, needing the vegetation not only for food but also for the moisture it contains, put the greatest grazing pressure on the land. The soil may again be stripped bare, with the resultant deterioration and reduced future growth of vegetation as previously described for cropland.

Natural drought cycles thus play a role in desertification. However, in the absence of intensive human land use, the degradation of the land during drought is typically less severe, and the natural systems in the arid lands can recover when the drought ends. On a human timescale, desertification—permanent conversion of marginal dry lands to deserts—is generally observed only where human activities are also significant. Were it not for extensive irrigation now, much of the Dust Bowl area of the 1930s could be subject to desertification; see box 17.1.

Desertification is cause for concern principally because it effectively reduces the amount of arable (cultivatable) land on which the world depends for food. An estimated 600 million people worldwide now live on the arid lands. All of those lands, in some measure, are potentially vulnerable to desertification. More than 10% of those 600 million people live in areas identified as actively undergoing desertification now. Some projections suggest that, by the end of this century, one-third of the world's once-arable land will be rendered useless for the culture of food crops as a consequence of desertification and attendant soil deterioration. The recent famine in Ethiopia may have been precipitated by drought, but it will be prolonged by desertification brought on by overuse of land incapable of supporting concentrated human or animal populations. Other regions in which desertification of semiarid lands is an actual or potential problem include the Sudan, parts of central China, and areas of the Great Plains and southwestern United States; note the locations of the semiarid lands in figure 17.2.

Box 17.1

The Dust Bowl

The Dust Bowl area proper, although never exactly defined, comprised close to 100 million acres of southeastern Colorado, northeastern New Mexico, western Kansas, and the Texas and Oklahoma panhandles. The farming crisis there during the 1930s resulted from an unfortunate combination of factors: clearing or close grazing of natural vegetation, drought (rainfall less than 50 centimeters [20 inches] per year for several years), sustained winds (averaging more than 15 kilometers/hour [10 miles/hour], with velocities ranging much higher during storms), and poor farming practices, including widespread disregard of wind erosion as a potential problem. There had been droughts in the area previously. But in the late 1800s and the early decades of this century, mechanization of farming made possible the rapid expansion of cultivated acreage, from about 12 million acres in 1879 to over 100 million acres by 1929, which, in turn, greatly increased the size of the area threatened by adverse conditions. When drought then killed the crops that had supplanted the prairie grasses, serious soil erosion was possible.

The action of the wind was most dramatically illustrated during the fierce dust storms that began in 1932. The storms were described as "black blizzards" that blotted out the sun (figure 1). Black rain fell in New York, black snow in Vermont, as windblown dust moved eastward. In May 1934, one 36-hour storm whipped up a dust cloud more than 2000 kilometers (1250 miles) long. People choked on the dust, some dying of suffocation or of a "dust pneumonia" similar to the silicosis miners develop from breathing rock dust.

As the dust from Kansas and Oklahoma settled on their desks, politicians in Washington, D.C., realized that something had to be done. In April 1935, a permanent Soil Conservation Service was established to advise farmers on land use, drainage, erosion control, and other matters.

By the late 1930s, concerted efforts by individuals and state and federal government agencies to improve farming practices to reduce wind erosion—together with a return to more normal rainfall—had considerably reduced the problems. However, drought struck again in the 1950s, and tens of millions of cropland acres were damaged by

FIGURE 1 Example of a 1930s dust storm in the Dust Bowl.
Photograph courtesy of USDA Soil Conservation Service.

FIGURE 2 Barren fields buried under windblown soil.
Photograph courtesy of USDA Soil Conservation Service.

wind erosion in 1954. In a dry period in the winter of 1965, high winds raised dust 10,000 meters (about 6 miles) into the air and carried some of it east as far as Pennsylvania. Millions of acres more were damaged in the mid-1970s. In recent years, the extent of wind damage has been limited somewhat by the widespread use of irrigation to maintain crops in dry areas and dry times. But, as was noted in chapter 15, some of the important sources of that irrigation water are rapidly being depleted. Future droughts may yet produce more scenes like those of figures 1 and 2.

SUMMARY

Wind arises from differential heating of the atmosphere, its movement complicated by topography and by the earth's rotation. Globally, it is a far less efficient agent of surface change than is water. Its effects are particularly prominent in regions of limited vegetation and precipitation. Wind erodes material through abrasion by wind-carried sediment, producing ventifacts, or through deflation, the removal of unconsolidated material. The principal depositional eolian landform is the dune. Dune shapes vary with the relative importance of sediment supply, wind action, and presence of vegetation. If winds blow consistently in one direction, dunes may also migrate. Where a source of abundant fine sediment exists, selective transport and redeposition of this material by wind produces loess, which often contributes to excellent farmland.

Deserts, defined as regions incapable of supporting enough vegetation to sustain significant human or animal populations, form in a variety of climates. The majority are in warm or temperate regions and are dry. Many deserts are characterized by internal drainage of ephemeral streams, which deposit alluvial fans of sediment and form playas within the basins into which they drain. The amount of land area classified as desert is presently increasing as a result of desertification accelerated by human activities in arid lands.

TERMS TO REMEMBER

abrasion 305
alluvial fan 312
barchan dune 308
deflation 305
desert 310
desertification 313
desert pavement 306

dune 307
eolian 307
ephemeral stream 311
flash flooding 311
internal drainage 312
loess 309
longitudinal dune 308

parabolic dune 308
pediment 312
playa 312
rain shadow 310
slip face 308
transverse dune 308
ventifact 305

QUESTIONS FOR REVIEW

1. What is the principal cause of wind? Briefly describe three factors that influence the direction of wind flow.
2. Compare and contrast the processes of sediment transport by streams and by wind, and the sizes of the materials moved.
3. Describe the nature of wind abrasion and the formation of ventifacts.

4. What is *deflation,* and how is it involved in the formation of desert pavement?
5. Summarize the process of dune formation and migration.
6. How do barchan and parabolic dunes differ? Why?
7. What is *loess?* Name three kinds of material from which loess deposits may be derived.
8. What is a *desert?* Are all deserts found in hot climates? Explain.

9. Describe how a mountain range can cast a rain shadow.
10. Explain the phenomenon of flash flooding and the formation of alluvial fans and playas in terms of common drainage characteristics in and near deserts.
11. What is *desertification?* Explain two ways in which it is accelerated by human activities.

For Further Thought

1. Obvious graded bedding is less common in wind-laid than in water-laid sediments; sorting of wind-deposited sediments is commonly very good. Suggest explanations for these observations.

2. If you live in a snowy area, make an inspection tour around your home or school buildings after a snowstorm with high winds. Note the distribution of snow, and try to relate it to the distribution of obstacles to wind flow. (Readers in sparsely vegetated areas might do the same after a windstorm in dry weather, although the sediment-distribution patterns may be more subtle and require closer scrutiny.)

Suggestions for Further Reading

Battan, L. J. 1979. *Fundamentals of meteorology*. Englewood Cliffs, N.J.: Prentice-Hall.

Brookfield, M. E., and Ahlbrandt, T. S., eds. 1983. *Eolian sediments and processes*. New York: Elsevier.

Cooke, R. U., and Warren, A. 1973. *Geomorphology in deserts*. Berkeley: University of California Press.

Ellis, W. 1987. "Africa's Stricken Sahel." *National Geographic* 172 (August): 140–79.

Goudie, A. 1983. *Environmental change*. 2d ed. New York: Oxford University Press.

Goudie, A., and Wilkinson, J. 1977. *The warm desert environment*. New York: Cambridge University Press.

Greeley, R., and Iversen, J. D. 1985. *Wind as a geological process*. Cambridge, England: Cambridge University Press.

Hurt, R. D. 1981. *The Dust Bowl*. Chicago: Nelson-Hall.

Hyde, P. 1987. *Drylands: the deserts of North America*. San Diego: Harcourt, Brace, Jovanovich.

Secretariat of the U.N. Conference on Desertification, Nairobi. 1977. *Desertification: Its causes and consequences*. New York: Pergamon Press.

Sheridan, D. 1981. *Desertification in the United States*. Washington, D.C.: Council on Environmental Quality.

Wells, S. G., and Haragan, D. R., eds. 1983. *Origin and evolution of deserts*. Albuquerque: University of New Mexico Press.

18

ICE AND CLIMATE

Icebergs and meltwater testify to summer shrinking of Angel Glacier in Jasper National Park, Alberta, Canada; global warming could lead to melting of glaciers on a global scale. © Doug Sherman/Geofile

While ice now covers only about 10% of the continental land area, features attributed to glacial episodes over the earth's history are found over about three-fourths of the continental surface. Only a few tens of thousands of years ago, sheets of ice covered major portions of North America, Europe, and Asia. Melting glaciers recharged aquifers used today for water; glacial sediments from older ice advances themselves make up some of the aquifers. In this chapter, we survey the nature of glaciers, the distinctive erosional and depositional features they produce, and other characteristics of glaciated or cold regions.

As we are becoming increasingly aware, human activities may be altering the climate of the future. Initial discussions of the effects of global warming tended to focus on the potential consequences of extensive melting of present remaining glacial ice, and indeed this may be significant. Studies of past ice ages can shed light on what might be expected in this respect, and we will explore this aspect of global warming in this chapter. It is important, however, to realize that there could be other effects of global warming with impacts at least as serious, not only for humans, but for earth's flora and other fauna as well.

ICE AND THE HYDROLOGIC CYCLE

Approximately 75% of the fresh water on earth is stored as ice in glaciers. Water enters this reservoir as precipitation (usually snowfall) and leaves it by evaporation or by melting. The world's supply of glacial ice is the equivalent of 60 years of precipitation over the whole earth, or, to put it another way, represents 900 years' flow of all the world's rivers at their present discharge.

Where glaciers are large or numerous, glacial meltwater may be the principal source of summer streamflow. By far the most extensive glaciers in the United States are those of Alaska, which cover about 3% of the state's land area. Summer streamflow from these glaciers is estimated at nearly 50 trillion gallons of water. Even in less extensively glaciated states like Washington, Montana, California, Idaho, Colorado, Oregon, and Wyoming, streamflow from summer meltwater amounts to tens or hundreds of billions of gallons.

Anything that modifies glacial melting patterns can profoundly affect regional water supplies. Dusting the glacial surface with a thin layer of dark material, such as coal dust, increases the heating of the surface, and hence the rate of melting and water flow. Conversely, cloud seeding over glaciated areas could increase precipitation and the amount of water stored in the glaciers. Increased meltwater flow can be useful not only in increasing water supplies but also in achieving higher levels of hydroelectric power production. Techniques for modifying glacial meltwater flow are not now being used in the United States, in part because the majority of U.S. glaciers are in national parks, primitive areas, or other protected lands. However, some of these techniques are being practiced in parts of the former Soviet Union and China.

THE NATURE OF GLACIERS

Fundamentally, a **glacier** is a mass of ice on land, formed by recrystallization of snow, that moves under its own weight. A quantity of snow sufficient to form a glacier does not fall in a single winter, so in order for glaciers to develop, the climate must be cold enough that some snow and ice persist year-round. This, in turn, requires an appropriate combination of elevation and latitude.

Glaciers tend to be associated with the extreme cold of polar regions. However, temperatures also generally decrease at high elevations (figure 18.1), so glaciers can exist in mountainous areas even in tropical or subtropical regions. Three mountains in Mexico have glaciers, all at elevations above 5000 meters (over 16,000 feet). Similarly, there are glaciers on Mount Kilimanjaro in East Africa (which is on the equator), but only at elevations above 4400 meters (14,500 feet).

Glacier Formation

For glaciers to form, there must be sufficient moisture in the air to provide the necessary precipitation. In addition, the amount of winter snowfall must exceed summer melting, so that the snow accumulates year by year. Most glaciers start in the mountains, as snow patches that survive the summer. Slopes that face the poles (north-facing in the Northern Hemisphere and south-facing in the Southern Hemisphere), protected from the strongest sunlight, favor this survival. So do gentle slopes, on which snow can pile up thickly instead of plummeting down periodically in avalanches.

As the snow accumulates, it is gradually transformed into ice. The weight of overlying snow packs it down, drives out much of the air, and causes it to recrystallize into coarser, denser, interlocking ice crystals. In its initial stages, the process is somewhat like what happens as one packs snow into a denser, icier snowball. Alternate thawing and freezing after deposition hasten the transformation. (Glacial ice could, in fact, be regarded as a very-low-temperature metamorphic rock.) The material intermediate in texture between freshly fallen snow and compact, solid ice is called **firn.** It is not unlike the coarse frozen material that forms along roadsides in snowy areas during winter and early spring, from piles of snow plowed off the roads. Complete conversion of snow into glacial ice may take from 5 to 3500 years, depending on such factors as climate and rate of snow accumulation at the top of the pile. Eventually, the mass of ice becomes large enough that, if there is any slope to the terrain, the ice begins to slide or flow downhill.

FIGURE 18.1 The snow line on a mountain range reflects the decrease in temperature with increasing elevation. Olympic Range, Washington.

Movement of Glaciers

The movement of a glacier may be a nearly imperceptible 20 meters (about 65 feet) per year or, for brief periods at least, a glacier may *surge* at up to 10 kilometers per year (about 30 meters, or close to 100 feet, per day). Glacial movements can occur in several ways.

A major mechanism of glacial movement involves internal deformation within the ice, including plastic flow. The internal deformation can be demonstrated, in part, by the fact that not all portions of a glacier move equally quickly. Typically, flow is fastest at the top and center, in part because friction at the base of the glacier (and the sides of a glacier in a valley) slows it down (figure 18.2). This is analogous to the effect of friction within the channel

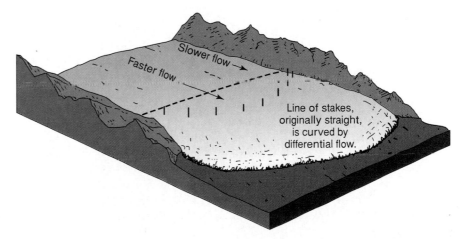

FIGURE 18.2 Differential rates of movement of different parts of a glacier indicate internal deformation.

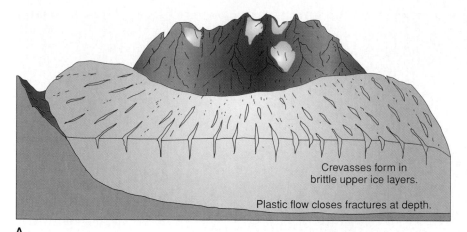

Crevasses form in
brittle upper ice layers.

Plastic flow closes fractures at depth.

A

on streamflow velocity. Also, the internal deformation is most pronounced in the deeper zones of the glacier, which are under the greatest confining pressure. The uppermost layers of the glacier, riding on more plastic layers below, behave rigidly or brittlely, fracturing to make the **crevasses,** or cracks, that are the peril of climbers who cross glaciers (figure 18.3). Like other rocks, then, glacial ice can be deformed by folds, joints, or faults. Like other rocks, glacial ice is most likely to behave brittlely at low pressure (near the ice surface), more plastically deep in the glacier where confining pressures are higher.

A secondary mechanism of movement is by sliding on basal meltwater. Because ice is less dense than water, the pressure on the ice at a glacier's base may be sufficient to melt a little of it (just as ice melts under a skater's blade even in cold weather). The ice mass as a whole may then slide on this meltwater.

The Glacial Budget

Matter is continually cycled through a glacier. Where it is cold enough for snow to fall, fresh material accumulates, adding to the weight of ice and snow that pushes the glacier downhill. At some point, the advancing edge of the glacier terminates, either because it flows out over water and breaks up, creating icebergs by a process known as **calving,** or, more commonly, because it has flowed to a place that is warm enough that ice loss **(ablation)** by melting, evaporation, wind erosion, or calving is at least as rapid as the rate at which new ice flows in to replace it (figure 18.4). The **equilibrium line** is the boundary at which there is a balance between material added and material lost.

Over the course of a year, the size of a glacier varies somewhat (figure 18.5). In winter, the rate of accumulation increases, and melting and evaporation decrease. In summer, snowfall is reduced or halted, and melting accelerates; ablation exceeds accumulation.

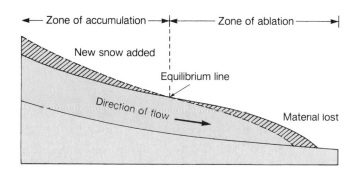

FIGURE 18.4 Longitudinal cross section of a glacier (schematic). In the zone of accumulation, the addition of new material exceeds loss by melting or evaporation; the reverse is true in the zone of ablation.

B

FIGURE 18.3 (A) Crevasses form in the brittle upper portion of the glacier (schematic). (B) Crevasse formation as this glacier in Denali National Park, Alaska, is stretched around a rocky peak.

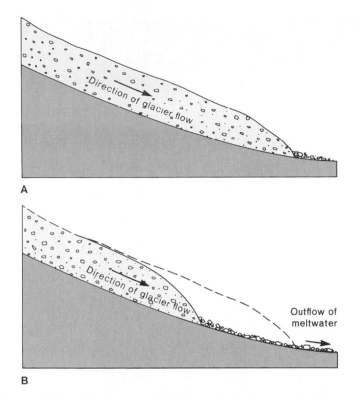

FIGURE 18.5 (A) Advance of glacier carries both ice and sediment downslope. (B) Rapid melting causes apparent retreat of glacier, accompanied by deposition from the melting ice.

If the climate remains stable, the glacier achieves a sort of dynamic equilibrium, and the average extent of the glacier remains constant from year to year. Over longer periods, if there is a sustained increase in net precipitation/accumulation, the glacier thickens and extends farther from its source; it is then said to be *advancing*. Conversely, if warmer or drier conditions over an extended period cause a net loss of ice year after year, the extent of the glacier will diminish, and it is described as *retreating*, although it does not, of course, flow backwards toward its source, or uphill. Noticeable variations in glacial extent have been observed in recent history. From the mid-1800s to the early 1900s, many mountain glaciers worldwide retreated far up into their valleys. Beginning about 1940, the trend seemed to reverse, and glaciers generally advanced. Now, retreating glaciers predominate again. Glaciers worldwide do not necessarily all advance or retreat simultaneously, however, because local climatic conditions may deviate from overall world trends.

Types of Glaciers

Glaciers are divided into two types on the basis of size and occurrence. The most numerous today are the **alpine glaciers,** typically found in mountainous regions, most often at relatively high elevations (figure 18.6). Many occupy valleys in the mountains and are termed, logically, *valley glaciers.* Most of the estimated 70,000 to 200,000 glaciers in the world today are alpine glaciers.

FIGURE 18.6 Alpine glaciers. College Fjord, Alaska. Note the streaks of sediment being carried along by the flowing ice.
Photograph by W. A. Montgomery.

FIGURE 18.7 Typical U-shaped glacial valley in Glacier National Park, Montana.

The larger and rarer **continental glaciers** are also known as *ice caps* (generally less than 50,000 square kilometers in area) or *ice sheets* (larger). They can cover whole continents and reach thicknesses of a kilometer or more. Though they are far fewer in number than the alpine glaciers, the continental glaciers collectively contain far more ice. At present, the two principal continental glaciers are the Greenland and the Antarctic ice sheets. (The Arctic polar ice mass is not a true glacier, as it is not based on land.) The Antarctic ice sheet is so large that it could easily cover the forty-eight contiguous United States, and it reaches depths of over 2 miles (3 kilometers). The geologic record indicates that, at several times in the past, even more extensive ice sheets existed on earth.

Given the size of continental glaciers, it is clear that to think of them as flowing "downhill" in the obvious sense is misleading; they cover, and flow across, many local ups and downs of terrain. Instead, they flow down a gradient in thickness and mass, from the regions of maximum accumulation, under the influence of gravity. Therefore, when the flow directions of past continental glaciers are determined, their source regions can also be identified.

GLACIAL EROSION

The mass and solidity of a glacier make it a very effective agent of both erosion and sediment transport, more so than either wind or liquid water.

Sediment Production and Transport

Rocks fall from valley walls onto glacial ice and are subsequently carried along with the glacier. Note the dark stripes of sediment along the margins of the alpine glaciers in figure 18.6. Additional material becomes frozen into the ice at the base and sides of the glacier. Ice itself is too soft to erode the rocks over which it flows, but these rock fragments frozen into the ice cause erosion by **abrasion** of the surrounding rocks at the base of the glacier or along valley walls, just as the grit of sandpaper can abrade wood or metal. The result of glacial abrasion is sediment, produced with little alteration of the original chemical and mineralogical character of the parent rock. Continued pulverizing into ever-finer fragments eventually produces a powdery, silt-sized sediment termed **rock flour.**

Because ice is solid, it transports sediments with equal efficiency, regardless of particle size. Everything is moved together and, when the ice melts, everything is deposited together, as shown in figure 18.5. Deposition of the resultant sediment is described later in the chapter.

The Glacial Valley

The differences in the character of erosion by water and by ice result in differences in shape between glacial and stream valleys. Glacial valleys are characteristically U-shaped in cross section (figure 18.7), in contrast to the V-shaped val-

FIGURE 18.8 Tribuary glaciers (entering from left). Denali National Park, Alaska.

leys of streams. They are broadened and deepened partly by abrasion by rocks frozen into the ice and partly by a process known as **plucking.** Plucking occurs as water seeps into cracked rocks at the base of the glacier and freezes, attaching the rocks to the glacier. As the glacier moves on, it may tug apart the fractured rock and pluck away chunks of it. The process is accelerated by the fact that water expands as it freezes in the cracks, driving apart rock fragments by **ice wedging.**

Just as streams may have tributaries, so may alpine glaciers, as valleys join and the ice of several glaciers merges (figure 18.8). The main glacier, larger and more massive, carves a larger and deeper-floored valley. When the main and tributary glaciers melt, the tributary glaciers leave **hanging valleys** (figure 18.9). These smaller, shallower valleys are abruptly truncated by the deeper valley of the main ice mass.

Plucking at the head of an alpine glacier, combined with weathering of the surrounding rocks, produces a rounded or bowl-like depression known as a **cirque** (figure 18.10A). The cirque becomes a hospitable setting for more snow accumulation to feed the glacier. When the glacier melts away, a cirque bottom may stay filled with water, making a small, rounded lake called a **tarn** (figure 18.10B).

FIGURE 18.9 Hanging valley (right) above the floor of Rock Creek Valley, Montana.

A

B

FIGURE 18.10 Cirques and tarns. (A) Cirque at head of glacier in Denali National Park, Alaska. (B) Tarns in old glacial cirques in Beartooth Mountains, Montana.

FIGURE 18.11 A horn formed by erosion by several glaciers around a single peak. Denali National Park, Alaska.

Other erosional features are formed when more than one glacier erodes a mountain. Where several alpine glaciers flow down in different directions from the same high peak, each chipping away at its head, the result may eventually be a pointed **horn** chiseled from the peak (figure 18.11). Glaciers flowing in parallel, each forming a steep-sided valley, may form an **arête,** a sharp-spined ridge between the valleys (figure 18.12).

Other Erosional Features of Glaciers

Gravel and boulders frozen into the base of the ice and dragged along as the ice moves act as a natural, coarse sandpaper. They scrape parallel grooves, or **striations,** in softer rocks over which they move (figure 18.13). Striations are more than evidence of the past presence of a glacier; they also indicate the direction in which the glacier flowed (parallel to the striations). Striations can be produced by both continental and alpine glaciers.

Many alpine glaciers begin by flowing down the valleys of mountain streams. Even a continental glacier may do so locally, widening, deepening, and tending to straighten the stream valley in the process. The Finger Lakes of upstate New York formed in this way, as an ice sheet gouged deeper into old stream valleys (figure 18.14). The lakes, too, are elongated in the direction of glacier flow.

FIGURE 18.12 An arête formed between two parallel glacial valleys. Denali National Park, Alaska.

FIGURE 18.13 Striations on a rock surface show direction of glacial flow. Notice that the grooves extend continuously across the fractures in the rock and flake off where the surface layer is weathered away. They are clearly a surface feature, not a characteristic of the rock itself.

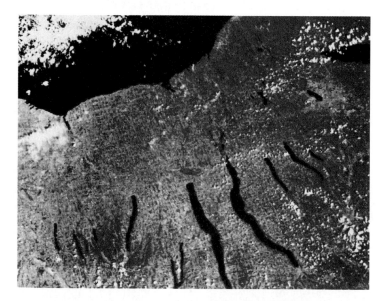

FIGURE 18.14 The Finger Lakes region, New York state (Landsat satellite image).
© *NASA.*

FIGURE 18.15 Till deposited by an Alaskan glacier shows characteristically poor sorting.

The basins now occupied by the Great Lakes were deepened by continental glaciation and filled with glacial meltwater as the glaciers retreated. The basic drainage network of the Mississippi River system was established at the same time, as it carried that meltwater to the sea. Even mountains can be scoured into elongated remnants by the immense mass of ice represented by a continental ice sheet.

DEPOSITIONAL FEATURES OF GLACIERS

Sooner or later, melting glacial ice deposits its load of sediment, either directly or through the action of meltwater. This can produce a variety of features. One of the simplest is the isolated large boulder that has been carried along in the ice far from its parent rock. If it cannot be identified as having been derived from the local bedrock, it is a glacial **erratic.**

The general term for glacially transported sediment is **drift.** Drift may be further subdivided, depending on whether it is stratified or unstratified. This, in turn, generally reflects whether it has been deposited directly from the melting ice or redistributed and deposited by glacial meltwater.

Direct Deposition: Till and Moraines

Sediment deposited directly by melting ice is called **till.** Till is unstratified and typically poorly sorted with respect to particle size or density (figure 18.15), in contrast to most water- or wind-deposited sediments. The rock formed when till is lithified is **tillite.** Many tills and tillites are very porous and permeable, and they make excellent aquifers.

A landform made of till is a **moraine.** Moraine comes in several forms. The concentrations of rock debris along the sides of a valley glacier, which fall onto the glacier from the valley walls above or are ground out by the ice, are **lateral moraines.** Where tributary glaciers join a valley glacier, the lateral moraines toward the inside join as a ribbon of

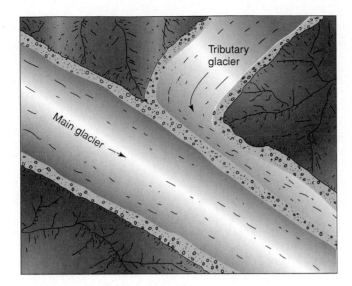

FIGURE 18.16 Lateral moraines join to form a medial moraine as a tributary glacier joins the main ice mass. See also figure 18.6.

moraine within the combined ice mass and thus become a **medial moraine** (figure 18.16). Merging of many tributaries eventually gives the resulting composite glacier a striped appearance (recall figure 18.6). When the glacier retreats, recognizable ridges of lateral or medial moraine may be left behind. A broad blanket of till may also be deposited by basal ice as **ground moraine.**

Glacial sediment transport continually brings a fresh supply of sediment to the glacier's end. If the extent of the glacier remains the same for several years, a ridge of till accumulates there that is known as an **end moraine.** Much end moraine is the consequence of thrust faulting of debris-rich basal ice to the front of the glacier. Formation of a ridge-like landform can also be enhanced to some extent by the advancing glacier acting like a bulldozer on previously deposited ground moraine. A single glacier may leave multiple end moraines. In such a case, the one marking the farthest advance of the tongue of ice is the **terminal moraine** (figure 18.17). End moraines deposited during glacial retreat, when the ice front was temporarily stationary, are **recessional moraines.**

Continental glaciers also may leave terminal moraines, by which advances of the great ice sheets can be mapped (figure 18.18). Alternatively, the ice may override previously deposited moraines, streaking them out into oval mounds elongated in the direction of ice flow. These are **drumlins** (figure 18.19).

Meltwater Deposition: Outwash

A melting glacier not only deposits sediment but also releases volumes of water that may, in turn, redistribute the sediment. The resultant water-deposited sediment is glacial **outwash.** As a water-laid sediment, outwash tends to be stratified and better sorted than till. The very fine rock flour can be transported especially readily, for its fine size tends to keep it in

FIGURE 18.17 This glacier, in Denali National Park, Alaska, shows concentric loops of moraine (center) left by retreat of glacier. The terminal moraine is overgrown by vegetation. Note the outwash also.

suspension. It is also easily redistributed later, by the action of wind, to make the distinctive fine sediment called *loess,* discussed in the previous chapter.

Several distinct landforms can occur in outwash, as in moraines. One is a winding, snakelike outwash ridge deposited by meltwater streams flowing within and beneath a melting glacier. The resultant feature—a stream deposit without a stream valley—is an **esker** (figure 18.20). Glacial retreat can also strand blocks of ice in thick outwash deposits. As the ice melts, it leaves behind a hole in the outwash, called a **kettle** (figure 18.21). Many of the lakes in Wisconsin, Minnesota, Alaska, and other previously glaciated states have formed when shallow water tables intersected kettle holes. The collective action of many meltwater streams can deposit a broad sheet of outwash—an **outwash plain**—in front of a glacier (figure 18.22). The outwash plain may be crisscrossed by braided streams of meltwater, as shown in

FIGURE 18.18 Terminal moraines of northeastern Illinois reflect the advances of many separate tongues of ice over hundreds of thousands of years.

Source: H. B. Willman and J. C. Frye, Glacial Map of Illinois, *Illinois State Geological Survey, 1970.*

figure 18.17. The sands and gravels of outwash deposits often provide essential raw material for construction—road-building, cement-making, and other applications.

Deposition in Glacial Lakes

End moraines can act as dams within a glacial valley, trapping meltwater to form a lake. Sediment deposition in such a lake typically shows an annual cyclicity. The principal

FIGURE 18.19 Drumlins are shaped by the flow of continental glaciers over sediment. Acadia National Park, Maine.
© Bruce F. Molnia/TERRAPHOTOGRAPHICS/BPS.

FIGURE 18.20 An esker formed by a stream flowing within or under a glacier, depositing sediment. Alaska.

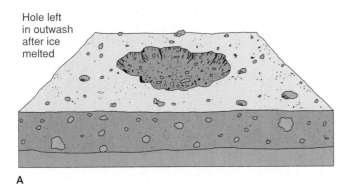

Ice block
stranded
in outwash

Hole left
in outwash
after ice
melted

A

B

FIGURE 18.22 Outwash plain below Exit Glacier, near Seward, Alaska. Note kettle holes filled with "glacial milk," water full of suspended rock flour.

FIGURE 18.21 Kettles. (A) Formation of a kettle as a stranded block of ice melts. (B) Examples of kettle lakes in Denali National Park, Alaska.

flushing of sediment into the lake occurs during the accelerated melting of summer; in winter, addition of sediment is minimal. The coarser sediments settle out rather quickly, but the finest rock flour may take months to settle. At the same time, microorganisms such as algae may flourish in the warmer, sunnier summer months, then die and settle to the bottom with the coming of winter. Each year, then, a distinct, two-part layer of sediment is deposited that is coarser at the bottom (spring/summer deposition), finer-grained and richer in organic matter at the top (fall/winter). Each such annual sediment couplet is a **varve.** Hundreds of these annual depositional layers may be preserved in varved sediments and can be counted, somewhat like tree rings, to determine the age of the sediments.

On a grander scale, the difference in climate associated with past continental glaciation left visible traces even beyond the area actively glaciated. The glaciated areas, which now have temperate climates, were icy; nearer the equator, where hot, dry conditions now prevail, the weather was more temperate and moister. Abundant rainfall led to the formation of **pluvial lakes** (so named from the Latin for "rain"). The most impressive of these, glacial Lake Bonneville, covered an area about one-third the size of Utah. As the climate dried out, the lake dried up, leaving behind remnant shorelines (by which its extent has been mapped) and, ultimately, the Bonneville salt flats. (The

Box 18.1

Permafrost

In temperate climates, where temperatures drop below freezing for only a few months of the year, the ground freezes shallowly during this period but thaws during warmer weather. In alpine and arctic climates, at high elevations or high latitudes, the winter freeze is so deep and pervasive that summer's thaw does not penetrate the whole frozen zone and may barely thaw the surface. Below lies a layer of more or less permanently frozen ground—**permafrost.** The top of the frozen zone is sometimes referred to as a *permafrost table,* and it behaves somewhat like a water table, rising and falling with seasonal climatic variations (figure 1).

One consequence of the persistent freezing is the *patterned ground* commonly observed in arctic regions (figure 2). Contraction of the soil during freezing causes it to break up into polygonal chunks somewhat analogous to basalt columns.

As long as permafrost stays frozen, it constitutes a fairly solid base for structures. If disturbed and warmed, whether by natural thawing or by construction, some of

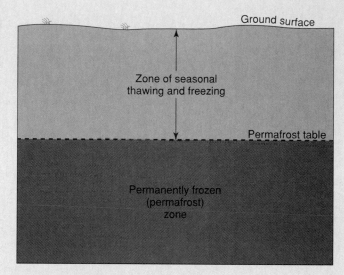

FIGURE 1   Permafrost and the permafrost table (schematic).

FIGURE 2   Patterned ground, broken into polygons, in an alpine climate.

Great Salt Lake of Utah seems to be a younger lake formed on the old salty bed of Lake Bonneville, deriving its high salt content by solution of that older salt.)

GLOBAL CLIMATE, PAST AND FUTURE

Climate is the result of the interplay of a number of factors. The main source of energy input to the earth is sunlight, which warms the land surface, which in turn radiates heat into the atmosphere. Globally, how much heating occurs is related to the sun's energy output and how much of the sunlight falling on earth actually reaches the surface. Incoming sunlight may be blocked by cloud cover or, as we have previously seen, by dust and sulfuric acid droplets from volcanic eruptions. Heat (infrared rays) radiating outward from earth's surface may or may not be trapped by certain atmospheric gases, as described below.

Locally, at any given time in earth's history a wide variety of local climates can obviously exist simultaneously—icy glaciers, hot deserts, steamy rain forests, temperate regions—

the ice melts. Below is still-frozen soil, so, depending on the topography, the water may drain slowly or not at all. The result is a mucky, sodden mass of saturated soil that is difficult to work in or with and that is structurally weak (figure 3). This was a major problem during the construction of the Trans-Alaska Pipeline (figure 4). In fact, where the pipeline passes underground, it actually had to be refrigerated in some places to keep the warm oil from melting the surrounding permafrost, which might cause the pipeline to sag and perhaps break.

FIGURE 3 Consequences of the melting of permafrost: differential subsidence of railroad tracks. Copper River region, Alaska. This track had to be abandoned in 1938. *Photograph by L. A. Yehle, USGS Photo Library, Denver, CO.*

FIGURE 4 The Trans-Alaska Pipeline, supported by upright posts that are cooled by metal "fins" dissipating heat

and a given region may also be subject to wide seasonal variations in temperature and rainfall. One of the challenges in determining present global climate trends, then, is deciding just how to measure global climate at any given time. Commonly, several different kinds of data are used to characterize present climate, including air temperatures, over land and sea, at various altitudes; snow and ice cover on land, and extent of sea ice; ocean-water temperatures; distribution of atmospheric moisture and precipitation. An understanding of

past climatic fluctuations would be helpful in developing models of possible future climate change, but the evidence available is generally less direct and comprehensive.

Evidence of Climates Past

Aspects of local climate may often be deduced from the geologic record, especially the sedimentary-rock record. For example, the now-vegetated dunes of the Nebraska Sand Hills

FIGURE 18.23 Aerial photograph of the Nebraska Sand Hills area, formerly a desert, shows a well-preserved barchan dune about 75 meters high, now covered by vegetation under moister modern conditions.
Photograph by T. S. Ahlbrandt, USGS Photo Library, Denver, CO.

(figure 18.23) are remnants of an arid "sand sea" in the region 18,000 years ago, when conditions must have been much drier than they are now. We have noted that glacial deposits may be recognized in now-tropical regions. As will be seen in the chapter on mineral and energy resources, coal deposits indicate warm, wet conditions, conducive to lush plant growth, perhaps in a swampy setting. Some idea of global climate at a given time may be gained from compiling such data from a variety of localities (taking into account changes in latitude related to continental drift); but such records are far from complete.

From marine sediments comes evidence of water-temperature variations. The proportion of calcium carbonate ($CaCO_3$) in Pacific Ocean sediments in a general way reflects water temperature, because the solubility of calcium carbonate is strongly temperature-related. The carbonate or silica skeletons of marine microorganisms contain considerable oxygen, and the relative proportions of the isotopes oxygen-16 and oxygen-18 in these skeletons are related to the (near-surface) water temperature at the time those "hard parts" were formed.

The longevity of the massive continental glaciers—sometimes hundreds of thousands of years—makes them useful in preserving evidence of both air temperature and atmospheric composition in the past. Tiny bubbles of contemporary air may be trapped as the snow is converted to dense glacial ice. The oxygen-isotope composition of the ice re-

flects the air temperature at the time the parent snow fell. If those temperatures can be correlated with other evidence, such as the carbon-dioxide content of the air bubbles or the presence of volcanic ash layers in the ice from prehistoric explosive eruptions, the climate-prediction models may be refined accordingly. The ice sheets themselves are remnants of a climatic shift known as an ice age.

Ice Ages and Their Possible Causes

The term **ice age** (uncapitalized) was used by Professor Louis Agassiz in the late nineteenth century to describe a period of widespread continental glaciation. When capitalized, the term *Ice Age* is generally used to refer to a time in the relatively recent geologic past when extensive continental glaciers covered millions more square kilometers of area than they now do. There were many cycles of advance and retreat of the ice sheets during this epoch, known to geologists as the Pleistocene, which spanned the time from about 2 million to 10,000 years ago. At its greatest extent, Pleistocene glaciation in North America covered essentially all of Canada and much of the northern United States. The ice was well over a kilometer thick over most of that area. Climatic effects extended well beyond the limits of the ice, with cooler temperatures and more precipitation common in surrounding lands, and worldwide sea levels lowered as vast volumes of water were locked up as ice in ice sheets (figure 18.24).

Pleistocene glaciation can be studied in some detail because it was so recent, and the evidence left behind is thus fairly well preserved. It is not, however, the only large-scale continental glaciation in earth's past. There have been at least half a dozen ice ages over the earth's history, going back a billion years or more. As previously noted, the pre-drift reassembly of the pieces of the continental jigsaw puzzle is aided by matching up features produced by past ice sheets on what are now separate continents.

An underlying question is what kinds of factors might be involved in creating conditions conducive to the formation of such immense ice masses, bearing in mind that ice a kilometer thick represents far more than a few winters' snow. The proposed causes of ice ages fall into two groups: those that involve events external to the earth and those for which the changes arise entirely on earth.

One possible external cause, for example, would be a significant change in the sun's energy output. Present cycles of sunspot activity cause variations in sunlight intensity, which should logically result in temperature fluctuations worldwide. However, the variations in solar-energy output would have to be about ten times as large as they are in the modern sunspot cycle to account even for short-term temperature fluctuations observed on earth. To cause an ice age of major proportions and of thousands of years' duration, any cooling trend would have to last much longer than the eleven years of the present sunspot cycle. Another problem with linking solar activity fluctuations with past ice ages is simply lack of evidence. Although means exist for estimating temperatures on earth in the past, scientists have yet to conceive of a way to determine the pattern of solar activity in ancient times. Thus there is no way to test the hypothesis, to prove or disprove it.

Another proposed cause of ice ages is the observed variation in the tilt of the earth's axis in space. This tilt varies over time relative to the earth's orbital plane. Changes in tilt do not change the total amount of sunlight reaching earth, but they do affect its distribution. If, for a time, the poles were tilted farther away from incident sunlight, the polar region might become sufficiently colder that a large ice sheet could begin to develop. Some evidence supports this theory. When glaciers are extensive and large volumes of water are locked up in ice, sea level is lowered. The reverse is true as ice sheets melt. It is possible to use ages of Caribbean coral reefs to date high stands of sea level over the last few hundred thousand years, and these periods of high sea level correlate well with periods during which maximum solar radiation fell on the mid–Northern Hemisphere. Conversely, then, periods of reduced sun exposure are apparently correlated with ice advance (lower sea level). However, this precise a record does not go back very far in time. It may also be that changes in tilt are related to changes in precipitation patterns over the oceans, and thus to changes in seawater

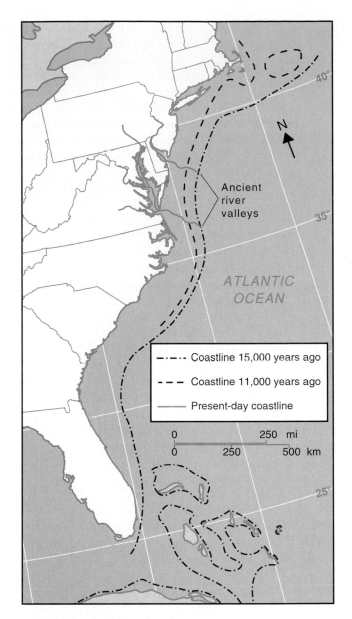

FIGURE 18.24 Shoreline of eastern North America extended much farther seaward at peak of last major ice advance of the Ice Age. By approximately 6000 years ago, melting had raised sea level approximately to its present level.
From Michael Bradshaw and Ruth Weaver, Physical Geography. *Copyright © 1993 Mosby-Year Book, Inc. Reprinted with permission of Times Mirror Higher Education Group, Inc., Dubuque, Iowa. All rights reserved.*

salinity and mixing patterns, which in turn would alter the distribution of organisms in the oceans and the oceans' effectiveness as a "sink" for atmospheric carbon. (To understand what this has to do with heating, see the later section on the greenhouse effect.)

Since the advent of plate-tectonic theory, a novel suggestion of a cause of ice ages has invoked continental drift. Prior to the breakup of Pangaea, all the continental landmasses were close together. This concentration of continents meant much freer circulation of ocean currents elsewhere on the globe and more circulation of warm equatorial waters to the poles. The breakup of Pangaea disrupted the oceanic circulation patterns. The poles could have become substantially colder as their waters became more isolated. This, in turn, could have made sustained development of a thick ice sheet possible. A limitation of this mechanism is that it only accounts for the Pleistocene glaciation, while considerable evidence exists for extensive glaciation over much of the Southern Hemisphere—India, Australia, southern Africa, and South America—some 200 million years ago, just *prior* to the breakup of Pangaea. Beyond that, not enough is known about continental positions prior to the formation of Pangaea to determine whether continental drift and resulting changes in oceanic circulation patterns could be used to explain earlier ice ages.

Another possible earth-based cause of ice ages might involve blocking of incoming solar radiation by something in the atmosphere. The resultant cooling might be adequate to induce the start of an ice age. We have observed that volcanic dust and sulfuric-acid droplets in the atmosphere can cause measurable cooling, but with the modern examples at least, the effects were significant only for a year or two. In none of these cases was the cooling serious enough or prolonged enough to cause an ice age. Even some of the larger prehistoric eruptions, such as those that occurred in the vicinity of Yellowstone National Park or the explosion that produced the crater in Mount Mazama that is now Crater Lake, would have been inadequate individually to produce the required environmental change, and again, the dust and acid droplets from any one of these events would have settled out of the atmosphere within a few years. However, there have been periods of more intensive or frequent volcanic activity in the geologic past. The cumulative effects of multiple eruptions during such an episode might have included a sustained cooling trend and, ultimately, ice-sheet formation.

The foregoing is a brief and incomplete sampling of the many processes and phenomena that may have caused climatic upheavals in the past. The possibilities are many, and, as yet, there is insufficient evidence to demonstrate any single theory correct. Of more immediate concern is the realization that present human activities may also have begun to alter the climate irrevocably in the other direction, toward global warming.

The Greenhouse Effect and Global Warming

On a sunny day, it is much warmer inside a greenhouse than outside it. Light enters through the glass and is absorbed by the ground, plants, and pots inside. They, in turn, radiate heat: infrared radiation, not visible light. Infrared rays cannot readily escape through the glass panes; the rays are trapped, and the air inside the greenhouse warms up. The same effect can be observed in a closed car on a bright day.

In the atmosphere, carbon dioxide molecules act similarly to the greenhouse's glass. Light reaches the earth's surface, warming it, and the earth radiates infrared rays back. But the infrared rays are trapped by the carbon dioxide molecules, and a portion of the radiated heat is thus trapped in the atmosphere. Hence the term **"greenhouse effect."** (See figure 18.25.) As a result of the greenhouse effect, the atmosphere stays warmer than it would if that heat radiated freely back out into space.

The evolution of a technological society has meant rapidly increasing energy consumption. Historically, we have relied most heavily on carbon-rich fuels—wood, coal, oil, and natural gas—to supply that energy. These probably will continue to be important energy sources for several decades at least. One combustion by-product that all of these fuels have in common is carbon dioxide gas (CO_2).

Excess water in the atmosphere readily falls out as rain or snow. Some of the excess carbon dioxide is removed by geologic processes, including photosynthesis and solution in the oceans to form carbonate, but in the past century or so, since the start of the so-called Industrial Age in the mid-nineteenth century, the amount of carbon dioxide in the air has increased by an estimated 25%—and its concentration continues to climb (figure 18.26). If the heat trapped by carbon dioxide is proportional to the concentration of carbon dioxide in the air, the increased carbon dioxide, it is feared, will gradually cause increased greenhouse-effect heating of the earth's atmosphere. In recent years, scientists have also recognized greenhouse-effect heating potential from other trace gases, including water vapor, methane (CH_4), nitrous oxide (N_2O), and chlorofluorocarbons, which collectively may pose at least as much of a warming threat.

One potential problem arising from increased greenhouse-effect heating is agricultural. In many parts of the world, agriculture is already only marginally possible because of hot climate and little rain. A temperature rise of only a few degrees could easily make living and farming in these areas impossible. Also, the warmer it gets, the faster the remaining ice sheets will melt. Significant decreases in the extent of ice caps could further alter global weather patterns, compounding the agricultural problems. Recent projections suggest that summer soil-moisture levels in the Northern Hemisphere could drop by up to 40% with a doubling in atmospheric CO_2. This is a sobering prospect for farmers in areas where rain is barely adequate now.

Another consequence on a local level might be more "killer heat waves" such as that experienced in the upper midwest in the summer of 1995. For people vulnerable to

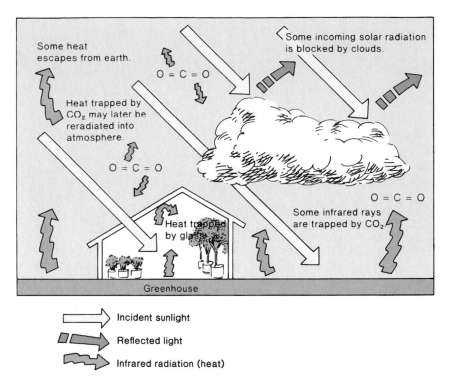

FIGURE 18.25 The greenhouse effect (schematic). Both glass and air are transparent to visible light. Like greenhouse glass, carbon dioxide molecules in the atmosphere trap infrared rays radiating from the sun-warmed surface of the earth.

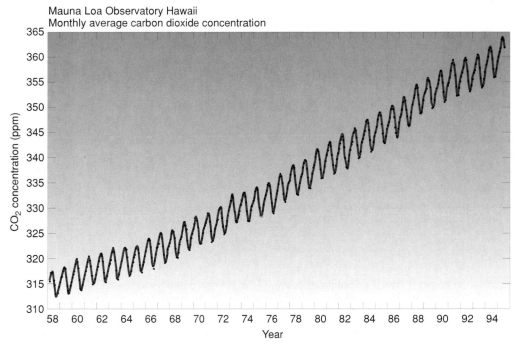

FIGURE 18.26 Rise in atmospheric CO_2 over past several decades is clear. (Zigzag pattern probably reflects seasonal variations in uptake by plants.)

Used by permission of Charles D. Keeling, Scripps Institution of Oceanography. The measurements were obtained in a cooperative program of NOAA and the Scripps Institution of Oceanography.

FIGURE 18.27 An 80-meter rise in sea level would flood many major cities in the United States and elsewhere.

extreme heat, a few degrees may mean the difference between survival and death. The fact that a region's average temperature might rise a few degrees may be unimportant, but an increase of a few degrees at the high end of the range could be critical.

Yet another potential problem is that meltwater from the ice caps would, in turn, be added to the oceans, and the net effect would be a rise in sea level. If *all* the ice melted, sea level could rise by close to 75 meters (about 250 feet). About 20% of the world's land area would be submerged. Many millions, perhaps billions, of people now living in coastal or low-lying areas would be displaced, since a large fraction of major population centers grew up along coastlines and rivers. The Statue of Liberty would be up to her neck in water. The consequences to the continental United States and southern Canada are illustrated in figure 18.27. Also, raising of the base levels of streams draining into the ocean would alter the stream channels and could cause significant flooding along major rivers.

Such large-scale melting of ice sheets would take time, perhaps several thousand years. On a shorter timescale, how-

ever, the problem could still be significant. Continued intensive fossil-fuel use could easily double the level of carbon dioxide in the atmosphere before the middle of the next century. This would produce a projected temperature rise of 4° to 8°C (7° to 14°F), which would be sufficient to melt at least the West Antarctic ice sheet completely in a few hundred years. The resulting 3- to 6-meter rise in sea level, though it sounds small, would nevertheless be enough to flood many coastal cities and ports, along with most of the world's beaches. This would be both inconvenient and extremely expensive. (The only consolation is that the displaced inhabitants would have decades over which to adjust to the changes.) Studies of the Ross Ice Shelf in Antarctica have recently raised another specter: sudden breakoff of huge masses of ice, which would plunge into the sea as monstrous icebergs, raising global sea level abruptly and catastrophically by about 6 meters. At present, far too little is known to assess the likelihood of this accurately.

One might reasonably ask how much heating results from particular increased CO_2 levels—in other words, how

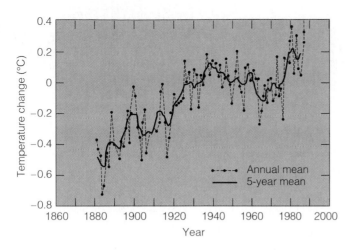

FIGURE 18.28 Fluctuations in global temperature over the last few decades do not precisely mirror CO_2 rise; compare with figure 18.26.

Source: Data from J. H. Steele, "The message from the oceans," Oceanus, 32, Summer, 1989, p. 6.

acute will greenhouse-effect heating become, how soon? Unfortunately, there is no simple answer to such questions. Too many variables are involved in global climate modeling. For example, if air temperatures rise, evaporation of water from the oceans would tend to increase. More water in the air means more clouds. More clouds, in turn, would reflect more sunlight back into space before it could reach and heat the surface. (Some speculate that over centuries, the world might then cool off, perhaps even enough to initiate another ice advance or ice age far in the future.)

In fact, NASA has identified cloud cover as the single largest source of uncertainty in projections of greenhouse-effect heating. Interactions between atmosphere and oceans, as with El Niño episodes, contribute further uncertainty.

Clouds aside, we cannot readily isolate the effects of increased CO_2 in the atmosphere, because other atmospheric chemicals and constituents vary over time too. The world actually seemed to be cooling from 1940 to the mid-1960s; it has been suggested that that cooling trend was a result of soot and other particulate air pollution from human activities, acting like volcanic ash to reflect sunlight. More recently, those pollutants have been reduced and temperatures have begun to rise somewhat (figure 18.28). Even so, we cannot

equate this temperature rise with greenhouse-effect heating from CO_2, for climate shows historic fluctuations over periods of years to centuries that equal or exceed the recent warming. Data for just the last 10,000 years suggest a range of up to 6°C (11°F) in average annual temperature, far beyond documented recent temperature fluctuations, yet over a period when human influence on atmospheric chemistry was certainly negligible. In short, it is difficult to separate the signal from the noise in the data.

It is important to bear in mind, too, that global climate changes do not affect all parts of the world equally. Associated changes in wind-flow patterns and amounts and distribution of precipitation will cause differential impacts in different areas, not all of which will be equally resilient. A region of temperate climate and ample rainfall may not be seriously harmed by a temperature change of a few degrees or several inches more or less rainfall. Places where temperature or moisture conditions already make living marginal are more vulnerable. And because of the many variables involved and the immense complexity of the calculations, global-climate computer models must limit the resolution at which effects are calculated. Typical models operate on areal units, or cells, that are hundreds of kilometers on a side, meaning that small local climatic anomalies may be missed.

While the magnitude and rate of anticipated heating are debated, there is consensus that increased CO_2 means increased trapping of heat in the atmosphere. In order to moderate that heating, either the rate of input of CO_2 into the atmosphere could be reduced, or the natural "sinks" for atmospheric CO_2 enhanced. A major sink for CO_2 is use by plants in manufacturing food by photosynthesis, and this is one reason for concern about the impacts of large-scale destruction of rain forests. In order to reduce CO_2 emissions, we could reduce dependence on fossil fuels; but this is likely to be a gradual process. Nor will its effects on global warming necessarily be wholly good: Reduction in burning of coal can reduce emissions of both CO_2 *and* sulfur gases (see chapter on mineral and energy resources). However, the sulfur gases, which make sulfuric acid in the atmosphere, may have a *cooling* effect, as we have learned from volcanoes. Such frustrating uncertainties about the precise impacts of various changes in atmospheric chemistry spur continued research to refine the models.

Glaciers past and present have sculptured the landscape not only in mountainous regions but over wide areas of the continents. Formed from many seasons of accumulated snow gradually converted to ice, glaciers flow downslope by internal deformation or on a basal meltwater layer, under the influence of gravity. Glaciers leave behind U-shaped valleys, striated rocks, and poorly sorted sediment (till) in a variety of landforms (moraines). Terminal moraines mark glaciers' maximum extent; the orientation of striations and drumlins indicates the direction of flow. Further outwash of sediments by meltwater produces water-laid sediments of glacial origin. Most present glaciers are alpine glaciers. The two major remaining continental glaciers (ice sheets) are on Greenland and Antarctica.

At several times in the past, sustained periods of colder temperatures have led to major advances of ice sheets over the continents (ice ages). The last of these was during the Pleistocene, within the past two million years. Ice advances were accompanied by corresponding eustatic lowering of sea level. Proposed causes of ice ages include plate tectonics, volcanism, changes in the tilt of the earth's axis in space, and changes in solar activity, but no single cause has been proven conclusively.

In modern times, the burning of fossil fuels has been increasing the amount of carbon dioxide in the atmosphere. The resultant greenhouse-effect heating may begin to melt the remaining ice sheets, causing a rise in global sea level and ultimately flooding many coastal areas. The extent and rate of global warming, however, are extremely difficult to predict, given the number of variables and magnitude of natural climatic fluctuations. It is important to remember, too, that local changes in temperature and/or precipitation patterns may differ greatly from average global trends.

TERMS TO REMEMBER

ablation 320
abrasion 322
alpine glacier 321
arête 325
calving 320
cirque 323
continental glacier 322
crevasses 320
drift 326
drumlins 327
end moraine 327
equilibrium line 320
erratic 326

esker 327
firn 318
glacier 318
greenhouse effect 334
ground moraine 327
hanging valley 323
horn 325
ice age 332
ice wedging 323
kettle 327
lateral moraine 326
medial moraine 327
moraine 326

outwash 327
outwash plain 327
permafrost 330
plucking 323
pluvial lakes 329
recessional moraine 327
rock flour 322
striations 325
tarn 323
terminal moraine 327
till 326
tillite 326
varve 329

QUESTIONS FOR REVIEW

1. What proportion of the earth's fresh water is currently stored as ice? Describe how this water cycles through a glacier.

2. Describe three conditions that encourage glacier formation.

3. What is the distinction between continental and alpine glaciers?

4. What is *rock flour,* and how is it produced?

5. Describe the origin of
   (a) hanging valleys and
   (b) striations.

6. How does glacial till differ, in terms of sorting, from many water-laid sediments? Why?

7. What is a *moraine?* Distinguish among the following: ground moraine, terminal moraine, medial moraine.

8. How do outwash deposits originate? What is an *esker?*

9. Describe the annual depositional cycle of a glacial lake that gives rise to varves.

10. Cite and briefly summarize any three proposed causes of ice ages.

11. What is the *greenhouse effect?* Why is it a subject of increasing concern in modern times?

12. An increase in atmospheric CO_2 may, in the short term, lead to global warming. Over the longer term, however, the result may be global *cooling*. Explain.

FOR FURTHER THOUGHT

1. How would you distinguish an esker from sediment deposited by a stream not associated with a glacier? From a natural levee?

2. Assume that, during the Pleistocene glaciation, one-third of the continental land area was covered by ice. If the total areal extent of the continents is 149 million square kilometers, that of the oceans is 361 million square kilometers, and sea level was lowered by approximately 130 meters, what was the average thickness of ice over the glaciated areas (ignoring isostatic effects)? You may also find it interesting to examine a bathymetric (depth) chart of the oceans to see how much new land would have been created by that much lowering of sea level.

SUGGESTIONS FOR FURTHER READING

Ausubel, J. H. 1991. A second look at the impacts of climate change. *American Scientist* 79 (May/June): 210–21.

Barth, M. C., and Titus, J. J., eds. 1984. *Greenhouse effect and sea-level rise.* New York:VNR Company.

Bentley, C. R. 1980. If West Sheet melts rapidly—What then? *Geotimes* (August): 20–21.

Berner, R. A., and Lasaga, A. C. 1989. Modeling the geochemical carbon cycle. *Scientific American* 260 (March): 74–81.

Broecker, W. S., and Denton, G. H. 1990. What drives global cycles? *Scientific American* 262 (January): 48–56.

Embleton, C., and King, C. A. M. 1975. *Glacial and periglacial geomorphology.* 2nd ed. London: Edward Arnold.

Environmental Protection Agency, 1983. *Projecting sea-level rise.* 2nd ed. Washington, D.C.: U.S. Government Printing Office.

———. 1990. The greenhouse effect: What can we do about it? *EPA Journal* 16 (March/April).

Eyles, N., ed. 1983. *Glacial geology.* New York: Pergamon Press.

Flint, R. F. 1971. *Glacial and quaternary geology.* New York: Wiley-Interscience.

Goldthwait, R. P., ed. 1975. *Glacial deposits.* New York: Dowden, Hutchinson, and Ross.

Goudie, A. 1983. *Environmental change.* 2nd ed. New York: Oxford University Press.

Imbrie, J., and Imbrie K. P. 1979. *Ice ages.* Short Hills, N.J.: Enslow.

John, B. S. 1977. *The ice age, past and present.* London: Collins.

Meier, M., and Post, A. 1980. *Glaciers: A water resource.* U.S. Geological Survey.

Perry, T. S. 1993. Modeling the world's climate. *IEEE Spectrum* 30 (July): 33–42.

Sharp, R. P. 1988. *Living ice: Understanding glaciers and glaciation.* New York: Cambridge University Press.

Woods Hole Oceanographic Institution. The oceans and global warming. *Oceanus* 32 (Summer).

Woodwell, G. M. 1978. The carbon dioxide question. *Scientific American* 238 (January): 34–43.

Zorpetle, G. 1993. Sensing climate change. *IEEE Spectrum* 30 (July): 20–27.

19

FOSSILS AND EVOLUTION

Ordovician-age trilobites from Criner Hills, County, Oklahoma.
© Doug Sherman/Geofile

What are fossils? How does an organism become a fossil? How are the various fossilized organisms related? These questions are addressed by the study of paleontology, taphonomy, and taxonomy, respectively. The answers to such questions invite speculation on the nature of the fossil record: Is the fossil record complete? Does it show any biases?

The fossil record also leads paleontologists to ask why assemblages of fossilized organisms change—or evolve—over time. Additional questions then arise: What is meant by evolution? What evidence is there of evolution? What causes evolution? How quickly does it occur?

In this chapter, we attempt to answer all of these questions by examining the physical evidence of past life, fossils, and the theories that arise from their study.

FOSSILS

Early Ideas

A characteristic feature of many sedimentary rocks is the occurrence of fossils. Even though fossils were observed and written about by early Greek philosophers as far back as 550 B.C., the true nature of fossils was not accepted by most geologists until the late 1700s and early 1800s. Only a very few of the early Greek philosophers, for example, Xenophanes, recognized fossils as the remains of organisms. Others, such as Aristotle and Theophrastus, believed that fossils were formed by "plastic forces" (whatever they are) in the earth. During the Middle Ages fossils were termed "figured stones," and confusion about their origin prevailed. A few scholars revived the idea of "plastic forces" forming fossils. Some said fossils were placed in rocks by Satan to confuse and torment humankind. Others maintained that fossils were "jokes of nature." Still others said that stars were responsible for fossil formation. During the Renaissance, three views about the origin of fossils predominated. One view was that fossils were the remains of plants and animals that had perished in the flood of Noah's time recorded in Genesis. Another view stated that fossils were the remains of monsters and giants. A third view, argued by Leonardo da Vinci, was that fossils were the remains of once-living organisms. Although it took a while for this latter view to become established, by the late eighteenth century the true nature of fossils was widely accepted.

Paleontology

Originally, a "fossil" was anything dug from the ground, whether organic or inorganic. However, in contemporary usage the term **fossil** is restricted to the remains of organisms, unaltered or altered, or evidence of organic activity.

Figures 19.1 and 19.2 show examples of the most commonly found fossils—those of shallow-water marine invertebrates. The study of fossils is called **paleontology,** an inclusive term for a number of disciplines that study the diverse aspects of past life. Various types of paleontology include *paleobiology,* the study of the biology of past life; *paleozoology,* the study of fossil animals; *paleobotany,* the study of fossil plants; *micropaleontology,* the study of fossil microorganisms; and *paleoecology,* the study of how fossil organisms related to their environment and to each other.

P aleontologists put an age restriction on fossils: fossils must be older than 10,000 years. This date marks the approximate end of the ice ages, the last retreat of glaciers from North America. Any "fossil" found in sediments or rocks younger than 10,000 years is called a subfossil. Ancient objects made by humans are not fossils. Archaeologists term these objects *artifacts.*

TAPHONOMY (PRESERVATION)

The study of the methods of fossilization is called **taphonomy.** The road to becoming a fossil is long and difficult and traveled by few organisms after their death.

Fossilization

Upon an organism's death and before its burial, biologic agents, such as scavengers and bacteria, are likely to destroy an organism, particularly its soft parts. Physical elements of the environment, such as waves and currents, can transport and rework the sediment, causing further destruction of the organism. After burial in the sediment and before lithification (the process in which sediment changes into rock), additional reworking, dissolution by ground water, and diagenesis (physical and chemical changes that occur after the sediment is deposited) can destroy an organism's remains. After the sediment lithifies, further dissolution, deformation, metamorphism, weathering, and erosion can alter or destroy the fossil. Even if the fossil survives, it is unlikely to be exposed at the earth's surface. And even exposure at the surface does not guarantee that an observant paleontologist will find it.

In general, for an organism to overcome the obstacles to fossilization, two conditions must be met. First, the organism must possess hard parts. Many invertebrates (animals without backbones) have shells, and vertebrates (animals with backbones) have bones and teeth, which are likely to be fossilized. Second, environmental conditions for fossilization must be favorable. Generally, this means rapid burial of

FIGURE 19.1 Common marine invertebrate fossils (I). (A) Sponges. (B) Corals. (C) Bryozoans. (D) Brachiopods. (E) Snails.
(A–E) © Times Mirror Higher Education Group, Inc./Bob Coyle, photographer.

A

B

C

D

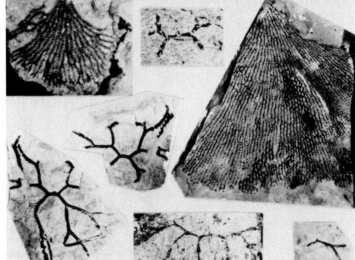

E

FIGURE 19.2 Common marine invertebrate fossils (II).
(A) Pelecypods. (B) Cephalopods. (C) Trilobites. (D) Echinoderms.
(E) Graptolites.

(A–D) © Times Mirror Higher Education Group, Inc./Bob Coyle, photographer.
(E) Photograph by R. J. Ross, Jr., USGS Photo Library, Denver, CO.

TABLE 19.1	*Summary of the Types of Fossil Preservation*

Preservation of all or part of the organism

A. Unaltered
   1. Soft parts
   2. Hard parts

B. Altered
   1. Permineralization
   2. Replacement
   3. Recrystallization
   4. Carbon films

Preservation of the organism's shape

A. Mold

B. Cast

Preservation of signs of organic activity (trace fossils)

A. Tracks

B. Trails

C. Burrows

D. Borings

E. Coprolites

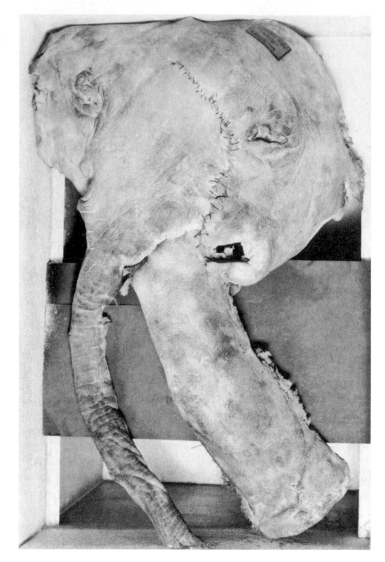

FIGURE 19.3 Unaltered remains of an organism: a mammoth that was frozen in permafrost.
Photograph courtesy of the American Museum of Natural History, Neg. #320496.

the organism upon death. In this way the organism is protected from many of the destructive physical and biologic processes mentioned previously.

Types of Fossilization

The many methods by which an organism can become a fossil can be divided into three broad categories (table 19.1): (1) preservation of all or part of the organism, either unaltered or altered; (2) preservation of the organism's shape; and (3) preservation of signs of organic activity (trace fossils).

In very rare circumstances, the unaltered remains of the soft parts of an organism are found (figure 19.3). Mammoths frozen in permafrost in Siberia, ground sloths mummified in caves in the U.S. Southwest, insects preserved in amber, and saber-toothed tigers and other vertebrates found in tar pits in California are all examples of this exceptional method of fossilization. More commonly, the unaltered remains of hard parts, including shell material of invertebrates and pollen and spores of plants, are found.

Most often, however, the organism's remains have been altered during the fossilization process. Three processes of alteration, which often overlap and are frequently difficult to differentiate in the preserved fossil, are recognized. **Permineralization** occurs when material has been added to the pore spaces in the object. **Replacement** occurs when the original material has been removed by dissolution and new material added on an atom-by-atom or molecule-by-molecule basis. "Petrified" wood is an example of these two processes (figure 19.4). The wood has been replaced by mineral matter that has also filled the internal pore spaces of the material. **Recrystallization** occurs when a mineral recrystallizes into another form. Many invertebrate shells that were originally made of aragonite have now recrystallized to calcite. Aragonite is chemically identical to calcite—both are $CaCO_3$ chemically—but has a different crystalline structure. One final type of alteration occurs in several types of organisms, particularly plants. When the volatiles in the plant tissue escape from the rock (carbonization), **carbon films** are left behind (figure 19.5).

FIGURE 19.4 Permineralization and replacement. Silicified wood.
Photograph courtesy of Carla W. Montgomery.

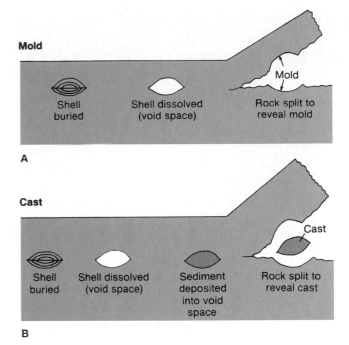

FIGURE 19.6 Origin of molds and casts. (A) Formation of a mold. (B) Formation of a cast.

FIGURE 19.5 Carbon films. A plant fossil.
Photograph by E. D. McKee, USGS Photo Library, Denver, CO.

In some rocks, all the organism's original remains have been destroyed, but the fossil's shape in the rock has been left intact (figure 19.6). For example, if a shell is buried in sediment that eventually lithifies, ground water moving through the rock may dissolve the shell, leaving a void in the rock. If the rock is then split, revealing this void, a **mold** is found (figure 19.6A). The empty space in the rock may also be filled with sediment (for example, fine mud) deposited by

the ground water. Over time, the sediment lithifies. The rock may then be split to reveal a **cast** of the organism (figure 19.6B). Molds and casts frequently preserve delicate structures and are an important source of information about an organism's anatomy and structure.

A large category of fossils consists not of actual remains of organisms but, rather, of indications of organic activity. These are termed **trace fossils** (figure 19.7). A wide variety of trace fossils have been cataloged. *Tracks* are footprints left in sediment as the organism passed. Dinosaur tracks and hominid footprints are good examples. *Trails* are imprints of parts of the body. Among the most easily recognized trace fossils are *burrows* made in sediment that, at the time of burrowing, was unlithified. Worm and crustacean (for example, shrimp) burrows are probably the most common type of trace fossil in the fossil record. In contrast to burrows, *borings* are made into solid rock (lithified sediment) or shell material (also hard). *Coprolites* (fossilized feces) are also trace fossils.

Trace fossils provide clues regarding organisms and organism behavior not found in other fossil types. Tracks and trails provide evidence of the organism's anatomical structure, size, and behavioral patterns, such as walking, grazing, running, digging, resting, crawling, dwelling, and feeding. Coprolites provide evidence of the organism's diet and internal morphology.

A

B

FIGURE 19.7 Trace fossils. (A) Worm burrows. (B) Reptile tracks.
(B) Photograph by E. D. McKee, USGS Photo Library, Denver, CO.

A

B

Virtually all fossils can fit into one of the previous categories. **Pseudofossils,** on the other hand, superficially (and sometimes quite convincingly) look like fossils (figure 19.8). Dendrites (features formed by water that seeped into cracks and deposited minerals), various sedimentary structures (such as cone-in-cone structures), and concretions may all resemble fossils.

The Fossil Record

Given the obstacles to fossilization, what is the nature of the fossil record? Is it a complete record of past life? Are there inherent biases in the fossil record? The answers to the last two questions are no and yes. The fossil record is an incomplete, biased sample. But to recognize the limits of the fossil record is not to imply that information deduced from it is worthless; rather, this recognition shows where restraint may be necessary in interpretations.

C

FIGURE 19.8 Pseudofossils. (A) Dendrites. (B) Cone-in-cone structures. (C) Concretions.
(B) Photograph by R. L. Rioux, USGS Photo Library, Denver, CO. (C) Photograph by R. Arnold, USGS Photo Library, Denver, CO.

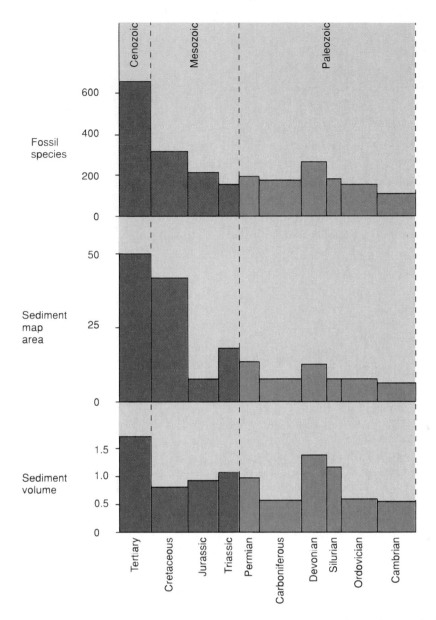

FIGURE 19.9 Sediment volume, sediment map area, and number of fossil species compared for each period of the Phanerozoic.

From Principles of Paleontology *by D. M. Raup and S. M. Stanley. Copyright © 1978 by W. H. Freeman and Company. Reprinted with permission.*

How complete is the fossil record? The number of fossil species described from rocks of all ages is about 250,000 (the term *species* is defined later in the chapter). Biologists estimate that the number of living species on earth may be as high as 4 million. While, clearly, the picture of past life is not complete, this is to be expected for two reasons. First, the difficult path of fossilization precludes many organisms from ever becoming fossils. In many environments, the organic material is recycled so quickly there is nothing left to fossilize! Second, the rock record itself is not complete—at no single location has sedimentation been continuous for long intervals of time. Derek Ager, in his book *The Nature of the Stratigraphic Record,* stated the situation succinctly by saying there is "more gap than record." Fossils cannot be found in rocks destroyed by erosion. Figure 19.9 shows that there appears to be a relationship between the number of fossil species described for a given period and the amount of sedimentary rock preserved for that period.

Compounding the completeness problem is the problem of bias. The fossil record is heavily biased toward organisms that had hard parts (they have a better chance of being preserved) and lived in shallow-water marine environments (these environments are the most commonly represented in the rock record because they experience less erosion than do terrestrial environments). Additional, and sometimes subtle, biases creep in, involving where rocks crop out, where geologists study, and the method(s) used to collect fossil samples.

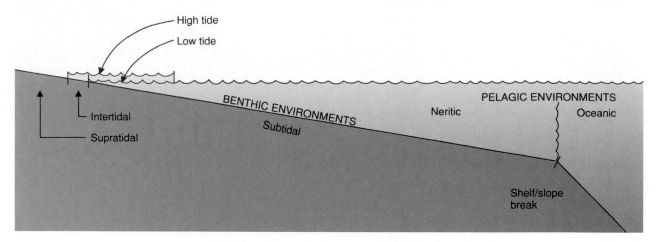

FIGURE 19.10 Classification of marine environments.

USES OF FOSSILS

Fossils have a variety of uses in a number of areas:

1. Correlation. Fossils are most often used in correlation—establishing a time equivalency between strata. Mapping a fossil's geographic distribution during one interval of time (*horizontal occurrence*) and determining the length of time that the species was alive (*vertical distribution*) allow correlation of rocks from different regions of the earth. Especially useful are *index fossils*.

2. Paleoenvironments. The types of fossils found at a particular locality help in determining the paleoenvironment. Quite commonly, it is possible to distinguish among various types of terrestrial environments and between shallow and deep marine environments by studying the types of fossils that occur and the nature of the enclosing sediment.

3. Paleoecology. Fossils contribute to paleoecology studies. Study of the fossil's morphology, the associated fossils, and the type of sediment in which the fossil was found can lead to deductions about how the animal adapted to the environment, its relationship to other organisms, and how ancient communities were structured.

4. Paleogeography. The geographic distribution of fossils also aids in paleogeography studies. From these studies it is possible to determine the distribution of land and sea areas, paleolatitudes, ancient shorelines, and land connections.

5. Paleoclimates. Geologists can also determine the paleoclimate(s) for a given region by studying the types of fossils and their morphologies. Certain types are restricted to certain climates. Other fossils show different morphologies in different climates. For example, leaves from plants growing in the tropics differ in shape from those of temperate-climate plants.

6. Evidence of evolution. The vertical distribution of fossils provides evidence of evolutionary processes that have occurred in the past.

7. Evidence of plate tectonics. The distribution of fossils provides evidence of plate tectonics by showing how fossil assemblages changed as continents moved. For example, if fossils of a terrestrial species are found on continents that now have an ocean between them, possibly the continents were connected at one time.

8. Determining the age and placement of rock strata. Index fossils are frequently used to determine the age of a rock unit(s).

MARINE ENVIRONMENTS AND LIFE HABITS

Two subjects of paleoecology are important to discussions of fossils: marine environments and life habits.

Marine Environments

Marine environments are shown in figure 19.10. Two regions are defined: (1) *benthic*, or bottom, environments, and (2) *pelagic* environments—those in the water column. Benthic environments are divided into three regions based on the positions of the high and low tides. Areas beyond high tide are called *supratidal*. These areas are only infrequently covered with water—for example, during storms. The bottom area between high and low tide is called *intertidal*. This area may be exposed to air once or twice a day depending on the local tidal conditions. The area from the low-tide mark all the way to the edge of the continent (the shelf/slope break) is called *subtidal*. This corresponds to the continental shelf. Pelagic environments are within the water column above the bottom. Two are defined: The water environment above the continental shelf (the subtidal benthic environment) is called the *neritic* environment. Beyond the shelf, above deep ocean basins, is the *oceanic* environment.

TABLE 19.2 *Taxonomic Classification of a Plant (Fern), Invertebrate (Brachiopod), Reptile (Dinosaur), and Human*

	Fern	Brachiopod	Dinosaur	Human
Phylum	Tracheophyta	Brachiopoda	Vertebrata	Vertebrata
Class	Filicineae	Articulata	Reptilia	Mammalia
Order	Flicales	Strophomenida	Saurischia	Primates
Family	Osmundaceae	Strophomenidae	Tyrannosauridae	Hominidae
Genus	*Osmunda*	*Strophomena*	*Tyrannosaurus*	*Homo*
Species	*claytoniana*	*planumbona*	*rex*	*sapiens*

Life Habits

Within marine environments, organisms have adopted three modes of life: Animals that float within the water column are called *planktonic*. Examples include jellyfish and plankton (typically, microscopic single-celled plants and animals). Animals that swim within the water column are called *nectonic* (sometimes spelled nektonic). This group includes fish, dolphins, and whales. *Benthonic* (or benthic) organisms live on the bottom. Some live within the bottom sediment (infauna), such as worms and other invertebrates that tunnel through the sediment. Others live on top of the bottom sediment (epifauna). Those that move about are called vagile or "vagrant" (for example, lobsters and starfish); those that remain stationary are called sessile (for example, corals and sponges).

Taxonomy (Classification)

Taxonomy is the process of classifying a group of objects according to some scheme. The purpose of classification is to discover any underlying patterns or themes. The goal is to determine how members in the group are related.

In 1758, Carolus Linnaeus (Carl von Linne) was the first to classify all the known organisms. Linnaeus constructed six levels of categories: kingdom, class, order, genus, species, and variety. The fundamental starting point for this classification is the species. Placement of individuals into species is based on internal anatomy and external form and shape. Similar species are then grouped into one genus, similar genera (the plural of genus) into one order, and so on. Thus, going up the scale from species to kingdom, more and more organisms are included in each group, with less and less similarity among group members.

The modern system of classification grew out of the Linnaean system. In the modern system, the idea of similarity and difference among organisms is retained, as is the fundamental starting point of species, but evolutionary relationships among organisms are also considered. Because many new species have been described, additional categories have been added to the Linnaean system. The major categories of the modern classification are kingdom, phy-lum, class, order, family, genus, and species. (This can be remembered by using Colin Fletcher's wonderful mnemonic: King Phillip, Come Out For God's Sake!—the first letter of each word corresponds to the first letter of the modern taxonomic categories.) Additional categories can be constructed when needed by adding the prefix *sub-* (for example, subclass), *infra-* (infraclass), or *super-* (superfamily) to the basic names. Table 19.2 shows several examples of taxonomic classification.

A lthough the modern system of classification is well known, it is not the only method by which organisms can be indexed. Two other, much more complicated, methods of taxonomy are *numerical taxonomy,* classifying organisms on the basis of some assigned numerical property, and *cladistics,* classifying organisms on the basis of "primitive" and "derived" characters.

Species Concept

As noted earlier, the basis of classification in both the Linnaean system and the modern system of taxonomy is the species. Newly described organisms are assigned to a species, or a new species is named according to the rules of the International Code of Zoological (or Botanical) Nomenclature.

Biologists define a species as a group of individuals that either can interbreed or has the potential to interbreed—a **biospecies.** This definition is based on genetics (production of viable offspring) and establishes that species are reproductively isolated from each other. In theory, but not always in practice, if a biologist has doubts about whether two populations are of the same species, he or she can perform experiments to see if interbreeding between members of the two populations in question can occur.

Paleontologists have difficulty applying biologists' definition of a species to fossils. Since many of the organisms in question are extinct, genetic compatibility is no longer possible to determine. Therefore, paleontologists define a species on the basis of morphologic similarity (external form and internal anatomy) between individuals—a **morphospecies.** Of course, there can be problems with this method. First, it is not always clear how genetic differences are expressed morphologically. Second, how much morphologic variation is required before a new species is defined? These concerns are not overwhelming but must be considered whenever discussing fossil species.

Taxonomic Problems

In addition to the species problem, paleontology also has two unique problems with taxonomic classification.

First, what is to be done with extinct, enigmatic life-forms? How can one classify something that is unlike anything living today? For example, in rocks of Late Precambrian age, there are animals that display a symmetry in body plan unlike that of any group today. How should these organisms be classified?

Another problem is transitional forms. A group may fall between two others morphologically. Is it really a separate group? Alternatively, to which of the other two groups is it more closely related?

EVOLUTIONARY THEORY

The ideas about evolution arise, in part, on the basis of taxonomy and, in part, on the basis of the principle of superposition. The classification of organisms shows similarities among them, implying an evolutionary relationship. The principle of superposition is also a basis of evolution. In a vertical outcrop of strata, such as the Grand Canyon, the oldest rocks are at the bottom and the youngest rocks are at the top, according to this principle. In this succession of strata, geologists observe that various types of sedimentary rocks are repeated over and over. This is because particular sedimentary environments, which produced the sedimentary rocks, occurred again and again throughout geologic time.

But the fossils in the sedimentary rocks do not repeat over and over in successive layers. Rather, the assemblage of fossils changes. Beginning at the bottom of the sequence is a distinct assemblage of fossils within the strata. Proceeding upward, some fossils disappear while new fossils appear. By the top of the sequence, the assemblage of fossils is different from that at the bottom. Why did certain organisms disappear and others appear, causing the assemblages to differ? Certain organisms disappear from the rocks either because they became extinct or because they migrated out of the area. But why do new organisms appear in the assemblage? Biologists and geologists postulate that these new organisms either evolved from earlier forms or were introduced from other areas by migration. Hence geologists can explain why the assemblages differ, and the fossil record shows a trend (from the time it began to the time it ended)—extinction and evolution acting together.

Evolution is defined simply as change through time. Two types of evolution are recognized based on the extent of the change. **Microevolution** involves changes within a species—for example, differences both within and between populations of the same species. This term is also sometimes used to describe the changes an organism undergoes within its lifetime (for example, from tadpole to frog and caterpillar to butterfly). **Macroevolution** involves large-scale changes of major lineages—for example, a line of amphibians evolving into reptiles.

Charles Darwin's Theory

Charles Darwin and Alfred Wallace are credited with independently developing the basic theory of evolution. (Both announced their results in 1858.) Darwin's theory of evolution can be summarized by four basic observations. First, the size of a population of organisms within a geographic area remains relatively constant over time. Second, even though the size of the population is constant, the number of offspring produced by any one individual within the population is large. Many more offspring are produced than can survive. Third, individuals within the population display variation. No two individuals are identical. Fourth, since relatively few offspring survive, each individual must experience a "struggle for life." The individuals are under pressure—from the environment, from predators, and from individuals within the same species. The conclusion drawn from these four observations is that those organisms that have characteristics that best allow them to carry on this struggle will survive and reproduce, passing along those traits that are genetically controlled to the next generation.

The concept of *natural selection* was Darwin's major contribution to science. In an oversimplified definition, natural selection is the unseen "force" or "forces" (or process) that weed out individuals unable to carry on successfully the struggle for life.

Darwin's theory was remarkable in explaining how evolution could operate, but Darwin recognized two problems with it. First, since he maintained that the process was gradual, where were all the supposed transitional forms in the fossil record? Darwin resorted to a long-held notion—that the rock record is incomplete—and argued that, if the rock record were more complete, these transitional forms would be seen. Second, Darwin could not explain how variation arose in populations or how the various characteristics were passed on from one generation to the next. This problem was solved by Gregor Mendel and his study of genetics.

Genetic Mutations

Gregor Mendel's work with the hereditary characteristics of peas provided an understanding of the mechanism by which characteristics of a parent are passed on to the offspring. DNA (deoxyribonucleic acid) is the code that carries the genetic information to the succeeding generation. Owing to various factors, such as ultraviolet light, cosmic rays, and chemicals, or simply spontaneously, the DNA code may not replicate itself exactly. This results in a mutation—a change in the DNA code sequence—which is expressed as a morphologic change. Mutations are random; that is, they arise by chance. Most mutations are harmful to the organism, if not fatal, but the effects of many mutations are environmentally dependent—a mutation in one environment may be harmful, but in another beneficial. Mutations supply the variation we see in individuals in a population and from generation to generation.

Modern Evolutionary Theory (Organic Evolution)

The current theory of organic evolution (also called Neo-Darwinism or the modern synthesis) was developed in the 1930s and 1940s and combines the ideas of genetic mutation and natural selection. Random mutations are the source of variation. They are "random" because there is no preferred direction—they arise by chance. Natural selection gives direction to the evolutionary process by favoring certain characteristics over others.

EVIDENCE FOR EVOLUTION

No theory should be presented without some evidence. Darwin substantiated his theory of evolution with abundant evidence accumulated during his five-year voyage as naturalist on the HMS *Beagle* and also from field studies in Britain and much library research. Since Darwin's time, additional evidence for evolution has been found. Macroevolutionary processes (for example, the change from fish to amphibian), however, may take millions of years. This sort of process is not seen in the laboratory, which further shows that the scientific method as applied to the geosciences must differ somewhat from the classic approach.

Anatomical evidence for evolution includes

1. Embryology. The close resemblance of embryos of certain organisms.
2. Homology (homologous structures). In some animals, the bones of the limbs have all been modified for different functions. The configuration of the bones arose from a common ancestor, and the modification through the process of evolution (figure 19.11).
3. Vestigial structures. More evidence of evolution is found in vestigial structures—organs inherited from an ancestor that now serve no purpose.

Biological evidence for evolution includes

1. Selective breeding. Microevolution has been demonstrated in selective breeding. Changes in plants and dogs through this process are common.
2. Ontogenetic changes. Evolutionary changes can occur during an organism's lifetime; for example, the dramatic change from a tadpole to a frog or a caterpillar to a butterfly.
3. Biochemistry. The similar biochemistry of organisms implies common ancestry.

Paleontological evidence for evolution is clearly documented in the fossil record. Well-documented examples of several types of evolution are known.

THE SPEED OF EVOLUTION

Evolution at the lowest level begins with speciation, the development of a new species. Answering the question, How long does it take for a new species to evolve? is the start of understanding how quickly evolutionary processes generally operate. After Darwin, the process of evolution was thought to be slow and gradual. Then, in 1972, a new model was proposed to explain speciation.

Phyletic Gradualism

Phyletic gradualism is a model that shows evolution occurring slowly and gradually over many, many generations (figure 19.12A). This was Darwin's view of how evolution operates. Figure 19.12A shows a population of a certain species (A) at time 1. Members of this population display some morphologic trait—for example, size—that clusters around an average value. Over time, the average value of the specified morphologic trait changes. (Perhaps, overall, individuals of the population get bigger.) Eventually, the morphologic change is so great that if the new population (B, at time 2) had been living during the time of the initial population (A), it would have been considered a different species. Species A has now evolved into species B. How this situation might appear in the rock record is shown in figure 19.12A. The geologic literature contains numerous well-documented examples of lineages displaying this type of evolution (for example, Jurassic ammonites and clams). However, most of the time the "transitional forms" occurring between population A and population B are not found. Darwin stated that the rock record is incomplete. This may be true in some cases but certainly not all. The problem of transitional forms has been both a continuing source of frustration and a point of contention.

Punctuated Equilibrium

A literal reading of the fossil record shows that new organisms appear quite suddenly in the strata, stay relatively

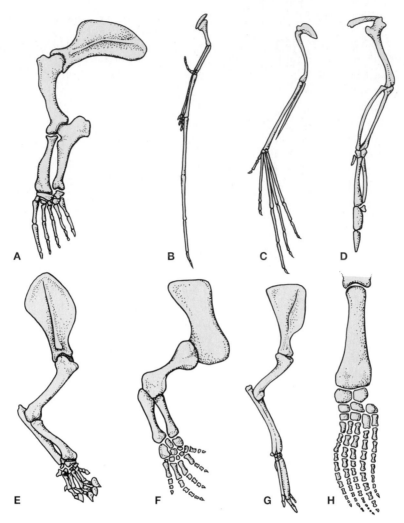

Figure 19.11 Homologous structures—anatomical evidence of evolution. (A) Seal. (B) Pterodactyl (Cretaceous). (C) Bat. (D) Bird. (E) Sabertoothed Cat (Quaternary). (F) Edaphosaur (Permian mammal-like reptile). (G) Horse (Oligocene). (H) Plesiosaur (Jurassic).

unchanged throughout their duration, and then disappear just as suddenly. In an attempt to reconcile the ideas of evolution with what is seen in the fossil record, Niles Eldredge and Steven J. Gould in the early 1970s proposed a model they termed **punctuated equilibrium.** According to their model, within the geographic distribution of a species, the individuals living on the extreme edge of the geographic range are under the most evolutionary pressure, certainly more than that experienced by individuals well within the geographic range. These "peripheral isolates" undergo evolution (change) at an accelerated rate. Speciation is very rapid and occurs in small regions at the edge of the geographic range of the main population (figure 19.12B). If the new species, which has evolved very rapidly, migrates back into the geographic range of the old species, in the rock record it will look as if a new species has suddenly appeared. In addition, if the new species is much better adapted to the environment, it could cause the extinction of the old species, for example, by taking all the nesting sites or food. Within this

framework, the rock and fossil record is more complete than we thought. The odds of preserving a rapid, geographically limited evolutionary event in the rock record are quite small. No wonder it seems that species appear suddenly. The geologic literature also contains numerous examples of this type of evolution (for example, Cambrian trilobites).

Which model of speciation—phyletic gradualism or punctuated equilibrium—is correct? Like most competing ideas, it is probably not a case of either/or, but rather of both/and. Both processes have operated in particular lineages in the past.

PATTERNS OF EVOLUTION

In looking at a large variety of fossil species over long periods of geologic time, one sees several broad patterns of evolution (figure 19.13). In figure 19.13, species A is undergoing a large amount of morphologic change over a brief time interval. This process is referred to as *orthogenesis*—change in

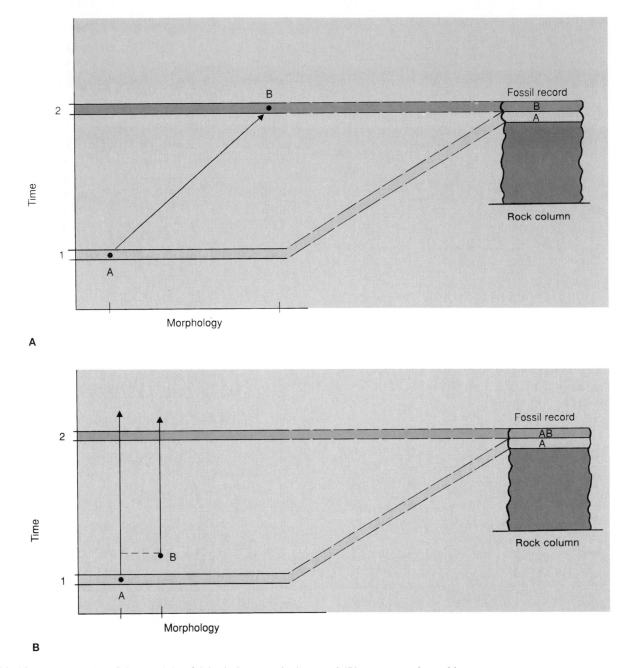

FIGURE 19.12 Comparison of the models of (A) phyletic gradualism and (B) punctuated equilibrium.

morphology over time. Species B is also undergoing morphologic change through time, but at a slower rate. Species A and B display *convergent evolution:* the two species are becoming more and more similar. Typically, convergent evolution occurs when different organisms adapt to the same environment. The similar shapes of some fish, Mesozoic plesiosaurs (reptiles), and dolphins (mammals) in the aquatic environment is one such example. Species C in figure 19.13 is changing in a similar manner as species B, an example of *parallel evolution.* Species D, however, is changing quite differently from species C, undergoing *divergent evolution.*

Finally, species E is evolving extremely rapidly over a very short time interval, with new additional species produced quickly. This pattern of evolution is called *adaptive radiation.* Adaptive radiation is common during certain times of geologic history. Past extinctions of large groups of organisms opened ecologic niches that had been filled by these animals before their extinction. The surviving groups of animals evolved rapidly over a short time to fill the vacated ecologic niches. The explosive radiation of mammals at the start of the Cenozoic to fill the ecologic niches left by the dinosaurs is one prominent example.

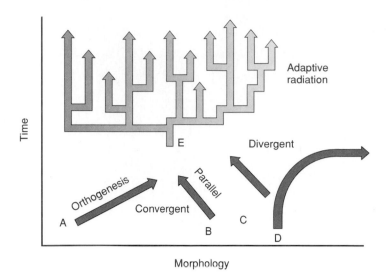

FIGURE 19.13 Patterns of evolution.

EXTINCTIONS

The extinction of a species is a natural process in the history of life. Analysis of the fossil record indicates that most species become extinct within 1 to 1½ million years of their first appearance.

Causes of Extinction

Any number of physical factors can cause a species to become extinct. Plate tectonics is one such factor. For example, the breaking apart of continents may modify the climate so much that organisms are not able to evolve fast enough to cope. Some geologists have correlated magnetic reversals with the extinctions of shallow-water marine protozoa. Evidently the reduced (or perhaps even absent) magnetic field during a reversal allows a greater concentration of cosmic radiation to enter the atmosphere. Another physical cause of extinctions is extraterrestrial meteorite/asteroid impacts. These events, which involve swarms of meteors striking the earth, disrupting the atmosphere and climate, are believed to have occurred at regular intervals in the earth's history.

Sometimes extinctions can be traced to a biologic cause. For example, a new predator may evolve that preys on a certain species. Before this species solves the problem of outwitting the predator, it may be decimated, eventually becoming extinct.

Mass Extinctions in Earth History

During geologic history, there have been times when large groups of organisms became extinct within a short period (a few million years). Some of these **mass extinctions** have been used to define period breaks (figure 19.14). Many plausible theories have been advanced to explain why mass extinctions occur. The terminal Paleozoic extinctions are thought to be a result of plate tectonics: the joining of all the continents into one supercontinent, Pangaea, destroyed much of the suitable shallow-water, marine, continental-shelf area. Studies have also shown a more general relationship between extinctions, large and small, and plate tectonics (figure 19.15). When continents join, faunal diversity appears to decrease. When continents fragment, faunal diversity increases. Recently, a controversial hypothesis for mass extinctions has been postulated, particularly for the extinction of the dinosaurs at the end of the Cretaceous—extraterrestrial meteor or asteroid impacts. While meteors and/or asteroids have certainly struck the earth in the past, the causal relationship between impacts and extinctions is still debated (figure 19.16; table 19.3).

FOSSILS AND THE GEOLOGIC TIME SCALE

The fossil record shows many evolutionary and extinction events. From an examination of the timing of these events, geologists have been able to divide geologic time into major units based on the history of life (figure 19.17). Phanerozoic time (from about 700 million years ago to the present) is divided into three eras—Paleozoic, Mesozoic, and Cenozoic—on the basis of the overall nature or predominant character of fossils found in the strata. The subdivision of eras into periods is done primarily on the basis of extinction events. In general, the Paleozoic fossil record is characterized by shallow-water marine invertebrates. Thus, the Paleozoic is sometimes called the Age of Invertebrates. The Mesozoic rock record contains a relatively large percentage of continental deposits with fossils of reptiles; hence, the Mesozoic has been termed the Age of Reptiles. The Cenozoic fossil record contains a large number of mammals and flowering plants that evolved during this time. The Cenozoic is informally named the Age of Mammals.

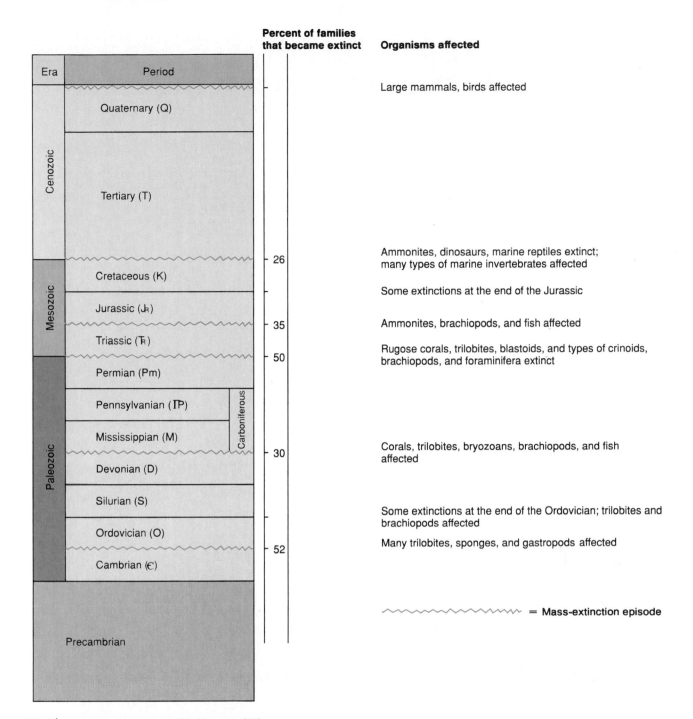

FIGURE 19.14 Major extinctions in the history of life.

Source: Data from A. Hallam, Facies Interpretation and the Stratigraphic Record, *1981, W. H. Freeman and Company, New York, NY.*

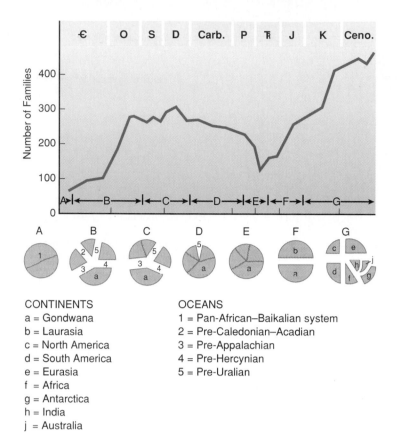

CONTINENTS
a = Gondwana
b = Laurasia
c = North America
d = South America
e = Eurasia
f = Africa
g = Antarctica
h = India
j = Australia

OCEANS
1 = Pan-African–Baikalian system
2 = Pre-Caledonian–Acadian
3 = Pre-Appalachian
4 = Pre-Hercynian
5 = Pre-Uralian

FIGURE 19.15 Plot of faunal diversity through time. Note the apparent correlation of low diversity when continents are joined and high diversity when continents are fragmented.

Reprinted with permission from Nature, *Vol. 228, #5272, figure 1, page 658, November 14, 1970, MacMillan Magazines Limited.*

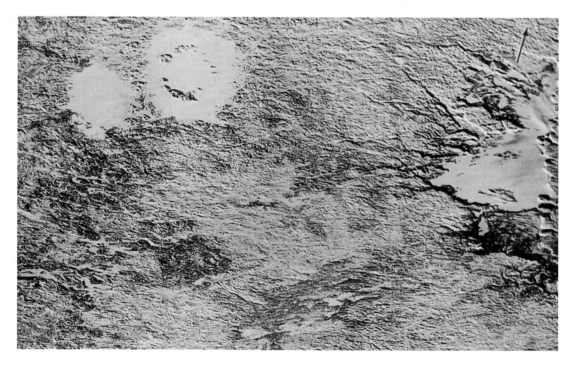

FIGURE 19.16 Satellite photographs reveal possible impact structures on the earth. Clearwater Lakes, Quebec.
© *NASA.*

TABLE 19.3 *Ages of Major Impact Structures*

Time Span	Structure	Age (Ma)	Diameter (km)
2.5 million years to present	Barringer Crater	0.01–0.04	1.2
	Sithlemenkat	0.02	12.4
	Lonar Crater	0.05	1.8
	Kofels	0.01–0.10	5.0
	Wilkes Land (?)	0.69	240.0
	Ashanti	1.0 ± 2	10.5
	Talemzane	1.0	1.8
	Pretoria Salt Pan	1.0	1.1
	New Quebec	2.0 ± 1.0	3.0
	Tenoumer	2.5 ± 0.5	1.8
About 15 million years ago	Reis	14.8	22.0
	Steinheim	15.1	3.5
	Pfahldorf	15.0 (?)	2.5
	Sornhul	15.0 (?)	1.5
	Mandelgrund	15.0 (?)	1.0
	Mendord	15.0 (?)	2.5
	Willenhofen	15.0 (?)	2.0
	Hemauer Pulk	15.0 (?)	2.0
	Wipfelsfurt	15.0 (?)	0.9
	Sausthal	15.0 (?)	1.0
	Haughton	15.0 (?)	17.0
About 42 million years ago	Lake Mistastin	35–40	20
	Lake Wanapitei	37±2	10
	Popigay	40–45	100
	El'gytkhyn	40	23
About 70 million years ago	Kamensk	65	25
	Gusev	65	3
	Boltyshsk	70 (?)	25
	Minsk	70	(?)
	Eagle Butte	74–78	10
	Dumas	74–78	2
About 100 million years ago	Mien	92±6	6
	Steen River	95±7	13.5
About 135 million years ago	Gosses Bluff	130±6	50
	Bolytsh	140 (?)	25
About 160 million years ago	Rochechart	154–173	15
About 180 million years ago	Puchezh-Katunksk	183±3	80
About 205 million years ago	Manicouagan	202–210	65
	Lake St. Martin	225±40	24
	Red Wing Creek	Triassic (?)	8
	Hartney	Triassic (?)	7
	Viewfield	Triassic (?)	2
About 360 million years ago	Flynn Creek	Late Dev.	3.2
	Kaluga	L–M Dev.	15
	Charlevoix	372–321	37
	Lake La Moinerie	350±50	11
About 450 million years ago	Brent	450±30	4
	Carswell	485±50	38
	Skeleton Lake	450±30	3.5

Table 19.3 modified from Earth History & Plate Tectonics: An Introduction to Historical Geology, 2E *by Carl K. Seyfert and Leslie A. Sirkin. Copyright © 1973, 1979 by Carl K. Seyfert and Leslie A. Sirkin. Reprinted by permission of Addison Wesley Educational Publishers.*

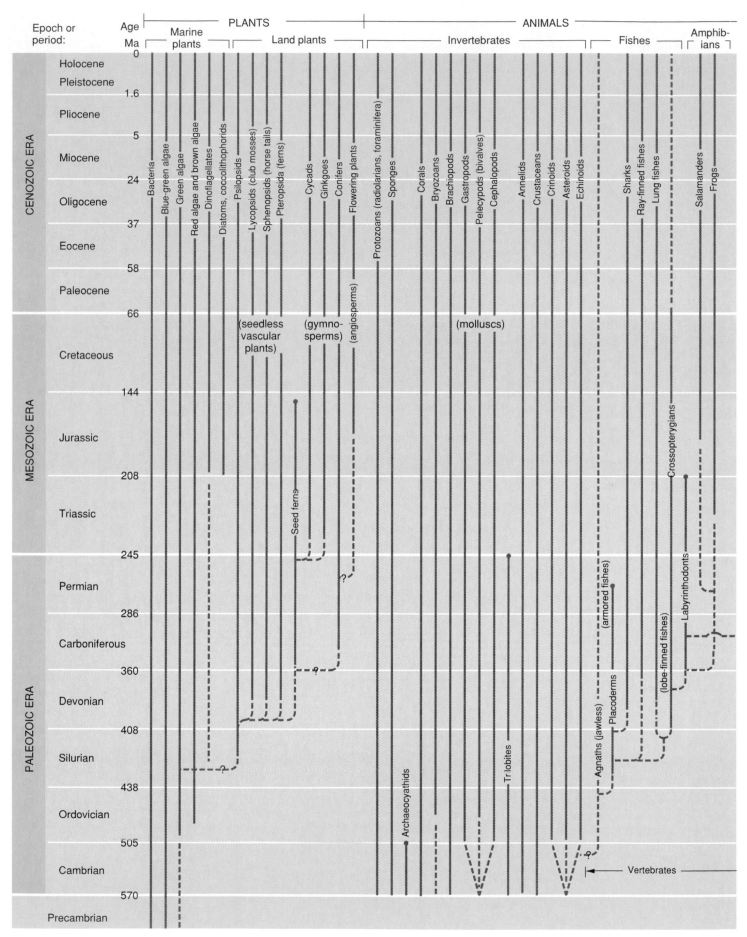

FIGURE 19.17 First appearances and extinctions of major plant and animal groups. See Appendix C for more detailed information and illustrations.
Copyright © 1971 by Arthur N. Strahler. Used by permission.

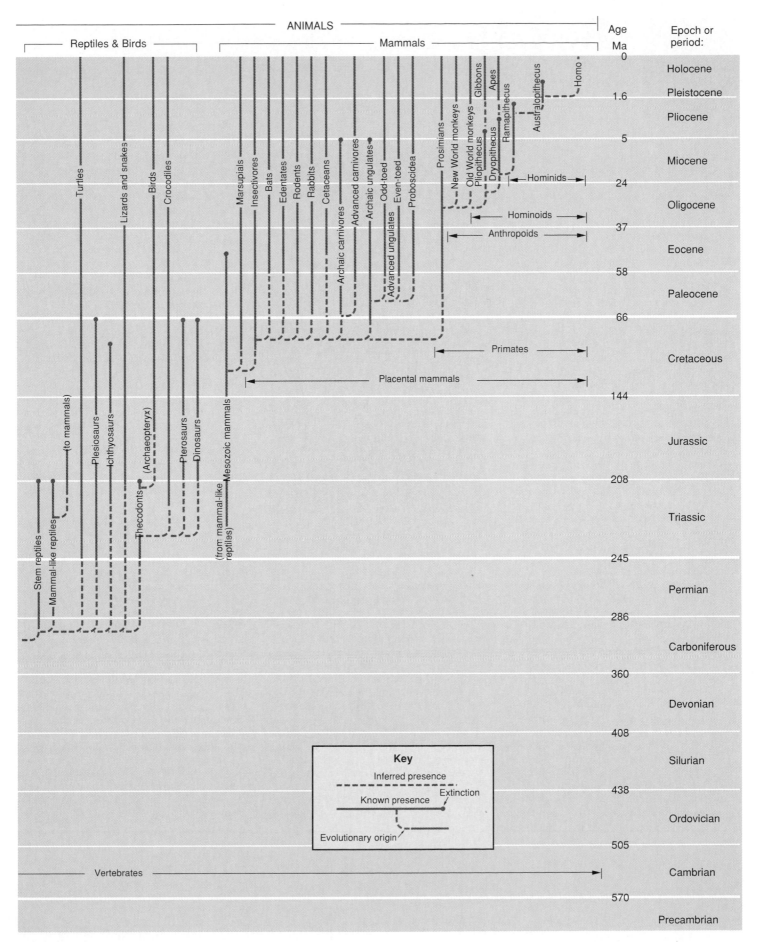

Figure 19.17 (continued)

Box 19.1

Fossil Collecting

Paleontology, like astronomy, is one field of science in which amateurs can make important contributions. Amateur fossil collectors help professional paleontologists by discovering new fossil species, finding new fossil localities, and collecting additional fossils from known localities.

The goals of any fossil collector are to illustrate or document some aspect of past life and to convey this information to another individual. A fossil collection should try to answer the following questions: (1) Why were these fossils collected? (2) Where were these fossils collected? (3) What do the fossils tell about life of the past?

Correct fossil collecting must be done with a purpose, ask the right questions, and use the appropriate equipment. Sometimes nonprofessionals err when collecting fossils, usually by not recording important information. The following brief guide should prevent the more common beginners' mistakes and ensure that the information gathered can be used by other paleontologists, thereby adding to our knowledge of past life.

The first, and most often overlooked, consideration is safety. Work gloves help protect the hands from sharp edges. A hard hat and safety glasses (or protective goggles if you wear glasses) are a must (figure 1). Any outcrop—particularly vertical walls in a quarry or along a roadcut—is potentially unsafe. Rocks fall without warning. And hammering on rock can turn the smallest fragments into dangerous projectiles.

A knapsack or backpack is helpful for carrying collected specimens. A knapsack with one large pocket for holding samples and numerous, smaller, side pockets for carrying equipment works best. Newspapers or cloth sample bags are used to wrap and protect fossils for the journey from outcrop to home. For smaller fossils, freezer storage bags with seals are ideal. A geologist's hammer or hammers (chisel-edge rock pick and crack hammer) and chisels (splitting

FIGURE 1 Geologist at an outcrop.

chisels, assorted sizes) are needed to extract the rock containing the fossils from the outcrop.

A notebook and pencil allow for a written description of the outcrop and fossils. Smaller, loose-leaf notebooks work best since they fit into pockets easily (freeing the hands when needed) and the pages with field notes can be removed after each field trip. Field labels can be made by cutting 3½- by 5-inch index cards in half. A camera should be used, if possible, because a visual record, not only of the outcrop but also of where the fossils were removed, enhances both field notes and faded memories. A camera case is worthwhile, since the camera can easily slip from around

the neck, particularly when one is leaning over to collect a sample, and fall against the rocks. For describing smaller specimens, a hand lens is useful.

Food and a bottle of water are essential miscellaneous equipment. Insect repellent is also a good idea, depending on the season, and a first-aid kit is necessary for the inevitable nicks and cuts.

Amateurs' most common, and most destructive, mistake is failing to record where they collected their samples. A box of fossils with no locations specified loses almost all its scientific value! The collector should record as accurately as possible the fossil's place of discovery.

Location description consists of two parts: (1) Recording the section, township, and range from the 7½-minute USGS topographic map that covers the area; and (2) writing a verbal description (for example: Roadcut on left side of county road H, 1.1 km south of its intersection with State Highway 2). The aim is to define the outcrop location so precisely that anyone unfamiliar with the area could find the site easily.

Frequently, no permission is required to collect along roadcuts. However, permission is almost always required to collect from quarries, even if they are inactive. Asking permission politely and leaving the area in the same state as when one arrived are important. A collector should spend some time looking over a fossil-collecting locality. If there is a published description of the outcrop, the collector should differentiate the major units (formations). If there is no published description, the collector can describe the geology, perhaps making a rough sketch. After this initial familiarization, the collecting should be easier. Once a collector finds a fossil, he or she should record information about the fossil in a field notebook. Typical questions include, What type of fossil is this? Is the fossil in life position; that is, is it attached? Does the fossil appear to be transported from somewhere else? Were other fossils

FIGURE 2 (A) Field label. (B) Catalog label.

The field label (A) reads:

DATE: 7/10/96

LOCATION: Spring Grove Underpass

FORMATION/MEMBER:

Platteville fm., McGregor mbr.

NOTES: Brachiopod collected from quarry wall 4′5″ above McGregor mbr.–Pecatonica mbr. contact.

A

The catalog label (B) reads:

IDENTIFICATION NO.: F-85-6
FOSSIL NAME: Opikina minnesotensis
DATE COLLECTED: 7/10/96
LOCALITY: Spring Grove Underpass (SE, sec. 17, TIOIN, R7W)
FORMATION/MEMBER: Platteville fm., McGregor mbr.
AGE: Ordovician
REMARKS: Typical bottom-dwelling brachiopod. Numerous individuals found in discontinuous layers. From quarry wall 4′5″ above McGregor/Pecatonica mbr. contact.

B

associated with this one? What type of sediment encloses the fossil? Only after these questions are answered does the collector hammer or chisel around the fossil to free the rock containing the sample. Having the fossil with some of its surrounding rock is preferable. Some fossils, however, lie loosely on the ground and should not be overlooked.

A field label for the fossil that includes the date, location, formation (member), and where on the outcrop the sample was collected should be made (figure 2A). Again, a rough sketch on the back of the label is helpful. Was the fossil a "grab sample" along the base of the outcrop? Did it come from a slab that had fallen from the outcrop? Was it taken from a bedding plane of the rock on the outcrop itself? If so, where on the outcrop? This information is as important as the fossil itself. The label and sample are then wrapped in a piece of newspaper and placed in a collecting bag. Wrapping the samples is necessary. Nothing is more maddening than to find that a beautiful sample was broken or scratched by another rock in the knapsack while being transported home.

Even though individual fossils are important, slabs containing many species (whatever their condition) are valuable as well, in some cases more so since a wealth of paleoecological data can be derived from such specimens.

As soon as possible after a visit to an area, the collector prepares the fossils for display. (Otherwise one tends to forget details.) The samples are washed in warm water and scrubbed with a toothbrush to remove excess dirt. Trimming the sample to remove excess rock is usually done only to make the size of the sample more manageable. In most cases, it is better not to free the fossil completely from the enclosing matrix. An important piece of information is lost (the type of sediment the fossil was buried in), and the risk of breaking the fossil is high (fossils represent sites of weakness in the rock). A catalog label card is made for each fossil using a 3½-by 5-inch file card (figure 2B). This label card should include the identification number, fossil name, date, locality, formation, age, and remarks. A small white dot is then painted on the corner of the fossil or on the sediment enclosing the fossil, and the matching identification number from the catalog card is written in permanent ink.

The geology department or library of a nearby university is an excellent place to begin the search for the names of collected fossils. The "Treatise on In-

vertebrate Paleontology" series (Boulder, Colorado, and Lawrence, Kansas: The Geological Society of America and the University of Kansas, 1979) is the most complete source for fossil identification. In addition, geologists and earth scientists are more than willing to help identify unknown specimens.

Once a collector has acquired a number of fossils, he or she compiles an index to keep track of the different types. Copies of the catalog label cards can be used to construct various indexes. For example, one copy of the catalog card can be filed in a "Taxonomy" index, where all the cards of various classes, orders, or genera are filed together. A second copy can be filed in an "Age" index, where all the cards of fossils of the same age are filed together. A third copy can be filed in a "Location" index, where all the cards of fossils found at one particular location are filed together. This may sound like busywork, but cataloging and indexing information turn a fossil collection from a pile of pretty stones into a collection with scientific value. Nowadays most paleontologists use personal computers to keep track of index information. This saves both time and space. As always, the warning "Back up your data" is applicable here.

SUMMARY

Fossils are the remains of organisms or the traces of organic activity. The study of fossils is called paleontology. The three major types of fossilization are (1) preservation either with or without alteration, (2) preservation of the organism's shape (molds and casts), and (3) preservation of signs of organic activity—trace fossils. Because of many physical, chemical, and biologic processes, few organisms become fossilized. Thus the fossil record is incomplete.

The modern taxonomic categories are kingdom, phylum, class, order, family, genus, and species. The starting point of modern taxonomy is the species. Biologists and paleontologists define species in different ways because of the nature of their respective samples (living versus fossilized organisms).

The theory of evolution explains how organisms change with time and why new lines of organisms appear. Two processes, acting together, cause a population of animals to exhibit a morphologic change: random mutations, which supply the variation, and natural selection, which determines the direction of the change. Evidence for the theory of evolution comes from many sources: anatomical evidence (embryology, homologous and vestigial structures), biologic evidence (selective breeding, ontogenetic changes, and biochemistry), and paleontological evidence (the fossil record).

The speed of evolutionary change in some lines of organisms may be slow and best explained by phyletic gradualism; in other lines, the evolutionary change is fast and best explained by punctuated equilibrium.

Extinction of a species is a natural process. Both physical and biologic factors may cause a species to become extinct. Mass extinctions are a frequent occurrence in the history of life. Many subdivisions of the geologic time scale are based on extinction events.

TERMS TO REMEMBER

biospecies 349
carbon films 344
cast 345
evolution 350
fossil 341
macroevolution 350
mass extinctions 354

microevolution 350
mold 345
morphospecies 349
paleontology 341
permineralization 344
phyletic gradualism 351
pseudofossils 346

punctuated equilibrium 352
recrystallization 344
replacement 344
taphonomy 341
taxonomy 349
trace fossils 345

QUESTIONS FOR REVIEW

1. What are the three basic categories of fossilization methods?
2. In what ways is the fossil record an incomplete and biased sample? What does this mean to the paleontologist?
3. How do the biologic and paleontological definitions of *species* differ? What problems arise from this difference?

4. What observations in the rock and fossil record lead paleontologists and biologists to conclude that evolution has operated in the past?
5. What contributions did Linnaeus, Darwin, and Mendel make to the theory of evolution?
6. List the evidence for the theory of evolution.

7. Contrast phyletic gradualism and punctuated equilibrium.
8. What are some hypothesized causes of mass extinctions in the earth's past?
9. What should be done before going into the field to collect fossils?

FOR FURTHER THOUGHT

1. The fossil record supports the theory of evolution, yet the record is incomplete. Compare and contrast what the fossil record shows versus what the theory of evolution predicts it should show. What aspects of the theory of evolution might need modification if our fossil record were more complete?

2. When a species disappears from the rock record, that means that it either became extinct or migrated out of the area. What evidence would allow a geologist to determine which of these two alternatives took place?

SUGGESTIONS FOR FURTHER READING

Ager, D. 1981. *The nature of the stratigraphic record*. New York: Halsted Press, John Wiley and Sons.

Beerbower, J. R. 1968. *Search for the past: An introduction to paleontology*. 2d ed. Englewood Cliffs, N.J.: Prentice-Hall.

Dowdeswell, W. H. 1984. *Evolution: A modern synthesis*. London: Heinemann.

Gould, S. J. 1980. *The panda's thumb: More reflections in natural history*. New York: W. W. Norton.

Hitching, F. 1982. *The neck of the giraffe*. New York: New American Library.

McKerrow, W. S., ed. 1978. *The ecology of fossils*. London: Gerald Duckworth.

Raup, D. M., and Stanley, S. M. 1978. *Principles of paleontology*. 2d ed. San Francisco: W. H. Freeman.

Schopf, T. J. M., ed. 1972. *Models in paleobiology*. San Francisco: Freeman, Cooper.

Seilacher, A. 1967. Fossil Behavior. *Scientific American* 217: 72–80.

20

THE ARCHEAN

A banded-iron formation, Jasper Knob Ispheming, Michigan.
© *Doug Sherman/Geofile*

INTRODUCTION

The Precambrian, from the start of the earth about 4.6 billion years ago to about 570 million years ago, covers approximately 88% of geologic time. The story of the Precambrian is really the story of how the earth (and life) formed. Fundamental questions about the earth's early history include, How did the earth's crust form? Where did all the seawater come from? How did the earth develop an atmosphere? The study of Precambrian rocks provides geologists with clues about the answers to these questions.

Precambrian geology also leads to a number of other questions, such as, Where are Precambrian rocks exposed at the surface? What are Precambrian rocks like? What do these rocks tell us about the development of the earth?

Most intriguing are Precambrian fossils, which allow paleontologists to speculate about the conditions necessary for organic life to arise from inorganic processes.

Origin of the Universe

The origin of the universe is a fundamental question astronomers have tried to answer since they first observed the sky. The currently accepted model for the origin of the universe is the **Big Bang theory,** first proposed by Abbé Georges Lemaître. The theory itself is rather straightforward—all energy and matter exploded from a primeval fireball (figure 20.1). Almost immediately, the subatomic particles that compose atoms began to form as the temperature cooled. During this stage, radiation was common and the universe was made up essentially of hydrogen and helium gases. Cooling and condensation of matter and gases continued for many millions of years, until the temperature was cold enough and densities high enough for gravity to coalesce the material into very large regions. This was the start of the formation of galaxies. Soon after, the solar system began to form. The heavier elements (those with masses greater than those of hydrogen and helium) were created later by nuclear reactions inside stars.

Two pieces of astronomical evidence strongly support the Big Bang theory. First, the *Doppler effect* shows that galaxies are moving away from each other. You can experience the Doppler effect by listening to an ambulance or fire truck pass by when you are standing stationary. The whistle or siren emits sound waves at a constant frequency. However, as the ambulance moves toward you, more sound waves reach your ear per unit time, creating a higher pitch. After the ambulance passes, fewer sound waves reach your ear per unit time, creating a lower pitch. Of course, stars do not emit sound; rather, it is light waves that are affected. When looking at the spectra of a star, astronomers see that the light waves have been shifted to the lower end (red end) of the spectrum. (Hence, the Doppler shift is sometimes termed the *red shift*.) This indicates that galaxies are moving away from each other.

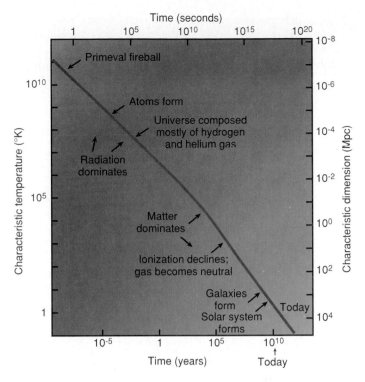

FIGURE 20.1 A highly schematic history of the universe after the Big Bang. The horizontal scale is logarithmic, thus the first year of early history of the universe occupies about the left half of the figure. As the temperature cooled (diagonal black line) atoms formed, then galaxies, then solar systems.

From William K. Hartman, Astronomy The Cosmic Journey. *Copyright © 1987 Wadsworth Publishing Company, Belmont, CA. Reprinted by permission.*

A second piece of evidence for the Big Bang was predicted before it was discovered. Physicists and astronomers predicted that there would be leftover radiation from the explosion of the Big Bang. Mathematical calculations showed that this cosmic background radiation (CBR) would be in the 1 to 5 K range. Radio astronomers using very sensitive equipment found a uniform background of radiation that corresponded to 3 K radiation, exactly within the predicted range. (This radiation, the remaining energy [heat] from the Big Bang, is called the 3 K radiation.)

Of course, with new discoveries the Big Bang theory continues to be modified; however, the overall scheme appears to be acceptable.

Origin of the Solar System

Since the late 1950s and early 1960s, the United States has sent many probes to the sun, moon, and other planets to study the members of our solar system (table 20.1). The data obtained from these probes, particularly the *Voyager* series, provide a wealth of information that will be studied for many years to come. Astronomical studies and probe data of our solar system show that the solar system displays many

TABLE 20.1 *Summary of U.S. Solar, Lunar, and Interplanetary Probes*

Probe	Launch Date	Destination	Encounter Date	Current Status
Mariner				
1	22 July 1962	Venus	—	Did not achieve orbit
2	27 Aug. 1962	Venus	14 Dec. 1962	Orbiting sun
3	5 Nov. 1964	Mars	—	Contact lost after launch; orbiting sun
4	28 Nov. 1964	Mars	14 July 1965	Orbiting sun
5	14 June 1967	Venus	19 Oct. 1967	Orbiting sun
6	24 Feb. 1969	Mars	31 July 1969	Orbiting sun
7	27 Mar. 1969	Mars	5 Aug. 1969	Orbiting sun
8	8 May 1971	Mars	—	Did not achieve orbit
9	30 May 1971	Mars	13 Nov. 1971	Contact lost 27 Oct. 1972; orbiting Mars
10	3 Nov. 1973	Venus	5 Feb. 1974	
		Mercury	29 Mar. 1974	
		Second flyby	21 Sept. 1974	
		Third flyby	16 Mar. 1975	Orbiting sun
Voyager				
1	5 Sept. 1977	Jupiter	5 Mar. 1979	
(= Mariner 11)		Saturn	12 Nov. 1980	Leaving solar system
2	20 Aug. 1977	Jupiter	9 July 1979	
(= Mariner 12)		Saturn	25 Aug. 1981	
		Uranus	24 Jan. 1986	
		Neptune	26 Aug. 1989	Will leave solar system
Pioneer				
1	11 Oct. 1958	Moon	—	Failed
2	8 Nov. 1958	Moon	—	Third-stage rocket did not ignite
3	6 Dec. 1958	Moon	—	Failed
4	3 Mar. 1959	Moon		Flew by moon
5	11 Mar. 1960	Sun	26 June 1960	Solar orbit
6	16 Dec. 1965	Sun		Solar orbit
7	17 Aug. 1966	Sun		Solar orbit
8	13 Dec. 1967	Sun		Solar orbit
9	8 Nov. 1968	Sun		Solar orbit
E	27 Aug. 1969	Sun	—	Failed
10	2 Mar. 1972	Jupiter	3 Dec. 1973	Leaving solar system
11	5 Apr. 1973	Jupiter	2 Dec. 1974	
		Saturn	1 Sept. 1974	Leaving solar system
12	20 May 1978	Venus	4 Dec. 1978	Landed on Venus
(= Venus 1)				
13	8 Aug. 1978	Venus	9 Dec. 1978	Main unit burned in atmosphere 9 Dec. 1978
(= Venus 2)				Four probes launched and landed on surface
Viking				
1	20 Aug. 1975	Mars	20 July 1976	Landed on Mars
2	9 Sept. 1975	Mars	3 Sept. 1976	Landed on Mars

TABLE 20.2 Characteristics of Our Solar System

1. All the planets' orbits lie roughly in a single plane.

2. The sun's rotational equator lies nearly in this plane.

3. Planetary orbits are nearly circular.

4. The planets and the sun all revolve in the same west-to-east direction, called prograde, or direct, revolution.

5. Planets differ in composition.

6. The composition of planets varies roughly with distance from the sun: dense, metal-rich planets lie in the inner system, whereas giant, hydrogen-rich planets lie in the outer system.

7. Meteorites differ in chemical and geologic properties from all known planetary and lunar rocks.

8. The sun and all the planets except Venus and Uranus rotate on their axis in the same direction (prograde rotation) as well. Obliquity (tilt between equatorial and orbital planes) is generally small.

9. Planets and most asteroids rotate with rather similar periods, about 5 to 10 hours, unless obvious tidal forces slow them (as in the earth's case).

10. Distances between planets usually obey the simple Bode's rule.

11. Planet-satellite systems resemble the solar system.

12. As a group, comets' orbits define a large, almost spherical cloud around the solar system.

13. The planets have much more angular momentum (a measure relating orbital speed, size, and mass) than the sun. (Failure to explain this was the great flaw of the early evolutionary theories.)

From William K. Hartman, *Astronomy The Cosmic Journey.* Copyright © 1987 Wadsworth Publishing Company, Belmont, CA. Reprinted by permission.

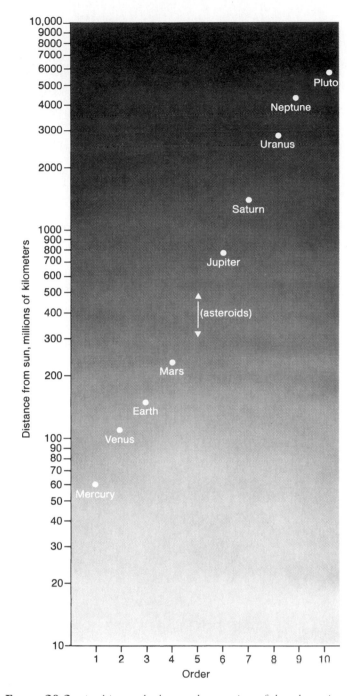

FIGURE 20.2 As this graph shows, the spacing of the planets' orbits exhibits a geometric regularity. However, notice that the vertical axis is a log scale. In reality, the planets are not regularly spaced from the sun.

fundamental characteristics (table 20.2). Any hypothesis of the origin of the solar system should account for most, if not all, of these features. One important characteristic of the solar system is the spacing of the planets from the sun; it exhibits a geometric regularity (figure 20.2). A second characteristic is the fundamental dichotomy between the terrestrial planets (Mercury, Venus, Earth, and Mars), which are small, dense, and rocky, and the Jovian planets (Jupiter, Saturn, Uranus, and Neptune) which are very large, low density, and gaseous (figure 20.3).

The **solar nebular hypothesis** describes the origin of the solar system and accounts for most of the characteristics. This hypothesis states that the solar system began to form about 5.0 to 4.6 billion years ago. Evidence for this age comes from the dating of meteorites—most show an age of about 4.6 billion years.

Step 1: The Solar Nebula. In this region of the Milky Way galaxy, a large cloud of gases and dust was rotating—the solar nebula. The cloud of gases and dust contracted to form a disk, drawn by gravitational forces. The isotope abundances of certain meteorites seem to indicate that an explosion of a nearby star may have helped to collapse the cloud and produce the protosun. Because of the contraction, the inner region of the cloud began to increase in temperature while the outer parts formed a disk in which the gases began to cool. In the outer regions of the disk, the cooling allowed elements to condense. Closer to the center of the disk, where temperatures were higher, silicates formed. In the outer regions, where temperatures were lower, ammonia and methane gas and ice formed. These substances could not form close to the center of the disk because the high temperatures would have vaporized the gases and melted the ice.

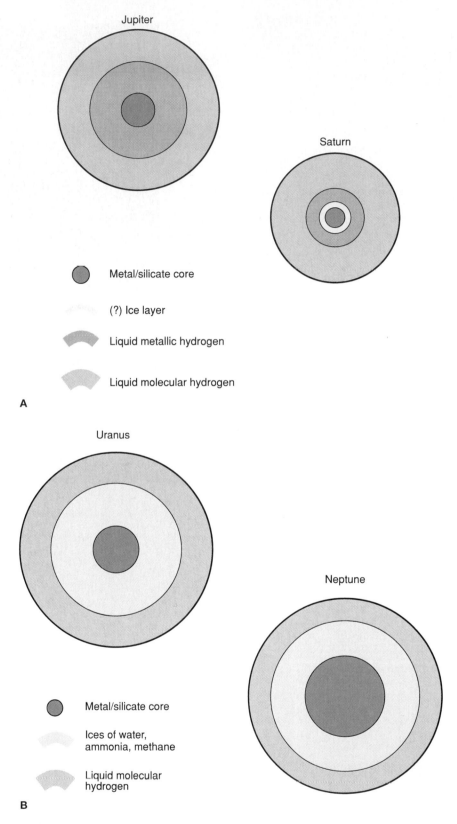

FIGURE 20.3 Cross sections of the Jovian planets. (A) Jupiter and Saturn have both liquid metallic hydrogen and liquid molecular hydrogen zones. (B) Uranus and Neptune lack the metallic hydrogen and have icy mantles.
Source: © NSSDC/Goddard Space Flight Center.

Step 2: The Protosun. As the nebula was contracting, temperatures at the center of the disk increased enough for nuclear reactions to occur. This led to the formation of the sun.

Step 3: Planetesimals. While the sun was forming at the center of the nebula, small to large (millimeters to kilometers) bodies composed of dust and gas were formed by collisions. Since dust and gases in the nebula were orbiting in the same direction, collision velocities were low. This meant that particles tended to accumulate rather than break apart. Once planetesimals were large enough, their gravitational fields pulled in more material and other, smaller planetesimals. This material was continuously accreted to form even larger planetesimals.

Step 4: Protoplanets. Finally, much of the gases, dust, and planetesimals accreted and collided to form the major planets. The major differences in composition between the terrestrial and Jovian planets are due to their distance from the sun when they formed. Only in the distant regions of the solar nebula, where temperatures were lower, could gases remain stable; hence, the Jovian planets' gaseous atmospheres.

Origin of the Moon

Understanding the origin of the moon presents a curious and, at present, much-debated challenge. Studies of the rocks obtained from the *Apollo* missions to the moon show that the moon's composition and structure are somewhat like the earth's (figure 20.4). The bulk composition of moon rocks is similar to that of rocks of the earth's mantle. However, there are several important differences in composition between lunar and terrestrial rocks. The lunar rocks are low in iron (and, consequently, quite unlike the earth's core) and have fewer volatiles and slightly more of a group of elements called refractories. Interestingly, the proportions of oxygen isotopes in lunar rocks are identical to those in terrestrial rocks. This would indicate that the moon formed at about the same distance from the sun as did the earth.

Several hypotheses have been proposed to explain the origin of the moon. Currently, the impact-trigger hypothesis is accepted by most workers in this area. The *fission hypothesis* states that the moon is composed of material that spun off the earth as the earth was accreting. This idea does not explain the source of energy needed to separate a volume of material from the main mass of accreting particles. The *co-accretion hypothesis* (or binary hypothesis) states that the moon formed in orbit around the earth at the same time that the earth was forming. The co-accretion idea has problems explaining certain compositional differences between the earth and the moon, for example, their differences in iron content. For a while, the *capture hypothesis* seemed reasonable. This hypothesis states that the moon formed somewhere in the solar system and was later captured by the gravitational attraction of the earth. Again, several problems

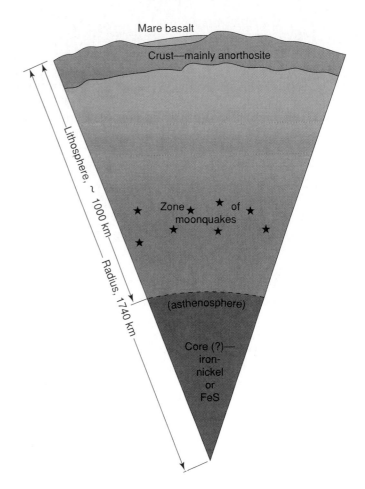

FIGURE 20.4 Cross section of the lunar interior. The existence and size of the core are uncertain.

arise; for example, capture of a planetary body by gravitational forces is a rare phenomenon because the "window" of capture is narrow. If the moon had come in at too steep an angle, it would have collided with the earth. If the angle were too shallow, the moon would have been merely deflected in its path, not captured. In 1984, a new hypothesis was proposed and today is the current working hypothesis of lunar origin. The *impact-trigger hypothesis* states that early in the earth's formation, after the core had formed, a very large body (planetoid, asteroid, or something else) struck the earth. The impact of this large body caused a great volume of mantle debris to be ejected. It was this debris that then accreted to form the moon. Chemical studies and computer models of the proposed impact provide evidence for this model. The impact-trigger hypothesis also explains the difference in iron content between the earth and the moon—only mantle material was ejected. Overall, the impact-trigger hypothesis seems to fit the available evidence.

The Precambrian

Duration

The Precambrian began about 4.6 billion years ago, with the formation of the earth, and ended about 570 million years ago, with the appearance of skeletonized fossil forms in the rock record. The Precambrian represents about 88% of geologic time.

Subdivisions

Geologic time may be divided into two broad intervals: (1) the *Cryptozoic* (meaning "hidden life"), the time represented by the Precambrian; and (2) the *Phanerozoic* (meaning "evident life"), the time represented by the Paleozoic, Mesozoic, and Cenozoic eras. The Precambrian consists of two eons: the Archean eon, from 4.6 to 2.5 billion years ago, and the Proterozoic eon, from 2.5 billion to about 570 million years ago. Some geologists further subdivide the Archean and Proterozoic into smaller units, but there is disagreement about the boundaries of these smaller units (figure 20.5).

One major problem with establishing a Precambrian chronology scheme, and also with correlating Precambrian-age rocks, is the absence of distinctive fossils, particularly index fossils. Most fossils found in Precambrian-age strata are of algae and bacteria. The absence of index fossils makes correlation of Precambrian-age rocks from one region to another extremely difficult. Because most Precambrian rocks are igneous or metamorphic in origin, direct isotopic dating of the rocks is often used to establish a Precambrian-chronology scheme. The dates obtained from isotopic dating represent the time of either the formation of the igneous rock or the metamorphism. Since igneous and metamorphic rocks typically form during orogenic (mountain-building) episodes, the dates obtained from Precambrian-age igneous and metamorphic rocks usually represent orogenies. However, unlike index fossils (which are geographically widespread), orogenic events are geographically restricted and do not occur worldwide. Therefore, the chronology scheme established for the Precambrian of North America may differ from that set up for other continents, because orogenies may have happened at different times on each continent. (Even the Precambrian chronology scheme for the United States differs somewhat from that for Canada.)

The Early Earth

During the earliest part of the Precambrian—the Early Archean—many of the earth's major features, including the continental and oceanic crust, seawater, and the atmosphere, developed.

Differentiation

The formation of the earth's major features resulted from the overall differentiation of the earth. During its formation, the earth became partially molten—hot enough so that elements and minerals became mobile. This heat came from several sources: the contraction of the planet by gravitational pull, the decay of radioactive isotopes, and extraterrestrial impacts. During the differentiation process, the heavier elements sank to the mantle and core, while the lighter elements rose to the surface; most of the very light elements were expelled from volcanoes.

Origin of the Crust

Continental crust and oceanic crust differ in both chemical composition and density. An average sample of continental crust is chemically and mineralogically equivalent to the plutonic igneous rock granite. The density of continental crust is about 2.6 grams per cubic centimeter. Oceanic crust is composed mostly of the volcanic igneous rock basalt, produced at the midocean ridges. The density of oceanic crust is about 3.3 grams per cubic centimeter.

The major chemical differences between granite and basalt are in the relative abundance of SiO_2 (silica) and of denser elements such as iron (Fe) and magnesium (Mg). Granite has more silica and less Fe and Mg than does basalt; thus granite is less dense than basalt.

The origin of oceanic and continental crust is related to the earth's overall differentiation. The first crustal material formed was *ultramafic* in composition—high in iron and magnesium, and relatively low in silica. Several processes might then have acted, singly or in unison, to decrease the amount of iron and magnesium and increase the amount of silica—that is, to progressively change the ultramafic magma into mafic magma and then into felsic magma (see next paragraph), which ultimately produced both the basaltic and the granitic types of crust.

One simple process that might have occurred is *partial melting* of rock to form a magma. If a mafic or ultramafic rock is slowly heated, the minerals with low melting temperatures melt first. According to Bowen's Reaction Series, these minerals are quartz, potassium feldspar, and mica—the minerals that make up felsic igneous rocks. Therefore, felsic-type magmas will form simply by the incomplete melting of a more mafic rock. If a mafic rock incompletely melts to form a felsic-type magma, and this magma cools, granite is formed. The partial melting process also probably contributed to the formation of mafic magmas and rocks from ultramafic magmas.

Another process that might have formed mafic- and felsic-type rocks is *fractional crystallization,* which occurs in magmas that contain minerals with different melting temperatures. As an ultramafic magma cools, minerals with the highest melting temperatures crystallize first. Minerals with just slightly lower melting temperatures crystallize next, and so on. At the last stages of cooling, a residual magma with a higher concentration of silica and a lower concentration of mafic minerals than that of the original magma is left. (To oversimplify, imagine that an ultramafic rock is put in a crucible, melted, and allowed to cool. The magma will solidify to form an ultramafic rock, along with a minor amount of

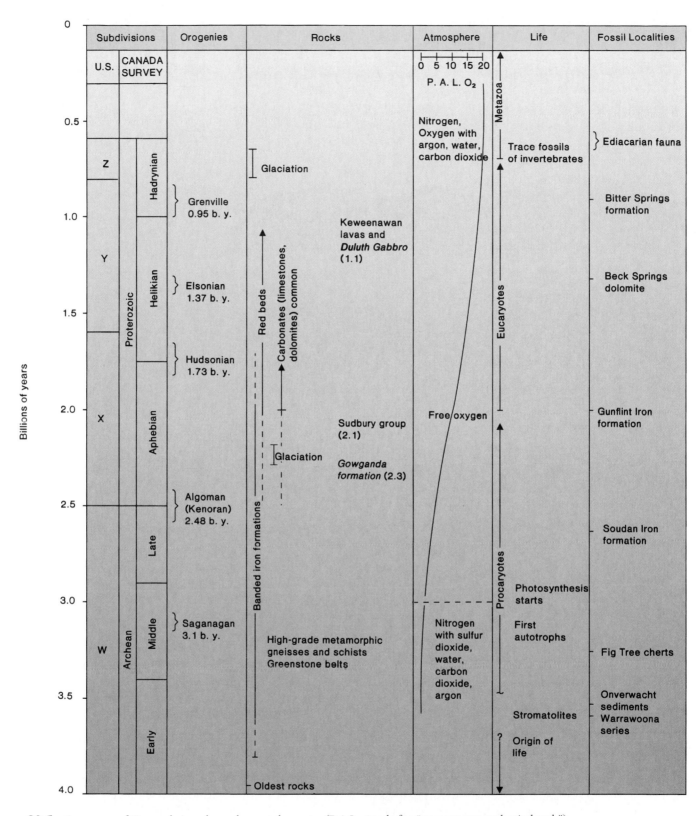

FIGURE 20.5 Summary of Precambrian chronology and events. (P.A.L. stands for "percent atmospheric level.")

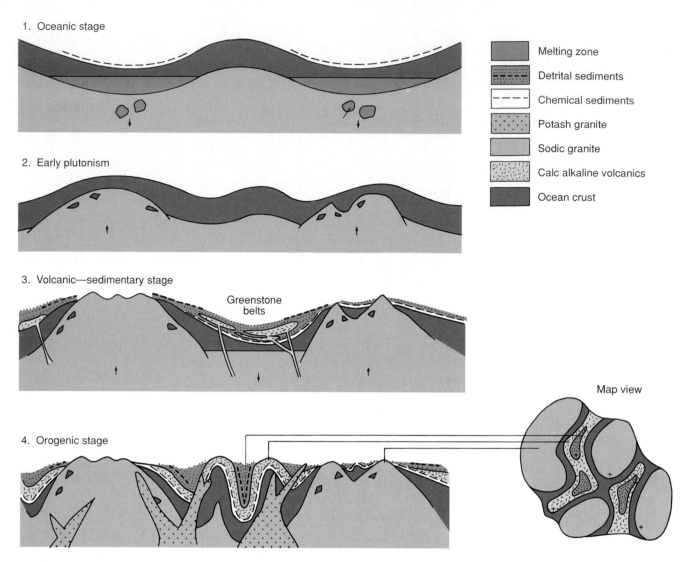

1. Oceanic stage

2. Early plutonism

3. Volcanic—sedimentary stage

Greenstone belts

4. Orogenic stage

Map view

Melting zone

Detrital sediments

Chemical sediments

Potash granite

Sodic granite

Calc alkaline volcanics

Ocean crust

FIGURE 20.6 Formation of the crust during the Archean. Stage 1: Minor sedimentation in topographic lows with associated partial melting. Stage 2: Magmas rise to form batholiths. Stage 3: Erosion of the uplifted areas (batholiths) and igneous intrusions form Greenstone belts. Stage 4: Aggregation of granitic nuclei into protocontinents.

After A. Y. Glikson, Geological Society of America Bulletin, *vol. 43, 1972, p. 3338. Used by permission of the author.*

mafic rock.) Fractional crystallization of the ultramafic magmas of the mantle explains the formation of basaltic oceanic crust during the early part of the Precambrian.

<blockquote>

Many geologists who work in Precambrian terranes and Archean- age rocks envision a magma ocean on the surface of the earth around four billion years ago, but this idea is disputed by some. The entire earth was not molten—just the outer part. In this ultramafic- and mafic-magma ocean floated thin and unstable, felsic-type crust that was constantly melted and recycled. Thus, the earliest felsic crust is thought to have been equivalent to islands floating not in an ocean of water, but in an ocean of molten rock.

</blockquote>

Although the formation of some of the granitic-type crust can be explained using the concepts just discussed, the growth and further chemical evolution (for example, the increase in the amount of silica) of the granitic continental crust was more involved. Additional mechanisms were needed, and there are several possibilities. One is plate tectonics. Although plate tectonics did not function exactly as it does today, some of the Archean ocean crust was very mobile. Because of the high density of the basalt, the oceanic crust was constantly undergoing subductionlike movement, being forced downward into the mantle. When this occurred, masses similar to island arcs were formed, most of which were more felsic than mafic in composition. Eventually, the collision of many of these arcs produced the first continental masses (figures 20.6 and 20.7).

Continental Accretion

After the smaller fragments of felsic-type continental crust collided with other fragments and island arcs to form larger continents, tectonic activity was confined to the margins of these protocontinents. The subduction and collision that occurred throughout the Precambrian added continental material to

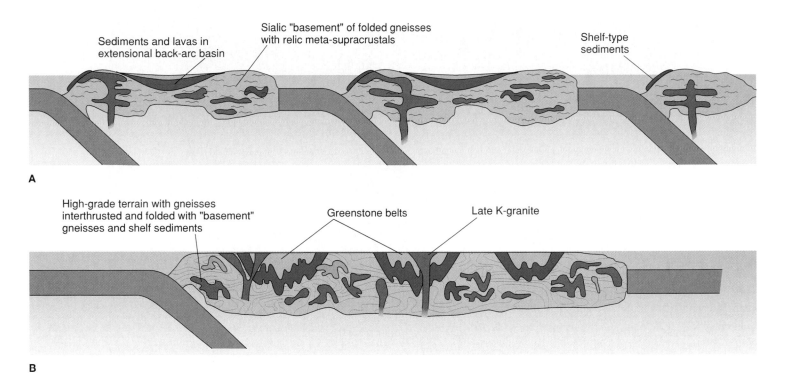

FIGURE 20.7 The growth of continents during the Archean according to a plate-tectonic model. (A) Isolated continental fragments undergo lateral movement. (B) Collision of isolated fragments produces larger, more stable continental masses.

TABLE 20.3 *Comparison of the Earth's Earliest Atmosphere with Its Present Atmosphere*

Time Frame	Major Components of the Atmosphere
0 to 3 billion years ago	Nitrogen, oxygen, with argon, carbon dioxide, water vapor
3.0 to 4.5 billion years ago	Nitrogen, with argon, water vapor, carbon dioxide, sulfur dioxide
Before 4.5 billion years ago	Hydrogen, methane, with ammonia, hydrogen sulfide, nitrogen, argon, water vapor

the margins of the continent, just as the size of a ball of clay is increased by sticking smaller pieces of clay onto the sides. This method of continental growth has been termed **continental accretion**. Continental accretion explains the increase in the sizes of the continents through time. Evidence for the process of continental accretion comes from dating the Precambrian rocks of North America. In general, these rocks show an age progression—the rocks decrease in age (become younger) away from the center of the continent.

Origin of the Oceans

The origin of the tremendous volume of water in the oceans also resulted from the overall differentiation of the earth. Ocean water came from the interior of the earth through **outgassing.** Outgassing refers to the transfer of volatiles (lighter elements, gases, and water) from the earth's interior to the surface through volcanic action. The volume of water trapped in the structure of the rocks in the interior of the earth was certainly enough to produce the oceans. The only significant questions remaining about the origin of the oceans are (1) When did the process of outgassing start? and (2) At what rate was water added to the oceans through outgassing? The oceans were fairly well established by 2 billion years ago, however, the presence of stromatolites in the Early Archean (3.5 billion years ago) indicates that water bodies existed very early in earth's history.

Origin of the Atmosphere

The earth's earliest atmosphere was not like our present atmosphere. It was produced through outgassing, in a manner similar to the development of the oceans. (Both water and gases were expelled from the numerous volcanoes present during the Early Archean.) Two lines of evidence help geologists construct the composition of the earth's earliest atmosphere: (1) study of the atmospheres of the Jovian planets, such as Jupiter; and (2) study of the gases emitted by present-day volcanoes. From these studies, geologists postulate that the earth's early atmosphere was composed principally of nitrogen, with lesser amounts of carbon dioxide (CO_2), water, helium, argon, and some sulfur dioxide (SO_2) (table 20.3).

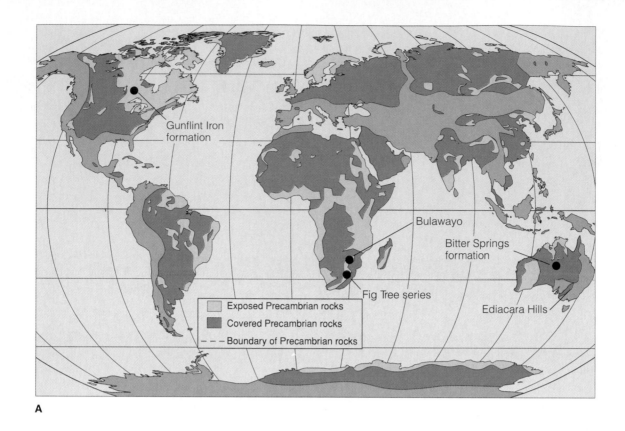

A

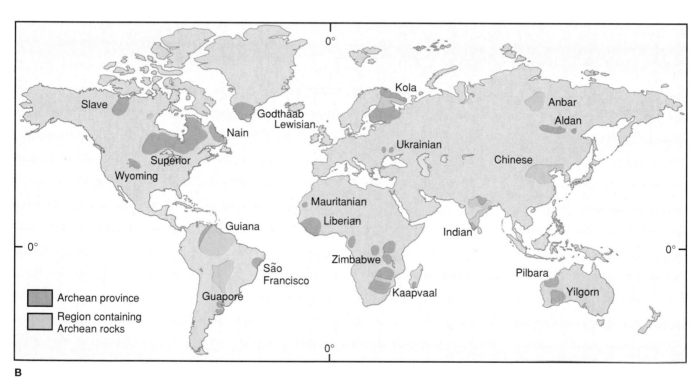

B

FIGURE 20.8 (A) Extensive outcrops of Precambrian rocks on the continents with important fossil localities noted. (B) Major Archean provinces.

(A) From A. M. Goodwin, in B. F. Windley, The Early History of the Earth. Copyright © 1976 John Wiley & Sons, Ltd., Chichester.

One important component missing from the early atmosphere was free oxygen (O_2—oxygen that is not combined with any other elements), although oxygen makes up about 20% of the modern atmosphere. Evidence that oxygen was absent from the early atmosphere comes from several sources. The first important line of evidence is the formation of a rock unique to Precambrian times, called banded iron formations. These rocks contain many very thin, interbedded layers of iron and silica. The iron is in the reduced state (not oxidized, that is, not combined with oxygen). If free oxygen had been present at the time of this rock's formation, the iron would have been oxidized (chemically joined with oxygen to produce FeO—similar to what happens when a car rusts). Second, oxygen is not emitted in large quantities from present-day volcanoes, so by analogy, the volcanoes of Archean time are unlikely to have emitted significant volumes of oxygen. Therefore, the oxygen found in the modern atmosphere must come from somewhere else. Two sources have been identified.

One source of oxygen was *photosynthesis* by plants. Once plants had evolved, their photosynthesis processes contributed great quantities of oxygen to the atmosphere through the breakdown of water. (Note in figure 20.5 that the addition of oxygen to the atmosphere was gradual. Although the earliest atmosphere was devoid of oxygen, by the end of Precambrian time, almost all the oxygen now in the atmosphere was present.) The other source of oxygen was inorganic chemical reactions that occur in the upper atmosphere and produce oxygen as an end product. These reactions are collectively called *photochemical dissociation* processes.

The absence of oxygen from the atmosphere had a major role in the evolution of life—an oxidizing environment would have precluded the formation of cells.

CRATONS AND SHIELDS

Cratons

Most extensive outcrops of Precambrian rocks occur in continental interiors (Greenland is an exception), because these areas generally have been stable over long periods of geologic time (figure 20.8). Plate-tectonic processes are much more likely to affect the margins of continents, which explains why the oldest rocks on earth are often found on continental interiors. The large, stable, interior region of a continent is called the **craton.** The craton is composed of two parts: the **shield** and the **platform** (figure 20.9). The interior part of the craton is made up of Precambrian igneous and metamorphic rocks that crop out at the surface. These rocks represent the ancient continental nucleus and form the shield. Shields are found on every continent. Surrounding the shield, and deposited on top of the Precambrian rocks, are Phanerozoic (Paleozoic, Mesozoic, and Cenozoic) strata that make up the platform. Platform rocks typically are folded in broad folds. Precambrian rocks that underlie the platform rocks are called *basement* rocks. To drill through the platform strata into the Precambrian rocks below is to "strike basement."

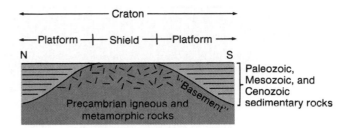

FIGURE 20.9 Cross section through the craton showing the relationship between the shield and the platform.

Shields

The interior region of the North American craton where Precambrian igneous and metamorphic rocks are exposed is called the *Canadian Shield*. The Canadian Shield includes much of the central and eastern part of Canada, some parts of the United States (northern Minnesota, northern Wisconsin, northern Michigan, and the Adirondacks of New York), and parts of Greenland.

The Canadian Shield is made up of a number of **structural provinces** (figure 20.10). Each structural province is defined by isotopic dates obtained from igneous and metamorphic rocks within that province. In general, dates within each province cluster around a 300-million-year interval, and dates obtained from rocks within the province differ markedly from those of adjacent provinces. Most provinces are linear—long in one dimension and short in another. Isotopic dates and the morphologies of the structural provinces provide evidence that each province is the result of an orogenic event that occurred on the margin of the ancient continent. These events may have been subduction related or, perhaps, the results of collisions with other small fragments of continents.

Structural provinces become progressively younger from the interior of the continent outward. This distribution pattern supports the model of continental accretion. (Just as significantly, however, this pattern of progressively younger ages from continent center to continent margin has not been shown in some shields of the southern continents.)

THE ARCHEAN

The Archean record of the Precambrian represents a time quite unlike today. Because of the high heat flow in the Archean, the crust was relatively thin and very mobile. The oldest crustal rocks found on earth today are 3.9 to 4 billion years old. Any crust formed before this time was recycled (remelted). The oceans were also forming throughout much of the Archean, as was the atmosphere.

Two significant problems facing geologists who study Archean-age rocks are that these rocks are often not exposed and that they are highly altered and intensely metamorphosed. Archean rocks are the oldest rocks and thus are very likely to have been covered by younger sediments or altered by tectonic events.

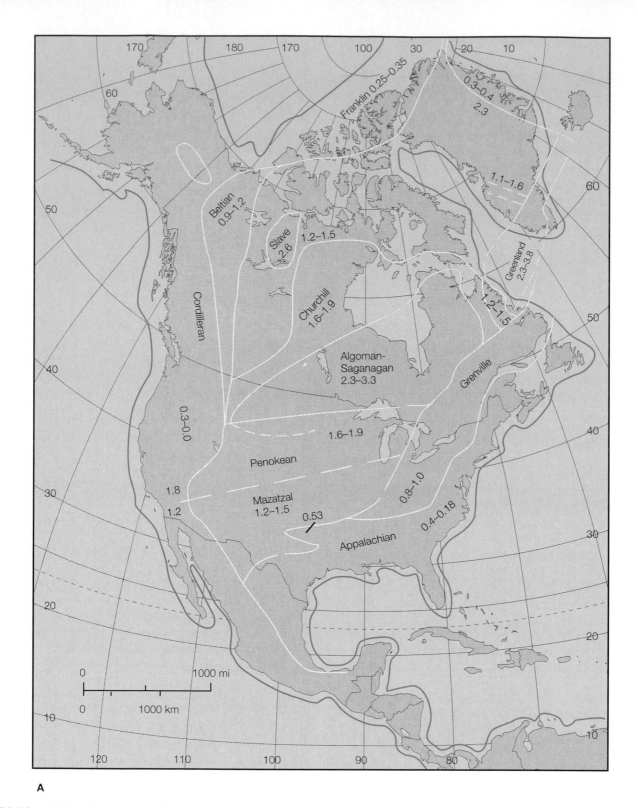

A

FIGURE 20.10 (A) The Canadian Shield and Precambrian structural provinces (in billions of years).

(A) Redrawn from Dott and Batten, Evolution of the Earth, 4th ed. Copyright © 1988, McGraw-Hill, Inc., New York, NY. Reprinted by permission of McGraw-Hill, Inc.

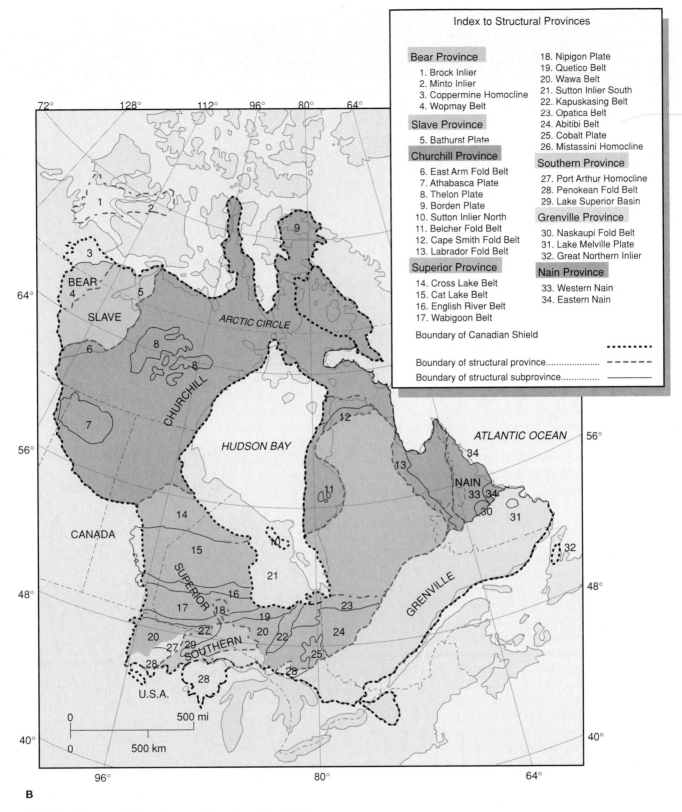

Index to Structural Provinces

Bear Province
1. Brock Inlier
2. Minto Inlier
3. Coppermine Homocline
4. Wopmay Belt

Slave Province
5. Bathurst Plate

Churchill Province
6. East Arm Fold Belt
7. Athabasca Plate
8. Thelon Plate
9. Borden Plate
10. Sutton Inlier North
11. Belcher Fold Belt
12. Cape Smith Fold Belt
13. Labrador Fold Belt

Superior Province
14. Cross Lake Belt
15. Cat Lake Belt
16. English River Belt
17. Wabigoon Belt

18. Nipigon Plate
19. Quetico Belt
20. Wawa Belt
21. Sutton Inlier South
22. Kapuskasing Belt
23. Opatica Belt
24. Abitibi Belt
25. Cobalt Plate
26. Mistassini Homocline

Southern Province
27. Port Arthur Homocline
28. Penokean Fold Belt
29. Lake Superior Basin

Grenville Province
30. Naskaupi Fold Belt
31. Lake Melville Plate
32. Great Northern Inlier

Nain Province
33. Western Nain
34. Eastern Nain

Boundary of Canadian Shield ....

Boundary of structural province ....

Boundary of structural subprovince ....

FIGURE 20.10 (B) Structural provinces of the Canadian Shield.

(B) C. H. Stockwell, et al., "Geology of the Canadian Shield" in Economic Geology Report, *1:44–150, 1968. Geological Survey of Canada, Natural Resources Canada. Reproduced with permission of the Minister of Supply and Services Canada, 1993.*

FIGURE 20.11 A Precambrian shield gneiss.
© Doug Sherman/Geofile.

Archean Rocks

Lines of evidence regarding the oldest rocks on the earth's surface—rocks of Archean age—include petrographic evidence, structural relations from field studies, isotopic evidence (potassium/argon, rubidium/strontium, uranium/lead, and thorium/lead), and geochemical evidence.

Almost all rocks of Archean age are either igneous rocks or metamorphosed igneous and sedimentary rocks. In general, Archean rocks are divided into two broad groups: (1) high-grade metamorphic terranes and (2) **greenstone belts.**

The high-grade metamorphic terranes of Archean age consist mostly of schists and gneisses (figure 20.11). The schists were probably originally shales; the gneisses were originally shales or granites. Rare quartzites also occur. In general, the rock types, associations, and occurrences indicate that the high-grade metamorphic terranes represent the very earliest continental-type material.

Greenstone belts are quite different in occurrence, rock type, and stratigraphic succession from the high-grade metamorphic terranes. Greenstone belts are linear regions of mostly metamorphosed basaltic and other volcanic rocks, with a high percentage of immature clastic sedimentary rocks. Most greenstone belts formed 3 to 3.8 billion years ago. The name *greenstone* comes from the widespread occurrence of the green, metamorphic mineral chlorite in the metamorphosed basalts. The sequence of strata in well-developed greenstones can be quite thick—anywhere from 10,000 to 25,000 meters. The stratigraphic sequence within a greenstone belt begins with volcanic rocks and ends with sedimentary rocks. The base of the greenstone belt is usually marked by flood basalts and

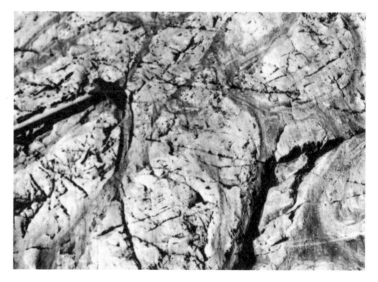

FIGURE 20.12 A Precambrian greenstone.
Photograph by W. P. Puffett, USGS Photo Library, Denver, CO.

pillow basalts (figure 20.12). Occurring stratigraphically higher are immature clastic sedimentary rocks, such as conglomerates, graywacke sandstones, and other immature and volcanic sandstones. Chert is also common. As a whole, the greenstone-belt sequence represents igneous and sedimentary rocks of the oceanic and island-arc type.

A well-studied Archean terrain is the Yellowknife greenstone belt in the southern part of the Slave Province, Canada. A geologic map of the region appears to show a

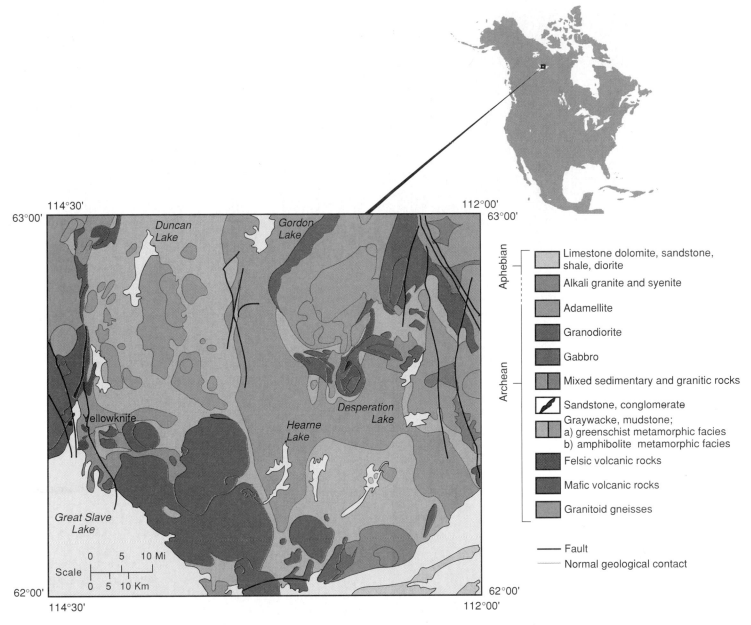

FIGURE 20.13 Geologic map of the Yellowknife-Hearne Lake area, southern Slave Province, Canada.
Source: Data from J. B. Henderson, Chemical Evolution of the Early Precambrian, *1977, Academic Press, Orlando, FL.*

Legend:

Aphebian
- Limestone dolomite, sandstone, shale, diorite
- Alkali granite and syenite

Archean
- Adamellite
- Granodiorite
- Gabbro
- Mixed sedimentary and granitic rocks
- Sandstone, conglomerate
- Graywacke, mudstone; a) greenschist metamorphic facies b) amphibolite metamorphic facies
- Felsic volcanic rocks
- Mafic volcanic rocks
- Granitoid gneisses

— Fault
— Normal geological contact

random mass of rock (figure 20.13). Yet careful study shows that several patterns occur in this fault-bounded basin. Most notably, mafic volcanic rocks are restricted to the edges of the basin, while graywackes and mudstones are quite common on the interior. This relationship, as well as extensive field studies of the structural relationships between rocks, has led geologists to postulate a basinal model to reconstruct the original depositional environment (figure 20.14). Volcanic rocks were erupted along the edges of the basin, while the graywackes and mudstones were deposited on the interior as the basin filled. Other geologists, however, favor a plate-tectonic model involving an arc-trench system. In this model, various regions of rock are the remnants of accreted terranes (see chapter 24 for a discussion of accreted terranes). Still other workers have said this region developed as a result of seafloor spreading in a marginal basin.

The various models proposed to explain the origin of the Yellowknife greenstone belt illustrate how the geologic history of a complex geologic region is deciphered by geologists. After much field work, mapping the types of rocks and their structural relations, and many hours of analyzing the rocks in the laboratory, a hypothesis explaining the original depositional environment and subsequent geologic history is proposed. Other geologists may view the evidence in another way, proposing a different hypothesis. Geologists will then determine what evidence will help support or refute the various hypotheses. Additional laboratory and field work is done to try to find this evidence. In some cases it may take years of searching for additional field evidence before enough is found to allow one hypothesis to be accepted.

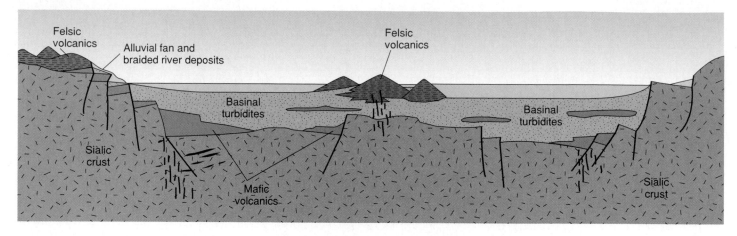

FIGURE 20.14 Basinal model of the development of the Yellowknife area, Slave Province, Canada.

Two other important Archean rock types are komatiites and kimberlites. Komatiites are volcanic lavas with an unusually high concentration of magnesium. All komatiites have been metamorphosed and in the field may not look at all like solidified lava flows. The importance of komatiites is twofold. They provide a means of studying the composition and temperature of the earth's early mantle, since it is thought that is where the magmas originated. They also have tremendous economic value; for example, gold is often associated with these rocks. Kimberlites are also an ultramafic type of rock (high in iron and magnesium), brought to the surface through "pipes." Most of the world's diamonds are mined from "kimberlite pipes." Almost all diamonds come from the kimberlite pipes of South Africa; however, an isolated "pipe" occurs in the United States—the diamond mine of central Arkansas.

Tectonics During the Archean

Plate tectonics was not functioning in the Archean in the same manner it does today for two reasons: (1) heat flow in the Archean was much higher than it is today, and (2) oceanic-type crust was not distinctly differentiated from continental-type crust. Because of the high heat flow, the crust was thin and mobile. The turnover rate of crust was high—that is, crust that solidified melted rapidly. In addition, most of the crust was mafic (basaltic) in composition. Any continental granitic crust that formed was thin and unstable. Because there was little distinction between oceanic and continental crust, many plate-tectonic processes, such as ocean–continent subduction and continent–continent collision, could not operate as they do today. Not until after the Algoman orogeny (2.5 billion years ago) did oceanic and continental crust become distinctly different.

The Algoman Orogeny

The **Algoman orogeny** (also called the Kenoran orogeny) occurred approximately 2.5 to 2.6 billion years ago on the Canadian Shield. The Algoman orogeny marks the boundary between the Archean eon and the Proterozoic eon of the Precambrian, and was only one of many orogenies that occurred during this period of worldwide granite formation. The Algoman and related orogenies were magmatic rather than tectonic in origin and marked a fundamental turning point in the crustal evolution of the earth. Much of the continental crust thickened, increased in volume, and became relatively stable at this time.

SUMMARY

The Precambrian represents approximately 88% of geologic time and is divided into two eons: the Archean and the Proterozoic. During the early part of the Archean, the crust, seawater, and atmosphere developed. The formation of all three was part of the overall differentiation of the earth at that time. The crust formed from partial melting and fractional crystallization of mafic and ultramafic magmas. The general progression of ultramafic mantle rocks forming mafic (basaltic) oceanic rocks that, in turn, formed granitic continental crust is seen throughout the Archean. The ocean water came from the interior of the earth through outgassing by volcanic activity. Outgassing also explains the origin of the atmosphere and its components, except for oxygen. The two sources that gradually supplied oxygen to the atmosphere were photosynthesis and photochemical-dissociation processes.

Precambrian rocks are typically exposed on continental-shield regions—the stable interior part of each continent where Precambrian-age rocks crop out. Each shield is surrounded by younger, Phanerozoic-age strata that make up the platform. Together, the shield and the platform comprise the continental craton.

The main characteristics of Precambrian rocks are that most have been intensely altered and highly metamorphosed and that they lack index fossils. The oldest rocks, which are of Archean age, can be divided into two groups: the high-grade metamorphic gneisses that represent ancient continental crust and the greenstone belts that represent ancient oceanic crust. Tectonic processes during the Archean were not functioning exactly as they do today because of higher heat flow. The Algoman orogeny, which occurred 2.5 billion years ago, produced large volumes of granitic crust and sharply differentiated, stable, granitic continental crust from basaltic oceanic crust. This allowed modern plate-tectonic processes to begin.

TERMS TO REMEMBER

Algoman orogeny 380
Big Bang theory 365
continental accretion 373
craton 375

greenstone belts 378
outgassing 373
platform 375
shield 375

solar nebular hypothesis 367
structural province 375

QUESTIONS FOR REVIEW

1. Describe the Big Bang theory. What evidence supports this theory?
2. Describe the Solar Nebular hypothesis.
3. How did the moon form? Compare and contrast at least three separate hypotheses.
4. Why are chronology schemes for Precambrian rocks different in the United States and Canada?
5. How did the crust form?
6. What was the origin of seawater?
7. How did the earth's earliest atmosphere differ from its present one?
8. Where did all the oxygen in our atmosphere come from?
9. Distinguish between the craton, the shield, and the platform.

FOR FURTHER THOUGHT

1. The early earth was certainly unlike the earth today. Contrast how the following processes were different in Archean time: weather; seawater composition; ocean circulation— tides, currents, and waves; rates and types of erosion; and types of sedimentation. Can you offer any evidence for your answers?
2. Using figure 20.5, calculate the percentages of geologic time represented by the Archean, the Proterozoic, and the Phanerozoic.

SUGGESTIONS FOR FURTHER READING

Cloud, P. E., Jr. 1968. Atmospheric and hydrospheric evolution on the primitive earth. *Science* 160: 729–36.

Nisbet, E. G. 1987. *The young earth: An introduction to Archean geology.* Boston: Allen and Unwin.

Silk, J. 1989. *The Big Bang.* New York: W. H. Freeman.

Stockwell, C. H., McGlynn, J. C., Emalie, R. F., et al. 1968. Geology of the Canadian Shield. In *Geology and economic minerals of Canada: Geological survey of Canada.* Economic Geology Report 1, pp. 44–150.

Walter, M. R. 1977. Interpreting Stromatolites. *American Scientist* 65: 563–71.

Windley, B. F., ed. 1976. *The early history of the earth.* Chichester, England: John Wiley and Sons.

Windley, B. F. 1984. *The evolving continents.* 2nd ed. Chichester, England: John Wiley and Sons.

21

THE PROTEROZOIC

Agawa Rock is composed of Algoman granite. Lake Superior Provincial Park, Ontario, Canada. © Doug Sherman/Geofile.

The Proterozoic marks a great turning point in the history of the earth and life. During the Proterozoic the crust stabilized and, by the end of the Algoman orogeny, two distinct types of crust had formed—continental (granitic) and oceanic (basaltic) crust. This development allowed for familiar plate-tectonic processes to begin. The oceans had also stabilized, and oxygen in the atmosphere increased. The evolution of life progressed. Prokaryotes and eukaryotes appeared in the fossil record. The end of the Proterozoic is marked by the appearance of metazoans in the fossil record.

THE PROTEROZOIC

Differences Between the Archean and the Proterozoic

The Proterozoic is sometimes termed the platform stage of the Precambrian, in contrast to the mobile-crustal stage of the Archean. The production of large volumes of granite and the stabilization of continents during the Algoman orogeny had a number of significant effects on the surface of the earth (table 21.1). First, the crustal differentiation between granitic continental crust and basaltic oceanic crust allowed modern plate-tectonic processes to begin. (For example, various di-

vergent processes, such as continental rifting, and various convergent processes, such as subduction and collision, could now operate in a manner similar to that of Phanerozoic time.) Second, the large, stable continents provided a site for the weathering of rocks. Erosion, deposition, and weathering cycles can be recognized in the changing chemistry of sedimentary rocks of this age. Third, widespread depositional environments (environments in which sediment is typically deposited) formed on the stable continental surfaces. Different terrestrial and shallow-marine environments can be recognized in the sedimentary rocks of this age. Fourth, detrital sediments (for example, sandstones) became texturally more mature (the grains became more rounded, and many unstable minerals were broken down and carried away), owing to changes in the rates and intensities of weathering. Fifth, the large continents also stabilized the environment somewhat, allowing for the widespread deposition of various chemical precipitates, such as limestones, to occur.

Distribution of Rocks

Most Precambrian rocks of North America are igneous and metamorphic and crop out on the Canadian Shield region (figure 21.1). Note in figure 21.1 the complex relationship of the shield rocks. Much of Precambrian (both Proterozoic and Archean) geology is deciphered by detailed study of

---

TABLE 21.1 *Geologic Differences Between the Archean and the Proterozoic of the Precambrian*

Feature	Archean	Proterozoic
Crust	Thin, mobile	Thick, stable
Plate tectonics	Few modern processes	Modern plate-tectonic processes present
Weathering	Not important	Significant—several weathering cycles distinguished in some rocks
Depositional environments	Unstable	Stable, widespread
Detrital sediments	Texturally immature	Texturally mature sediments found
Carbonates	Few	Many

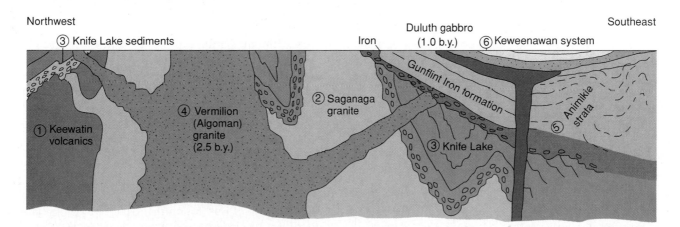

FIGURE 21.1 Geologic cross section of part of the Canadian Shield. Numbers indicate relative ages (1=oldest). Note how the principles of relative-age dating—crosscutting relationships and inclusions—can be applied to Precambrian rocks to develop a geologic history of a region.

Redrawn from Dott and Batten, Evolution of the Earth, *4th ed. Copyright © 1988 McGraw-Hill, Inc., New York, NY. Reprinted by permission of McGraw-Hill, Inc.*

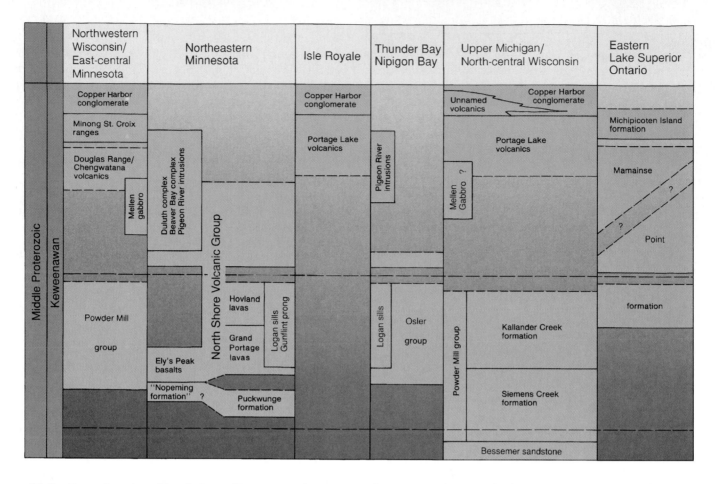

FIGURE 21.2 General stratigraphic relations of Proterozoic (Keweenawan) igneous rocks around Lake Superior.
After John C. Green, in Wold and Hinze, editors, "Geology and Tectonics of the Lake Superior Basin" in GSA Memoir 156, fig. 3, p. 50, 1982, Geological Society of America. Used by permission of the author.

the contacts between the different igneous and metamorphic rocks. The principles of crosscutting relations and inclusions, and sometimes of superposition, allow a great deal of Precambrian geology to be worked out in detail.

The shield area of North America is the best place to observe Precambrian strata cropping out at the surface. A representative stratigraphic column of Proterozoic strata of the shield is shown in figure 21.2. Important Canadian Archean rocks include the Vermilion and Saganaga granites and the Timiskaming, Knife Lake, and Yellow Knife groups. The Canadian Shield, however, is not the only place to observe Proterozoic strata. In certain places, tectonic processes may have brought Precambrian rocks to the surface (the Smoky Mountains of Tennessee and the Tetons of Wyoming), or the platform strata may have eroded away, exposing the underlying Precambrian basement rock (the Inner Gorge of the Grand Canyon in Arizona) (figures 21.3, 21.4, 21.5).

Proterozoic Rocks

Of all the Precambrian-age rocks, perhaps the most interesting and most enigmatic are the **banded iron formations**

(figure 21.6). Banded iron formations are alternating thin laminae (beds) of silica, in the form of chert, and iron. The iron in the rocks is hematite (Fe_2O_3). Banded iron formations formed in water. The source of the iron was either from volcanic emanations or weathering of rocks. (Evidence exists for both sources.) They indicate that the ocean was probably chemically stratified and that the ocean was free of free-oxygen (oxygen alone, not chemically combined with another element) until about two billion years ago. Many banded iron formations of the Mesabi, Vermilion, and Cuyuna iron ranges of Minnesota are mined for their iron content (figure 21.7). Other important Proterozoic rocks include the Early Proterozoic Sudbury group, the Middle Proterozoic Duluth gabbro, and the Middle and Late Proterozoic Keweenawan lavas.

Glaciations

Because of the high heat flow of Archean time, glaciations were impossible. But during the Proterozoic, the earth cooled enough for glaciation to occur in some places. Two, possibly three, glacial episodes have been recorded in sedimentary

FIGURE 21.3 The Great Smoky Mountains, Sevier County, Tennessee.
Photograph by W. B. Hamilton, USGS Photo Library, Denver, CO.

FIGURE 21.4 Folding of Precambrian-age strata. Lake Ellen Wilson area, Glacier National Park, Flathead County, Montana.
Photograph by P. Carrara, USGS Photo Library, Denver, CO.

FIGURE 21.5 Mt. Rushmore is sculpted in Precambrian rocks that are part of an isolated dome in southwestern South Dakota.
Photograph courtesy of Dennis J. Bebel.

rocks of Late Proterozoic time. One occurred about 2.3 billion years ago on the Canadian Shield in southern Ontario. The *Gowganda formation* includes tillites and other conglomerates, and it also has striated pebbles and overlies a striated surface (figure 21.8). All evidence points to a glacial origin for

these sediments. Other areas in Canada also show evidence of glaciations. In eastern Newfoundland, for example, rocks show additional evidence of Proterozoic glaciations.

THE EDIACARIAN PERIOD

Careful study of rocks of latest Precambrian age (570 to 700 million years old) has revealed an interesting suite of fossils. A complex fossil metazoan (multicelled organisms) community was first discovered in the Ediacara Hills of South Australia (figure 21.9). This fauna, the first evidence of metazoans in the rock record, is of moderate diversity, with numerous invertebrates that all lack shells. At least twenty different genera have been described, including organisms resembling jellyfish, segmented worms, and corals, and possible arthropods, molluscs, and echinoderms.

Since the discovery of the Ediacarian Fauna, similar fossils of this age have been found in strata on many continents. For example, in eastern Newfoundland, thick sequences of Late Precambrian (Ediacarian) rocks have been found. One locality in this region is being considered as the stratotype (that is, type section) for the Precambrian–Cambrian boundary. Many geologists have proposed that these rocks should be assigned to the **Ediacarian period.** Although some authors place the Ediacarian period as the first period of the Paleozoic, most place it as the last period of Precambrian time.

PRECAMBRIAN LIFE

The most prominent characteristic of Precambrian fossils is the lack of index fossils. Until Ediacarian time, almost all fos-

FIGURE 21.6 A Proterozoic banded iron formation.
Photograph courtesy of Clarence Casella, Northern Illinois University.

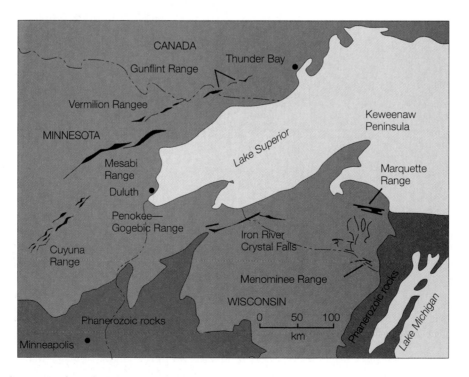

FIGURE 21.7 Iron-mining districts in the Lake Superior region.
From Stearn, Carroll, and Clark, Geological Evolution of North America, *3d ed. Copyright © 1979 John Wiley & Sons, Inc., New York, NY. Reprinted by permission of John Wiley & Sons, Inc.*

FIGURE 21.8 The Gowganda conglomerate, a Precambrian tillite.
From W. G. Miller, Annual Report 19 (2):74, 1913. Ontario Bureau of Mines.

FIGURE 21.9 Fossils of the Ediacarian Fauna.
Photographs courtesy of M. F. Glaessner.

sils of Precambrian age are of single-celled algae and bacteria. The lack of identifiable index fossils makes correlation of Precambrian rocks difficult.

Because most rocks of Precambrian age have been metamorphosed, the fossils within the rocks have been altered and in many cases destroyed. But chert bands and chert nodules, irregular masses of silica that formed on the sea floor, resist metamorphic changes. Therefore, the primary rock type to examine for Precambrian-age fossils is chert—either in beds or in nodules.

Evidence of Precambrian Life

Abundant evidence for Precambrian life exists. Fossil evidence includes microfossils (fossils of microscopic-sized organisms) and **stromatolites** (figure 21.10). Although the earliest organiclike structures in the rock record may not be true fossils—they may be inorganic structures—many localities of Archean age do have unequivocal microfossils. Stromatolites are not strictly fossils; they are a type of organosedimentary structure formed by algae. When algae grow in shallow water, they often form a mat on the sea floor. Frequently, the algal mat gets covered by fine-grained sediment. The algae then grow through the thin layer of sediment to form a new mat, and the process is repeated over and over. The end result is a moundlike structure of fine-grained sediment, typically carbonate in composition, that reflects the influence of the growing algae. Fossils of algae may be found within stromatolites, but the structure itself was not a living animal.

Chemical evidence for the existence of Precambrian life has also been found. Various types of organic molecules have been found in Precambrian-age rocks. Carbon-isotope ratios provide additional evidence of life. Animals may use certain isotopes of carbon preferentially in their metabolic

FIGURE 21.10 Precambrian algal stromatolites.
Photograph by R. G. McGimsey, USGS Photo Library, Denver, CO.

processes. Therefore, the carbon isotopes have a characteristic ratio (signal) when influenced by animals. Carbon isotopes with the right ratio, an organic signature, have been found in Precambrian-age rocks.

Origin of Life

One common misconception concerning the origin of life is the belief that, for billions of years, organic molecules floated around and collided in a primordial organic "soup," which, after trillions and trillions of collisions, resulted in a single-celled organism crawling out onto land. Not so. The oldest crustal rocks are approximately 3.9 billion years old. Although the rocks were subsequently metamorphosed, some had originally been sedimentary rocks. Fossils have been found in these earliest sedimentary rocks (although the organic nature of these structures is disputed by some). Thus, it appears that life did not begin slowly and gradually. Several lines of evidence and careful analysis indicate that life appeared on the earth just as soon as the earth was cool enough to support it.

Another misconception concerning the origin of life is that, to explain how life arose on earth, chemists and biologists must somehow be able to mix different chemicals in a test tube and have a living cell form. Again, not so. Creating cellular life in the laboratory is currently impossible. Instead, scientists try to postulate a reasonable set of conditions under which life could have arisen on the early earth. And for the most part, they have succeeded.

Biologists and paleontologists have defined five basic questions that need to be answered when discussing the origin of life: (1) Where did the raw materials for life come from? (2) How did monomers develop? (3) How did polymers develop? (4) How was an isolated cell formed? and (5) How did reproduction start?

The first question—Where did the raw materials for life come from?—has an easy answer: The early earth provided all the elements and chemicals needed for life to begin.

The second question—How did monomers develop?—addresses the question of how organic materials could form from inorganic processes. Monomers are simple organic compounds, such as amino acids and simple sugars. For a long time scientists believed that there was a barrier between inorganic processes and organic processes. Surely, inorganic chemical reactions could not produce organic molecules. But this idea was proved wrong by several very famous experiments performed in the late 1940s and early 1950s—the *Miller-Urey experiments,* named after co-workers Stanley L. Miller and Harold Urey. Through their experiments, Miller and Urey proved that organic molecules, amino acids in particular, could be produced by inorganic processes. They placed the ingredients thought to have been present at the start of the earth in a large flask. These ingredients included water, hydrogen, methane, and ammonia. The mixture was then heated to boiling. The steam and water droplets were subjected to electrical sparks, to simulate early storm events, and then the gas was cooled and recirculated. Miller and Urey continued the experiment for just over one week. At the end of that time, the mixture in the bottom of the flask contained abundant amino acids and many other simple organic compounds.

Since then, these types of experiments have been performed many times, with similar results. The one basic requisite for the success of the experiments is the absence of oxygen. Oxygen prevents the simple organics from forming. Abundant evidence exists to show that the atmosphere was devoid of free oxygen during the early Archean, when processes similar to this were occurring. In summary, the Miller-Urey experiments have shown that amino acids could have easily formed under postulated Precambrian conditions.

Another source for amino acids, excluding inorganic chemical processes, is meteorites. This idea is relatively new and highly controversial. Upon careful examination of the interiors of some meteorites, researchers have found amino acids and other organic structures. Perhaps when the earth was very young, the profusion of meteorite impacts "seeded" amino acids in the developing oceans. But there is one major problem with this idea. How do we know the amino acids were originally present in the meteorites? Perhaps the meteorites became contaminated with amino acids when they struck the earth or when a geologist handled the rock. Whether this idea proves to be possible or not, amino acids could have easily formed in the developing oceans by inorganic processes, which satisfactorily answers the question regarding the origin of monomers.

The third question—How did polymers develop?—concerns how simple organic molecules linked into longer chains of organic molecules, a process called *polymerization.* Examples of polymers include various types of proteins and nucleic acids (such as RNA). Again, numerous models have been developed to explain polymerization. One recent idea is that the simple organic molecules may have become concentrated on the surfaces of chemically complex minerals, such as clays, and grown into longer chains. Another idea is that water containing amino acids and other simple organic molecules may have totally evaporated near volcanic vents on the shores of ancient oceans. Experiments have shown that when water containing amino acids evaporates completely, the amino acids spontaneously form longer chains—polymers. Many other models of polymerization have also been used. Overall, the question of polymerization has also been answered.

The fourth question—How was an isolated cell formed?—concerns the development of the cell wall. Once complex organic molecules formed, how were they isolated into a cell-like structure? Only a few models for this process have been proposed, and they are not without flaws. To date, the question of the development of a cell wall has proved difficult to answer.

The fifth question—How did reproduction start?—remains, essentially, unanswered.

In summary, answering questions about the origin of life involves developing a reasonable set of conditions under which life could have arisen. Five basic questions are asked concerning the origin of life. So far, the first three have satis-

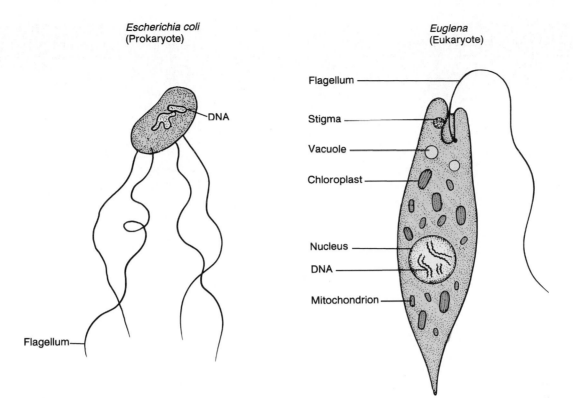

FIGURE 21.11 Comparison of prokaryotes and eukaryotes.

TABLE 21.2 *Comparison of Prokaryotes and Eukaryotes*

Feature	Prokaryotes	Eukaryotes
Organisms represented	Bacteria and cyanobacteria	Protists, fungi, plants, and animals
Cell size	Small; generally 10 to 100 micrometers	Large; generally 1 to 10 micrometers
Metabolism and photosynthesis	Anaerobic or aerobic	Aerobic
Motility	Nonmotile or with flagella made of the protein flagellin	Usually motile; cilia or flagella constructed of microtubules
Cell walls	Of characteristic sugars and peptides	Of cellulose or chitin, but lacking in animals
Organelles	No membrane-bounded organelles	Mitochondria and chloroplasts
Genetic organization	Loop of DNA in cytoplasm	DNA organized in chromosomes and bounded by nuclear membrane
Reproduction	By binary fission	By mitosis or meiosis
Cellular organization	Mainly unicellular	Mainly multicellular, with differentiation of cells

factory answers. Answering the last two, however, will require much additional research.

First Fossils

Most Precambrian fossils are of single-celled organisms of bacteria, algae, and similar life-forms. Two fundamental types of cells are found in Precambrian-age rocks, and three biologic processes are known to have been used by these cells.

The earliest fossils are of **prokaryotes**—cells without a nucleus (figure 21.11; table 21.2). The genetic material in a prokaryotic cell is concentrated in a single fiber rather than within a nucleus, and reproduction is by cell division. The cell wall is complex. Modern prokaryotes include bacteria and blue-green algae. The earliest undisputed prokaryotes are 3.5 billion years old. These cells used fermentation to produce energy.

Eukaryotes are cells with a nucleus (figure 21.11; table 21.2). These cells have their genetic material in many fibers, concentrated in the nucleus, and because of this may reproduce by sexual reproduction. The earliest eukaryotes are about 2 billion years old. Eukaryotes probably

TABLE 21.3 *Biological Processes Used to Obtain Energy*

Fermentation

$C_6H_{12}O_6 \longrightarrow 2CO_2 + 2C_2H_5OH + $ Energy

Sugar   Carbon   Ethyl
(glucose)   dioxide   Alcohol

Photosynthesis

$6CO_2 + 6H_2O + $ Sunlight $\xrightarrow{\text{(chlorophyll)}}$ $C_6H_{12}O_6 + 6O_2$ (Sugar used for energy)

Carbon   Water   Sugar   Oxygen
dioxide   (glucose)

Respiration

$C_6H_{12}O_6 + 6O_2 \longrightarrow 6CO_2 + 6H_2O + $ Energy

Sugar   Oxygen   Carbon   Water
(glucose)   dioxide

---

arose from prokaryotes, but the processes by which this happened are not known. One idea is that one or two types of prokaryotes invaded another prokaryote, thereby forming a more complex cell.

Three separate biologic processes were used by different early organisms to obtain energy: (1) fermentation, (2) photosynthesis, and (3) respiration (table 21.3). The first animals produced energy by fermentation, in which organic compounds (sugars) are used to release energy. Photosynthesis uses solar energy (sunlight) to change water and carbon dioxide into organic molecules and oxygen. Respiration uses oxygen and sugar to produce energy in addition to water and carbon dioxide.

Precambrian Fossil Record

Many different fossil localities have been found in Precambrian-age rocks. Two important and well-known Archean fossil localities are found in Australia and South Africa. The oldest sedimentary rocks that contain undisputed fossils belong to the *Warrawoona group* of northwestern Australia. In these 3.5-billion-year-old rocks, bacteria, blue-green algae, and possible stromatolites have been found. Another Archean fossil locality is the *Fig Tree group* of South Africa. These rocks, slightly younger than the Warrawoona group, are approximately 3.2 billion years old

and contain many varied rod-shaped and spherical fossil forms—probably of algal and bacterial organisms.

One of the better-known Proterozoic fossil localities is the **Gunflint Iron Formation,** 1.6 to 1.9 billion years old, of northern Minnesota. In the Gunflint, many diverse forms are found, including blue-green algae, algae, bacteria, definite stromatolites, and other enigmatic forms unlike anything living today (figure 21.12).

Rise of Metazoans Near the End of the Precambrian

The transition from the single-celled eukaryotes to the multicelled metazoans preceding the time of the Ediacarian Fauna is not shown in the fossil record. Two factors played a role in this transition: (1) oxygen levels in the atmosphere and (2) the presence of sexual reproduction. Abundant oxygen is needed for the complex metabolic processes that occur in metazoans. One interesting idea is that, just before Ediacarian time, oxygen became abundant enough in the water that a benthic community of organisms could develop. Rather than being restricted to the water column, organisms could now settle on the bottom. In the bottom environment shells are beneficial, so many organisms developed shells. The development of shelled animals defines the end of the Precambrian and the start of the Cambrian.

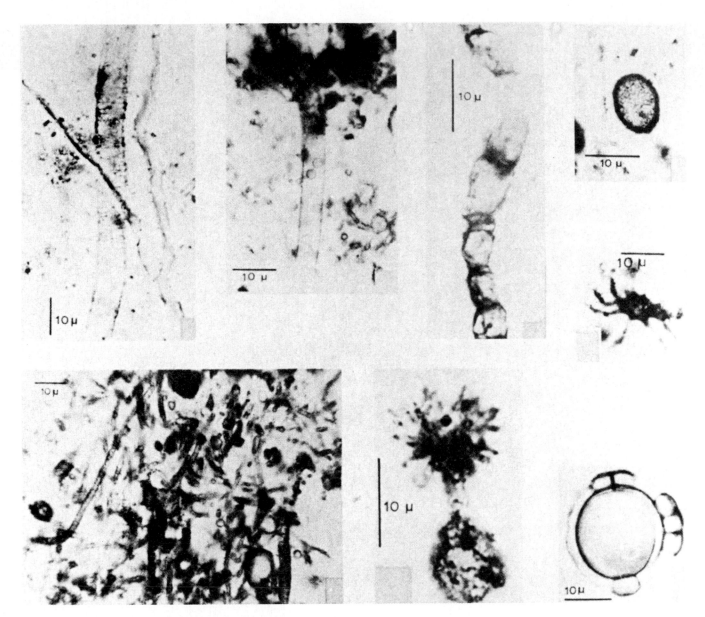

FIGURE 21.12 Fossils from the Gunflint Iron Formation.
Photographs courtesy of E. S. Barghoorn.

Box 21.1

Regional Geology

THE LAKE SUPERIOR REGION

The Proterozoic strata that underlie and crop out around Lake Superior form a syncline (see chapter 11) and are part of a Precambrian-age intracontinental rift called the Midcontinent rift system (figure 1).

Rocks of this rift system are exposed around Lake Superior—in Ontario, northern Michigan, northern Wisconsin, and northern Minnesota. The rift system continues southwestward into Kansas, although the igneous rocks that form it become buried beneath the Phanerozoic strata of the platform. The buried rift structure is recognized by a large positive-gravity anomaly shown on gravity maps of the midcontinent. The rift system also extends from Lake Superior southeastward into southern Michigan.

The rocks surrounding Lake Superior were deposited in a structural basin, now a topographic basin as well, and lie on a series of Archean-age and older Proterozoic igneous and metamorphic rocks.

The development of the midcontinent rift in the Lake Superior region can be summarized by three major events: (1) continental rifting and formation of a series of smaller, interconnected basins; (2) deposition of clastics in the basins; and (3) subsequent faulting of the rocks.

Initially, the region was rather stable, as is indicated by the sedimentary strata that appear to have been deposited into a basin by rivers. Then continental rifting began, and great quantities of both intrusive and extrusive igneous rocks were formed. Northwest of Lake Superior, in northern Minnesota, a series of ultramafic intrusions formed—the Duluth complex. These intrusions consist predominantly of gabbro. Rifting of the continent produced a basinal structure, extending northeast–southwest, into which both lavas and sediments were deposited. The major suite of extrusive lavas—the Keweenawan supergroup—formed between 1.1 and 1.2 billion years ago. These units are flood basalts, and over 10,000 meters of them have been measured. Sediments, consisting of flu-vial sandstones, were deposited in the basin during the rifting episode and are found interlayered with the volcanics.

Following the rifting and volcanism, sediments were also deposited along the basin margin. These strata consist primarily of conglomerates, sandstones, and shales.

The last stage in the history of these units is an episode(s) of faulting of the strata. Interestingly, there is evidence on the Slate Islands (northern Lake Superior) of a meteorite impact at this time (Middle Proterozoic) that may also have contributed to the structural history of these rocks.

The strata comprising the Midcontinent rift system of Lake Superior are important for a number of reasons. First, many minerals of economic importance have been found in these rocks, particularly native copper in the Keweenawan lavas. Second, the strata give geologists a detailed look at Precambrian plate-tectonic processes.

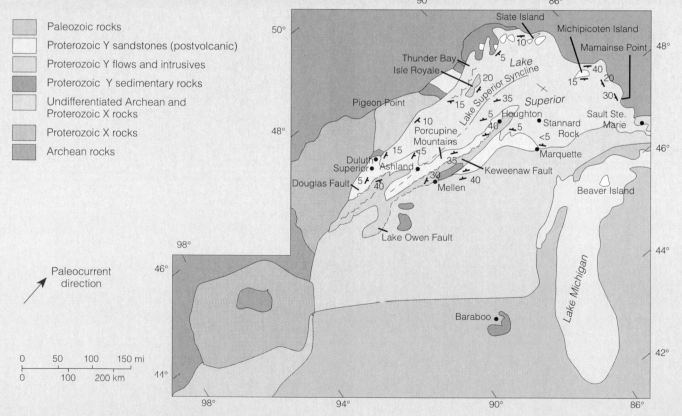

FIGURE 1 Generalized geologic map of the Lake Superior region. (Refer to figure 20.5 for the time represented by Proterozoic X and Proterozoic Y Precambrian rocks.)

After D. M. Davidson, in Wold and Hinze, editors, "Geology and Tectonics of the Lake Superior Basin" GSA Memoir 156, fig. 2, p. 7, 1982, Geological Society of America. Used by permission of the author.

SUMMARY

The most interesting Proterozoic rocks are the banded iron formations—alternating laminae of pure iron and chert. Most of these are mined for their iron.

Five separate questions need to be answered regarding the origin of life.

The first three, concerning raw materials, monomers, and polymers, all have satisfactory answers. The last two questions involve the development of cell walls and reproduction and lack satisfactory answers.

Most Precambrian fossils consist of prokaryotes and eukaryotes—bacteria and algae. Only at the very end of the Precambrian, during the Ediacarian interval, do complex metazoan communities appear.

TERMS TO REMEMBER

banded iron formation 384
Ediacarian period 386

eukaryote 389
Gunflint Iron Formation 390

prokaryote 389
stromatolites 387

QUESTIONS FOR REVIEW

1. Why are all banded iron formations Precambrian in age?
2. What is the Ediacarian period?

3. What major questions must be answered in solving the problem of the origin of life?

4. What problems are encountered in the study of Precambrian fossils?
5. Briefly outline the development of life during the Precambrian.

FOR FURTHER THOUGHT

1. The rise of the metazoans at the end of the Proterozoic is a major event in the history of life. Throughout the Archean and Proterozoic, life consisted primarily of algae and bacteria. After the appearance of metazoans, evolution proceeded rapidly. What physical or biological factors were important in metazoan evolution? Did any one parameter change to facilitate metazoan development?

SUGGESTIONS FOR FURTHER READING

Barghoorn, E. S., and Schopf, J. W. 1966. Micro-organisms three billion years old from the Precambrian of South Africa. *Science* 152: 758–63.

Glaessner, M. F. 1984. *The dawn of animal life: A biohistorical study*. Cambridge, England: Cambridge University Press.

Schopf, J. W. 1978. The evolution of the earliest cells. *Scientific American* 239 (3): 110–38.

22

THE EARLY PALEOZOIC

Cross-bedded Cambrian-age sandstone from the Wisconsin Dells.
© Glen Allison/Tony Stone Images

INTRODUCTION

The division between the Precambrian and Paleozoic is marked by a major event in the history of life—the widespread appearance in the rock record of organisms with hard parts.

The story of the Early Paleozoic in North America is really a story of widespread, shallow-water seas. Characteristic sediments, such as sandstones near the shore and limestones in deeper water, were deposited in uniform patterns over large geographic areas on the continent.

Most of the tectonic events occurred along the continent's eastern margin, with the first pulse of orogenic activity in the Appalachian area.

Diverse marine invertebrates lived in the shallow-water seas. At the start of the Early Paleozoic, these invertebrates increased drastically in both number and kind. By the end of the Early Paleozoic, they were ubiquitous on the shallow, limy bottoms. By the end of the Early Paleozoic, life took a step out of the marine realm as plants invaded the land.

Maps and Cross Sections

An earth historian uses various tools to decipher the history of the earth. Two of the most important, and most frequently used, tools are maps and cross sections.

All of the many types of maps that geologists use are designed to show some aspect of geology or geography, usually for a specific interval of time. *Paleogeologic* maps show the distribution of rock types of a selected region at a specific time. *Paleogeographic* maps show the distribution of land and sea in a selected region at a specific time. *Lithofacies* maps show the areas over which sediments of different lithologies are distributed. *Isopach* maps show the thickness of an interval of strata (perhaps of one or several formations) over a region.

Cross sections, in contrast to maps, are slices through the earth. They are used to show correlations of strata across a region and can reveal how rocks change in thickness from one location to another.

Together, maps and cross sections provide geologists with the horizontal and vertical distributions of the rocks in an area. This information is necessary for constructing the region's geologic history.

Arches and Basins

Arches and basins are characteristic features of the platform strata (the relatively undisturbed, cratonic strata surrounding the shield); each shows a distinctive pattern of sedimentation. Arches and basins influenced sedimentation on the craton throughout much of the Early Paleozoic (figure 22.1). To identify the presence and influence of arches and basins in rocks, geologists use (1) rock types; (2) isopach maps, to see how sediment thickens and thins around these features; and (3) unconformities, to determine the extent of the features.

Arches were topographic high areas that influenced surrounding sedimentation (figure 22.2A). Physically, arches

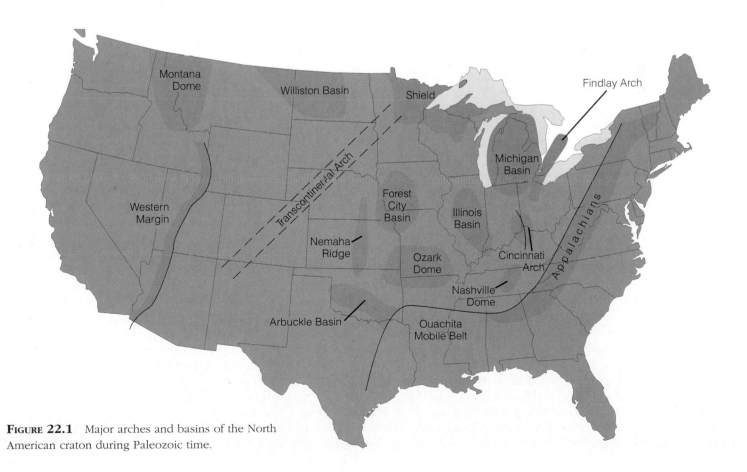

FIGURE 22.1 Major arches and basins of the North American craton during Paleozoic time.

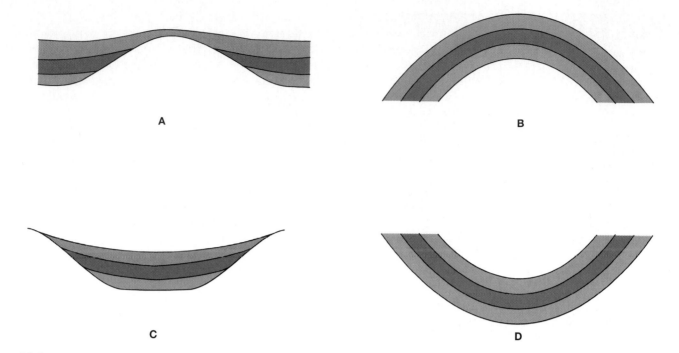

FIGURE 22.2 A comparison of (A) arch, (B) anticline, (C) basin, and (D) syncline. All are typical cratonic features of the Early Paleozoic.

were areas of extremely shallow water, or possibly even emergent islands, that supplied sediment in epeiric seas (see the next section). Usually, arches that formed on the craton were linear—long in one dimension, short in another. Because arches were topographic highs, very little sediment was deposited on the axis of the arch (the highest area), whereas during the same time interval, a greater amount of sediment was deposited on the flanks of the arch. Thus, strata thin as they approach and go over the arch. Frequently, the axis of the arch was eroded, and strata are not continuous over the arch, as is reflected by many unconformities in the rocks overlying the arch. Because of the shallow water depths near arches, sandstones and other coarse-grained detrital sediments were deposited near the arch; limestones and fine-grained sediments were deposited farther away.

In certain regions, arches that were more circular than linear formed. These features are called *domes*. Sediment types and characteristics of domes are very similar to those of arches.

Anticlines are structural features produced when rocks are folded into an upfold (figure 22.2B). The folding occurs long after sedimentation is completed. Superficially, anticlines resemble arches in cross section, but there are several important differences. As shown in figure 22.2B, the strata do not thin across the structure. Because the anticline forms after sedimentation is completed, the layers of sediment retain about the same thickness everywhere along the fold. Also, rock type is not influenced by the anticline. Near arches, sandstone is typically deposited, but anticlines occur after sedimentation is completed and thus do not influence rock type as arches do.

TABLE 22.1 *Comparison of Arches and Basins*

Characteristic	Arch	Basin
Shape	Linear	Circular
Lithologies	Sandstones	Fine-grained sediments
Thickness of strata	Thin near center	Thickest near center
Unconformities	Many	Few

Basins are topographic low areas that act as "pockets" or depressions to receive sediments (figure 22.2C). Usually, basins on the craton are circular to oval in plan view, in contrast to the typical linear plan of arches. The contrast in shape between arches and basins is due to the tectonic forces that form each feature. Within basins, the beds are thinnest near the edges of the basin but become thicker near the center. In addition, unconformities are less common than they are in arches, owing to the more continuous nature of sedimentation in the relatively deeper water.

Synclines, like anticlines, are structural features formed long after sedimentation is completed (figure 22.2D). They are downfolds in strata. As with anticlines, the thickness of a sediment layer remains approximately constant across the fold. Similar to anticlines and arches, structural basins—structural downwarps of strata formed after sedimentation is completed—are also common.

A summary of the principal characteristics of arches and basins is shown in table 22.1.

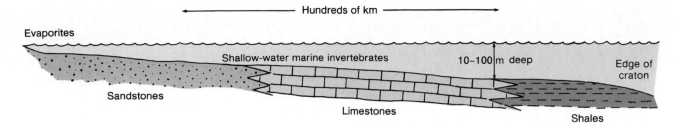

FIGURE 22.3 Sedimentary facies of an epeiric sea.

Epeiric Seas

Epeiric seas were relatively shallow, geographically widespread bodies of water that once covered large parts of the continent. *Epeiric* is from the Greek *epeiros,* meaning "continent." Thus, epeiric seas were "seas over continents." The main problem encountered by geologists studying epeiric-sea deposits is that no complete modern analog (a modern example that might serve as a model for ancient cases) of the shallow-water, tropical epeiric sea of the past exists on earth today. This is because continents stand much higher today than they did in the past. The best modern example of a tropical epeiric sea is the Bahamas Bank, off the west coast of the Bahamas. However, this area does not begin to compare in size with ancient epeiric seas. Hudson Bay closely approximates the dynamics of epeiric seas, but it is located in such high latitudes that tropical climate conditions are not found, and invertebrate life is scarce. Again, it is a close, but not an exact, analog.

By studying the sediments deposited in epeiric seas, geologists can be reasonably certain of the seas' principal characteristics (figure 22.3). Epeiric seas were geographically widespread. Because the continents in the past were so low, only a small rise in sea level flooded hundreds of square kilometers of a continent. An important, but unresolved, question is, How deep were epeiric seas? Most geologists agree that a water depth of less than 100 meters is a reasonable estimate. Some estimates, however, are as great as 300 meters. Throughout most of the Paleozoic, North America was near the equator. Therefore, the epeiric seas found on the North American continent were tropical, and the water temperature was warm. Since the water was relatively shallow, and the epeiric seas were so widespread, tides, waves, and currents must have been different from those occurring in the ocean today. How different? Again, the question is not resolved.

Only a few types of sediments were deposited in epeiric seas. In shallow-water and beach environments, sands were common. Typically these are now found as extremely pure quartz sandstones, called quartz arenites, that display cross-bedding and ripple marks. Farther out, limestones are found. In deeper-water deposits, shales predominate.

Cratonic Sequences

The periods are convenient subdivisions of the eras of geologic time. The boundaries between periods are usually placed at times of major extinctions—at biologic markers. A different way of looking at the rock record is to consider how sediment was—or was not—deposited on the craton and along the margins of the continent. During a *transgression* of seas (an advance of sea over the land surface caused by a rise in sea level), sediment was deposited on the craton, whereas during a *regression* of seas (a retreat of sea from land caused by a drop in sea level), some of this sediment was eroded, and unconformities formed. By mapping the large packages of sediment formed during transgressions, and using the unconformities produced during the regressions as boundaries, geologists can map **sequences** for cratonic sediments during the Phanerozoic. The term *sequence* is informally used as a lithostratigraphic designation (a subdivision of the rock record based on lithology) and applies to cratonic strata (table 22.2).

The cratonic sequences are shown in figure 22.4, with geographic extent plotted on the horizontal axis, time on the vertical axis. By looking at a selected horizontal line, geologists can determine the places where sedimentation was occurring on the margins and craton at a certain time. From oldest to youngest, the North American sequences are (1) Sauk, (2) Tippecanoe, (3) Kaskaskia, (4) Absaroka, (5) Zuni, and (6) Tejas. (All names were coined by the developer of the sequence concept, Lawrence Sloss, from the names of Indian tribes found in the Midwest.)

Sedimentation Patterns

Maps and cross sections of rocks of various ages reveal several patterns of sedimentation recurring throughout the Paleozoic, Mesozoic, and Cenozoic. These basic patterns provide a framework for thinking about regional sedimentation patterns in the past.

One sedimentation pattern is found on the craton, the stable interior part of the continent. Sediments deposited on the craton in epeiric seas are usually geographically extensive and very thin. This is because most cratonic sedimentation involved deposition in basins and around arches in epeiric seas. Typical lithologies include sandstones and limestones and, sometimes, shale.

Another sedimentation pattern is found in regions of mountain building, at convergent plate boundaries. Sediments deposited in these "mobile belts" are usually linear and extremely thick. Typical lithologies include immature sandstones (mostly graywackes, that is, poorly sorted sandstones that contain angular quartz grains and a large percentage of rock fragments) and shales.

TABLE 22.2 *Correlation Chart of the Cratonic Sequences and Subsequences to the Periods of the Paleozoic, Mesozoic, and Cenozoic.*

System	Series	Stage	End (Ma)	Subsequence (Ma)	Subsequence
Quaternary		Pleistocene	0		
Tertiary	Neogene	Pilocene	2		Tejas III
		Miocene	5.1		
	Paleogene	Oligocene	24.6	29	Tejas II
		Eocene	38	39	Tejas II
		Paleocene	54.9	60	Tejas I
Cretaceous	Upper	Maastrichitian	65		
		Campanian	73		
		Santonian	83		Zuni III
		Coniacian	87.5		
		Turonian	88.5		
		Cenomanian	91	96	
	Lower	Albian	97.5		
		Aptian	113		
		Barremian	119		Zuni II
		Hauterivian	125		
		Valanginian	131	134	
		Berriasian	138		
Jurassic	Upper	Portlandian	144		
		Kimmeridgian	150		
		Oxfordian	156		Zuni I
	Middle	Callovian	163		
		Bathonian	169		
		Bajocian	175		
	Lower	Aelenian	181	186	
		Toarcian	188		
		Pleinsbachian	194		
		Sinemurian	200		
		Hettangian	206		
Triassic	Upper	Rhaetian	213		Absaroka III
		Norian	219		
		Carnian	225		
	Middle	Ladinian	231		
		Anisian	238		
	Lower	Scythian	243	245	
Permian	Ochoa / Guadalupe	Tatarian	248		
		Kazanian	253		Absaroka II
	Leonard	Kungurian	258		
		Artinskian	263	268	
	Wolfcamp	Sakmarian	268		
	Virgin / Missouri	Stephanian	286		
Pennsylvanian	Des Moines	Westphalian (D C B A)	296		Absaroka I
	Atoka				
	Morrow	Numurian (C B A)			
Mississippian	Chester	Visean	333	330	
	Valmayer				Kaskaskia II
	Kinderhook	Tournaisian	352	362	
Devonian	Upper	Famennian	360		
		Frasnian	367		
	Middle	Givetian	374		Kaskaskia I
		Eifelian	380		
		Emsian	387		
	Lower	Siegenian	394	401	
		Gedinnian	401		
Silurian	Upper	Pridolian	408		
		Ludlovian	414		Tippecanoe II
	Middle	Wenlockian	421		
	Lower	Llandoverian	428		
Ordovician	Upper	Ashgillian	438	438	
		Caradocian	448		
	Middle	Llandeilan	458		Tippecanoe I
		Llanvimian	468		
	Lower	Arenigian	478		
		Tremadocian	488	488	
Cambrian	Upper	Trempealeauan	505		Sauk III
		Franconian		515	
		Dresbachian			
	Middle		523		Sauk II
	Lower		548	548	Sauk I
Precambrian	Upper Proterozoic	Ediacaran	590	600	

After L. L. Sloss, *Sedimentary Cover—North American Craton, The Geology of North America*, Volume D-2, 1988, Geological Society of America, Boulder, CO. Reprinted by permission.

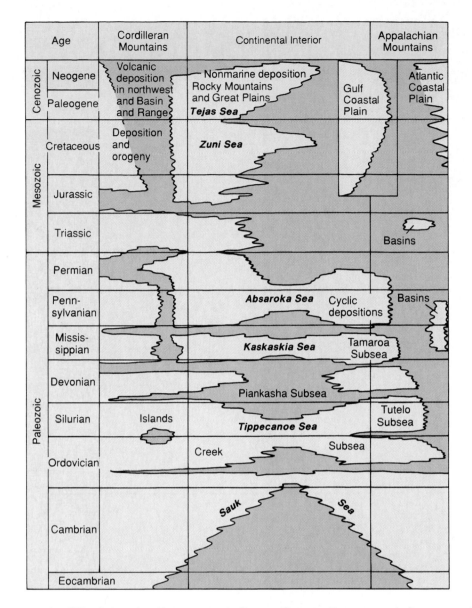

FIGURE 22.4 Cratonic sequences of North America. Brown areas indicate sediments. Green areas indicate no sediments of that age at that location.

From Historical Geology *by Leigh W. Mintz; adapted from* General Geology *by Robert J. Foster; Charles E. Merrill Pub. Co. Reprinted by permission of Leigh W. Mintz.*

CHARACTERISTICS OF THE EARLY PALEOZOIC

The Early Paleozoic consists of the Cambrian (505 to 570 million years ago), the Ordovician (438 to 505 million years ago), and the Silurian (408 to 438 million years ago) periods of the Paleozoic era.

The Lower Paleozoic rock record of North America is the result of a combination of three processes: (1) epeiric-sea sedimentation over and in (2) arches and basins during (3) transgressions and regressions. These controlling factors determined sediment type, distribution, and thickness. Most of the Lower Paleozoic rock record consists of sandstones,

limestones and dolomites, evaporites, and shales—typical epeiric-sea deposits. Most of the strata, particularly limestones, contain an abundant invertebrate fossil fauna.

G eologists look at rocks in a number of ways—for example, what kind of rock is it? When was the rock deposited? The distinction between rocks and time is fundamental. By convention, when referring to time (such as periods or eras), geologists use the terms *early, middle,* and *late;* when referring to rocks, geologists use the terms *lower, middle,* and *upper.* Thus, Lower Cambrian rocks were deposited during Early Cambrian time, and so on.

A

B

Cambrian and Early Ordovician

Figure 22.5 (A) Paleogeography of the United States during Early Paleozoic time. (B) Position of the continents during the Cambrian and Early Ordovician.

The world of the Early Paleozoic was quite different from our world today. The earth rotated faster, and consequently, the days were shorter. (Current estimates for Cambrian time are 21 hours per day, 420 days per year, based on growth rings of coral that lived at that time.) The moon was closer to the earth, making tidal forces stronger. Furthermore, there were no land plants or animals. The continents were barren.

In Cambrian time, North America was little more than a large landmass covered by epeiric seas and several small islands, all situated on the equator and rotated about 90 degrees clockwise from the continent's present orientation (figure 22.5). During the Phanerozoic, North America rotated 90 degrees counterclockwise and simultaneously moved north, by continental drift, to its present position.

From the Late Precambrian through the Cambrian, continents diverged. Between North America on one side and Europe and Africa on the other, the Proto-Atlantic Ocean was formed (sometimes referred to as the Iapetus Ocean). Then, in Ordovician time, a convergence started as North America began to close in on Europe in the north and Africa in the south, destroying the Proto-Atlantic Ocean between them. This movement continued into Late Paleozoic time (Devonian) when Africa and Europe finally collided with North America, completely destroying the Proto-Atlantic Ocean.

The eastern margin of North America was passive during the Cambrian. This situation changed in the Ordovician, when subduction of the sea floor began at the margin—the result of convergence. Subduction activity continued into the Silurian and set the stage for the continent–continent collision of North America with Europe and Africa during the Devonian.

The western margin of North America was a passive continental margin throughout the Early Paleozoic. No convergence or mountain building was experienced until Devonian time.

THE NORTH AMERICAN CRATON

On the North American craton, most Lower Paleozoic sediments were deposited in shallow-water epeiric seas that covered much of the continent. Consequently, most of the sedimentary rocks of the platform of this age are relatively thin and widespread.

A general relationship exists between the types of sediment deposited and the transgressions and regressions of the Early Paleozoic. In the Early Cambrian, a major transgression began—the start of the Sauk Sea. As the continent began to flood, clastic deposition (primarily of sandstones) dominated on and near the shores of the advancing sea. By the end of the Cambrian, most of this clastic material had been deposited. As a result, carbonate deposition became important. A brief regression in the Early to Middle Ordovician caused erosion of some sediments, and a few sandstones became exposed and eroded and were recycled. In the Middle Ordovician (through the Late Silurian), another transgression began—the start of the Tippecanoe Sea. Once this epeiric sea became larger, carbonate deposition again became very widespread. Finally, at the end of the Ordovician, shale was deposited over a large part of the craton, owing to mountain building in the east (figure 22.6).

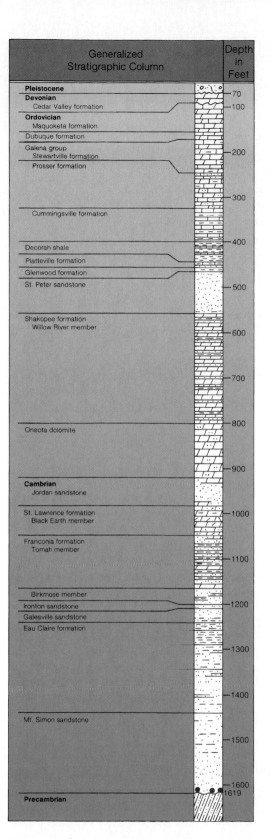

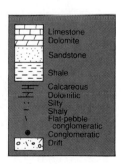

FIGURE 22.6 Geologic column of Cambrian-Ordovician strata of the Upper Mississippi Valley.

From G. S. Austin, Geology of Minnesota: A Centennial Volume, *p. 472. Copyright © 1972 Minnesota Geological Survey, St. Paul, MN. Reprinted by permission.*

FIGURE 22.7 Cross-bedded Cambrian sandstones (quartz arenites) of the Wisconsin Dells. Note the cross-bedding.
Photograph courtesy of H. H. Bennett Studio Foundation, Inc.

A n inverse relationship exists between the amount of sandstone deposited and the amount of carbonate deposited; the more sandstone, the less carbonate, and vice versa. This is because the sources of the two types of sediment are entirely different. The source of sediment for sandstones is commonly the exposed rocks on land that are being weathered. Rivers then carry the sediment to the marine environment. The source of carbonate sediment is organic—algae and invertebrates in the marine environment. These organisms cannot survive in turbid and muddy water, precisely the type of water produced when a lot of clastic material is introduced. Therefore, when sand is found in the water, the animals that produce carbonate sediment are not.

The Transcontinental Arch

One of the major influences on sediment deposition during the Early Paleozoic was the Transcontinental Arch, a topographic high. The arch extended roughly from Lake Superior to Arizona. It formed from minor folding of the platform. (Mild structural folding of platform strata is caused by epeirogenic movements.) In Cambrian time, the arch directly influenced sedimentation, but it also underwent a significant amount of folding later in its history.

Sedimentary Rocks of the Cambrian and Ordovician Ages

Quartz arenites are perhaps the most common cratonic sediment of Cambrian age (figure 22.7). Most of these sandstones are extremely mature, consisting of sand-sized, well-rounded quartz grains and very little else. (Some of the deeper-water sandstones have a greenish color because they contain a green mineral called glauconite.) The sandstones typically show cross-bedding and ripple marks, indicating that they were deposited in the beach environment and shallow waters of an epeiric sea. These quartz sandstones are probably the most mature sandstones in the world (that is, they consist of only quartz sand) for a number of reasons. First, the sources of these sandstones were the granitic rocks of the shield, which had been exposed to almost 500 million years of weathering activity. This long erosional interval contributed to the maturity of the source sand. Second, since no land plants existed during this time, the intensity and rate of erosion were greater than those of today. Third, the sandstones were deposited in an extremely high-energy environment—a beach. The constant pounding of waves removed most of the easily weathered minerals, leaving nothing but quartz. Because this combination of conditions was not repeated later in the Phanerozoic, the sandstones are characteristic of this time and only a few other times.

FIGURE 22.8 Typical Ordovician-age limestones of the craton. Quarry, Durand, Illinois.

FIGURE 22.9 Shelly limestones. Abundant brachipod shells are enclosed in the limestone matrix.
Photograph by E. B. Hardin, USGS Photo Library, Denver, CO.

The brief regression from Early to Middle Ordovician time meant that sediments previously deposited on the North American craton experienced erosion. Some of the Early Ordovician carbonates were eroded, exposing the underlying Cambrian sandstones (the sands had probably lithified by this time). These already-mature sandstones were again eroded and redeposited along the beaches of the transgressing Tippecanoe Sea. The "supermature" sandstone thus produced is called the *St. Peter sandstone* and is one of the more famous sandstone formations. The St. Peter sandstone can be traced from Kentucky (where it is younger) across the midcontinent—Ohio to Illinois—to Nebraska and north into Minnesota.

> T he St. Peter sandstone is an example of recycling of sediments. The sources of the sediment of the Cambrian sandstones were the Precambrian shield rocks. The sources of the sediment of the Ordovician-age St. Peter sandstone were the Cambrian sandstones. Recycling of sediments occurred again and again throughout the Phanerozoic.

The transgressing Tippecanoe Sea became one of the largest epeiric seas. Conditions were uniform across its extent. Organisms thrived in the epeiric seas. Algae (principally) and other invertebrates produced abundant carbonate sediment (figure 22.8). As this sediment was deposited, shells of the invertebrate animals were incorporated into it. These geographically widespread, relatively thin limestones containing abundant invertebrate fossils are called *shelly limestone facies* (figure 22.9). The shelly limestones are a common cratonic sediment of this time.

Although not an abundant epeiric-sea deposit, shales, interstratified with carbonates, occur on parts of the craton during this time. The source of the shale was sediment eroded from the Taconic Mountains of eastern New York. There was simply too much fine-grained material to be deposited along the margin, and so the excess was deposited in the epeiric sea along with the carbonate.

Other cratonic sediments of the Early Paleozoic that formed a minor amount of the rock record included conglomerates, oolites, and flat-pebble conglomerates (figure 22.10). Conglomerates are found in Lower Cambrian rocks near the Baraboo region of Wisconsin, which at that time was adjacent to the shield. The large cobbles and pebbles are probably storm-deposited clasts (fragments of rocks). Only in the high-velocity, storm-produced waves could such large clasts be transported and deposited. Oolites formed in shallow-water or intertidal conditions as the result of calcium carbonate precipitating from solution. Each flat-pebble conglomerate records a storm episode. When a storm swept over part of a shallow epeiric sea, some of the lithified sea floor was

A

B

C

FIGURE 22.10 Other Early Paleozoic cratonic sediments and sedimentary structures. (A) Conglomerates. (B) Flat-pebble conglomerates. (C) Recent mud cracks forming in a pool of water that has dried up and subsequently filled again.

broken up into chunks. These clasts, typically tabular in shape, were then deposited in a single layer. Thus flat-pebble conglomerates in these rocks record a single storm episode that occurred over 400 million years ago! Other typical sedimentary structures in these rocks include cross-bedding, ripple marks, and mud cracks (figure 22.10).

Sedimentary Rocks of the Silurian Age

In Silurian time, sedimentation patterns on the craton became more complex because of the mountain building occurring in the east. A large basin formed in the Michigan area (figure 22.11). The **Michigan Basin** structure developed as the bottom dropped out from under Michigan, partly literally and partly figuratively. The crust was depressed and isostatically lowered, perhaps as a large amount of very dense mantle rock was emplaced in the lower crust below Michigan.

Several factors acted simultaneously to produce different sedimentation types and patterns from those of Cambrian and Ordovician time. Widespread algae and invertebrates produced abundant carbonate sediment in the Michigan Basin area. The production of sediment just kept pace with the sinking of the crust. Therefore, although the region was sinking, it was filling with sediment at approximately the same rate, so all of the sediments were shallow-water deposits. Surrounding the basin were massive carbonate reefs. These barrier reefs prevented the continuous circulation of water between the surrounding epeiric sea and the waters in the Michigan Basin. Together, the rapid evaporation of the waters, due to the tropical climate conditions, and the restricted replenishment of water to the basin, due to the barrier reefs, led to the deposition of widespread evaporites, such as gypsum, anhydrite, and halite. Occasionally—for example, during times of minor transgressions—the waters were deeper, and reefs thrived. During minor regressions or times of significant restriction of water circulation, evaporites were precipitated, and the reefs died. (Corals could not tolerate the higher salinities.) This alternating cycle of carbonate and evaporites is characteristic of Silurian-age rocks of the Michigan area of the craton (figure 22.12).

The geologic setting in the Michigan Basin also led to the production of widespread dolomites. Dolomites are calcium-magnesium carbonates—$CaMg(CO_3)_2$—and differ from limestones ($CaCO_3$) chemically only by the addition of magnesium (Mg) to the rock. Algae, coral (which made the reefs), and invertebrates produced abundant carbonate sediment. During times of evaporite conditions, much of the water in the Michigan Basin was evaporated, leaving behind a tremendous concentration of ions, including magnesium, in the water. (When seawater evaporates, only the water is evaporated, not the salts.) The remaining waters, enriched in numerous elements, formed *brines*. These brines percolated down through the underlying carbonates, often changing the limestone to dolomite by adding magnesium. Therefore, dolomites are intimately associated with evaporites in the Michigan Basin region.

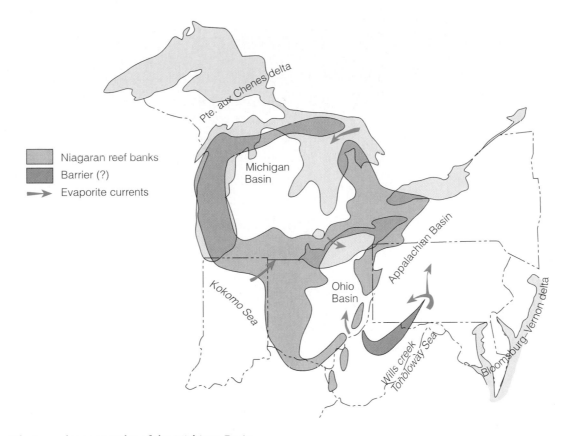

FIGURE 22.11 Silurian paleogeography of the Michigan Basin.

From Alling and Briggs, AAPG Bulletin, vol. 45, fig. 12, p. 544, 1961. Copyright © 1961 American Association of Petroleum Geologists, Tulsa, OK. Reprinted by permission.

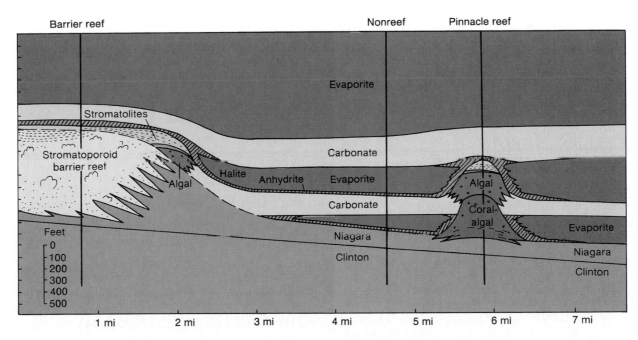

FIGURE 22.12 Generalized geologic column of Silurian evaporites and carbonates of the Michigan Basin.

From K. J. Mesolella et al., AAPG Bulletin, vol. 58, fig. 6, p. 40, 1974. Copyright © 1974 American Association of Petroleum Geologists, Tulsa, OK. Reprinted by permission.

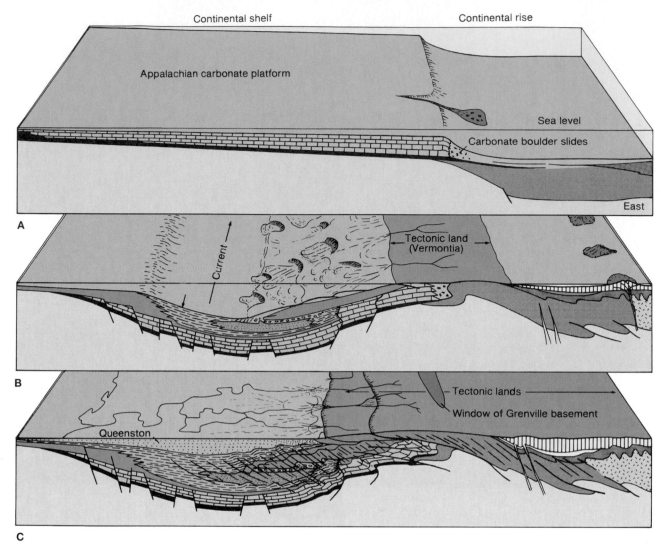

FIGURE 22.13 Tectonics of the Appalachian region during the Early Paleozoic. (A) Late Proterozoic through Early Ordovician. (B) Middle Ordovician. (C) Late Ordovician.

After John M. Bird, Geological Society of American Bulletin, *vol. 81, number 4, fig. 7, p. 1043, 1970, Used by permission of the author.*

THE EASTERN MARGIN

The Appalachians

The history of the eastern margin of North America during the Paleozoic is the history of the Appalachian Mountains. The Appalachians are the result of three orogenic episodes: (1) the Taconic orogeny (Ordovician), (2) the Acadian orogeny (Devonian), and (3) the Appalachian orogeny (Pennsylvanian). The Appalachians began to form in the Ordovician during the Taconic orogeny, a result of subduction of the proto-Atlantic plate due to convergent motion. Following this orogenic episode, in the Devonian, the mountains experienced their most intense period of deformation during the Acadian orogeny, when Africa finally collided with North America and destroyed the intervening ocean. At that time, the Appalachians were as high and as majestic as the Rockies are today. Finally, starting in the Pennsylvanian and continu-

ing into the Permian, the Appalachians experienced another active period during the Appalachian orogeny, when thrust faulting dominated.

The Cambrian

During the Cambrian, the eastern margin was a passive continental margin (figure 22.13A). Two general environments accumulated sediments: a continental-shelf region with shallow-water carbonate sediments and a deeper-water environment with deposits of shales.

The Ordovician and Silurian (the Taconic Orogeny)

In the Ordovician, the tectonic situation along the eastern margin changed dramatically as Europe (in the north) and Africa (in the south) started to converge on North America.

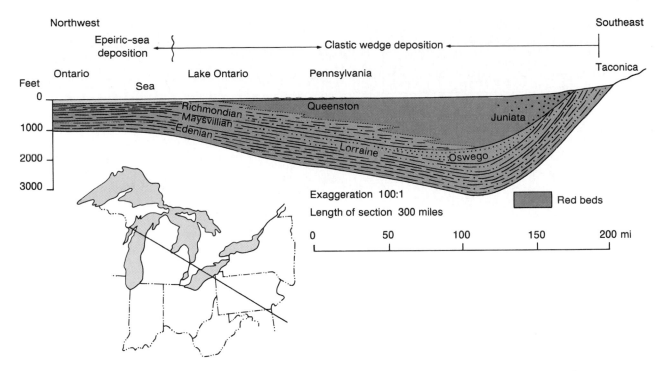

FIGURE 22.14 Cross section through the clastic wedge from the Appalachians to the North American craton. Note how a greater quantity of sediment was deposited in the appalachian region (southeast) during a given interval of time compared with the quantity of sediment deposited in the cratonic epeiric sea (northwest).

From Kay and Colbert, Stratigraphy and Life History. *Copyright © 1965 John Wiley & Sons, Inc., New York, NY. Reprinted by permission of John Wiley & Sons, Inc.*

(The direction of motion is relative—the continents were actually all moving toward each other.) Subduction began along the margin, and the Proto-Atlantic Ocean started to be destroyed. Because of the convergent plate motion, a complex series of tectonic events was initiated along the margin and on the extreme eastern edge of the craton. These events marked the start of the **Taconic orogeny.**

At first, subduction produced an island arc along the eastern margin. Initially, this arc was a series of volcanic islands; later it became a large, continuous, uplifted area of land. As the arc formed, the crust behind it (between the arc and the craton) was depressed, forming a deep-water basin (figure 22.13B). This basin filled primarily with immature sandstones (graywackes) and volcanics from the island arc, although the craton also contributed a minor amount of sediment. Owing to the amount of sediment shed off the arc and the uplifted land, Ordovician formations in the Appalachians are much thicker compared with the thin platform-cratonic deposits of the same age (figure 22.14).

In the Middle and Late Ordovician, thick, black shales (such as the *Martinsburg shale*) were deposited in the Appalachian region. One characteristic type of fossil found in the shales is graptolites (refer to figure 19.2E). In contrast to the shelly limestone facies of the craton, these sediments are called the *graptolite shale facies.*

When paleontologists first found graptolites—an extinct group of colonial, planktonic organisms—in the shales, they were puzzled about how graptolites lived in or on the muddy bottoms of the deep-water basins. Evidence seemed to indicate that essentially no organism could live in this hostile environment because of the anoxic conditions (absence of oxygen). However, additional fossil-collecting produced well-preserved samples of graptolites that showed that these organisms were connected to bladders or floats. Thus, graptolites were planktonic—they floated; they did not live in the hostile environment of the sea bottom. When the graptolite colony died, it sank to the bottom. The hostile environment prevented any scavengers and bacteria from destroying the graptolites, so they were preserved in great numbers.

In the Late Ordovician and into the Silurian, sediment types again changed as the basin behind the arc filled with sediments. The series of volcanic islands developed into a rather large, uplifted landmass that supplied great quantities of sediment to the basin and filled it (figure 22.13C). Deep-water sedimentation gave way to shallow-water marine and terrestrial sedimentation with representative alluvial, fluvial

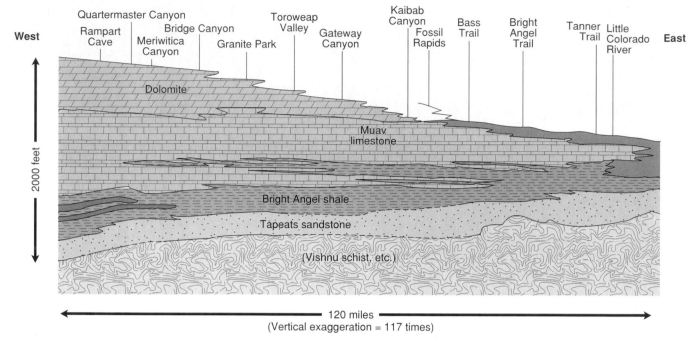

West | Quartermaster Canyon | Toroweap Valley | Kaibab Canyon | Tanner Trail | East
Rampart Cave | Bridge Canyon | Gateway Canyon | Fossil Rapids | Bass Trail | Bright Angel Trail | Little Colorado River
Meriwitica Canyon | Granite Park

Dolomite

Muav limestone

Bright Angel shale

Tapeats sandstone

(Vishnu schist, etc.)

2000 feet

120 miles
(Vertical exaggeration = 117 times)

Figure 22.15 Correlation chart of Cambrian rocks in the Grand Canyon and vicinity.
Source: J. S. Shelton, **Geology Illustrated,** *Copyright 1966 W. H Freeman and Company.*

(river), and deltaic sediments. Sandstones and shales are the resultant typical rock types. This association indicates that the basin behind the island arc was now filled.

Together, these sediments deposited along the eastern margin at this time represent a tremendous volume of material shed off the Taconic highlands. The large wedge of sediment formed has been termed the **Queenston clastic wedge,** or the Queenston delta. The wedge is thickest near the source of the sediments—the mountains—and thinnest near the edge bordering the epeiric sea. (A clastic wedge is a typical feature created whenever mountains are formed.)

In the Silurian, the situation remained essentially unchanged. Volcanics were still produced, but volcanic centers shifted northward as that part of the margin began to close. One very interesting Silurian formation is the *Clinton Iron formation,* found in the southern Appalachians, which consists of sand-sized oolites in which the calcium carbonate has been replaced by iron. The iron came from the intense weathering of the igneous rocks formed by the Taconic orogeny.

O ne interesting sidelight concerning the Clinton Iron formation shows how certain rock types have influenced history: During the last stages of the Civil War, President Abraham Lincoln ordered a naval blockade from the Atlantic into the Gulf of Mexico, thereby surrounding the South. Supplies to the South were cut off. The blockade was ineffective in the case of iron, however, because it was mined from the Clinton Iron formation. This source of iron for the South may have briefly prolonged the Civil War.

The Western Margin

The western margin of the continent was primarily a passive continental margin during the Early Paleozoic. This margin trailed behind the continent as North America and Europe and Africa closed in on one another.

The Cambrian rocks of the Grand Canyon show a classic transgressive-facies pattern (figures 22.15 and 22.16). In the Early Cambrian, the *Tapeats sandstone* was deposited on the shores and in the shallow waters of an advancing epeiric sea. As the sea transgressed, water depths deepened and the *Bright Angel shale* of the Middle Cambrian age was deposited on top of the sandstone. Finally, in the Late Middle Cambrian, the *Muav limestone* was deposited on top of the shale.

Many of these shallow-water, passive-margin deposits of the western margin have been exposed by much-later tectonic (orgenic) events. For example, many of the mountains in western Canada are formed from Early Paleozoic deposits (figure 22.17).

There is evidence of some very minor subduction activity along the western margin in Ordovician time. A well-known sandstone was deposited in Middle Ordovician time during the transgression marking the start of the Tippecanoe Sea. The *Harding sandstone,* which crops out along the Rocky Mountain Front Ranges, is the basal sandstone of the transgressive epeiric sea.

In Silurian time, the western margin was again a passive trailing margin. Shallow-water carbonates and deeper-water graywackes and shales were deposited. The first mountain-building processes to affect the western margin began in Devonian time with the Antler orogeny.

Figure 22.16 Early Paleozoic strata exposed in the Grand Canyon.
Photograph courtesy of Malcolm P. Weiss.

Life in the Early Paleozoic

Life in the Early Paleozoic was confined primarily to the shallow-water marine environment of the epeiric seas. Several major evolutionary events occurred: Animals with skeletons appeared at the start of the Cambrian and are used to define the base of the Cambrian. In Late Cambrian rocks, the first fossil traces of vertebrate skeletons (fish) are found. During the Silurian, plants moved out onto the barren land surface.

The Cambrian

Two major evolutionary events in the history of life occurred at the very start of the Cambrian: (1) a rapid increase in the number and types of phyla and (2) the advent of skeletons.

For several billion years, life on earth consisted primarily of algae and bacteria. Then, in a very short span of time, almost all of the major phyla (about seven, plus a few not represented today) evolved at the start of the Cambrian. Why? Some paleontologists hypothesize that the rapid increase is merely a reflection of skeletonization (the appearance of skeletons)—the organisms were around before this time, but they were not preserved, because they were soft-bodied. A careful examination of the fossil record, however, shows that not only do skeletonized forms appear at the start of the Cambrian, trace fossils (which, in this case, represent activities of soft-bodied animals) also undergo rapid

Figure 22.17 Early Paleozoic strata of the Temple Mountain group. Banff National Park, Canada.
Photograph by H. E. Malde, USGS Photo Library, Denver, CO.

FIGURE 22.18 Diorama of Cambrian epeiric-sea life. Trilobites and worms are crawling along the bottom in the foreground. Tall, branching sponges can be seen in the right foreground and the background.
Photograph courtesy of Field Museum of Natural History, Chicago, #GEO80872.

diversification. Apparently, soft-bodied animals were also undergoing rapid evolution. A more commonly held idea is the "petri-dish model"—an idea analogous to what happens when bacteria are introduced into a petri dish containing nutrient-rich material. At first, the bacteria are few in number. Then, because of the "unlimited" food available, they reproduce rapidly, and the number of bacteria increases over a very short period. Finally, as food becomes scarce and pollutants increase, the rate of growth slows and finally stops. At the start of the Cambrian, the entire epeiric-sea/ocean environment was opened, and any number of new ecologic niches were available. Organisms underwent a period of extremely rapid evolution to fill these niches. After this initial explosive-radiation event, the number of phyla evolving was reduced to zero. (Only one major phylum, the bryozoa, evolved after this time, in the Middle Ordovician.)

Why the advent of skeletons? Perhaps the most reasonable explanation is related to the fact that organisms require a large supply of oxygen to make skeletons, and that the start of the Cambrian may mark the time at which enough free oxygen (O_2 — oxygen not combined with anything else) had accumulated in the environment for organisms to finally produce skeletons. Whatever the reason, a wide variety of animals now had skeletons, which meant that the odds for their preservation increased dramatically.

The first animals to increase rapidly and become dominant were the *trilobites* (figure 22.18). Trilobites increased more in number and variety than did any other marine invertebrate during this time. About 50% of all Cambrian fossils are trilobites! They are the perfect index fossil for this period. Almost all trilobites were vagrant benthos, crawling along the sea bottom in search of food. They sucked sediment into

Figure 22.19 Outcrops of the Burgess shale.
Photograph courtesy of the Smithsonian Institution.

their mouths (they had no teeth!), digested the organic material, and expelled the waste.

Another important index fossil for Cambrian rocks is the *archaeocyathids*. These double-cupped animals either belong in their own phylum or are related to sponges. Apparently, archaeocyathids were not suited for the earth, because they evolved in the Early Cambrian and became extinct at the end of the Middle Cambrian. Other groups, however, found the marine environment suitable. *Brachiopods* appeared and came to dominate the epeiric-sea bottoms from Ordovician through Devonian time. Most brachiopods of Cambrian time were inarticulates, lacking a definite hinge line between valves. One common inarticulate is represented by the genus *Lingula*. *Lingula* brachiopods found today are unchanged in appearance from their Cambrian predecessors. Thus, *Lingula* is possibly the longest-living genus on earth! During the Ordovician, however, the articulates, which possessed a definite hinge line between valves, became dominant.

Of all the other groups to evolve, perhaps the most important were the molluscs. Molluscs are a tremendously diverse group of invertebrates. By the Middle Cambrian, three of the major types of molluscs had appeared—the gastropods (represented today by snails); the pelecypods (also called bivalvia, represented today by clams and oysters); and the cephalopods (represented today by nautiloids and squid).

One of the most important fossil localities ever discovered, the **Burgess shale** of British Columbia, contains rocks from Middle Cambrian time (figure 22.19). The Burgess shale was discovered in 1910 by Charles Doolittle Walcott (1850–1927) when, perhaps apocryphally, his pack mule slipped on a rock, uncovering some rather strange fossils. Walcott followed the talus up the slope to where the rocks were exposed and collected a large group of fossils. Since Walcott's time, numerous fossil-collecting expeditions have uncovered a host of additional fossils at the site. The Burgess shale fauna includes diverse soft-bodied animals that have not been preserved anywhere else in the world (mostly worms and arthropods, but also a host of unusual and strange animals). During the Middle Cambrian, in British Columbia, a large reef developed adjacent to a basin filled with rather deep water. Rock and sediment, and the animals living in and around the rocks adjacent to the reef, occasionally broke loose and slid into the deep water where, owing to hostile environmental conditions, no scavengers or predators lived. The organisms were buried and killed instantly. With no scavengers present, the soft-bodied animals were preserved. The Burgess shale fauna are like a snapshot of life in Cambrian time, giving a much more accurate representation of the life of the past than those found anywhere else. Recent discoveries in Greenland have also yielded many fossils similar to those found in the Burgess shale.

The first vertebrate fossils are Late Cambrian in age and consist of the exterior plates of the first fish, the agnathids. Agnathids were extremely primitive. They lacked true jaws and were covered with bony plates. Most were probably deposit-feeders, straining mud and sediment to obtain organic material. Agnathids were probably the only vertebrates from the Late Cambrian through the Silurian.

T he story of fish evolution is one of streamlining. The first fish, the agnathids, were little more than armored tanks moving slowly along the bottom. In Devonian time, the first jawed fish appeared, and the bony plates were reduced. This trend continued through Devonian time, as the plates were reduced to the scales familiar in modern fish.

The Ordovician

During the Ordovician, several events occurred in the history of life. The Middle Ordovician regression caused organisms to be restricted to the fringes of the craton. Evolutionary pressures were intense, and several new groups evolved. The transgression that followed produced the Tippecanoe Sea, one of the most extensive epeiric seas ever. Climate was uniform, and organisms thrived.

The most common animals living on the carbonate bottoms of the epeiric seas were brachiopods and *bryozoans* (figure 22.20). Together, they formed a community that extended into Devonian time. *Tabulate* and *rugose* corals first appeared. Tabulates were colonial corals, superficially similar to modern types. Rugose corals could be either solitary or colonial. Solitary forms often were shaped like horns, and the group is referred to as the horn-corals. One common pelagic

FIGURE 22.20 Diorama of Ordovician epeiric-sea life. A large, straight-shelled cephalopod is shown in the right foreground. Large, dome-shaped corals and branching bryozoans are in the background. Swimming forms are cephalopods.
Photograph courtesy of Field Museum of Natural History, Chicago, #GEO80820.

(floating) animal was the *graptolite*. Graptolite colonies floated all over the ocean surface. Thus, graptolites are an important index fossil for rocks of Ordovician age.

The Silurian

Two major events in the history of life occurred in the Silurian: the first coral reefs appeared, and the first land plants evolved.

Much of the life on the bottom of the epeiric seas remained relatively unchanged from Ordovician time (figure 22.21). Similar organisms were present; only the abun-

dance and relative importance of each group changed. Brachiopods and bryozoans were still abundant, as were other invertebrates.

One major change occurred when corals, particularly the colonial tabulates, began to form coral reefs. Although *reefs* (usually defined as wave-resistant structures) were common before this time, they had been made previously by other invertebrates, such as algae and sponges, and had not been very large structures. When tabulate corals started to form reefs, they changed the epeiric-sea environment entirely by opening up a new ecologic niche not seen be-

FIGURE 22.21 Diorama of Silurian epeiric-sea life. Several types of corals are shown in the left foreground. Trilobites, crinoids, and brachiopods are on the bottom in the right foreground. Stalked crinoids are shown on the left.
Photograph courtesy of Field Museum of Natural History, Chicago, #GEO80875.

fore—the back-reef area. The large, linear, *barrier reef* grew from a reef core consisting of massive limestone. Coral grew more quickly on the seaward side of the reef, where waves brought nutrients and carried away wastes faster than they did elsewhere. But during storms, waves also expended most of their energy and did most of their damage on the seaward side of the reef core. Thus, in front of the reef, large chunks of broken coral accumulated, referred to as reef talus. The importance of the barrier reefs is that they acted to block waves and currents coming in from seaward areas. (They are called barrier reefs for this reason.) Behind the barrier reefs, quiet, sheltered marine areas opened up.

Animals were quick to exploit this new habitat, and many organisms grew in the quiet marine waters of the back-reef areas. Also common in this environment was carbonate sand, produced from erosion of reefs.

The first unequivocal plant fossils occur in rocks of Silurian age. Probable plant fossils (fossil spores) occur in rocks of Ordovician age, but these fossils are still under debate. Plants had to overcome three major problems in making the transition to land: (1) dessication, (2) development of a fluid-transport system, and (3) reproduction. Plants developed tough outer-epidermal cells to prevent drying out in the nonmarine environment. Internal cells became elongated and

FIGURE 22.22 Psilopsids—the first land plants.

toughened to produce an internal system of fluid transport and also to provide strength for support. Spores were developed for reproduction purposes. The earliest plants are called *psilopsids* and look like no more than roots with a few stems (figure 22.22).

The importance of land-plant evolution cannot be overstated. For one thing, as soon as lowland forests developed (in the Devonian), the rates of weathering on land changed. In addition, chemical styles of weathering changed as plants took nutrients from rocks and produced soils. With photosynthesis occurring in land plants, the levels of atmospheric oxygen increased. Some believe, because the appearance of amphibians in the Devonian postdates the appearance of land plants by about forty million years, that land plants may have been influential in providing enough oxygen to produce an atmospheric ozone layer that inhibited the penetration of ultraviolet rays. The early amphibians were then protected from the ultraviolet rays, as a consequence of land-plant evolution.

BOX 22.1
Regional Geology
THE UPPER MISSISSIPPI VALLEY

FIGURE 1 A Middle Ordovician hardground from Minnesota.

The Upper Mississippi Valley and the surrounding area is a scenic region of the Midwest. Almost all the rocks that crop out at the surface, and also those that underlie the soils and glacial deposits, are Cambrian- and Ordovician-age sandstones and carbonates (limestones and dolomites). The Mississippi River has eroded through most of the Ordovician strata, thereby exposing the Cambrian sandstones—for example, along the cliffs near La Crosse, Wisconsin, and farther north. To the west (Iowa and Minnesota) and east (Wisconsin and Illinois) of the Mississippi River valley, coming out of the river valley, Ordovician carbonates are common.

The strata in the Upper Mississippi Valley are typical epeiric-sea sediments (sandstones, limestones, and dolomites),

and they display the typical tabular shape of sediment deposited in epeiric seas—long in two dimensions, short in the third.

During the Early Paleozoic, this region was inundated by two major epeiric seas. In the first, the Sauk Sea, sandstones were deposited during the Cambrian. The scenic sandstone cliffs of the Wisconsin Dells are composed of these sedimentary rocks. Following a subsequent regression and transgression (the Tippecanoe Sea), carbonates became common—the shelly limestones. Many interesting sedimentological features are found in these rocks, such as hardgrounds—synsedimentary, lithified sea floors (figure 1). Hardgrounds developed during an interruption in sedimentation, during which the sea floor became lithi-

fied and mineralized with iron sulfides. On many hardgrounds, a complex community of organisms lived.

Other sedimentary rocks were deposited on top of the Ordovician carbonates during the Middle and Late Paleozoic. However, during the Mesozoic, when the craton was exposed, most of these later rocks were eroded. Finally, in the Pleistocene, repeated glaciation deposited tills and other associated sediments.

The strata of the Upper Mississippi Valley have many economic uses: the sand is used in glassmaking, and the limestones and dolomites are used in building and road construction. These rocks also provide evidence about epeiric processes. On the craton there are few better places to study the deposits left by epeiric seas.

Distributions, types, and thicknesses of Lower Paleozoic sedimentary rocks are the results of epeiric-sea sedimentation over arches and in basins during marine transgressions and regressions. Cambrian and Ordovician sedimentary rocks consist primarily of quartz arenites and shelly limestones. In Silurian time, evaporites were deposited in the Michigan Basin area.

The Appalachians began to form in the Ordovician with the Taconic orogeny. This orogeny was the result of subduction of the proto-Atlantic oceanic plate as Africa and North America converged.

The western margin of North America was essentially a passive margin throughout most of the Early Paleozoic.

Several important advances in the history of life took place in the Early Paleozoic. The start of the Cambrian period was marked by the appearance of skeletonized organisms and a dramatic increase in the number of phyla. In Ordovician time, brachiopod/bryozoan bottom communities were common in epeiric seas. The first vertebrates are of Late Cambrian age. In Silurian time, the carbonate bottom communities remained relatively unchanged. On land, plants first appeared.

A summary of Early Paleozoic events is shown in figure 22.23.

Period	Western margin	Craton	Eastern margin	Life
Silurian	Passive margin —Shallow-water carbonates —Deep-water graywackes	Late Silurian —Regression Michigan Basin —Barrier reefs —Limestones and dolomites —Evaporites Early-Late Silurian —Transgression (Tippecanoe Sea)	Continued subduction —Volcanics —Ironstones in the southern Appalachian region	Brachiopods/bryozoan epeiric-sea communities First coral reefs —Back-reef environment opened up First land plants: psilopsids Index fossils: brachiopods
Ordovician	Subduction (minor) —Volcanics —Lavas Harding sandstone	Middle-Upper Ordovician —Transgression (Tippecanoe Sea) —Shelly limestones —St. Peter sandstone Early-Middle Ordovician —Regression	Taconic orogeny —Subduction —Queenston clastic wedge —Graptolite shales —Martinsburg shale —Trenton limestone	Articulate brachiopods and bryozoan epeiric-sea communities Tabulate and rugose corals Index fossils: graptolites First land plants?
Cambrian	Passive margin Transgressive sequence in the Grand Canyon —Muav limestone —Bright Angel shale —Tapeats sandstone	Transcontinental Arch Early-Late Cambrian —Transgression (Sauk Sea) —Quartz arenites	Passive margin —Continental shelf —carbonates —Deep-water shales —Potsdam sandstone	Major events —Development of skeletons —Increase in the number of phyla Trilobites Index fossils: trilobites, archaeocyathids First vertebrates: agnathids Burgess Shale Fauna

FIGURE 22.23 Early Paleozoic: summary diagram.

TERMS TO REMEMBER

anticline 396
arch 395
basin 396
Burgess shale 411

epeiric seas 397
Michigan Basin 404
quartz arenite 402
Queenston clastic wedge 408

sequence 397
syncline 396
Taconic orogeny 407

QUESTIONS FOR REVIEW

1. How are arches different from anticlines? How are basins different from synclines?
2. What are some of the characteristics of epeiric seas? What is the biggest problem facing geologists who study epeiric-sea processes and sedimentation?
3. What are cratonic sequences? How are they defined?
4. Where was North America situated in Early Paleozoic time?
5. How did sedimentation on the craton differ in Silurian time from sedimentation in Cambrian and Ordovician time? What caused this difference?
6. What tectonic events characterize the Taconic orogeny?
7. What was happening on the western margin of North America during Early Paleozoic time?
8. What two major events in the history of life occurred at the start of the Cambrian?
9. What types of animals were common in Early Paleozoic epeiric seas?
10. What problems did plants have to overcome in making the transition to land?

FOR FURTHER THOUGHT

1. The evolution and expansion of land plants in Silurian and Devonian time and the development of the back-reef area behind epeiric-sea barrier reefs in Silurian time are examples of "one-time" events in the history of the earth and life. Both events changed the earth in fundamental ways, and both involve the evolution and adaptation of organisms. Would the extinction of a species affect the earth in a similar manner? Can you think of any examples?

2. What types of geologic and biologic evidence do geologists look for in epeiric-sea deposits to help deduce the dynamics and life habits of organisms in ancient epeiric seas?

SUGGESTIONS FOR FURTHER READING

Bott, M. H. P. 1976. Mechanisms of basin subsidence—an introductory review. *Tectonophysics* 36:1–4.

Gould, S. J. 1990. *Wonderful life: The Burgess shale and the nature of history*. New York: W. W. Norton.

Irwin, M. L. 1965. General theory of epeiric clear water sedimentation. *American Association of Petroleum Geologists Bulletin* 49:445–59.

Klein, G. deV. 1982. Probable sequential arrangement of depositional systems on cratons. *Geology* 10:17–22.

Palmer, A. R. 1974. Search for the Cambrian world. *American Scientist* 62:216–24.

Rodgers, J. 1971. The Taconic orogeny. *GSA Bulletin* 82:1141–77.

Sloss, L. L. 1963. Sequences in the cratonic interior of North America. *GSA Bulletin* 79:93–113.

23

THE LATE PALEOZOIC

Permian-age strata (Dechelly Sandstone overlying the Cutler Formation) of Monument Valley, Arizona. © Doug Sherman/Geofile

In the Late Paleozoic, epeiric seas withdrew from the North American continent following the formation of the Appalachians, and epeiric-sea sedimentation was replaced with more transitional and nonmarine sedimentation.

Orogenies occurred on the continent's eastern and western margins. The Appalachians formed in the east and were a truly spectacular mountain chain at this time. On the western margin, an orogenic episode initiated the start of the Rocky Mountains.

During the Late Paleozoic, land plants became widespread, and amphibians and reptiles evolved. At the end of the Late Paleozoic, widespread mass extinctions wiped out many shallow-water, marine invertebrates, and this marked the end of the Paleozoic.

CHARACTERISTICS OF THE LATE PALEOZOIC

The Late Paleozoic consists of the Devonian (360 to 408 million years ago), the Mississippian (320 to 360 million years ago), the Pennsylvanian (286 to 320 million years ago), and the Permian (245 to 286 million years ago) periods. The Mississippian period is named for outcrops in Illinois along the Mississippi River; the Pennsylvanian period is named for outcrops in the state of Pennsylvania. These two periods are recognized only in the United States and Canada. In Europe and on other continents, the time span represented by the Mississippian and Pennsylvanian periods cannot be practically divided in two because the rock types are so similar. Therefore, European geologists call the period from 286 to 360 million years ago the **Carboniferous,** referring to the large amount of coal in deposits of this age.

During the Late Paleozoic, sedimentation on the craton and on the eastern margin changed from predominantly marine (Devonian and Mississippian marine sediments) to nonmarine (Pennsylvanian and Permian transitional and nonmarine sediments). This change was caused by mountain building in the east and the associated continental uplift. The epeiric seas, so common in Early Paleozoic time, became restricted to narrow regions of the continent during the Late Paleozoic.

North America was near the equator during the Late Paleozoic, although rotated slightly counterclockwise from its Early Paleozoic position (figure 23.1). Tropical conditions prevailed.

Convergence of the continents continued throughout the Late Paleozoic. The northern continents and landmasses—North America, Greenland, Europe, and Asia—collided to form a large northern continent called **Laurasia.** The southern continents—Africa, South America, India, Australia, and Antarctica—made up a larger southern continent called **Gondwanaland** (sometimes shortened to Gondwana). As Laurasia and Gondwanaland converged and collided in the Late Paleozoic, a huge supercontinent called **Pangaea** formed. Pangaea began to form around Pennsylvanian time and broke apart in the Mesozoic.

E vidence for the formation of Gondwanaland by the collision of the southern continents is incomplete. Probably, the southern continents were separate units until the Early Paleozoic, when all converged. Throughout the Early and Late Paleozoic, Gondwanaland was intact. In the Late Jurassic (about 150 million years ago), Gondwanaland began to break apart into the individual southern continents we see today.

THE NORTH AMERICAN CRATON

During the Late Paleozoic, sedimentation on the craton changed in character, from typical shallow-water, epeiric-sea sedimentation to nonmarine sedimentation, as the continent was uplifted in the east by orogenic processes.

Devonian Sedimentation

Sedimentation on the North American craton during the Early Devonian was sporadic. Because of the continent–continent collision of North America and Africa, much of the craton was folded and warped into a series of arches and basins. A minor regression occurred from the Middle Silurian to the Early Devonian and marked the end of the Tippecanoe Sea. The subsequent transgression in the Middle Devonian (the start of the Kaskaskia Sea) caused flooding only in the low basin areas and not along the topographic highs (arches). Consequently, Lower Devonian sediments are relatively thin and confined to basinal areas. Two prominent basins that received sediment were the Michigan Basin and the Williston Basin. Both contain abundant evaporites of this age. Only in Late Devonian time, when epeiric seas had become firmly established, were sediments deposited over large regions. Typical Late Devonian sediments include shelly limestones and reef limestones analogous to Early Paleozoic carbonates.

In the Late Devonian, and into the Early Mississippian, the *Chattanooga shale* was deposited along the eastern margin and throughout much of the cratonic epeiric sea. The Chattanooga shale is an unusually thick, black-shale deposit representing the introduction of a large volume of mud. A thick accumulation of black shale is an uncommon epeiric-sea deposit because conditions in epeiric seas were not conducive to shale deposition—black shale is usually deposited in anoxic environments in deeper water. Why was such a large volume of shale deposited in the Kaskaskia epeiric sea? Some geologists postulate that the transgression producing the Kaskaskia Sea led to unusually

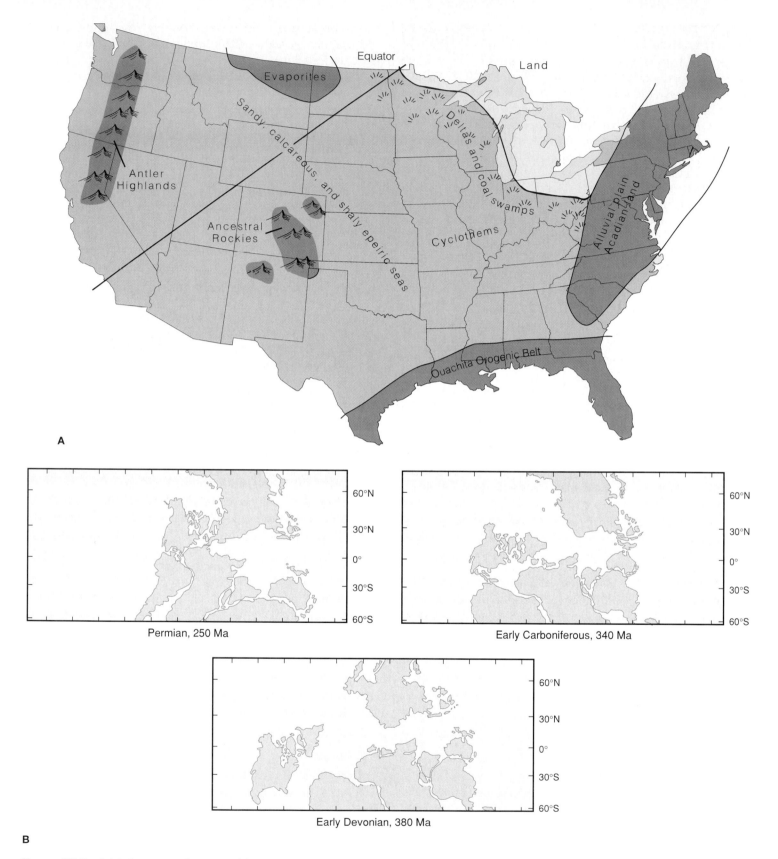

A

B

FIGURE 23.1 (A) Paleogeographic map of the United States during the Late Paleozoic. (B) Positions of the continents during the Late Paleozoic.

FIGURE 23.2 Crinoidal clastic limestones.
Photograph courtesy of Malcolm P. Weiss.

FIGURE 23.3 An outcrop of a Pennsylvanian cyclothem. The lowest bed is the coal layer. The strata above this layer, to the top of the outcrop, are limestones and shales and represent the marine section of the cyclothem.
Photograph courtesy of D. L. Reinertsen, Illinois State Geological Survey.

deep epeiric seas. Perhaps this epeiric sea was much deeper than those of the Early Paleozoic. Another idea is that the salinity levels of the Kaskaskia Sea were higher than normal, perhaps because of the abundant-evaporite conditions of the Silurian and Early Devonian, and that the unusually high salinity inhibited carbonate deposition. Another, more plausible, idea is that the Appalachians in the east shed a huge volume of sediment. Coarser-grained sediment was deposited nearer the mountains, but the tremendous amount of finer-grained sediment was carried out into the epeiric sea on the craton and deposited.

Mississippian Clastic Limestones

After the deposition of the Chattanooga shale, carbonate sedimentation again became dominant in Mississippian time. Most of the Mississippian carbonates are superficially similar to Lower Paleozoic epeiric-sea carbonates, but they display one important difference. In Mississippian time, echinoderms, particularly crinoids and blastoids, underwent rapid evolution. A tremendous number of crinoids lived at this time. As they died, their stems disarticulated on the bottom, leaving billions of tiny disks or hoops, which were incorporated into the sediment. The segments of crinoid stems acted dynamically just like clastic sand particles. Thus, some of these limestones display cross-bedding and ripple marks, just as sandstones do, and are therefore called **clastic limestones** (figure 23.2). However, unlike sandstones, the entire rock is carbonate, and the sand grains are represented by disarticulated crinoid-stem pieces.

Other Mississippian cratonic sediments include typical reefal limestones and minor evaporites.

At the very end of Mississippian time, along the eastern part of the craton, sands of various mixtures were deposited. These sands represent renewed erosion of the shield (the source of the sand) and are another example of recycling of sediments.

Pennsylvanian Cyclothems and Coal

In Pennsylvanian time a strange style of sedimentation occurred—sedimentation in cycles. A typical **cyclothem** consists of a series of nonmarine sediments overlain by a series of marine sediments (figure 23.3). The cycle of nonmarine/marine strata is repeated again and again in a vertical sequence. In an ideal cyclothem, the base of the cyclothem is marked by an unconformity (figure 23.4). The section of nonmarine strata begins with fluvial (river), cross-bedded sands and gravels. Above these are sandy shales, followed by freshwater limestones with terrestrial and freshwater fossils.

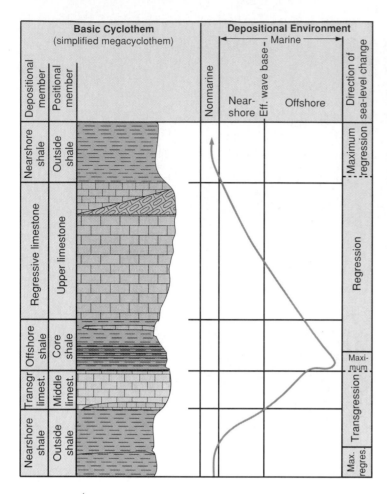

FIGURE 23.4 Generalized schematic of the ideal Pennsylvanian cyclothem.

FIGURE 23.5 Pennsylvanian-age cratonic limestones. Quarry, Fultanham, Ohio.
Photograph courtesy of Malcolm P. Weiss.

Next is a layer of *underclay* (a clay-rich layer representing an ancient soil) followed by a coal horizon, both deltaic deposits. The marine section consists of a shale/limestone/shale/limestone/shale sequence, with all of the strata containing marine fossils. Very few cyclothems contain all the layers—many are missing one or two—but the sequence is always the same: nonmarine sediments followed by marine sediments.

One prominent characteristic of cyclothems is that the thickness and development of the marine or nonmarine sections varies regionally. In Pennsylvania and the eastern United States, the nonmarine section of the cyclothem is thicker and more developed than the marine section, and coals are particularly well developed (which explains why much of the coal industry is centered in Pennsylvania, Kentucky, West Virginia, and Tennessee). The greater development of the nonmarine section indicates that these cyclothems represent the area near uplifted and exposed land. In Ohio and Illinois, on the midcontinent, the nonmarine and marine sections are approximately equal in thickness (figure 23.5). However, in Texas, along the western part of the North American craton, the marine sequence is thicker and more developed than the nonmarine sequence, and

coals are poorly developed or absent. These latter cyclothems must represent the region nearer to, or in, deeper-water conditions.

What caused the cyclic sedimentation pattern in Pennsylvanian time? The nonmarine section of the cyclothem is represented by sediments deposited in terrestrial (river) and transitional (deltaic) environments. The marine section of the cyclothem is represented by sediments deposited in the rapidly changing marine environments. Any hypothesis proposed to explain the nature of cyclothems must therefore explain rapid transgressions and regressions. Of the numerous processes suggested, two, acting together, are most plausible. First, the rapid transgressions and regressions appear to be the result of repeated glaciation in the southern continents, in Gondwanaland. Not only is the timing of glacial cycles right for the transgressions and regressions, but there is also supporting evidence—abundant glacial sediments, such as tillites, of this age are found on the southern continents. Because the craton was very close to sea level, small rises and falls of the level of the sea meant that thousands of square kilometers were either flooded or exposed. Therefore, prominent, geographically widespread cyclothem units could easily be produced. Second, acting on a smaller scale, and producing many of the minor cyclothems, were deltaic processes operating on the shores between the swamps and the sea.

An important unit in the cyclothem is coal, mined today for energy purposes. Coal forms when plant material sinks to the bottom of a lake or swamp and is buried under layers of additional plant material and silt, sand, or both. Normally, decay of organic material by scavengers and bacteria keeps pace with the amount of organic material introduced into the

FIGURE 23.6 Permian-age DeChelly sandstone. Monument Valley, Kayente, Arizona.
© Doug Sherman/Geofile.

FIGURE 23.7 The Appalachians.
© NASA.

environment. But in environments such as swamps, deltas, and some lakes, the amount of organic material produced is so great that the scavengers and bacteria cannot decompose it all. Therefore, much of the plant material remains undecomposed and is buried. Over time, this plant material changes to coal under the influence of pressure from overlying sediments and heat from within the earth (the geothermal gradient). Under the pressures and temperatures that are normal near the surface of the earth, the transformation of plant material to coal takes a long time.

Permian Red Beds

In Permian time, sedimentation styles and types again changed as a major regression occurred. Nonmarine sedimentation became dominant, while both the marine sedimentation that characterized Mississippian time and the transitional sedimentation of Pennsylvanian time declined. **Red beds** became a characteristic rock type on the continent (figure 23.6). Red beds are typically sandstones containing iron. In this case the amount of iron is trivial, probably less than 5%. Even this small amount of iron, however, stains the sandstone a deep red—hence the name. Although some questions remain concerning the environment of red-bed deposition and the timing of the introduction of iron into the sandstone, most geologists agree that the red beds were deposited in nonmarine environments. The general, nonmarine conditions that prevailed on all continents during Permian and Triassic time make red beds a characteristic rock type of this age all over the world.

THE EASTERN MARGIN

The Appalachians

The Appalachian Mountains, which began to form in the Ordovician during the Taconic orogeny, completed their development in Late Paleozoic time and have been undergoing erosion ever since (figure 23.7). Their peaks are not rugged and sharp like the Rockies, because they are an older mountain chain and have been undergoing erosion for a longer period.

The Devonian Acadian Orogeny

In Late Devonian to Early Mississippian time (a period of about twenty-five million years), a continent–continent collision

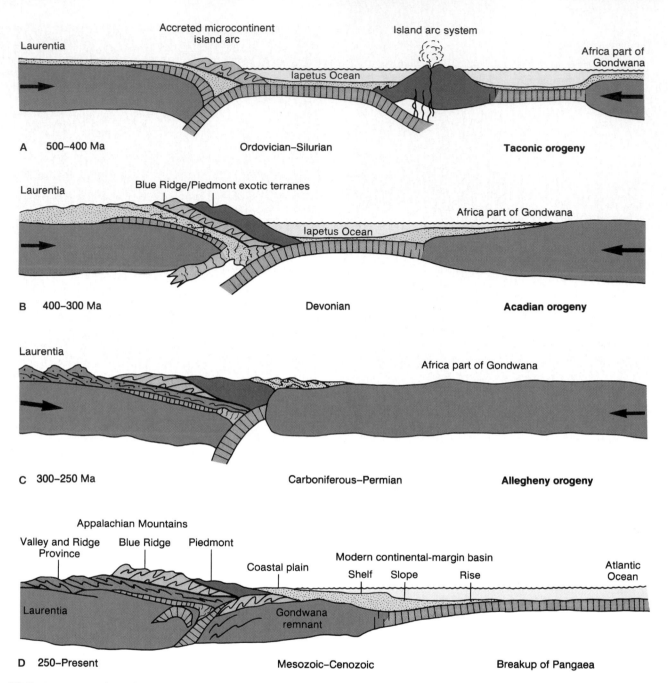

FIGURE 23.8 Tectonics of North America's eastern margin during the Late Paleozoic. (A) Taconic orogeny (Ordovician-Silurian, 400–500 million years ago). (B) Acadian orogeny (Devonian, 300–400 million years ago). (C) Allegheny orogeny (Carboniferous-Permian, 250–300 million years ago). (D) Breakup of Pangaea (Mesozoic-Cenozoic, 250 million years ago to present).

A Trip Through Time *by Cooper/Miller/Peterson, © 1990. Adapted by permission of Prentice-Hall, Inc., Upper Saddle River, NJ.*

finally occurred between Africa and North America. This produced the **Acadian orogeny** (named for a French-colonial term for a region in New England and adjacent Canada) (figure 23.8B). The Acadian-orogenic episode was much more intense than the Taconic orogeny of the Ordovician because it was collision related rather than subduction related. Large amounts of granite were produced, regional metamorphism was widespread, and some thrust faulting occurred. During Devonian time, the Appalachians were likely as prominent as the Rockies are today.

G eologists can estimate the height of the Appalachians during Devonian time using several pieces of evidence. First, the erosion rate of rock and sediment in mountains today can be measured, perhaps using the Rocky Mountains as an example. Next, the thickness and areal extent of Devonian formations in the Appalachian region can be used to estimate the amount of sediment that was shed off the newly created Appalachian Mountains during the Devonian. By knowing how much sediment was deposited over a given time at a given rate, geologists can estimate the height of the old mountain chain.

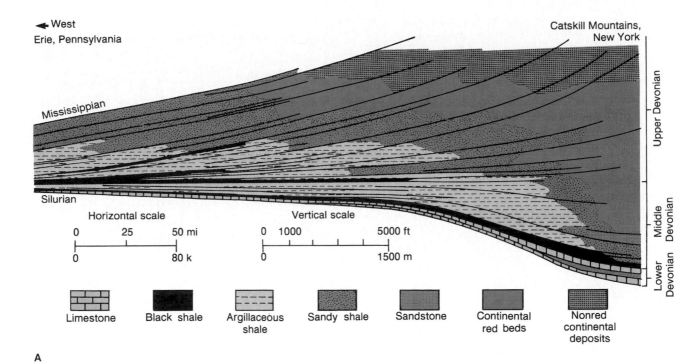

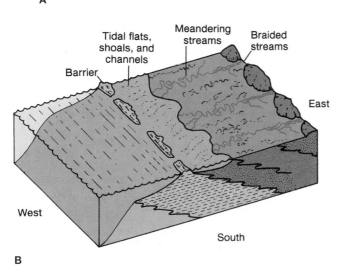

FIGURE 23.9 The Catskill clastic wedge. (A) Cross section of Devonion-age strata from Erie, Pennsylvania (west), to the Catskill Mountains, New York (east). (B) Inferred paleoenvironments.

Source: Moore, GSA Fiftieth Anniversary Volume, 1941, based on G. H. Chadwick and G. A. Cooper. (B) After John R. L. Allen and P. F. Friend, Geological Society of America Special Paper *106:21–74, 1968, Geological Society of America.*

Because of the uplift, erosion was intense and the mountains shed large volumes of sediment symmetrically on either side. Along the eastern margin, this large volume of sediment produced another clastic wedge, the **Catskill clastic wedge,** or Catskill Delta (figure 23.9). The Catskill clastic wedge was similar in shape to the Ordovician-age Queenston clastic wedge, but was much larger. Devonian sedimentary rocks in the Catskill clastic wedge are over 1800 meters thick, compared with the typical, 200- to 300-

meter-thick cratonic sedimentary rocks of equivalent age. All sedimentary rocks of this age along the eastern margin thicken and become coarser toward the east (the source area of the sediment). Coarse-grained sandstones, some conglomerates, and shales were deposited on an alluvial plain that extended all along the eastern margin. On the alluvial plain near the base of the mountains, conglomerates are common. Farther west, near the edge of the alluvial plain and the epeiric sea, finer-grained sandstones and shales were deposited in deltaic and marginal marine environments. Growing on the vast alluvial plain was the Gilboa Forest, one of the first widespread forests recognized in the fossil record.

In Europe and Africa, a mirror-image clastic wedge analogous to the Catskill clastic wedge formed. An important group of rocks, representing a part of this clastic wedge, is termed the *Old Red sandstone* and is found throughout most of Europe.

Other Devonian Orogenies

The Acadian orogeny was just one of several orogenies that occurred along the eastern and northern margins of North America during Devonian time. Farther north, as the eastern margin of Greenland collided with Europe, the *Caledonian orogeny* occurred. The Caledonian orogeny is equivalent in most aspects of tectonic style and pattern to the Acadian orogeny. Even farther north, in Arctic Canada, the *Ellesmerian orogeny* took place as part of a continental block collided with the northernmost part of Canada and Alaska. Tectonic events of the Ellesmerian orogeny appear to be similar to those of the Caledonian and Acadian orogenies, but little is known of this orogeny because the current location of the rocks (the Canadian Arctic) makes collecting and field mapping difficult.

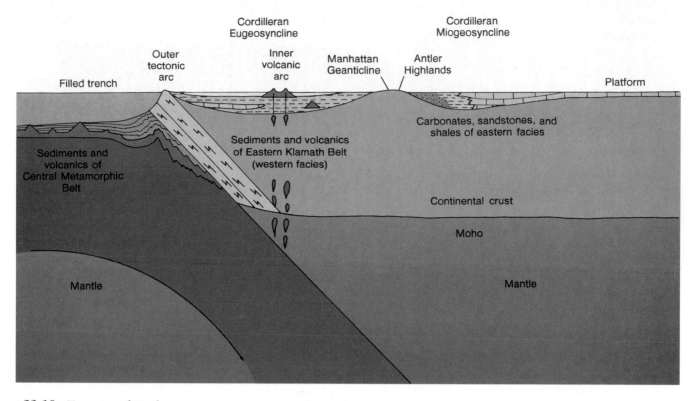

FIGURE 23.10 Tectonics of North America's western margin during the Late Paleozoic.

Figure rendered from the Earth History & Plate Tectonics: An Introduction to Historical Geology, *2E by Carl K. Seyfert and Leslie A. Sirkin. Copyright © 1973, 1979 by Carl K. Seyfert and Leslie A. Sirkin, Reprinted by permission of Addison Wesley Educational Publishers, Inc.*

The Pennsylvanian Appalachian Orogeny

In Mississippian time, large amounts of nonmarine shales and sandstones were deposited adjacent to the Appalachians. The *Pocono group,* consisting of these nonmarine clastics, can be seen today in Pennsylvania forming some of the vertical ridges of the Appalachians.

During the Pennsylvanian (and continuing through the Permian and into the Triassic), the final orogenic phase of the Appalachian Mountains took place—the **Appalachian (or Allegheny) orogeny** (figure 23.8C). The tectonics of the orogeny were characterized by intense folding and marked thrust faulting as blocks of rock were thrust west toward the craton. Associated with the thrust faulting was metamorphism. This orogenic phase produced another clastic wedge, the Dunkard clastic wedge, which was smaller than the Catskill clastic wedge. Geologists who have studied the rocks and structures produced by the Appalachian orogeny generally agree that the orogenic event represents the last phase of the continent–continent collision that began in the Late Devonian. However, other researchers maintain that the orogeny is the first phase of the breakup of North America and Africa, which was completed in Late Triassic time.

The Ouachita Orogeny

Mountain building also occurred in the Late Paleozoic along North America's southern margin, with the *Ouachita orogeny.* The Ouachita orogeny, which began in Mississip-

pian time, was subduction related and is considered by many to be the southern extension of the Appalachian orogeny. The orogenic phase continued into the Pennsylvanian, when the Marathon orogeny occurred, producing the Ouachita Mountains and the Wichita Mountains in the Oklahoma and Texas region. Some overthrusting of sediments to the north occurred, and nonmarine clastics are a common rock type.

THE WESTERN MARGIN

The western margin began to experience orogenic activity in the Devonian, and this activity continued into Mississippian and Pennsylvanian time. These orogenic events set the stage for the formation of the Rockies in Mesozoic time.

The Devonian

Starting in the Late Devonian, and continuing through the Mississippian and into the Pennsylvanian, the collision of a volcanic island arc with the western margin produced the **Antler orogeny,** named for Antler Peak, Nevada (figure 23.10). The collision caused widespread thrust faulting, with blocks overthrust to the east. Some metamorphism occurred, but little volcanism. Since no major mountains were formed, no clastic wedge was produced. However, clastic sediments (sandstones) are common. To the west of the *Antler Belt* (sometimes called the Antler Highlands) was deeper water; to the east, between the highlands and the craton, a very narrow epeiric sea formed in which carbonate was deposited.

FIGURE 23.11 Flat-lying Pennsylvanian- and Permian-age sedimentary rocks along the San Juan River. The Goosenecks, Utah.
Photograph courtesy of Stephen R. Mattox.

FIGURE 23.12 The DeChelly Sandstone (Permian) forms the monuments of Monument Valley, Utah-Arizona.
Photograph by H. E. Malde, USGS Photo Library, Denver, CO.

The Mississippian and Pennsylvanian

Carbonates continued to be deposited in the epeiric sea between the Antler Belt and the craton during the Mississippian. (Many mountain peaks in Utah, Wyoming, Idaho, Montana, and the Canadian Rockies are composed of these Mississippian-age limestones.) Interbedded sandstone is common—a result of erosion of the Antler Highlands. Sedimentary rocks are common on the western margin during this time (figure 23.11).

The *Ancestral Rockies* formed in Mississippian and Pennsylvanian time just east of the Antler Belt. They represent the first orogenic phase of the Rocky Mountains. The formation of these mountains is related to large-scale movement along near-vertical reverse faults, rather than to subduction or island-arc collision. Coarse gravels and immature sandstones were deposited in alluvial and deltaic environments near the mountains. The mountains shed large amounts of arkoses (feldspar-rich sandstones), which are seen today at the Red Rocks Amphitheater just outside Denver and the Flatirons (wedge-shaped layers of resistant sandstones) in Colorado.

The Permian

In the Permian, the extreme western margin (west of the Antler Belt) was adjacent to deep water. Shales were deposited. Upwelling from the deeper water brought up a tremendous amount of nutrients, including phosphates, which were deposited along with the shales. The *Phosphoria formation* is an important phosphate-bearing deposit in the western states today.

In southwestern Texas, a large organic-reef complex grew around an extremely deep water basin (perhaps over 300 meters deep). During the regression in Permian time, the reefs became exposed, and the corals died. Subsequently, the basin filled with evaporites and red beds. In contrast to the Michigan Basin, this basin was a deep-water basin that simply filled with sediment, rather than a shallow-water basin that isostatically accumulated a large thickness of sediment. The reef structure (called the Capitan Reef), now exposed in the Guadalupe Mountains of West Texas, has provided paleontologists with one of the best examples of ancient coral-reef structure and ecology.

In other areas, sandstone was deposited (figure 23.12).

LIFE IN THE LATE PALEOZOIC

The Late Paleozoic saw many advances in the history of life. Forests became common in the Late Devonian, and throughout Mississippian and Pennsylvanian time vast swamps were common. In the Devonian, amphibians evolved from fish; in the Late Pennsylvanian, one line of amphibians evolved into reptiles. Reptiles diversified in the Permian, and this set the stage for the Mesozoic, the "Age of Reptiles."

The Devonian

Invertebrates and the reef communities continued throughout the Devonian and were similar to those of the Early Paleozoic (figure 23.13). Differences from the Silurian period involved the relative abundances of the various invertebrate groups. Corals and coral reefs, brachiopods, and bryozoans were

FIGURE 23.13 Diorama of Devonian epeiric-sea life. Most of the tentacle forms are corals. Cephalopods are shown crawling along the bottom. An elaborate, spiny trilobite is in the center foreground.
Photograph courtesy of Field Museum of Natural History, Chicago, #GEO80821.

especially abundant and widespread. One important line of invertebrates evolved in the Devonian—the coiled *cephalopods*. The coiled cephalopods, a class of molluscs represented today by the squid, octopus, and nautilus, were to become an important index fossil for Mesozoic-age marine strata.

Devonian time is often called the "Age of Fish" because fish underwent rapid evolution both in kinds and in numbers (figure 23.14). Agnathids, the common jawless fish of Ordovician and Silurian time, gave rise to the first jawed fish, the *placoderms* ("plate-skinned" fish) in the Late Silurian. The armored group of agnathids became extinct in the Devonian. Placoderms had smaller bony plates; they became more streamlined. Fossils of placoderms include not only the smaller bony plates but also internal bones, indicating that a vertebrate skeleton had at least partially developed. Placoderm evolution proceeded rapidly in the Early Devonian and reached a peak in the Middle Devonian. By the Late Devonian, placoderms began to decline rapidly, finally becoming extinct in the Permian. However, in the Devonian, placoderms gave rise to two important groups of fish. First were the *Chrondrichtyes* ("kron-DRICK-tees"), or cartilaginous fish,

FIGURE 23.14 Typical Devonian-age fish.
Photograph courtesy of Field Museum of Natural History, Chicago, #82665.

modern representatives of which include the sharks, skates, and rays. The primitive Devonian sharks were similar to modern types. Probably no other animal has evolved to more perfectly fit its ecologic niche than the shark. It is the perfect killing machine! The second major group of fish to evolve in the Devonian from the placoderms were the *Osteichthyes* ("os-tee-ICK-thees"), or bony fish. While relatively primitive at first, this group also underwent explosive evolution in the Devonian and is the dominant type of fish in modern oceans, lakes, and streams.

In the Devonian, one line of fish, the rhipidistians, of the order Crossopterygii, evolved into the first amphibians, the **labyrinthodonts** (figure 23.15). Labyrinthodonts are so named because the structure of the surface of their teeth resembles a maze or labyrinth. Labyrinthodonts evolved rapidly and peaked in the Carboniferous. They became extinct in Triassic time. Several major evolutionary changes had to occur for a line of fish to evolve into amphibians (table 23.1). First, and most important, was the development of lungs. But other important structural changes included

(1) modification of the spinal column and limbs to support the animal on land, (2) development of a three-chambered heart to pump blood more effectively, (3) development of eyelids and a lubrication system for the eye, and (4) modification of the internal ear structure. The transition from fish to amphibian is one of the better-documented macroevolutionary changes in the fossil record. The rhipidistians differ little from the earliest amphibians.

I n Devonian time, the crossopterygians (an order of fish that includes air-breathing forms) evolved into two separate lines. One line was the rhipidistians, which evolved into the amphibians; the other line was the coelacanths ("SEEL-a-canths"), an entirely marine group of fish. Coelacanths were thought to have become extinct in the Mesozoic. Then, in late 1938, a fisherman off the southern coast of Africa caught a coelacanth! Apparently, coelacanths were still in existence and living in very deep water. Since then, several coelacanths have been captured, and a few have been photographed in their environment. Most are found near the Comores Islands, between Madagascar and Africa in the Indian Ocean. A current debate among scientists concerns capturing coelacanths for dissection and study. Since no one knows the size of the population of coelacanths, scientists are reluctant to capture individuals for fear of contributing to the extinction of a species.

FIGURE 23.15 A Devonian labyrinthodont.
From Historical Geology *by L.W. Mintz; Charles E. Merrill Pub. Co.; Reprinted by permission of L.W. Mintz.*

TABLE 23.1 *Summary of Major Evolutionary Changes from (A) Fish to (B) the Ancestral Labyrinthodonts to (C) Labyrinthodonts (Amphibians) to (D) the First Reptiles*

	A Rhipidistian Crossopterygians	B Ancestral Labyrinthodonts	C Labyrinthodonts	D Stem Reptiles
Feature	(Lobe-finned fishes) *Eusthenopteron*	(Ancient amphibians) *Ichthyostega*	Anthracosaurs	Seymourians *Seymouria*
Skull	Solidly roofed – – – – – – – – –	Basically similar to A – – – – – – –	Little changed from B; more flattened	– – – – Little changed from C
Appendages	Fins with small bones and large fin flaps	→ Limbs fully boned and articulated – – – – – – – –	Little changed from B; larger, more massive	– – – – Little changed from C
Girdles	No pelvic girdle	→ Strong pectoral and pelvic girdles – – – – – – –	Little changed from B; heavier	– – – – Moderate change from C
Vertebrae	Not interlocking – – – – – – – –	Little changed from A; arches touching	→ Different from B: interlocking vertebrae, strong vertebral column	– – – – – – – Little changed from C
Dorsal fin, tail	Fin rays on tail – – – – – – – –	Dorsal and caudal fin rays retained	→ No fin rays	
Body covering	Bony scales over – – – – – – – – all of body	Bony scales retained	→ Bony scales gone – – – – – –	Horny epidermis
Geologic age	Middle Devonian	Upper Devonian	Carboniferous	Lower Permian

——→ Major change   – – – – – – – – Little change

A. N. Strahler. Based on data of A. S. Romer, 1966, and E. H. Colbert, 1980. From Arthur N. Strahler, *Science and Earth History: The Evolution/Creation Controversy* (Buffalo, N.Y.: Prometheus Books). Copyright © 1987 by Arthur N. Strahler. Reprinted by permission of the author and publisher.

Figure 23.16 A Devonian forest. A lycopod (A). A horsetail (B). A tree fern (C).

Photograph courtesy of Field Museum of Natural History, Chicago; painting by C. R. Knight, #CK49T.

On land, plants underwent rapid diversification. Psilophytes evolved rapidly, and by the Late Devonian lowland forests were common (figure 23.16). A famous fossil forest was the Gilboa Forest, which grew on the alluvial plain of the Catskill clastic wedge adjacent to the Appalachian Mountains. The first ferns appeared in the Late Devonian, some remarkably similar to modern species.

Insects first appeared in the Silurian. Most fossils of the first insects are of various wingless species. Oddly, spiders (which are not insects, but arachnids, a type of arthropod) have been found in the Devonian and Carboniferous strata of Germany. Apparently, one line of arthropods rapidly found a perfect ecologic niche. (Spiders are another group, like insects, that is highly successful. Throughout most of our lives, we are probably never more than 10 feet away from a spider!)

The Mississippian

In the Mississippian, the number of *echinoderms* in the epeiric seas increased dramatically. Crinoids experienced a period of explosive evolution, so much so that the Mississippian has been termed the "Age of Crinoids."

The Pennsylvanian

Several changes in epeiric-sea invertebrate life occurred in the Pennsylvanian (figure 23.17). One group of brachiopods, the productids, developed spines on their shells to prevent them from sinking in mud. They then invaded the muddy bottom regions of the epeiric seas (common because of the input of clastic material from the Appalachians), an environment inaccessible to other brachiopods. Other brachiopods developed highly curved shells, an unusual adaptation, also for living on muddy bottoms.

Several groups of *foraminifera* (one-celled animals) became important, particularly the fusilinid foraminifera. Fusilinids grew quite large—many resemble wheat grains. They are used as index fossils for Pennsylvanian and Permian marine strata. This group of foraminifera became extinct at the end of the Permian.

S ome of the Mazon Creek fossils, found near the Mazon Creek area of Illinois, like those of the Burgess shale, are found nowhere else in the world. The fossils are found in concretions and consist of shallow-water marine animals (arthropods, crustaceans, and fish), terrestrial animals (insects and arachnids), and plants. There is even an unknown form called *Tullimonstrum* (Tully Monster), the state fossil of Illinois, named after the amateur paleontologist who discovered it.

The vast coal swamps of the Carboniferous were unique—an environment never paralleled after Carboniferous time (figure 23.18). Various types of plants such as **lycopsids, sphenopsids,** and ferns were numerous, and most grew to great heights. Lycopsids, sometimes called "scale trees" because of their scaly bark, were a common type of Pennsylvanian-age plant. (Today, the only living members of this group are the club mosses and *Lycopodium,* a ground pine.) The lycopsids that grew in Pennsylvanian time had branches only at the top of the trunk. On the ends of the branches were long, narrow leaves resembling spikes. Some lycopsids grew to 40 meters in height! Sphenopsids, represented today by the horsetails and scouring rushes, were also common. They developed leaves at each joint in their long, ribbed stems. Many of these sphenopsids grew to 7 meters in height; modern horsetails are typically less than 1 meter in height. Ferns were also common and grew to large sizes.

FIGURE 23.17 Diorama of Pennsylvania epeiric-sea life. Sponges, with snails crawling on them, are in the left background. Cephalopods, snails, and brachiopods are on the bottom. Note the large, coiled cephalopod on the right.
Photograph courtesy of U.S. National Museum.

FIGURE 23.18 A Pennsylvanian-age swamp. Large Lepidodendron trees are in the background. Ferns are also common. Note the amphibian and dragonfly in the foreground.
Mural by P. R. Haldorsen; photograph courtesy of the National Museum of Canada.

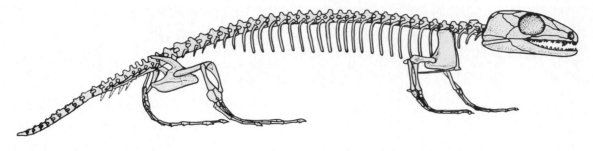

FIGURE 23.19   A cotylosaur—a Pennsylvanian reptile.

Amphibians (labyrinthodonts and other types) and insects lived in the widespread swamps. The insects diversified, and some grew to truly great proportions—cockroaches 10 centimeters in length and dragonflies with wingspans of over 1 meter! (The largest dragonfly today has a wingspan of about 10 centimeters.)

One interesting question, not yet fully resolved, is, Why did plants, insects, and animals grow to such large sizes in the past? (including the 40-meter-high lycopsids, the huge dragonflies of the Pennsylvanian, and the dinosaurs of the Mesozoic). Certainly there appears to be no structural or genetic reason why, for example, insects today are not larger. One idea is that predators for an individual group were not present in the Late Paleozoic. Predators of insects, such as birds, did not evolve until Cretaceous time. Perhaps if species are left unchecked by any natural force, greater size is a common evolutionary direction.

Forests and swamps were so abundant and widespread over the earth that two distinct floras can be recognized. The northern (or Euameric) flora, which consisted of mostly lycopsids and ferns, grew in North America, Europe, and China. The plant types indicate that the climate here was tropical to subtropical. In contrast, the southern or *Glossopteris* flora grew on the southern continents—Gondwanaland. *Glossopteris*, a type of seed fern, was the major constituent of this flora. The southern-flora fossils are found in sedimentary rocks associated with tillites, indicating that the climate was cool. This flora appears to have grown in lowland areas near glaciers.

One important event in plant evolution, the evolution of the gymnosperms, occurred in the Late Paleozoic. Gymnosperms are the nonflowering, seed-bearing groups of plants. Included in this group are the cycads, ginkgoes, seed ferns, and conifers. The gymnosperms became the dominant plant type of the Mesozoic.

In the Late Pennsylvanian, one line of labyrinthodont amphibians gave rise to the first reptiles, the **cotylosaurs.** This stem group of reptiles strongly resembles the amphibians from which it arose and is another well-documented macroevolutionary event (figure 23.19; See table 25.1). The structural modifications needed to form a reptile from an amphibian were much easier to achieve than those involved in the fish-to-amphibian transition (table 23.1). The most important develop-

ment in the evolution of reptiles was the development of the *amniotic egg.* Early amniotic eggs possessed a leathery outer shell (as opposed to the hard, calcium-carbonate shell of another amniotic egg, the chicken egg) that allowed the exchange of gases (for example, oxygen and carbon dioxide) but prevented internal fluids from escaping. Thus, dessication was no longer a problem, and this freed the reptiles from the necessity of being in water at the reproductive stage. The egg also contained albumen, which stored water during development, and various membranes, which the embryo used in different ways during development. (Some membranes contained yolk sacs; others allowed for the storage of waste products.) Other amphibian-to-reptile structural changes included modification of the limbs, spine, pelvis, and jaws.

The Permian

Nonmarine environments predominated during Permian time, as all the continents converged to form the supercontinent Pangaea. On land, the cotylosaurs evolved into several lines of reptiles, including two important groups of mammal-like reptiles—the *pelycosaurs* ("pel-ICK-oh-sawrs") and the *therapsids* (figure 23.20; see table 25.1). Pelycosaurs, commonly called "sailbacks," were quite common in the Permian but became extinct by the end of the period. The group of pelycosaurs that included *Dimetrodon* was active, fearsome, and predatory (figure 23.20A). The function of *Dimetrodon's* "sailback" was a point of debate for some time after its initial discovery. Some paleontologists thought it was for protection, but since few other animals were large enough to prey on *Dimetrodon,* this idea seems implausible. Another idea, only recently developed, is that the sail served as a heatregulating mechanism for body temperature. By pumping blood into its sail and standing broadside to the wind, a *Dimetrodon* could cool itself internally. This possible behavior leads to several interesting speculations about the physiology of these "reptiles." The carnivorous group of pelycosaurs gave rise to the therapsids in the Late Pennsylvanian.

Dimetrodon is often included in kits of plastic models of dinosaurs sold at many toy stores. *Dimetrodon,* however, lived in Permian time and became extinct before the appearance of true dinosaurs. *Dimetrodon* is also classified into a different subclass of reptiles from the dinosaurs because it was an entirely different type of reptile.

A

B

FIGURE 23.20 Permian reptiles. (A) Dimetrodon, a "sailback" pelycosaur. (B) Therapsids—mammal-like reptiles.
Photographs courtesy of Field Museum of Natural History, Chicago; paintings by C. R. Knight, #CK45T and #CK22T.

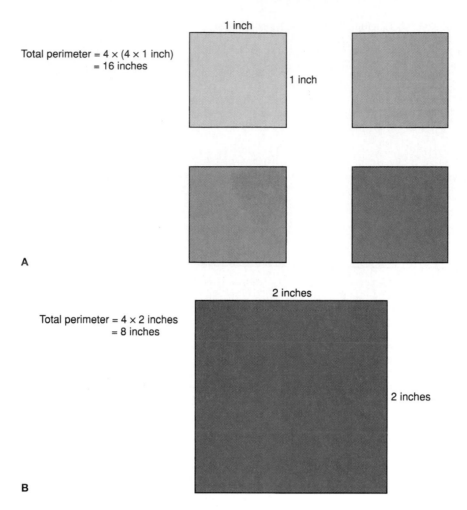

Figure 23.21 Reduction in continental-shelf area due to continent–continent collisions. (A) Before continental collisions. (B) After continental collisions.

The other group of mammal-like reptiles, the therapsids, was varied (figure 23.20B). There were two types: herbivores and carnivores. The carnivore group possibly gave rise to mammals in Triassic time.

Permian Mass Extinctions

At the end of Permian time, almost 50% of all families (not species!) of shallow-water marine invertebrates became extinct in one of the most profound mass extinctions in the history of life. Major groups to become extinct included rugose corals, blastoids, several groups of brachiopods and crinoids, fusilinid foraminifera, and the last of the trilobites. Other groups, such as bryozoans and ammonites, were also greatly affected.

For a long time, paleontologists could not determine the cause of this widespread catastrophe. At first it was proposed that the Permian mass-extinction event was an illusion created by an incorrect reading of the fossil record and that the extinctions may not have happened at all! Because of the nonmarine conditions prevalent on the continents during this time, sediment, generally speaking, was eroded rather than deposited. The actual volume of preserved sedimentary rocks

of Permian age is very small compared with the volume of sediment preserved for other periods. Since the volume of sediment is small, the number of fossils preserved in the sediment is also small. Therefore, it was theorized that the extinction event reflects only the smaller number of fossils found, owing to the absence of sediment preservation, and not to a great number of organisms dying out. At first, this seemed a good idea, but it has problems. For example, statistics can be employed to correct for the lack of preserved sediment (that is, to judge if a paucity of fossils would still occur even if more sediment were preserved). In addition, at exceptional Permian-fossil localities, the number of fossils found is still small. Thus, the extinction event is real and not a result of the incompleteness of the rock record.

Another idea, developed after plate-tectonic theory became well established, stated that all the continents had converged to form the supercontinent Pangaea, destroying in the process a tremendous amount of shallow-water, continental-shelf area. Figure 23.21 gives examples of the amount of shelf area destroyed by continent–continent collisions. Figure 23.21A shows four separate continents, scaled at 1 inch by 1 inch. The total shelf area of each continent is 4 inches

Box 23.1

Regional Geology

THE APPALACHIANS

The development of the Appalachian Mountains was a long and complex process. At their peak of development, the Appalachians were very rugged, similar to the present-day Rockies. Today, however, the Appalachians are a subdued mountain chain, having undergone erosion for at least 200 million years (figure 1). Traditionally, the Appalachians have been divided into five basic provinces: (1) the Appalachian Plateaus, (2) the Valley and Ridge Province, (3) the Blue Ridge Province, (4) the Crystalline Appalachians (the New England Uplands and the Piedmont Plateau), and (5) the Coastal Plains. The length of time of formation, the large geographic area affected, and the complex structural relationships that exist between the strata have made the Appalachians a fascinating area of geologic inquiry.

An often-overlooked aspect of the Appalachians is their important influence in the development of geologic thought. Many early geologists looked at the Appalachians, tried to decipher their history, and presented theories on how mountains formed. Most of these early theories were partially correct. Despite not having a coherent framework on which to base their ideas, early geologists carefully mapped the strata, establishing the stratigraphy and structure of the major units. Later geologists used these data in their interpretations. With the advent of the idea of plate tectonics, the history of the Appalachians underwent revision. Various ideas were bantered about, such as thin-skinned versus thick-skinned tectonics. The question

FIGURE 1 Aerial view of the Appalachians.
© *Nasa.*

here is whether the Precambrian basement rocks are involved in the deformation of the strata (thick-skinned) or not (thin-skinned). Currently, the Appalachians are again being reviewed to determine whether accreted terranes (see chapter 9) were involved in their formation. The Appalachian Mountains have presented enough questions to keep geologists busy for many, many years.

(4 sides × 1 inch). The total shelf area is then 4 × 4 inches = 16 inches. If the four continents collide to form one supercontinent, 2 inches on a side, the total shelf area available now is 4 × 2 inches = 8 inches (figure 23.21B). The shelf area is halved. This simple model clearly explains the Permian mass extinctions. Convergence of the continents destroyed the shallow-water marine habitat. Organisms became extinct for the simple reason that their environment was being destroyed and few widespread epeiric seas were available to inhabit.

Very recently, another alternative has been suggested to explain the widespread extinctions—climatic changes caused by volcanic eruptions. The Siberian Traps in Siberia, large deposits of volcanic basalts, represent large-scale volcanism. Isotopic dating of the rocks previously indicated that the eruptions took place after the terminal-Permian extinctions. However, recent dating suggests that the eruptions are coincident with the Permian mass extinctions. If the current results are valid, climatic changes associated with these eruptions could well explain the Permian mass extinctions.

SUMMARY

Late Paleozoic sedimentary rocks reflect a change from the typical Early Paleozoic epeiric-sea carbonates to Late Paleozoic transitional (deltaic) and nonmarine environments. On the North American craton and near the eastern margin, clastic limestones were common during Mississippian time. In Pennsylvanian time, cyclothems were deposited. Cyclothems consist of a section of nonmarine strata overlain by a section of marine strata, with the sequence repeated over and over again. Transgressions and regressions caused by glaciations in Gondwanaland appear to be the cause of the cyclothems. In Permian time, red beds were deposited in the nonmarine environments typical of the craton.

Africa collided with North America's eastern margin during the Devonian to produce the Appalachian Mountains in the Acadian orogeny. Other orogenies also occurred along the southern and northern margins of the continent during the Devonian. In the Pennsylvanian, thrust faults occurred as part of the Appalachian orogeny.

The western margin experienced some mountain-building processes in the Devonian, with the Antler orogeny, and in the Mississippian and Pennsylvanian with the creation of the Ancestral Rockies.

The Late Paleozoic saw many advancements in the history of life. In the epeiric seas, little changed during the first part of the Late Paleozoic. On land, however, plants became firmly established and produced widespread swamps and forests in the Pennsylvanian. Amphibians (labyrinthodonts) evolved from fish in the Devonian and inhabited the swamp environments. Reptiles (cotylosaurs) evolved from amphibians in the Late Pennsylvanian and diversified in the Permian.

At the end of the Permian, shallow-water marine invertebrates suffered a mass-extinction event as the continental shelf and the epeiric-sea environments were destroyed when all the continents converged to produce Pangaea.

A summary of Late Paleozoic events is shown in figure 23.22.

Period	Western margin	Craton	Eastern margin	Life
Permian	*Phosphoria formation* —Phosphate —Upwelling Organic-reef complex in southwest Texas —*Capitan Limestone* *DeChelly sandstone*	Craton mostly exposed —Regression Red beds (sandstones)	Appalachian orogeny continued	Mass extinctions of the end of Permian Diversification of reptiles —Pelycosaurs —Therapsids Index fossils: fusulinid foraminifera
Pennsylvanian	Widespread tectonic activity Ancestral Rockies —Movement along near-vertical faults —Gravels, sandstones	Cyclothems —Coal Transgression (Absaroka Sea) Numerous minor transgressions and regressions	Appalachian orogeny —Thrust faulting —Dunkard clastic wedge Ouachita orogeny in the south continued	Plants (swamps) —Lycopsids —Sphenopsids —Ferns First reptiles: Cotylosaurs
Mississippian	Ancestral Rockies? East of the Antler Belt —Carbonates —Sandstones *Redwall limestone*	Tilting of the continent in the east Late Mississippian —Regression Early Mississippian —Crinoidal limestones —*Salem limestone*	Nonmarine sandstones and shales *Pocono group* Ouachita orogeny in the south started —Subduction	"Age of Crinoids" —Crinoids —Blastoids First gymnosperms
Devonian	Antler orogeny —Island-arc collision —Thrust faulting —Klamath Arc	Extensive warping of the craton Middle–Late Devonian —Transgression (Kaskaskia Sea) —Shelly limestones —*Chattanooga shale* Early–Middle Devonian —Regression	Acadian orogeny —Collision of Africa and North America —Catskill clastic wedge —*Portage group*	Corals and brachiopods in epeiric seas Lowland forests (Gilboa forest) —First seeds "Age of Fish" First amphibians: Labyrinthodonts

FIGURE 23.22 Late Paleozoic: summary diagram.

Terms to Remember

Acadian orogeny 424
Antler orogeny 426
Appalachian orogeny 426
Carboniferous 419
Catskill clastic wedge 425

clastic limestones 421
cotylosaurs 432
cyclothem 421
Gondwanaland 419
labyrinthodonts 429

Laurasia 419
lycopsids 430
Pangaea 419
red beds 423
sphenopsids 430

Questions for Review

1. Which continents made up Laurasia? Gondwanaland?
2. What are two explanations for the deposition of the Chattanooga shale, an atypical epeiric-sea deposit?
3. In what ways are Mississippian clastic limestones similar to Cambrian sandstones? In what ways are they different?
4. Describe an idealized cyclothem. What caused the cyclic sedimentation?
5. Which strata in cyclothems have economic value?
6. How did the Devonian Acadian orogeny in the Appalachian region differ from the Ordovician Taconic orogeny?
7. What two orogenies occurred on the western margin during the Late Paleozoic?
8. What structural modifications were needed in the evolution of an amphibian from a fish?
9. Describe the plants that grew in Pennsylvanian-age swamps and forests.
10. What major evolutionary adaptation occurred in reptiles during the Late Paleozoic?
11. What caused the Permian mass extinctions?

For Further Thought

1. The term *swamp* includes a range of geologic environments. In what way were the Pennsylvanian swamps similar to present-day swamps? In what way were they different? What are some of the major "swamp" environments on earth today?
   Can you now more accurately define what is meant by "swamp"?
2. Calculate the time represented by the Mississippian period, by the Pennsylvanian period, and by the Permian period. Compare these with the time represented by the Cambrian period and by the Ordovician period. How does the length of the Late Paleozoic (Mississippian, Pennsylvanian, and Permian) compare with the length of the Early Paleozoic (Cambrian, Ordovician, and Silurian)?

Suggestions for Further Reading

Branson, C. C. 1962. *Pennsylvanian system in the United States: A symposium.* Tulsa, Okla.: American Association of Petroleum Geologists.

Colbert, E. H. 1969. *Evolution of the vertebrates,* 2d ed. New York: John Wiley and Sons.

Dineley, D. L. 1984. *Aspects of a stratigraphic system: The Devonian.* New York: John Wiley and Sons.

Heckel, P. H. 1980. Paleogeography of eustatic model for deposition of mid-continent Upper Pennsylvanian cyclothems. In Fouch, T. D., and E. R. Magathan, eds. Paleozoic paleogeography of the west-central United States, Rocky Mountain paleogeography, symposium 1: The Rocky Mountain Section, pp. 197–213. Society of Economic Paleontologists and Mineralogists.

Thomas, B. A. and Spicer, R. A. 1987. *The evolution and paleobiology of land plants.* London: Croom Helm.

Wanless, H. R. 1962. Pennsylvanian rocks of the Eastern Interior Basin. In Branson, C. C., ed. Pennsylvanian system in the United States, pp. 4–59. Tulsa, Okla.: American Association of Petroleum Geologists.

Williams, H., and Hatcher, R. D., Jr. 1982. Suspect terranes and accretionary history of the Appalachian orogeny. *Geology* 10: 530–36.

24

THE MESOZOIC

Cross-bedded Navajo sandstone, Zion National Park, Utah.
© Terry Donnelly/Tom Stack and Associates

In the Mesozoic, most of the North American craton was emergent, and many Early and Late Paleozoic strata were eroded from various parts of the craton.

Along the eastern margin, North America and Africa separated to form the present-day Atlantic Ocean, and the Appalachians underwent continuous erosion.

Widespread tectonic activity occurred along the western margin as several orogenic pulses formed various mountain chains, including the Rocky Mountains.

Of course, dinosaurs were the most interesting Mesozoic life-forms. Were they just large lizards, or something more? Why did they die out? Their disappearance at the end of the Cretaceous currently has paleontologists debating the cause of these mass extinctions. Mammals immediately occupied the ecologic niches vacated by the dinosaurs, and this set the stage for the Cenozoic.

CHARACTERISTICS OF THE MESOZOIC

The Mesozoic era consists of the Triassic (208 to 245 million years ago), the Jurassic (144 to 208 million years ago), and the Cretaceous (66 to 144 million years ago) periods. In the Mesozoic, the focus of orogenic events shifted from the eastern margin, and the orogenies that formed the Appalachians, to the western margin, and the orogenies that formed the western Cordillera.

North America was emergent throughout most of the Mesozoic (figure 24.1). The continent continued its counterclockwise rotation and approached its present orientation, but at more southern latitudes.

The tectonics of Pangaea played an important role in the geologic events of North America during the Mesozoic. Evidence for the existence of Pangaea from the Late Paleozoic through the Late Triassic includes (1) the fit of various continental margins, particularly of South America and Africa; (2) the distribution of land fossils found on different continents, particularly the seed fern *Glossopteris* and reptiles such as the cotylosaurs and *Lystrosaurus;* (3) the polar-wander curves calculated for each continent; and (4) the distribution of unique rock types, particularly the **Gondwana rock succession** (figure 24.2).

The Gondwana rock succession is a generalized stratigraphic sequence of rocks found on all the southern continents. (To a geologist, the southern continents are South America, Africa, India, Australia, and Antarctica.) Although there are many local differences in geology, the general sequence is the same. The base of the stratigraphic sequence consists of Carboniferous-age tillites and other evidence of glaciation—striated pebbles, striated pavements, and so on.

At some stratigraphically higher point, Permian-age nonmarine sedimentary rocks, such as fluvial, swamp, and deltaic sedimentary strata, are found. These deposits reflect the nonmarine conditions found on many continents during the Permian period. These are followed by Permian- and Triassic-age red beds and some eolian sandstones, again indicating nonmarine conditions. Finally, Late Triassic and Early Jurassic–age basalts are found, which indicates the initiation of continental rifting—the start of the breakup of Pangaea.

In the Mesozoic, almost all the tectonic events experienced by North America resulted from the breakup of Pangaea and the divergence of the continents. As Africa and North America began to diverge in the Late Triassic, a new ocean formed between them—the present-day Atlantic Ocean. Because North America moved relatively westward, away from Africa, the eastern margin turned into a passive trailing edge, whereas the western margin was now an active leading edge experiencing convergent plate motion with the oceanic Pacific Plate. Subduction occurred along the western margin throughout the Mesozoic and the Cenozoic. The convergent tectonics produced both the Rocky Mountains and many of the geologic features west of the Rockies.

THE NORTH AMERICAN CRATON

The majority of Mesozoic-age sediments of the craton reflect nonmarine environments on an emergent continent. Widespread epeiric seas, similar to those formed during Early Paleozoic time, were no longer common. Mesozoic epeiric seas were geographically restricted and confined principally to the craton's western edge. Because most of the continent was emergent, erosion of older sediments and rocks occurred. Thus, Mesozoic-age strata are missing from much of the central part of the craton. Most Triassic- and Jurassic-age sedimentary rocks of the craton are confined to the extreme western edge, where the epeiric seas formed between the highlands of the western margin and the craton.

The Triassic

Lower Triassic rocks of the craton consist of red beds and sandstones. Along the extreme western edge of the craton, where a restricted epeiric sea formed between the craton and the highlands to the west, limestones were formed. These limestones provide geologists with one of the few glimpses of Early Triassic marine life.

The Jurassic

Along the western edge of the craton, between the craton and the highlands, an epeiric sea began to form in the Early Jurassic. Just to the east of this epeiric sea, the

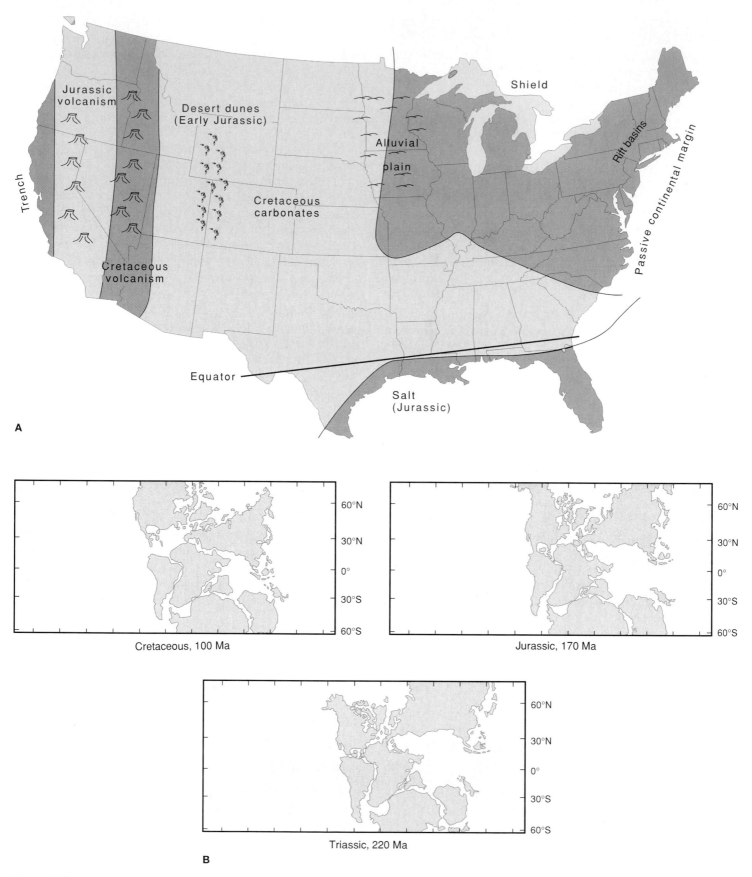

FIGURE 24.1 (A) Paleography of the United States during the Mesozoic. (B) The positions of the continents during the Mesozoic.

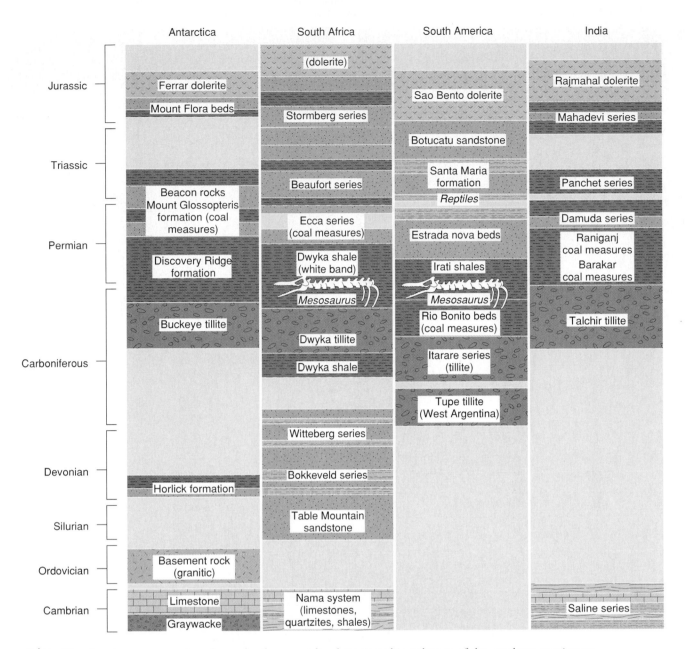

FIGURE 24.2 The Gondwana succession shown by the generalized stratigraphic columns of the southern continents.

Figure from Earth History and Plate Tectonics *An Introduction to Historical Geology 2nd Edition by Carl Seyfert and Leslie Sirkin. Copyright © 1973, 1979 by Carl A. Seyfert and Leslie A. Sirkin. Reprinted by permission of Addison Wesley Educational Publishers, Inc.*

Navajo sandstone was deposited (figure 24.3). The Navajo sandstone is another extremely pure quartz sandstone; it resulted from the recycling of older sediments that probably originated from the craton to the northeast. Although the origin of the Navajo sandstone has been disputed—some geologists argue a shallow-marine origin, others an eolian (wind-deposited) origin—the current consensus is that the sands comprising the Navajo sandstone were deposited as coastal sand dunes along the eastern edge of the epeiric sea. The large-scale cross-bedding typical of the formation indicates an eolian origin.

In Middle and Late Jurassic time, the *Sundance epeiric sea* continued to advance and became more widespread until it spread from Canada, along the western edge of the craton, to the Gulf of Mexico.

On the western edge of the craton, the *Morrison formation* was produced during the Late Jurassic. The Morrison formation is a famous sequence of sandstones and some conglomerates bearing terrestrial-vertebrate fossils. The sediments making up the Morrison formation were deposited on an alluvial plain as part of a clastic wedge that formed as the result of the Nevadan orogeny.

FIGURE 24.3 The many faces of the Navajo sandstone. (A) The Navajo sandstone. Note the large-scale cross-bedding. (B) The Navajo sandstone. Rainbow Bridge National Monument, San Juan County, Utah. (C) Navajo sandstone (Jurassic). Mountain of the Sun, Zion National Park, Utah.

(A) © Times Mirror Higher Education Group, Inc./Doug Sherman, photographer. (B) Photograph by W. R. Hansen, USGS Photo Library, Denver, CO. (C) Photograph by W. B. Hamilton, USGS Photo Library, Denver, CO.

The Cretaceous

At least four widespread epeiric seas formed over much of the western edge of the craton during the Cretaceous. Situated just east of the Rocky Mountain belt, some of these seas extended from Canada to the Gulf of Mexico. Great quantities of clastics were deposited in these epeiric seas (figure 24.4). The source of the clastics was the newly uplifted mountains in the west. Chalk, a type of carbonate, was also deposited. The *Niobrara chalk* of Kansas is one example. Abundant plant life on restricted parts of the craton led to the formation of coal. (Coal in North America is almost always of one of two ages—Pennsylvanian, most likely, or Cretaceous.)

FIGURE 24.4 The Dakota sandstone (Cretaceous). Garden of the Gods, El Paso County, Colorado.
Photograph by J. C. Maher, USGS Photo Library, Denver, CO.

THE EASTERN MARGIN
Triassic Rift Basins

The start of the breakup of Pangaea in Triassic time greatly affected the types, thicknesses, and distribution of sediments along the eastern margin.

During Late Triassic time, long, narrow *rift basins,* bounded by faults, formed along the eastern margin as North America and Africa began to diverge (figure 24.5). The rifting occurred to the east of the Appalachian Mountains, which marked the zone of earlier convergence, so America obtained a part of Africa as a result of the convergence–divergence sequence. The divergence produced not only basins but also intervening highland areas. The alternating series of long and narrow highland mountains and low basins is called horst-and-graben structure. Great quantities of nonmarine sediment were dumped into the basins (the grabens). In some places, over 5000 meters of sediment is found in vertical section. The *Newark series* is one example of these Triassic-age, rift-basin sedimentary series. Most of the sedimentary rocks found in the rift basins consist of poorly sorted, immature sandstones, such as arkoses, other types of sandstones, and conglomerates. The sediments came from the adjoining highland mountains, the horsts. Most sediments were deposited either by streams that drained into or ran through the basins or by lakes that formed in the basins.

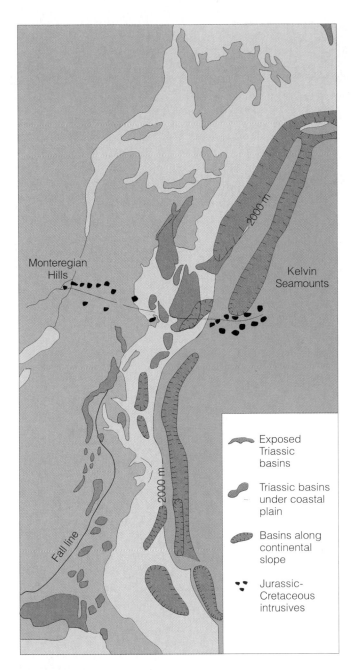

FIGURE 24.5 Triassic fault basins of North America's eastern margin.
From Stearn, Carroll, and Clark, Geological Evolution of North America, *3d ed. Copyright © 1979 John Wiley & Sons, Inc., New York, NY. Reprinted by permission of John Wiley & Sons, Inc.*

The sedimentary rocks found in the Newark series and other strata deposited in the rift basins of this age are important to paleontologists because they contain the earliest dinosaur tracks and other important vertebrate fossils, such as the first mammals.

Igneous rocks formed as flows, dikes, and sills. Magma flowed onto the surface in response to continental rifting. Numerous dikes (igneous rock bodies that cut across other rocks) also formed. The Palisades sill of New York, seen along the Hudson River, is a typical sill—an intrusive igneous structure (figure 24.6).

The Jurassic and Cretaceous Passive Margin

Rifting proceeded in the Jurassic and Cretaceous, and a fully developed, passive continental margin formed (figure 24.7). The Atlantic Coastal Plain (the region between the Appalachian Mountains and the Atlantic Ocean) and the passive eastern margin continued to develop until modern times. Sediments and rocks were eroded from the Appalachians and carried by rivers over the coastal plain and deposited on the continental shelf. Deposition along the shelf, together with the passive tectonic situation, produced a well-developed continental shelf, slope, and rise.

THE WESTERN MARGIN

The relative westward motion of North America as it diverged from Africa produced convergent tectonic motion along the western margin throughout most of the Mesozoic. Thus, many geologic events and thick sediment accumulations along the western margin are of Mesozoic age.

FIGURE 24.6 The Palisades sill.
© John Serra.

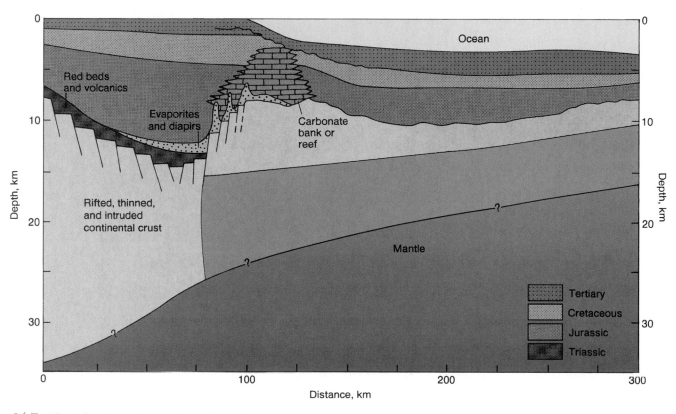

FIGURE 24.7 The Atlantic passive continental margin.
From J. A. Grow et al., AAPG Memoir. 29, fig. 13, p. 80, 1979. Copyright © 1979 American Association of Petroleum Geologists, Tulsa, OK. Reprinted by permission.

TABLE 24.1 *Summary of Cordilleran Orogeny*

Name	Age	Characteristics	Tectonic Cause
Laramide	Late Cretaceous–Paleocene	Mostly vertical uplift of rocks	Low-angle subduction
Sevier	Cretaceous	Overthrusting of blocks to the east	Low-angle subduction
Nevadan	Jurassic	Large batholiths of granite produced	Low-angle subduction
Sonoman	Triassic	Thrust faulting	Collision of island arcs

The Cordilleran Orogeny

The formation of the Rocky Mountains was a long and complicated process that began in the Triassic (some geologists extend the beginning back even further, to the Carboniferous-age Ancestral Rockies), continued through the Mesozoic, and finally ended in the Early Cenozoic (Eocene). The entire orogenic episode that produced the Rockies and other mountains of western North America is called the **Cordilleran orogeny** (pronounced "Cor-dee-AIR-an"). The Cordilleran orogeny is separated into several smaller orogenic episodes or pulses: the Triassic Sonoman orogeny, the Jurassic Nevadan orogeny, the Cretaceous Sevier orogeny, and the Late Cretaceous to Early Tertiary (Paleocene) Laramide orogeny (table 24.1). The Nevadan and Laramide orogenies were the major orogenic events.

Tectonics

The tectonics and geology of North America's western margin during the Mesozoic resulted from the convergence of the western margin and the oceanic Pacific Plate. This produced a prolonged period of subduction along the margin and formed an extensive ocean–continent subduction zone (figure 24.8).

Overall, the tectonic styles and patterns of the western margin were the result of subduction-related processes (shown in figure 24.8 and summarized in table 24.1). In the Triassic, the direction of subduction was toward the west, or toward and under the oceanic Pacific Plate. In Jurassic time, the direction of subduction changed—it was now toward the east, or toward and under the western margin of the continent. During Late Jurassic time, through the Cretaceous, and into the Cenozoic, the subduction angle decreased, so that the base of the continent and the subducting oceanic plate came into contact at the extreme western edge of the margin first. Over time, this zone of continent–ocean interaction moved eastward. The interaction of the subducting oceanic plate with the base of the continent resulted in orogenic pulses. Thus, the oldest orogenies on the western margin are farther west, and the younger orogenies are eastward, toward the craton. (Intuitively, this seems backward, particularly if one considers the idea of continental accretion and the ages of the structural provinces of the Canadian Shield.)

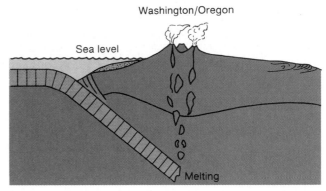

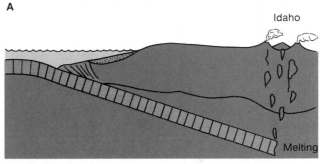

FIGURE 24.8 Tectonics of North America's western margin during the Mesozoic. (A) Late Jurassic. (B) Late Cretaceous. Note how the shallowing subduction angle through the Mesozoic caused younger orogenic episodes farther east, toward the craton.
From Earth and Life through Time, *2E by Steven M. Stanley. Copyright © 1989 W. H. Freeman and Company. Reprinted by permission.*

The Triassic

In Triassic time, rifting occurred along the eastern margin, and so compression occurred along the western margin as the continent moved in a relatively westward direction. A subduction zone formed along the western margin during early Triassic time and produced the first orogenic episode or pulse of the Cordilleran orogeny—the **Sonoman orogeny**—starting in Early Triassic time (possibly even the Late Permian) and ending in Late Triassic time. This pulse was the result of a series of collisions of island arcs, their associated sediments, and other accreted terranes along the western margin. Thrust faulting was common as large blocks were thrust eastward toward the craton. Volcanics and graywackes, deposited adjacent to the island arcs, are a common rock type. At this time, the angle of subduction was toward the west, under the Pacific Plate.

In the Triassic, on the eastern side of the orogenic belt near the craton, a series of varied nonmarine sediments were deposited that reflected the orogenic pulse and resulting environmental conditions. The Triassic-age formations shown in figures 24.9 and 24.10 are widespread throughout western North America. To the west of the orogenic belt, in deeper water, many immature graywacke sandstones and deepwater shales were deposited.

In general, the west-to-east progression of deep-water sediments adjacent to an orogenic belt (and associated nonmarine sediments), adjacent to an epeiric sea (and associated shallow-water carbonate sediments), and adjacent to the craton is a common theme throughout the Mesozoic.

Accreted Terranes

Accreted terranes are geologic provinces (that is, very large blocks) that are found in orogenic regions and are internally consistent in their geology but markedly different in geology from the adjoining terrane (figure 24.11). Geologic features that may be consistent within each accreted terrane include (1) the rock types, (2) the stratigraphy, (3) the types and ages of the fossils, (4) the calculated polar-wander curve, (5) the type or types of metamorphism, and (6) the type or types of igneous activity. Essentially, each accreted terrane shows its own unique geologic history. Accreted terranes are separated from each other or from the continental mass by major faults. Physically, many accreted terranes may be either the remains of island arcs or parts of oceanic plateaus that are remnants of continental material.

Accreted terranes are "suspect" (an older term is, in fact, "suspect terrain") because, in many cases, it is not known where they formed or exactly when they were emplaced along the margin. One accreted terrane emplaced on North America's western margin appears to have formed 30 degrees south of the equator. (Evidence for this comes from the calculated polar-wander path and fossils.) Certainly this terrane shows a remarkable northward movement! Figure 24.11 shows how some terranes have been stretched in a northward direction—sort of squished all along the western margin. This indicates that the direction of subduction along the western margin during the Mesozoic was not strictly eastward, but rather northeastward, with both an eastward and a northward component.

The Jurassic

In Jurassic time, the subduction angle along the western margin changed from westward to eastward. At that time, the Pacific Plate began to be subducted under the edge of the continent.

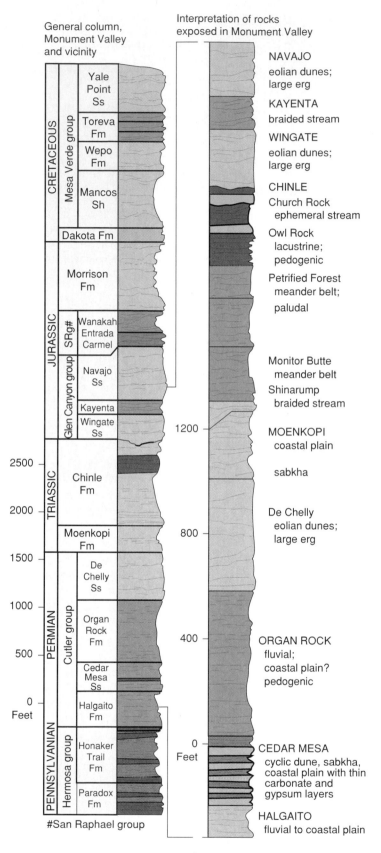

FIGURE 24.9 Stratigraphic column of rocks in Monument Valley, Arizona and Utah. Note that predominately terrestrial sedimentation (eolian [wind-deposited] and fluvial [rivers and streams]) has occurred in this region from the Late Paleozoic through the Mesozoic.

After R. C. Blakey and D. L. Baars, 1987. Rocky Mountain Section of the Geological Society of America, 1987, fig 2, p. 362, Geological Society of America. Reprinted by permission of the authors.

FIGURE 24.10 Windgate sandstone (Triassic) and other Triassic- and Jurassic-age sedimentary rocks. Capitol Reef National Monument, Wayne County, Utah.
Photograph by W. B. Hamilton, USGS Photo Library, Denver, CO.

The **Nevadan orogeny,** a more widespread orogenic event, began in the Late Triassic or Early Jurassic, ended in the Late Jurassic or Early Cretaceous, and produced large batholiths of granite from the Sierra Nevada mountains of California through Washington and the Rockies of Idaho, all the way to British Columbia. Igneous activity produced batholiths in northern British Columbia and the southern Yukon from Jurassic (Cassia batholith) into the Cenozoic (Kastberg intrusion and Sea Gull batholith).

Two major regions of sedimentation are recognized along the western margin during the Jurassic: a western belt and an eastern belt. The western belt of sedimentation was adjacent to an island arc and extended from California to British Columbia. Most of the sediments are volcanic in origin, although some limestones and reefal carbonates are also found. East of the western belt, and adjacent to the craton's western edge, sediment was deposited in an epeiric sea. Many limestones with marine fossils, and some minor sandstones, are found. (The Navajo sandstone was deposited just east of this belt of deposition.)

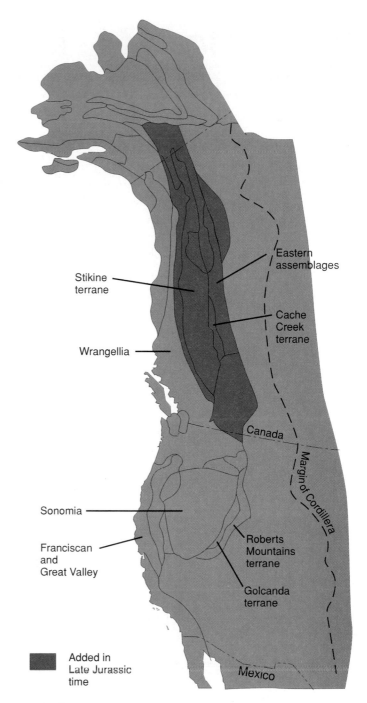

FIGURE 24.11 Accreted terranes.
Reproduced with permission, from the Annual Review of Earth and Planetary Sciences, *Vol. 11, © 1983 by Annual Reviews Inc.*

The Cretaceous

In Late Cretaceous time, continued orogenic activity—the **Sevier orogeny**—affected much of the Rocky Mountain region. Much of the tectonics of this orogeny involved overthrusting of blocks of rock toward the craton. In conjunction with the overthrusting, conglomerates and sandstones were deposited. Eastward of this orogeny, epeiric seas again developed.

Box 24.1

Regional Geology

The Gulf of Mexico

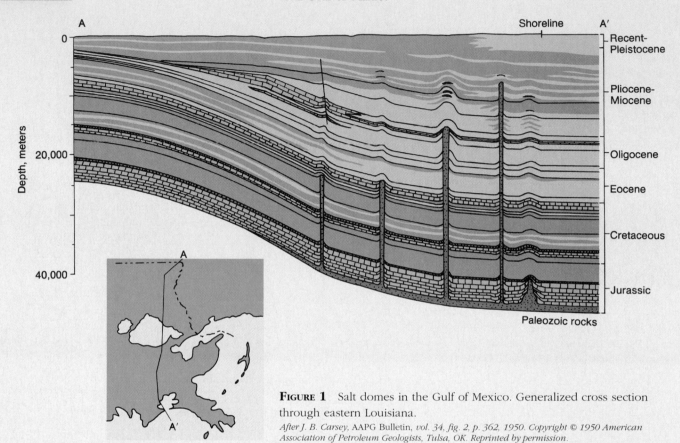

Figure 1 Salt domes in the Gulf of Mexico. Generalized cross section through eastern Louisiana.

After J. B. Carsey, AAPG Bulletin, vol. 34, fig. 2, p. 362, 1950. Copyright © 1950 American Association of Petroleum Geologists, Tulsa, OK. Reprinted by permission.

The Gulf of Mexico started to form in very Late Triassic time as continental rifting occurred—the beginning of the breakup of Pangaea. As North America began to separate from Africa, a major rift valley formed. Abundant terrestrial sediments were deposited in the rift valley.

The area then underwent subsidence, followed in the Early Jurassic by a marine invasion. In the Middle Jurassic, the shallow marine water, combined with a hot and arid climate, produced abundant evaporites. One major salt deposit—the Louann salt—eventually formed traps for the oil. Anhydrite also formed.

In the Late Jurassic, the middle of the Gulf basin experienced divergence— a minor episode of seafloor spreading.

Some oceanic-type crust (basalt) is now found in the middle of the Gulf, a remnant of this rifting episode. Also during the Late Jurassic, the evaporite conditions ended, and normal marine sedimentation dominated. One important rock formation, the Smackover limestone, was deposited. Today, this rock is a major petroleum reservoir.

In the Early Cretaceous, the seafloor spreading stopped. Reefs were extremely common in the Gulf, and many grew very large. In the Late Cretaceous, chalk was deposited. In the Early Cenozoic, sediments record cycles of sedimentation corresponding to several transgressions and regressions. During the transgressions, marine calcareous marls (fine-grained limestones) and

sands and clays were deposited. During the regressions, thick deltaic sediments were deposited.

The strata of the Gulf are extremely important because they contain oil. Oil is found in many of the carbonate strata—for example, the Smackover limestone—having migrated there from somewhere else. In addition, sometime after the evaporites were deposited, at least after several thousands of meters of other sediment were deposited on top of the evaporites, they began to form *salt domes* (figure 1). (Salt, like ice, is plastic over long periods.) As the salt flowed upward, it punctured younger beds and formed traps for subsequent oil migration.

Many of the Mesozoic tectonic events that occurred on North America are related to the breakup of Pangaea and the westward movement of North America away from Africa.

Because the North American craton was emergent throughout much of this period, Mesozoic-age strata are missing from much of the craton's central part. Most Mesozoic-age cratonic strata are confined to the craton's extreme western edge, since it was here that narrow epeiric seas formed. Only in Cretaceous time, when epeiric seas were more widespread, did sedimentation occur over significant parts of the craton.

As the continents diverged, rift basins formed along North America's eastern margin in the Triassic. Large volumes of nonmarine sediment were deposited in these long, narrow basins. As rifting progressed in Jurassic and Cretaceous time, a passive continental margin formed. In the Cretaceous, the Atlantic Coastal Plain formed. Since Mesozoic time, the Appalachians have been eroding and the sediment carried eastward to form a fully developed continental shelf, slope, and rise.

In Mesozoic time, several orogenic pulses occurred along North America's western margin, all as part of the Cordilleran orogeny. The later orogenic pulses occurred farther eastward than the older orogenies owing to the shallowing of the angle of subduction as the oceanic Pacific Plate coupled with the underside of the North American continent.

A summary of Mesozoic geologic and biologic events is shown in figure 24.12.

Period	Western margin	Craton	Eastern margin	Life
Cretaceous	Sevier orogeny —Overthrusting —Conglomerates —Sandstones *Dakota sandstone* *Mesaverde formation*	Several transgressions and regressions along the western edge of the margin —Some coal —*Niobrara chalk*	Atlantic Coastal Plain Mature passive margin	Mass extinctions at the end of Cretaceous First angiosperms
Jurassic	Nevadan orogeny —Granite batholiths —Subduction flip-flop —Local uplifts —*Morrison formation*	Very western edge of craton: Middle Jurassic —Transgression —Sundance Sea Early Jurassic —*Navajo sandstone* Widespread evaporites in the Gulf of Mexico	Fault-block mountains eroded —Broad, low-lying surface produced	Dinosaurs First birds: *Archaeopteryx* Index fossils for marine strata: ammonites Gymnosperms common —Cycads and ginkgoes —"Age of Cycads"
Triassic	Sonoman orogeny —Subduction/collision —Volcanics —Graywackes —Thrust faulting *Chinle formation*	General emergence of the craton —Nonmarine environments —Red beds Gulf Coast starts to form	Rifting —Grabens —Palisades sill *Newark series* —Arkose —Sandstone	Reptiles dominant on land First mammals Mostly clams and ammonites in marine strata Scleractinian corals evolve (modern corals)

FIGURE 24.12 Mesozoic: summary diagram.

TERMS TO REMEMBER

accreted terranes 446
Cordilleran orogeny 445

Gondwana rock succession 439
Nevadan orogeny 447

Sevier orogeny 447
Sonoman orogeny 445

QUESTIONS FOR REVIEW

1. What is the evidence for the existence of the supercontinent Pangaea?
2. How do cratonic sedimentary rocks of Mesozoic age differ from those of Early and Late Paleozoic age?

3. Why were epeiric seas confined to the western edge of the craton during the Mesozoic?
4. Why are sedimentary rocks that were deposited in Triassic rift basins along the eastern margin important to paleontologists?

5. Describe the tectonics of the western margin during Mesozoic time.
6. What are accreted terranes? Why are they sometimes called "suspect"?

FOR FURTHER THOUGHT

1. What evidence would indicate the change in angle of subduction from west to east along North America's western margin in the Jurassic? Can you think of several possible causes for this subduction "flip-flop?"

SUGGESTIONS FOR FURTHER READING

Churkin, M., Jr., and Eberlein, G. D. 1977. Ancient borderland terranes of the North American Cordillera: Correlation and microplate tectonics. *GSA Bulletin* 88:769–86.

Coney, P. J. 1972. Cordilleran tectonics and North America plate motion. *American Journal of Science* 272: 603–28.

Klein, G., deV. 1969. Deposition of Triassic sedimentary rocks in separate basins, eastern North America. *GSA Bulletin* 80:1825–32.

McPhee, J. 1981. *Basin and range.* New York: Farrar-Straus-Giroux.

25

MESOZOIC LIFE

Petrified logs (Triassic), Petrified Forest National Park, Arizona.
© Doug Sherman/Geofile

INTRODUCTION

The Mesozoic is known as the "Age of Reptiles" because of the dominance of the dinosaurs from the Middle Triassic to the end of the Cretaceous. A tremendous amount of work has gone into studying dinosaurs in the past few years. What kind of animals were they? Why did they become extinct? The resolution of one question leads to the asking of two more. Debates rage. Hypotheses are proposed. Rocks are broken and fossils found. It's an exciting time to be a geologist.

The dinosaurs seem to overshadow other major events in the history of life during the Mesozoic, including the evolution of mammals in the Triassic. A very important event in plant evolution also occurred during the Mesozoic—flowering plants evolved.

THE TRIASSIC

Early Triassic–age invertebrate faunas are not well known, because epeiric seas were not widespread until Late Triassic time. Organisms were recovering from the terminal-Permian mass extinctions. Particularly on the western edge of the North American craton, however, some Early Triassic–age invertebrate fossils have been found. These fossils consist mostly of ammonites (see the next section) and clams. Strangely, many of the ammonites suffered another extinction episode at the end of the Triassic, although one family survived to give rise to the abundant ammonites of Jurassic time.

An important group of corals—the scleractinians—evolved from rugose corals in the Early Triassic. Initially, only solitary forms were produced, but colonial forms soon appeared. By the end of the Triassic, reefs were made by scleractinian corals. The scleractinians are the major coral group of present-day oceans.

The Mesozoic is often referred to as the "Age of Reptiles." One line of reptiles, the *thecodonts,* gave rise to the dinosaurs in the Middle Triassic, and dinosaurs dominated the land surface until the end of the Cretaceous. But not all reptiles were confined to land. Several groups invaded the marine environment—for example, the *ichthyosaurs,* the *mosasaurs,* and the *plesiosaurs.* Ichthyosaurs and plesiosaurs were the dominant marine reptiles (figure 25.1).

Some authors claim that the Loch Ness Monster is a plesiosaur or a family of plesiosaurs trapped in Loch Ness since the last ice age or some other earlier time. The story of the search for "Nessie" is long and complicated and, for the most part, involves nonscientific methods. Despite the sonar and pictorial evidence that something big is indeed living in the lake, persuasive scientific evidence has not yet been obtained. If there are large creatures living in Loch Ness, they are almost certainly not plesiosaurs. (But it would be marvelous if they were!)

Reptiles also took to the air during the Mesozoic. The best known of the Mesozoic-age flying reptiles is *Pteranodon* ("tair-AN-oh-don"; figure 25.2). The pterosaurs were the first flying vertebrates. Although debated, it appears that members of this group included both gliders and active flyers. Some apparently had funny bodies! And some

FIGURE 25.1 Mesozoic marine reptiles. The long-necked plesiosaurs and ichthyosaurs are fighting for a meal.
Photograph courtesy of Field Museum of Natural History, Chicago; mural by C. R. Knight, #CK34T.

FIGURE 25.2 Mesozoic flying reptiles. Various pterosaurs glide through the air in the background. Archaeopteryx, the first bird, is shown in the top center.

Photograph courtesy of Field Museum of Natural History, Chicago; painting by C. R. Knight, #CK39T.

were huge—the largest animals ever to fly. Pterosaurs arose in the Early Jurassic but became extinct in the Late Cretaceous.

On land, the first mammals appeared in the Late Triassic (figure 25.3). Small skull fragments, jaws, and teeth are the only fossil remains of these animals yet found. Many of the jaws show reptilian characteristics, indicating that mammals evolved from mammal-like reptiles in the Late Triassic. The first mammals were small and probably nocturnal. Throughout the Jurassic and Cretaceous, they remained so and did not become dominant until after the extinction of dinosaurs, sixty-five million years ago.

Plant life on land continued to diversify in the Triassic and Jurassic. Ferns, sphenopsids, and some lycopsids were abundant in the Triassic. **Gymnosperms,** several seed-bearing, nonflowering groups of plants, were also common. Two groups of gymnosperms dominated the Triassic and Jurassic landscapes—the cycads and the ginkgoes (figure 25.4). Cycads probably arose from the seed ferns in the Carboniferous and became very abundant in the Jurassic, but they declined greatly in the Late Cretaceous. Several species of cycads are alive today. Ginkgoes evolved in the Permian, probably also from a seed-fern ancestor. They display a very characteristic, fan-shaped leaf. One species of ginkgo is alive today. Conifers, which arose in the Carboniferous, diversified greatly in the Early Mesozoic. Many forms similar to modern types, such as pines, cedars, and firs, were abundant.

FIGURE 25.3 Diagrammatic representation of the first mammal.
Source: Crompton, 1968.

One very important fossil locality for Triassic-age conifers is the Petrified Forest in Arizona (figure 25.5). During Triassic time, conifers were growing next to rivers. When the trees died, their trunks fell into the water and were carried to bends and meanders in the river, where they became jammed against each other. Eventually they became waterlogged and sank. Covered by river sediment, they were preserved when silica-rich water seeped through the sediment, replacing wood with silica, atom by atom. Thus, the petrified logs of the Petrified Forest still show growth rings and other features but are composed of silica instead of the original woody material.

FIGURE 25.4 Mesozoic flora. Various conifers are in the background. Cycads and ferns dominate the foreground.
Drawing by Z. Burian under the supervision of Professor J. Augusta.

FIGURE 25.5 Triassic-age mudstones form the Painted Desert. Petrified Forest National Monument, Arizona.
Photograph by W. B. Hamilton, USGS Photo Library, Denver, CO.

THE JURASSIC

During the Jurassic, two groups of invertebrates in the epeiric seas underwent rapid evolution and diversification—the ammonites and the pelecypods (bivalves) (figure 25.6). The more important of the two were the **ammonites,** a group of coiled cephalopods. Essentially, the evolution of the ammonites involved two trends: an increase in coiling, and an increase in the complexity of the suture pattern. (Sutures are the lines of intersection between the shell and the previously used, internal living chambers.) Cephalopods were (and many of the living representatives still are) active predators. The fossil record of ammonites shows the shells coiled, so that the center of gravity was nearer the front, where the animal lived. Because of this, the ammonites could move quickly and still retain the shell's protective benefits. To support the shell, the suture pattern became more complex and structurally stronger, covering more surface area. Because of the coiling and the shell's abundant and varied suture patterns, many genera and species of ammonites can be recognized. In addition, because ammonites were nectonic (swimmers), their fossils are found all over the world in marine strata of this age. Consequently, almost all Jurassic-age marine strata are indexed on the basis of ammonites. Ammonites are, perhaps, the best index fossil.

FIGURE 25.6 An ammonite.
Photograph by W. A. Cobban, USGS Photo Library, Denver, CO.

FIGURE 25.7 Archaeopteryx—the first bird.
Photograph courtesy of American Museum of Natural History, Neg. #325288.

On land, birds arose from an ornithopod-dinosaur ancestor. The dinosaurs that gave rise to birds were very small, perhaps about chicken size. The first bird, *Archaeopteryx* ("ar-kee-OP-ter-icks"), was little more than a reptile with feathers (figure 25.7). It had a reptilelike skeleton, a reptilian tail with feathers, and teeth. (Modern birds lack teeth.) In a very real sense, dinosaurs did not all become extinct; some evolved into the birds.

One of the few places *Archaeopteryx* fossils have been found is in the Solnhofen limestone of southern Germany. Like the Burgess shale, the Solnhofen limestone is an exceptional fossil locality, containing many fossils found nowhere else in the world. The limestone is an extremely fine-grained—termed *lithographic*—limestone that was deposited in quiet lagoons next to an epeiric sea. *Archaeopteryx* apparently settled to the bottom of the lagoon after they died. Jellyfish, insect, and various reptile fossils are also found in the Solnhofen limestone.

DINOSAURS

Dinosaurs were an incredibly successful group of animals, ruling the earth for 150 million years. They evolved from a stem line of small reptiles, the thecodonts, in the Middle Triassic (about 215 million years ago) and became extinct at the end of the Cretaceous (66 million years ago). The name *dinosaur*, which means "terrible lizard" or "fearfully great lizard," was first proposed in 1841 by Sir Richard Owen and is an informal term for a very diverse group of animals. "Dinosaur" also implies that these animals were large reptiles, and dinosaurs did in fact display several reptilian characteristics. They laid shelled eggs like those of modern reptiles. In addition, their jaws, ears, teeth, scaly skin, and pelvic structure were all reptilelike.

The first finds of dinosaur skeletal remains in the western United States provided the context for a colorful and controversial episode of vertebrate paleontology in the mid- to late-1800s. Rivalries were common among scientists, each of whom was eager to be the first to discover, name, and publish a new dinosaur species. The most famous rivalry involved Edward Drinker Cope (1840–1897) and Othniel Charles Marsh (1831–1899). Their hatred for each other was well known. Sometimes their field workers even fired guns at each other in an effort to claim prime collecting localities! In fact, each started his own scientific journal so he could be the first to publish his discoveries. In this intense effort to outdo each other, the two men frequently made mistakes—for example, fitting a skull to nearby vertebrae even if they had not occurred in association with each other. Unfortunately for Cope, he died first, which allowed Marsh to have the last laugh. Marsh named a fossil after Cope—the coprolite, which is fossilized excrement.

Dinosaur Classification

Dinosaurs are classified as reptiles. A generalized classification scheme for reptiles is shown in table 25.1. A brief review of reptile history provides a context for the understanding of

TABLE 25.1 *Summary of Reptile Classification*

Class Reptilia

Subclass Anapsida "without arch"	
Order Cotylosauria	First reptiles
Order Chelonia	Turtles
Order Mesosauria	Freshwater mesosaurs

Subclass Synapsida "with arch"	
Order Pelycosauria "basin lizards"	Mammal-like reptiles
Order Therapsida "mammal arch"	Advanced mammal-like reptiles

Subclass Euryapsida "broad arch"	
Order Protorosauria "first lizards"	Terrestrial euryapsids
Order Ichthyosauria "fish lizards"	Aquatic reptiles
Order Placodontia	Aquatic euryapsids
Order Sauropterygia	Marine euryapsids

Subclass Diapsida "double arch"	
Infraclass Lepidosauria	
Order Eosuchia	Primitive lepidosaurs
Order Rhynchocephalia "beaked lizards"	Beaked lepidosaurs
Order Squamata	Lizards, snakes
Infraclass Archosauria	
Order Thecodontia "socket-toothed"	Primitive archosaurians
Order Crocodilia	Crocodiles
Order Pterosauria "winged lizards"	Flying reptiles
Order Herrerasauria "Herrera lizards"	First dinosaurs
Order Saurischia "lizard-hipped"	Lizard-hipped dinosaurs
Suborder Theropoda "beast feet"	

Subclass Diapsida ("double arch") continued	
Infraorder Coelurosauria "hollow-tail lizards"	
Infraorder Carnosauria "flesh lizards"	
Infraorder Deinonychosauria "terrible-claw lizards"	
Infraorder Ornithomimosauria "bird-mimic lizards"	
Infraorder Oviraptorosauria "egg-thief lizards"	
Suborder Sauropodomorpha "lizard-feet forms"	
Infraorder Prosauropoda "before sauropods"	
Infraorder Sauropoda "lizard feet"	
Suborder Segnosauria "slow lizards"	
Order Ornithischia "bird-hipped"	Bird-hipped dinosaurs
Suborder Ornithopoda "bird feet"	
Suborder Scelidosauria "limb lizards"	
Suborder Stegosauria "roof lizards"	
Suborder Ankylosauria "fused lizards"	
Suborder Pachycephalosauria "thick-headed lizards"	
Suborder Ceratopsida "horned faces"	

Reptile classification from Colbert, 1980; Dinosaur classification from Lambert, 1990.

dinosaurs as a group. Reptiles are classified according to the number and position of openings in their skull. An *apse* is a bony arch that occurs between openings in the skull. Anapsida ("without arch") reptiles have no opening in the skull and, consequently, no arch. Synapside ("with arch") reptiles have a single opening low on the side of the skull. Euryapsida ("broad arch") reptiles have a single opening high on the skull. Diapsida ("double arch") reptiles have two openings in the skull, and hence, two arches.

The subclass Anapsida (first reptiles/turtles) includes three orders. The cotylosaurs ("ko-TILL-oh-sawrs") (Pennsylvanian–Triassic) were the first reptiles and evolved from labyrinthodonts in Pennsylvanian time (see figure 23.19). The group includes both carnivores and herbivores. *Seymouria* (Early Permian, Texas) is an example. Turtles, of the order Chelonia (Early Triassic–Recent), were the direct descendants of the cotylosaurs. The third group, the mesosaurs (Late Pennsylvanian and Early Permian),

were freshwater reptiles. Mesosaur fossils have been found in freshwater deposits in southern Africa and southern Brazil and have been used as evidence to support the theory of continental drift.

The subclass Synapsida (mammal-like reptiles) consists of two orders. Pelycosaurs (Pennsylvanian–Late Permian) include both herbivores and carnivores. Members of this group had sails—for example, *Dimetrodon* (Permian, Texas; see figure 23.20A). Carnivorous pelycosaurs gave rise to the therapsids. The therapsids (Carboniferous–Early Jurassic) had limbs that extended from below the body, rather than from the sides of the body (see figure 23.20B). This group included both herbivores and carnivores. The carnivore group gave rise to the mammals in Early Triassic time.

The subclass Euryapsida (mainly marine reptiles) includes four groups. Protorosauria (Permian) were primitive, land-dwelling euryapsids of Permian time. Ichthyosaurs (Jurassic–Cretaceous) were active, marine predators. Pla-

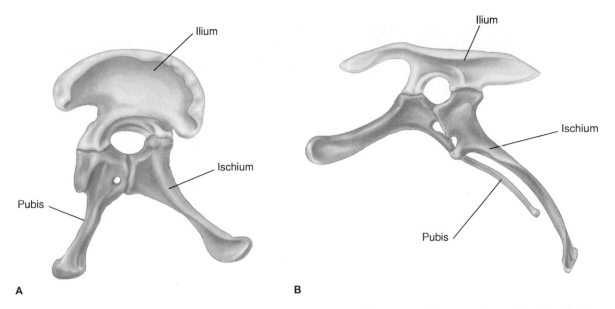

FIGURE 25.8 Structural differences in the pelvic region of the two main orders of dinosaurs. (A) Saurischian pelvis (reptile-hipped). (B) Ornithischian pelvis (bird-hipped).

codonts (Triassic) were also marine euryapsids. The order Sauropterygia (Triassic–Cretaceous) includes the plesiosaurs. (This group is said to live on in the "monster" of Loch Ness!)

The largest subclass, Diapsida (ruling reptiles), includes the modern-day lizards and snakes, and the extinct dinosaurs. Diapsids are generally divided into two broad groups, or infraclasses—lepidosaurs and archosaurs. The lepidosaurs include the order Eosuchia, primitive lepidosaurs; the order Rhynchocephalia, the beaked lepidosaurs; and the order Squamata, lizards (Late Triassic–Recent) and snakes (Cretaceous–Recent). The archosaurians are characterized by a skull opening in front of each eye and include the thecodonts, the crocodiles, the flying reptiles, and the dinosaurs.

Thecodonts ("THEE-koh-dahnts") (Permian–Triassic) were rather small, agile reptiles—for example, *Hesperosuchus*. Some thecodonts showed bipedal tendencies, and various types gave rise to the rest of the archosaurians. The order Crocodilia (Triassic–Recent) includes the modern crocodile. Crocodiles were contemporaries of the dinosaurs. The order Pterosauria (Early Jurassic–Late Cretaceous), is composed of the first vertebrates to fly, for example, *Pteranodon*.

The dinosaurs include three orders of the archosaurians. (Note that the term "dinosaur" does not occur in the formal classification of reptiles.) The two main orders of dinosaurs are differentiated on the basis of their respective pelvic structures (figure 25.8). **Saurischian dinosaurs** have a pelvic structure similar to that of reptiles. **Ornithischian dinosaurs** have a pelvic structure similar to that of birds.

Members of the order Herrerasauria ("her-AIR-uh-sawrs") were the first known predatory dinosaurs—for example, *Herrerasaurus*. It is thought that this group gave rise to the other two orders of dinosaurs.

The classification of dinosaurs undergoes constant revision. This is not surprising, since about one-quarter of all known species of dinosaurs have been discovered within the last twenty years. Because of the similarities between dinosaurs and birds, Robert Bakker and Peter Galton proposed in 1974 that dinosaurs be removed from the class Reptilia and placed in their own class, Dinosauria. This new class would include three orders: Ornithischia, Saurischia, and Aves (birds). (Birds are currently placed in their own class, Aves.) This proposal has not been widely accepted, but it does indicate a very natural grouping of the animals involved.

The dinosaur order Saurischia includes a wide variety of herbivores and carnivores (figure 25.9).

The members of the suborder Theropoda were all bipedal carnivores. Traditionally, the order Coelurosauria ("see-LOOR-oh-sawrs") included only small, predatory dinosaurs, but larger members of this group are also known. Examples of this group include *Coelurus* and *Compsognathus* ("KOMP-so-NAY-thus"). Recent classifications have placed the first bird, *Archaeopteryx*, in this group as well. This placement indicates the close link between smaller members of the dinosaur group and the birds. The order Carnosauria includes the larger predatory dinosaurs. *Tyrannosaurus* ("ty-RAN-oh-SAW-rus") and *Allosaurus* ("AL-oh-SAW-rus") are two well-known examples. *Tyrannosaurus* is often portrayed in books and films as an active predator (an animal that kills and eats other animals), effectively hunting down its prey, usually *Triceratops*. The *Tyrannosaurus–Triceratops* battles

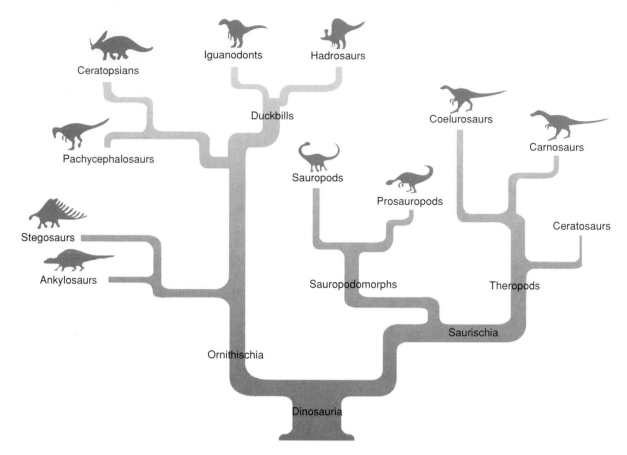

FIGURE 25.9 A current classification of dinosaurs.

Adapted and reprinted with permission by Geotimes, *from Kevin Padian, February 1991.*

look good on television or the movie screen, but in reality *Tyrannosaurus* may have also scavenged a good deal. (A scavenger eats only dead animals and does not kill them.) The order Deinonychosauria is composed of very agile, active predators of Cretaceous time. *Deinonychus* ("DY-noh-NICK-us") and *Velociraptor* ("VELL-oh-sih-RAP-tor") are two examples. *Deinonychus* was equipped with a large claw on the middle toe of its three-toed foot that it used to stab out at its prey (figure 25.10). Fossil evidence (tracks) indicates that *Deinonychus* hunted in packs. Certainly a paraphrase of an old paleontology saying is appropriate here: If a pack of *Deinonychus* finds you, "it's all over but the screaming." The order Ornithomimosauria (Late Cretaceous) is composed of bipedal, birdlike theropods. *Ornithomimus* ("OAR-nith-oh-MY-mus"), an example of this group, closely resembled the modern-day ostrich. The order Oviraptorosauria (Late Cretaceous) is similar to the Ornithomimosauria. *Oviraptor* ("OH-vih-RAP-tor") is an example.

Members of the suborder Sauropodomorpha include small, large, and very large quadrupedal herbivores. Prosauropoda (Triassic–Jurassic) include small and large, bipedal and quadrupedal herbivores. *Plateosaurus* ("PLAT-tih-oh-SAW-rus") is an example. The Sauropoda (Early Jurassic–Late Cretaceous) include large, quadrupedal herbivores. The best-known example is *Apatosaurus* ("ay-PAT-oh-SAW-rus) (formerly known as *Brontosaurus*), which displays the familiar characteristics of this group: a relatively small head,

a very long neck, a large body, and a very long tail. (Dino, from the "Flintstones" cartoon, is a sauropod.) *Apatosaurus* is almost always pictured living in and near swamps. Supposedly, it needed water to support its great weight. But if *Apatosaurus* had lived in water, and in particular, deep water (say, up to its neck), the pressure of the water on the outside of its body would have prevented it from breathing! Almost all available evidence shows that *Apatosaurus* lived in herds on land. *Brachiosaurus* ("BRAY-kee-oh-SAW-rus"), *Diplodocus* ("DIP-loh-DOH-kus"), and *Supersaurus* ("SUE-per-SAW-rus") are other examples of these huge dinosaurs.

A recent study of the paleontological literature showed that *Brontosaurus* was first named *Apatosaurus*, and thus, according to the strict rules of nomenclature, the name *Brontosaurus* has been changed back to *Apatosaurus*. *Brontosaurus*, however, is still in popular usage in many books.

The suborder Segnosauria (Late Cretaceous) includes a group of dinosaurs that have both saurischian and ornithischian features. *Segnosaurus* is an example.

The other major order of dinosaurs, Ornithischia, is composed of herbivores of both bipedal and quadrupedal types (figure 25.9). The suborder Ornithopoda (Middle

FIGURE 25.10 Velociraptor—an agile predator.

Jurassic–Late Cretaceous) includes both small and large herbivores. Some members of this group were bipedal. The duck-billed dinosaurs, typical of this group, were the most numerous and diverse of all dinosaurs. Many had strange, bony skulls. Examples include *Camptosaurus* ("KAMP-toh-SAW-rus"), *Iguanodon* ("ih-GWAH-no-don") (the first skeletal bones found of a dinosaur were of an *Iguanodon*), *Hadrosaurus,* and *Corythosaurus.* The suborder Scelidosauria is also composed of bipedal herbivores. *Scutellosaurus* ("SKEW-tell-oh-SAW-rus") is an example. The suborder Stegosauria (Middle Jurassic–Late Cretaceous) includes the plated dinosaurs. All were quadrupedal herbivores. *Stegosaurus* ("STEG-oh-SAW-rus"), which lived only during the Jurassic, is characterized by a row of bony plates along its back. While protection would seem the logical explanation for the bony plates, they almost never show the teeth marks of predators. Possibly the plates were used as body-temperature regulators, similar to the "sailbacks" of pelycosaur reptiles of Permian time. If so, this indicates a fairly sophisticated physiology. The suborder Ankylosauria (Middle Jurassic–Late Cretaceous) includes the armored dinosaurs. All, like the stegosaurs, were quadrupedal herbivores. *Ankylosaurus* ("AN-kee-loh-SAW-rus") is an example. The suborder Pachycephalosauria (Late Jurassic–Late Cretaceous) is composed of bipedal, bipedal/quadrupedal herbivores. *Pachycephalosaurus* ("PAK-kih-SEFF-ah-loh-SAW-rus") is well known. The final suborder, Ceratopsida, includes all the horned dinosaurs. Common mostly in the Late Cretaceous, they were all quadrupedal herbivores. Examples are *Protoceratops* ("PROH-toh-SAIR-ah-tops"); *Triceratops* ("try-SEER-ah-tops"), with its familiar three-horned face, bony neckshield, and parrotlike beak; and *Torosaurus* ("TOR-o-SAW-rus"), which had the largest skull of any land animal.

Dinosaur Fact and Fiction

Unfortunately, much of the way dinosaurs have been depicted in popular books and movies is not entirely correct. The fossil record allows paleontologists to learn a great deal about dinosaurs, and many, many reasonable inferences from the available evidence are possible. Let's examine the fossil evidence and plausible inferences derived from this evidence.

Skeletons

Most of the information about dinosaurs comes from their fossilized skeletons. Starting at the top of the dinosaur skeleton and proceeding downward:

Skulls range from relatively small, in sauropods, to relatively large, in theropods. Although some skulls are solid bone, many of the larger skulls have large holes. The purpose of the holes was probably to lighten the skull, provide room for muscles, and possibly serve as a location for glands. Some skulls are very large and bony (for example, in the pachycephalosaurians and some duck-billed dinosaurs). Paleontologists have inferred that these types of dinosaurs probably engaged in mating rituals involving the butting of heads, similar to the rituals of modern-day mountain goats. Brain size can be deduced by studying dinosaur skulls. Theropods had relatively big brains; sauropods and stegosaurs had very small brains. The placement of the eyes on the skull varied from group to group. Some dinosaurs had eyes on the sides of their skull, others had eyes that faced forward. Modern reptiles can distinguish color, and there is no reason to believe dinosaurs did not also see color. All dinosaurs lacked external ears, but hadrosaur (ornithopod) skulls indicate that these animals may have heard sound. Some ornithopod skulls show large nasal openings. One

FIGURE 25.11 A scene from Alberta, Canada, during Late Cretaceous. From left to right: a *Kristosaurus,* an *Edmontonia* (ankylosaur), and two *Hypacrosaurus* (duckbills) flee from a *Tyrannosaurus* that seems concerned about the horned *Monoclonius* coming from behind. Meanwhile, Pterosaurs and birds fly away.

reconstruction even showed a dinosaur with a trunk(!), although this interpretation is disputed. Rather, these nasal structures suggest that some dinosaurs had vocal capabilities. Ceratopsian and other skulls have beaks, which were used for cropping vegetation.

The teeth of dinosaurs vary according to life habit. The theropods had long, bladelike teeth. (Calvin, of the comic strip "Calvin and Hobbes," was not wrong in calling *Tyrannosaurus* teeth "six inch chisels of death.") *Allosaurus* even had a loose jaw construction similar to that of modern snakes, which allowed it to open its jaws extremely wide to swallow large quantities of flesh. The sauropod and ornithischian teeth were composed of a series of ridges, which indicates cropping and chewing of vegetation.

The rib cages of most dinosaurs are large, indicating that the heart/lung apparatus was large and sophisticated.

The pelvic structure provides the basis of classification of dinosaurs. The ornithischian pelvic structure is certainly unlike that of modern reptiles. The pelvis also clearly shows that dinosaur limbs extended directly downward from the body.

The limbs of the dinosaurs varied, depending on the group. The bipeds (for example, theropods) had claws, indicating that they were active predators. Other groups, such as sauropods, ornithopods, stegosaurs, and ankylosaurs, had nails instead of claws.

Despite what virtually all books and movies show, dinosaur tails were long, stiff structures held aloft behind the dinosaur (figure 25.11). Dinosaur tracks rarely show a drag-

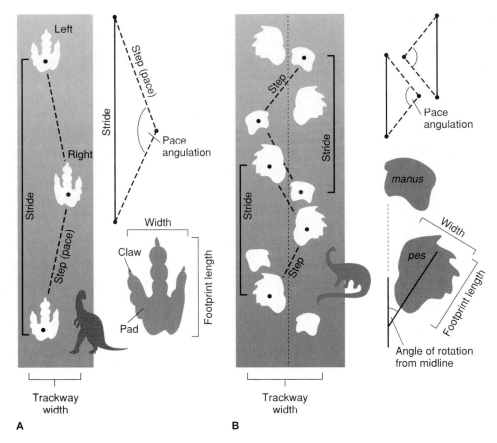

FIGURE 25.12 Important characteristics of dinosaur tracks. (A) Tracks of a bipedal dinosaur. (B) Tracks of a quadrupedal dinosaur.
Adapted from Martin Lockley, Tracking Dinosaurs: A New Look At An Ancient World. Copyright © 1991 Cambridge University Press, New York. Reprinted with permission of Cambridge University Press.

ging tail (see below). In theropods, the tail probably acted as a counterbalance when the animal was running. Some paleontologists have recently suggested that the tail in large sauropods and also in stegosaurs may have been used as a "third leg" to balance on as the animal reared up to eat leaves. This is an interesting idea, although there is little supporting evidence. Other tails were clearly used as defensive mechanisms—the spiked tail of *Stegosaurus,* for example, and the club tail of *Ankylosaurus.*

The study of the dinosaur bones themselves shows several interesting features. Many dinosaur bones look remarkably mammalian. In mammal bones, large "holes" mark the positions of blood vessels. Surrounding these canals are rings of bony tissue. This structure has been termed the Haversian system. Dinosaur bones also show this Haversian system. Stress tests on dinosaur bones (corrected, of course, for the fact that they are now stone) indicate that it is unlikely that some dinosaurs, such as the giant sauropods, ran. If they had, they would have broken their leg bones! Limb bones also show evidence of disease and predation, and healed broken bones indicate that dinosaurs were not immune to accidents. Like other reptile bones, dinosaur bones display growth rings. From these growth rings it is possible to determine both the rate of growth of dinosaurs and their life expectancy. It appears that many dinosaurs lived as long as 120 years, with the life expectancy of the large sauropods (for example, *Apatosaurus*) approaching 200 years.

Skin/Scales

Amazingly, fossilized skin imprints of dinosaurs have been preserved in the fossil record. The imprints show that dinosaur skin was composed of scales. The scales, called tubercles, are similar in form to those of the Gila monster. Some dinosaurs also had feathers! All information about the color of dinosaurs is lost to paleontologists. Many of the colorful patterns shown in dinosaur picture-books look great, but all the colors and color patterns are inferred. We simply do not know how dinosaurs were colored. Alternatives to coloration in dinosaurs include monochrome coloring like that of the modern elephant or crocodile (probably the best inference) or a camouflage pattern of spots, stripes, or shading.

Tracks

Dinosaur tracks permit many anatomical inferences and reveal a wide range of behaviors (figure 25.12). Comparison of dinosaur tracks to skeletons clearly shows that dinosaur limbs extended directly down from the body and were not splayed out from the side of the body as they are in modern reptiles. The depths of dinosaur footprints give some indication of the weights of the dinosaurs. One set of tracks shows that one individual limped! Gait (walking or running) is easily shown by track pattern. Running tracks of the active predators show that some of the large theropods ran at least 25 miles per hour. In some areas, hundreds of dinosaur tracks have been found, showing that some dinosaurs (some

FIGURE 25.13 A scene from a watering hole in the western United States during Late Jurassic time. From left to right: an *Allosaurus* watches *Camarasaurus* walk away while *Diplodocus* drink and a *Stegosaurus* departs. The foliage of conifers is typical for this time.

of the large sauropods) moved in herds (figure 25.13). Paleontologists infer that this pattern may be due to seasonal migrations. (The sight of hundreds or thousands of *Apatosaurus* walking across the landscape must have been tremendous!) Close examination of these "herd tracks" have indicated to a few researchers that the herds were structured—larger animals walked on the outside, while smaller, immature individuals walked on the inside. This interpretation is highly controversial. Clearly, however, the large sauropods were land-dwellers and were not confined to swamps, as once believed. Since the large sauropods could not run, and proba-

bly were not camouflaged, herding behavior would have been an effective alternative to discourage attack by theropods. Other tracks show that dinosaurs (particularly *Deinonychus*) hunted in packs. This type of behavior is certainly unlike that of modern reptiles. Perhaps the most interesting observation of dinosaur tracks is what is missing— tailprints! Only rarely have dinosaur tailprints been found. Contrary to most representations, dinosaurs did not drag their tails behind them. (Note: Dragging a tail requires a lot of energy and is also quite harmful to the scales.) One set of *Apatosaurus* tracks has raised quite a debate. The tracks show only the front footprints of the *Apatosaurus*. From this it has been inferred that the dinosaur was swimming, using only its front legs to push its body along. However, other explanations have been put forth; it now appears that the "swimming *Apatosaurus*" was not, in fact, swimming at all. The most recent reading of the tracks indicates that they are incomplete—some were not preserved.

A well-known locality for dinosaur tracks is the Paluxy River locality in Texas. The site is famous for showing many different types of dinosaur tracks. For a long time, local residents claimed that other strange-looking tracks resembling human footprints also were found in the strata. "Scientific creationists" were quick to point out that this proved that humans and dinosaurs lived at the same time. However, careful examination of the controversial footprints has shown that they are the incomplete tracks of a bipedal dinosaur, while other human-looking footprints were hoaxes, carved in the rock. The "scientific creationists" have recently conceded that no unequivocal human footprints occur in the Paluxy River strata.

Eggs/Nesting Sites

Dinosaurs hatched from eggs. Studies of the eggs have revealed fossilized embryos, which allow paleontologists to study the growth and development of dinosaurs. Exceptional nesting-site localities have been found in Mongolia and Montana. Careful study of the nesting sites (eggs and young dinosaur skeletons) has shown that the young stayed in the nest for some time, perhaps several years. This finding reveals that dinosaurs cared for their young for long periods of time.

Other Evidence

Other evidence of dinosaur behavior and ecology comes from a variety of sources. Gizzard stones with smooth, concave surfaces have been found, providing evidence of dinosaur digestion. Plant fossils found in what would have been the stomach region of a fossilized animal provide clues to dinosaur diet. Pine needles, for example, were found in the stomach region of an *Apatosaurus*, providing additional evidence that these large sauropods lived on land, not exclusively in swamps. Assemblages of dinosaur fossils, particularly from Alberta, give clues to the diversity and number of dinosaurs and to predator/prey ratios. The stratigraphic position (vertical distribution of fossils in the rock column) of dinosaur remains indicates that evolution was rapid among dinosaur groups.

Warm-blooded Dinosaurs?

Since their discovery, dinosaurs have been classified as reptiles, and indeed, they display many reptilian characteristics.

Recently, however, some paleontologists have questioned whether all dinosaurs were cold-blooded, as are modern reptiles. Ectothermic ("external heat"), or cold-blooded, animals such as fish, amphibians, and reptiles do not generate internal heat. Their body temperature is not really cold; rather, it closely approximates the environmental temperature. For example, snakes and lizards often sun themselves on rocks in the morning to elevate their body temperature. Later, if the temperature rises too much, they seek shade. Endothermic ("internal heat") animals—all mammals and birds—generate internal heat. Most of the food you eat, for example, goes to maintain your body temperature of 98.6°F.

Several lines of evidence suggest that dinosaurs displayed endothermic characteristics. This evidence includes the following: (1) *Erect posture.* Several groups of dinosaurs—the theropods, ornithopods, and pachycephalosaurians—were bipedal, with the limbs extending straight down from the body, a characteristic of dinosaurs and mammals. Yet no modern reptiles show this structure or behavior. (2) *Bone structure.* If a bone cross section of a crocodile (reptile), a dinosaur, and an elephant (mammal) are compared, the dinosaur bone looks very much like the elephant bone, including its system of Haversian canals. However, dinosaur bones do show growth rings (characteristic of some reptiles), which do not occur in mammal bones. (3) *Predator/prey ratios.* In areas where abundant dinosaur bones have been found, the numbers of predator and prey individuals have been counted. The predator-to-prey ratio closely approximates that of modern mammal predator-prey communities (for example, lion-gazelle communities of Africa). Of course, problems arise with this evidence, due to fossilization and preservation biases. (4) *Distribution.* Some large dinosaurs apparently lived in rather cold climates (even correcting for the positions of the continents in the Mesozoic), far outside the temperature range of modern reptiles. Crocodiles, common in Florida, are not found in Wisconsin because they could not survive the winter. Yet large dinosaurs apparently survived in rather cold climates. (5) *Specialized dentition.* Some dinosaurs have dentition patterns that indicate a varied diet, implying a complicated metabolic system. (6) *Brain size.* It makes no sense to compare absolute brain sizes of animals; whales have larger brains than mice because whales are larger animals. Rather, the brain-size/body-size ratio is compared. When this is done, humans are found to have the largest brain size compared to body size. Certain dinosaurs—for example, theropods—had very large brains for their body size. In general, the smallest to largest progression in brain size of dinosaurs is as follows: sauropods—armored and plated dinosaurs—horned dinosaurs—large ornithopods—theropods—coelurosaurs. (7) *Inferred heart/lung structure.* Studies of the skeletons of dinosaurs show that dinosaurs probably had large hearts and lungs. Some paleontologists have suggested that these hearts may have been quite different from those of modern reptiles. (8) *Sails and bony plates.* Sail-like structures on some dinosaurs (*Spinosaurus*) and bony plates on others

FIGURE 25.14 Angiosperm evolution. Note how some of the earliest angiosperm leaves resemble modern leaves.

From Origin and Early Evolution of the Angiosperms, *by J. A. Doyle and L. J. Hickey, edited by C. B. Beck, © 1976, Columbia University Press, New York. Used by permission of James A. Daley.*

(*Stegosaurus*) suggest that dinosaurs regulated their body temperatures. By flushing blood into these structures, the animals could cool or warm their body temperature rather quickly. (9) *Inferred behavior patterns.* Theropod dinosaurs, particularly the carnosaurs and deinonychusians, were agile, active predators. This behavior is not found in modern reptiles. The nesting behavior—caring for the young for long periods of time—and herding of dinosaurs is also not characteristic of modern reptiles. (10) *Oxygen isotopes.* Recent work with oxygen isotopes in the bones of living animals shows that in reptiles there is a great difference in temperature between the bones of the main body and those of the limbs. In a reptile, the limbs cool more rapidly than does the main body. However, in mammals there is little temperature difference between bones because the whole body is maintained at a constant temperature. These temperature differences in bones are reflected in the oxygen-isotope ratios of particular bones. The oxygen-isotope ratios in dinosaur bones showed, in seven of nine groups studied, little temperature difference between main body bones and limb bones. Some dinosaurs, particularly *Tyrannosaurus,* were very much like mammals in their body-temperature distribution.

In summary, evidence indicates that several groups of dinosaurs were not simply cold-blooded, like reptiles. Alternatively, they may not have been warm-blooded, like mammals. A third alternative is that some dinosaurs were homeothermic, meaning that their large bodies could store heat for long periods without losing body temperature. (This is similar to water in a large lake that does not freeze over in winter.) Thus, they may have been cold-blooded but, because of the heat-storing capacity of their bodies, acted as warm-blooded animals do. However, problems arise with this interpretation. If dinosaurs could not maintain a uniform body temperature, how could they survive cold nights, periods of extended cloudiness, and seasonal changes?

THE CRETACEOUS

On land, an important event in plant evolution occurred during the Cretaceous—the evolution of angiosperms (figure 25.14). **Angiosperms** are flowering plants that have a protected seed structure; many have fruit. The earliest undisputed angiosperm fossils are Cretaceous in age, but possible spores of angiosperm origin are found in Jurassic-age strata. Angiosperms rapidly diversified in the Cretaceous and into the Cenozoic; today they represent over 95% of all land plants.

O ne interesting aspect of angiosperm evolution is the impact it had on another group living on the land—the insects. For example, paralleling the evolution of flowering plants in the Cretaceous, there was a rapid expansion in insect evolution, especially in butterflies and bees. These two radically different groups have had a long, complex, and intimate history.

CRETACEOUS EXTINCTIONS

The mass extinction that occurred at the end of the Cretaceous period is one of the most intriguing, puzzling, and controversial events in the history of life. About 25% of all families of organisms became extinct at this time. The most important aspect of the terminal-Cretaceous extinction is its selectivity. A relatively high proportion of animals, both terrestrial and marine, experienced little change at this time. Among microscopic marine organisms that declined were planktonic foraminifera and coccolithophorids (both carbonate-producers). Some marine invertebrates such as ammonites, belemnites, and inoceramids experienced gradual decline before becoming extinct at the end of the Cretaceous. Several marine vertebrates died out, including mosasaurs, ichthyosaurs, and plesiosaurs. Two important terrestrial vertebrate groups that became extinct were the flying reptiles (pterosaurs and *Pteranodons*) and the dinosaurs. Just as important, however, fish, amphibians, and reptiles (such as crocodiles, lizards, snakes, and turtles) all appear to have been unaffected by the terminal-Cretaceous events. Plants show a gradual increase in extinction rate, which has been attributed to cooling climates.

Extinction Theories/Hypotheses/Guesses

Many ideas have been proposed to explain the terminal-Cretaceous extinctions. The selectivity of the extinctions is one reason why paleontologists have difficulty accepting only one extinction cause.

Plate Tectonics

Plate tectonics, and associated continental drift and mountain building, is one of the leading models for the Cretaceous extinctions. Plate tectonics could have caused any number of different physical parameters to change, which in turn may have caused various groups to become extinct. Regressing seas and continental fragmentation and joining could have changed the climate, both in average temperature and contrast of seasons, and decreased habitat diversity.

Climate Change

Climate change is a leading hypothesis among paleontologists. A change in climate, particularly a cooling, could have affected the dinosaurs in a number of ways; their body temperature could have been lowered, their reproduction cycles disrupted, and their sex determination changed (recent studies of alligator nests show that when the temperature is less than 86°F, more females hatch; when the temperature is greater than 93°F, more males hatch). Throughout the early and middle Mesozoic, climates were fairly uniform—warm, with little change between seasons. Estimates of Cretaceous climates range from 6° to 12°C warmer than today. Plant fossils from the end of Cretaceous time show that not only was the climate cooling, there were also greater seasonal temperature variations. Additional evidence of deteriorating climate comes from microscopic marine organisms from deep-sea sediments of the Weddel Sea, Antarctica. Studies show that about 200,000 years before the end of the Cretaceous, the waters warmed briefly, then cooled greatly (by 1° to 2°C).

Extraterrestrial Impact

The most controversial explanation—and presently probably the most studied aspect of earth history—is the extraterrestrial-

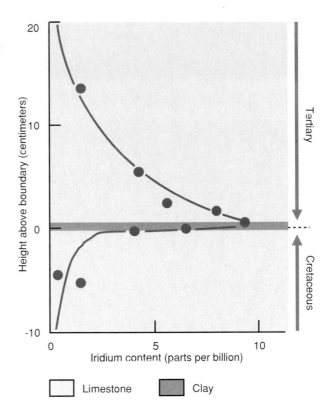

FIGURE 25.15 Plot of iridium concentration (parts per billion) versus stratigraphic position (height in centimeters) at Gubbio, Italy. The iridium "spike" (area of highest concentration) occurs in the clay marking the end of the Cretaceous.

Source: Data from R. McNeill Alexander, Dynamics of Dinosaurs and Other Extinct Giants, *1989, Columbia University Press, New York, NY.*

impact hypothesis. Originally proposed by Luis Alvarez and others in 1980, the hypothesis states that sixty-six million years ago, one or several asteroid or meteor bodies collided with the earth. The collision sent large quantities of dirt, dust, and ash into the atmosphere, blocking sunlight. Temperatures fell and the climate cooled drastically. Even if the climate-cooling effect has lasted only for several years, it might have been enough to cause the mass extinctions.

Evidence for an extraterrestrial impact seems convincing. Initially, the model of an extraterrestrial impact was proposed after the discovery of a worldwide concentration of iridium in shales deposited at this time (figure 25.15). (Because of the widespread extinction of microscopic marine organisms, which produced carbonate, the terminal-Cretaceous marine rocks are invariably shales rather than limestones.) Iridium is a rare element, not found in significant quantities on the earth's surface. During the formation of the earth, iridium either sank to the mantle and core or was lost to space. Hence, the only two sources of iridium at the surface are lavas produced from magmas deep within the earth or meteors. The iridium concentration seen in shales of the Late Cretaceous may be due to dust from extraterrestrial bodies that struck the earth. An alternative source for the iridium concentration, however, might be the extremely large volumes of

Figure 25.16 Evidence for an extraterrestrial impact at the end of the Cretaceous. Shocked quartz grain.
Photograph courtesy of G. Izett.

bodies struck the earth at different places within a brief time.) Glass spherules up to 6 millimeters in diameter have been found in rock layers near Haiti. These spherules are virtually crystal-free and would have formed in a manner similar to that of microtektites. The composition of the glass spherules also points to an impact on land. Recently, very small (3 to 5 nanometers in diameter) diamonds have been found in rocks near Alberta, Canada. Such diamond dust would form as a result of the extremely high pressures created during the collision of an asterioid body with the earth. A soot layer has also been found in rocks of the same age as those that show the iridium concentration. This soot layer has led some researchers to postulate a global firestorm set off by the impact. In Late Cretaceous sediments from New Zealand, Spain, and Denmark, carbon was deposited 10,000 times faster than in the rocks immediately above and below. The carbon also contains retene, a hydrocarbon formed by the burning of coniferous trees. In terminal-Cretaceous rocks from Denmark, geologists have found extraterrestrial amino acids (isovaline and alpha-aminoisobutyric acid). These amino acids have been found in meteorites but are rare on earth. The use of amino-acid evidence to support an impact event is controversial. How do amino acids survive the high temperatures of the impact event? There is also some question as to the stratigraphic placement of the rock layers that contain these amino acids; they apparently occur centimeters above and below the stratigraphic position of the iridium concentration.

The idea of an extraterrestrial impact has led geologists to look for the "smoking gun"—the impact crater. Unusually thick shale deposits, with abundant, large "shocked" quartz grains, found in the Caribbean Sea led researchers to first postulate this area as a likely impact site. An area south of Cuba (with the impact centered in the Isle of Pines) was examined first, but recent work has discounted this area. At present, the most likely impact site is on the northern end of the Yucatan Peninsula. In the rocks beneath the town of Chicxulub, Mexico, a buried crater structure of approximately the right age has been mapped using geophysical methods. A series of *cenotes* (sinkholes) on the surface outline the buried crater's rim. Of course, the impact crater(s) may never be found. If the impact occurred in the ocean, that crust may have been subducted since then; if so, all the rocks would now be lost.

Perhaps the best summary of how paleontologists view the impact hypothesis is by Robert Sloan of the University of Minnesota, who said, concerning the impact, "[it] surely did not help, although it is hardly the sole reason" for the extinctions.

Volcanism—Deccan Flood Basalts
A strong case can be presented for a terrestrial cause of the mass extinctions. Widespread volcanism would have had an effect similar to that of an extraterrestrial impact—ash and

lava that were extruded from volcanoes and fissures in India at this time—the Deccan flood basalts. Recent work has shown that microorganisms, bacteria, and fungi can increase and decrease the iridium concentration in rock. Perhaps there has been an organic effect. Evidence from a number of microscopic rock and mineral grains is used to support the impact hypothesis. "Shocked quartz grains" are formed from the stress and pressure of an impact event (figure 25.16) and would form if an asteroid struck continental rock. Microtektites are formed when masses of dust and rock particles are thrown into the atmosphere during the impact, cool, and fall to earth. These microscopic grains have been found in deep-sea sediments off the coast of Japan. Some contain the mineral clinopyroxene, which forms only at high temperatures (over 800°C). This mineral composition would indicate that the asteroid struck basaltic rock. (Because geologists find evidence that the impact occurred both on land and in the ocean, they postulate that either several asteroid bodies struck the earth at different places at the same time, or two asteroid dust blocking the sunlight and, in turn, cooling the climate. Large-scale volcanic basalts of the correct age have been found in India—the Deccan flood basalts. Volcanism can also account for both the iridium anomaly (the magmas, from deep within the earth, were rich in iridium) and the widespread ash content (burning of forests). Although some models of such events have predicted widespread warming of the climate, most researchers agree that cooling would occur instead. The ash would block the sunlight, and production of sulfate aerosols would deplete the ozone. Volcanism can also produce glass spherules and microtektites.

Change in the Atmospheric O_2:CO_2 Ratio

Perhaps the most interesting proposal for the extinctions is that there was a change in the atmosphere. Great quantities of chalk and limestone were laid down in epeiric seas during the Cretaceous. These rocks are composed of the mineral calcite ($CaCO_3$), which is made when plants (algae) and animals (corals) take CO_2 out of the water (CO_2 that ultimately came from the atmosphere) and produce calcite. Perhaps this tremendous depletion of CO_2 affected the atmosphere in some way: Climate change? Temperature extremes? While there is little direct evidence for the consequences of this reduction of CO_2, the idea remains intriguing.

Migration

Due to continental drift, some populations of dinosaurs may have migrated to new areas unavailable before. Such migrations may have introduced new predators to species that were unable to cope and became extinct. While this idea may sound plausible, it doesn't explain why other groups became extinct as well.

Diseases/Epidemics

Closely associated with the migration idea is that of diseases and epidemics. Perhaps migrations of dinosaurs to new areas inadvertently caused diseases to be introduced to other populations, creating epidemics and mass death. But if that were true, why were other groups affected as well? Unfortunately, there is no evidence to support this idea.

Supernova

Other researchers have postulated that a supernova near our solar system bathed the earth in lethal doses of radiation, or perhaps affected the climate somehow, causing the extinctions. This idea has one glaring fault—there is no astronomical evidence of a supernova near our solar system.

Mammals

This idea seems rather comforting for us as mammals—for 150 million years, mammals were outcompeted by the dinosaurs. Finally, according to this idea, the mammals attacked and ate the dinosaur eggs or in some way successfully outcompeted the dinosaurs. No evidence exists for this speculation. Certainly mammals became the dominant land animals *after* the extinction of the dinosaurs, but they probably had little direct effect on dinosaur extinctions.

Increase in Radiation

During a reversal of the earth's magnetic field, there is a relatively short period of time (thousands of years) when the magnetic field is virtually nonexistent. This allows increased radiation and other cosmic rays to strike the earth's surface. Perhaps this is what happened to the dinosaurs. However, although extinctions of microscopic marine organisms in the Cenozoic have been related to magnetic reversals, large animals (for example, Cenozoic mammals and hominids) have all survived magnetic reversals in the past.

Plants

According to another theory, the extinction of dinosaurs might have been related to plants. Perhaps new plants evolved that lacked certain nutritional elements dinosaurs needed—or that the dinosaurs could not digest. Or, alternatively, perhaps plants developed poisons that caused the death of dinosaurs. Unfortunately, there is no evidence to support this theory. The dinosaurs were unaffected by the evolution of flowering plants in the Mesozoic. In addition, the evolution of new plant types does not appear to have coincided with dinosaur extinctions and decline.

Conclusions

On balance, the iridium anomaly and the shocked quartz grains tend to support the impact event. But (and this is an important question to consider) if an extraterrestrial impact did occur, did it contribute to the extinction of dinosaurs and other organisms? The question is hotly debated, and no definitive answer exists. Proof of an extraterrestrial impact does not necessarily prove that the impact contributed to any extinctions.

To further complicate the whole issue, recent detailed, meticulous collecting of dinosaur fossils from the western United States in rocks just below the iridium-enriched layers (that is, just before the extinction event) apparently show that dinosaurs were thriving quite well right up until their extinction. If so, then it is likely that an impact caused the extinctions. The debate rages.

In summary, the cause or causes of the terminal-Cretaceous extinctions remain a fascinating area of scientific inquiry. Currently, continental drift and its associated climatic change are considered to be the most probable causes of the demise of the dinosaurs and other organisms, but the extraterrestrial-impact hypothesis may yet prove valid.

SUMMARY

Ammonites are the most useful index fossil for marine strata of Mesozoic age. On land, dinosaurs ruled the earth. Dinosaurs are divided into two basic orders: (1) the saurischians, or reptile-hipped dinosaurs, which include the theropods (*Tyrannosaurus*) and the sauropods (*Apatosaurus*); and (2) the ornithischians, or bird-hipped dinosaurs, which include ornithopods, ankylosaurs, stegosaurs, and ceratopsians

(*Triceratops*). Abundant evidence exists to indicate that dinosaurs were not merely large-scale lizards. This evidence includes their erect posture, bone structure, predator-prey ratios, distribution, specialized dentition, brain size, inferred heart/lung structure, sails and bony plates, inferred behavior patterns, and oxygen-isotope ratios in bones. Dinosaurs appear to have been very complex animals.

The breakup of Pangaea, and the resulting cooler climates, probably led to the decline of the dinosaurs. An extraterrestrial-impact event at the end of the Cretaceous, or large-scale volcanism, may have acted as an extinction catalyst for a process already under way.

Mammals arose in the Triassic but remained mostly small, nocturnal forms until the Early Cenozoic.

TERMS TO REMEMBER

ammonite 454
angiosperm 464

gymnosperm 453
ornithischian dinosaurs 457

saurischian dinosaurs 457

QUESTIONS FOR REVIEW

1. What important events in the history of life took place in the Jurassic?
2. List the two major orders of dinosaurs, and give examples of each.

3. What evidence leads paleontologists to believe that dinosaurs were not simply large lizards?

4. Why did dinosaurs become extinct at the end of the Cretaceous?

FOR FURTHER THOUGHT

1. Mammals arose in the Triassic yet did not compete successfully with the dinosaurs throughout the rest of the Mesozoic. Why were the dinosaurs more successful than the mammals?

SUGGESTIONS FOR FURTHER READING

Alexander, R. McN. 1989. *Dynamics of dinosaurs and other extinct giants.* New York: Columbia University Press.

Alvarez, L., et al. 1980. Extraterrestrial cause for the Cretaceous-Tertiary extinctions. *Science* 209: 1095–1108.

Bakker, R. T. 1972. Anatomical and ecological evidence of endothermy in dinosaurs. *Nature* 238: 81–85.

Bakker, R. T. 1975. Dinosaur renaissance. *Scientific American* 232(4): 58–78.

Bakker, R. T. 1989. *The dinosaur heresies.* New York: William Morrow.

Colbert, E. H. 1980. *Evolution of the vertebrates. A history of the backboned animals through time.* 3d ed. New York: John Wiley and Sons.

Lambert, D. 1990. *The dinosaur data book.* New York: Avon.

Lockley, M. 1991. *Tracking dinosaurs. A new look at an ancient world.* New York: Cambridge University Press.

Ostrom, J. 1984. *Dinosaurs.* 2d ed. Carolina Biology Readers 98. Burlington, N.C.: Carolina Biological Supply.

Paul, G. S. 1988. *Predatory dinosaurs of the world.* New York: Simon and Schuster.

Thomas, R. D. K., and Olson, E. C., eds. 1980. *A cold look at the warm-blooded dinosaurs.* Boulder, Colo.: American Association for the Advancement of Science.

THE CENOZOIC

Eocene-age strata, Bryce Canyon National Park, Utah.
© Doug Sherman/Geofile

Despite the Cenozoic's long time span (sixty-six million years), geologists have concentrated primarily on only the last one or two million years, the time of the Pleistocene glaciations. Much of North America's present topography and geography is the result of these widespread continental glaciers.

Cenozoic tectonics along North America's western margin produced many of the features we see today—the Grand Canyon, the sharp and rugged topography of the Rocky Mountains, the Basin and Range structure of Nevada, the Columbia Plateau of Washington and Oregon, and the San Andreas fault of California.

On land during the Cenozoic, the mammals experienced rapid evolution, increasing both in size and in numbers. The last part of the Cenozoic is marked by an important evolutionary event—the development of humans!

CHARACTERISTICS OF THE CENOZOIC

The Cenozoic era began sixty-six million years ago, at the end of the Mesozoic era, and continues to the present. During the Cenozoic, most of the craton was exposed. The most prominent Cenozoic-age sediments of the craton are the tills and other glacial sediments left by the continental glaciers of Pleistocene time. On North America's eastern margin, the Appalachians continued to erode, and the Atlantic continental margin developed into a fully passive margin with a well-developed shelf, slope, and rise. The main focus of geologic events was North America's western margin as many of the geologic features west of the Rocky Mountains formed, such as the Grand Canyon, the Basin and Range province of Nevada, the San Andreas fault of California, and the Cascade Range of northern California, Oregon, Washington, and British Columbia.

Most North American geologists divide the Cenozoic into two periods: (1) the Tertiary period and (2) the Quaternary ("qua-TER-na-ry") period (figure 26.1). The Tertiary is then subdivided into the Paleocene, Eocene, Oligocene, Miocene, and Pliocene epochs. (Most of the epochs were defined by Charles Lyell in 1833 on the basis of the percentage of extant, or still-living, fauna found as fossils.) The Quaternary is subdivided into the Pleistocene and the Holocene (sometimes called Recent) epochs. Other geologists, however, particularly European geologists, use marine planktonic (floating) and benthic (bottom-dwelling) microorganisms to divide the Cenozoic era into the Paleogene period and the Neogene period. The Paleogene includes the Paleocene, Eocene, and Oligocene epochs. The Neogene period includes the Miocene, Pliocene, Pleistocene, and Holocene epochs. Either scheme is correct, and both are useful to remember. The first scheme is the division historically used in North America; the second scheme is a more natural subdivision of the rocks of Europe.

North America completed both its counterclockwise rotation and its northward movement in the Cenozoic. The continent moved to its present orientation and position (figure 26.2).

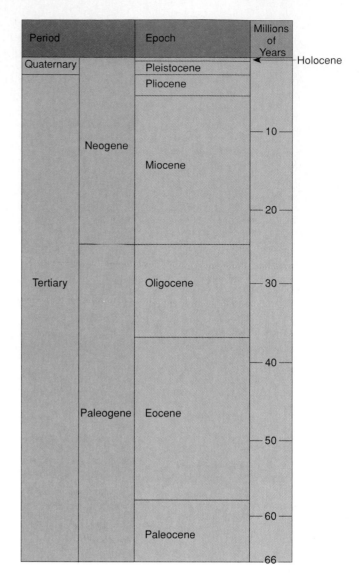

FIGURE 26.1 Cenozoic subdivisions.

During the Cenozoic, the continents continued to diverge and separate, as Pangaea broke apart, and to move to their present positions (figure 26.3). The timing of the continental separations was responsible for many of the diverse geologic and biologic features of Cenozoic age.

Overall, the general tectonic motion of the continents was one of divergence as North and South America, Europe, and Africa all moved away from the Atlantic midocean ridge. Consequently, the Atlantic Ocean grew larger. This meant that subduction was occurring somewhere, and it was—along most of the western, northwestern, and northern margins of the Pacific Ocean. As the Atlantic Ocean grew, the Pacific Ocean shrank.

One interesting feature of the separation of the continents was the remarkable journey of India as it diverged from Gondwanaland. After India separated from Gondwanaland, it moved very rapidly northward, where it collided with the Asian continent to form the Himalayas.

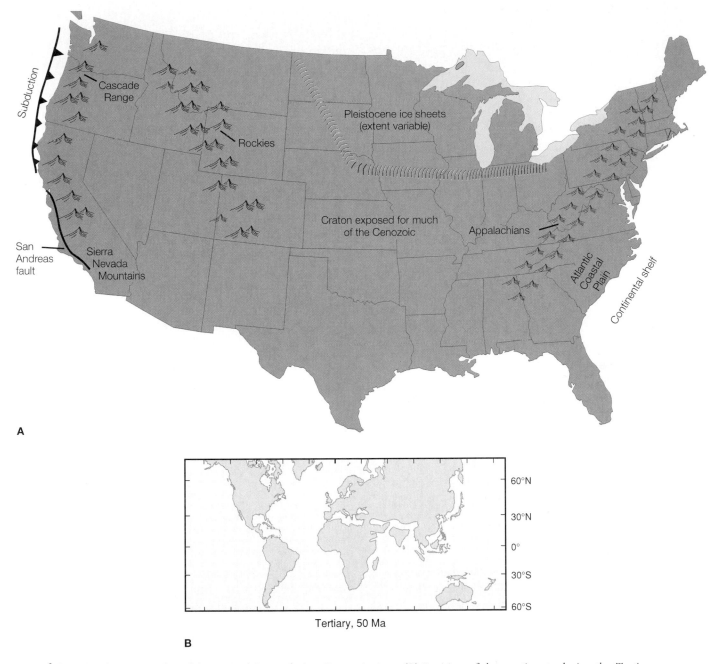

FIGURE 26.2 (A) Paleogeography of the United States during Cenozoic time. (B) Position of the continents during the Tertiary.

THE NORTH AMERICAN CRATON
Terrestrial Environments

During almost all of the Cenozoic, the North American craton was exposed. By now, the continent was riding relatively high, owing to the Appalachian Mountains in the east and the Rockies in the west. The epeiric seas, common in the Early Paleozoic and rare in the Mesozoic, were absent from the continent's interior during the Cenozoic. Worldwide transgressions and regressions occurred, but because the craton was topographically higher than before, only the very edges of the continent (for example, the shelf) were affected.

Because of the craton's exposed nature, erosion was the dominant process, and sediment was deposited only in terrestrial environments. Abundant sediments were shed eastward onto the craton from the newly uplifted Rockies, forming a large clastic wedge similar to the Paleozoic clastic wedges of the Appalachians.

In the Pleistocene, abundant sediments also were deposited on the craton—not by epeiric seas but by glaciers.

Pleistocene Glaciations

The Pleistocene epoch of the Quaternary period was a time of widespread continental glaciations throughout the northern

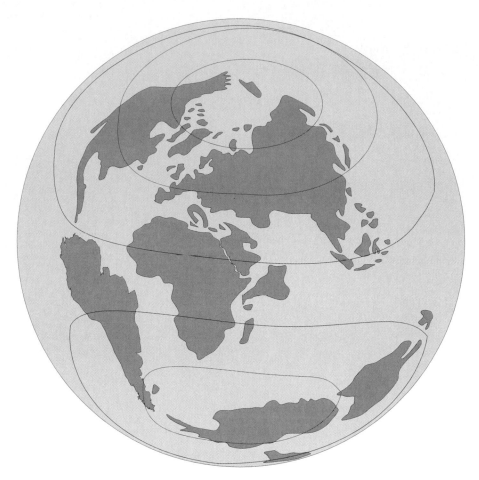

FIGURE 26.3 Paleogeography of the world about sixty million years ago.

Figure rendered from Earth History and Plate Tectonics: An Introduction to Historical Geology, *2E by Carl K. Seyfert and Leslie A. Sirkin. Copyright © 1973, 1979 by Carl K. Seyfert and Leslie A. Sirkin. Reprinted by permission of Addison Wesley Educational Publishers, Inc.*

latitudes (figure 26.4). Geologists debate the exact beginning of the Pleistocene because, originally, the Pleistocene was defined not on the basis of fossils, as were the other epochs, but rather on the presence of tills and other evidence of glaciation. Because the glaciations did not occur simultaneously worldwide, the base of the Pleistocene is not synchronous throughout the world. Some geologists place the start of the Pleistocene at 1.6 million years ago; others place it at approximately 1 million years ago. Geologists have since redefined the base of the Pleistocene using fossils. On land, vertebrate fossils, such as the appearance of the modern horse and elephant, define Lower Pleistocene strata. In marine sediments, the disappearance (extinction) of the discoasters (planktonic microorganisms) defines the start of the Pleistocene.

Whichever date is used for the start of the Pleistocene, widespread continental glaciations did not begin until approximately 1 million years ago. (Climate deterioration, however, started much sooner—cooling began about the mid-Oligocene, 35 to 40 million years ago.) The end of the Pleis-

tocene is well defined, placed at 10,000 years ago, when the continental glaciers disappeared from the midlatitude region.

Evidence of Pleistocene continental glaciations shows that, in North America, many separate glacial advances occurred (figure 26.5). The repeated cycle of advances (glaciations) and retreats (interglacial warming periods) is typical of continental glacial episodes.

Although, historically, geologists have recognized four different glacial cycles in North America, each with two or more subcycles, anywhere from three to six separate cycles have been recognized in Pleistocene-age rocks from Europe. And evidence from the marine realm indicates that possibly as many as eight distinct cycles occurred. The exact number of cycles varies from place to place, but it is the cyclic nature of continental glaciations that is important, not the actual number of cycles.

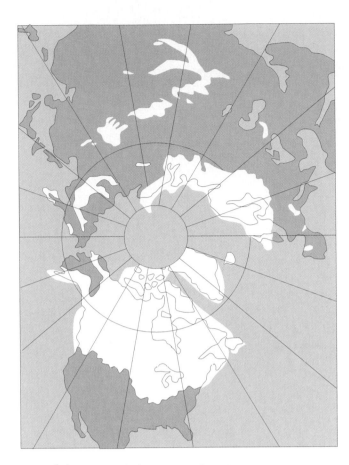

FIGURE 26.4 Extent of Pleistocene glaciations.

From M. Gray, Update: The Quaternary Ice Age. *Copyright © 1988 Cambridge University Press, New York, NY. Reprinted by permission.*

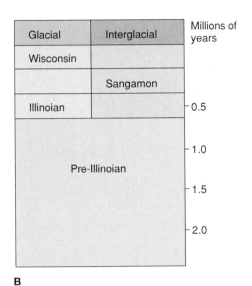

Glacial	Interglacial	Millions of years
Wisconsin		
	Sangamon	
Illinoian		0.5
	Yarmouth	
Kansan		1.0
	Aftonian	1.5
Nebraskan		
	Pre-Nebraskan	2.0

A

Glacial	Interglacial	Millions of years
Wisconsin		
	Sangamon	
Illinoian		0.5
		1.0
Pre-Illinoian		
		1.5
		2.0

B

FIGURE 26.5 Glacial and interglacial episodes during the Pleistocene.

The continental glaciations profoundly affected many of the physical and biologic aspects of the northern continents. Some of the physical terrestrial effects included (1) eustatic sea-level changes, (2) isostatic rebound of the continents, (3) glacial erosional and depositional landforms, (4) glacial lakes, (5) major drainage (river) changes, and (6) glacial sediments.

During a major glacial episode, water must be taken from the oceans and put onto the land in the form of ice. This causes a eustatic (worldwide) sea-level drop—a regression. When the climate warms, the ice melts, and water is put back into the oceans, causing a worldwide transgression. Numerous features indicate that the continents have recently experienced several transgressions and regressions. Wave-cut benches, submerged beaches, exposed coral reefs, and other fossils all indicate that sea level was different in the recent past. Rarely, a fisher or oceanographer finds a fossil of a mammoth (extinct Pleistocene elephant) or some other extinct, terrestrial vertebrate far out on the continental shelf, under many meters of water. The animal was living on land that is now underwater—the continental shelf—and when it died, it was buried in the sediment. The continental-shelf area is perfectly habitable, except that now it is underwater.

Another consequence of continental glacial episodes is that, when the ice forms and large glaciers are widespread, the land is depressed—physically lowered. (This is caused by *isostatic equilibrium,* the tendency for continents to float in a buoyant equilibrium on top of denser mantle material.) After a glaciation episode, the land rises as the weight (the ice) is removed. The isostatic rebound of many regions, particularly Scandinavia, can be directly measured.

Glacial erosional and depositional features are recognized throughout the northern United States and Canada. Erosional features, such as striated pavements, are found throughout New England and even in Central Park in New

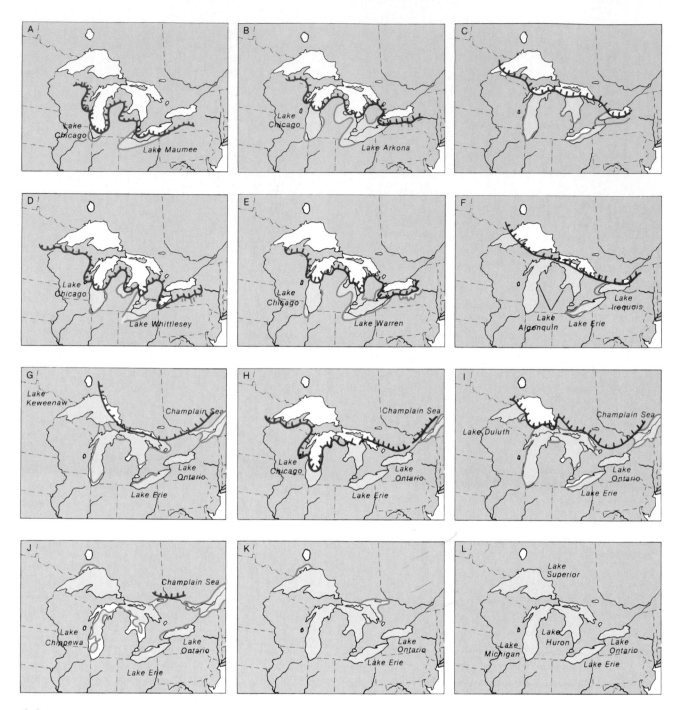

FIGURE 26.6 Phases in the development of the Great Lakes. (A–C) Glacial retreat. (D) Port Huron advance. (E) Port Huron advance, later phase. (F) Glacial retreat after Port Huron. (G) Two Creeks. (H) Valders advance. (I) Glacial retreat. (J) Transition, Wisconsin Holocene. (K) Holocene. (L) Modern Great Lakes.

Reprinted by permission of American Scientist, *journal of Sigma XI, The Scientific Research Society.*

York City. Striated pebbles, which form only by the grinding action of glaciers, are found in many midwestern sediments. Numerous glacial-depositional features, such as drumlins and eskers, are common in Washington, Montana, North Dakota, Minnesota, Wisconsin, and Illinois.

Lakes are a common result of continental glaciations (figure 26.6). The Great Lakes were formed by the erosive action of continental glaciers during the last ice age. The ad-vancing glacial ice scoured out large valleys, which were further eroded by successive glacial advances. After the last glacial advance, the land tilted upward with the removal of the ice, and the large valleys filled with water from the melting ice, forming the Great Lakes. Another example of glacial lakes is the Finger Lakes of New York. From an aerial view, they show the north-to-south path of the glacial movement as the ice advanced.

Some glacially formed lakes of Pleistocene age are now dried up. Glacial Lake Agassiz of northwestern Minnesota, northeastern North Dakota, and southern Canada (the Red River valley region), was a prominent glacial lake during the last ice age. Sediment deposited on the bottom of the lake eventually formed the region's rich soils. Another large lake formed by glaciers was glacial Lake Missoula, which formed in western Montana. The lake was frequently filled with glacial meltwater and drained rapidly through a series of catastrophic events. The lake's western edge was dammed by southward-flowing glacial ice that blocked off a westward-flowing river. Runoff filled the deep valleys upstream from the ice dam to form an extensive lake that was up to 300 meters deep. As water depth increased, the ice dam either began to float upward or failed owing to great lateral pressure; in either case, a torrent of water rushed westward over the Columbia Plateau to the Pacific Ocean. These repeated large-scale floodings formed the **Channeled Scablands,** a geologic province in western Idaho, eastern Oregon, and eastern Washington.

FIGURE 26.7 Typical glacial deposits of the upper Midwest.
Photograph courtesy of Ross D. Powell, Northern Illinois University.

One interesting interaction between the drainage pattern of the Great Lakes and the process of isostasy occurred during the Pleistocene. When the glaciers retreated for the last time and the meltwater filled the valleys, the drainage was to the north into Canada. This was because the glaciers first melted in the south, and due to isostatic rebound, the land there was rising. The land to the north of the Great Lakes was still isostatically depressed. After the glaciers melted in the north, the land rebounded even higher than it did in the south because the thickness of the ice in the north had been greater. This changed the direction of drainage in the region from north to south.

An abundance of glacially deposited sediment is found throughout North America's upper Midwest (figure 26.7). **Tills,** which are poorly sorted, mixed-lithology deposits emplaced by a glacier as it melts, are extremely common. Many of the rich soils of the Midwest developed on Pleistocene-age glacial tills. Another common deposit is *loess,* windblown, silt-sized sediment. As winds blew across the widespread tills, they picked up the finer-sized sediment fraction and deposited it elsewhere. Uniform deposits of loess were commonly produced by this process. *Glacial erratics,* common in many tills and soils of the Midwest, are rocks brought into an area from someplace else by a glacier. Deposited along with tills and other glacial sediments, erratics are often seen in piles on the edges of fields—put there by farmers who must yearly "harvest" a crop of erratics. Frequently, the glacial erratics in the Midwest are northern-shield rocks, because the glacial movement was from north to south.

Other major terrestrial consequences of the continental glaciations were biologic in nature rather than physical. The repeated glacial advances and retreats caused plants and animals to migrate, evolve, and become extinct. Plant and insect life was particularly susceptible to minor climatic changes. When glaciers advanced and retreated, so did the plant life and the insects. Because of this, some types of fossils can be used to measure the various glacial advances and retreats. Two fossil groups frequently used for determining glacial positions at various times are (1) plant spores and pollen and (2) beetle communities. Plant spores and pollen found in lake sediments, and the types of beetles found in plant remains (peat) from Canada, are two sensitive indicators of the repeated glacial advances and retreats.

Many effects of the glaciations were also felt in the marine realm. Two of these were (1) varying oxygen-isotope ratios and (2) changes in foraminifera coiling directions.

Oxygen isotopes (both oxygen-16 and oxygen-18) are used preferentially by organisms in making their shells. (The oxygen is used in the carbonate radical of the calcium carbonate—$CaCO_3$—shells.) Stated simply, the oxygen isotope used by certain organisms for shell construction depends on the water temperature. Determining the ratio of oxygen-16 to oxygen-18 in certain invertebrate shells, then, can indicate whether the water was warm or cold, with cold water indicating a glacial period and warm water indicating an interglacial episode.

Strangely, the direction in which some planktonic foraminifera coil their shells also depends on the water temperature. (Coiling of foraminifera produces a "top" and "bottom" to the foraminifera shell, so it is possible to determine the coiling direction.) The direction of the coiling is species dependent. In one abundant, cold-water species, coiling of the shell to the right is common in cold-water conditions; coiling to the left is common in very cold-water conditions. In deep-sea sediment cores, sediments with foraminifera coiling in one direction alternate with sediments showing foraminifera coiling in the opposite pattern, reflecting changes in water temperature due to glacial/interglacial episodes.

Canadian geologists are currently using evidence from a wide variety of sources to determine the number and extent of Pleistocene glaciations. On eastern Baffin Island, evidence is collated from the following sources: (1) mapping moraine units, (2) mapping relative sea level and its changes, (3) determining radiometric dates, (4) examining the remains of marine bivalves, (5) looking at micro- and macrofossils in peat, (6) examining the weathering of boulders, and (7) mapping soils that developed during interglacial episodes.

THE EASTERN MARGIN

The Appalachians

The Appalachians experienced a prolonged and continuous period of erosion during the Cenozoic. Essentially, the mountains were eroded, and the sediment was carried eastward by rivers and deposited on the continental-shelf region. Periodically, the Appalachians were rejuvenated; that is, they experienced mild uplifts. At these times, the rate of erosion was increased as streams and rivers cut deeper into the rocks. At least three separate episodes of rejuvenation/erosion can be recognized along the Atlantic coastal plain (the region between the Appalachians and the Atlantic Ocean).

The Eastern Continental Margin

The sediments carried eastward from the Appalachians by rivers were dumped along the coast on the continental-shelf region. The sediments were reworked and transported into deeper water to produce not only a shelf but also a continental rise. The well-developed continental shelf, slope, and rise of the eastern seaboard are features of a mature passive continental margin. In the future, should convergent plate motion begin along the eastern margin, these sediments will form the flanks of future mountains in the east.

THE WESTERN MARGIN

The tectonic style of North America's western margin that began in the Mesozoic continued into the Cenozoic and is responsible for forming many of the major landforms west of the Rockies. The western margin of North America is divided into three general regions for Cenozoic time (figure 26.8). The first region is the Rocky Mountain Region, which includes the Rocky Mountains proper from Idaho and Montana

to Colorado and southward. The second region, the Intermontane Region, includes part of Idaho, the extreme eastern parts of Washington and Oregon, and the state of Nevada—essentially everything between the first and third regions. The Basin and Range is an important province in this region. The third region, the Pacific Mountain System, includes, approximately, the states of Washington, Oregon, and California and the Pacific Coast proper.

Tectonics

The types of tectonics experienced by North America's western margin in the Early Cenozoic differed from the types of tectonics active during the Late Cenozoic (figure 26.9). Throughout most of the Paleogene, the oceanic Pacific plate continued to be subducted under the continent's western margin, as it had been since Mesozoic time (figure 26.9A). Compression was the major tectonic force on this convergent plate boundary. Then, in Oligocene time, approximately thirty million years ago, part of the Pacific midocean ridge (though at this time it was hardly in midocean) collided with North America's western margin—along California—and was subducted (figure 26.9B). The collision caused diverse effects, some of which were felt far inland. In the Pacific Ocean, the direction of plate motion changed from principally north to northwest. (This is shown by the "kink" in the Hawaiian island chain.) Along California, the tectonic motion changed from convergent to shear as the plates began to slide past each other, rather than one being subducted under the other. However, along Washington and Oregon, the Pacific midocean ridge was not subducted—the ridge was still west of the subduction zone—and subduction continued along Oregon and Washington. This singular point explains the difference in tectonics experienced by California and Washington/Oregon.

The collision of the Pacific midocean ridge with North America's western margin also produced profound effects farther eastward, in the Intermontane Region. In essence, the tectonic motion changed from convergent stress to divergent stress in this area. Much of the region began to rift apart, apparently in response to the change in tectonic motion along California from convergent to shear. Of course, the rifting appears to be incomplete; that is, the continent has not broken apart completely in this region, but divergent stress is still occurring today, as evidenced by several major earthquakes in the Basin and Range region during this century.

The Rocky Mountains

The Rockies continued their development in earliest Cenozoic time with the **Laramide orogeny.** The Laramide orogeny was the final pulse of the ongoing Cordilleran

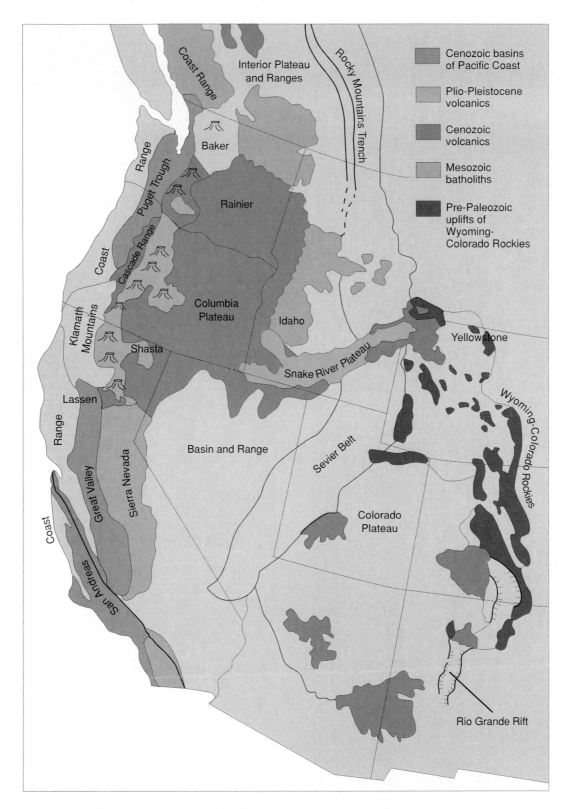

FIGURE 26.8 Physiographic provinces of the western United States.

From Stearn, Carroll, and Clark, Geological Evolution of North America, *3d ed. Copyright © 1979 John Wiley & Sons, Inc., New York, NY. Reprinted by permission of John Wiley & Sons, Inc.*

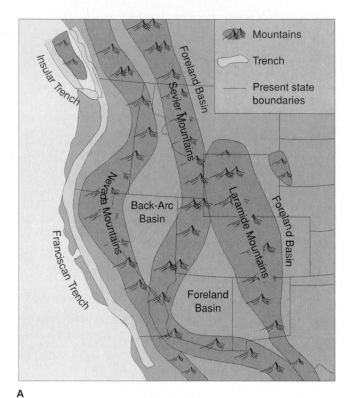

A

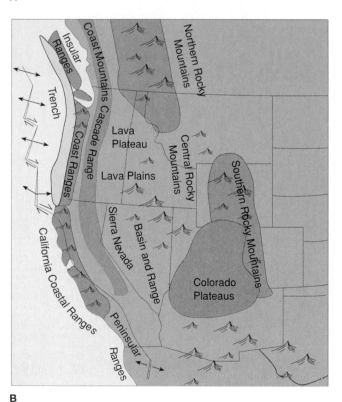

B

Figure 26.9 Western North American plate boundary during the (A) Early and (B) Late Cenozoic. Subduction occurred along the western margin during the Early Cenozoic, producing the various orogenies and related features. Subduction of the midocean ridge in the Late Cenozoic along California produced the San Andreas fault and probably caused the development of the Basin and Range.

From Historical Geology *by Leigh W. Mintz; Charles E. Merrill Pub. Co. Reprinted by permission of Leigh W. Mintz.*

Figure 26.10 Devil's Tower is a Paleocene-age igneous volcanic neck exposed by erosion. Crook County, Wyoming.
Photograph by D. E. Trimble, USGS Photo Library, Denver, CO.

orogeny; it affected a region extending all the way from Alaska through the Yukon Territory, Alberta, and British Columbia and from Montana southward to Texas. The tectonics of the orogeny involved mostly vertical uplift of rocks, producing the pronounced topographic relief of many of the individual mountains in the Rocky Mountain chain. Some volcanoes were also produced (figure 26.10). A clastic wedge—the *Fort Union clastic wedge*—formed in association with the orogeny, and many types of terrestrial-mammal fossils are found in its strata.

In addition to the clastic-wedge strata, uplift of many of the mountains produced *intermontane basins*, which are steep sided, rectangular, and fault bounded (figure 26.11). A variety of terrestrial environments formed in these basins, including alluvial fans, rivers, and lakes. Because the sedimentation rate was rapid, large volumes of sediment were preserved. The strata contain a host of terrestrial fossils of mammals, plants, insects, and fish.

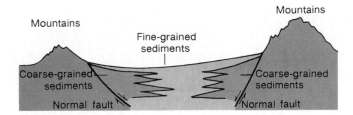

FIGURE 26.11 Intermontane basins.

In Eocene time, large lakes formed just to the west of the Rockies in southeast Wyoming, northeast Utah, and northeast Colorado. Many of these lakes were shallow, and they frequently dried up almost completely. The sediments composing the *Green River formation* were deposited in a large lake of this type (figure 26.12). Because of the fluctuating water level and the rapid influx of sediments, many fossilized fish were preserved in the strata. In many cases, the fish were obviously stranded in shallow pools by the receding waters and died as the water in the small pools evaporated. Algae were a prominent component of the life of the lake. In the lake's interior, fine-grained carbonate was deposited. Today this combination of organics and fine-grained sediment has produced oil shale, which has much economic potential because of its petroleum content.

The "oil shale" of the Green River formation is not a shale, nor does it contain oil. The rock type that composes much of the Green River formation is a marl—a fine-grained limestone. The petroleum product in the marl is called *kerogen*, a crude form of petroleum. With further refining, kerogen can be turned into oil products. To be accurate, the kerogen-marl deposits, rather than the oil-shale deposits, of the Green River formation are a possible future energy resource.

In addition, in Eocene time, the intermontane basins that had formed from the orogenic uplifts in Paleocene time filled with sediment. (Subsequent erosion of these types of sediments has produced many attractive features of western North America, such as the multicolored strata of Bryce Canyon, figure 26.13.)

Throughout much of the remainder of the Cenozoic, the Rockies and adjacent regions experienced repeating cycles of uplift (with perhaps some associated faulting and volcanism) and erosion of the uplifted region (and infilling of basins) (figures 26.14 and 26.15). Evidence of volcanism is still seen today in such areas as Yellowstone National Park and its geysers (figure 26.16).

The Intermontane Region

Strangely, the central Cordillera experienced divergent motion during the mid-Cenozoic. What caused divergent activity at a predominantly convergent (subduction) plate boundary?

FIGURE 26.12 The Green River formation.
Photograph courtesy of Stephen R. Mattox.

FIGURE 26.13 The brightly colored, highly dissected Wasatch formation (Eocene). Bryce Canyon National Park, Utah.
Photograph by W. B. Hamilton, USGS Photo Library, Denver, CO.

FIGURE 26.14 The Tetons were produced during Cenozoic uplift of older strata along nearly vertical faults.
Photograph by W. B. Hamilton, USGS Photo Library, Denver, CO.

FIGURE 26.15 Sculpted Tertiary-age (White River group, Oligocene) sediments. Badlands National Monument, South Dakota.
Photograph by D. E. Trimble, USGS Photo Library, Denver, CO.

FIGURE 26.16 Geysers in the Norris geyser basin. Yellowstone National Park, Wyoming.
Photograph by W. B. Hamilton, USGS Photo Library, Denver, CO.

Because the divergence coincides with the subduction of part of the Pacific ridge system, it probably reflects the change in plate motion at the western margin.

In the Pliocene (five to ten million years ago), the Colorado Plateau experienced a period of prolonged and gradual uplift owing to a change in the thermal properties of the underlying crust—the crust became hotter. The cause of the abundant heat flow, however, is not known. The uplift caused the Colorado River to downcut into its channel and produce a large exposure of rocks—the **Grand Canyon** (figures 26.17 and 26.18). Frequently, authors attribute the Grand Canyon only to the downcutting of the Colorado River. The Grand Canyon, however, is the result of two processes acting simultaneously—regional uplift and stream downcutting.

The **Basin and Range** province in Nevada was also formed in the southern part of the Intermontane Region during Oligocene time. The Basin and Range region consists of a series of horsts and grabens (similar to the Triassic fault basins that formed when North America diverged from Africa in the Early Mesozoic). The alternating horst-and-graben

structure could only have been formed as a result of divergent stress. Some plutonic rocks are also found in the region, indicating spreading.

In the north, similar divergent plate motion was occurring during the Miocene. The continent experienced rifting as flood basalts poured from fissures as lava flows. The flood basalts covered an extensive region of the Northwest—Washington, Oregon, and Idaho. The many flows, one on top of another, produced the **Columbia Plateau.** Overall, over 2500 meters of basalt is found in vertical section, yet each flow is quite thin (5 to 10 meters). Some of these basalts were eroded and sculpted by violent floods caused by the sudden releases of water from Lake Missoula when the ice dams that impounded the water were broken.

The Pacific Coast System

Today, the tectonic motion along the Pacific Coast is of two types: (1) convergent along the edge of British Columbia, Washington, Oregon, and northern California; and (2) shear or transverse along central and southern California. The difference in tectonic motion is due to the subduction of part of the Pacific ridge system along California in mid-Cenozoic time.

The subduction zone along Washington and Oregon is an ocean–continent subduction zone. As at all such convergent plate boundaries, a series of volcanoes was produced on the land—the **Cascade Range**—during the Cascade orogeny. The Cascade Range includes such volcanoes as Mount St. Helens, Mount Lassen, Mount Rainier, Mount Hood,

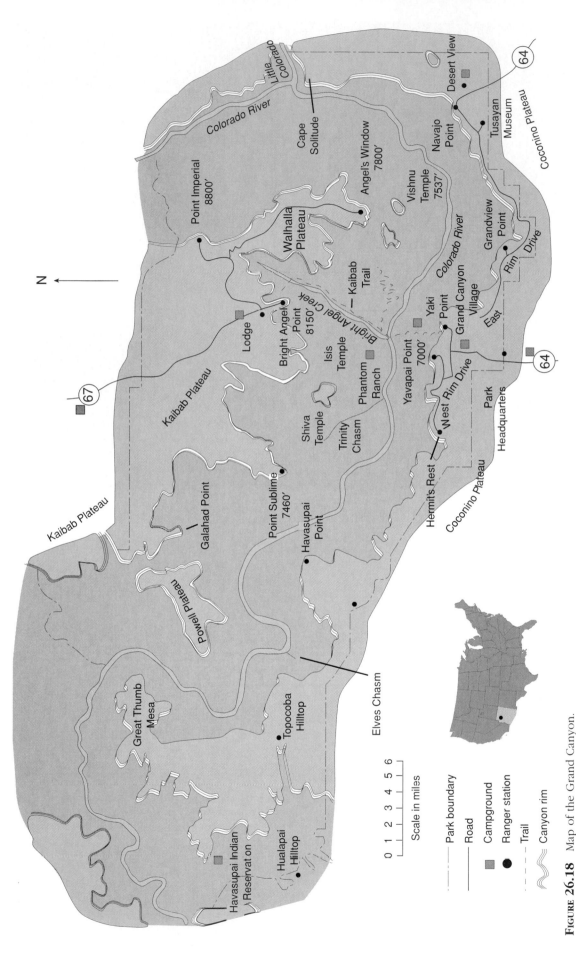

FIGURE 26.18 Map of the Grand Canyon.

From Harris and Tuttle, Geology of National Parks, 3d ed. Copyright © 1990, Kendall/Hunt Publishing Company.

FIGURE 26.19 Mount St. Helens of the Cascade Range.
Photograph by Cascade Volcano Observatory, USGS Photo Library, Denver, CO.

and Mount Baker (figure 26.19). The eruption of Mount St. Helens in May 1980 shows dramatically that subduction continues and that the melting plate still produces magmas for the periodic volcanic eruptions.

The change in tectonic motion along California from convergent to shear produced the **San Andreas fault** system (figure 26.20). The Pacific Plate is moving relatively northward, the North American Plate southward.

A common myth is that movement along the San Andreas fault will somehow cause California to "fall off" into the ocean. Owing to the relative north/south motion of the Pacific and North American plates, however, divergence will not occur; that is, part of California will not break apart and move into the Pacific Ocean. Rather, over time, the tectonic motion of the Pacific Plate will cause part of Los Angeles to move northward, while motion of the North American Plate will cause part of San Francisco to move southward until the two fragments are joined as one large city!

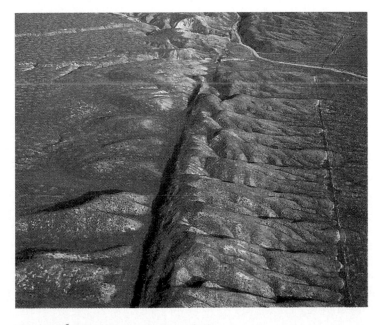

FIGURE 26.20 The San Andreas fault.
Photograph by R. E. Wallace, USGS Photo Library, Denver, CO.

LIFE IN THE CENOZOIC

Invertebrates

Shallow-water marine life during the Cenozoic was restricted to the continental shelves at the continental margins. The epeiric seas, which formerly had been so abundant, became rare because the continents were much higher in Cenozoic time than during the Paleozoic and the Mesozoic eras.

In the shallow-water marine environments of the continental shelves, invertebrates thrived. Many of the groups common in modern oceans continued, and overall, the modern ecosystem developed. Fossils of corals, bryozoans, echinoderms, brachiopods, and other invertebrates of Cenozoic age all show decidedly modern appearances.

In oceanic areas, the water environment above deep-ocean basins, both phytoplankton (microscopic plants) and zooplankton (microscopic animals) were abundant in shallow waters. Some more important plant groups were *coccolithophores,* composed of calcium carbonate ($CaCO_3$), and *diatoms,* composed of silica (SiO_2). Important protozoan plankton included *foraminifera,* composed of calcium carbonate, and *radiolarians,* composed of silica (figure 26.21). When the microscopic plants and animals died, they sank to the bottom and were incorporated into the deep-ocean sediments. Because their distinctive remains are well preserved and widely distributed, they serve as extremely useful index fossils for correlating Cenozoic marine sediments.

Life on Land After the Mesozoic

Throughout virtually all of the Mesozoic, dinosaurs occupied the ecologic niches of the large vertebrates. However, at the end of the Mesozoic, climates changed slowly from warm to cool and temperate. Mammals, which were warm-blooded and possessed hair, were better able to adapt to the cooler and more seasonally variable conditions. Thus, after the extinction of the dinosaurs at the end of the Cretaceous, mammals underwent explosive evolution (adaptive radiation), and their numbers and diversity increased to take over the ecologic niches left vacant by the dinosaurs and to invade new ones. By the end of the Paleocene epoch, more than fifteen orders of mammals had appeared.

Vertebrates

Reptiles of many types continued into the Cenozoic, but they were not nearly as successful or as abundant as during the Mesozoic. Snakes, crocodiles, alligators, turtles, and lizards all survived the terminal-Cretaceous extinctions, and all are found today.

The Cenozoic is the "Age of Mammals." The earliest mammals were similar to the Mesozoic types—small herbivores. By at least the end of the early Paleocene, however, two important groups had arisen, the marsupials and the insectivores.

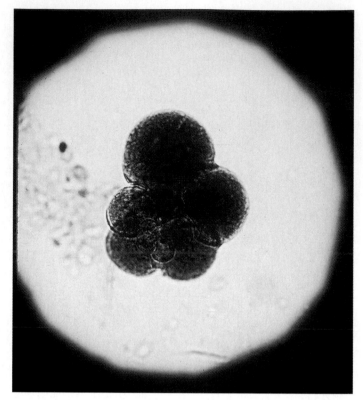

A

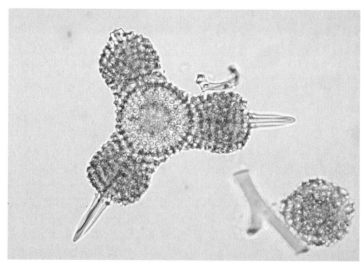

B

FIGURE 26.21 (A) Foraminifera. (B) Radiolarians.
Photographs courtesy of Dr. Hsin Yi Ling, Northern Illinois University.

Marsupials are pouched mammals. The young marsupial spends part of its early developmental stage within a pouch attached to the mother's abdomen. Many types of marsupials are alive today—kangaroos, opossums, wombats, and wallabies. Although widespread on certain continents today,

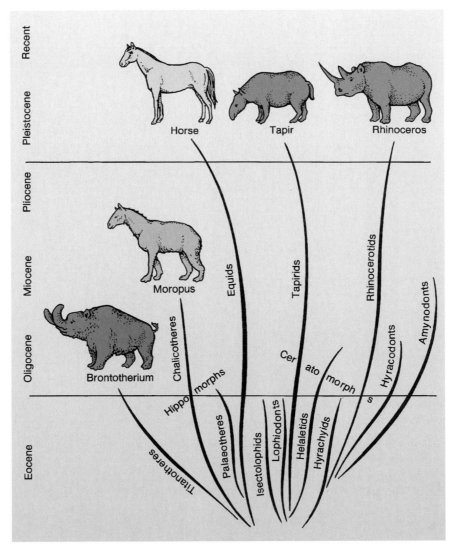

FIGURE 26.22 Ungulates: the perissodactyls.

From E. H. Colbert, Evolution of the Vertebrates. *Copyright © 1966 John Wiley & Sons, Inc., New York, NY. Reprinted by permission.*

marsupials are not as successful as the placental mammals, in which the young develop entirely within the mother's body.

The earliest placental mammals were the **insectivores,** which evolved in the Cretaceous. The best illustrative examples of an early insectivore are the modern shrew and mole. In the Early Paleocene, insectivores were rather primitive and unspecialized. Soon, however, the insectivores gave rise to an abundance of other mammals, and by the Middle Paleocene, one fundamental group of mammals had arisen, the ungulates (figure 26.22). The term **ungulates** refers to the large group of plant-eating, hoofed mammals. The earliest ungulates were an abundant group called the *condylarths.*

In the Eocene, carnivorous mammals evolved from an ungulate ancestor. At first, the carnivorous mammals had hoofs, but this group became extinct at the end of the Eocene and was replaced by clawed carnivorous mammals—

for example, cats and dogs—in Early Oligocene time. Other ungulates continued to evolve. Two of the more important groups are the perissodactyls and the artiodactyls.

Perissodactyls are the odd-toed hoofed mammals. Five different groups of perissodactyls are known: the extant horses, tapirs, and rhinoceroses; and the extinct chaliocotheres ("kol-EYE-oh-koh-THER-eez") and titanotheres. Perissodactyls arose from condylarths and were very abundant during the Miocene.

The development of the horse is often used as evidence of evolution because the changes involved are so well documented in the fossil record (figure 26.23). The earliest horse is *Hyracotherium* (formerly called *Eohippus*), fossils of which are found in Late Paleocene–age strata in North America and Europe. This horse was very small, the size of a modern dog, and had four hoofed toes on the front feet

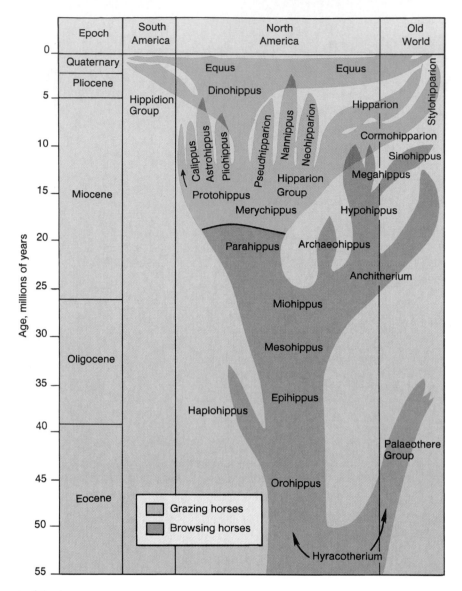

FIGURE 26.23 Development of the horse.

From MacFadden, Paleobiology, *vol. 11, fig. 1, p. 247, 1985. Copyright © 1985 Paleontological Society, Ithaca, NY. Reprinted by permission.*

and unspecialized, primitive dentition. From associated fossils, it is believed *Hyracotherium* lived in forests. Throughout the Eocene, Oligocene, Miocene, and Pliocene epochs, horse evolution followed a very definite trend. The horses got bigger, lost several toes (the modern horse stands on one toe, the middle toe), and changed their diets. Grasslands developed in Middle Miocene time, which opened up a new environment not seen before. Horses successfully adapted to this new environment, feeding on the extensive grasses. Because the grasses contained a large amount of silica, horses' molars changed and their jaws lengthened. In addition, because living on the open grasslands afforded little protection in the way of camouflage, horses adapted to running and evolved fewer toes.

Artiodactyls, or even-toed hoofed mammals, include cattle, camels, pigs, deer, hippopotamuses, sheep, goats, and giraffes. Artiodactyls evolved from condylarths in Eocene time and spread throughout the continent (figure 26.24).

Continental Drift and Mammal Distribution

The timing of the breakup of certain continents and their subsequent reconnection during the Cenozoic has controlled the present distributions of mammal fossils on the continents. Two examples that illustrate this controlling influence are the mammal types found in North and South America and Australia.

During much of the Early Cenozoic, North and South America were separated. The land bridge between the two continents—Central America—had not yet formed. Each

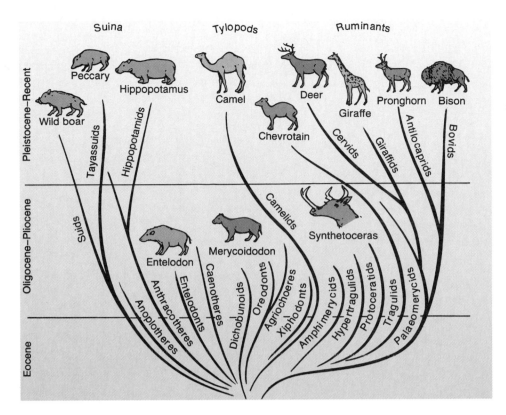

FIGURE 26.24 Artiodactyls.

From E. H. Colbert, Evolution of the Vertebrates *fig. 109, p. 378. Copyright © 1966 John Wiley & Sons, Inc., New York, NY. Reprinted by permission.*

continent developed its own separate and distinct fauna. North American mammals were predominantly placental and carnivorous, South American mammals predominantly marsupial. When Central America formed as a series of volcanic provinces in response to subduction of the Pacific Plate during the Middle Cenozoic, North and South America became connected. Mammals migrated from one continent to the other. Carnivorous mammals invaded South America and rapidly became dominant. Most of the marsupial mammals of South America suffered mass extinctions, although a few escaped and migrated north—for example, the North American opossum.

In Australia, another marsupial-mammal population flourished (for example, kangaroos). Because Australia separated from Southeast Asia before the carnivorous mammals could migrate to Australia, no large carnivores exist in Australia today. Therefore, in Australia the marsupial mammals did not have to compete with carnivorous mammals and today remain a stable population.

Plants

Plant communities of earliest Cenozoic time differed from those of Mesozoic time by displaying a much greater percentage of cooler-climate species.

P lants are excellent types of fossils for reconstructing ancient climates because they are very sensitive to temperature changes. Differences in the abundance of various species or in the leaf morphology of one species can signal a climatic change. Fossil plant leaves cannot always be used in climate reconstructions, however, because of poor preservation. Therefore, plant spores and pollen are used and can be very helpful in determining changing climatic patterns of the past.

Angiosperms, the flowering plants, rapidly outnumbered gymnosperms in the Cenozoic. Angiosperm evolution was paralleled by the evolution and increase in the number of kinds of insects.

One of the more important evolutionary events in plant life during the Cenozoic was the appearance of grasslands in the Miocene. Many animals, especially mammals such as the horses, were quick to exploit this new grassland environment.

Fossil Humans

The evolution of primates, and the subsequent evolution of humans, occurred during the Cenozoic. Primates differ from other mammals in several respects: The primate skeleton is

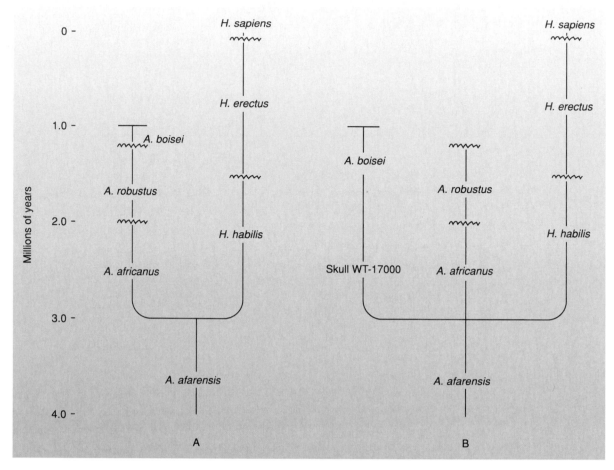

FIGURE 26.25 Current models of human evolution. (A) The conventional view. (B) A newer view, based on recent discoveries.

generalized, which allows for a great range of behavior. The lower limb bones, the vertebrae, and the clavicle are all examples of skeleton adaptations. Primates have mobile digits with nails rather than claws at the end. Their visual apparatus is complex; their tooth types are distinct. Most primates have an upright body trunk. Primates also have a very large brain relative to body size. Finally, the gestation process of primates is longer than that of most other mammals.

The primates are a varied and diverse order of mammals. Some of the many types of primates include (1) the prosimians (lower primates), such as the tree shrews, tarsiers, slender lorises, and lemurs; (2) the anthropoids (higher primates), such as the New World Monkeys (for example, the howler monkey, spider monkey, and capuchin), the Old World Monkeys (for example, the baboons, macaques, and mandrills), the pongids (the great apes, such as the orangutan, the chimpanzee, and the gorilla), and the family **Hominidae,** or humans.

Primate evolution leading to humans is complex. One important question is, Why did humans split from the other primate line? The following four answers have been proposed, and all are probably somewhat correct:

1. The availability of grasslands as a different environment for hominids.
2. The development of bipedalism in hominids.
3. The development of tool use in hominids.
4. Competition with other animals.

Currently, there is much debate in the field of paleoanthropology about the evolution of the various hominid groups. Recent finds have led to new interpretations of how humans evolved (figure 26.25). The new controversy and debate is very healthy for the science and will, in the long run, lead to a more accurate picture of human evolution.

The start of human evolution begins with the first line of hominids, the Australopithecine group, ***Australopithecus***

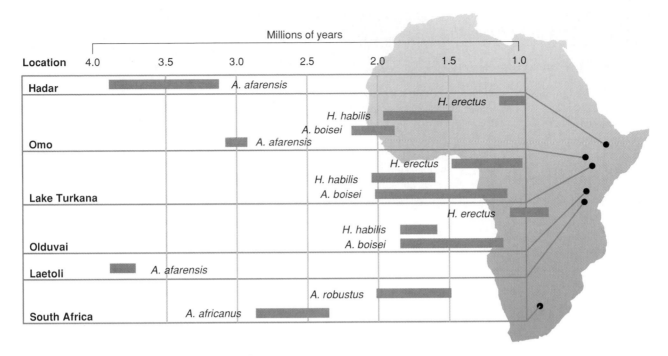

FIGURE 26.26 Important hominid sites in Africa. Hominids are plotted by type, location, and age.

Stephen Misencik, 1980 Copyright, "Lucy."

("southern ape"), which evolved in Africa approximately four million years ago from a Dryopithecine ape ancestor.

The oldest Australopithecus is *A. afarensis* ("southern ape of Afar"). The first *A. afarensis* was found in 1972 by Donald Johanson at Hadar (figure 26.26) in the Afar Triangle of Ethiopia, in Africa, and was named "Lucy" (from the Beatles' song "Lucy in the Sky with Diamonds," which was playing on the radio when she was discovered). The importance of Lucy is that her skeleton is over 40% complete. Previously, interpretations of the earliest hominids were based only on fragments of pelvic bones, skulls, and teeth. The fact that Lucy's skeleton was fairly complete meant that these interpretations could now be tested, and they were found to be substantially correct. Additional finds of *A. afarensis* skeletal fragments have indicated that these hominids were small (table 26.1), with the males larger than the females. Their faces were decidedly apelike, with a low forehead, a flat nose, a brow ridge, and no chin. Most important, *A. afarensis* was fully bipedal. This discovery helped answer a hotly debated question: Which came first in human evolution, bipedalism or larger brains? Lucy showed that hominids were first bipedal.

The next group, *A. africanus* ("southern ape of Africa"), was slightly larger (table 26.1). *A. africanus* was widespread, fully bipedal, although the arms were relatively long, and had human dentition. There is some question about whether these hominids had the ability to manufacture and use tools. *A. africanus* gave rise to two, possibly three, other groups of hominids.

A. robustus ("robust southern ape"; once called *Paranthropus,* "beside man") appears to have evolved from *A.*

TABLE 26.1 *A Comparison of Hominid Groups (Heights and Weights are Approximate)*

Hominid Group	Height	Weight (pounds)
A. afarensis	3–4'	65
A. africanus	3–4'	50–90
A. robustus	5–5½'	100–150
A. boisei	5–6'	125–175
H. habilis	4–5'	110
H. erectus	5–6'	90–160
H. sapiens neanderthalensis	5'7"	150
H. sapiens sapiens	5'6"–5'8"	150

africanus about 2 to 2.5 million years ago. Hominids in this group were again larger (table 26.1). *A. robustus* also differed from *A. africanus* in skull shape (larger and flatter) and in dentition (massive cheek teeth, smaller front teeth). The dentition clearly indicates that *A. robustus* was an herbivore.

The conventional view of human evolution held that *A. robustus* then gave rise to another group, *A. boisei* ("Boise southern ape"; once called *Zinjanthropus,* "East Africa man") (see figure 26.25A). This group was named after a Mr. Boise, a British businessman who financed the expedition that found the first skeletal fragments in 1959. Once again these hominids were larger than the previous group. *A. boisei* are characterized by immense molars and premolars, indicating

Evolutionary Step	Time (years)
Major advances in tool making (Mousterian technology)	150,000
Use of fire	700,000
Migration from Africa to Eurasia	1.0 million
Settlement, meat eating	1.5 million
Start of brain expansion	2.1 million
First stone tools	2.5 million
Bipedalism	5–10 million

that they chewed great quantities of leaves. *A. boisei* were fully bipedal and had large skulls with massive brow ridges. Possibly these hominids made stone tools—chipped stone and pebble fragments have been found in association with the skeletal remains. A recent discovery of a skull by Alan Walker (and Richard Leakey) west of Kenya's Lake Tarkana has forced a reexamination of this type of hominid. The skull (referred to by its museum number, WT-17000) is clearly *A. boisei* but was found in sediments about 2.5 million years old—much, much older than other deposits where *A. boisei* skulls have been found. Thus the conventional idea that *A. boisei* evolved from *A. robustus* may be incorrect. *A. boisei* may have evolved from *A. africanus* or *A. afarensis* much earlier (see figure 26.25B).

The **Homo** ("man") line also evolved rather early and represents an entirely different line from that of the Australopithecines. One idea proposed is that both the advanced Australopithecines and the *Homo* line evolved from a common ancestor in Late Pliocene time.

The first *Homo* group, *H. habilis* ("handy man"), was named by Louis Leakey, Philip Tobias, and John Napier in 1964. Relatively small, hominids in this group had a larger brain case than that of the Australopithecines. *H. habilis* also made stone tools and built simple shelters. Other examples of this group include "*Telanthropus*," from South Africa, and "*Meganthropus*," from southeast Asia. Two conflicting views state that *H. habilis* is either a distinct species or simply a more primitive form of *H. erectus*. This is a matter of interpretation and does not detract from the marked differences between the Australopithecine line and the *Homo* line.

H. erectus ("upright man") evolved from *H. habilis* about 1.6 million years ago and died out about 200,000 years ago. These hominids were larger, with long, low skulls. *H. erectus* used tools and fire, had good building methods, and certainly engaged in hunting both large and small game. This group was also very widespread throughout the Middle Pleistocene. Skeletal fragments have been found in Africa; in east Asia (Peking Man, "*Sinanthropus*"), southeast Asia (Java Man, "*Pithecanthropus*"), and China; and in Germany (Heidelberg Man), southwest France (Tautavel skull), Hungary (Vertesszöllös skull), and Greece (Petralona skull).

The term "Archaic sapiens" refers to a group of incomplete skulls and skeletal fragments, approximately 300,000 years old, from Europe and Africa that appear to be primitive forms of *Homo sapiens*.

H. sapiens, our species, evolved from *H. erectus* in the Late Pleistocene. Two variants are found in the fossil record: *H. sapiens neanderthalensis*, the Neanderthals, and *H. sapiens sapiens*, modern humans. The Neanderthals (named for the Neander Valley, near Dusseldorf, West Germany, where the first skeletons were found), are an often-misunderstood group. Previously pictured as hulking, brutish cave-dwellers (this misinterpretation resulted partly from the fact that one Neanderthal reconstruction was based on the skeleton of an arthritic individual), they are now believed to have been much more like modern humans. Neanderthals used a variety of stone tools, were capable hunters (particularly of mammoths), constructed shelters, used caves, wore clothes, and were the first group to bury their dead. Neanderthals were also adapted for cold climates. The reason for their disappearance is something of a mystery. Various ideas proposed include their inability to cope with the drastically cooling climates that marked the start of another ice age (unlikely, since Neanderthals were well equipped for cold climates and could always migrate to other areas), and changing ecological patterns in which *H. sapiens sapiens* successfully outcompeted the Neanderthals.

Our line, *H. sapiens sapiens*, "Cro-Magnons," coexisted with the Neanderthals. Cro-Magnons used complicated tools made from both stone and bone, migrated seasonally, used many different hunting methods, buried their dead, and developed culture and art.

In summary, the basic split of hominids into two different groups, the Australopithecine group and the *Homo* group, is well established. Finer points, such as the relationship of *A. boisei* to other members of the Australopithecine group, and the naming of *H. habilis*, continue to be debated. A summary of the major evolutionary steps in human evolution is shown in table 26.2.

In the study of human evolution, Africa has been considered the "cradle," or starting point. Important hominid sites in Africa are shown in figure 26.26. Hadar is in the Afar Triangle in Ethiopia. It was here that Donald Johanson discovered many *A. afarensis* skeletal fragments. Several important sites are also found around Lake Rudolf in eastern Africa. Omo is on the banks of the Omo River, just north of Lake Rudolf. Koobi Fora, just to the east of Lake Rudolf, is where Richard Leakey found the famous "1470" skull of a 2.6-million-year-old *Homo*. Lothagam, near the southern part of Lake Rudolf, is a site where a river has cut into 5 to 6-million-year-old sediments. An important find from this site is Bryan Patterson's 5.5-million-year-old Australopithecine jaw. Olduvai Gorge, in Tanzania, is about 400 miles south of Lake Rudolf. Louis and Mary Leakey, beginning in 1960, and Donald Johanson and Tim White, beginning in 1986, have uncovered many *H. habilis* skeletal fragments. Laetoli, nearby, is the place where a complete set of hominid footprints was discovered. The tracks show several individuals walking across what was then volcanic ash.

Box 26.1

Regional Geology

TETONS OF WYOMING

The Teton Range in northwestern Wyoming is typical of a series of large, northwest-trending mountain ranges that occur in Wyoming (figure 1). Two prominent peaks in Grand Teton Park are Mount Teton (13,747 feet high) and Mount Moran. The core of the Tetons is composed of Precambrian igneous and metamorphic rocks. The Tetons' majestic appearance is due, in part, to the large fault scarp that forms their eastern face—a fault scarp that rises over 2 kilometers from the surrounding plains.

The earliest phase in the formation of the Tetons occurred during the Precambrian as the igneous and metamorphic rocks that make up the core of the mountains were formed around the edge of the shield. The mountains were formed in the Late Tertiary and were associated with a number of other uplifts that occurred during this time (for example, the Beartooth uplifts of Montana, the Black Hills uplifts of South Dakota, and the Uncompahgre uplifts of Colorado). During the Late Tertiary, large blocks of rocks were thrust upward along normal and reverse faults. Displacements of over 6000 meters occurred. Importantly, in most cases, the faults were nearly vertical.

The Tetons (and associated mountains) are important for two reasons. First, Precambrian igneous and metamor-

FIGURE 1 The Tetons.
Photograph courtesy of Diane Carlson.

phic rocks are exposed. Here, geologists can obtain another view of Precambrian petrologic and tectonic processes away from the Canadian Shield. Second, the tectonic style of formation of the Tetons is different from that previously seen in the Cordilleran region. For much of the Cordilleran history, the stresses were dominantly horizontal and compressive, typical of subduction-zone margins. During the formation of the Tetons, stress was relieved by motion that was nearly vertical, as expressed by the faults. This represents a different style of tectonics that must be considered at convergent plate margins.

Perhaps the most famous hoax in the history of science is the infamous Piltdown Man controversy of the early 1900s. Charles Dawson, an amateur anthropologist, "discovered" a modern human skull and an ape-like jaw fragment in a gravel pit on an old farm at Piltdown Manor, Sussex. Dawson reported his discovery to Arthur Smith Woodward of the British Museum, and together they announced their findings to the Geological Society of London on 12 December 1912. Termed *Eoanthropus dawsoni* ("Dawson's dawn man"), the reconstructed skull and jaw were proclaimed to be a major find in the field of paleoanthropology. Questions regarding the authenticity of the reconstruction were raised almost immediately. The part of the jaw where it fit to the skull was missing (quite convenient). This raised doubts as to whether the skull and jaw were in fact joined together. Finally, in 1953, a team of three British scientists, Kenneth P. Oakley (chemistry), J. S. Weiner (anatomy), and Wilfred E. Le Gros Clark (anatomy) settled the Piltdown question once and for all. Extensive chemical and anatomical tests showed convincingly that the skull was a modern human skull, the jaw that of an ape with the teeth filed down. The question immediately arose, Who committed this fraud? Dawson? Woodward? Evidence points to Dawson, but this question is still not resolved.

An important lesson can be found in why this rather obvious inconsistency was not discovered earlier. The reason is twofold. First, because the skeletal fragments were considered valuable, few scientists were allowed to examine them directly. Plaster casts, which omitted vital clues, were used instead. Second, English scientists wanted badly to find ancient hominid fossils in England. Initially, scientists were perhaps reluctant to prove that this find was a fake. Piltdown Man *was* a fake. Yet, in the end, it was scientists who determined that it was a fake. This shows that science is, above all, a self-correcting discipline.

SUMMARY

During the Cenozoic, the continents continued to drift apart as the Atlantic Ocean grew in size and the Pacific Ocean shrank.

On the North American craton, terrestrial environments dominated throughout the Cenozoic. During the Pleistocene, advances and retreats of the glaciers had several profound effects. The glacial episodes caused regressions along continental margins as water left the oceans to become glacial ice; the intervening interglacial warming periods caused transgressions. After the glaciers retreated from the continents for the final time, the continents experienced isostatic rebound. Many types of glacial erosional features, such as striated pavements and pebbles, and depositional features, such as drumlins and eskers, were formed, as were numerous glacial lakes. Many of these lakes—for example, the Great Lakes and New York's Finger Lakes—are still present today; other lakes formed by glaciers have dried up completely—for

example, glacial Lakes Agassiz and Missoula.

North America's eastern margin experienced several episodes of uplift followed by erosion, which produced recognizable erosional surfaces along the Atlantic coastal plain. The eastern margin of the continent developed into a fully mature, passive continental margin as rivers carried sediments from the eroding Appalachians to the continental-shelf region.

On North America's western margin, compressional tectonics progressed as subduction of the oceanic Pacific Plate continued. Uplift of the Rockies during the Laramide orogeny of the Paleocene epoch produced much of the mountains' rugged topography. In the Middle Cenozoic, a fundamental tectonic change occurred along the western margin as the oceanic ridge of the Pacific Plate was subducted along the California coast. This resulted in tectonic motion changing from convergence to shear along

California and produced the San Andreas fault system. In the Intermontane Region, other effects of this change in tectonic motion were felt as the region experienced an episode of uplift—the Colorado Plateau—and incomplete rifting and divergence. The divergence produced the Basin and Range province of Nevada and the flood basalts of the Columbia Plateau of Washington and Oregon. Along the coasts of Washington and Oregon, subduction continued—the Pacific midocean ridge system was not subducted here—and produced the volcanic Cascade Range.

During the Cenozoic, phytoplankton and zooplankton were common in the shallow waters of oceanic basins. These microscopic plants and animals are used as index fossils for correlating Cenozoic marine sediments. On land, the mammals experienced rapid evolution, increasing in both size and number.

A summary of Cenozoic geologic and biologic events is shown in figure 26.27.

Period	Western margin	Craton	Eastern margin	Life
Neogene	Pliocene —Colorado Plateau —Grand Canyon —Teton Range —Cascade Range Miocene —Regional uplifts —Volcanism	Pleistocene —Continental glaciation —Glacial lakes —Till —Loess	Erosion of the Atlantic Coastal Plain Periodic rejuvenations —Uplift and erosion	Hominid evolution Grasses evolve in Miocene
Paleogene	Oligocene —Basin and Range —Columbia Plateau —Basalts Eocene —*Green River formation* —Intermontane basins Paleocene Laramide orogeny —Mostly vertical uplifts —Fort Union clastic wedge	Craton exposed	Erosion of the Atlantic Coastal Plain Periodic rejuvenations —Uplift and erosion	Mammals expand and diversify —Ungulates: perissodactyls and artiodactyls Index fossils for marine strata: Foraminifera

FIGURE 26.27 The Cenozoic: summary diagram.

TERMS TO REMEMBER

artiodactyls 486
Australopithecus 488
Basin and Range 480
Cascade Range 481
Channeled Scablands 475
Columbia Plateau 481

Grand Canyon 480
Hominidae 488
Homo 490
insectivores 485
Laramide orogeny 476
marsupials 484

perissodactyls 485
San Andreas fault 483
till 475
ungulates 485

QUESTIONS FOR REVIEW

1. What were several physical and biologic effects of the Pleistocene-age continental glaciations?
2. Compare and contrast tectonics of the eastern margin during the Mesozoic era and the Cenozoic era.

3. Why were epeiric seas absent from the North American craton during Cenozoic time?
4. Describe the differences in tectonics on the western margin between the Early Cenozoic and the Late Cenozoic.
5. What is the reason for divergent plate motion in the central Cordillera during the mid-Cenozoic?

6. What important groups of marine plankton are used for the correlation of marine strata?
7. Briefly outline mammal evolution during the Cenozoic.
8. How did continental drift control mammal distribution during the Cenozoic?

FOR FURTHER THOUGHT

1. Two examples of the influence of plants on the evolution of animals are the evolution of angiosperms and the development of insects and the evolution of grasses and the development of the horse. Can you think of any other examples of how plants may have influenced the development of animals?
2. The entire Cenozoic era is about sixty-six million years long. Compare the evolutionary events that have occurred during the Cenozoic with those that occurred during a comparable length of time in the past—for example, during the Cambrian period.

SUGGESTIONS FOR FURTHER READING

Atwater, T. 1970. Implications of plate tectonics for the Cenozoic tectonic evolution of western North America. *GSA Bulletin,* 81: 3513–35

Campbell, B. 1985. *Human evolution.* 3d ed. New York: Aldine.

Johanson, D. C., and Edey, M. A. 1981. *Lucy: The beginnings of humankind.* New York: Simon and Schuster.

Kurten, B. 1969. Continental drift and evolution. *Scientific American* 220(3): 54–64.

———. 1971. *The age of mammals.* London: Weidenfeld and Nicolson.

Redfern, R. 1983. *The making of a continent.* New York: Times Books.

Steiner, J., and Grillmair, E. 1973. Possible galactic causes for periodic and episodic glaciations. *GSA Bulletin,* 84: 1003–18.

Tattersall, I., and Eldredge, N. 1977. Fact, theory, and fantasy in human paleontology. *American Scientist* 65: 204–11.

27

MINERAL AND ENERGY RESOURCES, TODAY AND TOMORROW

Aerial view of Lake Mead behind Hoover Dam emphasizes the volume of water and size of reservoir that may be involved in hydropower/water-supply projects such as this; for these reasons, among others, relatively few sites are suitable for large-scale hydropower generation.
© Peter/Stef Lamberti/Tony Stone Images

We are all affected by issues relating to geologic resources—their availability, their cost, the consequences of their use. The United States has a particularly resource-hungry lifestyle; we in this nation use a portion of world resources far out of proportion either to our population or to our own resource supplies. This chapter offers a brief introduction to our mineral and energy resources, including a look at what may lie ahead.

The chapter begins with a look at the occurrences of a variety of rock and mineral resources and then briefly considers the U.S. and world supply-and-demand picture. Aspects of that picture suggest a need to develop additional sources of mineral materials for the future, and several possibilities are examined. Environmental impacts of mining activities are also noted. We then take a similar look at the energy sources currently in most common use and finally survey briefly some additional energy sources that may become increasingly important in the future.

BASIC CONCEPTS: RESERVES AND RESOURCES

The term *resources* is often used, informally, to designate some natural materials that are useful and/or economically valuable to present society. In the context of minerals and fuels, the term has a more precise definition and is distinguished from the related term *reserves*.

The **reserves** represent that quantity of a given material that has been found and could be exploited economically with existing technology. The term is usually further restricted to apply only to material not yet consumed. (The term *cumulative reserves* encompasses that quantity of the material already used, together with remaining reserves.) The category of **resources** is a much larger one: it includes reserves, plus deposits of the material that are known but that cannot be exploited economically given current prices and technology, plus deposits that have not yet been found but are believed to exist, based on extrapolation from the abundance and geologic occurrence of known deposits. Clearly, the reserves are the most conservative estimate of the amount of a given mineral or fuel available, especially in the near future.

Because economic considerations enter into the definition of reserves, changing economic factors can cause particular mineral or fuel deposits to be reclassified as reserves that were previously classified only as part of the resources. For example, increases in demand, or changes in other factors that drive up prices, make some previously uneconomic mineral deposits profitable to mine. Improvements in technology that decrease the cost of extracting the particular material of interest likewise affect the profitability of developing particular deposits.

In the context of energy, a further distinction is made between *renewable* and *nonrenewable* resources. **Nonre-**newable resources are those that are not being produced at present or that are being produced at rates much slower than current consumption rates—fossil fuels, for example. On a human timescale, the supplies of nonrenewable fuels are finite. As these resources become depleted, there is increasing interest in more extensive development of **renewable** resources, those that either can be replenished on a human timescale (such as wood) or those that can be used without actual depletion of supply (like solar energy).

TYPES OF MINERAL DEPOSITS

The bulk of the earth's crust is composed of fewer than a dozen elements. In fact, eight chemical elements make up more than 98% of the crust. Many of the elements *not* found in abundance in the earth's crust, including industrial and precious metals, essential components of chemical fertilizers, and elements like uranium that serve as energy sources, are vitally important to society. Some of these are found in very minute amounts in the average rock of the continental crust: copper, 0.006%; tin, 2 parts per million (ppm); gold, 4 parts per *billion* (ppb). Clearly, then, many useful elements must be mined from very atypical rocks.

Ore Deposits Defined

An **ore** is a rock in which a valuable or useful metal occurs at a concentration sufficiently high to make it economically worth mining. A given ore deposit may be described in terms of the **enrichment** or **concentration factor** of a particular metal:

$$\text{Concentration factor (of a given metal in an ore)} = \frac{\text{Concentration of the metal in that ore}}{\text{Concentration of that metal in average continental crust}}$$

The higher the concentration factor, the richer the ore (by definition), and the less of it needs to be mined to extract a given amount of metal.

Because ores are unusual rocks, it is not surprising that the known economic mineral deposits are very unevenly distributed around the world. See, for example, figure 27.1.

Magmatic Deposits

Magmatic activity gives rise to several kinds of deposits. Certain igneous rocks, just by virtue of their compositions, contain high concentrations of useful silicates or other minerals. These deposits may be especially valuable if the rocks are coarse-grained *pegmatites,* from which individual minerals are easily recovered. In some pegmatites, crystals may be over 10 meters (30 feet) long. Many pegmatites are also enriched in uncommon elements. Pegmatites commonly crystallize from the residual fluids left after most of a body of magma has solidified; many rare or trace elements not readily incorporated into the crystal structures of the common silicates are likewise left over, concentrated in the residual fluids at this

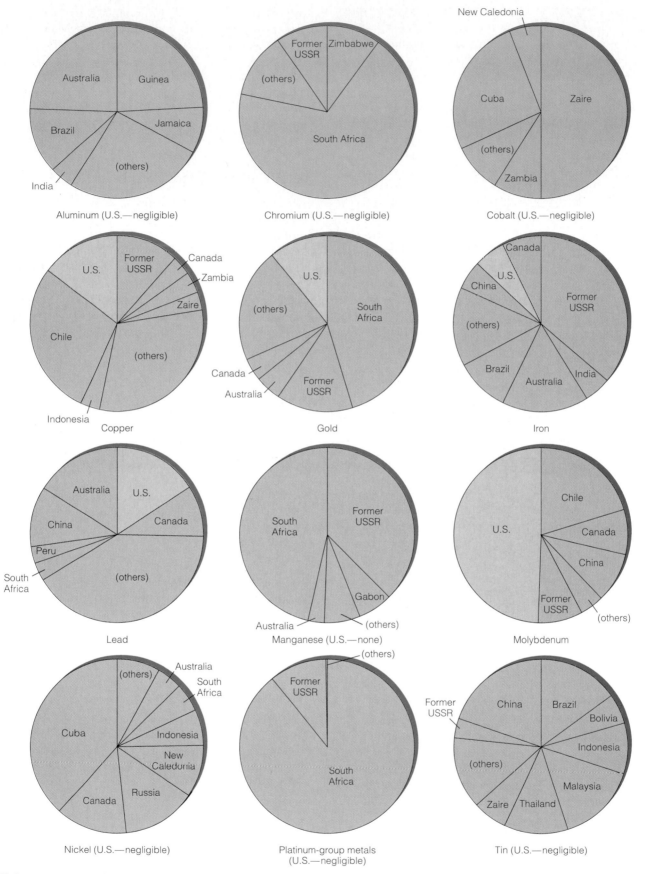

FIGURE 27.1 Proportions of world reserves of some nonfuel minerals controlled by various major producers. Although the United States is a major consumer of most metals, it is a major producer of very few.

Source: Data from Mineral Commodity Summaries 1993, *U.S. Bureau of Mines.*

stage. Among these are lithium, boron, beryllium, and uranium. Rarer minerals mined from pegmatites include tourmaline and beryl. (Tourmaline is used as a gemstone and for crystals in radio equipment and is also mined for the element boron that it contains. Beryl is mined for the metal beryllium when the crystals are of poor quality, or for the gemstones aquamarine and emerald when found as large, clear, well-colored crystals.)

Other useful minerals may be concentrated within a cooling, crystallizing magma chamber by gravity. If they are more or less dense than the magma, they may sink or float as they crystallize, instead of remaining suspended in the solidifying silicate mush, and accumulate in thick layers that are easily mined. This, then, is a form of fractional crystallization (chapter 3) that produces economic mineral deposits. Chromite (an oxide of chromium) and magnetite (Fe_3O_4) are both quite dense. In a magma of suitable bulk composition, rich concentrations of these minerals may form in the lower part of the magma chamber by gravitational settling during crystallization.

Hydrothermal Ores

Not all mineral deposits related to igneous activity form within igneous rock bodies. Magmas have water and other fluids dissolved in or associated with them. Particularly during the later stages of crystallization, the fluids may escape from the cooling magma, seeping through cracks and pores in the surrounding rocks, carrying with them dissolved salts, gases, and metals. These warm fluids can leach additional metals from the rocks through which they pass. In time, the fluids cool and deposit their dissolved minerals, creating a **hydrothermal** (literally, "hot water") ore deposit (figure 27.2).

The particular minerals deposited vary with the composition of the hydrothermal fluids, but, worldwide, a great variety of metals occur in hydrothermal deposits: copper, lead, zinc, gold, silver, platinum, uranium, and others. Because sulfur is a common constituent of magmatic gases and fluids, the ore minerals are frequently sulfides.

The hydrothermal fluids need not all originate within the magma. Sometimes, circulating subsurface waters are heated sufficiently by a nearby cooling magma to dissolve, concentrate, and redeposit valuable metals in a hydrothermal ore deposit. The fluid involved may also be a mix of magmatic and nonmagmatic fluids.

As noted in chapter 12, recent investigations of underwater spreading ridges have revealed hydrothermal fluids gushing from the sea floor at vents along these ridges, depositing sulfides and other minerals as they cool (figure 27.3). Similar activity accounts for the metal-rich muds at the bottom of the Red Sea, which is also a spreading-rift zone. Seawater seeps into the sediments and fractured sea floor at ridges, is warmed, reacts with the seafloor rocks, and dissolves metals. The warmed waters rise and cool, encounter and react with unaltered seawater, and deposit a variety of metallic minerals.

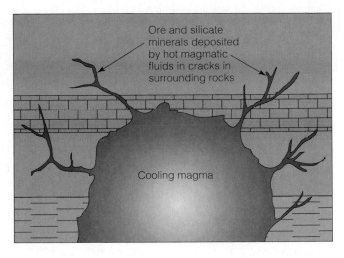

FIGURE 27.2 Hydrothermal ore deposits form in various ways; here, ore deposition occurs in veins around a magma chamber.

FIGURE 27.3 Effects of hydrothermal activity along the East Pacific Rise: an active "black smoker chimney" vent. The black is mainly fine-grained iron sulfide.
Photograph by W. R. Normark, USGS Photo Library, Denver, CO.

Sedimentary Deposits

Sedimentary processes can also produce economic mineral deposits. Some such ores have been deposited directly as chemical sedimentary rocks. Layered sedimentary iron ores are an example (figure 27.4). These large deposits, which may extend for tens of kilometers, are, for the most part, very ancient. Their formation is believed to be related to the development of the earth's atmosphere. Under the oxygen-free conditions believed to have prevailed initially on earth, iron from the weathering of continental rocks would have been very soluble in the oceans. As photosynthetic organisms began producing oxygen, that oxygen would have reacted with the dissolved iron and caused it to precipitate. If the majority of large iron ore deposits formed in this way, during a time long

FIGURE 27.4 Massive sample of layered sedimentary iron ore, slightly folded. Displayed outside the Smithsonian Institution, Washington, D.C.

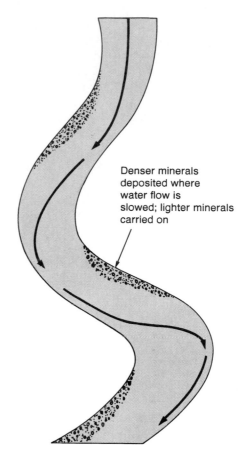

Denser minerals deposited where water flow is slowed; lighter minerals carried on

FIGURE 27.5 Formation of placer deposits.

past when the earth's surface chemistry was very different from what it is now, it follows that additional similar deposits are unlikely to form now or in the future.

Other sedimentary mineral deposits can form from seawater, which contains a variety of dissolved salts and other chemicals. When a body of seawater trapped in a shallow sea dries up, it deposits these minerals in **evaporite** deposits. Some evaporites may be hundreds of meters thick. Ordinary table salt (NaCl, known mineralogically as halite) is one mineral commonly mined from evaporite deposits. Others include gypsum and salts of the metals potassium and magnesium.

Most large evaporites form in a marine setting with restricted water flow and an evaporation rate relatively rapid by comparison with the influx of additional water. Under these conditions, water in the restricted basin becomes progressively more concentrated in its dissolved salts until the salts begin to precipitate out of the oversaturated solution. One situation in which conditions may be appropriate for evaporite deposition is in the early stages of continental rifting, when the sea first begins to invade the thinned rift zone. Additional sites of possible evaporite deposition include large lakes and also shallow inland seas that become isolated from the oceans.

Other Low-Temperature Ore-Forming Processes

Streams also play a role in the formation of mineral deposits, although they are rarely the sites of primary formation of ore minerals. As noted in chapter 14, streams often deposit sediments that are well sorted by size and density. The sorting action can effectively concentrate certain weathering-resistant, dense minerals in places along the stream channel. Such deposits are termed **placers.** The minerals of interest

are typically weathered out of the rocks of the stream's drainage basin and then transported, sorted, and concentrated while other minerals are dissolved or swept away (figure 27.5). Gold, diamonds, and tin oxide (cassiterite) are examples of minerals that have been mined from the sands and gravels of placer deposits. (The process of panning for gold makes use of the fact that dense minerals can be selectively concentrated in moving water.)

Even weathering alone can produce useful ores by leaching away unwanted minerals and leaving a residue enriched in a valuable metal. For example, the extreme leaching of tropical climates gives rise to lateritic soils from which nearly everything has been dissolved except for aluminum and iron oxide and hydroxide compounds. Most commercial aluminum deposits are *bauxite,* an aluminum-rich laterite in which the aluminum is found primarily as hydroxides. (The iron content of the laterites is generally not as high as that of the richest sedimentary iron ores described earlier. However, as the richest of those iron deposits are mined out, it may become profitable to mine laterites for iron as well as for aluminum.) Weathering has also enriched the shallower zones of certain copper deposits in the western United States.

Metamorphic Deposits

Finally, the mineralogical changes caused by the heat or pressure of metamorphism can produce secondary minerals

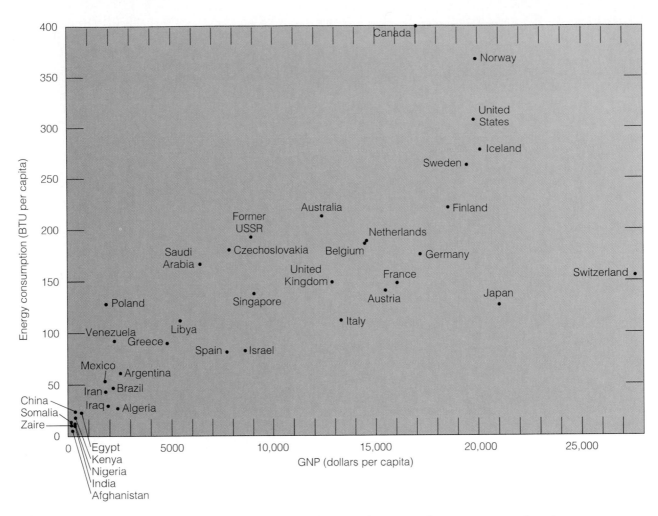

FIGURE 27.6 There is a variable but generally positive correlation between the GNP and energy consumption: the more energy consumed, the higher the value of goods and services produced, and generally, the higher the level of technological development as well. Similar relationships are commonly observed with respect to mineral resources.

Source: Data from 1992 Information Please Enviromental Almanac, *World Resources Institute.*

of value. Graphite, used in "lead" pencils, in batteries, as a lubricant, and for many applications where its very high melting point is critical, is usually mined from metamorphic deposits. Graphite consists of carbon, and one way in which graphite can be formed is by the metamorphism of coal, which is also carbon-rich. Asbestos, used less in insulation now but still prized for its heat- and fire-resistant properties, is a hydrous ferromagnesian silicate formed by the metamorphism of mafic igneous rocks, with the addition of water. Talc is formed similarly, at very low temperatures. Garnets, which are used as abrasives as well as semiprecious gems, are also common metamorphic minerals, especially in rocks of moderate to high metamorphic grade.

U.S. AND WORLD MINERAL SUPPLY AND DEMAND

The very uneven global distribution of mineral resources has, from time to time, meant that political disruption in one or a few countries has seriously disrupted the supply of a particular commodity. From this viewpoint, then, it is of some interest to examine the mineral resource supplies of the United States alone and relate these supplies to the nation's consumption of mineral materials. From a broader perspective, leaving aside the unanswerable question of the extent to which each country's mineral wealth may be considered available to all, we can also look at the world mineral-supply picture.

The other side of the supply/demand picture is demand, including predictions of future demand. World population continues to climb—past 5 billion already, swelling toward an estimated 8.5 billion by the year 2025. More people naturally demand more use of mineral and energy resources. But the problem is more complex and more acute than a simple matter of population growth. *Per capita* mineral and energy consumption grows as standard of living improves and societies become more technological; see, for example, figure 27.6.

Projection from historical trends may or may not be appropriate. Between 1947 and 1974, the production rates of most minerals grew at between 2 and 10% per year. The effects of such exponential growth in demand can be seen in

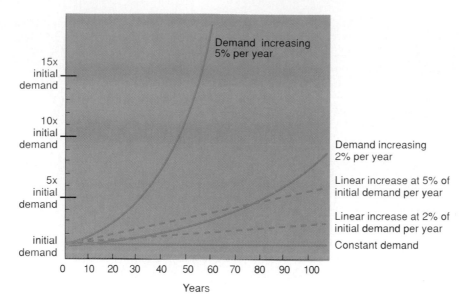

figure 27.7. If demand increases at 2% per year, it will double not in 50 years but in 35. An increase of 5% per year leads to a doubling of demand in 14 years and a tenfold increase in 47 years! On the other hand, depressed world economic conditions since the mid-1970s caused demand for many resource materials to level off or even to decline over the short term. Data for the early and mid-1980s showed that demand on the upswing again, at least in the industrialized countries, such as the United States. In looking at predictions of the longevity of mineral reserves, one should bear in mind the underlying assumptions about future demand.

The reserves are the amounts of materials most often used to make supply projections, as these deposits can be counted upon to be available in the near term. Even those figures may be somewhat imprecise, especially on a worldwide basis. Many mining and energy companies and some nations prefer to reveal as little about their unexploited assets as possible. In some countries, firms are taxed, in part, on the size of their remaining reserves, so their estimates may be on the low side. Still, over past decades, scientists have combined published information on reserves and mineral-production data with basic geologic knowledge to make what are believed to be reasonably good estimates of worldwide reserves. Estimates of U.S. reserves are, presumably, even more accurate.

All told, the United States consumes huge quantities of mineral and rock materials, relative to both its population and its production of most minerals. Figure 27.8 illustrates U.S. per capita consumption of many mineral and rock resources. Table 27.1 gives a global perspective by showing 1992 U.S. primary production and consumption data and the percentage of world totals that these figures represent. Keep in mind when looking at the consumption figures that fewer than 5% of the people in the world live in the United States.

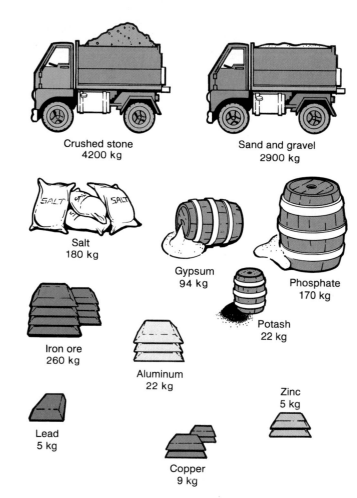

FIGURE 27.8 Per capita consumption of mineral and rock resources in the United States, 1992, based on population of approximately 250 million persons. (1 kg = 2.2 lb).

TABLE 27.1 *U.S. Production and Consumption of Rock and Mineral Resources, 1992**

Material		U.S. Primary Production	U.S. Primary Production as % of World	U.S. Consumption	U.S. Consumption as % of World†	U.S. Production as % of U.S. Consumption‡	
						Primary	Total
Metals	aluminum	4,000	22.7	5,500	31.2	73	102
	chromium	0	0	435	3.3	0	26
	cobalt	0	0	7.2	29.0	0	24
	copper	1,720	19.3	2,300	25.8	75	93
	iron ore	56,200	6.6	66,100	7.8	85	85
	lead	410	12.8	1,220	38.1	34	102
	manganese	0	0	610	3.2	0	0
	nickel	5.5	0.01	145	15.8	4	10
	tin	negligible	0	48.0	24.0	0	27
	zinc	520	7.2	1,225	16.6	42	62
	gold	320	14.7	100	4.6	320	520
	silver	1,800	13.1	3,900	28.5	46	49
	platinum group	7.8	2.6	113	38.4	7	74
Nonmetals	clay	42,151	§	38,044	§	111	111
	gypsum	16,300	15.1	26,000	24.1	63	65
	phosphate	47,000	33.3	42,000	30.1	112	112
	potash	1,760	7.0	5,390	21.5	33	33
	salt	39,800	19.4	45,800	22.3	87	87
	sulfur	10,600	20.1	12,900	24.5	82	82
Construction	sand and gravel, construction	805,700	§	806,000	§	100	100
	stone, crushed	1,155,000	§	1,158,000	§	100	100
	stone, dimension and facing	1,086	§	‖	§	‖	‖

Source: *Mineral Commodity Summaries 1993*, U.S. Bureau of Mines.

*All production and consumption figures in thousands of metric tons, except for gold, silver, and platinum-group metals, for which figures are in metric tons, and construction materials, for which figures are in thousands of tons.

†Assumes that overall, world production approximates world consumption. This may not be accurate if, for example, recycling is extensive.

‡"Primary" and "Total" headings refer to U.S. production.

§World data not available.

‖Consumption not available in tonnage. However, $415 million worth of dimension stone was imported (value of domestic production, $183 million, of which $65 million worth was exported).

As noted earlier, world demand for minerals (and energy) has slackened somewhat since the mid-1970s relative to earlier decades. Demand is unlikely to remain depressed over the longer term, however. Such a situation would imply, among other things, no significant economic growth or improvement in standard of living for the billions of people living in less-developed countries.

Table 27.2 presents projections of the lifetimes of selected U.S. and world mineral and rock reserves, assuming constant demand at 1992 levels, which are generally depressed somewhat from those of prior years by the residual effects of world recession. Note that, for many of the metals, the present reserves are projected to last only a few decades, even at constant 1992 consumption levels. Consider the implications if consumption were to resume the rapid growth rates of the mid-twentieth century.

Such projections also presume unrestricted distribution of minerals, so that the minerals can be used as needed and where needed, regardless of political boundaries or such economic factors as a nation's purchasing power. It is instructive to compare global projections with corresponding data for the United States only. Even at 1992 consumption rates, the United States has less than half a century's worth of reserves of many of the materials listed. Of some metals, notably aluminum, chromium, manganese, platinum, and tin, the United States has virtually no reserves at all. The resultant dependence of the United States on imports is illustrated in figure 27.9.

TABLE 27.2 *World and U.S. Production and Reserves Statistics, 1992**

Material	Production	World Reserves	Projected Lifetime (years)	U.S. Reserves	Projected Lifetime (years)
bauxite	105,000	23,000,000	219	20,000	4.1
chromium	12,800	1,400,000	109	0	0
cobalt	25	4,000	160	0	0
copper	8,900	310,000	35	45,000	20
iron ore	844,500	150,000,000	178	16,100,000	244
lead	3,200	63,000	20	10,000	8.2
manganese	18,800	800,000	42	0	0
nickel	916	47,000	51	23	0.2
tin	200	8,000	40	210	4.4
zinc	7,365	140,000	19	16,000	13
gold	2,170	44,000	20	4,770	47
silver	13,700	280,000	20	31,000	7.9
platinum group	294	56,000	190	250	2.2
gypsum	108,000	2,600,000	24	800,000	31
phosphate	141,000	12,000,000	85	1,230,000	29
potash	25,035	9,400,000	375	83,000	15
sulfur	52,700	1,400,000	27	140,000	11

Source: *Mineral Commodity Summaries 1993*, U.S. Bureau of Mines.

Reserves include only currently economic deposits.

*All production, reserve, and resource figures in thousands of metric tons, except for gold, silver, and platinum-group metals, for which figures are in metric tons.

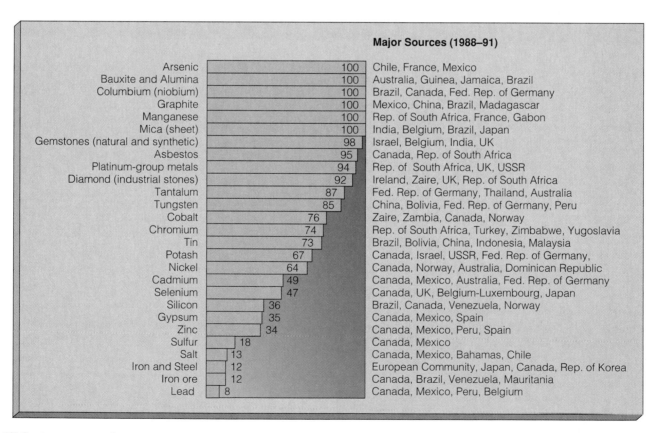

Major Sources (1988–91)

Arsenic	100	Chile, France, Mexico
Bauxite and Alumina	100	Australia, Guinea, Jamaica, Brazil
Columbium (niobium)	100	Brazil, Canada, Fed. Rep. of Germany
Graphite	100	Mexico, China, Brazil, Madagascar
Manganese	100	Rep. of South Africa, France, Gabon
Mica (sheet)	100	India, Belgium, Brazil, Japan
Gemstones (natural and synthetic)	98	Israel, Belgium, India, UK
Asbestos	95	Canada, Rep. of South Africa
Platinum-group metals	94	Rep. of South Africa, UK, USSR
Diamond (industrial stones)	92	Ireland, Zaire, UK, Rep. of South Africa
Tantalum	87	Fed. Rep. of Germany, Thailand, Australia
Tungsten	85	China, Bolivia, Fed. Rep. of Germany, Peru
Cobalt	76	Zaire, Zambia, Canada, Norway
Chromium	74	Rep. of South Africa, Turkey, Zimbabwe, Yugoslavia
Tin	73	Brazil, Bolivia, China, Indonesia, Malaysia
Potash	67	Canada, Israel, USSR, Fed. Rep. of Germany,
Nickel	64	Canada, Norway, Australia, Dominican Republic
Cadmium	49	Canada, Mexico, Australia, Fed. Rep. of Germany
Selenium	47	Canada, UK, Belgium-Luxembourg, Japan
Silicon	36	Brazil, Canada, Venezuela, Norway
Gypsum	35	Canada, Mexico, Spain
Zinc	34	Canada, Mexico, Peru, Spain
Sulfur	18	Canada, Mexico
Salt	13	Canada, Mexico, Bahamas, Chile
Iron and Steel	12	European Community, Japan, Canada, Rep. of Korea
Iron ore	12	Canada, Brazil, Venezuela, Mauritania
Lead	8	Canada, Mexico, Peru, Belgium

FIGURE 27.9 Proportions of U.S. mineral needs supplied by imports as a percentage of apparent consumption. Principal sources of imports indicated at right.

Source: U.S. Commodity Summaries 1993, *U.S. Bureau of Mines.*

Some relief can be anticipated from the economic component of the definition of reserves. That is, as currently identified reserves are depleted, the law of supply and demand will drive up minerals' prices. This, in turn, will make some of what are currently subeconomic resources profitable to mine, and they will effectively be reclassified as reserves.

The technological advances necessary for the development of some of the subeconomic resources, however, may not be achieved rapidly enough to help substantially. Also, if the move is toward developing lower- and lower-grade ore, then, by definition, more and more rock will have to be processed and more and more land disturbed to extract a given quantity of mineral or metal. Mining an ore that is 1% copper means processing five times as much ore for a given yield of copper as mining that copper from a 5%-copper ore.

Minerals for the Future:
Some Options Considered

If demand for minerals cannot realistically be reduced, supplies must be increased or extended. New sources of minerals must be developed in either traditional or nontraditional areas, or minerals must be better conserved.

A variety of new methods are being applied in the search for the less easily located ore deposits. Geophysics, for example, provides some assistance. Rocks and minerals vary in density and in electrical and magnetic properties, so changes in rock types or distribution below the earth's surface can cause gravitational and magnetic anomalies that can be measured at the surface, as well as small variations in the electrical conductivity of rocks. Some ore deposits may be detected in this manner. As a simple example, because many iron deposits are strongly magnetic, large magnetic anomalies may indicate the presence and the extent of a subsurface body of iron ore. Radioactivity is another readily detected property, and uranium deposits can be located with the aid of a Geiger counter or scintillation counter.

Geochemical prospecting adds other methods that are increasingly widely used. Some studies are based on the recognition that soils reflect, to a degree, the chemistry of the materials from which they formed. The soil over a copper-rich ore body, for instance, may itself be enriched in copper relative to surrounding soils. Occasionally, plants can be sampled instead of soil. Certain plants tend to concentrate particular metals, making the plants very sensitive indicators of locally high concentrations of those metals. (For example, the "locoweed" that causes strange symptoms in grazing cattle does so because it concentrates the toxic element selenium.) Water or stream sediments may likewise be sampled and analyzed for high concentrations of the metals being sought. Even soil gases can supply clues to ore deposits. Mercury is a very volatile metal, and high concentrations of mercury vapor have been found in gases filling soil pore spaces over mercury ore bodies.

Remote sensing, which includes the use of aerial photographs and satellite imagery, is another useful tool for mineral exploration. Landsat satellites provide images cover-

FIGURE 27.10 Landsat satellite images may reveal details of the geology that will aid in mineral exploration. Vegetation, also sensitive to geology, may enhance the image. In this view of South Africa in the rainy season, recognizable geologic features include a granite pluton (round feature at top center) and folded layers of sedimentary rock (below).
© *NASA.*

ing the world every 18 days. This gives at least a preliminary look at many areas that might otherwise be too remote or difficult to reach on the ground. Moreover, the satellite images can be processed through computers to sharpen the images or to focus on unusual features in the geology or vegetation that might bear further investigation (figure 27.10). Remote sensing must still be followed by ground-based geologic studies eventually. Perhaps its greatest usefulness at present is in limiting the scope of these slower and more costly detailed studies, by pinpointing areas particularly likely to yield new mineral deposits and thus allowing further exploration efforts to be concentrated in the most promising areas.

Advances in geologic understanding also play a role in mineral exploration. Development of plate-tectonic theory has helped geologists to recognize the association between particular types of plate boundaries and the occurrence of certain kinds of mineral deposits. This association may direct the search for new ores. For example, because geologists know that molybdenum deposits are often found over existing subduction zones, they can logically explore for more molybdenum deposits in other present or past subduction zones. Also, the realization that many of the continents were once united suggests likely areas to look for

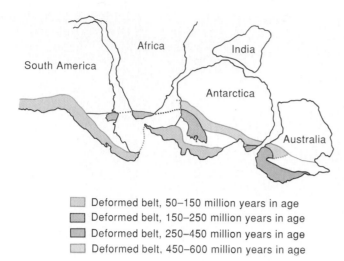

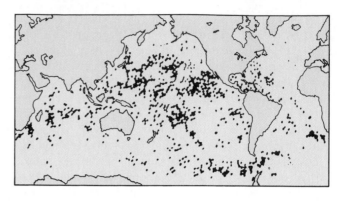

Deformed belt, 50–150 million years in age
Deformed belt, 150–250 million years in age
Deformed belt, 250–450 million years in age
Deformed belt, 450–600 million years in age

FIGURE 27.11 Pre-continental-drift reassembly of landmasses suggests locations of possible ore deposits in unexplored regions by extrapolation from known deposits on other continents.
Adapted with premission from C. Craddock et al., 1970, Geologic Maps of Antarctica, Antarctic Map Folio Series, Folio 12, *American Geographical Society.*

FIGURE 27.12 Manganese-nodule distribution on the sea floor.
Adapted from G. R. Heath, "Manganese Nodules: Unanswered Questions" in Oceanus 25 (3): 37–41, 1982. Copyright © 1982 Woods Hole, MA. Reprinted by permission.

more ores (figure 27.11). If mineral deposits of a particular kind are known to occur near the margin of one continent, similar deposits might be found at the corresponding edge of another continent that was once connected to the first. If a mountain belt on one continent is rich in some kind of ore and seems to have a counterpart range on another continent that was once linked to the first, the same kind of ore may occur in the matching mountain belt. Such reasoning is partly responsible for the supposition that economically valuable ore deposits may exist on Antarctica. The most successful exploration efforts may integrate several of the methods just described, making use of the particular advantages of each.

Marine Mineral Resources

Virtually every chemical element is dissolved in seawater. However, most of the dissolved material is halite. Vast volumes of seawater would have to be processed to extract small quantities of such metals as copper and gold. Moreover, current technology is not adequate to extract, selectively and efficiently, a few specific metals of interest from all that salt water on a routine basis, although the Japanese have been experimenting with extraction of uranium from seawater. Other types of underwater mineral deposits have greater potential.

As noted in chapter 18, during the last ice age, when a great deal of water was locked in ice sheets, eustatic sea levels were lower than at present. Much of the now-submerged continental-shelf area was dry land, and streams running off the continents flowed out across the exposed shelves to reach the sea. Where the streams' drainage basins included appropriate source rocks, those streams might have concen-

trated valuable placer deposits. With the melting of the ice and rising sea levels, any placers on the continental shelves were submerged. Finding and mining these deposits may be more costly and difficult than land-based mining, but as land deposits are exhausted, placer deposits on the continental shelves may well be worth seeking.

The hydrothermal ore deposits forming along some seafloor spreading ridges are another possible source of needed metals. In many places, the quantity of material being deposited is too small, and the depth of water above would make recovery prohibitively expensive, at least over the near term. However, the metal-rich muds of the Red Sea contain sufficient concentrations of such metals as copper, lead, and zinc that some exploratory dredging is underway, and several companies are interested in the possibility of mining those sediments. Along a section of the Juan de Fuca Ridge, off the west coast of Oregon and Washington, hundreds of thousands of tons of zinc- and silver-rich sulfides have already been deposited, and the hydrothermal activity continues.

Perhaps the most widespread undersea mineral resources, and the most frequently and seriously discussed as near-term resources, are the *manganese nodules*. Ranging in size up to about 10 centimeters in diameter, these are lumps composed mostly of manganese oxides and hydroxides. They also contain lesser amounts of iron, copper, nickel, zinc, cobalt, and other metals. Indeed, the value of the minor metals may be a greater financial motive for mining manganese nodules than the manganese itself. The nodules are found over much of the deep-sea floor, in regions where sedimentation rates are slow enough not to bury them (figure 27.12). At present, the costs of recovering these nodules are high compared with the costs of mining the same metals on land, and the technical problems associated with recovering the nodules from beneath several kilometers of water remain to be worked out, as does the practical question of who owns seabed resources in international waters and who has the right to exploit them (see box 27.1).

Box 27.1

Exclusive Economic Zones

Traditionally, nations bordering the sea claimed territorial limits of 3 miles (5 kilometers) outward from their coastlines. Realization that valuable minerals might be found on the continental shelves led some individual nations to assert their rights to the whole width of the adjacent continental shelf. As noted in chapter 9, the widths of continental shelves vary widely. Nations bordered by narrow shelves found this approach unfair. Some consideration was then given to equalizing claims by extending territorial limits to 200 miles offshore, which would encompass not only all of the continental shelves but, in most places, a considerable expanse of sea floor as well.

The eight years of intermittent negotiations collectively described as the "Third U.N. Conference on the Law of the Sea" produced a Law of the Sea Treaty that attempted to bring some order out of the chaos of oceanic territorial claims. Among other provisions, the Treaty established **Exclusive Economic Zones** (EEZs) extending up to 200 miles from the shorelines of coastal nations, within which those nations have exclusive rights to mineral-resource exploitation. These areas are not restricted to continental shelves only. Some deep-sea manganese nodules fall within Mexico's EEZ; Saudi Arabia and the Sudan have the rights to the metal-rich muds on the floor of the Red Sea. Included in the EEZ off the west coast of the United States is a part of the East Pacific Rise spreading-ridge system, with its active hydrother-

mal vents and sulfide deposits. Areas of the ocean basins outside any nation's EEZ are under the jurisdiction of an International Seabed Authority, and the treaty includes some provision for financial and technological assistance to developing countries wishing to share in resources there.

The United States did not sign the treaty, but in March 1983, President Reagan unilaterally proclaimed a U.S. Exclusive Economic Zone that extends to the 200-mile limit offshore. This move expands the undersea territory under U.S. jurisdiction to some 3.9 billion acres, more than 1½ times the land area of the United States and its territories (figure 1). So far, however, exploration of this new domain has been minimal.

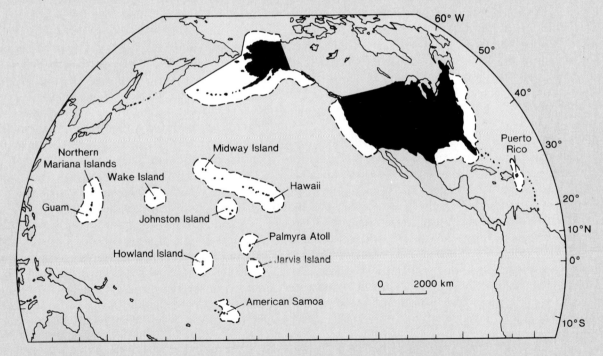

FIGURE 1 The Exclusive Economic Zone of the United States.
Source: U.S. Geological Survey, 1983 Annual Report.

Metal	1985	1986	1987	1988	1989	1992
aluminum	16.4	15.2	15.6	19.5	21.6	30
chromium	24.9	20.7	24.0	22.5	20.9	26
cobalt	5.7	15.2	14.1	14.0	14.5	25
copper	23.5	22.3	22.7	23.4	24.9	42
lead	50.3	50.1	54.7	56.4	60.3	62
manganese	0	0	0	0	0	0
nickel	27.2	25.3	24.6	32.3	34.4	30
platinum group	38.6	42.7	54.2	52.1	72.1	67
zinc	5.5	7.0	8.1	8.8	9.5	29

Source: *Mineral Commodity Summaries 1990*, and *Mineral Commodity Summaries 1993*, U.S. Bureau of Mines.

Recycling

The most effective way to extend some mineral reserves may be through recycling, which reduces the need for additional primary production from reserves. Some metals are already extensively recycled, at least in the United States (see table 27.3). Worldwide, recycling is less widely practiced, in part because the less-industrialized countries have accumulated fewer manufactured products from which materials can be retrieved. Two additional benefits of recycling are a reduction in the volume of waste-disposal problems and a decrease in the extent to which more land must be disturbed by new mining activities.

Unfortunately, not all materials lend themselves equally well to recycling. Among those that work out best are metals that are used in pure form in large chunks—copper in pipes and wiring, lead in batteries, and aluminum in beverage cans. The individual metals are relatively easy to recover and require minimal effort to purify for reuse. Recycling aluminum is also appealing from an energy (and cost) standpoint: it takes only one-twentieth as much energy to produce aluminum by recycling old scrap as it does to extract aluminum metal from bauxite. Sometimes, it is even economical to recover and recycle some of these discrete metallic items from old dump or landfill sites.

Where different materials are intermingled in complex manufactured objects, it is more difficult and costly to extract individual metals. Consider trying to separate the various metals of a refrigerator, a lawn mower, or a television set. Only in a few rare cases are metals valuable enough that the recycling effort may be worthwhile; for example, there is interest in recovering the platinum from catalytic mufflers. Alloys, mixtures of different metals in application-specific pro-

portions, present additional problems. Also, some materials are not used in discrete objects at all. The potash and phosphorous in fertilizers are strewn across the land and cannot be recovered. Road salt washes off of streets and into storm sewers and soil. Clearly, these things cannot be recovered and reused. For the foregoing and other reasons, it is unrealistic to expect that all mineral materials can ever be completely recycled. However, conservation—reducing the use of nonrenewable resources—and reuse of manufactured objects instead of often-wasteful indulgence in disposable versions (using a refillable pen or razor, a ceramic mug instead of plasticized-paper cups, and so on) can also contribute toward reducing the need for new mineral resources, when recycling is not a viable option.

IMPACTS OF MINING AND MINERAL-PROCESSING ACTIVITIES

Mining and mineral-processing activities can modify the environment in a variety of ways. Most obvious is the presence of the mine itself. Both underground mines and surface mines have their own sets of associated impacts.

Underground mines are relatively inconspicuous. They disturb a relatively small land area close to the principal shaft(s). Waste rock dug out of the mine may be piled close to the mine's entrance, but in most underground mines, the tunnels follow the ore body as closely as possible to minimize the amount of non-ore rock to be removed. When mining activities are complete, the shafts can be sealed, and the area often returns very nearly to its pre-mining condition. However, near-surface underground mines occasionally have collapsed years after abandonment, when supporting timbers have rotted away or ground water has enlarged the underground cavities through solution (figure 27.13). In some cases, the collapse occurred so long after the mining had ended that the existence of the old mines had been entirely forgotten.

Surface-mining activities consist of either open-pit mining (including quarrying) or strip-mining. Open-pit mining is practical when a large, three-dimensional ore body is located near the surface. Most of the material in the pit is the valuable commodity and is extracted for processing. Thus, this procedure permanently changes the topography, leaving a large hole in its wake. The exposed rock may begin to weather and, depending on the nature of the ore body, may release pollutants into surface runoff water.

Strip-mining, more often used to extract coal than mineral resources, is practiced most commonly when the material of interest occurs in a layer near and approximately parallel to the surface. Overlying vegetation, soil, and rock are stripped off, the coal or other material is removed, and the waste rock and soil cover is dumped back as a series of

FIGURE 27.13 Collapse of land surface over old abandoned copper mine in Arizona. Note roads, trees, and houses for scale.
Photograph by F. L. Ransome, USGS Photo Library, Denver, CO.

spoil banks. The finely broken-up material of spoil banks, with its high surface area, is very susceptible to both erosion and chemical weathering. Strip-mining is discussed further later in this chapter, in the context of coal.

The **tailings,** quantities of crushed and ground waste rock left over from mineral processing, may end up heaped around the processing plant to weather and wash away, much like spoil banks. Often, traces of the ore are left behind in tailings. Depending on the nature of the ore, rapid weathering of the tailings may leach out such harmful elements as mercury, arsenic, cadmium, and uranium, contaminating surface and ground waters.

The chemicals used in processing—for example, the cyanide used on gold ore—are often toxic also. Smelting to extract metals from ores may release not only sulfurous gases from sulfide ores but also lead, arsenic, mercury, and other potentially toxic, volatile elements along with exhaust gases (including sulfurous gases) and ash, unless emissions are tightly controlled.

The situation is improving, thanks to a combination of increased environmental awareness and strengthened environmental-protection laws. Old quarries can be converted to recreational lakes, old strip-mines reclaimed (as noted in the next section). Much greater care is being taken in disposal of acid mineral-processing waste (which can be neutralized by reaction with another rock material, powdered limestone) and control of toxic runoff from mining sites. And as long as we continue to enjoy a standard of living that involves high consumption of energy and manufactured goods, mining and some attendant environmental impacts are inevitable.

OUR PRIMARY ENERGY SOURCES: FOSSIL FUELS

The term *fossil* refers to any remains or evidence of ancient life. The **fossil fuels,** then, are those energy sources that formed from the remains of once-living organisms. These include oil, natural gas, coal, and fuels derived from oil shale

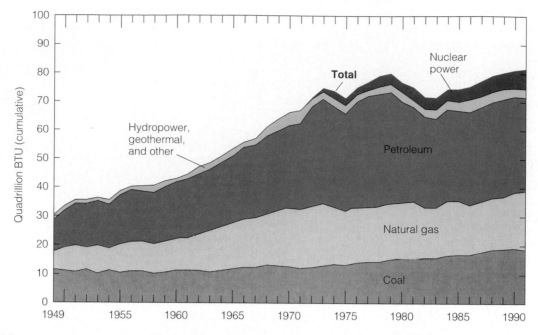

FIGURE 27.14 U.S. energy consumption over the past four decades. Note the consistent rise, except following the Arab oil embargo of the early 1970s, and in the early 1980s, when high prices and a sluggish economy combined to depress energy consumption.
Source: U.S. Energy Information Administration, Annual Energy Review 1991, *Department of Energy.*

and tar sand. The differences in the physical properties of the various fossil fuels arise from differences in the starting materials from which they formed and in what happened to those materials after the organisms died and were buried in the earth.

Oil and Natural Gas

These two fuels are currently our principal energy sources (figure 27.14). *Oil,* or petroleum, is not a single chemical compound. Petroleum comprises a variety of liquid hydrocarbon compounds (compounds made up of various proportions of the elements carbon and hydrogen). There are also gaseous hydrocarbons (so-called natural gas), of which the compound methane (CH_4) is the most common. How organic matter is transformed into liquid and gaseous hydrocarbons is not fully understood, but the transformation is believed to occur somewhat as follows:

Microscopic life is abundant over much of the oceans. When these organisms die, their remains settle to the sea floor. Near shore—for example, on many continental shelves—terrigenous sediments accumulate rapidly. In such a setting, the starting requirements for the formation of oil are satisfied: there is an abundance of organic matter, rapidly buried by sediment to prevent it from being destroyed by reaction with oxygen. Oil and most natural gas deposits are be-

lieved to form from such accumulations of marine microorganisms. (Additional natural gas deposits may be derived from the remains of terrestrial plants.)

As burial continues, the organic matter begins to change. Pressures increase as the weight of the overlying pile of sediment or rock increases; temperatures increase according to the geothermal gradient; and slowly, over long periods of time, chemical reactions occur. These reactions break down the large, complex organic molecules into simpler, smaller hydrocarbon molecules. The nature of the hydrocarbons changes with time and with continued heat and pressure. In the early stages of petroleum formation, the deposit may consist mainly of larger hydrocarbon molecules ("heavy" hydrocarbons), which have the thick, nearly solid consistency of asphalt. As the petroleum matures, and the breakdown of large molecules continues, successively "lighter" hydrocarbons are produced. Thick liquids give way to thinner ones, from which are derived lubricating oils, heating oils, and gasoline. In the final stages, most or all of the petroleum is further broken down into very simple, light, gaseous molecules—natural gas. Most of the maturation process occurs in the temperature range of 50° to 100°C (approximately 120° to 210°F). Above these temperatures, the remaining hydrocarbon is almost wholly methane; with further temperature increases, methane can be broken down and destroyed in turn (see figure 27.15).

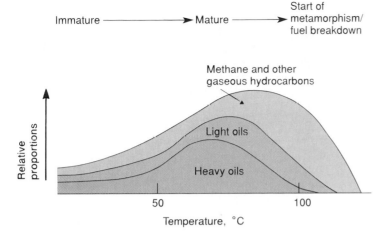

FIGURE 27.15 The process of petroleum maturation, in simplified form.

The amount of time required for oil and gas to form is not known precisely. Since virtually no petroleum is found in rocks younger than one to two million years old, geologists infer that the process is comparatively slow. Even if it took only a few tens of thousands of years (a geologically short period), the world's oil and gas are being used up far faster than significant new supplies could be produced.

Once the solid organic matter is converted to liquids and gases, the hydrocarbons can migrate out of the rocks in which they formed, if the surrounding rocks are permeable. The pores and cracks in rocks are commonly saturated with water at depth. Most oils and all natural gases are less dense than water, so they tend to rise as well as to migrate laterally through the water-filled pores of permeable rocks.

Commercially, the most valuable deposits are those in which a large quantity of oil and/or gas has been concentrated and confined by relatively impermeable rocks (commonly shales) in **traps** (figure 27.16). The **reservoir rocks** in which the oil or gas has accumulated should be relatively

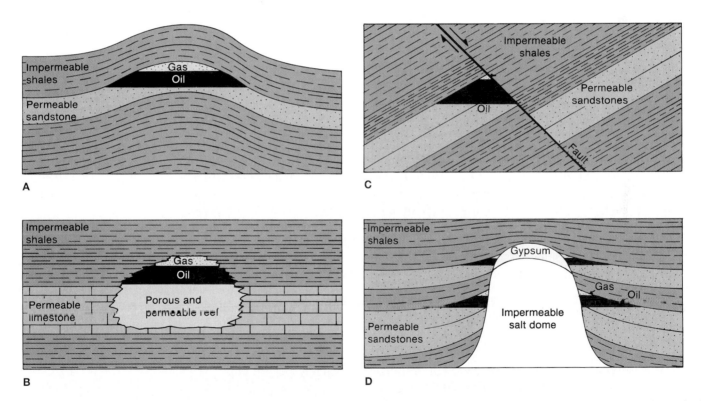

FIGURE 27.16 Types of petroleum traps. (A) A simple fold trap. (B) Petroleum accumulated in a fossilized ancient coral reef. (C) A fault trap. (D) Petroleum trapped against an impermeable salt dome that has risen up from a buried evaporite deposit.

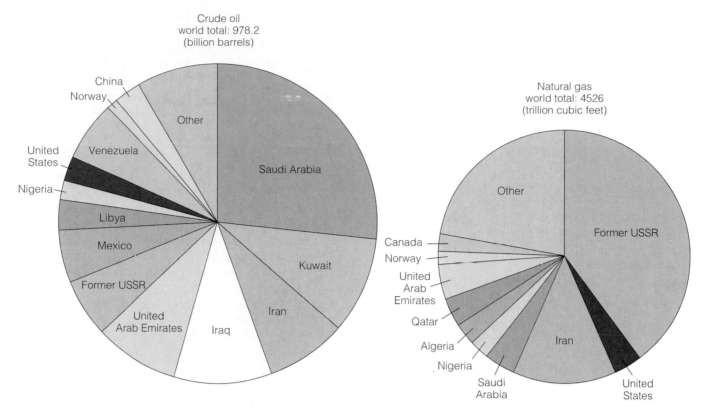

Crude oil
world total: 978.2
(billion barrels)

China
Norway
Other
United States
Venezuela
Nigeria
Libya
Saudi Arabia
Mexico
Former USSR
Kuwait
United Arab Emirates
Iraq
Iran

Natural gas
world total: 4526
(trillion cubic feet)

Other
Former USSR
Canada
Norway
United Arab Emirates
Qatar
Algeria
Nigeria
Iran
Saudi Arabia
United States

FIGURE 27.17 Estimated proven world reserves of crude oil and natural gas, January 1992.
Source: Averages of estimates of Oil and Gas Journal *and* World Oil, *as summarized in* International Energy Annual 1991, *U.S. Energy Information Administration.*

porous if a large quantity of petroleum is to be found in a small volume of rock, and should also be relatively permeable so that the oil or gas flows out readily once a well is drilled into the reservoir. Sandstones and limestones are common reservoir rocks. If the reservoir rocks are not naturally very permeable, it may be possible to fracture them artificially with explosives or high-pressure fluids to increase the rate at which oil or natural gas flows through them.

Oil is commonly quantified in terms of units of *barrels* (1 barrel = 42 gallons). Worldwide, estimated remaining reserves are close to 980 billion barrels (figure 27.17). The United States alone consumes nearly 30% of the oil used worldwide. Initially, U.S. oil resources probably amounted to about 10% of the world's total. Cumulative U.S. oil resources are estimated to have been not much above 200 billion barrels. Already close to half of that has been produced and consumed; remaining U.S. *resources* are estimated at about 120 billion barrels. Out of that, present proven *reserves* are under 30 billion barrels. For more than a decade, the United States has been using up somewhat more domestic oil each year than the amount of new domestic reserves discovered, or proven, so that net U.S. oil reserves have been decreasing year by year. Domestic production has likewise been declin-

ing (figure 27.18). Furthermore, without oil imports, U.S. reserves would dwindle still more quickly. The United States consumes nearly 6 billion barrels of oil per year and currently, over 40% is imported.

The production/consumption picture for conventional natural gas is similar to that for oil. Natural gas presently supplies nearly 25% of the energy used in the United States. The United States has proven natural gas reserves of nearly 200 trillion cubic feet. However, roughly 20 trillion cubic feet are consumed per year, and each year, less is found in new domestic reserves than the quantity consumed. As with oil, the U.S. supply of conventional natural gas is expected to be exhausted within decades, although the gas supplies are less severely depleted than the oil.

While decreasing reserves have prompted exploration of more areas, most regions not yet explored have been neglected precisely because they are unlikely to yield appreciable amounts of petroleum. The high temperatures involved in the formation of igneous and most metamorphic rocks would destroy organic matter, so oil would not have formed or been preserved in these rocks. Nor do these rock types tend to be very porous or permeable, so they generally make poor reservoir rocks as well, unless fractured. The

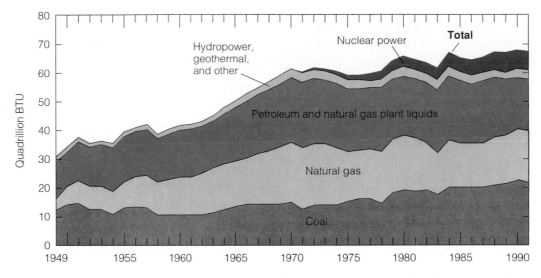

FIGURE 27.18 U.S. energy production since the mid-twentieth century by energy source. Note that domestic production began to decline despite sharp rises in fossil-fuel prices. (Natural gas plant liquids are liquid fuels produced during refining/processing of natural gas.)
Source: U.S. Energy Information Administration, Annual Energy Review 1991, *Department of Energy.*

large cratonic regions underlain predominantly by igneous and metamorphic rocks are simply very unpromising places to look for oil.

Despite a quadrupling in oil prices between 1970 and 1980 (*after* adjustment for inflation), U.S. proven reserves continued to decline. This is further evidence that higher prices do not automatically lead to proportionate, or even significant, increases in fuel supplies. And each time a temporary excess of production over demand causes petroleum prices to plummet, as in early 1986, exploration (as well as the development of new energy sources) comes to a virtual standstill.

Two decades ago, there was great interest in developing **enhanced-recovery** methods by which more of the oil and gas in a given deposit could be extracted—fracturing the rocks with explosives to increase permeability, or warming the oil with steam to decrease its viscosity, for instance. But such methods are expensive, so when oil and gas prices drop, interest in developing such methods likewise declines. Similarly, there is at present little research underway aimed at locating and developing **geopressurized natural gas** deposits, in which gas is dissolved in deep ground water at high pressure several miles deep in the crust: large quantities of gas may be present in such deposits, but they would be both technologically difficult and expensive to develop. Enhanced-recovery methods and exploitation of unconventional petroleum resources (see the following) were major reasons why, in 1995, the U.S. Geological Survey significantly increased its petroleum resource projections; but these remain nonrenewable energy sources, use of which increases atmospheric CO_2.

Coal

Coal is formed not from marine organisms but from the remains of land plants. A swampy setting, in which plant growth is abundant and where fallen trees, dead leaves, and other debris are protected from decay (either by standing water or by rapid burial under later layers of plant debris), is especially favorable to the initial stages of coal formation. The Carboniferous period of geologic time, named for its abundant coal deposits, was apparently characterized over much of the world by just the sort of tropical climate conducive to the growth of lush, swampy forests in lowland areas.

Given a suitable setting, the first combustible product formed is *peat*. Peat can form at the earth's surface, and there are places, like North Carolina's Dismal Swamp or the Malaysian Peninsula, where peat can be seen forming today. Peat can serve as an energy source—it is so used in Ireland—but it is not a very efficient one. Further burial, with more heat, pressure, and time, gradually dehydrates the organic matter and transforms the spongy peat into soft brown coal (*lignite*) and then to the harder coals (*bituminous* and *anthracite*). As the coals become harder, their carbon content increases, and so does the amount of heat released by burning a given weight of coal. As with oil, however, the heat to which coals can be subjected is limited: excessively high temperatures lead to metamorphism of coal into graphite.

Like oil, the higher-grade coals seem to require formation periods that are long compared to the rate at which coal is being used. Consequently, coal, too, can be regarded as a nonrenewable resource. However, the world supply of coal, and that in the United States also, represents a total energy resource far larger than that of petroleum (figure 27.19).

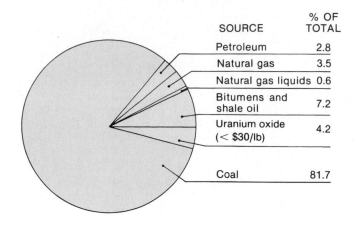

SOURCE	% OF TOTAL
Petroleum	2.8
Natural gas	3.5
Natural gas liquids	0.6
Bitumens and shale oil	7.2
Uranium oxide (< $30/lb)	4.2
Coal	81.7

Figure 27.19 Comparison of the total recoverable energy from various U.S. fossil-fuel reserves. The key word is *recoverable,* which reflects limitations of present technology.
Source: The Coal Data Book *(1980), The President's Commission on Coal.*

Coal resource estimates are less uncertain than are corresponding estimates for oil and gas. Coal is a solid, so it does not migrate. It is therefore found in the sedimentary rocks in which it formed; one need not seek it in igneous or metamorphic rocks. It occurs in well-defined beds that are easier to map than underground oil and gas concentrations. And because it formed from land plants, which did not become widespread until several hundred million years ago, one need not look for coal in more ancient rocks.

The estimated world reserve of coal is about 650 billion tons; total resources are estimated at over 10 trillion tons. The United States controls over 30% of the world reserves, some 200 billion tons. Total U.S. coal resources may approach ten times the reserves. Furthermore, most of that coal is yet unmined. At present, coal provides about 20% of the energy used in the United States. While the United States has consumed close to 50% of its petroleum resources, it has used up only a few percent of its coal. Even if only the *reserves* are counted, the U.S. coal supply could satisfy U.S. energy needs for several centuries, at current levels of energy consumption, if coal could be used for all purposes. As a supplement to other energy sources, coal could last correspondingly longer.

Coal can be converted to liquid or gaseous hydrocarbon fuels by causing the coal to react with steam or with hydrogen gas at high temperatures. The conversion processes are termed **liquefaction** (when the product is a liquid fuel) and **gasification** (when the product is gaseous). Both processes are intended to transform the coal into a cleaner-burning, more versatile fuel, thereby expanding its range of possible applications. Currently, however, these coal-derived fuels are economically uncompetitive with oil and natural gas.

Oil Shale and Tar Sand

Oil shale is very poorly named. The rock, while always sedimentary, need not be a shale, and the hydrocarbon in it is not oil! The potential fuel in oil shale is a waxy solid, *kerogen,* which is formed from the remains of plants, algae, and bacteria. The rock must be crushed and heated to over 500°C to distill out the "shale oil," which is then refined, somewhat as crude oil is, to produce various liquid petroleum products. The United States has about two-thirds of the world's known supply of oil shale (see locations in figure 27.20). The total estimated resource could yield 2 to 5 trillion barrels of shale oil. For a number of reasons, the United States is not yet using this apparently vast resource to any significant extent, and it may not be able to do so in the near future.

One reason for this is that much of the kerogen is so widely dispersed through the oil shale that huge volumes of rock must be processed to obtain moderate amounts of shale oil. Even the richest oil shale yields only about three barrels of shale oil per ton of rock processed. The cost of extraction is not presently competitive with that of conventional petroleum. Another problem is that a large part of the oil shale is located at or near the surface and logically would be exploited by strip-mining. Most of the richest oil shale, however, is located in dry areas of the western United States (Colorado, Wyoming, Utah; see figure 27.21), so land reclamation after mining would be correspondingly difficult. The water shortage presents a further problem. Current processing technologies require large amounts of water—about three barrels of water per barrel of shale oil extracted. In the western states, there is no obvious abundant source of water to process the oil shale. Finally, since the volume of the rock actually increases during processing, it is possible to end up with a 20 to 30% larger volume of waste rock than the volume of rock mined. Aside from problems caused by accelerated weathering of the crushed material, there remains the basic question of where to *put* it. Clearly, it will not all fit back into the space mined out.

Tar sands are sedimentary rocks containing a very thick, semisolid, tarlike petroleum. The heavy petroleum in tar sands is believed to have been formed in the same way and from the same materials as lighter oils. Tar-sand deposits may represent very immature petroleum deposits, in which the breakdown of large molecules has not progressed to the production of the lighter liquid and gaseous hydrocarbons. Alternatively, and perhaps more likely, the lighter compounds may simply have migrated away, leaving this dense, viscous material behind. Either way, the tarry petroleum is too thick to flow out of the rock. Like oil shale, tar sand presently must be mined, crushed, and heated to extract the petroleum, which can then be refined into various fuels.

Many of the environmental problems associated with oil shale likewise apply to tar sand. The tar is disseminated

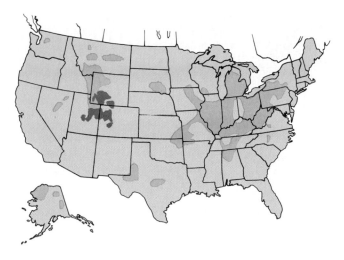

FIGURE 27.20 Distribution of U.S. oil-shale deposits. The richest of these is the Green River formation (brown).
Source: Data from J. W. Smith, "Synfuels: Oil Shale and Tar Sands" in Ruedisili and Firebaugh, eds., Perspectives on Energy, 3d ed., 1982, Oxford University Press.

FIGURE 27.21 The Green River formation in outcrop, Colorado. Note the sparse vegetation in this area, indicating dry conditions.
Photograph by R. L. Elderkin, Jr., USGS Photo Library, Denver, CO.

through the rock, so large volumes of rock must be mined and processed to extract appreciable petroleum. Many tar sands are near-surface deposits, so the mining method is commonly strip-mining. The processing requires a great deal of water, and the amount of waste rock after processing may be larger than the original volume of tar sand.

The United States has very limited tar-sand resources, so it cannot look to tar sand to solve its domestic energy problems, even if the environmental and developmental difficulties could be overcome. Canada, however, possesses tar sand and other "unconventional" petroleum resources that dwarf its "conventional" oil and gas deposits (figure 27.22). Chief among these "unconventional" petroleum resources are the Athabasca and related tar sands. The total petroleum estimated to exist in these tar sands amounts to 1243 billion barrels. The near-surface deposits, which actually crop out at the surface, are already being strip-mined and the oil extracted and refined, at costs comparable with those of developing new, conventional oil deposits.

Environmental Impacts of Fossil-Fuel Use

The primary negative impact of petroleum extraction and transportation relates to oil spills. These occur in two principal ways: from accidents during the drilling of oil wells and from wrecks of oil tankers at sea.

When an oil spill occurs, the oil, being less dense than water, floats. The lightest, most volatile hydrocarbons start to evaporate immediately, decreasing the volume of the spill somewhat (and polluting the air). Then a slow decomposition process sets in, due to sunlight and bacterial action. After several months, the spill may be reduced to about 15% of its original volume, and what is left is usually thick asphalt lumps. These can persist for years.

In calm seas, if a spill is small, it may be contained by floating barriers and picked up by specially designed ships that can skim up to fifty barrels of oil per hour off the water surface. This is clearly a slow process: three weeks after the 1989 *Exxon Valdez* spill of 10.2 million gallons of oil into the waters of Prince William Sound near the Trans-Alaska Pipeline port of Valdez (figure 27.23), only half a million gallons of oil had been recovered from the sea, and weather conditions threatened to breach the containing barriers at any time; huge quantities of oil had already washed up on and polluted the shore. Some attempts have been made to soak up other oil spills with peat moss, wood shavings, and even chicken feathers. The larger spills or spills in rough seas are a particular problem. Perhaps the best prospect for dealing with future oil spills is the development of "oil-hungry" microorganisms that will eat the spill for food and thus get rid of it. Scientists are currently developing suitable bacterial strains. Some experiments were made on the *Exxon Valdez* spill, and four months later, there was some good news on the microbe front: experiments involving fertilizing the native bacteria to encourage their growth showed much more rapid disappearance of oil from the fertilized shores. However, it is not yet known whether these bacteria were really "eating" the oil or just loosening it and thus hastening its washing away. (Even oil-eating microbes may not

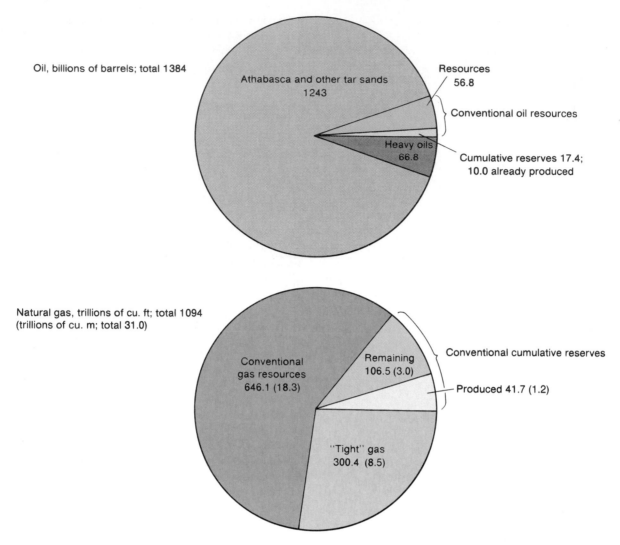

Oil, billions of barrels; total 1384

Athabasca and other tar sands
1243

Resources
56.8

Conventional oil resources

Heavy oils
66.8

Cumulative reserves 17.4;
10.0 already produced

Natural gas, trillions of cu. ft; total 1094
(trillions of cu. m; total 31.0)

Conventional
gas resources
646.1 (18.3)

Remaining
106.5 (3.0)

Conventional cumulative reserves

Produced 41.7 (1.2)

"Tight" gas
300.4 (8.5)

FIGURE 27.22 Canadian oil and gas resources. Note the significance of the tar-sand deposits.
Source: Data from R. M. Procter, et al., 1984, Oil and Natural Resources of Canada, Energy, Mines, and Resources Canada.

FIGURE 27.23 The oil-tanker port at Valdez, Alaska, at the southern end of the Trans-Alaska Pipeline. Since the *Exxon Valdez* wreck, all tankers in the sound must be escorted by tugboats.

be a perfect solution, however. What happens if these creatures make their way into tanks and oil-storage facilities?) For the time being, no good solution exists to the problems posed by a major oil spill.

A major problem posed by coal is the pollution associated with its mining and burning. Like all fossil fuels, coal produces carbon dioxide (CO_2) when burned. The additional pollutant that is of special concern with coal is sulfur. The sulfur content of coal can be over 3%, some in the form of pyrite (FeS_2), some bound in the organic matter of the coal itself. When the sulfur is burned along with the coal, sulfur gases—notably sulfur dioxide (SO_2)—are produced. These gases are toxic and can be severely irritating to lungs and eyes. The gases also react with water in the atmosphere to produce sulfuric acid, which is removed from the air as acid rain. Some of the sulfur in coal can be removed prior to burning, but the process is expensive and only partially successful, especially with organic sulfur. Alternatively, sulfur gases can be trapped by special devices ("scrubbers") in exhaust stacks, which use limestone or other materials to neutralize the acid vapor; but again, the process is expensive (in terms of both money and energy) and not perfectly efficient.

Coal use also produces a great deal of solid waste. The ash residue left after coal is burned typically amounts to 5 to 20% of the original volume. The ash, which consists mostly of noncombustible silicate minerals, also contains toxic metals, commonly including trace amounts of uranium. It presents a substantial waste-disposal problem.

Coal mining raises additional concerns. Underground mining of coal is notoriously dangerous, as well as expensive. After mining, too, there is potential for surface subsidence of the surface as supporting timbers rot, or even for underground fires (figure 27.24). The rising costs of providing a safer working environment for underground coal miners are largely responsible for a steady shift in coal-mining methods, from about 20% surface mining around 1950 to over 60% surface mining by 1980. No reversal of this trend is in sight. The good news, from the miners' perspective, is how much safer surface mining is (figure 27.25).

Most of the surface mining of coal is strip-mining. A particular problem with strip-mining coal is the associated sulfur. Some of this sulfur is left in the spoil banks, either in the small amount of coal left behind or in the pyritic shales often interbedded with the coal. This sulfur can react with water and air to produce acid runoff water, which slows revegetation of the area and may also cause a water-pollution problem (figure 27.26). (Even underground coal mines may have acid drainage, but the associated water circulation is generally more restricted.)

Mine reclamation is complicated by the need to restore a relatively low-sulfur surface layer in which the vegetation can become established. As with any reclamation, water availability is also a consideration. This is particularly true in the western United States, where half of the U.S. coal reserve (and much of the low-sulfur coal) is found; the dryness of the climate makes reclamation a major concern, as it would

FIGURE 27.24 Out-of-control fire in abandoned underground coal mine. Sheridan County, Wyoming.
Photograph by C. R. Dunrud, USGS Photo Library, Denver, CO.

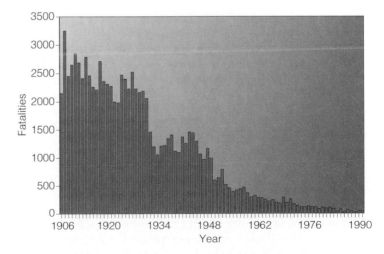

FIGURE 27.25 Drop in coal-mine fatalities reflects partly safer mining practices and partly the shift from underground mining to strip-mining.
Source: U.S. Bureau of Mines, Minerals Today, *October 1991, p. 19.*

FIGURE 27.26 Abandoned coal strip mine. Fulton County, Illinois. *Photograph by Arthur Greenberg, courtesy of EPA/National Archives.*

FIGURE 27.27 Reclaimed portions of Indian Head coal strip mine, Mercer, North Dakota, one year after seeding. *Photograph by H. E. Malde, USGS Photo Library, Denver, CO.*

be with development of oil shale. On the other hand, where water is ample, reclamation can produce a well-restored landscape (figure 27.27); it needn't leave a scarred and barren wasteland, as was once the case.

NUCLEAR POWER

Nuclear power actually comprises two different types of processes—*fission* and *fusion*—each with different advantages and limitations. Currently, only nuclear fission is commercially feasible.

Fission

Fission is the splitting-apart of atomic nuclei into smaller ones, with the release of energy (figure 27.28). Very few isotopes (some 20 out of the more than 250 naturally occurring isotopes) can undergo fission spontaneously and do so in nature. Some additional nuclei can be induced to split apart, and the naturally fissionable nuclei can be made to split more rapidly, thus increasing the rate of energy release. The fissionable nucleus of most interest for modern nuclear power reactors is the isotope uranium-235.

A uranium-235 nucleus can be induced to undergo fission by the firing of another neutron into the nucleus. The nucleus splits into two lighter nuclei (not always the same two) and releases additional neutrons as well as energy. Some of the newly released neutrons can induce fission in other nearby uranium-235 nuclei, which, in turn, release more neutrons and more energy in breaking up, and so the process continues in a **chain reaction.** A controlled chain reaction, with continuous, moderate release of energy, is the

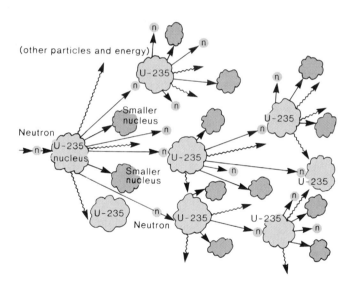

FIGURE 27.28 Nuclear fission and chain reaction involving U-235 (schematic). Neutron capture by U-235 causes fission into two smaller nuclei plus additional neutrons, other subatomic particles, and energy. Released neutrons, in turn, cause fission in other U-235 nuclei.

basis for fission-powered reactors. The energy released heats cooling water that circulates through the reactor's core. The heat removed from the core is transferred, through a heat exchanger, to a second water loop in which steam is produced. The steam, in turn, is used to run turbines to produce electricity (figure 27.29).

Only 0.7% of natural uranium is uranium-235. Natural uranium must be processed (enriched) to increase the con-

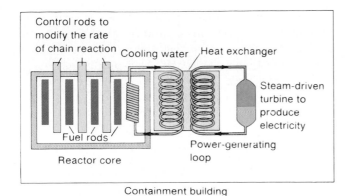

FIGURE 27.29 Conventional nuclear-fission electricity-generating plant (schematic).

centration of this isotope to several percent of the total uranium to produce reactor-grade uranium capable of sustaining a chain reaction. As the reactor operates, the uranium-235 atoms are split and destroyed, so that, in time, the fuel is so depleted ("spent") in this isotope that it must be replaced with a fresh supply of enriched uranium. With the type of nuclear reactor currently in commercial operation in the United States, nuclear electricity-generating capacity in the United States probably could not be increased to much more than four times present levels (to supply about 20% of total energy needs) by the year 2010 without serious shortages of uranium-235.

However, uranium-235 is not the only possible fission-reactor fuel. When an atom of the nonfissionable uranium-238 absorbs a neutron, it is converted to plutonium-239, which *is* fissionable. During the chain reaction inside the reactor, as freed neutrons move about, some are captured by the abundant uranium-238 atoms, making plutonium. Spent fuel could be *reprocessed* to extract this plutonium, as well as to reenrich the uranium in uranium-235. Fuel reprocessing with recovery of both plutonium and uranium could reduce the demand for new uranium by an estimated 15%.

A **breeder reactor** can maximize production of new fuel. Breeder reactors produce useful energy during operation, just as conventional reactors do, by fission in a sustained chain reaction within the reactor core. In addition, they are designed so that surplus neutrons not required to sustain the chain reaction are used to produce more fissionable fuels from suitable materials, like plutonium-239 from uranium-238, or thorium-233 from the common isotope thorium-232. Breeder reactors can synthesize more nuclear fuel for future use than they are actively consuming while generating power.

Breeder-reactor technology is more complex than that of conventional water-cooled reactors. The core coolant in a breeder is liquid metallic sodium, not water; the reactor operates at much higher temperatures. The costs are substantially higher than for present commercial reactors and might be close to $10 billion per reactor by the time any breeders could be completed in the United States. The breeding process is slow, too; the break-even point after which fuel produced would exceed fuel consumed might be several decades after initial operation.

Even conventional reactors are falling out of favor in the United States. At the end of 1991, there were 111 nuclear power plants operating in the country, with 8 more in various stages of construction. However, over the last two decades, nuclear plant cancellations have far exceeded new orders, and several utilities have revised their plans and decided to make new power plants coal-fired, not nuclear. The 111 nuclear plants accounted for 22% of electricity generation in the United States in 1991, or 8% of total energy consumed in this country.

The small but finite risk of damage to nuclear reactors through accident or deliberate sabotage is worrisome to many. One of the more serious possibilities is the so-called loss-of-coolant event, in which the flow of cooling water to the reactor core is interrupted. Resultant overheating of the core might lead to *core meltdown,* in which the fuel and core materials would deteriorate into a molten mass that might or might not melt its way out of the containment building (scientific opinion is divided on this possibility). A partial loss of coolant occurred at the Three Mile Island facility in Pennsylvania in March 1979. No full-scale loss-of-coolant accident with core meltdown has occurred in the United States, but the Chernobyl accident in 1986 involved extensive core meltdown, and release of considerable radioactive material, which drifted across much of northern Europe.

A Chernobyl-style accident cannot occur at a commercial nuclear reactor in the United States because of some fundamental differences in design. In any fission reactor, some material must be used as a *moderator* to slow the neutrons streaming through the core enough that they can interact with atomic nuclei to sustain the chain reaction. In the Chernobyl reactor, the moderator was graphite; the core was built largely of graphite blocks. In commercial U.S. reactors, the moderator is water. If the core of a water-moderated reactor overheats, the water vaporizes. Steam is a poor moderator, so the rate of chain reaction slows down. Graphite, however, remains solid up to extremely high temperatures, and hot graphite and cold graphite are both effective moderators. In a graphite reactor, therefore, a runaway chain reaction is quite possible, leading to increasingly intense and rapid heat production. At Chernobyl, the graphite eventually became so hot that it began to burn, turning the reactor core into the equivalent of a giant block of radioactive charcoal. (The whole reactor had to be smothered in concrete to put the fire out.)

The radioactive wastes from the production of fission power are another concern. These wastes are classified, somewhat imprecisely, into "low-level" and "high-level" wastes. The low-level wastes—volumetrically about 90% of radioactive wastes—are relatively low-radioactivity wastes that require minimal special handling for disposal. Examples include filters, protective clothing, and waste materials from medical and research laboratories. Low-level wastes commonly have been consigned to landfills for disposal. Currently, there are six commercial disposal sites for low-level wastes. Recent federal legislation requires states to take responsibility for their own low-level wastes by the early 1990s, but compliance is lagging.

Spent reactor fuel rods and by-products from their fabrication and reprocessing are examples of high-level wastes, for which various more-or-less elaborate disposal schemes have been proposed over the years. These have ranged from waste disposal in (under) polar ice caps or in subduction zones, to rocketing wastes into the sun, to placement of solid or liquid wastes in old abandoned mines. An international commission is considering the feasibility of radioactive-waste disposal in abyssal clays. In the United States, the practicabil-

ity of placing wastes in caverns in bedrock is being studied. For decades, one of the most promising solutions was considered to be disposal in old salt mines. Salt has a high melting point, the better to contain heat-producing wastes; it flows plastically under sustained stress, so it can self-seal if fractured; and salt is relatively impermeable.

I n late 1987, five years after the Nuclear Waste and Policy Act of 1982 mandated development of two high-level radioactive-waste disposal sites in the United States, Congress selected Yucca Mountain, Nevada, as the intended location of the first such site. The disposal unit at Yucca Mountain would be ash-flow tuff. Current plans call for intensive investigation of the site's suitability and apparent geologic stability, to be followed (if the results are satisfactory) by construction of the waste repository, starting in 1998, with the first waste disposal to occur in the year 2003. However, questions are still being raised about the long-term seismic stability of the site, groundwater circulation, and other safety issues; see figure 27.30.

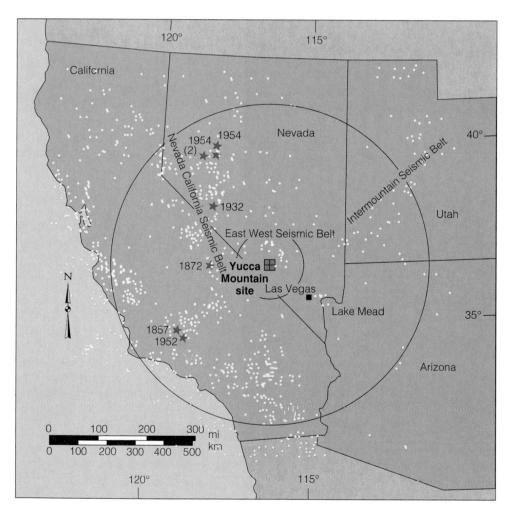

FIGURE 27.30 Seismicity in the southwestern United States, 1969–1978, showing earthquakes of magnitude 4 or higher. Stars denote quakes of magnitude 6.5 or higher.
Source: After Site Characterization Plan Overview, U.S. Department of Energy, 1988.

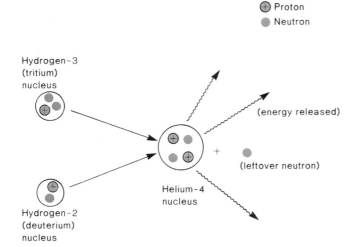

Proton
Neutron

Hydrogen-3 (tritium) nucleus

Hydrogen-2 (deuterium) nucleus

Helium-4 nucleus

(energy released)

(leftover neutron)

FIGURE 27.31 The nuclear fusion process (schematic).

Radioactive waste is a particular problem because radioactive materials cannot be treated by chemical reaction, heating, or other physical processes to make them nonradioactive. In this respect, they differ from many toxic chemical wastes that can be broken down by appropriate treatment. Radioactive materials have nuclei that are inherently unstable, and chemical reactions or changes in physical state do not affect the atomic nucleus. And whether nuclear power is utilized more extensively in the future or not, acceptable waste-disposal methods must be identified and implemented to dispose of wastes already accumulated.

Fusion

Nuclear **fusion** is the opposite of fission. Fusion is the process by which two or more smaller atomic nuclei combine to form a large one, with an accompanying release of energy (figure 27.31). It is the process by which the sun generates its vast amounts of energy. In the sun, simple hydrogen nuclei containing only a single proton are fused to produce helium. For technical reasons, fusion of the heavier hydrogen isotopes deuterium (nucleus containing one proton and one neutron) and tritium (one proton and two neutrons) would be easier to achieve on earth.

Hydrogen is plentiful since it is a component of water; the oceans contain, in effect, a vast reserve of hydrogen, an essentially inexhaustible supply of the necessary fuel for fusion. This is true even considering that deuterium is only 0.015% of natural hydrogen. The product of the projected fusion reactions—helium—is a nontoxic, chemically inert, harmless gas. There could be some mildly radioactive, light-isotope by-products of fusion reactors, but they would be much less hazardous than many of the products of fission reactors.

The principal reason for failure to use fusion power is lack of technology. To bring about a fusion reaction, the reacting nuclei must be brought very close together at extremely high temperatures (millions of degrees at least). The necessary conditions can be achieved in the interior of the massive sun, but no known physical material on earth can withstand such temperatures. The techniques being tested in laboratory fusion experiments are elaborate and complex, involving containment of the fusing materials with strong magnetic fields or using lasers to heat frozen pellets of hydrogen very rapidly. At best, the experimenters have been able to achieve the necessary conditions for fractions of a second, and the energy required to bring about the fusion reactions has exceeded the energy released thereby. Decades of research and development will be required before fusion becomes a commercial reality as an energy source.

In early 1989, news reports hailed the achievement of "cold fusion," fusion at room temperature in a beaker. Many subsequent attempts to verify the fusion reaction(s) or to replicate the original results were unsuccessful. The scientific community has generally concluded that, whatever process(es) occurred in those early experiments, it was not "cold fusion" and does not represent the solution to anticipated shortages of our current energy sources.

RENEWABLE ENERGY ALTERNATIVES

Fossil fuels and uranium are exhaustible resources. Over the longer term, development of renewable energy sources will be essential. A sampling of these follows.

Solar Energy

The sun can be expected to continue shining for approximately five billion years, so this resource is effectively inexhaustible, in sharp contrast to uranium or the fossil fuels. Sunlight falls on the earth without any mining, drilling, or other disruption of the land. Sunshine is free, under the control of no company or cartel and subject to no political disruption. The use of solar energy is essentially pollution-free, producing no hazardous wastes, air or water pollution, or noise. All of these features make solar energy an attractive option for the future.

The total solar energy reaching the earth is more than ample to provide for all human energy needs, both now and for the foreseeable future, but it is also spread over a broad surface area. Where large quantities of solar energy are used, solar-energy collectors must likewise cover a wide area. Sunlight is also variable in intensity, both from region to region, and from day to day as weather conditions change (figure 27.32). For various reasons, the two areas in which solar energy can make the greatest immediate contribution are in space heating and in the generation of electricity, uses that together account for about two-thirds of U.S. energy consumption. In each case, the low intensity and/or inconstancy of sunlight presents challenges.

While solar heating may be adequate by itself in mild and sunny climates, a conventional backup heating system is generally needed in areas subject to prolonged spells of

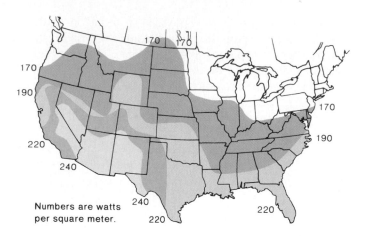

FIGURE 27.32 Distribution of solar energy over the contiguous United States. Maximum insolation occurs over the southwestern part of the country.

From S. W. Kendall and S. J. Nadis (eds.), Energy Strategies: Toward a Solar Future, *Union of Concerned Scientists, Copyright © Ballinger Publishing Company, Cambridge, MA.*

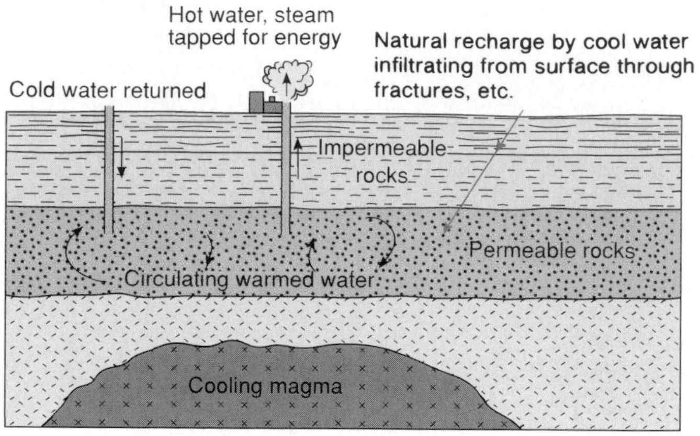

FIGURE 27.33 Extraction of geothermal energy using warmed, circulating ground water.

cloudiness or extreme cold. In such areas, use of solar energy can reduce, but not wholly eliminate, the consumption of conventional fuels.

Direct production of electricity using sunlight is accomplished through **photovoltaic cells,** also known simply as *solar cells.* For years, they have been the principal power source for satellites and for a few remote areas on earth difficult to reach with power lines. However, the low efficiency of present solar cells, together with the diffuse character of sunlight, make photovoltaic conversion an inadequate option for energy-intensive applications, such as factories. Even in the areas of strongest sunlight in the United States, incident radiation is on the order of 250 watts per square meter. The best solar cells are only about 20% efficient, meaning power generation of only about 50 watts per square meter. In other words, to keep one 100-watt light bulb burning would require at least 2 square meters of collectors, with the sun always shining. A 100-megawatt power plant would require 2 square kilometers of collectors (nearly 1 square mile)! This represents a large commitment both of land and of the mineral resources from which the collectors are made. There is also the problem of energy storage, especially difficult for large-scale applications.

Geothermal Energy

Magma rising into the crust from the mantle brings unusually hot material nearer the surface. Heat from the cooling magma heats any ground waters circulating nearby (figure 27.33).

This is the basis for **geothermal** power. The magma-warmed waters may escape at the surface in geysers and hot springs, signaling the existence of the shallow heat source below. More subtle evidence of the presence of hot rocks at depth comes from measurements of heat flow at the surface. High heat flow and recent magmatic activity go together and, in turn, are most often associated with plate boundaries. Therefore, most areas in which geothermal energy is being tapped extensively are along or near plate boundaries (figure 27.34).

How geothermal energy is used depends partly on how hot the system is. Where the ground water is warmed to temperatures comparable to what a home heating unit produces, the water can be circulated directly through homes to heat them. This is being done in Iceland and in parts of the former Soviet Union.

Other geothermal areas may be so hot that the water is turned to steam. The steam can be used, like any boiler-generated steam produced using more conventional fuel, to run electric generators. The largest U.S. geothermal-electricity operation is The Geysers, in California, which has operated since 1960 and now has a generating capacity of close to 2 billion watts, with additional power plants planned for the complex. Other steam systems are being used in Larderello (Italy), and in Japan, Mexico, New Zealand, the Philippines, Hawaii, and elsewhere.

In favorable cases, geothermal power generation is quite competitive economically with conventional methods of generating electricity. The use of geothermal steam is also largely pollution-free, provided that any associated sulfur gases are

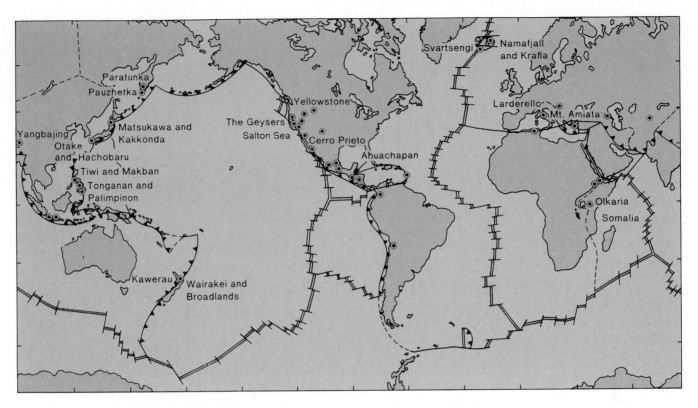

FIGURE 27.34 Distribution of geothermal power plants and major geothermal areas worldwide.

After map prepared by L. J. Patrick Muffler and Ellen Lougee, U.S. Geological Survey; plate tectonic boundaries supplied by Charles DeMets of the University of Wisconsin, Madison.

controlled. There are no ash, radioactive-waste, or carbon-dioxide problems with geothermal power as there are with other fuels. As with withdrawal of any subsurface water, however, there is danger of surface subsidence as geothermal waters are extracted, and disruption of the groundwater system may also affect water availability in wells. Use of geothermal power is restricted, primarily by three factors. First, each geothermal field can be used only for a limited period of time—a few decades, on average—before the rate of heat extraction is seriously reduced. This is so because rocks conduct heat very poorly. As hot water or steam is withdrawn from a geothermal field, it is replaced by cooler water that must be heated before use. Initially, the heating may be rapid, but in time, the permeable rocks become chilled to such an extent that water circulating through them heats too slowly or too little to be useful. The heat of the magma has not been exhausted, but it isn't transmitted efficiently to the water.

A second limitation of geothermal power is that, like geothermal power plants themselves, the resource is stationary, and most large cities are far removed from major geothermal resources.

The total number of sites suitable for geothermal power generation is the third limitation. Clearly, plate boundaries cover only a small part of the earth's surface, and many of them are inaccessible (seafloor spreading ridges, for instance). Not all have abundant associated subsurface water, either. For this reason, there is interest in developing **hot-dry-rock** geothermal areas, areas with young, hot rocks near the surface, but little water; these would be developed by the introduction of circulating fluids and would be correspondingly relatively expensive.

Conventional Hydropower

The principal requirements for the generation of substantial amounts of hydroelectric power are a large volume of water and the rapid movement of that water. Modern commercial generation of hydropower typically involves damming up a high-discharge stream, impounding a large reserve of water, and releasing it through a restricted outflow as desired, rather than operating subject to great seasonal variations in discharge (figure 27.35).

Figure 27.35 Glen Canyon Dam hydroelectric project.

Hydropower is a very clean energy source. The water is not polluted as it flows through the generating equipment. No chemicals are added to it, nor are any dissolved or airborne pollutants produced. The water itself is not consumed during power generation; it merely passes through the generating equipment. Hydropower is renewable as long as the streams continue to flow. Its economic competitiveness with other sources is demonstrated by the fact that nearly one-third of the electricity-generating plants in the United States in 1983 were hydropower plants.

The Federal Power Commission has estimated that the potential energy to be derived from hydropower in the United States, if tapped in every possible location, is about triple current hydropower use. In principle, then, hydropower could supply one-third to one-half of U.S. electricity, if consumption remained near present levels. However, ultimate development may not occur on such a scale.

Earlier chapters have already noted some of the problems posed by dam construction, including silting-up of reservoirs, habitat destruction and loss of farmland flooded by reservoirs, water loss by evaporation, and even earthquakes. Some sites with considerable power potential may not be available or appropriate for development. Other sites, in which the water flows along fault zones, may simply not be safe for dam construction. Many potential sites—in Alaska, for instance—are too remote from population centers to be practical unless power-transmission efficiency is improved.

An alternative to the development of many new hydropower sites would be to add hydroelectric generating facilities to dams already in place for flood control or recreational or other purposes. Although the release of impounded water for power generation alters streamflow patterns, it is likely to be less disruptive than either the original dam construction or the creation of new dam/reservoir systems.

Like geothermal power, conventional hydropower is also limited by the stationary nature of the resource. In addition, hydropower is somewhat more susceptible to natural disruptions than are other sources considered so far. Just as 100-year floods are rare, so are prolonged droughts, but they do happen.

FIGURE 27.36 Modern windmill array for power generation. Techapi Pass, California.
© *Times Mirror Higher Education Group, Inc./Doug Sherman, photographer.*

For various reasons, then, it is unlikely that many new hydropower plants will be developed. This clean, economical, renewable energy source can continue indefinitely to make a modest contribution to energy use, but it cannot be expected to supply much more energy in the future than it now does.

All large bodies of standing water on the earth, including the oceans and large lakes like the Great Lakes, have tides. Unfortunately, the energy represented by tides is too dispersed, in most places, to be useful. The difference in mean high-tide and low-tide water levels on the average beach is about 1 meter. A commercial tidal-power electricity-generating plant requires at least 8 meters difference between high and low tides for efficient generation of electricity, and a bay or inlet with a narrow opening that could be dammed to regulate water flow in and out. The proper conditions exist in very few places in the world. Tidal power is being used at a small plant at Passamaquoddy, Maine, and also at locations in France, the Netherlands, and the former Soviet Union. Worldwide, the total potential of tidal power is estimated at only 2% of the energy potential of conventional hydropower, which is itself limited.

Wind Energy

Because the winds are ultimately powered by the sun, wind energy can be regarded as a derivative of solar energy. It is clean and, like sunshine, renewable indefinitely. Windmills currently are used for pumping ground water and for the generation of electricity, usually on individual homesites.

Wind energy shares certain limitations with solar energy. It is dispersed not only in two dimensions but in three. Wind is also erratic and highly variable in speed, both regionally and locally. The regional variations in potential power supply are even more significant than they might appear from comparing average wind velocities because windmill power generation increases as the cube of wind speed. So, if wind velocity doubles, power output increases by a factor of 8 (2 × 2 × 2). But most of the consistently windy places in the United States are rather far removed physically from most of the high-population and heavily industrialized areas. Even where average wind velocities are great, strong winds do not blow constantly. This presents a storage problem that has yet to be solved satisfactorily. In the near future, wind-generated electricity might most effectively be used to supplement conventionally generated power when wind conditions are favorable.

Most schemes for commercial wind-power generation of electricity involve *wind farms,* concentrations of many windmills in a few especially favorable sites (figure 27.36). The Great Plains region is one of the most promising areas for initial sizeable efforts. The limits to wind-power use would then include the area that could be committed to windmill arrays, as well as the distance the electricity could be transmitted without excessive loss in the power grid.

About 1000 1-megawatt windmills would be required to generate as much power as a moderately large, conventional electric power plant. The windmills have to be spread out so as not to block each other's wind flow. However, the land would not have to be devoted exclusively to windmills. Farming and the grazing of livestock, common activities in the Great Plains area, could continue on the same land concurrently with power generation. The storage problem, however, would remain.

Biomass

The term **biomass** technically refers to the total mass of all the organisms living on earth. In an energy context, the term has become a catchall for various ways of deriving energy from organisms or from their remains.

The possibilities for biomass-derived fuels are many. Wood is a biomass fuel. There were eight wood-fueled electricity-generating plants in the United States in 1983, with a collective capacity ten times that of the wind-powered plants; by 1987, the plants fueled by burning wood and waste had a generating capacity nearly twenty times that of the wind-powered plants. Over 25% of U.S. households burn some wood for heat, and 4% rely on wood as their primary heat source.

The production of alcohol from plant materials for use as fuel—either by itself or as an additive to gasoline—is another present use of biomass fuel. Sometimes, using biomass energy sources means burning waste plant materials after a crop is harvested. Because certain plants produce flammable, hydrocarbon-rich fluids, the possibility of "raising" liquid fuel on farms is being evaluated. A variation on this theme involves microorganisms that produce oil-like substances: the organisms could be cultivated on ponds and their oil periodically skimmed off. Such possibilities represent potential future options.

A biomass fuel increasingly being utilized could be called "gas from garbage." Organic wastes, when decomposed in the absence of oxygen, yield a variety of gaseous products, among them methane (CH_4). The decay of organic wastes in sanitary-landfill operations makes these landfills suitable sites for the production of methane. Gas straight from the landfill is too full of impurities to use alone, but landfill-derived gas can be blended with purer, pipelined natural gas to extend the gas supply. The city of Los Angeles is among several now making some use of landfill gas. Methane can also be produced by decay of manures, and on some livestock feedlots, this is proving to be a partial solution to energy-supply and waste-disposal problems simultaneously.

All biomass fuels are burned to release their energy. They are carbon-rich and thus share the carbon-dioxide-pollution problem of fossil fuels. However, unlike the fossil fuels, they have the very attractive feature of being renewable on a human timescale.

SUMMARY

Economically valuable mineral deposits occur in a variety of geologic settings—igneous, metamorphic, and sedimentary. Valuable minerals may be found disseminated in igneous rocks, concentrated in plutons by gravity, crystallized as coarse-grained pegmatites, or deposited in hydrothermal veins associated with igneous activity. Sedimentary deposits include evaporites, placers, laterites, and other ores concentrated by weathering, and sediments and sedimentary rocks themselves (for example, limestone, sand and gravel, or sedimentary iron ores). Metamorphism plays a role in the formation of deposits of certain minerals, like graphite.

In assessing mineral or fuel supplies, one can speak of *reserves* (that quantity of the material that has been found and can be exploited economically with existing technology) or the larger quantity *resources* (reserves, plus additional deposits expected to be found, plus known, but uneconomic, deposits). Energy resources may be further classified as *renewable* or *nonrenewable,* depending on whether or not they are capable of replenishment on a human timescale.

Both the occurrence of and the demand for mineral and rock resources are very unevenly distributed worldwide. Projections for mineral use, even with conservative estimates of consumption levels, suggest that present reserves of most metals and other minerals could be exhausted within decades, both within the United States and globally. Price increases would lead to the development of some deposits that are presently uneconomic, but these are also limited. Strategies for averting further mineral shortages include applying new exploration methods to find more ores, looking to undersea mineral deposits not exploited in the past, and recycling metals to reduce the demand for newly mined material.

The world's major energy sources are the fossil fuels, formed from the remains of ancient plants and animals, modified by heat and pressure through time in the earth. All of the fossil fuels are nonrenewable energy sources. Petroleum and conventional natural gas deposits may be exhausted within decades. Coal resources could potentially last for centuries, but significant negative environmental impacts are associated with extensive coal use, including acid rain and acid mine drainage, land disturbance by mining activities, and the pollution and disposal problems posed by coal ash. Oil-shale deposits, containing the waxy hydrocarbon kerogen, are plentiful in the United States. However, the kerogen is thinly dispersed through the rocks, which would have to be strip-mined. Oil shale also requires a great deal of water to process, and most U.S. deposits occur in dry areas. U.S. tar-sand resources are negligible.

Conventional nuclear reactors consume the rare isotope uranium-235, the supply of which is severely limited. Significant expansion of the use of nuclear fission power would require the use of breeder reactors. All fission reactors present waste-disposal problems and raise concerns about radiation safety. Fusion power, which would be far cleaner, is technologically unfeasible for the present.

The inexhaustible or renewable energy sources include solar energy, geothermal power, hydropower, tidal and wind energy, and various biomass fuels. The principal practical applications of solar energy are for space heating and the generation of electricity; there is a storage problem in both contexts, and it is not yet possible to generate solar electricity efficiently for energy-intensive applications. Present use of geothermal energy, whether for hot-water heat or steam-generated electricity, is limited to a few sites along modern plate boundaries with recent volcanism. The thermal conductivity of rocks also limits the lifetime of each individual geothermal field, typically to several decades. The potential for additional development of either conventional hydropower or tidal power is restricted by the number of suitable sites. Optimum sites for wind-energy development are mostly far removed from major population centers, and wind shares with sunlight the problem of variable power generation with changing weather. Biomass fuels can be replenished on a human timescale but share with fossil fuels the problem of carbon-dioxide pollution. There are significant technological, supply, or environmental limitations associated with each of the various alternative energy sources.

TERMS TO REMEMBER

1. What is the distinction between *reserves* and *resources?* Under what conditions might some resources be reclassified as reserves? What is a *nonrenewable* resource?

2. What is an *ore?* How is its concentration factor defined?

3. Describe two kinds of magmatic ore deposits, and name one mineral mined from each.

4. What are *hydrothermal* ore deposits? Why are they especially associated with plate boundaries?

5. Under what conditions do evaporites form?

6. What is a *placer* deposit, and what kinds of minerals are concentrated in it?

7. Name and describe two kinds of exploration techniques used in the search for mineral deposits.

8. How has the development of plate-tectonic theory aided in finding more ore deposits?

9. Is seawater a potential source of essential metals? Explain.

10. Describe one marine mineral resource, not presently exploited significantly, that might become important in the future.

11. Why are some old mines and tailings piles being considered as future metal sources?

12. Describe one potential environmental problem or hazard associated with (a) underground mining, (b) surface mining, and (c) mineral processing after mining.

13. What is an Exclusive Economic Zone? List three kinds of mineral deposits that might be encompassed by an EEZ.

14. What are the two basic initial requirements for forming a fossil-fuel deposit?

15. Briefly outline the process of petroleum formation and maturation.

16. Sketch or describe any two types of petroleum traps.

17. Approximately what proportion of estimated U.S. oil resources has been produced and consumed?

18. Describe any enhanced-recovery method for oil. What are the attractions of such methods?

19. What is *geopressurized* natural gas, and where is it found?

20. From what materials is coal formed? How does coal's quality change with progressive heating?

21. Why is there interest in coal gasification and liquefaction processes?

22. Name and describe one environmental problem associated with (a) coal mining and (b) coal use/burning.

23. What is *oil shale?* How is fuel produced from it?

24. Cite three environmental or developmental problems associated with both oil shale and tar sand.

25. Compare and contrast nuclear fission and fusion processes.

26. If fission power is to be used extensively in the future, breeder reactors will be needed. Explain.

27. Cite two potential advantages of fusion power over fission. Why are fusion reactors not now in use?

28. Briefly describe the nature of a natural geothermal system. Why are geothermal areas restricted geographically?

29. What are the principal applications of geothermal energy? What are two important limitations on its usefulness?

30. What is a *hot-dry-rock* geothermal resource, and how could it be used?

31. For what is conventional hydropower now used? Discuss the extent to which additional hydropower-generation sites might be developed.

32. What areas have the greatest potential for wind-power development? Name at least two factors that now limit the potential contribution of wind power.

33. Describe any two biomass fuel sources. What drawback do they share with the fossil fuels? What advantage do they have?

FOR FURTHER THOUGHT

1. Select one of the metals aluminum, manganese, or tin, of which the United States controls very little or none of the identified world reserves. Investigate its uses in more detail; consider the implications of a cessation of imports. Has the United States any subeconomic deposits of this metal? Of what kind(s) are they, and how much metal do they contain relative to domestic demand?

2. Choose any one metallic mineral resource and investigate its geologic occurrence, its distribution (worldwide and, if applicable, within the United States), present consumption rates and past trends, and reserve and resource estimates. Evaluate the impact of the mining and processing methods customarily used to extract it.

3. Increased concern about energy conservation has led many homeowners to insulate their homes more and more snugly, to minimize heat loss by the escape of warmed air. Investigate the problems of "indoor air pollution" that have developed as a result.

4. Compare the cost estimates for shale oil, petroleum from tar sand, and gas from coal gasification with contemporary prices for conventional oil and natural gas. Do this for several different times—for example, 1972, 1978, 1986, and 1990. Determine the extent of oil and gas exploration and of development of the alternative fuels at the same times, and see what patterns, if any, emerge.

5. What energy sources are used by your local electric utility? Has the company changed its fuels, power-plant types, or plans for additional power plants over the last two or three decades? If so, to what extent have (a) economic factors and (b) environmental considerations entered into the decisions?

SUGGESTIONS FOR FURTHER READING

Broadus, J. M. 1987. Seabed materials. *Science* 235:853–60.

Brookins, D. G. 1981. *Earth resources, energy, and the environment.* Columbus, Ohio: Charles E. Merrill.

Craig, J. R., Vaughan, D. J., and Skinner, B. J. 1988. *Resources of The Earth.* Englewood Cliffs, N.J.: Prentice-Hall.

Deep ocean mining. 1982. *Oceanus* 25 (3).

Dick, R. A., and Wimpfen, S. P. 1980. Oil mining. *Scientific American* 243 (April): 182–88.

Earle, S. A. 1992. Assessing the damage one year later. *National Geographic* 179 (February): 122–34. (Pertains to the *Exxon Valdez* oil spill.)

Energy. San Francisco: W. H. Freeman. (A collection of offprints from *Scientific American*, 1970–1979.)

Fischett, M. A. 1986. The puzzle of Chernobyl. *IEEE Spectrum* 23 (July): 34–41.

Golay, M. W. 1990. Longer life for nuclear plants. *Technology Review* (May/June): 25–30.

Hoyle, F., and Hoyle, G. 1980. *Commonsense in nuclear energy.* San Francisco: W. H. Freeman.

Hunt, J. M. 1979. *Petroleum geochemistry and geology.* San Francisco: W. H. Freeman.

Hunt, V. D. 1982. *Handbook of energy technology.* New York: Van Nostrand Reinhold.

International Atomic Energy Agency, Vienna, Austria. (This agency publishes a quarterly bulletin on many aspects of nuclear power and other applications of radioactive materials.)

International Energy Agency. 1987. *Renewable sources of energy.* Paris: OCED.

Jensen, M. L., and Bateman, A. M. 1981. *Economic mineral deposits.* 3d ed. New York: John Wiley and Sons.

Kendall, H. W., and Nadis, S. J., eds. 1980. *Energy strategies: Toward a solar future.* Cambridge, Mass.: Ballinger.

Macauley, G., Snowdon, L. R., and Ball, F. D. 1985. *Geochemistry and geological factors governing exploitation of selected Canadian oil-shale deposits.* Geological Society of Canada Paper 85–13.

McGowan, J. G. 1993. Tilting toward windmills. *Technology Review* (July) 39–46.

National Academy of Sciences. 1979. *Energy in transition, 1985–2010.* San Francisco: W. H. Freeman.

National Geographic Society. 1981. *Special report on energy.* Washington, D.C.: National Geographic Society.

Petrick, A. 1986. *Energy resource assessment.* Boulder, Colo.: Westview Press.

Procter, R. M., Taylor, G. C., and Wade, J. A. 1984. *Oil and natural gas resources of Canada 1983.* Geological Society of Canada Paper 83–31.

Ruedisili, L. C., and Firebaugh, M. W. eds. 1982. *Perspectives on energy.* 3d ed. New York: Oxford University Press.

Sawkins, F. J. 1984. *Mineral deposits in relation to plate tectonics.* New York: Springer-Verlag.

Shusterich, K. M. 1982. *Resource management and the oceans.* Boulder, Colo.: Westview Press.

U.S. Bureau of Mines. 1993. *Mineral commodity summaries 1993.* (Published annually.)

———. *Minerals yearbook.* (Published annually.)

U.S. Department of Energy, Energy Information Administration. (Publishes a wide variety of energy-related data. Sources used in this chapter include *Annual energy outlook 1983* (published in 1984); *Annual energy review 1989* (1990); *Coal data book* (1980); *Commercial nuclear power: Prospects for the U.S. and the world* (1983); *Inventory of power plants in the U.S. 1983* (1984); *Nuclear plant cancellations: Causes, costs, and consequences* (1983); *Domestic uranium mining and milling industry* (1986); and *U.S. crude oil, natural gas, and natural gas liquids reserves* (1983).

Watson, J. 1983. *Geology and man*. Boston: Allen and Unwin.

Whitmore, E. C., Jr., and Williams, M. E., eds. 1982. *Resources for the twenty-first century*. U.S. Geological Survey Professional Paper 1193.

World Resources Institute. 1992. *World Resources 1992–93*. New York: Oxford University Press.

Zumberge, J. H. 1979. Mineral resources and geopolitics in Antarctica. *American Scientist* 67 (January): 68–76.

APPENDIX A
Unit Conversions

Common Prefixes

deci-	:	one-tenth	1 deciliter (dl)	=	0.1 liter
centi-	:	one-hundredth	1 centimeter (cm)	=	0.01 meter
milli-	:	one-thousandth	1 milliliter (ml)	=	0.001 liter
kilo-	:	one thousand	1 kilogram (kg)	=	1000 grams

Units of Length

1 centimeter	=	0.394 inches
1 meter	=	39.37 inches = 1.09 yards
1 kilometer	=	0.621 miles
1 inch	=	2.54 centimeters
1 yard	=	0.914 meters
1 mile	=	1760 yards = 1.61 kilometers

Units of Area

1 square centimeter	=	0.155 square inches
1 square meter	=	1.20 square yards = 1550 square inches
1 square kilometer	=	0.386 square miles
1 square inch	=	6.45 square centimeters
1 square yard	=	1296 square inches = 0.836 square meters
1 square mile	=	2.59 square kilometers
1 acre	=	4840 square yards = 4047 square meters

Units of Volume

1 cubic centimeter	=	0.061 cubic inches
1 cubic meter	=	1.31 cubic yards
1 cubic kilometer	=	0.240 cubic miles
1 cubic inch	=	16.4 cubic centimeters
1 cubic yard	=	0.765 cubic meters
1 cubic mile	=	4.17 cubic kilometers

Units of Liquid Volume

1 milliliter	=	0.0338 fluid ounces
1 liter	=	1.06 quarts
1 fluid ounce	=	29.6 milliliters
1 quart	=	0.946 liters
1 gallon	=	4 quarts = 3.78 liters
1 acre-foot	=	326,000 gallons = 1220 cubic meters

Units of Weight or Mass

1 gram	=	0.0353 ounces
1 kilogram	=	2.20 pounds
1 metric ton	=	1000 kilograms = 2200 pounds
1 ounce (avoirdupois)	=	28.4 grams
1 pound	=	454 grams = 0.454 kilograms
1 ton	=	2000 pounds = 909 kilograms
1 (troy) ounce	=	1.10 ounce (avoirdupois) = 31.2 grams

Temperature Conversions, Celsius (°C)/Fahrenheit (°F)

$$(°C) = ((°F) - 32)/1.8$$
$$(°F) = (1.8 \times (°C)) + 32$$

Energy Equivalents and Conversions

1 calorie = amount of heat required to raise the temperature of 1 milliliter of water by 1°C

1 BTU (British thermal unit) = amount of heat required to raise the temperature of 1 pound of water by 1°F

1 BTU = 252 calories

Average Energy Contents of Various Fossil Fuels

Fuel	Calories	BTU
1 barrel crude oil	1,460,000,000	5,800,000
1 ton coal	5,650,000,000	22,400,000
1 cubic foot natural gas	257,000	1020

APPENDIX B
Mineral and Rock Identification

MINERAL IDENTIFICATION

Table B.1 lists many of the more common minerals. Representative chemical formulas are provided for reference. Some appear complex because of opportunities for solid solution; some have been simplified by limiting the range of compositions represented, although additional elemental substitutions are possible.

A few general identification guidelines and comments:

Minerals showing metallic luster are usually sulfides (or native metals, but these are much rarer). Native metals have been omitted from the table. Those few that are likely to be encountered, such as native copper or silver, may be identified by their resemblance to household examples of the same metals.

Of the nonmetals, the silicates are generally systematically harder than the nonsilicates. Hardnesses of silicates are typically over 5, with exceptions principally among the sheet silicates; many of the nonsilicates, such as sulfates and carbonates, are much softer.

Distinctive luster, cleavage, or other identifying properties are listed under the column "Other Characteristics." In a few cases, this column notes restrictions on the occurrence of certain minerals as a possible clue to identification; for example, "found only in metamorphic rocks," or "often found in pegmatites."

ROCK IDENTIFICATION

One approach to rock identification is to decide whether the sample is igneous, sedimentary, or metamorphic and then look at the detailed descriptions in the corresponding chapter. How does one identify the basic rock type? Here are some general guidelines:

1. Glassy or vesicular rocks are volcanic.

2. Coarse-grained rocks with tightly interlocking crystals are likely to be plutonic, especially if they lack foliation.

3. Coarse-grained sedimentary rocks differ from plutonic rocks in that the grains in the sedimentary rocks tend to be more rounded and to interlock less closely. A breccia does have angular fragments, but the fragments in a breccia are typically rock fragments, not individual mineral crystals.

4. Rocks that are not very cohesive, that crumble apart easily into individual grains, are generally clastic sedimentary rocks. One other possibility would be a poorly consolidated volcanic ash, but this should be recognizable by the nature of the grains, many of which will be glassy shards. (Note, however, that extensive weathering can make even a granite crumble.)

5. More cohesive, fine-grained sedimentary rocks may be distinguished from fine-grained volcanics because sedimentary rocks are generally softer and more likely to show a tendency to break along bedding planes. Phenocrysts, of course, indicate a (porphyritic) volcanic rock.

6. Foliated metamorphic rocks are distinguished by their foliation (schistosity, compositional banding). Also, rocks containing abundant mica, garnet, or amphibole are commonly metamorphic rocks.

7. Nonfoliated metamorphic rocks, like quartzite and marble, resemble their sedimentary parents but are harder, denser, and more compact. They may also have a shiny or glittery appearance on broken surfaces, due to recrystallization during metamorphism.

Once a preliminary determination of category (igneous/sedimentary/metamorphic) has been made, table B.2 can be used in conjunction with the appropriate text chapter to identify the rock type. (Keep in mind, however, that the table and the text focus on relatively common rock types.)

Mineral	Formula	Color	Hardness	Other Characteristics
Amphibole (e.g., hornblende)	$(Na,Ca)_2(Mg,Fe,Al)_5Si_8O_{22}(OH)_2$	Green, blue, brown, black	5 to 6	Often forms needlelike crystals; two good cleavages forming 120-degree angle
Apatite	$Ca_5(PO_4)_3(F,Cl,OH)$	Usually yellowish	5	Crystals hexagonal in cross section
Azurite	$Cu_3(CO_3)_2(OH)_2$	Vivid blue	3½ to 4	Often associated with malachite
Barite	$BaSO_4$	Colorless	3 to 3½	High specific gravity, 4.5 (denser than most silicates)
Beryl	$Be_3Al_2Si_6O_{18}$	Aqua to green	7½ to 8	Usually found in pegmatites
Biotite (a mica)	$K(Mg,Fe)_3AlSi_3O_{10}(OH)_2$	Black	5½	Excellent cleavage into thin sheets
Bornite	Cu_5FeS_4	Iridescent blue, purple	3	Metallic luster
Calcite	$CaCO_3$	Variable; colorless if pure	3	Effervesces in weak acid
Chalcopyrite	$CuFeS_2$	Brassy yellow	3½ to 4	
Chlorite	$(Mg,Fe)_3(Si,Al)_4O_{10}(OH)_2$	Light green	2 to 2½	Cleaves into small flakes
Cinnabar	HgS	Red	2½	Earthy luster; may show silvery flecks
Corundum	Al_2O_3	Variable; colorless in pure form	9	Most readily identified by its hardness
Covellite	CuS	Blue	1½ to 2	Metallic luster
Dolomite	$CaMg(CO_3)_2$	White or pink	3½ to 4	Powdered mineral effervesces in acid
Epidote	$Ca_2FeAl_2Si_3O_{12}(OH)$	Green	6 to 7	Glassy luster
Fluorite	CaF_2	Variable; often green or purple	4	Cleaves into octahedral fragments; may fluoresce in ultraviolet light
Galena	PbS	Silver-gray	2½	Metallic luster; cleaves into cubes
Garnet	$(Ca,Mg,Fe)_3(Fe,Al)_2Si_3O_{12}$	Variable; often dark red	7	Glassy luster
Graphite	C	Dark gray	1 to 2	Streaks like pencil lead
Gypsum	$CaSO_4 \cdot 2H_2O$	Colorless	2	
Halite	$NaCl$	Colorless	2½	Salty taste; cleaves into cubes
Hematite	Fe_2O_3	Red or dark gray	5½ to 6½	Red-brown streak regardless of color of handsample
Kaolinite	$Al_2Si_2O_5(OH)_4$	White	2	Earthy luster
Kyanite	Al_2SiO_5	Blue	5 to 7	Found in high-pressure metamorphic rock; often forms bladelike crystals

Mineral	Formula	Color	Hardness	Other Characteristics
Limonite	$Fe_4O_3(OH)_6$	Yellow-brown	2 to 3	Earthy luster; yellow-brown streak
Magnetite	Fe_3O_4	Black	6	Strongly magnetic
Malachite	$Cu_2CO_3(OH)_2$	Green	3½ to 4	Often forms in concentric rings of light and dark green
Molybdenite	MoS_2	Dark gray	1 to 1½	Cleaves into flakes; more metallic luster than graphite
Muscovite (a mica)	$KAl_3Si_3O_{10}(OH)_2$	Colorless	2 to 2½	Excellent cleavage into thin sheets
Olivine	$(Fe,Mg)_2SiO_4$	Yellow-green	6½ to 7	Glassy luster
Phlogopite	$KMg_3AlSi_3O_{10}(OH)_2$	Brown	2½ to 3	Mica closely resembling biotite
Plagioclase feldspar	$(Na,Ca)(Al,Si)_2Si_2O_8$	White to gray	6	May show fine striations on cleavage surfaces
Potassium feldspar	$KAlSi_3O_8$	White; often stained pink or aqua	6	Two good cleavages forming a 90-degree angle; no striations
Pyrite	FeS_2	Brassy yellow	6 to 6½	Metallic luster; black streak
Pyroxene (e.g., augite)	$(Na,Ca,Mg,Fe,Al)_2Si_2O_6$	Usually green or black	5 to 7	Two good cleavages forming a 90-degree angle
Quartz	SiO_2	Variable; commonly colorless or white	7	Glassy luster; conchoidal fracture
Serpentine	$Mg_3Si_2O_5(OH)_4$	Green to yellow	3 to 5	Waxy or silky luster; may be fibrous (asbestos)
Sillimanite	Al_2SiO_5	White	6 to 7	Occurs only in metamorphic rocks; often forms needlelike crystals
Sphalerite	ZnS	Yellow-brown	3½ to 4	Glassy luster
Staurolite	$Fe_2Al_9Si_4O_{20}(OH)_2$	Brown	7 to 7½	Found in metamorphic rocks; elongated crystals may have crosslike form
Sulfur	S	Yellow	1½ to 2½	Sulfurous odor
Sylvite	KCl	Colorless	2	Cleaves into cubes; salty taste, but more bitter than halite
Talc	$Mg_3Si_4O_{10}(OH)_2$	White to green	1	Greasy or slippery to the touch
Tourmaline	$(Na,Ca)(Li,Mg,Al)(Al,Fe,Mn)_6^-$ $(BO_3)_3Si_6O_{18}(OH)_4$	Black, red, green	7 to 7½	Elongated crystals, triangular in cross section; conchoidal fracture

TABLE B.2 *A Key to Aid in Rock Identification*

Igneous Rocks	Sedimentary Rocks	Metamorphic Rocks
I. Extremely coarse-grained: rock is a pegmatite. (Most pegmatites are granitic, with or without exotic minerals.)	I. Rock consists of visible shell fragments or of oolites: *limestone*.	I. Nonfoliated; compact texture with interlocking grains: identified by predominant mineral(s).
II. Phaneritic (coarse enough that all grains are visible to the naked eye).	II. Rock consists of interlocking grains with texture somewhat like that of igneous rock and is light in color: probable chemical sedimentary rock.	A. If quartz-rich, perhaps with a sugary appearance: *quartzite*.
A. Significant quartz visible; only minor mafic minerals: *granite*.	A. Tastes like table salt: *halite*.	B. If calcite or dolomite (identified by effervescence, hardness): *marble*.
B. No obvious quartz; feldspar (light-colored) and mafic minerals (dark) in similar amounts: *diorite*.	B. No marked taste; hardness of 2 (if grains are large enough to scratch); does not effervesce: *gypsum*.	C. Rock consists predominantly of amphiboles: *amphibolite*.
C. No quartz; rock consists mostly of mafic minerals: *gabbro*.	C. Effervesces in weak HCl: *limestone* (calcite).	II. Foliated: classified mainly by texture.
D. No visible quartz or feldspar: rock is ultramafic.	D. Effervesces weakly in HCl, only if scratched: *dolomite*.	A. Very fine-grained; pronounced rock cleavage along parallel planes, to resemble flagstones: *slate*.
III. Porphyritic with fine-grained groundmass: go to part IV to describe groundmass (using phenocryst compositions to assist); adjective "porphyritic" will preface rock name.	III. Rock consists of grains apparently cemented or compacted together: probable clastic sedimentary rock.	B. Fine-grained; slatelike, but with glossy cleavage surfaces: *phyllite*.
IV. Aphanitic (grains too fine to distinguish easily with the naked eye).	A. Coarse grains (several millimeters or more in diameter), perhaps with a finer matrix: *conglomerate* if the grains are rounded, *breccia* if they are angular.	C. Coarser grains; obvious foliation, commonly defined by prominent mica flakes, sometimes by elongated crystals like amphiboles: *schist*.
A. Quartz is visible or rock is light in color (white, cream, pink): probably *rhyolite*.	B. Sand-sized grains; gritty feel: *sandstone*. If predominantly quartz grains, *quartz sandstone;* if roughly equal proportions of quartz and feldspar, *arkose;* if many rock fragments, and perhaps a fine-grained matrix, *graywacke*.	D. Compositional or textural banding, especially with alternating light (quartz, feldspar) and dark (ferromagnesian) bands: *gneiss*.
B. No visible quartz; medium tone (commonly gray or green); if phenocrysts are present, commonly plagioclase, pyroxene, or amphibole: probably *andesite*.	C. Grains too fine to see readily with the naked eye: *mudstone*. If rock shows lamination and a tendency to part along parallel planes, *shale*.	
C. Rock is dark, commonly black; any phenocrysts are olivine or pyroxene: *basalt*.	IV. Relatively dense; compact, dark, no visible grains; massive texture, conchoidal fracture: *chert* (silica).	
D. Rock is glassy and massive: *obsidian* (regardless of composition).		
E. Rock consists of gritty mineral grains, ash and glass shards: *ignimbrite* (welded tuff).		

APPENDIX C

Introduction to Topographic and Geologic Maps and Satellite Imagery

MAPS AND SCALE

Many kinds of information can be presented in map form. Topographic and geologic maps are the two kinds most frequently used by geologists. This section presents a brief introduction to both types of maps.

A point to note at the outset is that, in the United States, topographic and geologic maps commonly carry English, rather than metric, units. Many of these maps were drawn before there was any move toward adoption of metric units in the United States. The task of redrawing maps to convert to metric units is formidable, particularly with topographic maps. (A metric map series is being prepared by the U.S. Geological Survey, but it will be many years before its completion.)

A basic feature of any map is the map **scale,** a measure of the size of the area represented by the map. Map scales are reported as ratios—1:250,000, 1:62,500, and so on; or equivalently in words, for example, "one to two hundred fifty thousand." On a 1:250,000 scale map, a distance of 1 inch on the map represents 250,000 inches (almost 4 miles) in reality; on a 1:62,500 scale map, 1 inch equals about 1 mile of actual distance. The larger the scale factor, the more real distance is represented by a given distance on the map. As a result, the fineness of detail that can be represented on a map is reduced as the scale factor is increased. The choice of scale factor often involves a compromise between minimizing the size of the map (for convenience of use) or the number of maps required to cover the area, and maximizing the amount of detail that can be shown.

TOPOGRAPHIC MAPS

Topographic maps primarily represent the form of the earth's surface. Selected other features, both natural and artificial, may be included for information. Once one becomes accustomed to reading them, topographic maps can make excellent navigational aids.

Contour Lines, Contour Intervals

The problem of representing three-dimensional features on a two-dimensional map is addressed through the use of **contour lines.** A contour line is a line connecting points of equal elevation, measured in feet or meters above or below sea level. The **contour interval** of a map is the difference in elevation between successive contour lines. For instance, if the contour interval is 10 feet, contour lines are drawn at elevations of 1010 feet, 1020 feet, 1030 feet, and so on (in whatever range of elevations is appropriate to that map). If the contour interval is 20 feet, contours would be drawn at elevations of 1000 feet, 1020 feet, 1040 feet, and so on. The contour interval chosen for a particular map depends on the overall amount of relief in the map area. If the terrain is very flat, as in a midwestern floodplain, a 10-foot or even 5-foot contour interval may be appropriate to depict what relief there is. In a rugged mountain terrain, like the Rockies, the Cascades, or the Sierras, with several thousand feet of vertical relief, using a 10-foot contour interval would make for a very cluttered map, thick with contour lines; a 50-foot or even 100-foot contour interval may be used in such a case.

The relationship between the spacing of contours and the steepness of a slope can be seen in figure C.1. A comparison of the actual relief, as shown in figure C.1A, and the resultant arrangement of contour lines on the corresponding topographic map, shown in figure C.1B, illustrates that, on a given map, the more closely spaced the contours, the steeper the corresponding terrain. In other words, if the map shows many closely spaced contours, this means that there is a great deal of vertical relief over a limited horizontal distance. Note also that contours run across the face of a slope; the upslope and downslope directions are perpendicular to the contours.

Contours do close, although they may not do so on any one given map. A series of concentric closed contours indicates a hill. If there is a local depression, so that the same contour is encountered twice, once going up in elevation and again within the depression, the repetition of the contour within the depression is marked by a hachured contour, as shown is figure C.2.

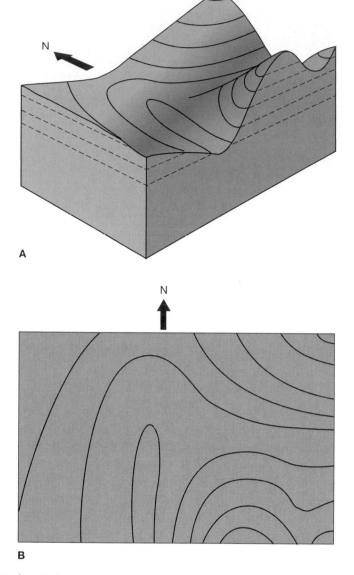

FIGURE C.1 Contours and relief. (A) Perspective view of a hilly area, with contours superimposed for reference. (B) The same contours as they would appear on a topographic map.

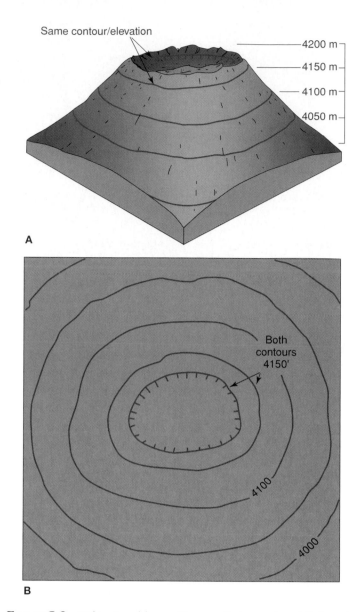

FIGURE C.2 Volcanic caldera with summit depression. (A) Actual relief. (B) As the volcano would appear on a topographic map.

Where contour lines cross a stream valley, they point upstream, toward higher elevations (figure C.3). Contour lines corresponding to different elevations should not ordinarily cross each other (that would imply that the same point has multiple elevations). The only exception to this would occur in the case of an overhanging cliff, where a range in elevations exists at one spot on the map.

Other Features on Standard Topographic Maps

The most extensive topographic maps of the United States have been compiled by the U.S. Geological Survey (USGS). The USGS has adopted a uniform set of symbols for various kinds of features, which facilitates the reading of topographic maps. Some of these symbols are illustrated in figure C.4.

Contour lines are drawn in brown. Major contours (typically, those contours corresponding to multiples of 100 feet, except on maps with large contour intervals) are drawn more boldly and labeled with the corresponding elevation. This is especially convenient in rugged terrain, where it would otherwise be easy to lose count of numerous closely spaced contours. Roads are drawn in black and/or red. Town limits are shown in light red, and the names of towns, cities, and mountains, or any other labels, are printed in black. Watery features are drawn in blue: lakes and perennial streams in solid blue, intermittent streams in dashed blue lines, swampy areas represented by blue symbols resembling tufts of low grasses. The background color for wooded areas is green; for open fields, brushy areas, deserts, bare rock, or any area lacking overhead vegetative cover, the background color is white. Revisions based on aerial photography ("photo-revisions") are shown in purple.

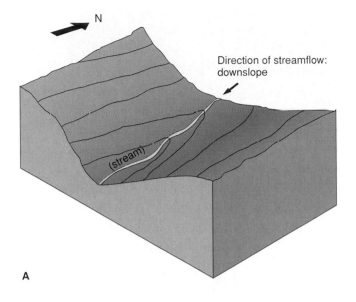

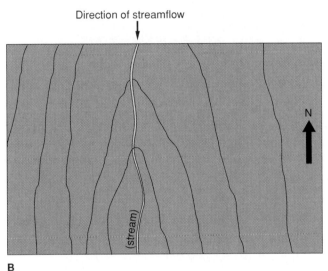

FIGURE C.3 Contours crossing stream valleys. (A) Relief with superimposed contours. (B) On topographic map, the contours point upstream.

Obtaining Topographic Maps

U.S. Geological Survey maps are sometimes identified as "15-minute" or "7½-minute" *quadrangle* maps. Such a designation means that the map area is a rectangle corresponding to so many minutes of latitude and longitude. Because it covers a smaller area, a 7½-minute quadrangle map shows more detail than a 15-minute quadrangle. Many parts of the United States have been mapped at both scales; for others, only 15-minute quadrangles are available. Individual maps are named for a town, mountain, or other prominent geographic feature within the map area: for example, "Cooke City 15-minute quadrangle," "Cutoff Mountain 7½-minute quadrangle." There are also special maps covering areas of particular interest, such as individual national parks.

Topographic maps may be ordered directly from the U.S. Geological Survey at nominal cost (variable with map size and scale). If you are uncertain of what map(s) you want, you can request free *index maps* of the corresponding state(s), along with price lists, before ordering. For example, if you were interested in obtaining quadrangle maps of selected parts of Yellowstone National Park, you would request a topographic index map of Wyoming and then choose the appropriate quadrangle(s) from that. The index map shows what part of the state is covered by which quadrangle map, the name by which each map is identified, and the date of the map or last revision thereof. A cautionary note: The same place name may be used both for a 15-minute quadrangle and for a 7½-minute quadrangle within it. Be alert to this when ordering, and specify quadrangle size if necessary.

Topographic maps can be obtained from two principal locations. For maps of areas east of the Mississippi River, contact:

> Branch of Distribution
> U.S. Geological Survey
> 1200 S. Eads Street
> Arlington, Virginia 22202

For areas west of the Mississippi River, maps can be obtained from:

> Branch of Distribution
> U.S. Geological Survey
> Box 25286 Federal Center
> Denver, Colorado 80225

GEOLOGIC MAPS

Topographic maps show the form of the land surface; geologic maps are one way of representing the underlying geology. The most common kind of geologic map is a map of **bedrock geology,** which shows the geology as it would appear with soil stripped away. (Other maps may show distribution of different soil types, glacial deposits, or other features of surface geology.) From a well-prepared geologic map, aspects of the subsurface structure can often be deduced.

Basic Concepts Related to Geologic Maps

Fundamental to making a geologic map is identifying a suitable set of **map units.** These may, for example, be individual sedimentary rock formations, distinguishable lava flows, or metamorphic rock units. The main requirement for a map unit is that it be identifiable by the presence or absence of some characteristic(s), and thus distinct from other map units chosen. The mapper then marks which of the map units are found at each place where the rocks are exposed.

Additional information, such as the orientation of beds or the location of contacts between map units, may also be recorded. Where obvious, contacts are drawn as solid lines; where only inferred, as dashed lines. (If one finds granite at point *A* and limestone at point *B* a short distance away, one can infer a contact between the granite and limestone somewhere between *A* and *B,* even if the exact spot is covered by soil.)

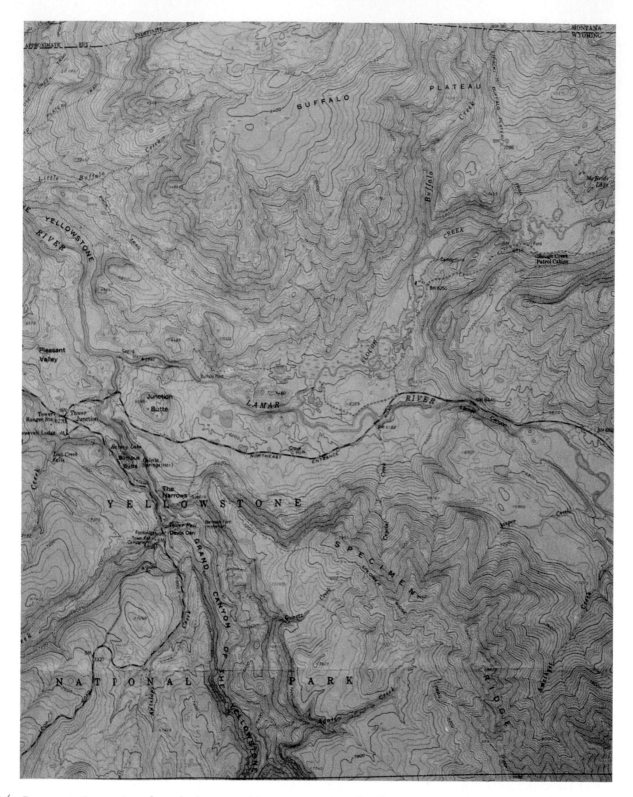

FIGURE C.4   Representative section of standard topographic map: a portion of Yellowstone National Park. Note the closely spaced contours along the Grand Canyon of the Yellowstone and edges of plateaus, the shapes of contours that cross streams, the closed contours around hills (for example, Junction Butte), and the swampy flat area (along Slough Creek).
Source: U.S. Geological Survey.

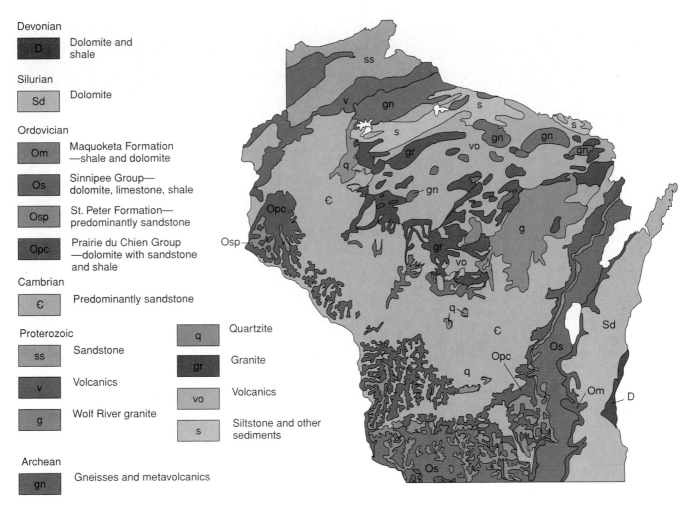

FIGURE C.5 Sample of geologic map with key: simplified bedrock geology of Wisconsin.
Source: After M. E. Ostrom, Wisconsin Geological and Natural History Survey, April 1981.

How easy it is to produce an accurate and complete geologic map depends on several factors (aside from the competence of the mapper!). Bedrock exposure is one. Thick soil, water or swamps, and vegetation can all obscure what lies below. In some areas, the geology is well exposed only along stream valleys; in others, the rocks are completely exposed, and mapping is greatly simplified. In glaciated areas, too, one need not only deal with the cover of glacial till (assuming that is not the material of interest) but also be cautious about mistakenly identifying a large, partially buried bit of glacial debris as a small exposure of bedrock. And some areas are simply much more complicated, geologically, than others.

On a geologic map, different units are represented in different colors for clarity (figure C.5). Units of similar age may be shown in different shades of the same color. The map is accompanied by a key showing all the map units, arranged in chronological order (insofar as their ages are

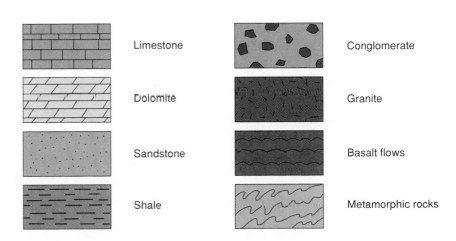

FIGURE C.6 Selected standard map symbols for various rock types.

known), with the youngest at the top. Ordinarily, a brief description of each map unit is given; alternatively, standard patterns may be used to indicate the general rock type (figure C.6). Each unit is also assigned a symbol. The first part of

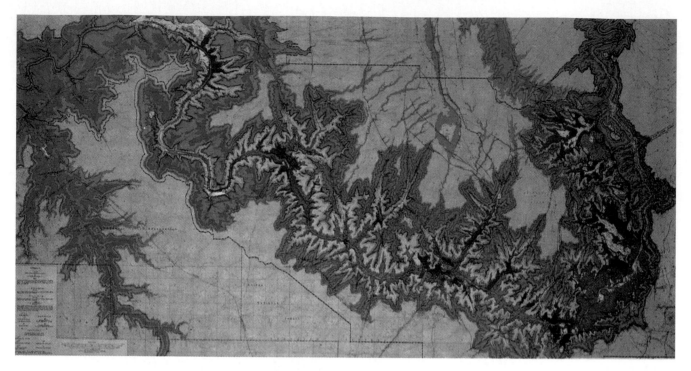

FIGURE C.7 The presence of the Grand Canyon allows much better knowledge of the area's subsurface geology. Geologic map shows different rock units in different colors. Note monotony outside canyon.
Used with permission of Grand Canyon Natural History Association.

the symbol consists of one or two letters corresponding to the unit's age (generally, the geologic era or period). This is usually followed by one to three lowercase letters corresponding to the rock type or unit's name (if any). For example, "p€qm" might be used for an unnamed Precambrian quartz monzonite, "Dl" for the Devonian Littleton Formation, "Qa" for Quaternary alluvium along a stream channel. These symbols are useful for distinguishing units mapped in similar colors, as well as for providing a general indication of the age of each unit directly on the map.

Interpretation from Geologic Maps—Examples

Interpretation of geologic maps likewise varies in difficulty, depending on the fundamental complexity of the geology and the extent of exposure (completeness of map). Sometimes, too, how much of the geology can be seen depends not only on cover or lack of it, but on topography. That is, in rugged terrain with considerable vertical relief, one has more of a three-dimensional look at the geology than in flat terrain. For example, in the vicinity of the Grand Canyon, the Paleozoic sedimentary sequence is quite flat-lying. So, for the most part, is the land surface. Outside the canyon, only the Kaibab limestone is exposed, forming a nearly level plateau and giving no evidence of what lies below. The cutting of the canyon has exposed many more rock units, down to the Precambrian. This is apparent on a geologic map (figure C.7).

The Grand Canyon has a regular, layer-cake geology that has a straightforward interpretation. Where the geology is more complex, patterns of repetition of units, and orientation of beds, must be used to interpret the structure. For example, the map pattern of figure C.8 shows circular exposures of rock units. All units dip away from the center of the circle, and the key shows that the oldest rocks are at the center. This, then, is a dome.

When sets of rocks are repeated, this may indicate the presence of a fault. In the map pattern shown in the top block in figure C.9, the same sequence of units is repeated; moreover, all are dipping in the same direction. It is difficult to create such a pattern by folding. A fault, however, can account for the result (bottom diagram in figure C.9). Offset of features that are otherwise continuous is another sign of possible faulting (figure C.10).

Cross Sections

Interpreting structure from geologic map patterns and rock orientations requires some practice and the ability to visualize in three dimensions. Often, the maker of a geologic map assists the map reader by supplying one or more geologic **cross sections.** A cross section is a three-dimensional interpretation of the geology seen at the surface. The line along which the cross section is drawn is indicated on the map. The cross section uses the same map units and symbols as the map proper and attempts to show the geometric relationships inferred to exist among those units—folds, faults, intrusive relationships, and so on.

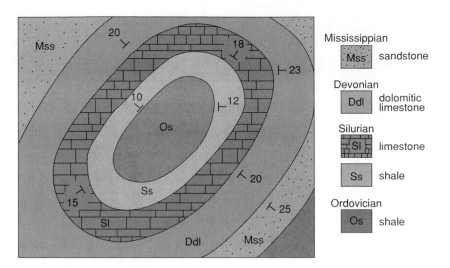

FIGURE C.8 A dome as it would appear on a geologic map.

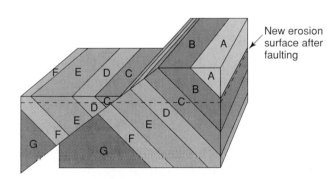

Faulting of tilted sediments

New erosion surface after faulting

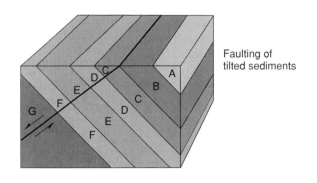

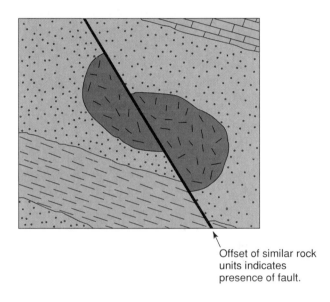

Offset of similar rock units indicates presence of fault.

FIGURE C.10 Igneous body and surrounding rocks offset by faulting.

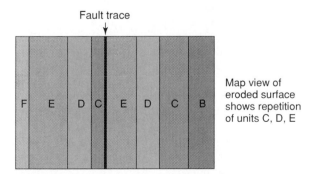

Fault trace

Map view of eroded surface shows repetition of units C, D, E

FIGURE C.9 Faulting can produce repetition of map units. Top surface of top block is original surface exposure of units; bottom diagram is map result after faulting and erosion to the dashed surface of center diagram.

A cross section is drawn by starting with a topographic profile along the chosen line and marking on it the geology as seen from the surface (figure C.11A). A structural interpretation that is consistent with all the known data is then devised (figure C.11B). Depending on the complexity of the geology and the completeness of exposure on which the original geologic map has been based, it may or may not be possible to develop a unique structural interpretation for the observed map pattern. If it is not, several plausible alternatives might be presented. Cross sections can be very valuable in evaluating site suitability for construction or other purposes.

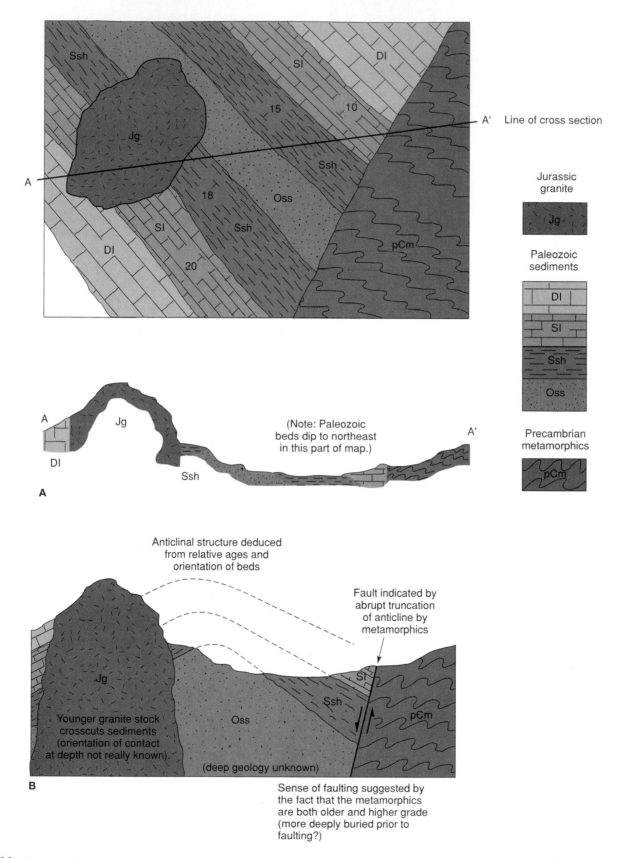

FIGURE C.11 Constructing a cross section. (A) Geology as seen at surface is sketched onto a topographic profile. (B) The pattern is interpreted, and a set of structures consistent with the pattern seen at the surface is sketched in.

Obtaining Geologic Maps

A variety of geologic maps are available from the U.S. Geological Survey, through the distribution offices listed earlier, in the section "Obtaining Topographic Maps." As with the latter, index maps are available to assist in selecting the geologic map(s) of interest.

Geologic mapping is more time-consuming than topographic mapping, and the U.S. Geological Survey has not had the personnel or funding to produce geologic maps of every part of the United States. However, many states have very active geological surveys that have undertaken extensive mapping programs within their home states. The information office of the state of interest should be able to supply information about the state geological survey. Contact information for U.S. state and Canadian province surveys can be found in Appendix E.

Once one has developed an interest in geology, one may begin asking, "What's that?" when observing rocks while traveling, especially in unfamiliar areas. For a quick answer, it may be helpful to consult a *geological highway map*. The American Association of Petroleum Geologists has produced a set of such maps that collectively cover all of the contiguous United States. Given that each map covers several states, the level of detail possible on each map is necessarily limited, but the maps do provide an overview of a region's geology. Information, prices, and maps can be obtained from:

American Association of Petroleum Geologists
P.O. Box 979
Tulsa, Oklahoma 74101

REMOTE SENSING AND SATELLITE IMAGERY

Remote sensing methods encompass all of those means of examining planetary features that do not involve direct contact. Instead, these methods rely on detection, recording, and analysis of wave-transmitted energy—visible light, infrared radiation, and others. Aerial photography, in which standard photographs of relatively large regions of the earth are taken from aircraft, is one example. Radar mapping of surface topography, using airplanes or spacecraft, is another. Still another involves analyzing the light reflected from the surface of a body. In the case of many planets, remotely sensed data may be the only kind readily available. In the case of the earth, remote sensing, especially using satellites, is a quick and efficient way to scan broad areas, to examine regions having such rugged topography or hostile climate that they cannot easily be explored on foot or with surface-based vehicles, and to view areas to which ground access is limited for political reasons. Probably the best-known and most comprehensive earth satellite imaging system is the one initiated in 1972, known as Landsat while under the control of the federal government, renamed Eosat when later "privatized."

The Landsat satellites orbit the earth in such a way that images can be made of each part of the earth. Each orbit is slightly offset from the previous one, with the areas viewed on one orbit overlapping the scenes of the previous orbit. Each satellite makes fourteen orbits each day; complete coverage of the earth takes eighteen days. Therefore, images of any given area should be available every eighteen days, although in practice, shifting distributions of clouds obscure the surface some part of the time at any point. Five Landsat satellites have been launched; a sixth is planned.

The sensors in the Landsat satellites do not detect all wavelengths of energy reflected from the surface. They do not take photographs in the conventional sense. They are particularly sensitive to selected green and red wavelengths in the visible light spectrum and to a portion of the infrared (invisible heat radiation, with wavelengths somewhat longer than those of red light). These wavelengths were chosen particularly because plants reflect light most strongly in the green and the infrared. Different plants, rocks, and soils reflect different proportions of radiation of different wavelengths. Even the same feature may produce a somewhat different image under different conditions. Wet soil differs from dry; sediment-laden water looks different from clear; a given variety of plant may reflect a different spectrum of radiation depending on what trace elements it has concentrated from the underlying soil or how vigorously it is growing. Landsat images can be powerful mapping tools.

Landsat Images and Applications

A common format for Landsat imagery is photographic prints at 1:1,000,000 scale. At that scale, a 23-centimeter (9-inch) print covers 34,225 square kilometers (13,225 square miles). The smallest features that can be distinguished in the image are about 80 meters (250 feet) in size, which gives some idea of the quality of the resolution. Multiple images can be joined into mosaics covering whole countries or continents.

Images are typically presented either in black and white or as *false-color composites*. The latter are produced by projecting the data for individual spectral regions through colored filters and superimposing the results. The false-color images are so named because the resulting pictures, though superficially resembling color photographs, do not present all features in the colors they would appear to the human eye. The most striking difference is in vegetation, which appears in shades of red, not green. Rock and soil usually show as white, blue, yellow, or brown, depending on composition. Water is blue to bluishblack; snow and ice are white. Examples of false-color Landsat images are used throughout this text. Landsat image data can also be further processed by computer to produce images in more "realistic" (expected) colors or to enhance particular features by emphasizing certain wavelengths of radiation.

Dozens of applications of Landsat imagery exist in the natural sciences—for example, basic geologic mapping, identification of geologic structures, environmental/ecological monitoring, and resource exploration. Because Landsat scans the same area repeatedly over time, seasonal changes (figure C.12) and the progress and extent of occasional events such

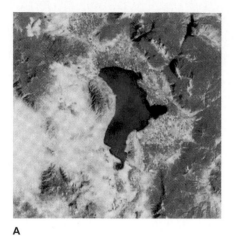 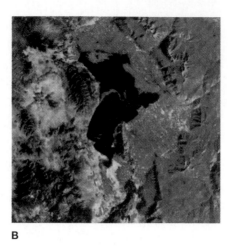 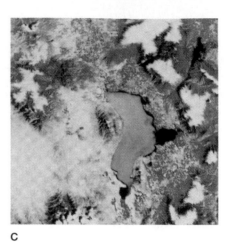

A   **B**   **C**

FIGURE C.12   Seasonal variations in Utah Lake. (A) In early August, some incoming sediment clouds the lake. (B) By mid-September, dry conditions have greatly reduced the extent of sediment input; water appears dark and clear. (C) The high runoff of spring, and the corresponding increased sediment input, is reflected in the very turbid water in this image, taken in late May.
© *NASA.*

FIGURE C.13   False-color Landsat image of the Imperial Valley of California; red in image is green in reality. Intensive agricultural development, supported by irrigation, shows as checkered red expanse along valley axis; outside irrigated area, land is dry and barren. The U.S./Mexican border (lower right) is sharply defined as a result of differences in land-use and agricultural practices.
© *NASA.*

as flooding (as shown in chapter 15) can be monitored. Landsat imagery also can be used to monitor development of and changes in surface features such as stream channels and currents or land-use patterns (figure C.13). It is helpful in identifying patterns of land use and in monitoring the progress of crops and the extent of damage to vegetation from fires, insects, or disease.

In all cases, Landsat imagery is especially useful when some *ground truth* can be obtained. Ground truth is information gathered by direct surface examination (best done at the time of imaging, if vegetation is involved), which can provide critical confirmation of interpretations based on remotely sensed data.

An overview of global diversity as seen through Landsat imagery and a survey of many applications of remotely sensed data are presented in *Mission to Earth: Landsat Views the World* by N. M. Short, P. D. Lowman, Jr., S. C. Freden, and W. A. Firch, Jr., published by NASA in 1976, and available through the U.S. Government Printing Office. Canadian features are analyzed in more detail in *Landsat Images of Canada—A Geological Appraisal* by V. R. Slaney, Geological Survey of Canada Paper 80-15, published in 1981, and available from the Canadian Government Printing Office. Individual Landsat images can be ordered from EROS Data Center, Sioux Falls, South Dakota, 57198.

APPENDIX D
Fossil Phyla

Key

R = Recent
Jr = Jurassic
Tr = Triassic
Pm = Permian
P = Pennsylvanian
M = Mississippian

D = Devonian
S = Silurian
O = Ordovician
C = Cambrian
PC = Precambrian

MAJOR GROUP	EXAMPLES

Kingdom: Animalia

Phylum: Protozoa
    Class: Sarcodina
        Order: Foraminiferida (O–R) . . . . Foraminifera (figure D.1)
        Order: Radiolaria (PC–R) . . . . Radiolaria (figure D.2)

Phylum: Archaeocyatha (C) . . . . Archaeocyathids

Phylum: Porifera (Late PC–R) . . . . Sponges (figure D.3)

Phylum: Cnidaria (= Coelenterata)
    Class: Hydrozoa . . . . Stromatoporoids
    Class: Scyphozoa . . . . Jellyfish
    Class: Anthozoa . . . . Corals, sea anemones (figure D.4)
        Order: Rugosa (O–Pm)
        Order: Tabulata (O–Jr)
        Order: Scleractinia (Tr–R)

Phylum: Bryozoa (= Ectoprocta) (O–R) . . . . Bryozoans (figure D.5)

Phylum: Brachiopoda . . . . Brachiopods (figure D.6)
    Class: Inarticulata (C–R)
    Class: Articulata (C–R)

Phylum: Mollusca
    Class: Gastropoda (C–R) . . . . Snails, limpets, slugs, pteropods (figure D.7)
    Class: Bivalvia (= Pelecypoda) (C–R) . . . . Clams, oysters, mussels, scallops (figure D.8)
    Class: Cephalopoda (Late C–R) . . . . Nautiloids, ammonoids, squid, octopi, cuttlefish, chambered
                                                                           nautilus (figure D.9)

MAJOR GROUP	EXAMPLES

Phylum: Annelida . . . . Segmented worms

Phylum: Arthropoda
    Class: Ostracoda (O–R). . . . . Ostracods
    Class: Trilobita (C–Pm). . . . . Trilobites (figure D.10)
    Class: Crustacea (D–R). . . . . Crabs, shrimp, crayfish, lobsters, barnacles
    Class: Insecta . . . . Insects
    Class: Arachnida (D–R). . . . . Spiders, scorpions, ticks, mites

Phylum: Echinodermata
    Class: Blastoidea (S–Pm). . . . . Blastoids
    Class: Crinoidea (C–R) . . . . Crinoids (figure D.11)
    Class: Asteroidea (O–R) . . . . Starfish (figure D.12)
    Class: Echinoidea (O–R). . . . . Sea urchins, sand dollars (figure D.13)
    Class: Cystoidea (O–D). . . . . Cystoids (figure D.14)
    Class: Holothuroidea . . . . Sea cucumbers

Phylum: Hemichordata (= Protochordata)
    Class: Graptolithina (C–M) . . . . Graptolites (figure D.15)

Phylum: Unknown
    Conodonts (C–Tr) . . . . Conodonts

Phylum: Chordata
  Subphylum: Vertebrata
    Class: Agnatha (C–R). . . . . Jawless fish; lamprey, hagfish
    Class: Placodermi (S–Pm). . . . . Early jawed fish
    Class: Chondrichthyes (D–R) . . . . Cartilagenous fish; sharks, rays, skates
    Class: Osteichthyes (D–R) . . . . Bony fish
    Class: Amphibia (D–R). . . . . Labyrinthodonts, frogs, toads, salamanders
    Class: Reptilia (P–R). . . . . Lizards, snakes, turtles, crocodiles, dinosaurs
    Class: Aves (Jr–R) . . . . Birds
    Class: Mammalia (Tr–R) . . . . Mammals

Kingdom: Plantae
    Division: Psilophyta. . . . . Rootless vascular plants; whisk ferns
    Division: Lycophyta. . . . . Lycopods; lycopsids (scale trees), club mosses, quillworts
    Division: Sphenophyta. . . . . Sphenopsids, horsetails
    Division: Pterophyta . . . . Ferns
    Division: Coniferophyta. . . . . Conifers
    Division: Cycadophyta. . . . . Cycads
    Division: Ginkgophyta. . . . . Ginkgoes
    Division: Anthophyta. . . . . Angiosperms (flowering plants)
      Class: Monocotyledoneae. . . . . Monocots
      Class: Dicotyledoneae . . . . Dicots

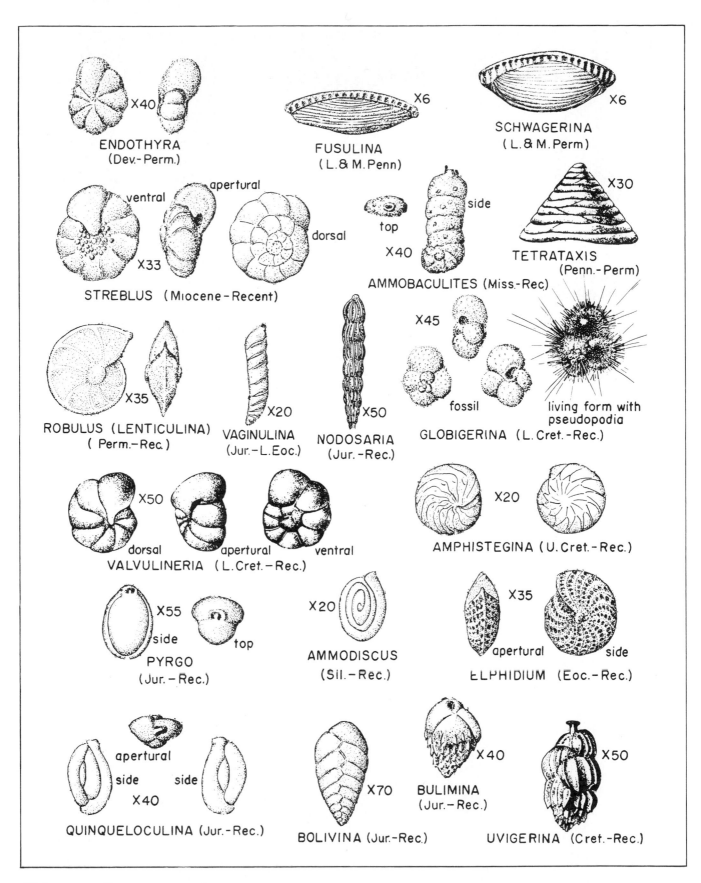

FIGURE D.1 Foraminifera.

From James C. Brice, et al., Laboratory Studies in Earth History, *5th ed. © Copyright 1993 Times Mirror Higher Education Group, Inc., Dubuque, Iowa. All Rights Reserved. Reprinted by permission.*

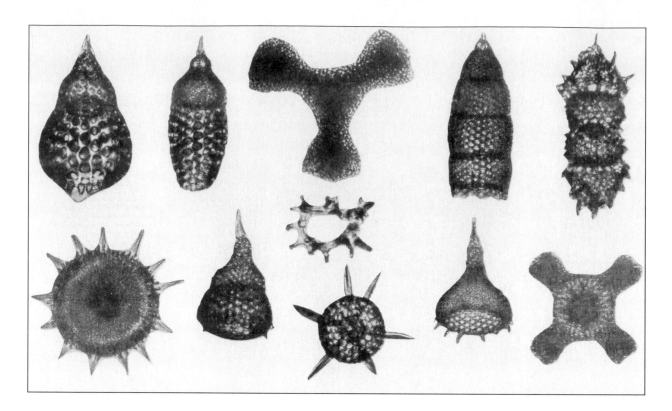

Figure D.2 Radiolaria.

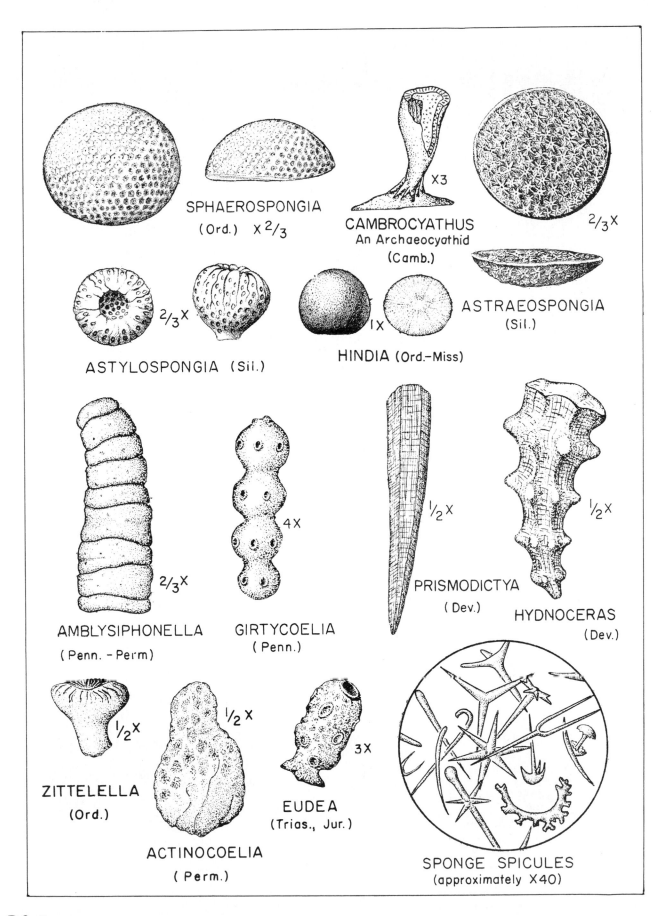

FIGURE D.3 Sponges.

Source: Collinson, C. W., 1959, Guide for Beginning Fossil Hunters, *Illinois State Geological Survey, Educational Series 4, 39 p.*

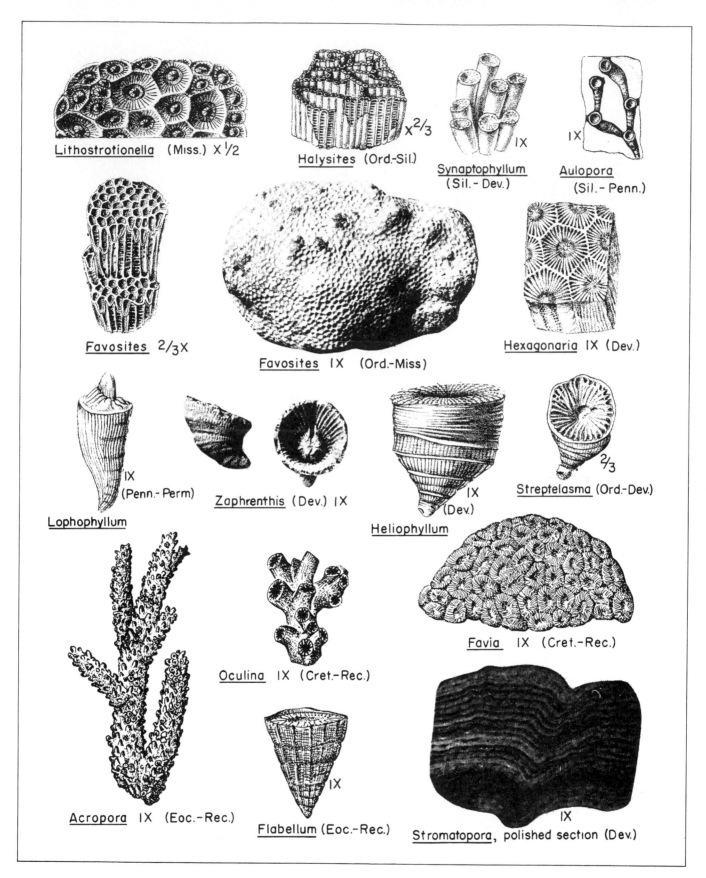

FIGURE D.4 Corals.

Source: Collinson, C. W., 1959, Guide for Beginning Fossil Hunters, *Illinois State Geological Survey, Educational Series 4, 39 p.*

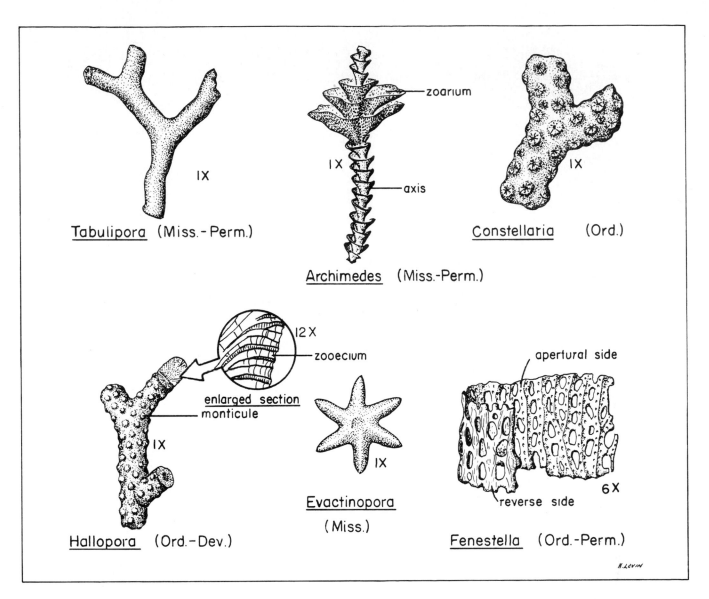

Figure D.5 Bryozoans.

From James C. Brice, et al., Laboratory Studies in Earth History, *5th ed. © Copyright 1993 Times Mirror Higher Education Group, Inc., Dubuque, Iowa. All Rights Reserved. Reprinted by permission.*

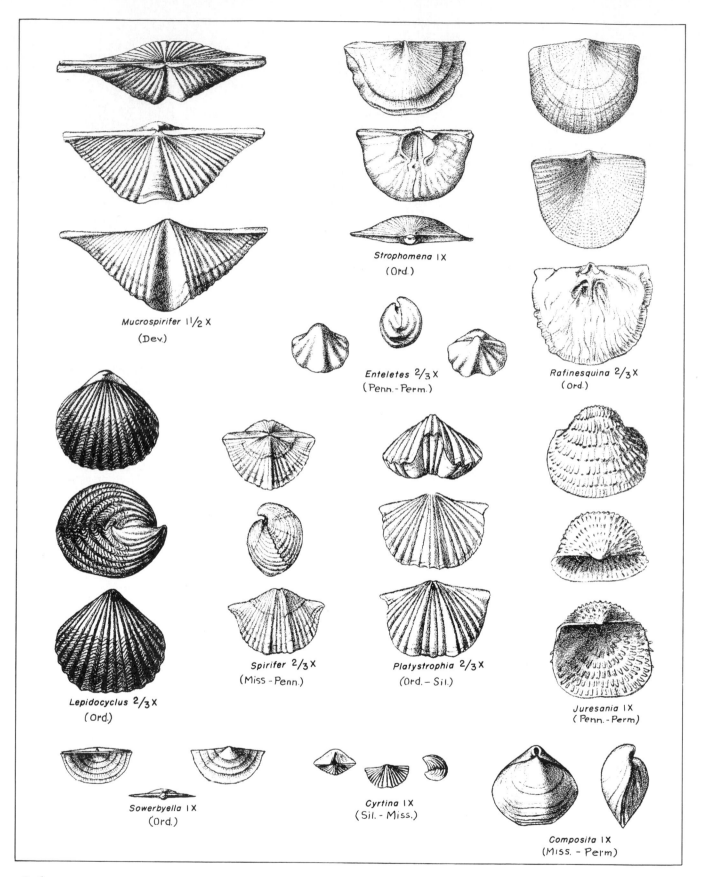

FIGURE D.6 Brachiopods.

Source: Collinson, C. W., 1959, Guide for Beginning Fossil Hunters, *Illinois State Geological Survey, Educational Series 4, 39 p.*

Figure D.7 Gastropods.

Source: Collinson, C. W., 1959, Guide for Beginning Fossil Hunters, *Illinois State Geological Survey, Educational Series 4, 39 p.*

Venus (Jur.–Rec.)

Grammysia (Sil.–Miss.)

Trigonia (Jur.–Rec.)

Cardiopsis (Miss.–Penn.)

Exogyra (Jur.–Cret.)

Allorisma (Miss.–Penn.)

Aviculopecten (Sil.–Perm.)

Ctenodonta (Ord.–Sil.)

Ostrea (Trias.–Rec.)

Pecten (Miss.–Rec.)

Gryphaea (Jur.–Cret.)

Cardium (Trias.–Rec.)

Nuculana (Sil.–Rec.)

Inoceramus (Jur.–Cret.)

FIGURE D.8 Bivalves.

Source: Collinson, C. W., 1959, Guide for Beginning Fossil Hunters, *Illinois State Geological Survey, Educational Series 4, 39 p.*

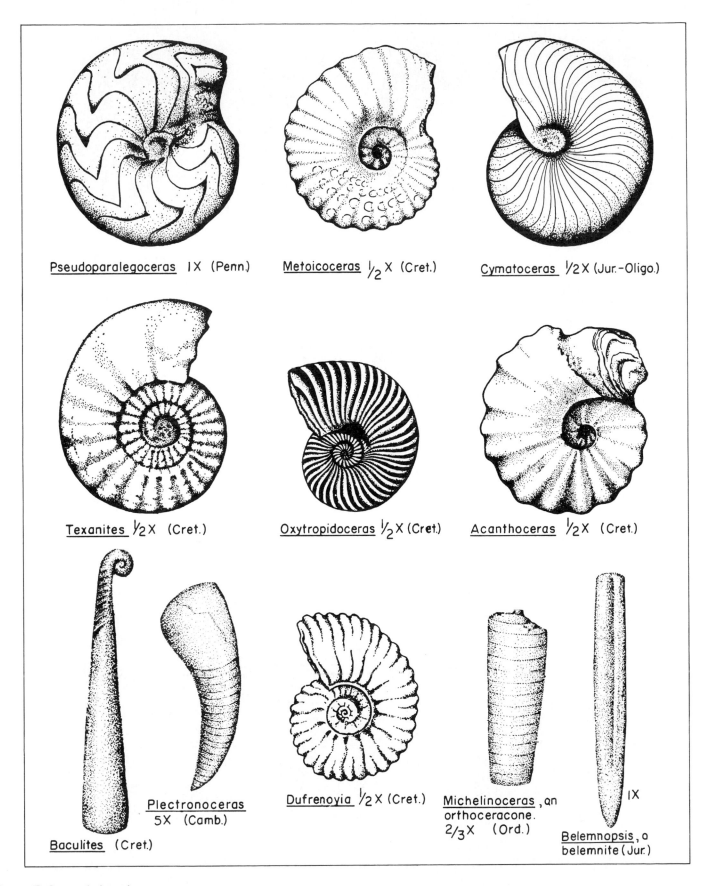

Pseudoparalegoceras IX (Penn.)   Metoicoceras ½X (Cret.)   Cymatoceras ½X (Jur.–Oligo.)

Texanites ½X (Cret.)   Oxytropidoceras ½X (Cret.)   Acanthoceras ½X (Cret.)

Plectronoceras 5X (Camb.)   Dufrenoyia ½X (Cret.)   Michelinoceras, an orthoceracone. 2/3X (Ord.)   Belemnopsis, a belemnite (Jur.)

Baculites (Cret.)   IX

FIGURE D.9 Cephalopods.
Modified from W. H. Matthews, III, 1960, Texas Fossils: An Amateur's Collector's Guidebook 2, *plates 32 and 33, University of Texas, Austin, Bureau of Economic Geology Guidebook 2.*

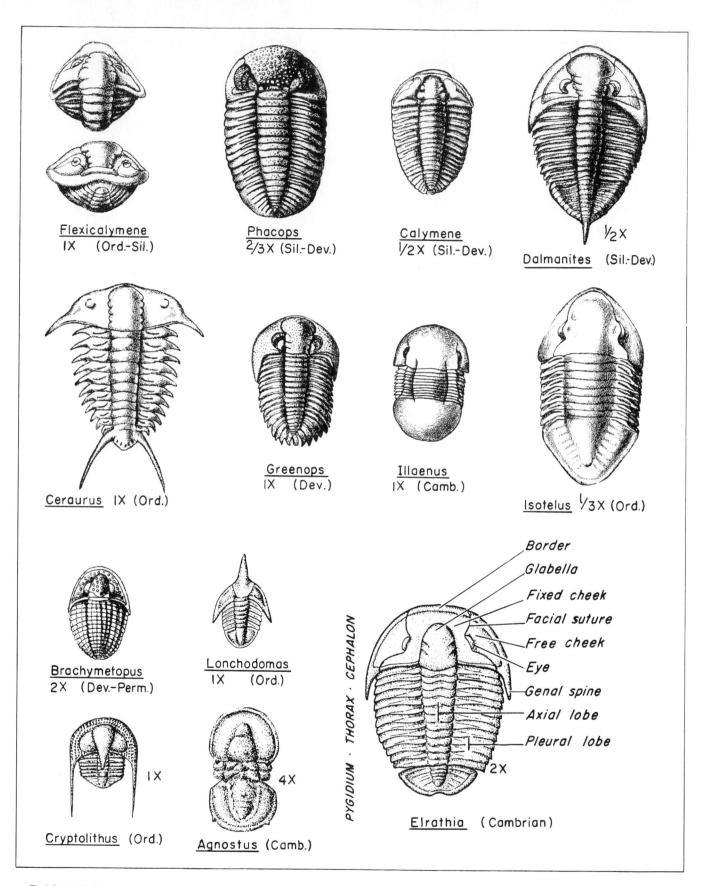

FIGURE D.10 Trilobites.

Source: Collinson, C. W., 1959, Guide for Beginning Fossil Hunters, Illinois State Geological Survey, Educational Series 4, 39 p.

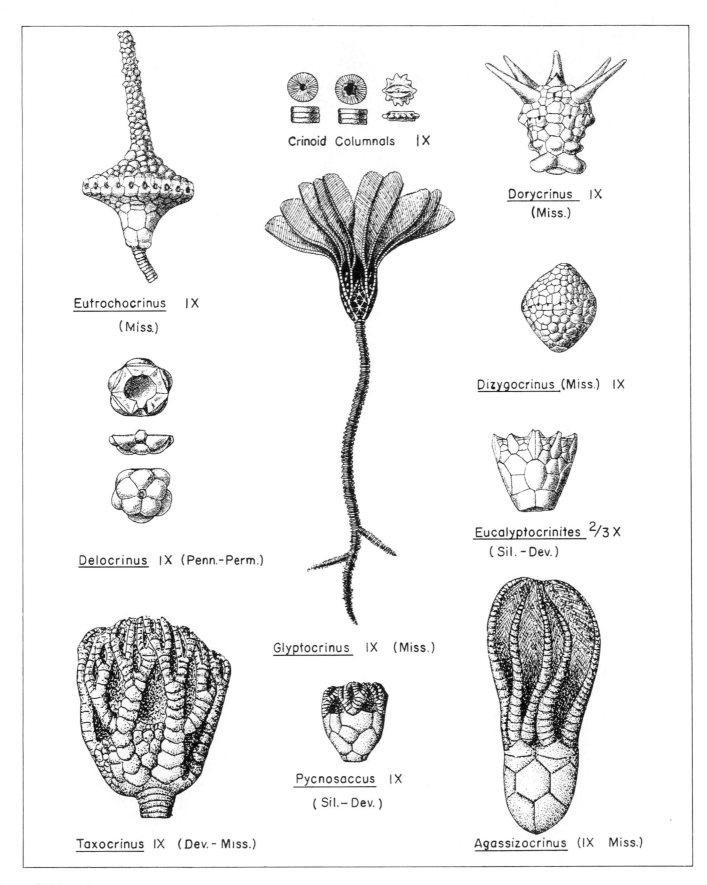

Crinoid Columnals IX

Eutrochocrinus IX
(Miss.)

Dorycrinus IX
(Miss.)

Dizygocrinus (Miss.) IX

Delocrinus IX (Penn.-Perm.)

Eucalyptocrinites 2/3 X
(Sil.-Dev.)

Glyptocrinus IX (Miss.)

Pycnosaccus IX
(Sil.-Dev.)

Taxocrinus IX (Dev.-Miss.)

Agassizocrinus (IX Miss.)

FIGURE D.11 Crinoids.
Source: Collinson, C. W., 1959, Guide for Beginning Fossil Hunters, *Illinois State Geological Survey, Educational Series 4, 39 p.*

Hudsonaster
5✕   (Ord.)

FIGURE D.12   Starfish.

From James C. Brice, et al., Laboratory Studies in Earth History, *5th ed. © Copyright 1993 Times Mirror Higher Education Group, Inc., Dubuque, Iowa. All Rights Reserved. Reprinted by permission.*

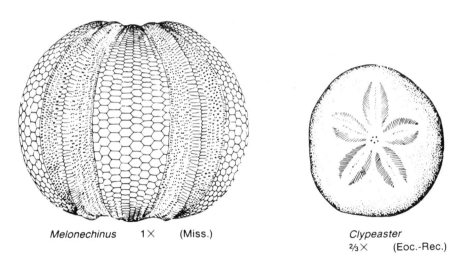

Melonechinus   1✕   (Miss.)

Clypeaster
⅔✕   (Eoc.-Rec.)

FIGURE D.13   Sea urchins and sand dollars.

From James C. Brice, et al., Laboratory Studies in Earth History, *5th ed. © Copyright 1993 Times Mirror Higher Education Group, Inc., Dubuque, Iowa. All Rights Reserved. Reprinted by permission.*

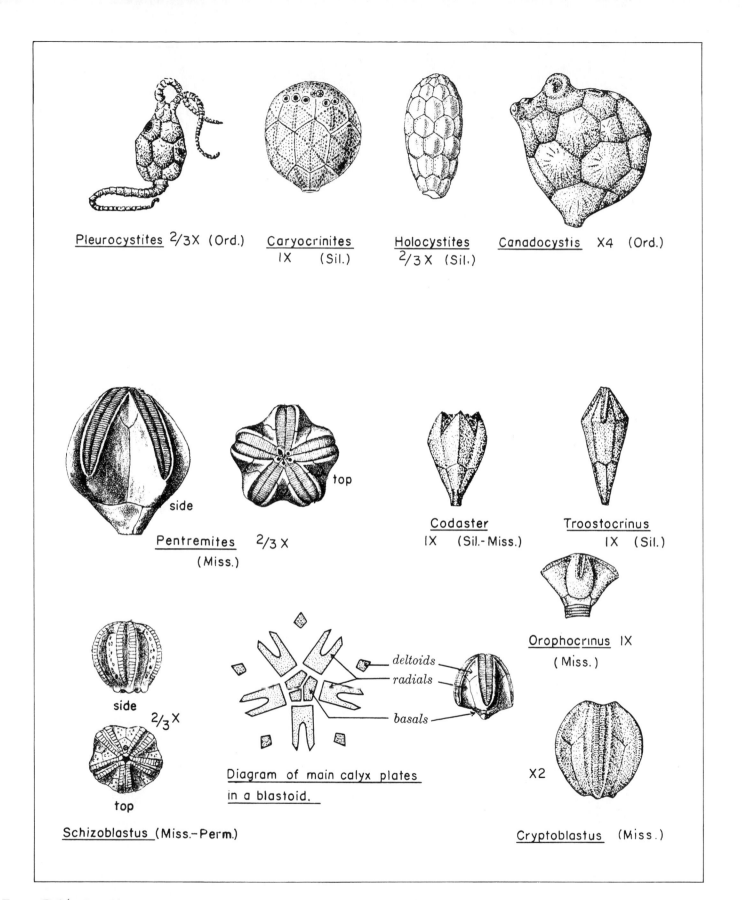

FIGURE D.14 Cystoids.

Source: Collinson, C. W., 1959, Guide for Beginning Fossil Hunters, *Illinois State Geological Survey, Educational Series 4, 39 p.*

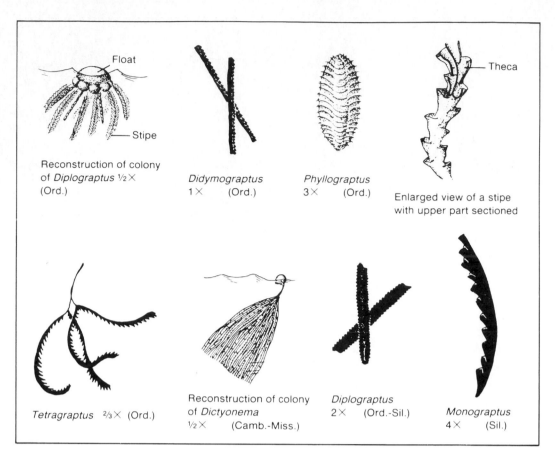

FIGURE D.15 Graptolites.

APPENDIX E

State, Province, and Territory Geological Surveys

Geologic surveys offer a wide variety of services to the reader. One important function of the survey is the publication of pamphlets, circulars, and bulletins about the geology and geologic history of the state, province, or territory. Most surveys have a catalog they will send upon request that lists the prices and availability of their publications. Typically included in the catalog are free publications that can be obtained. Make use of your local geologic survey!

UNITED STATES

AL   Geological Survey of Alabama, 420 Hackberry Lane, P.O. Box 0, Tuscaloosa, Alabama 35486, (205)349-2852 http://www.gsa.tuscaloosa.al.us

AK   Alaska Division of Geological & Geophysical Surveys, 794 University Avenue, Suite 200, Fairbanks, Alaska 99709-3645, (907)451-5000

AZ   Arizona Geological Survey, 416 West Congress, Suite 100, Tucson, Arizona 85701, (520)770-3500

AR   Arkansas Geological Commission, 3815 West Roosevelt Road, Little Rock, Arkansas 72204, (501)663-9714

CA   California Division of Mines and Geology, 801 K Street, Mailstop 1433, Sacramento, California 95814-3532, (916)445-5716 http://www.consrv.ca.gov/

CO   Colorado Geological Survey, 1313 Sherman Street, Room 715, Denver, Colorado 80203, (303)866-2611

CT   [Connecticut] Natural Resources Center, 79 Elm Street, Hartford, Connecticut 06106, (203)424-3555

DE   Delaware Geological Survey, DGS Building, University of Delaware, Newark, Delaware 19716-7501, (302)831-2833 http://www.udel.edu/dgs/dgs.html

FL   Florida Geological Survey, 903 West Tennessee Street, Tallahassee, Florida 32304-7700, (904)488-9380

GA   Georgia Geologic Survey, 19 Martin Luther King Jr. Drive Southwest, Room 400, Atlanta, Georgia 30334, (404)656-3214

HI   Hawaii Department of Land and Natural Resources, Division of Water and Land Development, Box 373, Honolulu, Hawaii, 96809, (808)587-0230

ID   Idaho Geological Survey, Morrill Hall, Room 332, University of Idaho, Moscow, Idaho 83844-3014, (208)885-7991 http://www.uidaho.edu/igs/igs.html

IL   Illinois State Geological Survey, Natural Resources Building, 615 East Peabody Drive, Champaign, Illinois 61820, (217)333-4747 http://www.isgs.uiuc.edu/isgshome.html

IN   Indiana Geological Survey, 611 North Walnut Grove, Bloomington, Indiana 47405, (812)855-7636 http://www.indiana.edu/~igs/index.html

IA   Iowa Geological Survey Bureau, Iowa Department of Natural Resources, 109 Trowbridge Hall, Iowa City, Iowa 52242-1319, (319)335-1575 http://www.igsb.uiowa.edu/

KS   Kansas Geological Survey, 1930 Constant Avenue, West Campus, University of Kansas, Lawrence, Kansas 66047, (913)864-3965 http://www.kgs.ukans.edu/

KY   Kentucky Geological Survey, 228 Mining and Mineral Resources Building, University of Kentucky, Lexington, Kentucky 40506-0107, (606)257-5500 Main Office; (606)257-3896 Publications http://www.uky.edu/KGS/home.htm

LA   Louisiana Geological Survey, University Station, Box G, Baton Rouge, Louisiana 70893, (504)388-5320 http://www.dnr.state.la.us

ME [Maine] Department of Conservation, Maine Geological Survey, 22 State House Station, Augusta, Maine 04333-0022, (207)287-2801

MD Maryland Geological Survey, 2300 St. Paul Street, Baltimore, Maryland 21218, (410) 554-5500
http://mgs.dnr.md.gov

MA [Massachusetts] Richard Foster, State Geologist, Executive Office of Environmental Affairs, 100 Cambridge Street, Boston, Massachusetts 02202, (617)727-9800 ext. 305 or (617)727-5830
http://www.magnet.state.ma.us/mepa

MI [Michigan] Department of Natural Resources, Geological Survey Division, P.O. Box 30256, Lansing, Michigan 48909-7756, (517)334-6907
http://www.deq.state.mi.us/gsd

MN Minnesota Geological Survey, 2642 University Avenue, St. Paul, Minnesota 55114-1057, (612)627-4780
http://www.geo.umn.edu/mgs/

MS Mississippi Office of Geology, P.O. Box 20307, Jackson, Mississippi 39289-1307, (601)961-5500

MO Missouri Division of Geology and Land Survey, Department of Natural Resources, Box 250, Rolla, Missouri 65401, (314)368-2100

MT Montana Bureau of Mines and Geology, Montana Tech of the University of Montana, 1300 West Park Street, Main Hall, Butte, Montana 59701-8997, (406)496-4174

NE Nebraska Geological Survey, Conservation and Survey Division, 113 Nebraska Hall, University of Nebraska-Lincoln, Lincoln, Nebraska 68588-0517 (402)472-3471
http://ianrwww.unl.edu/ianr/csd/index.htm

NV Nevada Bureau of Mines and Geology, University of Nevada, Mailstop 178, Reno, Nevada 89557-0088, (702)784-6691
http://www.nbmg.unr.edu

NH New Hampshire Geological Survey, Department of Environmental Services, Box 2008, 64 North Main Street, Concord, New Hampshire 03302, (603)271-3406 or (603)271-6482

NJ New Jersey Geological Survey, 29 Arctic Parkway, CN-427, Trenton, New Jersey 08625-0427, (609)292-1185
http://www.state.nj.us/dep/njgs/njgs.html

NM New Mexico Bureau of Mines & Mineral Resources, Campus Station, Socorro, New Mexico 87801, (505)835-5420
http://geoinfo.nmt.edu

NY New York State Geological Survey, Room 3140, Cultural Education Center, Empire State Plaza, Albany, New York 12230, (518)474-5816

NC North Carolina Geological Survey Section, 512 North Salisbury Street, Suite 527, P.O. Box 27687, Raleigh, North Carolina 27611-7687, (919)733-2423
http://www.ehnr.state.nc.us/EHNR/DLR/JEFF/rock1.htm

ND North Dakota Geological Survey, 600 East Boulevard Avenue, Bismarck, North Dakota 58505-0840, (701)328-9700

OH Ohio Department of Natural Resources, Division of Geological Survey, 4383 Fountain Square Drive, Building B, Columbus, Ohio 43224, (614)265-6576

OK Oklahoma Geological Survey, Energy Center, 100 East Boyd, Room N-131, Norman, Oklahoma 73019-0628, (405)325-3031
http://www.uoknor.edu/special/ogs-pttc

OR Oregon Department of Geology and Mineral Industries, 800 Northeast Oregon Street, No. 28, Suite 965, Portland, Oregon 97232, (503)731-4100
http://sarvis.dogami.state.or.us/homepage/

PA Pennsylvania Geological Survey, Department of Environmental Resources, P.O. Box 8453, Harrisburg, Pennsylvania 17105-8453, (717)787-2169

RI [Rhode Island] Office of the State Geologist, Department of Geology, The University of Rhode Island, Kingston, Rhode Island 02881-0807, (401)874-2265

SC South Carolina Geological Survey, 5 Geology Road, Columbia, South Carolina 29210-4089, (803)896-7708

SD South Dakota Geological Survey, Akeley Science Center, University of South Dakota, 414 East Clark Street, Vermillion, South Dakota 57069-2390, (605)677-5227
http://www.sdgs.usd.edu

TN Tennessee Division of Geology, Department of Environment and Conservation, L & C Tower, 13th Floor, 401 Church Street, Nashville, Tennessee 37243-0445, (615)532-1500

TX [Texas] Bureau of Economic Geology, The University of Texas at Austin, University Station, Box X, Austin, Texas 78713, (512)471-1534
http://www.utexas.edu/research/beg

UT Utah Geological Survey, 2363 South Foothill Drive, Salt Lake City, Utah 84109, (801)467-7970
http://utstdpwww.state.ut.us/~ugs

VT Vermont Geological Survey, 103 South Main Street, Center Building, Waterbury, Vermont 05671-0301, (802)241-3496
http://www.state.vt.us/anr/geology/vgshmpg.htm

VA   Virginia Division of Mineral Resources, P.O. Box 3667, Charlottesville, Virginia 22903, (804)293-5121

WA   Washington Department of Natural Resources, Division of Geology and Earth Resources, Mailstop 47007, Olympia, Washington 98504-7007, (360)902-1450

WV   West Virginia Geological and Economic Survey, P.O. Box 879, Morgantown, West Virginia 26507-0879, (304)594-2331

WI   Wisconsin Geological and Natural History Survey, 3817 Mineral Point Road, Madison, Wisconsin 53705, (608)262-1705 Main Office; (608)263-7389, Map Sales

WY   Wyoming State Geological Survey, P.O. Box 3008, University Station, Laramie, Wyoming 82071-3008, (307)766-2286
http://www_wwrc.uwyo.edu/wrds/wsgs/wsgs.html

CANADA

Alberta Geological Survey, 9945 108th Street, 6th Floor, North Petroleum Plaza, Edmonton, Alberta T5K 2G6 Canada, (403)422-1927

British Columbia Geological Survey Branch, Ministry of Energy, Mines and Petroleum Resources, 5th Floor, 1810 Blanshard Street, Victoria, British Columbia V84 1X4 Canada, (604)952-0429

Manitoba Energy and Mines, W. D. McRitchie, Geological Services Branch, Suite 360, 1395 Ellice Avenue, Winnipeg, Manitoba R3G 3P2 Canada, (204)945-6559

New Brunswick Minerals and Energy Division, Geological Surveys Branch, Department of Natural Resources and Energy, Box 6000, Fredericton, New Brunswick E3B 5H1 Canada, (506)453-2206

[Newfoundland] Department of Natural Resources Geological Survey, Box 8700, 95 Boneventure Avenue, St. John's, Newfoundland A1B 4J6 Canada, (709)729-2769 or (709)729-2301 or (709)729-3359

Northwest Territories Geology Division, Box 1500, Yellowknife, Northwest Territories X1A 2R3 Canada, (403)920-8210

Nova Scotia Department of Natural Resources, P.O. Box 698, Halifax, Nova Scotia B3J 2T9 Canada, (902)424-8633

Ontario Geological Survey, Ministry of Northern Development and Mines, Willet Green Miller Center, 933 Ramsey Lake Road, Sudbury, Ontario P3E 6B5 Canada, (705)670-5757

[Prince Edward Island] Energy and Mineral Section, Economic Development and Tourism, Carlottetown, Prince Edward Island C1A 7N8 Canada, (902)368-5010

Quebec Ministry of Natural Resources, 5700 4e Avenue Quest, Charlesbourg, Quebec G1H 6R1 Canada, (418)646-2727

Saskatchewan Geology and Mines Division, 1914 Hamilton Street, Regina, Saskatchewan S4P 4V4 Canada, (306)787-2476

[Yukon Territory] Exploration and Geological Services Division, 345-300 Main Street, Whitehorse, Yukon Territory Y1A 2B5 Canada, (403)667-3202

APPENDIX F
National Parks, National Monuments, and Other Places of Geologic Interest

This appendix lists reasonably well-known outcrop locations showing good exposures of rocks from the various geologic periods. The outline is by period and then by location (eastern margin, craton, western margin). Parenthetical notes after an entry indicate an important rock formation or geologic feature at that locality.

One of the very best places to observe the geology of North America is in our National Parks. Two outstanding guidebooks for reading about the National Parks are:

• *National Geographic's Guide to the National Parks of the United States* (1989, National Geographic Society, Washington, D.C., 432 pages) for tourist, camping, and general information.

• *Geology of National Parks* (Ann G. Harris and Esther Tuttle, 4th ed., Kendall/Hunt Publishing Co., Dubuque, Iowa, 652 pages) for the geologic features and geologic history of the parks.

Abbreviations

NP	=	National Park	NM =	National Monument
SP	=	State Park		

Hol	=	Holocene	Pm	=	Permian
Ple	=	Pleistocene	P	=	Pennsylvanian
Pli	=	Pliocene	M	=	Mississippian
Mio	=	Miocene	D	=	Devonian
Oli	=	Oligocene	S	=	Silurian
Eoc	=	Eocene	O	=	Ordovician
Pal	=	Paleocene	C	=	Cambrian
K	=	Cretaceous	PC	=	Precambrian
Jr	=	Jurassic	Pr	=	Proterozoic
Tr	=	Triassic	Ar	=	Archean

HOLOCENE
Eastern Margin
    FL
        Everglades NP (Hol Carbonate sands)
        Biscayne NP (Hol Carbonate sediments)
Western Margin
    WY   Yellowstone NP
        • Norris Geyser Basin; Old Faithful (Hol Geysers)
    AZ   Meteor Crater, near Winslow (Hol Meteorite impact)
    OR   Crater Lake NP (Hol Volcanic lavas, ash, cinder cones)
    CA   Lassen Volcanic NP (Hol Volcanic rocks)

PLEISTOCENE
Eastern Margin
    NY   Finger Lakes, western NY (Ple Glacial lakes)
    FL
        Everglades NP (Ple Miami Limestone)
        Biscayne NP (Ple Miami Limestone, Key Largo Limestone)
Craton
    WI   Kettle Moraine SP (Ple Tills)
Western Margin
    ID   Craters of the Moon NM (Ple Volcanoes)
    WY   Yellowstone NP (Ple Yellowstone Group)
    UT   Lake Bonneville, near Great Salt Lake
        • Bonneville Salt Flats (Ple Lake)
    WA Mount Rainier NP (Ple Volcanic lavas and ash)
    OR   Crater Lake NP (Ple Volcanics and glacial deposits)
    CA
        Lassen Volcanic NP (Ple Volcanic lavas, flows, pyroclastics)
        La Brea Tar Pits (Ple Tar)

PLIOCENE
Western Margin
- WY Grand Teton NP (Pli Teewinot Fm)
- AZ Pertrified Forest NP (Pli Bidahochi Fm)
- WA Olympic NP (Pli Marine sedimentary rocks)
- CA Lassen Volcanic NP (Pli Willow Lake Basalt)

MIOCENE
Eastern Margin
- FL Everglades NP (Mio Tamiami Fm)

Craton
- NE
  - Agate Fossil Beds NM (Mio Strata)
  - Court House Butte, Jail Rocks Butte, Smokestack Rock (Mio Arikaree Fm)

Western Margin
- CO Florissant Lake Basin, Florissant (Mio Strata)
- WA
  - Olympic NP (Mio Tectonic melange rocks)
  - Mount Rainier NP (Mio Fifes Peak Fm, Tatoosh Granodiorite)
  - Dry Falls SP
    - Columbia River Plateau (Mio Basalts)
- CA
  - Lighthouse Point, Conception (Mio Monterey Shale)
  - Channel Islands NP (Mio Volcanic rocks)
  - Death Valley NM (Mio Furnace Creek Fm)

OLIGOCENE
Craton
- SD Badlands NP (Oli Chadron Fm, Brule Clay, Sharps Fm)
- NE Scotts Bluff NM (Oli Brule Clay)

Western Margin
- CO Florissant Fossil Beds NM (Oli Tuffs)
- UT Zion NP (Oli Lava flows)
- NV Great Basin NP (Oli Volcanic lavas and ash)
- WA
  - Olympic NP (Oli Marine sedimentary rocks)
  - Mount Rainier NP (Oli Volcanic lavas and ash)
- CA Channel Islands NP (Oli Marine and nonmarine clastic sediments)

EOCENE
Craton
- TX White Bluff, Rio Grande (Eoc Midway Limestone)

Western Margin
- WY
  - Bitter Creek Valley (Eoc Green River Fm)
  - Yellowstone NP (Eoc Absaroka Volcanic Supergroup)
- UT Bryce Canyon NP
  - Bryce Amphitheater (Eoc Wasatch Fm)
- CA Channel Islands NP (Eoc Marine sedimentary rocks)

PALEOCENE
Western Margin
- ND Theodore Roosevelt NP (Pal Fort Union Group)
- WY Grand Teton NP (Pal Pinyon Conglomerate)
- C Little Red Deer River, north of Calgary, Alberta (Pal Paskapoo Sandstone)

CRETACEOUS
Craton
- AL Jones Bluff, Tombigbee River, Sumter Co. (K Selma Chalk)
- KS
  - Castle Rock, Gove Co. (K Niobrara Chalk)
  - The Sphinx (K Chalk deposits)
- TX
  - Big Bend NP (K Limestones)
  - Paluxy River, near Glen Rose, central TX (Lower C Glen Rose Limestone)

Western Margin
- CO
  - Front Range, Rocky Mountains, Boulder (K Dakota Sandstone)
  - Rocky Mountain NP (K Shales)
  - Mesa Verde NP (K Dakota Sandstone, Mancos Shale, Mesaverde Group)
- WY Rock Springs Dome (K Mesaverde Fm)
- UT
  - Bryce Canyon NP (K Dakota Sandstone, Tropic Shale)
  - Capitol Reef NP (K Dakota Sandstone, Mancos Shale, Mesaverde Group)
  - Arches NP (K Cedar Mountain Fm, Dakota Sandstone, Mancos Shale)
  - Zion NP (K Dakota Sandstone)
- WA North Cascades NP (K Intrusive igneous rocks and metamorphic rocks)
- CA Yosemite NP (K Intrusive igneous rocks)
  - El Capitan (K El Capitan Granite)

JURASSIC
Western Margin
- CO Dinosaur NM (Jr Morrison Fm)
- NM Navajo Church, northwestern NM (Jr Zuni Sandstone)
- UT
  - Zion NP
    - Checkerboard Mesa (Jr Navajo Sandstone)
    - The Great White Throne (Jr Navajo Sandstone)
  - Capitol Reef NP (Jr Morrison Fm)
  - Arches NP (Jr Entrada Sandstone, Morrison Fm)
    - Balanced Rock (Jr Entrada Sandstone)
  - Canyonlands NP (Jr Carmel Fm, Entrada Sandstone)

TRIASSIC

Eastern Margin
- CT
  - West Rock, New Haven (Tr Sandstones)
  - The Hanging Hills of Meriden (Tr Sediments and lava flows)
- NJ Palisades SP, along the Hudson River, Edgewater (Tr Basalts—Palisades Sill)

Western Margin
- AZ
  - Petrified Forest NP (Tr Moenkopi Fm, Chinle Fm)
  - Painted Desert (Tr Chinle Fm)
- UT
  - Zion NP (Tr Moenkopi Fm, Chinle Fm, Moenave Fm)
    - The Narrows (Tr Moenave Fm)
  - Capitol Reef NP (Tr Moenkopi Fm, Chinle Fm, Wingate Sandstone, Kayenta Fm)
    - Chimney Rock (Tr Moenkopi Fm, Shinarump Conglomerate)
  - Canyonlands NP (Tr Moenkopi Fm, Chinle Fm, Wingate Sandstone, Kayenta Fm)
  - Arches NP (Tr Moenkopi Fm, Chinle Fm, Wingate Sandstone, Kayenta Fm)
- WA North Cascades NP, WA (Tr Schists)

PERMIAN

Craton
- SD Wind Cave NP (Pm Minnekahta Limestone)

Western Margin
- CO
  - The Garden of the Gods, Colorado Springs (Pm Fountain Fm)
  - Red Rocks Ampitheatre, Denver (Pm Fountain Fm)
- TX Guadalupe Mountains NP
  - El Capitan (Pm Capitan Limestone)
- NM Carlsbad Caverns NP (Pm Goat Seep Dolomite, Capitan Limestone)
- AZ
  - Grand Canyon NP (Pm Coconino Sandstone, Toroweap Fm)
  - Canyon de Chelly NM (Pm Coconino Sandstone)
  - Walnut Canyon, near Flagstaff (Pm Coconino Sandstone)
- UT
  - Monument Valley
    - Mitten Butte (Pm De Chelly Sandstone)
  - Canyonlands NP (Pm Cutler Group)
    - The Land of Standing Rocks (Pm Cedar Mesa Sandstone)
    - Monument Basin (Pm Redbeds)
  - Natural Bridges NM (Pm Cedar Mesa Sandstone)
  - Capitol Reef NP (Pm Cutler Fm, Kaibab Limestone)
  - Arches NP (Pm Cutler Fm)

PENNSYLVANIAN

Eastern Margin
- KY Mammoth Cave NP (P Caseyville Fm)
- C Joggins, Nova Scotia (P Sandstones)

Craton
- SD Wind Cave NP (P Minnelusa Sandstone)

Western Margin
- CO Maroon Peak, southwest of Aspen (P Maroon Fm)
- AZ Grand Canyon NP (P Supai Group)
- UT
  - Canyonlands NP (P Hermosa Group)
  - Goosenecks, San Juan River, southeastern UT (P Strata)
- NV Carlin Canyon, Humboldt River, west of Elko (P Conglomerate)

MISSISSIPPIAN

Eastern Margin
- ME Acadia NP (M Granites)
- PA Second Mountain, near Harrisburg (M Pocono Sandstone)

Craton
- IN Bedford (M Salem limestone/Bedford Oolite)
- KY Mammoth Cave NP (M Limestones)
- IL Mississippi River bluffs, Alton (M Clastic limestones)
- AR Hot Springs NP (M Stanley Shale)
- SD Wind Cave NP (M Englewood Limestone, Pahasapa Limestone)

Western Margin
- C Banff NP, Alberta
  - Cascade Mountain (M Limestones)
  - Mt. Rundle (M Banff Fm, Rundle Fm)
- AZ Grand Canyon NP (M Redwall Limestone)
- CA Klamath Mountains, northern CA (M Bragdon Fm, Baird Fm)

DEVONIAN

Eastern Margin
- ME Acadia NP (D Diorite)
- NY
  - The Rocks, Conesville (D Oneonta Sandstone)
  - Letchworth SP, south of Rochester
    - Genesee River (Upper D Sandstones and shales)
  - Eighteen Mile Creek, western NY (D Shale and limestones)
  - Watkins Glen SP (Upper D Sandstones)
  - Riverside Quarry, Gilboa (D Strata)

Craton
- MI Thunder Bay Quarry, Alpena (D Alpena Limestone)
- AR Hot Springs NP (D Arkansas Novaculite)

Western Margin
- C Mt. Devon, Canadian Rockies, Alberta (D Limestones)
- AZ Grand Canyon NP (D Temple Butte Limestone)
- NV Toquima Range, central NV
  - Ikes Canyon (Lower D Limestone)

SILURIAN

Eastern Margin
- NY Niagara Falls, western NY (S Whirlpool Sandstone, Grimsby Sandstone, Rochester Shale, Lockport Dolomite)
- WV "Fluted Rocks," Great Cacapon River (S Limestones)

Craton
- C Georgian Bay Islands NP, Lake Huron, Ontario
  - The Flowerpots (S Lockport Dolomite)
- IL Thornton Reef Complex, Chicago (S Limestones)
  Des Plaines River, Joliet, southwest of Chicago (Lower S Edgewood Dolomite)
- AR Hot Springs NP (S Sandstone and shale)

ORDOVICIAN

Eastern Margin
- C St. Pauls' Inlet, west coast of Newfoundland (O Cow Head Brecccia)
- NY Trenton Falls, north of Utica, central NY (O Trenton Limestone)
  Deepkill Stream, north of Troy (Lower O Shales)

Craton
- IL Starved Rock, near La Salle (O St. Peter Sandstone)
- AR Hot Springs NP (O Marine, deep-water shales and chert)

Western Margin
- C Robson Peak, British Columbia (O Limestones)
- MT Bighorn Canyon (O Bighorn Dolomite)
- NM Franklin Mountains, southern NM (Upper O Limestones)
- NV Toquima Range, central NV
  - Willow Canyon (O Pillow lavas)

CAMBRIAN

Eastern Margin
- C Chapel Arm, Trinity Bay, eastern Newfoundland (Lower C Shales)
- NY Ausable Chasm, northeastern NY (Upper C Potsdam Sandstone)
- VA Shenandoah NP (C Chilhowee Group)

Craton
- MI Pictured Rocks (C Sandstones)
- WI Wisconsin Dells (Upper C Sandstones)
- SD Wind Cave NP (C Deadwood Sandstone)

Western Margin
- C Banff NP, Alberta
  - Mt. Assiniboine (C Limestone and sandstone)
  - Mt. Eisenhower (C Limestones and dolomites)
  - Lake Louise, Mt. Victoria (C Limestones and dolomites)
  Mt. Bosworth, Alberta (C Strata)

Yoho NP, near Field, British Columbia
  - Mt. Stephen; Mt. Wapta (C Burgess Shale)
  Mt. Robson, British Columbia (C Strata)
- CO Glenwood Canyon (Upper C Sandstones)
- AZ Grand Canyon NP (C Tapeats Sandstone, Bright Angel Shale, Muav Limestone)
- UT House Range, western UT
  - Notch Peak (C Limestone)
- NV Egan Range, south of Ely (C Limestone and dolomite)
  Great Basin NP (C Marine sedimentary rocks)

PRECAMBRIAN

Eastern Margin
- VA Shenandoah NP
  - Blue Ridge Mountains (Pr Igneous rocks)
- TN Great Smoky Mountains NP, NC/TN (Pr Ocoee Supergroup)

Craton
- C Kakabeka Falls, Ontario (Pr Huronian Supergroup)
  Sault Sainte Marie, Ontario (Pr Huronian Quartzite)
  Sudbury Basin, Sudbury, Ontario (PC Whitewater Group)
- MI Isle Royale NP, MI
  - Mott Island (Pr Keweenawan Series)
  Porcupine Mountains SP, MI
  - Lake of the Clouds (Pr Keweenawan lavas)
- WI Van Hise Rock, northwest of Baraboo (Pr Quartzite)
- MN Voyageurs NP (Ar Granites and lavas)
  Mesabi Iron Range, Virginia (Pr Iron formation)
  Soudan Iron Range, Soudan (Pr Iron formation)
  shores of Lake Superior, Duluth (Pr Keweenawan lavas)
- MO Silver Mine SP
  - St. Francois River (PC Igneous rocks)

Western Margin
- SD Mount Rushmore National Memorial, Black Hills (PC Igneous rocks)
- CO Rocky Mountain NP (Pr Basaltic dikes, granites, pegmatites)
  - Hallett Peak (PC Granites and gneisses)
  Black Canyon of the Gunnison NM (PC Schists)
- WY Grand Teton NP (PC Igneous rocks)
- MT Glacier NP
  - Mount Gould (Pr Belt Supergroup)
  - Chief Mountain (Pr Belt Supergroup)
  - Garden Wall (Pr Belt Supergroup)
- AZ Grand Canyon NP (Pr Unkar Group, Chuar Group)
  - Inner Gorge (Ar Vishnu Schist, Zoroaster Granite)

APPENDIX G
The Relative Geologic Time Scale

ERA/PERIOD/EPOCH	DERIVATION OF NAME	TYPE LOCALITY	REPRESENTATIVE U.S. LOCALITIES
CENOZOIC J. Phillips (1840)	"New" or "Recent" life		
QUATERNARY Paul G. Desnoyers (1829)	An addition to Arduino's threefold classification	northern France	
Holocene Portuguese Com. Inter. Geol. Congr. (1885)	(altogether recent)		Yellowstone NP, WY
Pleistocene Charles Lyell (1839)	*pleistos*, Greek "most" or "much" (most recent) 90–100%*	southern Italy	Kettle Moraine SP, WI
TERTIARY Giovanni Arduino (1760)	Third of Arduino's threefold classification of rocks	Apennine Mts. of northern Italy	
Pliocene Charles Lyell (1833)	*pleion*, Greek "more" (more or very recent) 50–90%*	northern Italy	Grand Teton NP, WY
Miocene Charles Lyell (1833)	*meion*, Greek "less" (middle or moderately recent) 20–40%*	western France	Agate Fossil Beds NM, NE
Oligocene Heinrich Ernst von Beyrich (1854)	*oligos*, Greek "little" (few or slightly recent) 10–15%*	Hanover Basin, northern Germany	Badlands NP, SD
Eocene Charles Lyell (1833)	*eos*, Greek "dawn" (dawn of recent) 1–5%*	Paris Basin Europe	Bryce Canyon NP, UT
Paleocene Wilhelm Philipp Schimper (1874)	*palaios*, Greek "ancient" (early dawn of recent) 0%*	northern France	Theodore Roosevelt NP, ND
MESOZOIC J. Phillips (1840)	"Middle" life		
CRETACEOUS J. J. d'Omalius d'Halloy (1822)	*creta*, Latin "chalk"	Margins of the British Channel	Big Bend NP, TX Mesa Verde NP, CO

*Percentage of remains of modern animals present in fossil faunas.

ERA/PERIOD/EPOCH	DERIVATION OF NAME	TYPE LOCALITY	REPRESENTATIVE U.S. LOCALITIES
JURASSIC Alexander von Humboldt (1799)	Jura Mountains (Alps), on the boundary of France and Switzerland	Jura Alps, Switzerland	Zion NP, UT Dinosaur NM, CO
TRIASSIC Friedrich von Alberti (1834)	Threefold division of rocks in Germany	Germany	Palisades SP, NJ Petrified Forest NP, AZ
PALEOZOIC Adam Sedgwick (1838) Rev. J. Phillips (1840)	"Ancient" life		
PERMIAN Roderick Impey Murchison (1841)	Perm, a Russian province	Province of Perm, Russia	Monument Valley, UT
(CARBONIFEROUS) Wm. D. Conybeare and William Phillips (1822)	Named for abundant coal-bearing strata in north central England	north central England	
PENNSYLVANIAN Henry Shaler Williams (1891)	Pennsylvania, the state	State of Pennsylvania	Canyonlands NP, UT
MISSISSIPPIAN Alexander Winchell (1869)	Mississippi, the river	Mississippi Valley	Mammoth Cave NP, KY
DEVONIAN Sedgwick and Murchison (1839)	Devonshire, England	Devonshire, England	Watkins Glen SP, NY
SILURIAN Roderick Impey Murchison (1835)	Silures, an ancient Celtic tribe that inhabited Wales	Border of Wales and England	Niagara Falls, NY
ORDOVICIAN Charles Lapworth (1879)	Ordovices, an ancient Celtic tribe that inhabited Wales	Wales and England	Trenton Falls, Utica, NY
CAMBRIAN Rev. Adam Sedgwick (1835)	*Cambria,* Latin name for Wales	Wales and England	Wisconsin Dells, WI

GLOSSARY

A

aa Rough, blocky lava flow.

ablation Loss of material from a glacier by melting, evaporation, or calving.

abrasion Grinding erosion by rocks entrained in glacial ice or by windblown sand.

absolute age Old name for *radiometric age.*

abyssal hills Low hills, several hundred meters high, found in the abyssal plains.

abyssal plains The flat areas of the deep ocean basins.

Acadian orogeny The orogeny in the Appalachian region caused by the continent–continent collision of North America and Africa during Devonian time.

accreted terrane A suspect terrane confirmed to be a crustal fragment that was transported and then added onto a continent.

acid rain Rainfall that is more acidic than typical precipitation, especially rainfall that is more acidic due to sulfur pollution in the atmosphere (forming sulfuric acid).

active margin A continental margin at which there is significant volcanic and earthquake activity; commonly, a convergent plate margin.

active volcano A volcano that is fresh-looking and has erupted within recent history.

aftershocks Smaller earthquakes that follow a major earthquake.

A horizon The topmost soil horizon, including topsoil; also known as the *zone of leaching.*

Algoman orogeny The orogeny that occurred on the Canadian Shield 2.5 billion years ago; characterized by the widespread production of granite.

alluvial fan A wedge-shaped sediment deposit left where a tributary flows into a more slowly flowing stream or where a mountain stream flows into a desert.

alluvium Stream-deposited sediment.

alpine glacier A small glacier found in a mountainous region; commonly, a valley glacier.

ammonites A group of coiled cephalopods; particularly useful as index fossils for Mesozoic-age marine strata.

amphiboles Hydrous ferromagnesian double-chain silicates.

amphibolite A metamorphic rock rich in amphiboles.

amphibolite facies A medium-grade, regional-metamorphic facies, commonly characterized by the production of abundant amphiboles.

andesite Volcanic rock intermediate in composition between basalt and rhyolite.

angiosperms The broad category of flowering plants.

angle of repose The steepest angle at which a slope of unconsolidated material is stable.

angular unconformity An unconformity at which the bedding of rocks above and below is oriented differently.

anion A negatively charged ion.

anticline An antiform in which the oldest rocks occur at the center of the fold, with progressively younger rocks found outward from the core.

antiform An arching fold in which the limbs dip away from the axis.

Antler orogeny The Devonian-age orogeny on North America's western margin; probably caused by the collision of an island arc with the margin.

aphanitic Rock texture in which grains are too fine to be distinguished readily with the naked eye.

Appalachian orogeny The Pennsylvanian-age orogeny in the Appalachian region; characterized by widespread thrust faulting.

aquiclude Rock that is effectively impermeable on a human timescale.

aquifer Rock sufficiently porous and permeable to be useful as a source of water.

aquitard Rock of low permeability, through which water flows very slowly.

arch A topographic high, often linear in shape, that has influenced sedimentation on and around it.

arête A sharp-spined ridge left by erosion by parallel valley glaciers to either side of the ridge.

artesian A system in which ground water in a confined aquifer is under extra hydrostatic pressure, so that the water can rise above the aquifer containing it.

artiodactyls The even-toed hoofed mammals; for example, pigs, hippopotamuses, deer, and cows.

ash Fine pyroclastic material.

assimilation The process by which magma incorporates and melts bits of country rock.

asthenosphere Weak, plastic, partly molten layer of the upper mantle directly below the lithosphere.

atom The smallest particle into which a chemical element can be subdivided.

atomic mass number The sum of the number of protons and the number of neutrons in a particular atomic nucleus.

atomic number The number of protons characteristic of a particular element.

aureole The contact-metamorphic zone around a pluton.

Australopithecus An early genus of hominids, best typified by *A. africanus.*

axial surface The plane dividing the two limbs of a fold.

axial trace The intersection of the axial plane of a fold with the land surface.

axis The line around which a fold is curved.

B

backwash The return flow of swash to the sea down the beach face.

banded iron formation A sedimentary rock formed only during the Precambrian, consisting of very thin, alternating layers of iron and chert.

bankfull stage The condition in which stream stage just equals stream bank elevation.

barchan dune A crescent-shaped transverse dune, with arms pointing downwind.

barrier island A long, low, narrow island parallel to a coast.

basalt Mafic volcanic rock; the volcanic equivalent of gabbro.

base flow Streamflow supported by ground water in adjacent rock or soil.

base level Ordinarily, the level of the water surface at a stream's mouth; the lowest level to which a stream can cut down.

basin A syncline in which rocks dip inward on all sides; also, any depression in which sediments are deposited.

Basin and Range A western geologic province, mostly in the state of Nevada, characterized by alternating linear mountains and valleys and formed by a pulse of divergent motion during the Middle Cenozoic.

batholith A massive, discordant pluton, often produced by multiple intrusions.

beach A gently sloping shore covered with sediment and washed by waves and tides.

beach face That portion of the beach exposed to direct wave and swash action.

bedding Depositional layering, as in sedimentary rocks.

bed load The amount of material moved by a stream along its bed, by rolling or by saltation.

bedrock geology The geology as it would appear with the overlying soil and vegetative cover stripped away.

Benioff zone The zone of earthquake foci dipping into the mantle away from a trench, resulting from subduction of lithosphere.

berm A flat or gently sloping zone behind a beach face.

B horizon The soil layer at intermediate depth; also known as the *zone of accumulation.*

Big Bang theory The currently accepted theory for the origin of the universe in which all matter "exploded" from a single point about 15–20 billion years ago.

biological sediments Sediments formed by the action of, or from the remains of, organisms.

biomass The total mass of all living organisms; also, fuel derived from modern organisms.

biospecies A group of individual organisms defined as a species on the basis of genetic compatibility.

blueschist facies A high-pressure metamorphic facies, not encountered under conditions of normal geothermal gradient and burial pressures.

body waves Seismic waves that travel through the earth's interior (P waves and S waves).

Bowen's Reaction Series The predicted sequence of crystallization of principal silicates from a magma.

braided stream A stream with multiple channels that divide and rejoin.

breccia Clastic sedimentary rock consisting of angular fragments in a finer-grained matrix.

breeder reactor A nuclear fission reactor designed to produce additional fissionable fuel while it operates to generate energy.

brittle Describes material that tends to rupture rather than deform under stress.

Burgess shale A Cambrian-age sedimentary rock unit found in British Columbia and noted for its exceptionally preserved fossils, particularly of soft-bodied animals.

C

caldera A large, bowl-shaped summit depression in a volcano; may be formed by explosion or collapse.

caliche A crusty, near-surface soil layer, cemented by calcite and other soluble salts; develops in arid climates.

calving The formation of icebergs as chunks of ice break off a glacier that terminates in water.

capacity The maximum total load a given stream can move.

capillary action The movement of water toward drier soil through fine pores in the soil and along the grain surfaces.

carbonate A nonsilicate mineral containing carbon and oxygen in the proportion of one atom of carbon to three atoms of oxygen (CO_3).

carbon film A fossil formed by a process in which volatiles in plant and animal tissues escape, leaving only traces of carbon.

Carboniferous The period of time represented by the Mississippian and the Pennsylvanian periods (360 to 386 million years ago).

Cascade Range The volcanic mountain chain in Washington and Oregon; Mount St. Helens is one member.

cast A replica of a fossil form created by the filling of a *mold.*

catastrophism A now-discredited theory that explained the earth's history as a static one, punctuated by global catastrophes that were the only agents of change.

cation A positively charged ion.

Catskill clastic wedge The clastic wedge formed by the Acadian orogeny in the Appalachian region during Devonian time.

cementation The process by which sediments are stuck together through the deposition of mineral material between grains.

Cenozoic The most recent era of geologic time, from sixty-six million years ago to the present.

chain reaction The sequence of fission events in which each fission event triggers further fission of atomic nuclei.

chain silicates Silicates in which silica tetrahedra are linked in one dimension by the sharing of oxygen atoms.

Channeled Scablands A geologic province in western Idaho, eastern Oregon, and eastern Washington; formed during the Pleistocene by extensive flooding and channeling of older flood basalts of the Columbia Plateau.

channelization Modification of a stream channel—for example, by straightening of meanders or dredging of the channel to deepen it.

chemical maturity A measure of the extent to which sediment has been depleted in soluble or easily weathered minerals.

chemical sediment Sediment precipitated directly from solution.

chemical weathering Solution or chemical breakdown of minerals by reaction with water, air, or dissolved substances.

chilled margin Fine-grained rock at the margin of a pluton; shows the effects of rapid cooling.

C horizon The deepest soil layer, consisting mainly of coarse chunks of bedrock.

cinder cone A volcanic structure built of pyroclastic materials.

cinders Pyroclastic material of intermediate size, with fragments ranging up to several centimeters across.

cirque A bowl-shaped depression formed at the head of an alpine glacier.

clastic Describes rock or sediment made of fragments of preexisting rocks and minerals.

clastic limestones Limestones composed of abundant crinoid columnals; particularly common on the North American craton during Mississippian time.

clay minerals A loosely defined group of aluminum silicates, with sheet-silicate structures, which form very fine crystals.

cleavage (mineral) The tendency of a mineral to break preferentially along planes in certain directions in the crystal structure.

cleavage (rock) The tendency of a rock to break along parallel planes, corresponding to planes along which platy minerals are aligned.

closed system A system that neither gains nor loses matter; an isolated system with respect to mass transfer.

coal Solid, carbon-rich fossil fuel formed from the remains of land plants.

coastline The zone at which land and water meet; also, the geometry of this zone.

Columbia Plateau A geographic province in western Idaho, eastern Oregon, and eastern Washington; formed by flood basalts of a rifting episode in the Middle Cenozoic.

column (geologic) Diagrammatic representation of vertical/chronologic sequence of rock units in a given region, shown in a column with the oldest unit at the bottom.

columnar jointing The development of polygonal columns in a lava flow during cooling.

compaction The compression and consolidation of sediment under compressive stress.

competence The largest size of particle a stream can move as bed load.

composite volcano A volcanic cone formed of interlayered lava flows and pyroclastics.

compound A chemical combination of two or more elements, in specific proportions, with a distinct set of physical properties.

compressive stress Stress tending to compress or squeeze an object.

concentration (enrichment) factor Enrichment of an ore in a metal of interest, relative to that metal's concentration in average crustal rock.

concordant Having contacts parallel to the structure in adjacent rocks.

cone of depression A conical depression of the water table or potentiometric surface caused by pumped extraction of ground water.

confined aquifer An aquifer overlain by an aquiclude or aquitard.

confining pressure The directionally uniform pressure to which rocks at depth are subjected.

conglomerate Clastic sedimentary rock consisting of rounded fragments in a finer-grained matrix.

contact metamorphism The metamorphism characteristic of wallrocks surrounding a pluton that are subjected to locally increased temperature only.

continental accretion The hypothesis that continents grew through time by material being accreted to the margins, probably by tectonic processes such as subduction.

continental divide A topographic high on a continent, such that water drains toward one side of the continent (or one ocean) on one side of the divide and toward another side (or ocean) on the other side of the divide.

continental drift The concept that the continents have shifted in position over the earth.

continental glacier A thick, extensive ice cap or ice sheet covering a significant portion of a continent.

continental rise The gently sloping region between the foot of the continental slope and the abyssal plains.

continental shelf The nearly level, shallowly submerged zone immediately offshore from a continent; water depths on the shelf are typically less than 100 meters.

continental slope The continental-marginal zone extending from the outer edge of the continental shelf down to the more gently sloping ocean depths (continental rise or abyssal plains).

continuous reaction series The crystallization series in which early crystals react with the melt without changes in mineralogy; the plagioclase branch of Bowen's Reaction Series.

contour interval The difference in elevation between successive contour lines on a map.

contour line A line joining points of equal elevation on a map.

convection cell A circulating mass of material (in air, water, or asthenosphere) in which warm material rises, moves laterally, cools, sinks, and is reheated, cycling back to rise again.

convergence (circulation) A zone in which opposing water currents meet.

convergent plate boundary A plate boundary at which two plates are moving together: a subduction zone or zone of continental collision.

coral atoll A ring-shaped coral reef structure, formed around an island that is now submerged or eroded away.

Cordilleran orogeny The general name for the Mesozoic and Early Cenozoic orogenies that formed the Rockies and other mountains in western North America; includes the Sonoman, Nevadan, Sevier, and Laramide orogenies.

core The innermost, iron-rich zone of the earth; the outer core is molten, the inner core solid.

correlation Determination that two or more distinct rock units are of the same age and/or are related in origin or history.

cosmic abundance curve A graph depicting the relative abundances of the chemical elements as a function of atomic number.

cotylosaurs The first reptiles.

country rock Rock into which a pluton is intruded; also known as wallrock.

covalent bond A bond formed by the sharing of electrons between atoms.

craton The stable continental interior.

creep (fault) A slow, gradual, more or less continuous slippage along a fault zone.

creep (mass movement) Very slow mass movement; not noticeable during direct observation.

crest (flood) The maximum stage reached during a flood event.

crest (wave) The highest point on a wave.

crevasses Deep, vertical cracks in brittle glacier ice.

cross-bedding A sequence of inclined sedimentary beds deposited by flowing wind or water.

cross section (geologic) An interpretation of geology and structure in the third (vertical) dimension commonly based on rock exposures and attitudes at the surface; drawn in a particular vertical plane.

crust The outermost compositional shell of the earth, 10 to 40 kilometers thick, consisting predominantly of relatively low-density silicates.

crystal form The external shape of crystals; distinguished from internal crystal structure.

crystalline Describes a solid having a regular, repeating geometric arrangement of atoms.

Curie temperature The temperature above which a magnetic material loses its magnetization; different for each such material.

cut bank Steep stream bank being eroded by lateral migration of meanders.

cyclothem A sequence of marine and nonmarine strata that formed on the North American craton during the Pennsylvanian; one of many such repetitive cycles.

D

daughter nucleus A nucleus produced by radioactive decay.

debris avalanche A flow involving a wide range of types and sizes of materials.

deep layer The deepest water in the oceans, with temperatures often near freezing; circulates very slowly.

deflation The removal of sediment by wind.

delta A sediment wedge deposited at a stream's mouth.

dendritic drainage An irregular, branching drainage pattern.

desert A region having so little vegetation that it is incapable of supporting a significant population.

desertification The rapid conversion of marginally habitable arid land into true desert, typically as accelerated by human activities.

desert pavement A surface of coarse rocks protecting finer sediment below; formed by the selective removal of fine surficial material.

de-watering The release of water from pores and/or by breakdown of hydrous minerals under conditions of increasing pressure or temperature.

diagenesis The set of processes by which lithification is accomplished; occurs at lower temperatures than does metamorphism.

dike A tabular, discordant pluton.

diorite A plutonic rock of intermediate composition, consisting of ferromagnesian minerals and feldspar, with little quartz and no olivine.

dip The angle made by a line or plane with the horizontal.

dip-slip fault A fault along which movement is vertical (parallel to the dip).

directed stress Stress that is not uniformly intense in all directions.

discharge The volume of water flowing past a point along a stream in a given period of time; equal to the flow velocity multiplied by the cross-sectional area of the channel.

disconformity An unconformity at which the bedding of rocks above and below is parallel.

discontinuous reaction series Crystallization sequence in which early crystals are transformed into new minerals by reaction with the melt, with abrupt changes in crystal structure; the ferromagnesian branch of Bowen's Reaction Series.

discordant Having contacts that cut across or are set at an angle to the structure of the adjacent rocks.

dissolved load The quantity of material carried in solution by a stream.

divergent plate boundary A plate boundary at which the plates are moving apart: an ocean spreading ridge or continental rift zone.

divide A topographic high separating two drainage basins.

dolomite A carbonate mineral, $CaMg(CO_3)_2$, or the chemical sedimentary rock made predominantly of that mineral.

dome An anticline dipping radially in all directions.

dormant volcano A volcano presently inactive but believed capable of future eruption.

downcutting The downward erosion by a stream toward its base level.

drainage basin The area from which a stream system draws its water.

drift Any glacially transported sediment.

drowned valley A stream or glacial valley partially flooded by seawater; occurs on a submergent coastline.

drumlins Elongated mounds of till oriented parallel to ice flow.

dune A low mound or ridge of sediment deposited by wind.

dynamic equilibrium The condition in which two opposing processes are in balance; for a stream, the condition in which erosion and deposition in the channel are equal.

dynamothermal metamorphism Regional metamorphism.

E

earthquake Ground displacement associated with the sudden release, in the form of seismic waves, of built-up stress in the lithosphere.

earthquake cycle The concept that there is a periodic quality about occurrence of major earthquakes on a given fault zone, with repeated cycles of stress buildup, rupture, and relaxation of stress through smaller aftershocks.

Ediacarian period A proposed name for the last part of the Precambrian; characterized by the appearance of metazoans in the fossil record.

elastic deformation Deformation in which strain is proportional to applied stress and the material returns to its original dimensions when the stress is removed.

elastic limit The stress beyond which material no longer behaves elastically.

elastic rebound The phenomenon whereby stressed rocks behave elastically before and after an earthquake, returning afterward to an undeformed, unstressed condition.

electron A negatively charged subatomic particle found outside the atomic nucleus.

element The simplest kind of chemical substance; elements cannot be decomposed further by chemical or physical means.

end moraine A ridge of till accumulated at the end of a glacier.

enhanced recovery Any of several methods used to increase the proportion of oil and/or gas extracted from a petroleum reservoir.

eolian Formed by or related to wind action.

epeiric seas Geographically widespread, shallow seas that covered large parts of the craton in the earth's past.

ephemeral stream A stream that flows only occasionally, in direct response to precipitation.

epicenter The point on the earth's surface directly above an earthquake's focus.

equilibrium line On a glacier, the boundary at which addition of material by precipitation just balances ablation loss.

era A major subdivision of the geologic time scale.

erratic An isolated large boulder not derived from local bedrock; a depositional feature of glaciers.

esker A winding ridge of till deposited by a stream flowing in and under a melting glacier.

estuary A coastal body of brackish water, open to the sea.

eukaryote A cell with a nucleus.

eustatic Describes the simultaneous worldwide rise or fall of sea level.

evaporite   A mineral deposit formed by the evaporation of water in a restricted basin; also, the minerals of such deposits.

evolution (biologic)   The change of a population or a species with time.

Exclusive Economic Zone   The territory extending 200 miles outward from a nation's shoreline, within which that country has the exclusive right to exploit marine resources.

exfoliation   The breakup of exposed plutonic rocks in concentric sheets due to the release of stress by unloading.

expansive clay   Clay that expands when wet, contracts as it dries out.

extinct volcano   A volcano expected never to erupt again.

extrusive   Igneous rock erupted onto the earth's surface.

F

facies (metamorphic)   The set of pressure-temperature conditions that leads to a particular distinctive metamorphic mineralogy or rock type.

facies (sedimentary)   The set of conditions that leads to the formation of a particular type of sediment or sedimentary rock; also, the rock or sediment so formed.

fall   A free-falling mass movement in which the moving mass is not always in contact with the land surface.

fault   A planar break in rock, along which there is movement of one side relative to the other.

fault breccia   Breccia formed by mechanical breakup of rocks during displacement along a fault zone.

feldspars   One group of framework silicates, containing aluminum and calcium, sodium, or potassium; collectively, the most abundant minerals in the crust.

felsic   Rock rich in feldspar and silica (quartz).

ferromagnesian   Silicate containing significant iron and/or magnesium.

firn   Dense, coarsely crystalline snow partially converted to ice.

fission   The process by which large atomic nuclei are split into smaller ones.

fissure eruption   An eruption of lava from a crack rather than from a pipelike vent.

flash flooding   A rapid rise of stream stage, common with ephemeral streams or after intense local precipitation events.

floodplain   The nearly flat area around a stream channel, into which the stream overflows during floods.

flood stage   The condition in which stream stage is above channel bank elevation, so that the stream overflows its banks.

flow   A mass movement of unconsolidated material in which the material moves in a chaotic or disorganized fashion, rather than as a coherent unit.

fluid injection   A proposed means of increasing pore fluid pressure and decreasing shear strength along locked faults to release built-up stress.

focus   The point of first break along a fault during an earthquake.

foliation   A texture, usually metamorphic, involving parallel alignment of linear or planar minerals, or compositional banding.

formation   A distinctive, mappable rock unit representing deposition under a uniform set of conditions and at one time.

fossil   The remains or evidence of ancient life.

fossil fuels   Any of the carbon-rich fuels produced from the remains of organisms through the action of heat and pressure over time.

fractional crystallization   The crystallization of magma with early crystals removed or isolated from later reaction with the remaining melt.

fracture   Irregular breakage of a crystal (or glass) without regard to particular planes in the crystal structure.

framework silicate   A silicate in which silica tetrahedra are linked in three dimensions by shared oxygen atoms.

frost wedging   The breakup of rock by the expansion of water freezing in cracks.

fumarole   A steam vent caused by subsurface water being heated by shallow magma or hot rock.

fusion   The process by which small atomic nuclei are combined to form larger ones.

G

gabbro   A mafic plutonic rock rich in ferromagnesians and plagioclase feldspar.

gasification   The conversion of coal to a gaseous fuel, such as methane.

geopressurized natural gas   Deeply buried regions in which natural gas is dissolved in pore waters under pressure.

geosyncline   A large syncline, of regional scale.

geothermal   Related to the heat of the earth's interior; use of geothermal power involves extraction of that heat through circulating subsurface water.

geothermal gradient   The rate of increase in temperature with depth in the earth.

geyser   A feature characterized by intermittent ejection of hot water and steam, heated by shallow magma or hot rocks.

glacier   A mass of ice, on land, that moves under its own weight.

glass   A solid lacking a regular crystal structure, in which atoms are randomly arranged.

gneiss   A metamorphic rock showing banded texture, usually defined by differences in mineralogy between bands.

Gondwanaland (or Gondwana)   A large continent of the Southern Hemisphere, formed by the five southern continents—Africa, Antarctica, Australia, India, and South America—and existing from the Early Paleozoic to the Mesozoic.

Gondwana rock succession   A generalized sequence of Permian- to Jurassic-age strata found on all five southern continents.

graben   A downdropped block bounded by steeply dipping faults.

graded bedding   Vertical progression of grain sizes within a sediment layer, from coarse to fine, or vice versa.

graded stream   A stream in dynamic equilibrium.

gradient   The steepness or slope of a stream channel along its length.

Grand Canyon   The famous canyon found in northwestern Arizona; formed by the Colorado River cutting into the Colorado Plateau.

granite   A plutonic rock rich in quartz and potassium feldspar.

granulite facies   The highest-grade regional-metamorphic facies.

gravity anomaly   A local or regional variation in the force of gravity as measured near the earth's surface.

greenhouse effect   Atmospheric heating resulting from the trapping of heat by carbon dioxide (CO_2) and other gases in the atmosphere.

greenschist facies   A low-grade regional-metamorphic facies, named for the common presence of the green minerals chlorite and epidote.

greenstone belts   Large regions of metamorphosed basalts of Precambrian age found on the shield.

groundmass   The finer-grained matrix of a porphyritic rock.

ground moraine A sheet of moraine left by a melting glacier.

ground water Water in the saturated zone, below the water table.

group A set of related sedimentary rock formations, usually having a common history.

gullying The formation by water of large erosional channels on a sloping soil surface.

Gunflint Iron Formation A Proterozoic-age formation famous for its fossils.

guyot A flat-topped seamount.

gymnosperm A general term for nonflowering plants that bear seeds; for example, conifers.

gyre A nearly closed, oval or circular water-circulation pattern.

H

half-life The length of time required for half of an initial quantity of a radioisotope to decay.

halide A nonsilicate containing a halogen element (Cl, F, Br, I).

hanging valley The valley of a tributary glacier, smaller than the main glacier valley and with a higher floor.

hardness The ability to resist scratching; measured on the Mohs scale of relative hardness.

hard water Water containing high concentrations of dissolved calcium, magnesium, and iron.

headward erosion The cutting back of a stream channel at its source.

height (wave) The difference in elevation between a wave's crest and trough.

hinge The most sharply curved part of a fold.

historical geology The study of the history of the earth and the evolution of its physical and biological features over time.

Hominidae The family of humans.

Homo The genus of humans.

horn A peak formed by headwall erosion by several alpine glaciers diverging from the same topographic high.

hornblende-hornfels facies A moderate-temperature contact-metamorphic facies characterized by abundant amphiboles.

hornfels A contact-metamorphic rock formed under conditions of low to intermediate temperature.

horst An uplifted block bounded by high-angle faults.

hot dry rock A potential geothermal resource; an area characterized by above-average heat flow but lacking abundant subsurface water.

hot spot An isolated area of active volcanism not associated with a plate boundary.

hot springs Springs heated by shallow magma bodies or young, hot rocks.

hydrograph A graph of stream stage or discharge as a function of time.

hydrologic cycle The cycle of precipitation, evaporation, infiltration, and migration of the water in the hydrosphere.

hydrosphere All water that is at and near the earth's surface and not chemically bound in rocks.

hydrostatic pressure Fluid pressure.

hydrothermal Literally, "hot water"; describes processes or ore deposits related to circulating subsurface water warmed by shallow magma or hot rock; hydrothermal fluids commonly contain dissolved minerals and gases.

hydrothermal vents Areas along spreading-ridge systems where waters heated by reaction with new lithosphere emerge into the colder ocean.

hydrous Containing water or hydroxyl (OH^-) ions.

hypothesis A conceptual model or explanation for a set of data, measurements, or observations.

I

ice age A period of very extensive continental glaciation; when capitalized ("Ice Age"), it refers to the last such episode, 2 million to 10,000 years ago.

ice wedging See *frost wedging*.

igneous Formed from or related to magma.

incised meanders Meanders cut deeply into rock, with little or no floodplain at channel level.

index fossil A fossil form useful in correlation; typically one that is widely distributed in area but narrowly restricted in time of occurrence.

index minerals Minerals stable over a restricted range of pressure and/or temperature conditions; useful in evaluating metamorphic grade attained.

inert Not tending to bond or form compounds with other elements.

infiltration The process by which water percolates into the ground.

insectivores A taxonomic order of mammals consisting of the most primitive placental mammals; for example, shrews.

intensity The size of an earthquake as measured by its effects on structures; one earthquake may have several intensities that decrease with increasing distance from the epicenter.

internal drainage Stream drainage into an enclosed, landlocked basin.

intrusion A body of igneous rock emplaced in preexisting rock; also, the process of such emplacement.

ion An electrically charged atom.

ionic bond A bond formed by electrostatic attraction between cations and anions.

island arc A line or arc of volcanic islands formed over, and parallel to, a subduction zone overlain by oceanic lithosphere.

isograd A line on a map connecting points of equal metamorphic grade, as determined by index minerals.

isostasy The tendency of crust and lithosphere to float at an elevation consistent with the density and thickness of the crustal rocks relative to the underlying mantle.

isostatic equilibrium The condition in which the mass of rock above a given level in the earth is everywhere the same.

isotopes Atoms of the same chemical element that differ in numbers of neutrons in the nucleus.

J

joint A planar break in rock without relative movement of rocks on either side of the break.

joint set A set of parallel joints.

K

karst topography Topography characterized by abundant sinkholes and other solution features.

kettle A hole in glacial outwash, formerly occupied by a block of stranded ice.

kimberlite An igneous rock, tending to occur in discordant pipelike plutons, originating up to 200 kilometers deep in the mantle, in which diamonds may be found.

knickpoint An abrupt change in streambed elevation—for example, at a waterfall.

L

labyrinthodonts The first amphibians; named for the complex structures on the surfaces of their teeth.

laccolith A concordant pluton with a flat bottom and domed country rock above.

lahar A volcanic mudflow deposit.

lamination Very fine or thin bedding.

landslide Any rapid mass movement; contrasted with *creep*.

Laramide orogeny The Early Cenozoic orogeny that affected the Rockies.

lateral moraine Moraine deposited at the sides of a valley glacier.

lateritic soil Extensively leached soil; characteristic of tropical climates.

Laurasia A large continent in the Northern Hemisphere that consisted of North America, Europe, Greenland, and Asia and existed from the Late Paleozoic to the Mesozoic.

lava Magma that flows out at the earth's surface.

law A basic concept or mathematical relationship that is invariably found to be true.

Law of Faunal Succession The concept that organisms, and thus fossil forms, change through time, each specific life-form corresponding to a unique period of earth history.

Law of Lateral Continuity The concept that when a sediment is laid down, it will laterally (1) become thinner and thinner, to zero thickness, (2) terminate abruptly against the basin of deposition, or (3) change into some other rock type (a facies change).

Law of Original Horizontality The concept that sedimentary rocks are generally deposited in horizontal layers, so that deviations from the horizontal reflect postdepositional disturbance.

Law of Superposition The concept that, in an undisturbed sedimentary section, the rocks on the bottom are the oldest, with the overlying rocks progressively younger toward the top of the sequence.

leaching The removal of soluble chemicals by infiltrating or percolating water.

levees Ridges along the bank of a stream; may be natural or artificial.

limbs The two sides of a fold, on either side of its axial plane.

limestone A carbonate-rich (especially calcite-rich) chemical sedimentary rock.

lineation Any linear structure in rock, including, but not limited to, parallel alignment of elongated minerals.

liquefaction (coal) The conversion of coal to a liquid hydrocarbon fuel, such as gasoline.

liquefaction (seismic) A quicksandlike condition with loss of soil strength; occurs when water-saturated soil is shaken by seismic waves.

lithification The conversion of sediment into rock.

lithofacies Literally, "rock facies"; rocks corresponding to a particular set of conditions of formation.

lithosphere The rigid outermost layer of the earth, 50 to 100 kilometers thick, encompassing the crust and uppermost mantle.

littoral drift Sand movement along the length of a beach, which occurs in the presence of longshore currents.

load The total amount of material moved by a stream.

locked fault An active fault where friction is preventing stress release through creep; no displacement is occurring.

loess Silt-sized sediment deposited by wind.

longitudinal dunes Dunes elongated parallel to the direction of wind flow.

longitudinal profile A diagram of the elevation of a stream bed along its length.

longshore current The net current parallel to a coastline, caused when waves approach the shore at an oblique angle.

lopolith A concordant pluton with a sagging floor that is concave upward.

low-velocity zone The zone within the upper mantle characterized by lower seismic velocities than those of layers immediately above and below it, a result of plastic behavior and/or partial melting of rocks in this zone.

luster The surface sheen exhibited by a mineral.

lycopsids The scale trees of Pennsylvanian time; represented today by the club mosses.

M

macroevolution The change of one line of plants or animals into another; involves large-scale changes—for example, one line of fish evolving into amphibians.

mafic A rock, magma, or mineral rich in iron and magnesium.

magma A silicate melt, usually containing dissolved volatiles and often crystals.

magma mixing The process by which two compositionally dissimilar magmas are combined into one.

magnitude The size of an earthquake as measured by ground displacement near the epicenter.

manganese nodules Lumps of manganese and iron oxides and hydroxides, along with other metals, found on the sea floor.

mantle The zone of the earth's interior between crust and core, rich in ferromagnesian silicates.

map unit A distinct, identifiable rock unit used in preparing a geologic map.

marble Metamorphosed limestone.

marsupials The pouched mammals; for example, the kangaroo and opossum.

mass extinctions Events in the fossil record characterized by the disappearance of many types of organisms within a geologically short time.

mass spectrometer An instrument used in radiometric dating to measure relative quantities of different isotopes.

mass wasting The downslope movement of material under the influence of gravity.

matrix The fine-grained sediment filling spaces between coarse grains in poorly sorted clastic sediment or rock.

meanders Lateral bends in a stream channel.

mechanical weathering The physical breakup of rocks, without change in their composition.

medial moraine Moraine formed by the joining of lateral moraines as tributary glaciers flow into a valley glacier.

member A subdivision of a formation.

Mesozoic The middle era of the Phanerozoic; the time from about 245 to 66 million years ago.

metamorphic grade A measure of the intensity of metamorphism to which a metamorphic rock was subjected.

metamorphism Literally, "change in form" of rocks, brought about particularly through the application of heat and pressure.

metasomatism The introduction of ions in solution into a rock, and the resulting alteration of that rock.

micas A group of sheet silicates characterized by the tendency to cleave well between sheets of silica tetrahedra.

Michigan Basin The structural basin underlying the state of Michigan.

microevolution Differences among different populations of the same species, or the changes an organism goes through during its life cycle.

migmatite "Mixed rock," partly melted during metamorphism, having a mix of igneous and metamorphic characteristics.

milling Erosion by water-borne sediment.

mineral   A naturally occurring, inorganic, solid element or compound, with a definite composition or compositional range and a regular internal crystal structure.

mineralogical maturity   The extent to which a sediment has been depleted in easily weathered minerals and enriched in resistant ones.

mineraloid   A material satisfying the definition of a mineral except that it lacks a regular internal crystal structure.

Mohorovičić discontinuity (Moho)   The boundary between crust and mantle, defined using seismic velocities.

mold   An impression of a fossil form preserved in rock or sediment.

moraine   A landform composed of till.

morphospecies   A definition of species applied to fossils—based on similarities in morphology rather than on genetic compatibility.

mouth   Where a stream reaches its base level or terminates.

mud cracks   Cracks in fine sediment caused by shrinkage during dehydration.

mudstone   A very fine-grained clastic sedimentary rock, such as siltstone or claystone.

N

native element   A mineral consisting of a single chemical element.

neap tides   The least extreme tides, which occur when the sun and moon are at right angles relative to the earth.

neck (volcanic)   A pipelike, discordant pluton.

neutron   An electrically neutral subatomic particle, generally found in an atomic nucleus.

Nevadan orogeny   The Jurassic-age orogeny along North America's western margin.

nonrenewable   Describes a resource that is not being produced at a rate comparable to that at which it is being consumed.

normal fault   A dip-slip fault in which the hanging wall moves down relative to the footwall.

nucleus   The central unit of an atom, containing protons and neutrons.

nuée ardente   A hot, glowing cloud of volcanic ash and gas, so dense that it flows down the volcano's slopes.

O

obduction   The process by which a segment of lithosphere is emplaced atop another.

oblique-slip fault   A fault involving both strike-slip and dip-slip movements.

obsidian   Massive volcanic glass.

oil shale   Sedimentary rock containing a waxy solid hydrocarbon, kerogen.

olivine   A ferromagnesian silicate with a structure consisting of individual silica tetrahedra.

oolites   Spheroidal carbonate grains formed in concentric layers.

ooze   A fine-grained, water-rich, siliceous or calcareous pelagic sediment of biogenic origin.

ophiolite   A complex assemblage of marine sediments, mafic and ultramafic rocks, found on a continent; generally believed to be a piece of obducted oceanic lithosphere.

order (stream)   The hierarchical rank of a stream based on the existence and complexity of tributaries: a first-order stream has no tributaries, a second-order stream has only first-order streams as tributaries, and so on.

ore   A rock in which a valuable or useful metal occurs in sufficient concentration to be economic to mine.

ornithischian dinosaurs   The order of dinosaurs that have a pelvic structure similar to that of modern birds.

orogenesis   The set of plutonic, metamorphic, and tectonic processes involved in mountain building.

orogeny   The period of time over which orogenesis occurs in a particular mountain range.

outgassing   The process by which gases were released to form the atmosphere and oceans as heavier elements sank to the core of the earth and lighter elements rose to the surface, during the differentiation of the earth in the early Archean.

outwash   Glacial sediment moved and redeposited by meltwater.

outwash plain   A flat sheet of outwash deposited by the collective action of many meltwater streams.

overbank deposits   Alluvium deposited outside a stream channel during flooding.

overturned fold   The extreme of a recumbent fold, in which the axial plane is tilted past the horizontal.

oxbows   Cut-off meanders.

oxide   A nonsilicate containing oxygen with one or more metals.

P

pahoehoe   A lava flow with a smooth, ropy appearance.

paleogeographic map   A map depicting a region's geography at some time in the geologic past, as indicated by the rock record.

paleomagnetism   "Fossil magnetism" preserved in rocks, reflecting the relative orientation of the earth's magnetic field at the time magnetization was acquired.

paleontology   The study of fossils.

Paleozoic   The oldest era in the Phanerozoic; the time from approximately 570 to 245 million years ago.

Pangaea   A single supercontinent that formed when Laurasia and Gondwanaland collided in the Late Paleozoic (approximately 200 million years ago).

parabolic dunes   Crescent-shaped dunes with arms pointing upwind.

parent nucleus (radioactive)   An unstable, decaying nucleus.

parent rock   The original rock from which a metamorphic rock was formed.

partial melting   The melting of only a portion of a rock.

passive margin   A geologically quiet continental margin, lacking significant volcanic or seismic activity.

peak lag time   The time lapse between a precipitation event and the peak stage or discharge of a flood event.

pedalfer   A moderately leached soil rich in iron and aluminum oxides and hydroxides.

pediment   A gently sloping bedrock surface at the foot of mountains bordering a desert.

pedocal   A soil retaining many of its soluble minerals, especially calcite.

pegmatite   A very coarse-grained igneous rock.

pelagic sediments   Fine-grained sediments of the open ocean.

period (wave)   The length of time between the passage of two successive wave crests or troughs by a fixed point.

periodic table   The regular arrangement of chemical elements that reflects patterns of chemical behavior related to the electronic structure of atoms.

perissodactyls   The odd-toed hoofed mammals; for example, horses and rhinoceroses.

permafrost   The permanently frozen soil found in many cold regions.

permeability   A measure of the ease with which fluids move through rocks or sediments.

permineralization   A fossilization process in which mineral matter is added to the pore spaces in the original organism; for example, mineral matter filling pore space in wood or bone.

phaneritic   Rock texture in which grains are coarse enough to be distinguished with the naked eye.

Phanerozoic   A major division (eon) of the geologic time scale, 570 million years ago to the present, spanning the time over which complex life-forms have been abundant.

phase change   A change in mineralogy, crystal structure, or physical state, with no gain or loss of chemical elements.

phenocryst   A coarse crystal in a porphyritic rock.

photovoltaic cell   A device for the direct conversion of sunlight to electricity; also known as a *solar cell*.

phreatic eruption   A volcanic steam explosion caused when magma heats subsurface water.

phreatic zone   The zone of saturation, in which pores in rock or soil are filled with water.

phyletic gradualism   The slow, gradual change of one species into another.

phyllite   A fine-grained metamorphic rock, formed by progressive metamorphism of slate, in which cleavage planes shine with light reflected from small mica flakes.

physical geology   That branch of geology concerned particularly with the materials of the earth and the physical processes that shape it.

pillow lava   A lava flow with a bulbous surface, made up of "pillows" or lobes with a glassy, quenched rind; formed from lava extruded under water.

pipe   See *neck.*

placer   A deposit of dense or resistant minerals concentrated by stream action.

plastic deformation   Behavior in which deformation is not proportional to applied stress and is permanent; the material stays deformed when stress is removed.

plateau (oceanic)   A broad topographic high, shallowly submerged or slightly exposed, within the ocean basins; often underlain by continental-type crust.

plate tectonics   The theory according to which the lithosphere is broken up into a series of rigid plates that can move over the earth's surface.

platform   The relatively undeformed Phanerozoic (Paleozoic, Mesozoic, and Cenozoic) strata that surround the shield.

playa   A "dry lake," floored by fine sediment, formed in a desert having internal drainage.

plucking   Glacial erosion caused as water freezes onto rock and the glacier then moves on, tearing away rock fragments.

plume   A rising column of magma in the asthenosphere.

plunging fold   A fold with a dipping (nonhorizontal) axis.

pluton   A body of plutonic rock.

plutonic   Describes igneous rock crystallized at depth.

pluvial lakes   Lakes formed through abundant rainfall during ice ages.

point bar   A sedimentary feature built in a stream channel, on the inside of a meander or anywhere the water slows.

polar-wander curve   A curve mapping past magnetic pole positions relative to a given region or continent.

polymorphs   Minerals having the same chemical composition but different crystal structures.

porosity   The proportion of void space (cracks, pores) in a rock.

porphyry   An igneous rock with coarse crystals in a fine-grained groundmass.

potentiometric surface   A theoretical surface indicating the elevation corresponding to hydrostatic pressure in a confined aquifer; analogous to the water table in an unconfined aquifer.

Precambrian   The eon spanning the time from the formation of the earth to the start of the Phanerozoic.

precursor phenomena   Detectable changes that occur prior to volcanic eruptions or earthquakes and might be used in prediction efforts.

prokaryote   A cell without a nucleus.

proton   A positively charged subatomic particle, generally found in an atomic nucleus.

pseudofossils   Sedimentary structures and other inorganic features that resemble fossils.

punctuated equilibrium   The rapid evolution of one species into another; usually occurs in geographically isolated areas away from the main geographic range of the species.

P waves   Compressional seismic body waves.

pyroclastic flow   A hot, fast-moving, denser-than-air flow composed of volcanic ash and other pyroclastics mixed with gas.

pyroclastics   Fragments of rock and lava emitted during an explosive volcanic eruption.

pyroxene-hornfels facies   A high-temperature contact-metamorphic facies, characterized by the production of abundant pyroxene.

pyroxenes   Single-chain silicates, mostly ferromagnesian.

Q

quartz   The simplest framework silicate, with the formula SiO_2.

quartz arenite   A sandstone consisting almost entirely of quartz sand grains.

quartzite   Metamorphosed quartz-rich sandstone.

Queenston clastic wedge   The clastic wedge formed by the Taconic orogeny in the Appalachian region during Ordovician time.

quick clay   An unstable, failure-prone clay sediment, derived from glacial rock flour deposited in a marine setting and weakened by later flushing with fresh water.

R

radial drainage   The drainage pattern resulting from streams radiating outward from a topographic high, such as a mountain.

radioactivity   The spontaneous decay or breakdown of unstable atomic nuclei.

radiometric age   A numerical date determined by the use of radioisotopes.

rain shadow   A dry zone landward of a coastal mountain range that is caused by loss of moisture from air passing over the mountains.

recessional moraine   An end moraine deposited by a retreating glacier during stationary periods.

recharge   The set of processes by which ground water is replenished.

recrystallization   The restructuring of a mineral into another form, often into coarser crystals.

rectangular drainage   A fracture-controlled drainage pattern in which streams make right-angle bends.

recumbent fold   A fold in which the axial plane is close to horizontal.

recurrence interval   The average length of time between floods of given severity on a given stream.

red beds   Nonmarine or marginal-marine sandstones with a low iron content, commonly of Permian or Triassic age.

refraction The deflection or change in direction of seismic body waves as they move across a boundary between two materials of different properties.

regional metamorphism Metamorphism on a large, or regional, scale; may be associated with mountain building; involves increases of both pressure and temperature.

regolith Surficial sediment deposit in place, analogous to soil but not capable of supporting plant life; generally lacks organic matter.

regression A long-term, seaward retreat of the shoreline.

relative dating Determining the sequence of rocks or events indicated by a particular rock section.

remote sensing Investigation or examination using light or other radiation rather than by direct contact; examples include the use of aerial photography, satellite imagery, or radar.

renewable Describes a resource that is capable of being replaced on a human timescale (for example, biomass fuels) or that is not significantly depleted by human use (for example, solar energy).

replacement A fossilization process in which a substance is replaced on an atom-by-atom basis by another substance; for example, "petrified" wood.

reserves The quantity of a mineral or fuel that has been located and can be exploited economically with existing technology.

reservoir rock Rock in which oil and gas deposits are found.

resources Reserves, plus that quantity of a useful mineral or fuel known or believed to exist but not exploitable economically with existing technology.

retention pond A basin used to hold water back from a stream temporarily after rain or a melting event, to reduce the risk of flooding.

reverse fault A dip-slip fault in which the hanging wall moves upward relative to the footwall.

rhyolite Silicic volcanic rock; the volcanic equivalent of granite.

rift valley A depression formed by grabens along the crest of a seafloor-spreading ridge or on a continent.

rill erosion Soil erosion on sloping land by water forming very small channels.

ripple marks The rippled surface formed on sediment by wind or water.

rock A solid, cohesive aggregate of one or more minerals or mineral materials.

rock cycle The concept that all rocks are continually subject to change and that any rock can be transformed through appropriate geologic processes into another type of rock.

rock flour Silt-sized sediment produced by glacial abrasion.

rupture Breakage or failure under stress.

S

saltation The process by which sediment is transported in a series of short jumps along the ground or stream bed.

salt dome A stocklike feature formed from salt rising from buried evaporite beds.

saltwater intrusion The replacement of fresh pore water by saline water as the fresh water is depleted.

San Andreas fault The Californian fault extending through Los Angeles north to San Francisco; a transform fault reflecting the opposite movement of the Pacific and North American Plates.

sandstone Clastic sedimentary rock made up of sand-sized particles.

sanidinite facies The highest-temperature contact-metamorphic facies.

saurischian dinosaurs The order of dinosaurs that have a pelvic structure similar to that of modern reptiles.

scale (map) The ratio of a unit of length on the map to the corresponding actual horizontal distance represented.

scarp A steep cliff resulting from vertical displacement along a fault; also, a similar feature formed by mass movement.

schist A medium- to coarse-grained metamorphic rock displaying schistosity.

schistosity The growth of coarse, platy minerals, especially micas, in parallel planes, due to directed stress.

scientific method The means of discovering scientific principles by formulating hypotheses, making predictions from them, and testing the predictions.

seafloor spreading The process through which plates diverge and new lithosphere is created at oceanic ridges.

seamounts Volcanic hills rising a kilometer or more above the sea floor.

section (geologic) The geologic column in a particular locality.

sediment An unconsolidated accumulation of rock and mineral grains and organic matter that has been transported and deposited by wind, water, or ice.

seismic gap A seismically quiet section of an active fault zone, where the fault is presumed to be locked.

seismic shadow An effect of the earth's liquid outer core, which blocks S waves and partially deflects P waves originating on one side of the earth from reaching the opposite side of the earth.

seismic tomography Method of investigating the earth's interior using lateral as well as vertical variations in seismic velocity.

seismic waves The form in which energy is released during earthquakes; divided into *body waves* and *surface waves*.

seismograph An instrument for detecting and measuring ground motion.

sensitive clay A weak, failure-prone clay sediment similar in behavior to quick clay but derived from other materials (for example, weathered volcanic ash).

sequence An informal, lithostratigraphic term referring to exceptionally large packages of sedimentary strata deposited during transgressions and regressions on the craton and the margins and bounded by widespread unconformities.

Sevier orogeny The orogeny that occurred on North America's western margin during Cretaceous time.

shale A clastic sedimentary rock made of clay-sized particles and having a tendency to break along parallel planes.

shearing stress Stress tending to cause different parts of an object to slide past each other across a plane.

shear strength The ability of a solid material to resist shearing stress.

sheet silicate A silicate in which tetrahedra are linked in two dimensions by shared oxygen atoms.

sheet wash Water flow over a sloping land surface, not confined to a channel.

shield A large, stable, continental region consisting of exposed Precambrian igneous and metamorphic rocks.

shield volcano A volcano with a low, flat, broad shape, formed by the buildup of many thin lava flows.

shock metamorphism Metamorphism characteristic of impact events, in which very high pressures are imposed abruptly and briefly; the extent of accompanying heating is variable.

shoreline The line along which the land and water surfaces meet.

silicate A mineral containing silicon and oxygen, with or without other elements.

silicic Rich in silica (SiO_2).

sill A tabular, concordant pluton.

sinkhole A circular depression formed by ground collapse into a solution cavity.

slate A low-grade, fine-grained, metamorphic rock exhibiting cleavage along parallel planes, due to alignment of clays, micas, and other sheet silicates.

slaty cleavage The rock cleavage characteristic of slate.

slide The movement of a coherent mass of rock or soil along a well-defined plane or surface.

slip face The downwind side of a dune; assumes a slope equal to the dune sediment's angle of repose.

slump A short-distance slide, commonly accompanied by a rotational movement of the slump block.

soil A surface accumulation of weathered rock and organic matter overlying the bedrock from which it formed; generally also defined as capable of supporting plant growth.

soil moisture Water in the vadose zone.

solar nebular hypothesis The hypothesis that the planets of the solar system coalesced from a rotating cloud of gas and dust (the solar nebula) about five billion years ago.

solid solution The phenomenon of substitution of one element for another in a mineral, within some compositional limits; also, a mineral in which this occurs.

solifluction A flow in wet soil above the permafrost layer, in alpine terrain.

Sonoman orogeny The orogeny that occurred on North America's western margin during Triassic time.

sorting The separation of minerals in a sediment by grain size; also, a measure of the extent to which this has occurred.

source The point at which a stream originates.

specific gravity The density of a mineral divided by the density of water.

sphenopsids A group of plants that have long, unbranched, ribbed stems; modern examples include horsetails and scouring rushes.

spheroidal weathering The chemical weathering of a rock into a spheroidal shape, in a series of concentric layers.

spoil banks Piles of waste rock and soil displaced during strip-mining.

spring A site where the water table intersects the ground surface so that water flows out at the surface.

spring tides The most extreme tides, which occur when the sun, moon, and earth are aligned.

stage The elevation of the water surface of a stream at a given point along the channel.

stock A pluton similar to, but smaller than, a batholith.

strain Deformation resulting from stress.

stratovolcano See *composite volcano*.

stratum A visually distinct layer in sedimentary rock that corresponds to a particular period of deposition (plural, *strata*).

streak The color of a mineral when powdered.

stream Any body of flowing water confined within a channel.

stream piracy The process by which a stream undergoing headward erosion cuts through a divide and begins to drain part of an adjacent drainage basin.

stress Force applied to an object.

striations (glacial) The parallel grooves cut in rock by rock fragments frozen into glacial ice.

strike The compass orientation of a line or plane as measured in the horizontal plane.

strike-slip fault A fault along which movement is horizontal only (parallel to strike).

strip-mining Mining by stripping off overlying rock, soil, and vegetation to expose the desired mineral or fuel; used for shallow, tabular bodies, especially coal beds.

stromatolite An organosedimentary structure formed by algae.

structural province A large region of rocks of Precambrian age that have similar radiometric ages.

subduction zone A convergent plate boundary at which a slab of oceanic lithosphere is being pushed beneath another plate (continental or oceanic) and carried down into the mantle.

submarine canyon A V-shaped canyon cut in the continental slope.

subsurface water Any water below the ground surface.

sulfate A nonsilicate mineral containing sulfur and oxygen in sulfate (SO_4) groups.

sulfide A nonsilicate mineral containing sulfur but lacking oxygen.

surface waves Seismic waves that travel along the earth's surface.

surf zone The region between the most seaward breakers and the farthest landward advance of swash.

surge A localized increase in the water level of an ocean or large lake, commonly associated with storms.

suspect terrane A region or set of rocks apparently unrelated to the adjacent regions in geology or history.

suspended load The quantity of material moved in suspension by a stream.

suture The zone along which two continental landmasses become joined.

swash The rush of water up the beach face following the breaking of waves.

S waves Seismic body waves characterized by shearing motion or displacement.

syncline A synform in which the youngest rocks are found in the center of the fold, with progressively older rocks exposed away from the center.

synform A trough-shaped fold in which the limbs dip toward the axis.

T

Taconic orogeny The orogeny in the Appalachian region during Ordovician time caused by the subduction of the proto-Atlantic plate due to the convergence of North America and Africa.

tailings Crushed rock waste from ore/mineral processing.

talus Coarsely broken rock debris from rockfalls or slides.

taphonomy The study of fossilization processes.

tarn A lake occupying a cirque.

tar sand Sand or sandstone containing a deposit of viscous, asphaltic petroleum.

taxonomy The process of classifying organisms according to some scheme.

tectonic Relating to large-scale movement and deformation of the earth's crust; *tectonics* is the study of tectonic phenomena.

tensile stress Stress tending to pull an object apart.

terminal moraine The end moraine marking the farthest advance of a glacier.

terraces (stream) Steplike plateaus surrounding a stream's present floodplain.

terrane A region or group of rocks with similar geology, age, or deformational style.

terrigenous sediment Clastic sediment derived from the continents.

theory A hypothesis that has been tested sufficiently against further observations or experiments that it has gained general acceptance.

thermocline A layer in the oceans of rapidly decreasing temperature with increasing depth, extending from the surface mixed layer to 500 to 1000 meters in depth.

thin section A thin slice of rock, mounted on a glass slide; typically so thin that light passes readily through most of the minerals in it.

thrust fault A reverse fault with a shallowly dipping fault plane.

tides The slow rise and fall of water level at a point, as the earth rotates through bulges of water caused by the earth's rotation and the gravitational pull of the moon.

till Glacial sediment deposited directly by melting ice.

tillite A rock formed from till.

trace (fault) The line of intersection of a fault plane with the earth's surface.

trace fossils Fossils in which evidences of organisms, rather than the organisms themselves, are preserved; for example, dinosaur tracks.

transform fault A strike-slip fault between offset segments of an oceanic spreading ridge.

transgression Landward encroachment of the sea.

transition zone The zone of the upper mantle below the low-velocity layer, characterized by several abrupt (though small) increases in seismic-wave velocities; extends to a depth of about 700 kilometers.

transverse dunes Dunes elongated perpendicular to the direction of wind flow.

traps Sites of localization or concentration of migrating oil and/or natural gas.

trellis drainage A rectilinear drainage pattern in which tributaries join the main stream at right angles.

trench An elongated, steep-walled valley on the sea floor, characteristic of and parallel to a subduction zone.

tributary stream A stream that flows into a larger stream.

trough The lowest point of a wave.

tsunami A seismic sea wave set off by a major earthquake in or near an ocean basin.

tuff Rock formed of consolidated volcanic ash.

turbidity Cloudiness of water, caused by suspended sediment.

turbidity current A density current of sediment-laden water that flows along the ocean bottom.

U

ultramafic Igneous rock extremely rich in ferromagnesians and poor in silica.

unconfined aquifer An aquifer overlain by permeable rocks or soil.

unconformity A surface within a sedimentary sequence at which there has been a period of nondeposition or erosion.

ungulates The broad category of plant-eating, hoofed mammals.

uniformitarianism The theory that the same physical laws have always operated and that, by observing present processes, we can understand the past history and development of the earth.

upconing A saltwater intrusion in a conical pattern below (and in reflection of) a cone of depression in an overlying freshwater lens.

upwelling The rising of deep, cold waters to shallower depths in response to reduced surface pressures.

V

vadose zone An unsaturated zone below the ground surface, in which pores are filled partly with water, partly with air.

varve A sediment couplet representing one year's deposition in a glacial lake.

ventifact A rock sculpted by wind abrasion.

vesicles Gas bubbles in igneous rock.

volcanic Pertaining to volcanoes; also, describes an igneous rock crystallized at or near the earth's surface, as from lava.

volcanic breccia Rock formed of a mix of ash, cinders, and coarser, angular, volcanic blocks.

volcanic dome A compact, steep-sided volcanic structure formed of very viscous lava.

volcano A vent from which magma, gas, and ash are erupted; also, the usually conical structure built by such eruption.

W

wallrock See *country rock*.

water table The top of the saturated (phreatic) zone.

wave (water) Periodic undulatory motion of the water surface.

wave base Depth at which the movement of water molecules in waves becomes negligible.

wave-cut bench A flat, erosional surface in rock, cut at water level by wave action and exposed by coastline emergence.

wavelength The horizontal distance between two successive wave crests or troughs.

wave refraction Deflection (change in direction) of a wave; for water waves, caused by variable water depth near shore.

weathering The set of physical, chemical, and biological processes by which rock is broken down in place.

X

xenolith A rock caught up in a magma as an inclusion.

Z

zeolite facies The lowest-temperature contact-metamorphic facies.

INDEX

A

aa lava, 53, 54
ablation, 320
abrasion, 305, 322
absolute ages, 125–26
abyssal hills, 206
abyssal plains, 206, 208
Acadian orogeny, 424, 425
accreted terranes, 201–2, 446, 447
accretion, continental, 372, 373
acid rain
    groundwater pollution from, 276, 277
    and weathering, 72–73, 74
active margins, 197–98, 209–10, 211
active volcanoes, 61, 62
adaptive radiation, 353, 354
advancing glaciers, 321
aftershocks, 165
Agassiz, Louis, 332
Age of Crinoids, 430
Age of Invertebrates, 354
Age of Mammals, 354, 484
Age of Reptiles, 354, 452
Ager, Derek, 347
agnathids, 411, 428
agriculture
    desertification and, 313, 314
    greenhouse effect and, 334
    water use, 280
A horizon, 77, 78
Alaska earthquake (1964), 156, 164, 166, 167,
       173, 293
albite
    crystallization, 37
    melting, 34, 35, 36, 37
    phase diagram, 36
algae
    Precambrian, 370, 387, 389
    as source of carbonate sediment, 402, 403
Algoman orogeny, 380
Allegheny orogeny, 424
Allosaurus, 457, 460, 462
alluvial fans, 213, 251, 312
alluvium, 249
alpha particles, 124
alpine glaciers, 321
aluminum
    as mineral resource, 496, 498
    recycling, 506
Alvarez, Luis, 465
amethyst, 22

amino acids, 388
ammonia, in solar system formation, 367
ammonites, 452, 454, 455
amniotic eggs, 432
amphibians, 429, 432
amphiboles, 27
    crystallization, 37
    metamorphism and, 105, 107
amphibolite, 111, 112
amphibolite facies, 114
Anapsida, 456
Ancestral Rockies, 427
andesite, 40, 42
angiosperms, 464, 487
angle of repose, 285, 286
angular unconformity, 120, 121
anhydrites, 404
animals. *See also* fauna
    behavior, as precursor phenomenon, 168–69
    diversity through time, plot of, 356
    invertebrates, Cenozoic, 484
    mammals. (*see* mammals)
    vertebrates, Cenozoic, 484–86
anion, 15
Ankylosaurus, 459, 461
annual rainfall, 254
anorthite, 37
anthracite coal, 511
anticlines, 186, 187, 396
antiform, 186
Antler Belt, 426
Antler orogeny, 408, 426
Antler Peak (Nevada), 426
Apatosaurus, 458, 461, 462–63
aphanitic texture, 41
Apollo missions, 369
Appalachian Mountains (North America)
    Cenozoic erosion, 476
    development, 435
    folded structure of, 197
    formation, 406, 407
    influence in development of geologic
        thought, 435
    Paleozoic, 406, 407, 423, 424
    as result of continent-continent collision, 199
Appalachian orogeny, 426
apse, 456
aquamarine, 497
aquiclude, 267
aquifers, 267–68, 271
aquitard, 267
aragonite, 344

archaeocyathids, 410, 411
Archaeopteryx, 455, 457
Archean eon
    boundary with Proterozoic eon, 380
    compared to Proterozoic, 383
    continents, growth of, 372, 373
    crustal formation, 201, 372
    environment, 375
    fossils, 390
    in geologic time scale, 130, 131, 370
    plate tectonics in, 380
    provinces, 374
    rocks, 375, 378–80
arches, 395–96
architecture, earthquakes and, 165–66
archosaurians, 457
arête, 325
argon-40, 126–27
arid lands. *See also* deserts
    defined, 313
    map of world's, 305
Aristotle, 341
arkoses, 427, 443
arsenic, 507
artesian system, 268
arthropods, 411
artifacts, 341
artiodactyls, 486, 487
asbestos, 499
ash, volcanic
    characteristics, 43
    hazards from, 57–59, 66
assimilation by magma, 37–38
asteroid impacts, 354, 356–57, 392, 465–66. *See*
    also meteorites
asthenosphere
    characteristics, 136–37
    lithosphere/asthenosphere boundary, 176
    seismic wave studies, deduced from, 175
asymmetric ripple marks, 93–94
Atlantic basin, 210
Atlantic Ocean, formation, 149
Atlantic passive continental margin, 444
atmosphere
    components, 373
    Cretaceous extinctions and changes in, 467
    origin, 373, 375
atmospheric circulation
    in ocean circulation, 219–21
    as origin of wind, 304–5
atolls, 214, 215
atomic mass number, 14

atomic number, 14
atoms, 14
augite, 27
aureole, contact, 112
Australopithecus, 488–89, 490
avalanches
    debris, 289, 293
    protection, 298
    rate of movement, 293
    recognizing previous areas of, 293
    snow, 292, 293, 298
axial surface, 186
axis, fold, 186

B

backwash, 229
Bacon, Francis, 135
bacteria, Precambrian, 370, 387, 389
bajada, 312
Bakker, Robert, 457
banded iron formations, 384, 386
banded ironstones, 90
Bandelier Tuff (New Mexico), 43
bankfull stage, 252
barchan dunes, 308, 332
barrels, oil, 510
barrier islands, 235, 236–38
barrier reefs
    Great Barrier Reef, 214, 228
    Paleozoic, 413
basalt
    Archean, 378
    Cenozoic, 481
    Deccan flood basalts, 466, 467
    formation, 39, 42
    gas bubbles trapped in, 34
    in Gulf of Mexico, 448
    in oceanic crust, 141, 142, 370, 372
    in ophiolites, 216, 217
    Proterozoic, 392
    spheroidal weathering, 77
    use of term, 42
base, wave, 225
base flow, 266
base level, 251
basement rocks, 375
Basin and Range (Nevada), 478, 480–81
basins
    defined, 187, 396
    Michigan Basin, 404–5, 419
    Williston Basin, 419
batholiths
    defined, 45
    granitic, 46
    in Nevadan orogeny, 447
bauxite, 498
beaches
    as clastic sedimentary environment, 95
    defined, 227
    profile, 228
    replenishment, 237
    sediment transport, 229–31
beach face, 228
Beagle, HMS, 351
Beartooth Mountains (Montana), 324
Becquerel, Henri, 124
bedding, sedimentary rock structures related
    to, 91–93
bedding plane, 91
bed load, 248
beetles, used to determine glacial positions, 475
belts, mountain, 194
Benioff zone, 210, 211
benthic environments, 348
benthonic organisms, 349

bentonite, 290
berm, 228
beryl, 497
beryllium, 497
beta particles, 124
B horizon, 78
Big Bang theory, 365
biochemistry, 351
biological sediments, 90
biomass as energy source, 524
biospecies, 349
biotite, 27
biotite mica, 37
birds
    classification, 457
    evolution, 455
    Mesozoic, 460
bituminous coal, 511
bivalvia, evolution, 411
blastoids, 421, 434
block faults, 189
blocks, volcanic, 43
blueschist facies, 115
body waves
    interior of earth, investigating, 174–78
    locating epicenters with, 160–61
    as precursor phenomenon, 168
    types, 160
bombs, 54
bonding in minerals, 15, 17
bones, dinosaur, 455, 461
borings, 345
boron, 497
Bowen, Norman L., 36
Bowen's Reaction Series
    chemical weathering and, 75
    defined, 36–37
brachiopods
    fossils, 342
    Paleozoic, 403, 411, 412, 427, 430
    Permian extinction, 434
Brachiosaurus, 458
brackish water, 232
braided stream, 250
breakers, 225, 226
breccia, 87
breeder reactor, 517
Bright Angel shale, 408
brines, 404
brittle materials, 155, 157
Brontosaurus, 458
bryozoans
    fossils, 342
    Paleozoic, 411, 412, 427
Buffon, Georges L. L. de, 123
buildings, earthquakes and, 165–66
Burgess shale, 411
burrows, 345

C

cadmium, 507
calcareous reefs, 218
calcite. *See also* calcium carbonate
    as carbonate, 28
    cementation of, 91
    chemical weathering, 70, 74
    dissolution in oceans, 218
    in fossilization process, 344
    in limestone, 89
    metamorphism, breakdown during, 105, 106
    properties, 23, 24, 218
    sedimentation, 95
    stalagmite/stalactite formation, 264
    structure, 21, 22

calcium, 276
calcium carbonate. *See also* calcite
    in desert soils, 313
    oolite formation from, 94, 95
    in pedocal soils, 79
    reefs, 95, 404, 447
    sedimentation, 95
    water temperature variations, tracking, 332
Caledonian orogeny, 425
caliche, 79
calving, 320
Camarasaurus, 462
Cambrian period, 399
    eastern margin of North America, 406
    in geologic time scale, 128, 129, 130
    life in, 409–11
    North American craton during, 401
    rocks, 394, 414–15
    start of, 390, 395
    time, 400
    western margin of North America, 408
Camptosaurus, 459
Canadian Shield
    Algoman orogeny, 380
    geologic cross section, 383
    glaciation, 386
    location, 375
    structural provinces, 375, 376, 377
capacity, stream, 248
capillary action, 79
Capitan Reef (Texas), 427
capture hypothesis, 369
carbon
    Cenozoic deposits, 466
    covalent bonding to form diamond, 17
    in diamond and graphite, 20–21
    ion formation, 15
    isotopes, 14–15
    in oil, 508
    Precambrian, 387–88
carbon-14, 124, 127
carbonate compensation depth, 218
carbonate minerals
    characteristics, 28
    effervescence of, 24
carbonate platforms, 95
carbonates
    biological contributions to, 90
    hard water and, 276
    metamorphic reactions in, 106
    of Michigan Basin, 404, 405
    Paleozoic, 401, 404, 405, 414–15, 421
    sedimentation, 95
    sources of, 402
    in stromatolites, 387
carbonate sand, 413
carbon dioxide
    Cretaceous extinctions and changes in, 467
    cycle, 10
    in early atmosphere, 373
    gas, in greenhouse effect, 334–37
carbon films, 344, 345
Carboniferous period
    coal formation, 430, 431–32, 511
    in geologic time scale, 129, 130, 419
Carlsbad Caverns (New Mexico), 264
Cascade orogeny, 481
Cascade Range (North America), 49, 62, 63, 481
cassiterite, 498
casts in fossilization, 345
catastrophism, 4–5
cation, 15
Catskill clastic wedge, 425
cave formation, 269
cementation, 91
cenotes, 466

crest
  stream, 254
  wave, 225
Cretaceous period
  dinosaurs. (*see* dinosaurs)
  extinctions, 354, 355, 464–67
  in geologic time scale, 128, 129, 130
  Laramide orogeny, 445
  life, 464–67
  North American craton, 442–43
  passive continental margin formation, 444
  Sevier orogeny, 445
crevasses, 320
crinoids, 421
  Age of Crinoids, 430
  Permian extinction, 434
cross-bedding, 92, 93
crossopterygians, 429
cross sections, 395
crust
  composition, 7, 19, 24, 33, 495
  formation, 7, 370, 372
  in lithosphere, 137
  mantle/crust boundary, 174–75
  oceanic. (*see* oceanic crust)
  origin of, 370, 372
  thickness of, measuring, 175
crustal roots, 194
Cryptozoic eon, 370
crystal form in minerals, 22
crystalline materials, 19
crystallization of magmas, 35, 36–40
crystals, 19–20
cumulative resources, 495
Curie temperature, 140, 178
currents
  coastal, 229–31
  ocean, 219–21
cut bank, 245
Cutler Formation, 418
Cuvier, Georges, 4
cyanide, 507
cycads, 453, 454
cycles, in environment, 9–10
cyclothem, 421–23

D

Dakota sandstone, 443
dam(s)
  flood control, 260
  Hoover, 260
  hydropower and, 521–22, 523
  large-scale, 494
  sediment, effect of dams on, 231
  stream profile, effect of dam on, 252, 253
  Vaiont, Italy, disaster (1963), 294
Darwin, Charles, 121, 350, 351
dating
  and geologic process rates, 131
  radiometric. (*see* radiometric dating)
  relative, 119–23, 128, 129
daughter nucleus, 124, 125, 126, 127
da Vinci, Leonardo, 341
Dawson, Charles, 491
dead volcanoes, 61
debris avalanche, 289, 293, 298
Deccan flood basalts, 466, 467
DeChelly sandstone, 418, 423, 427
deep layer, oceanic, 220
deflation, 305
deflation armor, 306
deformation
  glacial, 319–20
  resulting from stress, 155, 157–58

Deinonychus, 458, 462
delta
  as clastic sedimentary environment, 95
  formation, 250, 251
dendrites, 346
dendritic drainage pattern, 243, 244
density
  and continental buoyancy, 191–92
  in isostasy, 192–94
Denver, Colorado, earthquake (1960s), 168
deposition
  glacial, 326–29, 330
  in streams, 249–51
depth, in magma formation, 34, 35
desalination, 281
desertification, causes and impacts of, 313–14
desert pavement, 306, 307
deserts
  causes of natural, 310–11
  as clastic sedimentary environment, 94
  defined, 310, 311
  desertification, causes and impacts, 313–14
  dunes, 307–9
  landforms, 307–9, 311–13
  map of world's, 305
  streams in, 311–12
deuterium, 15
Devonian period
  Acadian orogeny, 423, 424
  Antler orogeny, 426
  Caledonian orogeny, 425
  Ellesmerian orogeny, 425
  in geologic time scale, 129, 130, 419
  life in, 427–30
  sedimentation, 419, 421
dewatering, 51
diagenesis, 90–91
diamond(s)
  covalent bond formation, 17
  dust, 466
  industrial, 22
  in kimberlite pipes, 380
  as mineral resource, 498
  physical properties, 20–21, 24
Diapsida, 456, 457
diatoms, 90, 95, 484
dikes
  in oceanic crust, 216, 217
  in relative dating, 120–21
  volcanic, 44
Dimetrodon, 432, 433, 456
dinosaurs, 118
  bones, 455, 461
  classification, 455–59
  defined, 455
  eggs, 463
  extinction, 354, 464–67
  fossils, 443
  nesting sites, 463
  scales, 461
  skeletons, 455, 459–61
  skin, 461
  tracks, 461–63
  warm-bloodedness, debate over, 463–64
diorite, 41, 42
dip, 189–91
Diplodocus, 462
dip-slip fault, 191
directed stress, 104, 105, 106, 107
discharge, stream, 244, 245
disconformity, 119, 120
discontinuous reaction series, 37
discordant plutons, 43, 44
diseases, extinctions and, 467
displacement along the fault, 164
dissolved load, 249

divergent evolution, 353, 354
divergent plate boundaries, 50, 51, 148–49
divergent stress, 476, 479
divide, 243
DNA (deoxyribonucleic acid), 351
dolomite
  as carbonate, 28
  formation, 89
  Paleozoic, 399, 414–15
  Silurian, 404
dolomitization, 91
domes, 187, 396
Doppler effect, 365
dormant volcanoes, 61, 62
downcutting, 245
drainage basin, 243
drift, 326
drowned valleys, 232, 233
drumlins, 327, 328
Duluth complex, 392
Duluth gabbro, 384, 392
dunes, 307–9
dunite, 41
Dust Bowl (United States), 306, 314
dynamic equilibrium, 10, 252
dynamic factors, 285
dynamothermal metamorphism, 112

E

Early Paleozoic era
  arches, 395–96
  basins, 395, 396
  characteristics, 399–401
  cratonic sequences, 397, 398
  eastern margin of North America, 406–7, 408
  epeiric seas, 397
  life in, 409–14, 416
  North American craton, 401–5
  sedimentary rocks of, 402–5
  summary diagram, 416
  transcontinental arch, 402
  western margin of North America, 408–9
earth
  age, 123–24, 130–31
  differentiation, 370
  early, 370, 372–75
  history, 5, 6–9
  humans, impact of, 11
  interior of the. (*see* interior of the earth)
  modern, 9–11
  moon, origin of, 369
  present as key to past, 4, 5
  solar system, origin of, 365–69
earthflows, 292, 293
earthquake cycle, 169
Earthquake Prediction Panel, 171
earthquakes
  basic principles, 155, 157–58
  control, 167–68
  at convergent plate boundaries, 150
  epicenter, 158, 159, 160–61
  focus, 158
  hazards from, 164–67, 172–74, 288–90
  historic, data on, 156
  intensity, 162–64
  landslides, earthquakes as triggers of, 288–90
  locations, 137, 156, 158–59
  magnitude, 162
  nuclear waste disposal and seismicity, 518
  periodicity of, 169–70
  plate movement and deformation as cause of, 6
  prediction, 168–71
  public response to predictions, 171

solubility, 269
as sulfate, 28
gyre, 220

H

Hadrosaurus, 459
Haicheng, China, earthquake (1975), 169
half-life, 124
halides, 29
halite
in evaporites, 89
formation, 17
as halide, 29
as mineral resource, 498
properties, 23, 24
Silurian formation, 404
solubility, 269
structure, 20, 22
Halley, Sir Edmund, 124
hanging valleys, 323
hanging wall, 191
Harding sandstone, 408
hardness in minerals, 22
hardpan, 79
hard water, 276
harmonic tremors, 61
Haversian system, 461
Hawaiian Islands
coastal erosion, 234
formation, 51, 145, 146
Loihi, 145
as shield volcanoes, 52–53, 64
volcanic hazards, 62, 64–65
wave refraction at, 226
Hayward fault, 172
hazards
coastal, 234–38
from earthquakes, 164–67, 172–74, 288–90
flood. (*see* flooding)
landslide, 285, 293–300
headward erosion, 252
height, wave, 225
Heimaey (Iceland), 56–57
helium, 365
Helmholtz, Hermann L.F. von, 123
hematite, 29
Herrerasaurus, 457
Hess, Harry, 141, 147
hiatus, 119
Himalaya Mountains (Asia), 150, 199, 470
hinge, 186
historical geology, 2
Holocene epoch, 129, 130, 470
homebuilding. *See* residential development
Hominidae, 488–90
Homo, 490
homology, 351, 352
Homo sapiens, 9, 488, 489, 490
Hoover Dam (Nevada), 260, 494
horn, 325
hornblende, 27
hornblende-hornfels facies, 114
hornfels, 114
horse, evolution and development, 485–86
horst, 196
horst-and-graben structure, 443, 480–81
hot-dry-rock areas, 521
hot spot
rates of plate movement with hot spots, monitoring, 145–46
volcanoes, 50, 51
hot springs, 269
humans
evolution, 8–9, 487–90, 491

fossil, 487–90, 491
impact of, 11
hurricanes
Gilbert, Hurricane (1988), 227
Saffir-Simpson hurricane scale, 227
Hutton, James, 4
hydrocompaction, 310
hydrogen
in early universe, 365
in fusion, 519
isotopes of, 15
in oil, 508
hydrograph, 254, 255
hydrologic cycle, 9, 242–43, 318
hydropower, 521–22, 523
hydrosphere, 242, 265
hydrostatic pressure, 268
hydrothermal activity, 105
hydrothermal ores, 497, 504
hydrothermal vents, 211, 212, 497, 505
hydrous minerals, 37, 105
hydrous silicates, 28
Hypacrosaurus, 460
hypocenter, earthquake's, 158
hypotheses, 4
Hyracotherium, 485

I

ice
and the hydrologic cycle, 318
in solar system formation, 367
ice ages, 332–34. *See also* glaciation
causes, 333–34
defined, 332
Pleistocene, 332, 333, 471–76
Proterozoic, 384, 386
ice caps, 232, 322
ice sheets, 322
ice wedging, 323
icthyosaurs, 452
igneous rocks
classification, 40–43
grain size, 38–40
intrusions, 43–46
as mineral resource, 495, 497
porosity, 266
in rock cycle, 30
textures, 38–40
Iguanodon, 459
illite, 75
impact structures, 113, 354, 356–57, 465–66
impact-trigger hypothesis, 369
incised meanders, 247, 248
index fossils
defined, 122–23
Mesozoic, 454
Paleozoic, 430
Precambrian, 370, 386–87
index minerals, 115
industry, water use, 278, 279, 280
inert elements, 17
infauna, 349
infiltration, 242
influent stream, 266
infra-, 349
inorganic, minerals as, 19
insectivores, 485
insects, 430, 432, 464
intensity, earthquake, 162–64
interference colors, 40
intergranular porosity, 265
interior of the earth
chemistry, 178–80
diagram of, 182

pressure, estimating, 180
seismic wave analysis of, 174–78
temperature, estimating, 180–81
intermontaine basins, 478, 479
internal drainage, 312
International Code of Zoological Nomenclature, 349
International Seabed Authority, 505
intertidal environments, 348
intraplate volcanism, 50, 51
intrusive rock structures, 43–46
invertebrates, 484
ions
bonding, 15, 17
charges and relative sizes of common, 19
defined, 15
formation, 15, 16
mass spectrometer counting of, 127
iridium, 465, 466, 467
iron
Clinton Iron formation, 408
in core, 177–78
as mineral resource, 496
Precambrian, 384, 386
iron ore, 89, 90, 497–98
irrigation water, 278, 280
island arc, 150
Appalchian region, 407
isograds, 115
isopach, 395
isostasy, 192–94
isostatic equilibrium, 192, 473
isotopes
characteristics, 14–15
fission and, 516–17
isotopic system, choice of, 126, 127
radioactive, 124–25, 126

J

jetties, 230
Johanson, Donald, 489, 490
joints, 188, 189
joint set, 188, 189
Joly, John, 124
Jovian planets, 366, 367, 368, 369
Juan de Fuca Ridge, 504
Jupiter, 366, 367, 368
Jurassic period
dinosaurs. (*see* dinosaurs)
in geologic time scale, 128, 129, 130
life, 453, 454–55
Nevadan orogeny, 445, 447
North American craton, 439, 441–42
passive continental margin formation, 444
western margin of North America, 446–47

K

Kant, Immanuel, 123
kaolinite, 75
karst topography, 270
Kaskaskia Sea, 419, 421
Kelvin, Lord William T., 123
Kenoran orogeny, 380
kerogen, 479, 512
kettle, 327, 329
Keweenawan lava, 384, 392
Kilauea (Hawaii), 9
predictors of eruptions, 62, 63
recent activity, 64
residential development and, 57, 64–65
kimberlites, 178, 380
knickpoints, 252, 253

T

tabulate corals, 411, 412
Taconic Mountains (New York), 403
Taconic orogeny, 406–8, 424
tailings, 507
talc, 24, 499
talus, 291
Tambora (Indonesia), 60
Tapeats sandstone, 408
taphonomy, 341, 344–47
tarn, 323, 324
tar sands, 512–13
taxonomy, 349–50
tectonics. *See also* plate movements; plate
      tectonics
   Archean, 380
   Cenozoic era, western margin during, 476, 478
   defined, 6, 135
   magmatism and, 50–51
   Mesozoic era, western margin during, 445
temperature
   atmospheric circulation and, 304–5
   for diagenesis, 90, 91
   internal, estimating, 180–81
   in magma formation, 34
   in metamorphism, 104, 105, 106
tensile stress, 155, 157
tensional stress, 107, 186, 189
terminal moraine, 327, 328
terraces, 247
terracing, 82–83
terrane, 201
terrestrial planets, 367, 369
terrigenous sediments, 217, 218
Tertiary period, 129, 130, 470. *See also individual
      epochs*
Teton Range (Idaho, Wyoming), 197, 480, 491
tetrahedron, silica, 25–28
texture, soil, 78, 79
thecodonts, 452, 457
Theophrastus, 341
theory, 4, 200
therapsids, 433, 434
thermocline, 220
thin sections, 40
thorium-232, 127, 517
thorium-233, 517
Three Mile Island, Pennsylvania, nuclear accident
      (1979), 517
thrust fault, 189, 190
tidal flat, 95
tidal waves, 166–67
tides, 226–27, 228, 523
till, 326, 475
tillite, 326, 387
time
   in crystal formation, 19
   dating. (*see* dating)
   geologic time scale, 128–31
   geology and, 3
   and rocks, geologic distinction between, 399
tin, 35, 495, 496
Tippecanoe Sea, 401, 403, 408, 411, 419
Tobias, Philip, 490
Tokyo, Japan, earthquake (1923), 156, 171
Torosaurus, 459
total dissolved solids (TDS), 275–76
tourmaline, 24, 497
toxic chemicals, soil erosion and, 82
trace, fault, 158
trace fossils, 345, 346
tracks
   dinosaur, 461–63
   as trace fossils, 345
trails, 345

Trans-Alaska Pipeline, 165
transcontinental arch, 402
transform faults, 149
transgressions
   in cyclothems, 422
   defined, 97, 397
   in Gulf of Mexico, 448
transgressive sequence, 97, 98
transition zone, 176
transported soil, 78
transverse
   dunes, 308, 309
   ridges, 308
traps, petroleum, 509
"Treatise on Invertebrate Paleontology," 361
trellis drainage pattern, 243–44
trench, 209, 210
Triassic period
   dinosaurs. (*see* dinosaurs)
   in geologic time scale, 129, 130, 439
   Gulf of Mexico formation, 448
   life, 452–54
   North American craton, 439
   rift basins, 443–44
   Sonoman orogeny, 445–46
   western margin of North America, 445–46
tributary stream, 243
Triceratops, 457, 459
trilobites
   in Cambrian, 410–11
   extinction, 434
   fossils, 343
tritium, 15
trough, wave, 225
Tsunami Early Warning System, 167
tsunamis, 166–67
tuff, 43
Tullimonstrum, 430
turbidity, 231
turbidity currents, 213
Twain, Mark, 131
type section, 128
Tyrannosaurus, 457–58, 460, 464

U

ultramafic, 41, 370
unconfined aquifer, 268, 271
unconformity, 119
underclay, 422
ungulates, 485
uniformitarianism, 4, 5
universe, origin, 365
unloaded, rock, 70
upconing, 275
upwelling, 221
uranium, 124, 497, 507
uranium-235, 125, 126, 127, 516–17
uranium-238, 127
Uranus, 366, 367, 368
urbanization
   filling wetlands for, consequences of, 274
   flood severity and, 256, 257
   soil erosion and, 82, 83
Urey, Harold, 388
U.S. Coast and Geodetic Survey, 167
U.S. Geological Survey, stream assessment, 243
U.S. Soil Conservation Service, 81
Ussher, Archbishop, 123

V

vadose zone, 265
vagile organisms, 349
Vaiont, Italy, disaster (1963), 294
valley glaciers, 321

varve, 329
vegetation. *See also* flora; plants
   biomass fuel and, 370
   desert, 310, 311
   dunes and, 308, 309
   effect of desertification on, 313
   and flooding, 254
   growth pattern as indication of creep, 295, 296
   and landslides, 287–88, 298
   wind erosion retarded by, 306–7
Velociraptor, 458, 459
ventifacts, 305, 306
vents, hydrothermal, 211, 212, 497, 505
Venus, 366, 367
Venus missions, 366
vertebrates, Cenozoic, 484–86
vesicles, 38
vestigial structures, 351, 352
Viking, 366
Vine, F. J., 141
vitreous luster, 23
volatiles, 34, 35
volcanic breccia, 43
volcanic glass, 23, 39, 42
volcanic hazards
   climate, impact on, 60–61
   lahar, 58
   lava, 56–57, 58, 59
   materials and eruptive style, 56–60, 66
   nuée ardente, 58–59
   predicting eruptions, 61–62
   pyroclastics, 57–59
   in U.S., current and future, 62–66
volcanic rocks, 41, 42–43
volcanism
   activity, classification of, 61
   Archean rock formed from, 380
   calderas, 55–56
   cinder cones, 54–55, 56
   composite volcanoes, 55
   and Cretaceous extinctions, 466–67
   divergent plate boundaries, magmatism at, 50,
      51
   dome, volcanic, 53
   fissure eruptions, 51–52
   hazards from. (*see* volcanic hazards)
   intraplate, 51
   magma generation, 50–51
   Mesozoic sediments and, 447
   microscopic examination of rocks, 40
   mountain formation by, 194–95
   plate movements and, 6
   porphyritic rock formation, 39, 40, 41
   predicting eruptions, 61–62
   rifting and, 149
   shield volcanoes, 52–53
   solidification of lava, 9
   at spreading ridges, 148
   subduction-zone, 50–51, 150
volcanoes
   El Chichón (Mexico), 61
   Heimaey (Iceland), 56–57
   Kilauea. (*see* Kilauea (Hawaii))
   Krakatoa (Indonesia), 60
   location of modern, 137
   Mauna Kea (Hawaii), 53
   Mauna Loa (Hawaii), 53
   Mont Pelée (Martinique), 58–59, 60
   Mount Etna (Italy), 57
   Mount Pinatubo (Philippines), 57, 61, 221
   Mount St. Helens. (*see* Mount St. Helens)
   Mount Vesuvius, 57–58
   Nevado del Ruíz (Colombia), 58
   rates of plate movement, volcanoes used to
      monitor, 145–46
   Tambora (Indonesia), 60
Voyager, 365, 366

W

Walcott, Charles D., 123–24, 411
Walker, Alan, 490
Wallace, Alfred, 350
wallrock, 43
Warrawoona group (Australia), 390
Wasatch formation, 479
waste disposal
  ground water pollution from, 276–77
  radioactive, 518–19
  at subduction zones, 151
water
  as compound of carbon and oxygen, 17
  quality, 275–77
watershed, 243
water table
  characteristics, 267
  defined, 265
  lowcring the, consequences of, 271–75
water use and supply
  categories of use, 278, 279
  extending the water supply, 280–81
  industrial versus agricultural use, 280
  quality, water, 275–77
  regional variations in water use, 278–79, 280
  surface water versus groundwater supply, 277–78, 279
wave base, 225
wave-cut benches, 232
wavelength, 225
waves
  benches, wave-cut, 232
  breakers, 225, 226
  defined, 225
  protection from, attempts at, 234–37
  refraction, 226

structure, 225–26
and surges, 227
and tides, 226–27, 228
weathering, 69
  and acid rain, 72–73, 74
  chemical, 70, 71–75, 76, 77
  defined, 70
  desert, 312–13
  mechanical, 70, 71, 75
  ore formation from, 498
  soil composition and, 79
Wegener, Alfred, 135, 136
Weiner, J. S., 491
wells, upconing, 275
West Antarctic ice sheet, 336
western margin of North America
  accreted terranes, 446, 447
  Cenozoic era, 476–83
  Mesozoic, 444–47
  Paleozoic, 408–9, 426–27
  physiographic provinces of, 477
wetlands, filling in, 274
White, Tim, 490
Wichita Mountains (Oklahoma, Texas), 426
Williston Basin, 419
wind
  deposits, 307–10
  dunes, 307–9
  erosion, 305–7
  loess, 309–10
  in ocean circulation, 219–21
  origin, 304–5
windbreaks, 82
wind energy, 523–24
wind farms, 523–24
Windgate sandstone, 447
Woodward, Arthur Smith, 491

Working Group on California Earthquake
  Probabilities, 170
worms, 410, 411

X

xenoliths, 46, 178
Xenophanes, 341

Y

100-year flood, 258–59
Yellowknife greenstone belt, 378–80
Yellowstone National Park (Idaho, Montana,
  Wyoming)
  geysers in, 2, 269, 480
  volcanic processes occurring at, 66
Yellowstone River (North America), 245
Yosemite National Park (California), 1
  granite dome formation, 195, 196
  weathering of rocks, 69, 70, 71
Yucca Mountain (Nevada), 518

Z

zeolite facies, 114
zeolites, 114
zinc, 497, 504
zone of accumulation, 78
zone of aeration, 265
zone of convergence, 198–99
zone of deposition, 78
zone of leaching, 77
zone of saturation, 265